CHEMICAL SIGNALS IN VERTEBRATES 4

Ecology, Evolution, and Comparative Biology

CHEMICAL SIGNALS IN VERTEBRATES 4

Ecology, Evolution, and Comparative Biology

Edited by

David Duvall
University of Wyoming
Laramie, Wyoming

and

Dietland Müller-Schwarze *and*

Robert M. Silverstein
State University of New York
Syracuse, New York

Plenum Press · New York and London

Library of Congress Cataloging in Publication Data

Satellite Symposium on Chemical Signals in Vertebrates (1985: Laramie, Wyo.)
Chemical signals in vertebrates 4.

"Proceedings of a Satellite Symposium on Chemical Signals in Vertebrates, of the Tenth International Comparative Endocrinology Conference, held July 27–29, 1985, in Laramie, Wyoming"—T.p. verso.

Includes bibliographies and index.

1. Vertebrates—Behavior—Congresses. 2. Chemical senses—Congresses. 3. Pheromones—Congresses. 4. Animal communication—Congresses. I. Duvall, David. Muller-Schwarze, Dietland. III. Silverstein, Robert Milton, 1916– . IV. International Comparative Endocrinology Conference (10th: 1985: Laramie, Wyo.) V. Title. VI. Title: Chemical signals in vertebrates four.

QL750.S28 1985 596'.051 86-17013

ISBN-13:978-1-4612-9310-1 e-ISBN-13:978-1-4613-2235-1

DOI: 10.1007/978-1-4613-2235-1

Proceedings of the Fourth International Conference on Chemical Signals in Vertebrates, held July 27–30, 1985, in Laramie, Wyoming

Softcover reprint of the hardcover 1st edition 1986

A Division of Plenum Publishing Corporation
233 Spring Street, New York, N.Y. 10013

PREFACE

This volume reviews recent developments in our understanding of chemical signaling in vertebrates. After sections dealing with general principles and chemical aspects of vertebrate pheromones, it follows a taxonomic approach, progressing from fish to mammals. The editors asked a diverse, international group of leading investigators, working on a wide array of vertebrate taxa and specific issues, to consider their efforts from comparative, evolutionary, and ecological viewpoints. The relative number of manuscripts in each part does not necessarily reflect current intensity of research, since the editors invited speakers who together would provide a balanced and comprehensive overview, while avoiding duplication. Still, the part on mammals is the longest.

Fourth in a series dating from 1977, this volume illuminates current trends and likely future developments in the field of chemical signaling in vertebrates. Going back even farther, the first chapter, a personal account of the past quarter century by Dr. Mykytowycz recalls the most important milestones, such as symposia, or the founding of societies and journals. He also credits those investigators who stand out by their seminal studies.

Some new trends can be discerned. First, more taxa, even those with hitherto only microsmatic reputation, such as birds (papers by Wenzel and by Mason and Clark) are examined for their olfactory involvement. This wider coverage allows application of evolutionary lines of thought that may aid in further hypothesis testing (e.g., Smith; Houck; Dawley; Gosling; Graves et al.; and Keverne and Rosser). Second, the primer pheromones have emerged as the vertebrate pheromones with the most consistent effects. For this reason, a number of laboratories worldwide are attempting to exploit these effects for livestock breeding (e.g., Albone et al.). After decades of effort, the first primer pheromone has been chemically identified (Jemiolo et al.). Third, the search for applications is taking new routes. For example, Sullivan and Crump explore the responses of pest rodents to predator odors.

Finally, we now know that the isolation and identification of biologically active vertebrate odor components, while in many ways similar in procedure to that so successfully applied in insects, poses a greater challenge. As in insects, several constituents usually must act together for the effect to occur. However, in vertebrates, odors functioning in recognition of species, group, individual or physiological state may consist of dozens of compounds. Instead of the "response-guided strategy" (Albone, 1984), where the bioassay follows successive fractionation steps, investigators of mammalian pheromones now speak of "odor images," "patterns," or "profiles," meaning that active constituents may occur in many or all fractions and that fractionation actually destroys the biological effect. The effect of various components of such an odor image may be additive, synergistic, or redundant (Muller-Schwarze et al.). An example of processing of chemical signals by humans (Doty) concludes the review. Needless to say, this field is more important than a single contribution can convey.

As this collection of essays testifies, the exploration of chemical signals has become more firmly embedded in behavioral ecology and sociobiology. Current research has been stimulated by the studies of insect pheromones and is guided by evolutionary concepts.

These papers were presented at the Fourth International Symposium on Chemical Signals in Vertebrates, held at the University of Wyoming, Laramie, from 27-30 July 1985. Our meeting was affiliated with the 10th International Symposium on Comparative Endocrinology, organized by Charles Ralph and held the previous week at Copper Mountain, Colorado. We thank the Union Carbide Corporation and the University of Wyoming Office of Research and College of Arts and Sciences for financial support. Richard Doty, Robert Jenkins, Walter Eggers, and Joan Wadlow deserve greatest thanks for their help in obtaining this support. The helpful counsel of John G. Vandenbergh, Ben S. Goodrich, and D. Michael Stoddart is also gratefully acknowledged.

Laramie "locals" who helped immeasurably with the actual presentation of the conference include Brent Breithaupt, Matt Goode, Al Redder, Brent Graves, Steve Buskirk, Bill Gern, Charles Ralph, Mike King, Geoff Carpenter, Mary David, Ruby Sanchez, Pat Beintema, Chuck Cooper, and our dedicated "army" of student volunteers. Brent Graves, Matt Goode, Al Redder, and Jeanne Trupiano-Duvall helped with proofreading. Mary David deserves special thanks for typing the entire camera-ready volume with so much patience and talent. Finally, our thanks to Jeanne Trupiano-Duvall for help with the indices and for her patience and support.

Laramie, Wyoming
Syracuse, New York

D. Duvall
D. Muller-Schwarze
R. M. Silverstein

Summer, 1986

CONTENTS

Part One: Some General Considerations

A Quarter of a Century of Studies of
Chemical Communication in Vertebrates
R. Mykytowycz 1

The Tetrapod Epidermis:
A System Protoadapted as a Semiochemical Source
Paul F. A. Maderson 13

Mammalian Semiochemistry:
Issues and Futures, with some Examples from a
Study of Chemical Signalling in Cattle
Eric S. Albone, Nicholas B. Blazquez, Jane French,
Susan E. Long, and Graham C. Perry 27

Motor Patterns Dedicated to Sensory Functions
David Chiszar 37

Part Two: Chemistry

Chemical Control of Feeding in Herbivorous
and Carnivorous Fish
Michael A. Adams and Peter B. Johnsen 45

Characteristics of Earthworm Washings
Detected by the Vomeronasal System of Snakes
Mimi Halpern, Nancy Schulman, and
Donald M. Kirschenbaum 63

Chemical and Biological Investigations
of Female Mouse Pheromones
Bozena Jemiolo, Franca Andreolini, and
Milos Novotny 79

Behavioral Function and Chemistry of Volatiles
From Feces and Anal Secretions of House Mice,
Mus musculus - A Preliminary Report
B. S. Goodrich, S. Gambale, P. R. Pennycuik,
and T. D. Redhead 87

Part Three: Fishes

The Evolution of Chemical Alarm Signals in Fishes
R. Jan F. Smith 99

Fish Reproductive Pheromones
N. E. Stacey, A. L. Kyle, and N. R. Liley 117

Chemosensory Orientation Mechanisms of Fish
Peter B. Johnsen 135

Release of Spawning Pheromone(s) by Naturally-ovulated and Prostaglandin-injected, Nonovulated Female Goldfish
P. W. Sorensen, N. E. Stacey, and P. Naidu 149

The Structural and Functional Relationships Between Olfactory and Reproductive Systems From Birth to Old Age in Fish
M. P. Schreibman, H. Margolis-Nunno, and L. Halpern-Sebold 155

Part Four: Amphibians

The Evolution of Salamander Courtship Pheromones
Lynne D. Houck 173

Pheromonal Markers as Territorial Advertisement By Terrestrial Salamanders
Robert G. Jaeger 191

The Recognition and Use of Chemical Signals By a Nesting Salamander
Don C. Forester 205

Evolution of Chemical Signals as a Premating Isolating Mechanism in a Complex of Terrestrial Salamanders
Ellen M. Dawley 221

Chemical Ecology of Kin Recognition in Anuran Amphibians
Bruce Waldman 225

Part Five: Reptiles

Pheromones and Reproduction in Garter Snakes
W. R. Garstka and D. Crews 243

The Role of Pheromone Trails in the Sociobiology Of Snakes
Neil B. Ford 261

Pheromone Mimicry in Garter Snakes
Robert T. Mason and David Crews 279

Initial Den Location by Neonatal Prairie Rattlesnakes: Functions, Causes, and Natural History in Chemical Ecology
Brent M. Graves, David Duvall, Michael B. King, Stan L. Lindstedt, and William A. Gern 285

Exploratory Tongue Flicking by Green Iguanas In Laboratory and Field
Gordon M. Burghardt, Barbara A. Allen, and Hannah Frank 305

Lizard Pheromones: Behavioral Resnonses and Adaptive Significance in Skinks of the Genus Eumeces
William E. Cooper, Jr. and Laurie J. Vitt 323

Chemoreception in the Homing and Orientation Behavior of Amphibians and Reptiles, with Special Reference To Sea Turtles
David Owens, Diana Crowell Comuzzie, and Mark Grassman 341

Part Six: Birds

The Ecological and Evolutionary Challenges Of Procellariiform Olfaction
Bernice M. Wenzel 357

Chemoreception and the Selection of Green Plants as Nest Fumigants by Starlings
J. Russell Mason and Larry Clark 369

Part Seven: Mammals

Economic Consequences of Scent Marking in Mammalian Territoriality
L. M. Gosling 385

Chemical Signals and Kin Recognition in Spiny Mice (Acomys cahirinus)
Richard H. Porter 397

Genetic Basis for Individual Discriminations: The Major Histocompatibility Complex of the Mouse
G. Beauchamp, A. Gilbert, K. Yamazaki, and E. A. Boyse 413

The Suppression of Ovarian Function By Chemosignals
John G. Vandenbergh 423

The Evolutionary Significance of the Olfactory Block to Pregnancy
Eric B. Keverne and Anne E. Rosser 433

Puberty-Influencing Chemosignals in House Mice:
Ecological and Evolutionary Considerations
Lee C. Drickamer 441

The Puberty Delaying Pheromone of the House Mouse:
Field Data and a New Evolutionary Perspective
David M. Coppola 457

Effects of Genotype, Social Interaction, and
Testosterone on Esterases in Male Mouse Urine
Jay Labov, Daniel Marra, Philip Allen, and
Diane Zavotsky 463

The Mammalian Vomeronasal System:
Its Role in Learning and Social Behaviors
Charles J. Wysocki, N. Jay Bean, and
Gary K. Beauchamp 471

Olfactory Communication Among Rats:
Information Concerning Distant Diets
Bennett G. Galef, Jr. 487

Alarm Chemosignals in a _Meriones unguiculatus_:
Prey-Predator Interactions
Del D. Thiessen and Robert Cocke 507

Avoidance Response of Pocket Gophers
(_Thomomys talpoides_) to Mustelid Anal Gland
Compounds
Thomas P. Sullivan and Douglas R. Crump 519

The Comparative Responses of Mice and Voles
To Conspecific and Heterospecific Odors in
The Field - A Lesson in Social Behavior
D. Michael Stoddart 533

Morphometric Patterns Among Microtine Rodents. I.
Sexual Selection Suggested by Relative Scent
Gland Development in Representative Voles (_Microtus_)
Frederick J. Jannett, Jr. 541

Female-induced Delay of Puberty in
Bank Vole and European Pine Vole Females
Anna Marchlewska-Koj and M. Kruczek 551

Familial Chemosignals Interfere with
Reproductive Activation in Female Pine Voles,
Microtus pinetorum
John J. Lepri 555

Odor Images: Responses of Beaver to
Castoreum Fractions
D. Muller-Schwarze, L. Morehouse, R. Corradi,
Cheng-hua Zhao, and R. M. Silverstein 561

Scent-marking in Pikas (_Ochotona princeps_):
Test of a Breeding-facilitation Hypothesis
Carron A. Meaney 571

Histology and Gross Morphology of the Sexually Dimorphic Sternal Gland in the North American Opossum, Didelphis virginiana Kerr
Donna Holmes Meisner 579

Chemical Communication in Gray Short-tailed Opossums (Monodelphis domestica) with Comparisons to other Marsupials and with Reference to Monotremes
Barbara H. Fadem 587

Observations on Chemical Communication and its Glandular Sources in Selected Insectivora
Walter Poduschka and Chris Wemmer 609

Plantar Glands in North American Mustelidae
Steven W. Buskirk, P. F. A. Maderson, and Robin M. O'Connor 617

The Marking Behavior of Male Przewalski's Horses
Lee Boyd and Lonnie Kasman 623

Chemical Communication Among Asian Elephants
L. E. Rasmussen, M. J. Schmidt, and G. D. Daves 627

Territorial Scent Marking by Two Sympatric Lemur Species
Anne Millhollen 647

Chemical Signals in Callitrichid Monkeys - A Comparative Review
G. Epple, A. M. Belcher, and A. B. Smith, III 653

Cross-cultural Studies of Taste and Smell Perception
Richard L. Doty 673

Author Index 685

Subject And Taxonomic Index 709

A QUARTER OF A CENTURY OF STUDIES OF CHEMICAL COMMUNICATION IN VERTEBRATES

R. Mykytowycz

Division of Wildlife and Rangelands
Research, CSIRO
P.O. Box 84, Lyneham, A.C.T. 2602
Australia

The purpose of this article is to give a brief outline of the events and processes which have taken place over the last twenty five years which have led to the emergence of a new biological discipline - namely that of chemical communication in vertebrates.

The simplest way of doing this is to recall my own involvement in it. Naturally, in adopting this approach, there is always the chance that the final picture may be somewhat biased.

References are made to the more important meetings and publications and to the formation of societies devoted to the promotion of interest in the field. In conclusion, the needs for future research are suggested.

THE NEED TO STUDY COMMUNICATION IN ANIMALS

In the early '50s we in Australia, for practical reasons, undertook a series of studies of various aspects of the biology of free-living European wild rabbits, Oryctolagus cuniculus. The hope was that the knowledge gained would enable us to control more efficiently the numbers of these animals which represent a serious agricultural problem. My personal task, originally, was to study the diseases of the rabbit. In the course of this work, however, I developed an interest in behaviour and became convinced of its importance in the population dynamics of free-living animals. At that time, the behaviour of animals ceased to be a trivial, theoretical subject; ethology became acknowledged as a separate discipline. This provided a legitimate excuse for me to enter into this new realm of biology.

In the course of our ethological studies, it became obvious that to preserve their characteristic pattern of behaviour and to assure an orderly coexistence, rabbits have to communicate with one another.

Like other mammals, rabbits probably make use of all modalities to exchange information. However, such behavioural activities as regular olfactory exploration of their surroundings and conspecifics, "chinning," the formation of "dung-hills," "tail-flagging" and the enurination of other individuals suggested that for wild rabbits, odour signals derived from the anal, chin, and inguinal glands and from urine are particularly important means of communication.

A search through the literature showed that the existence of the odour-producing glands mentioned above had already been recorded by anatomists and histologists, but in view of a complete absence of information on the behaviour of the species, the question of their functions was wrongly interpreted or completely ignored. Thus, there was an obvious need to clarify the role of the odoriferous skin glands in the European wild rabbit.

The scenario just outlined was not confined to the study of rabbits. A similar necessity appeared at about the same time for other species, as interest in mammalian behaviour, and the role of olfaction in it, grew rapidly.

FACTORS STIMULATING EARLY STUDIES OF OLFACTORY COMMUNICATION

Apart from the rise of ethology, there were other factors which stimulated interest in olfactory communication in vertebrates.

First of all, the early success of chemists in their research on insect pheromones, and the discussion and speculation which developed from it, provided strong encouragement and hope that "mammalian pheromones" may be just as easy to study. This would be of benefit to man in his efforts to manage populations of both free-living and domestic animals.

We now know that the early expectations of the practical importance of insect pheromones proved to be too optimistic. In any case, direct comparison of the behaviour of insects and higher animals leads to a misleading simplification. Also, modern analytical techniques of gas chromatography and mass spectrometry, which became commonly available during the mid-'60s, were soon found to be inadequate to solve the complexity of mammalian odours. The lack of suitable behavioural assays was another hampering factor.

On the other hand, it was perhaps fortunate that at this early stage of interest in olfactory communication of mammals those involved were not aware of all these problems; otherwise, the enthusiasm of many of them would surely have been dampened. Incidentally, the belief that the responses of mammals to odour signals are stereotyped and genetically controlled prevailed for a long time. This oversimplification, even today, often confuses many workers without previous experience in behavioural studies.

Exciting observations on the effects of odour on the timing of oestrus and the blocking of pregnancy in laboratory mice, which came to light around the mid-'50s, was another factor which inspired interest in olfactory communication.

Another stimulus which generated interest was the progress in understanding the physiology of olfactory systems for which we are indebted to such workers as E. D. Adrian, C. Pfaffmann and Y. Zotterman. The availability of electrophysiological techniques, which allowed the measurement of the electrical responses of the olfactory bulbs and peripheral olfactory systems to odour stimulation, promised easier experimentation.

EARLY SOURCES OF INFORMATION

In this early period of marshalling of ideas which eventually formed the foundation for the new field of interest, R. W. Moncrieff's book

"Chemical Senses" (1951, 2nd edition) contained all the information on olfaction available at that time and was of great help to the beginner.

An early interest shown by perfumers and natural product chemists in the secretions from mammalian skin glands did not stimulate attention as their studies were completely unrelated to the animal's behaviour. Nevertheless, the paper by Kingston (1965), summarising their achievements, proved a useful reference for further chemical studies.

Naturally, one could not afford to ignore the fairly numerous reports on the occurrence and histology of odour-producing skin glands throughout the animal kingdom. Some of them hinted at the possible role of skin glands in the communication of information related to reproduction. Subsequently, our own histological studies showed very clearly that social status - i.e., behavioural status of an individual, is reflected in the histological appearance of odoriferous glands which function for communication purposes (Mykytowycz, 1970).

FIRST STEPS TOWARDS THE ORGANIZATION OF GROUP MEETINGS

Even scientists cannot afford to live and work in complete isolation. As a result of the increasing interest in chemical senses generally, a number of meetings of workers in this field took place in the '50s (McCartney, 1968). Of special importance to us, however, was the First International Symposium on Olfaction and Taste which was held in Stockholm in 1962 as a satellite meeting to the International Congress of Physiology organised by Y. Zotterman. The program of this and subsequent triennial meetings, known as ISOT, although centered around physiology, also provided an opportunity to those interested in behaviour to meet and exchange ideas (Zotterman, 1963).

There were, however, few who were ready to take advantage of this opportunity. Only two or three contributions on the program of the first ISOT were directly related to the problems of the communicatory role of smell. In subsequent meetings, this area was somewhat more strongly represented. Nevertheless, what was important was the indication of the emerging awareness that there was a need to study chemical signaling in the olfactory behaviour of animals.

In 1962, sensory physiology itself was still in an early stage of development and this was very clearly expressed in the opening address by E. D. Adrian. He remarked that ... "Our subject has reached the right stage for serious discussion and we must get down to it. It concerns fundamental problems where physiology and psychology meet (at that stage behavioural matters were still the domain of psychology, not ethology), in fact we are in one of those borderlands which are the most fertile regions for the scientific advance...".

Although Adrian directed his remarks mainly toward his fellow physiologists, they were and still are appropriate to all those interested in olfactory communication in mammals.

During the first ISOT, only Jacques LeMagnen, the founder of the field of olfactory-endocrine relationships, whose later work linked animal and human studies, hinted at future developments. In his paper he remarked that ..."Electrophysiological investigation of olfactory and gustatory afferent fibres does not allow us to analyse the discriminatory process itself. Only behavioural data can actually reveal the differential effects of the given stimuli on the olfactory system. Thanks however to the work of Adrian, Pfaffman and of the Swedish and other workers, modern electrophysiological evidence has given us an idea of both the physiochemical and the intrinsic nervous mechanisms which at different levels account for the phenomena demonstrated by behavioural methods."

While the first ISOT in 1962 merely indicated the trends, the end of the '60s saw the appearance of separate groupings of workers interested in the study of chemical communication in animals.

It was most timely that the only known mammalian semiochemicals which fit nicely into the concept of insect pheromones - two steroids present in the saliva of the domestic boar (5α-androst-16-en-3-one and 3α-hydroxy-5α-androst-16-ene) - were isolated by Patterson (1969), just at the time when interest in olfactory communication was rising.

The first ever conference organised at the international level, which gathered together all those interested specifically in the chemical communication of animals, took place in June (27-28) 1968 at Auburn, Massachusetts. It was organized mainly by D. J. Moulton in cooperation with J. W. Johnston, and A. Turk (Johnston et al., 1970). The Auburn symposium brought together chemists, biologists, insect physiologists, mammalian endocrinologists, experimental psychologists, animal behaviourists and zoologists, as well as other specialists. This cocktail of different specialists was the best indication that a new, important biological discipline, chemical communication in animals, had emerged.

Most appropriately, E. O. Wilson, a founder of sociobiology, was invited to give the introductory speech. He already had at least ten years experience in the field of olfaction, having published a number of papers in which he not only summarised the achievements in the field of entomology, but also ventured into the mammalian world. Wilson's (1963) paper, published in "Scientific American," served as a key reference for many who contemplated entering the new niche of ethology.

During the '60s a further two ISOT and Gordon Conferences on the Chemical Senses were held. The Japanese Association of Taste and Smell (JUST) was formed in 1968. In the same year, the Commission on Chemoreception of the International Union of Physiological Sciences was organised. Its status and nature were formalised in 1972. The traditionally physiologically oriented program of the third ISOT held in New York in 1969 was augmented by papers on olfaction and behaviour of fish, birds and mammals. The speakers included J. E. Bardach, Bernice M. Wenzel, W. K. Whitten and T. G. Schultze-Westrum.

In 1968, another important event took place when the Ambrose Monell Foundation provided funds to create, in Philadelphia, a unique research and teaching center devoted exclusively to studies of the chemical senses. M. R. Kare was appointed as its Director and the late D. J. Moulton, who devoted his entire research career to the study of olfaction, contributed significantly towards its initial organisation.

The Monell Chemical Senses Center, originally a division of the University of Pennsylvania, was an exercise which proved to be highly successful. It has grown rapidly into a complex multidisciplinary institute in which major emphasis is placed on studies of the structure and behavioural function of chemical signals as well as on the physiology of their reception in mammals, including humans. Apart from M. R. Kare and D. J. Moulton, many others who figure prominently in the field of chemical communication became associated with the Monell Center - notably G. Epple, R. L. Doty, G. K. Beauchamp, G. Preti and A. B. Smith, III.

THE EXCITING '70s

While the main activities unfolded on the American continent, colleagues in Europe did not lag behind. In 1970, at the International Summer Course on Odour Perception, held in Utrecht, Netherlands, the European Chemoreception Research Organisation (ECRO) was inaugurated due to the efforts and support of M. Beets - a successful scientist and industrialist - and the enthusiasm of E. P. Köster. Although started as a European venture, ECRO opened its membership to colleagues from other parts of the world. In 1982, ECRO had over 400 members from more than 30 countries. Its function is to promote and support research and exchange information on all aspects of chemical senses. ECRO's activities include organising mini-symposia and congresses. Although it is oriented more towards physiology, there are also opportunities for those interested in chemical communication in animals to use its forum.

Thus, at the start of the '70s, students of various aspects of olfaction found themselves well organised into separate groups with fairly discrete, strongly defined mutual interests. The formation of these groups was aided not only by common professional backgrounds, but also by personal contacts and affinities. Those eager to solve the questions related to the behavioural role of olfaction in vertebrates no longer had to feel that they worked in isolation. Their numbers grew constantly. This was well demonstrated by frequent, smaller local and international meetings devoted specifically to chemical communication and by the appearance of special journals, newsletters and new books. The interest in olfactory communication widened beyond the laboratory species and those few which were the subjects of investigation initially.

There was always interest in the role of olfaction in man. In the past, physicians, psychiatrists, criminologists, anthropologists and others periodically touched upon this question, but their statements were not based on adequate systematic research data. This has been emphasised by McCartney (1968) in his useful publication, already mentioned above.

The advances in the studies of the effect of "primer pheromones" on oestrous cycles, cycles among women living in close proximity to one another. others, F. H. Bronson and W. K. Whitten, also inspired interest in olfaction in humans. The works of J. G. Vandenbergh and L. C. Drickamer added considerably to this.

The likelihood of the existance of human pheromones became the subject not only of more frequent speculation, but also of more systematic observation after Martha McClintock (1971) reported on the synchronisation of menstrual cycles among women living in close proximity to one another. Subsequently, specifically designed studies substantially strengthened the supposition that olfactory signals perceived subconsciously are responsible for this (Russell et al., 1980).

The studies of R. P. Michael and associates of the importance of the odour of vaginal discharge in communicating the reproductive states of rhesus monkeys, *Macaca mulatta*, also contributed towards the growth of interest in the possible role of olfaction in the behaviour of humans. W. Montagna was one of those who reinforced this trend through his studies of the histology of skin glands in primates, through his leadership at the Oregon Regional Primate Research Center and as an organiser of the Annual Symposia on Biology of the Skin. For example, during the 22nd Symposium of this series, a talk by myself and B. S. Goodrich as coauthor on the behavioural role of skin glands in animals, was meant to acquaint the participants, predominantly dermatologists, with a possible new function of the integument. Among other speakers, was F. J. Ebling who, through his own work and that of his associates on the hormonal control of skin glands, contributed significantly towards a better understanding of chemical communication.

Those concerned with the reproduction of animals were amongst the very first to pay attention to the importance of odour. No wonder reports on the effect of urine odour on the timing of oestrus and blocking of pregnancy in laboratory mice were inducive.

In 1972, during the 3rd International Symposium on Comparative Reproduction entitled "Environment and Reproduction in Mammals and Birds" held in Edinburgh, one session chaired by Hilda M. Bruce was devoted to olfaction. The decision to include this topic into the program was made under the influence and advice of A. S. Parkes, the noted British reproductive physiologist who, right from the very early stages, generated the awareness that olfaction was an important extrinsic factor influencing reproduction.

Information on the importance of the odour of the salivary steroid in the reproduction of domestic pigs, which J. P. Signoret effectively continued to disseminate, also provided a powerful stimulus among those with a practical approach to the study of reproduction in animals.

On the program of the 8th International Congress on Animal Reproduction and Artificial Insemination, which was held in Krakow, Poland in June 1976, there was a session entitled "Role of Pheromones in Reproduction" (Mykytowycz, 1976). Among the participants were colleagues representing Australia, Canada, France, India, U.K., Poland and U.S.A.

The first ever gathering of researchers interested exclusively in the behavioural role of olfaction in vertebrates, predominantly in mammals, took place in 1976 at Saratoga Springs, New York (see Muller-Schwarze and Mozell, 1977). This first Symposium on Chemical Signals in Vertebrates was attended by 140 participants. It objective was to bring together the physical and the biological scientists so that they might develop a greater understanding of one another's theoretical concepts and practical problems. Almost all laboratories involved in work on vertebrates were represented. This meeting gave a much needed impetus to an evolving interaction between researchers who were just beginning to approach chemical reception in higher animals, rather as integrated teams than as specialists, in separate disciplines working in parallel.

The Saratoga Springs Symposium provided a forum for the clarification of some newly emerging fundamental questions such as the suitability of the concept of "pheromones" in work with mammals, the question of learning in chemical communication, of diet in shaping odour profiles and the effects upon chemical communication of the multisensory context in which it must operate. Also, the question of the chemical complexity of odour signals received attention.

The Symposium in Saratoga Springs, as well as subsequent ones of this series held in Syracuse in 1979 and in Sarasota in 1982, were organised mainly by D. Muller-Schwarze in association with R. M. Silverstein. The organisation of this current symposium in Laramie, Wyoming in July 1985 has already involved colleagues of the younger generation, particularly David Duvall.

Physiologists, acting on the suggestion made by Lord Adrian referred to earlier, asked Wes Whitten to organise a symposium on olfaction during the International Physiological Congress held in Paris in July 1977. F. H. Bronson, M. Novotny, G. Raisman, J. P. Signoret and J. G. Vandenbergh participated as well as F. A. Beach, one of the very first contributors to the study of the role of smell in sex attraction. Chemical communication at last became "respectable" physiology.

Another important opportunity to take stock of the field of chemical communication occurred when the conference "Chemical Ecology: Odour Communication in Animals" took place at Noordwijkerhout, Netherlands in September 1978. It was sponsored by the 35 invited participants, 12 NATO and a number of non-NATO countries were represented.

F. J. Ritter of the Netherlands Organisation for Applied Scientific Research, was asked to organise an Advanced Research Institute to include higher animals as well as insects into the programme. Also on the organising committee were R. M. Silverstein, M. S. Blum, D. A. Evans and E. Priesner.

The conclusions and recommendations formulated by the separate panels set up during this meeting summarised the state of research in the field towards the end of the '70s (Ritter, 1979). Reference to some of the more important points is made elsewhere in this article.

In the same year as the NATO sponsored conference, a symposium entitled "Olfaction in Mammals" was held by the Zoological Society of London. Its objective, as with all other meetings, was to collect up-to-date information on progress in the field of chemical communication in vertebrates. This symposium was organised by D. M. Stoddart and the chairmen of the different sessions were E. S. Albone, R. D. Martin, R. A. Mugford and E. J. Neal. The proceedings edited by Michael Stoddart (1980) provide a valuable source of information to all interested in mammalian behaviour and biology.

In 1979, the Association for Chemoreception Sciences (AChemS) was formed in the U.S.A. through the enthusiastic efforts of Maxwell Mozell. Its aim is to provide an opportunity for closer interaction among the chemosensory community. Simultaneously, this association was meant to offer expert assistance whenever public or private bodies require consultation on broad questions related to chemosensory sciences.

Although the majority of members of AChemS are Americans, membership is open to colleagues from other countries.

Interest continued to spread world-wide. In the USSR it was initiated by V. E. Sokolov with the cooperation of E. P. Zinkevich and others. In 1979, the Zoological Society of Southern Africa held a symposium on "Animal Communication" in Capetown with an extensive session on chemical communication to which colleagues from other countries were invited.

Thus, towards the end of the '70s and at the beginning of the new decade, most conferences or meetings centered around the topic of the behaviour of mammals had to accommodate separate sessions on olfactory behaviour and communication.

PUBLICATIONS

The number of journals dealing with the behaviour of animals has multiplied rapidly in parallel with the growth of ethology. Most of them accept for publication contributions dealing with chemical communication in vertebrates. Naturally, the proceedings of all the above-mentioned symposia and conferences were helpful. However, with the steadily growing volume of research data, the need appeared to have special vehicles for the dissemination of information.

The newsletters of ECRO, initiated in 1972, and of AChemS, the first issue of which appeared in 1983, are very useful sources of information about current activities in the field of vertebrate olfaction, including chemical communication. A development of the utmost importance took place in 1973 when the first issue of "Chemoreception Abstracts" appeared for which E. P. Köster and P. MacLeod, two enthusiastic foundation members of ECRO, served as editorial advisers.

One year later, in 1974, "Chemical Senses and Flavour," which in 1978 changed its title to "Chemical Senses," was initiated. It was a useful journal for those interested in the physiological aspects of odour communication.

The closest to our interests, however, is the "Journal of Chemical Ecology" devoted specifically to promoting the study and understanding of the interactions of organsims with their physical and social environments through the mediation of chemical signals. The first issue of this journal appeared in 1975.

For this achievement, we are indebted to R. M. Silverstein and J. B. Simeone who have already served for a decade as the chief editors. On the original editorial board were 32 colleagues from 8 different countries. Initially, only four issues of the Journal of Chemical Ecology appeared each year; in 1984, the journal was published monthly.

This expansion of the "JCE" however, is not due to an increase in the number of submissions dealing with vertebrates. Unfortunately, the lack of significant chemical results in the studies of vertebrate odours inhibits authors from publishing their reports in a journal with a chemical connotation. No doubt, however, with a deeper involvement of chemists in the future, this attitude will change. It is comforting to have the "JCE" for it signifies our field of interest is mature, independent and self supporting.

In 1983, the policy of the "JCE" was further strengthened through the formation of the International Society of Chemical Ecology whose first president was Lincoln P. Brower. The "JCE" became the official journal of this Society.

BOOKS

Twenty years ago, when somebody wished to find some general information on the behavioural role of olfaction, the obvious and only source was R. W. Moncrieff's volume "The Chemical Senses" (1951) which, as mentioned earlier, contained mainly references to anecdotal, speculative earlier observations.

Much more detailed, although still on the whole based on the same speculative information, was McCartney's "Olfaction and Odours" (1968) which appeared at a time of heightened interest in this field. It contains over 1000 valuable references to early publications written in different languages.

In 1974, a volume edited by M. C. Birch entitled "Pheromones" appeared. The need for such a volume was explained by the publishers and editors in the following words: "...Unifying concepts have led to a situation in which the sharp boundaries between the various classical biological disciplines are rapidly disappearing...Thus while scientists are becoming increasingly specialised in their techniques, to an increasing extent they need an intellectual and conceptual approach on a wide and non-specialized basis... The study of pheromones demands an involvement with the behaviour of animals but that involvement has usually been kept to a minimum...."

The book is concerned with the biology of pheromones, avoiding techniques, chemistry and the cataloguing of identifications, all of which are indispensable, but tend to crowd out the biology. The text contains information on a wide range of animal species. The biology of vertebrates is well covered in it by separate articles.

H. H. Shorey's volume "Animal Communication by Pheromones" (1976) is claimed to be the first single-authored monograph to deal with pheromones. It minimises attention to selected species, discussing the same form of behaviour in less and more developed animals. It is a fairly condensed text containing over 700 useful references.

"Mammalian Olfaction, Reproductive Processes and Behavior" (1976) which was edited by one of the wave of younger researchers, R. L. Doty, deals specifically with mammals, as the title shows. It contains up-to-date information on the role of smell in the reproductive behaviour and endocrinology of mammals, and in addition, accommodates chapters on the anatomy, physiology and development of the nasal chemosensory pathways and an important critique of the pheromone concept in mammalian chemical communication by J. R. Beauchamp, R. L. Doty, D. G. Moulton and R. A. Mugford. A review of reproductive endocrine influences on human nasal chemoreception emphasizes the pressing need for more intensive critical investigation of the behavioural role of olfaction in humans, a line of research which, in fact, Doty subsequently followed up.

While most of the volumes mentioned above were still oriented strongly towards insect research and Doty's book dealt with only one aspect of the behavioural function of olfaction in mammals, two books by D. M. Stoddart, namely, "Olfaction in Mammals" being the edited proceedings of the Symposium of the Zoological Society of London and "The Ecology of Vertebrate Olfaction," both published in 1980, focused exclusively on mammals. Taken together, the two books provide a comprehensive picture of all aspects of chemical communication in laboratory, domestic and wild mammals.

In 1983, J. G. Vandenbergh published "Pheromones and Reproduction in Mammals" in which up-to-date research findings on the role of chemical communication in mammals are reviewed and the historical and theoretical perspectives are given. Among other topics, the ecological consequences of chemical communication are discussed.

Both the biologist's and the chemist's views concerning developments in the field of olfactory communication are expressed in the volume "Mammalian Semiochemistry: The Investigation of Chemical Signals between Mammals," published in 1983 by E. S. Albone. This is a very useful, long awaited source of information especially for biologists engaged in the investigation of olfactory communication, since it helps them to better understand the methods available to chemists in their studies of the composition of odour signals and the immense problems which chemists have to face.

CONCLUSIONS

A quarter of a century or so after the first steps were taken, the study of chemical communication in vertebrates has reached the stage where the new discipline, with all its separate facets can be clearly seen. Now a newcomer into the field does not need to search frantically for the necessary background information and directives.

This progress could not have been achieved but for the enthusiasm and dedication, in some instances even sacrifice, of a few individuals, most of whom have been mentioned already throughout the text of this article.

There is no doubt that the contributions of those who, for a couple of decades or so, stubbornly stuck to the field were most important. Thus, the names of many authors continued to reappear in the literature throughout the whole of this period over new reports which not only consolidated the basis, but also widened the horizon, of the field of chemical communication in vertebrates. Good examples are D. D. Thiessen, whose study subject was the Mongolian gerbil, Meriones unguiculatus, Bernice M. Wenzel, who practically in isolation pursued the question of the behavioural role of olfaction in birds, or G. M. Burghardt who sustained studies of chemical communication in reptiles. Wes Whitten probably stayed in the field longer than anybody else. Unfortunately, one cannot refer to many others since there is the danger that some just as notable may be omitted.

But one must mention once more those who, through the organisation of symposia and meetings, helped to develop the continuous interest and to maintain close personal contacts among researchers studying the chemical communication of vertebrates. The group of colleagues based in Syracuse headed by R. M. Silverstein and D. Muller-Schwarze particularly should be singled out. But for their efforts we would not be here today.

As emphasised repeatedly above, a lot has been achieved, but even more still has to be done. To conclude this review, it is appropriate to quote at least some recommendations which were formulated at the conclusion of the NATO sponsored meeting in 1978. Although a few years have elapsed, not much has changed in the meanwhile.

With regard to odour communication in mammals, the panel consisting of Gisela Epple, E. B. Keverne, D. Muller-Schwarze and R. Mykytowycz with F. H. Bronson as chairman put forward the following recommendations amongst others:

1. A much broader taxonomic basis is needed and a variety of wild animals should be used.
2. Much better behavioural assessments are needed emphasising the full behavioural involvement.
3. To determine what olfactory cues mean to mammals, we have to study the integration of these cues and others operating at the same time.
4. Chemical isolation techniques should be used more frequently to produce relevant cues with which one can study neural processing. Instead of the rather irrelevant chemicals commonly used in olfactory bulb physiology, one could utilize the cues that the animal actually uses.
5. There is a need for field experiments to complement general field observations and laboratory experiments on olfaction in mammals. Field experiments could include studying olfaction within the wider framework of ecological variation and adaptation.
6. There is a need to exploit the possibility of using olfaction in the regulation of the behaviour of domestic animals; to investigate the possibility of using primers for population regulation; to investigate the possibility of using odours for pest control.
7. Greater emphasis should be placed on interdisciplinary cooperation.
8. There is a particular need to involve people trained in chemical ecology.
9. There is a need for training people to acquire techniques and for exchange programs.

With all the up-to-date experience and the recommendations quoted above, we look forward to a bright future. It may be hoped that after another quarter of a century, we will be ready to talk to the animals using their own silent language and perhaps also better understand our own emotions which so frequently seem to be regulated by odours.

REFERENCES

Albone, E. S., 1984, "Mammalian Semiochemistry," John Wiley and Sons Ltd., Chichester, New York, Brisbane, Toronto, Singapore.

Birch, M. D., ed., 1974, "Pheromones," Am. Elsevier, New York.

Doty, R., ed., 1976, "Mammalian Olfaction, Reproductive Processes, and Behavior," Academic Press, New York, San Francisco, London.

Johnston, J. W., Moulton, D. G., and Turk, A., eds., 1970, "Advances in Chemoreception," Appleton-Century-Crofts, New York.

Kingston, B. H., 1965, The chemistry and olfactory properties of musk, civet, and castoreum, Int. Proceed. Sec. Int. Cong. Endocrinol., 209.

McCartney, W., 1968, "Olfaction and Odours," Springer-Verlag, Berlin, Heidelberg, New York.

McClintock, M., 1971, Menstrual synchrony and suppression, Nature (London) 229:244.

Moncrieff, R. W., 1951, "The Chemical Senses," 2nd edition. Leonard Hill, London.

Muller-Schwarze, D. and Mozell, M. M., eds., 1977, "Chemical Signals in Vertebrates," Plenum Press, New York and London.

Mykytowycz, R., 1970, The role of skin glands in mammalian communication, in: "Advances in Chemoreception, Vol. 1, Communication by Chemical Signals," J. W. Johnston, D. G. Moulton, and A. Turk, eds., Appleton-Century-Crofts, New York.

Mykytowycz, R., 1976, Role of pheromones in reproduction, in: "Proceedings VIIIth Intern. Congress on Animal Reprod. Artif. Insemin. Krakow, Vol. V, Final report.

Patterson, R. L. S., 1969, Boar taint: Its chemical nature and estimation, in: "Meat Production From Entire Male Animals," D. H. Rhodes, ed., J. and A. Churchill Ltd., London.

Ritter, F. J., ed., 1979, "Chemical Ecology: Odour Communication in Animals, Scientific Aspects, Practical Uses and Economic Prospects," Proc. Adv. Res. Inst. Chem. Ecol., Noordwijkerhout, Elsevier/North Holland Biomed. Press, Amsterdam, New York, Oxford.

Russell, M. J., Switz, M. G., and Thompson, K., 1980, Olfactory influences on the human menstrual cycle, Pharmacol. Biochem. and Behav., 13:737.

Shorey, H. H., 1976, "Animal Communication by Pheromones," Academic Press, New York, San Francisco, London.

Stoddart, D. M., ed., 1980, "Olfaction in Mammals," Proc. Symp. Zool. Soc., London.

Stoddart, D. M., 1980, "The Ecology of Vertebrate Olfaction," Chapman and Hall, London and New York.

Vandenbergh, J. G., 1983, "Pheromones and Reproduction in Mammals," Academic Press, New York, London, Paris, San Diego, San Francisco, Sao Paulo, Sydney, Tokyo, Toronto.

Wilson, E. O., 1963, Pheromones, Sci. Am., 208:100.

Zotterman, Y., ed., 1963, "Olfaction and Taste," Proc. 1st Intern. Symp., Stockholm, Pergamon Press, Oxford, London, New York, Paris.

THE TETRAPOD EPIDERMIS: A SYSTEM PROTOADAPTED AS A SEMIOCHEMICAL SOURCE

Paul F. A. Maderson

Biology Department
Brooklyn College of C.U.N.Y.
Brooklyn, NY 11210

INTRODUCTION

It is one thing to emphasize to an undergraduate class the importance of "communication" between organisms; it is quite another to comprehend how the organisms perceive and use the signal(s). Of cetacean vocalization it has been written: "This system is something we terrestrial mammals cannot imagine and for which we have no empathy," (Bateson, 1966). Researchers who pursue mechanisms of communication which involve vibrational or electromagnetic energy have long had access to technologies which could record and measure the signal(s) objectively. Until relatively recently no such tools were available for studies of chemical communication, and experiments relied largely on observation and bioassays. However, advances are being made at the molecular level (Lancet, 1984), and these have arrived at a most opportune time. The relative lack of objective measurement produced some extraordinary examples of circular reasoning. General texts often give the implicit or explicit impression that among vertebrates only mammals really use olfaction. However, the relatively poor performance of *Homo sapiens* in this context distorts our conceptualization of other classes. A brief review of available data on chemoreception in birds was summarized: "The best conclusion from all this is probably that the sense of smell in birds is comparable to that of man. It is present, but in most species is little used." (Yapp, 1970; cf. Wenzel, this symposium).

In his classic, "The Origin of Terrestrial Vertebrates," Schmalhausen (1968) devotes some attention to modifications of the nasal apparatus in early tetrapods, but offers no discussion of its biological role. "The Colonisation of Land," (Little, 1983) deals with all terrestrial animals, but scarcely mentions the evolution of chemoreceptor mechanisms.

Has our antiseptic (though highly-perfumed), 20th century culture, aided and abetted by Nature's having reduced the relative importance of olfaction in primates, precluded our ability to consider the importance of chemical communication in vertebrate evolution? Perhaps there are other factors which are important.

As discussed in the context of the history of study of developmental processes in reptiles (Billet et al., 1985), for the first two thirds of the 20th century, lower tetrapods were in a curious situation. The emergence of the diverse disciplines of contemporary zoology, coincident with the "filling in of the details of vertebrate evolution" (mainly by palaeontologists), forced many investigators into a typological mode of thinking. Once it was realized that all extant amphibians and reptiles represent the remnants of great Palaezoic and Mesozoic radiations, "comparative" disciplines concentrated primarily on "representative species."

Not until the 1960s did data emerging from the new discipline of physiological ecology reveal that amphibians and reptiles were not insensate brutes operating but sluggishly at the mercy of ambient temperature, a role cast for them by the vestiges of the 19th century "progressive evolutionary," i.e., mammal-directed, thinking. As their complex behaviors were studied and recorded in detail it became necessary to reassess their standing as highly specialized organisms in their own right (Greenberg, 1977). Furthermore, the possibility arose that they could serve as models for furthering our understanding of the patterns of evolutionary change which produced birds and mammals. Going beyond mere description of skeletal structure, as recorded in the fossil record, produced "palaeo-disciplines" which, while inherently untestable and therefore demanding great caution, force us into interdisciplinary thinking, an exercise valuable in and of itself. Such approaches have been applied to the integument _in toto_ (Maderson, 1972a,b) and to the evolution of pheromonal communication in synapsid reptiles (Duvall et al., 1983; Graves and Duvall, 1983a).

My studies of the skin of modern reptiles have involved, among other things, consideration of epidermal specializations which appear to be holocrine glands (see below). Those studies led me to consider the structure and biological role (Bock and von Wahlert, 1965) of apparently similar specializations in other tetrapods. When I was invited to participate in this symposium, I realized that a general question had been emerging in my mind for some years: Is there some common denominator that can explain the _enormous_ diversity of holocrine epidermal glands in tetrapods? Previous correspondence with a number of workers engaged in studies of chemical communication in reptiles had suggested the probable primacy of olfaction as a sensory modality in tetrapod evolution (Duvall et al., 1983; Maderson and Chiu, 1985). Widening the question to the context of the evolution of the sources of signals has proved an exciting challenge.

SOME DEFINITIONS

Interdisciplinary communication is often inhibited by semantic misunderstanding. Given my position as a skin biologist who has never _specifically_ undertaken a research project in the field of chemical communication, it behooves me to use _a_ terminology derived from the appropriate literature. I feel that it is preferable to use the term _semiochemical_ as defined in Albone (1984). My caution in this matter is predicated by two major factors. First, for none of the "skin glands" in amphibians and reptiles to which I shall refer is there a sufficient body of data to state categorically that _this_ signal produced by _this_ source has _this_ significance (Albone, 1984: fig. 2.1 and accompanying text). Second, some anatomists might question whether many of the reptilian structures should be properly termed "glands." This caveat is so fundamental that it deserves further explication.

Basic texts in histology, indeed even reviews aimed at specialists (Kurosumi et al., 1984), imply that an integumentary gland is an acinar structure produced by invagination of epidermal tissues into the dermis. From the gland, through a discreet orifice, there emerges a fluid secretion, usually with an obviously "oily" (implying lipidaceous), but often a serous, "watery," texture. While the topographic location of such glands is often specified e.g., axillary, inguinal etc., such is rarely necessary for the _definition_ of the structure as a semiochemical source. Such anatomical definitions, while based primarily on mammalian examples, describe the uropygial (preen) gland of birds (Jacob and Ziswiler, 1982) and the various multicellular derivatives of the amphibian epidermis (see Madison, 1977), but there are few similar examples in reptiles. I define structures in reptiles as "glands," and infer a possible semiochemical source, if they meet two or more of the following criteria: (1) any region on the body surface, stomodeum, or proctodeum where the cytological structure obviously differs from the surrounding tissue, (2) structures that occur with sufficient regularity in terms of number and/or topographic location (allowing for sexual dimorphism and/or circannual changes) as to be useful as taxonomic characters, (3) any integumentary specialization, whose structure does not suggest a sensory function, (4) any structure that gives rise to mature cells or materials in a form which might be termed a "secretion," perhaps having an odor detectable to the human olfactory sense. Discussion of inferable semiochemical sources based on observable behavioral interactions will appear later.

Finally, I use the term "protoadaptation" instead of "preadaptation" to avoid the teleological overtones of the latter. The principle of "excessive construction" proposes that systems or organisms are often capable of exhibiting biological roles which, while not the subject of selection in one environment, acquire significance in a changed environment (Gans, 1979). In the context of integumentary evolution, this principle emphasizes that fundamental metabolic activities of keratinocytes may be readily modulated (Flaxman and Maderson, 1976), and makes it easier to understand the multiple parallel and/or convergent trends seen in distantly related taxa. Possible scenarios concerning the way in which developmental pathways could be modulated so as to achieve gradualistic change in evolution (Maderson, 1981), including discussions of "neomorphs," appear elsewhere (Maderson, 1972a,b, 1975, 1983).

THE ANIMALS

For any animal, from protistan through vertebrate grades of organization, the non-living, external environment is a potential source of vibrational, electromagnetic, or chemical signals. Students of chemical communication seem to agree that this sensory mode must be the most ancient in the context of inter-individual communication, having its origins at the cell-cell level. Freeman (1978) mentions that J. B. S. Haldane alluded to the antiquity of chemical communication in eukaryote evolution over 20 years ago! Research over the past 25 years in developmental biology has revealed much about cell-cell communication (Yamada, 1983) and the techniques which have been developed in that context are now being used by researchers in olfaction (Lancet, 1984). For most animals it seems parsimonious to suggest that most semiochemicals would originate from the skin or the gastrointestinal tract, as these two regions of the body have essentially direct communication with the external environment. Available data concerning fish (Hara, 1982; see also this symposium) confirm these predictions for piscine vertebrates.

Why are semiochemical sources in terrestrial vertebrates mainly integumentary? In any taxon, the transition from an aquatic to a terrestrial environment demands the most drastic changes in the greatest number of organ systems (Little, 1983). For the integument two requirements must be met. First, whereas aquatic life offers little danger of mechanical damage which would breach the barrier between the milieu interieur and the physical environment, whenever a terrestrial animal moves about, such damage assumes primary import. Thus, terrestrial vertebrates show epidermal cornification, a feature seen but rarely in fish (Mittal and Banerjee, 1980). Second, while fish have "osmotic problems," few risk danger from immediate desiccation as do tetrapods, so that some sort of "water-proofing" system is necessary.

The molecular basis for water-proofing the tetrapod integument has only recently been clarified. It was long-supposed that keratins were responsible for both mechanical strength and water-proofing. Over the past decade, research has revealed that the cornified tissues play the greater role of mechanical protection, and simultaneously provide a framework for housing non-polar lipids which are <u>the</u> water-proofing agents (Lillywhite and Maderson, 1982; Maderson, 1984; Hadley, 1985). A broad view of the end-result of the invasion of the terrestrial environment would be as follows. Patterns of epidermal cell differentiation in our rhipidistian ancestors were modulated to form an outer, dead, corneous tissue; such changes might have begun in shallow waters (Schmalhausen, 1968) where mechanical damage impinged from the substrate and/or aquatic plants. Selection would have favored regular epithelial replacement controlled by intrinsic factors to cope with the probabilistic nature of the occurrence of mechanical trauma. There is indirect evidence for the presence of corneous tissues in fossil amphibia (Maderson, 1972a; Lillywhite and Maderson, 1982), but understanding the origin of waterproofing raises problems.

The amniote epidermis is characterized by the presence of extracellular lipids. These originate from membrane-bound organelles which appear in the keratinocytes as part of the normal differentiative process (see refs. in Maderson, 1984, 1985). Such structures are absent from the epidermis of modern amphibians, most species of which show but a single layer of cornified cells, and which are subject to desiccation. Some xeric-adapted species show (1) more extensive cornification and (2) behavioral repertoires where waxes deriving from specialized epidermal glands are spread over the skin surface (see refs. in Hadley, 1985). As it is widely accepted that the integument in modern amphibians is secondarily modified in association with cutaneous respiration, should we assume that the absence of intracellular liposomes is a derived character? This question cannot be answered directly. In the present context it suffices that data exist which imply that a cornified epidermis with cells with <u>some</u> capacity for the processing of water-proofing molecules, <u>could</u> have characterized the earliest tetrapods. It seems reasonable to assume that such a system was well-established by the time the amniotes arose. I will now discuss "typical" modern lizards as possible models for understanding the origin of epidermal semiochemical sources.

THE EARLIEST TERRESTRIAL ANIMALS IN THEIR TERRESTRIAL ENVIRONMENT

New data continually appear which imply earlier and earlier origins of "reptiles" (Carroll, 1982). Reptiles are amniotes possessing the characteristic embryonic membranes which are also found in mammals and birds. As data accumulate emphasizing the convergent evolution of "reptilian" osteological features in anthracosaurian amphibians, we may speculate that "amniote" anatomical and physiological features probably began to evolve in taxa which are <u>osteologically</u> amphibians!

The first "true" terrestrial vertebrates, whether <u>taxonomically</u> amphibians or reptiles, were definitely <u>not</u> lizards. However, there is good reason to imagine that they behaved like lizards (Maderson, 1972a: 164). Data are available that suggest that a superficial corneous layer of the epidermis, wherein the cells synthesized β-keratin, was an adaptation protecting the living cells from the deleterious effects of ultraviolet radiation (Maderson et al., 1984, and in preparation).

The flora of the first "upland" environment comprised primitive pteridophytes growing from between lichen-covered rocks (Edwards, 1980). Thus, our terrestrial ancestors probably scurried about on substrates physically similar to many contemporary microenvironments. Our knowledge of their body form and habitat implies that the preaxial and ventral aspects of the sprawled limbs, and the ventral body surface, would contact a hard substrate so that corneous materials would be "scraped off." A trail would result through what I will term uncontrolled semiochemical release. This scenario might imply that such a trail would attract predators and would therefore be selectively disadvantageous, but such an interpretation advertises our "mammalocentric" view of predator-prey relationships. Apart from the fact that all the animals living in this environment were leaving such trails, palaeoecological studies pioneered by Olson (1975) suggest that "terrestrial" food chains did not become established until the early Mesozoic.

Hard data and common sense seem to exclude reception of vibrational and electromagnetic energy as significant modes of intraspecific communication in the early days of terrestrial existence (Maderson and Chiu, 1985). Armed with an accumulated body of data showing (1) the ability of piscine vertebrates to use semiochemicals and their possession of sophisticated receptor-integrative neuronal pathways, and (2) the complex behaviors of modern reptiles, some of which are mediated by semiochemicals, permit the following proposition. The abraded regions of epidermis were protoadaptations for the discreet "glands" so characteristic of amniotes. In "typical lizards" such structures usually occur on the ventral integument. In serpentiform lizards and snakes, most semiochemical sources tend to be housed in the cloaca (see refs. in Price and LaPointe, 1981). This derived character state reflects the fact that the fossorial habits in such lineages preclude glands remaining on the body surface.

From our knowledge of mammalian behavior, it seems that sniffing of cloacal regions, urine, and fecal pellets is uncommon in reptiles. This may be a false impression based on inadequate data (see review, Madison, 1977; Crews and Greenberg, 1981; Fergusson et al., 1985), but if it proves to be a valid observation with, as one might expect, exceptions, an explanation is available. As semiochemicals emanating from the GI tract in fish are soon disseminated into an external environment of essentially infinite volume, such release probably does not attract predators. Non-controlled voiding of feces and urine in a terrestrial environment would have been relatively harmless during the earliest millenia of terrestrial life when such would merely have reinforced the integumentary trail. However, with the acquisition of more sophisticated, discriminatory, olfactory behavior during the initial reptilian radiations, such activities would have become selectively disadvantageous, at least in some lineages.

I do not wish to imply that the existence of GI semiochemicals and complex behavioral patterns pertaining to their deposition in mammals is necessarily unique in an evolutionary sense. Dr. Crews (pers. comm.) has suggested that in this, and many other contexts of conspecific communication, mammals merely "reinvented the wheel." There are data which support this contention. Grassman et al. (1984) have documented olfactory imprinting in marine turtles, a fact which suggests that integumentary glands in these forms (see most recently, Solomon, 1984) could be semiochemical sources in addition to materials emanating from the GI tract. Turtles, the most ancient of extant reptilian lineages, have, in their secondary marine habitats, reassumed the olfactory habits of their piscine ancestors. Gunderson (1976: 55) refers to data from Russian studies on cetaceans suggesting that these secondarily aquatic tetrapods have reverted to similar habits. A more striking pair of examples of reappearances of "old sensory habits" could scarcely be imagined.

The preceding leads to a consideration of what I will term *facultative, controlled semiochemical release*. Any vertebrate is a possible *source* of vibrational, electromagnetic, and semiochemical signals. Having been the source of either of the first two types of signal, the responsible individual can then stop vocalizing and/or moving, or hide or rely on camouflage. Whichever signals are given out, they are soon dissipated, and there are strategies for avoiding further transmission and the possible, resultant, unneeded attention. In this respect, semiochemical communication is a two-edged sword. It may be much more specific and therefore of greater use at fairly long distances than either audition or vision, but by the same token, the signal is longer-lived. If, as I assume, integumentary trails were initially selected for to facilitate intraspecific communication, it is reasonable to conclude, that by the time terrestrial predation evolved, uncontrolled semiochemical release would have become selectively disadvantageous. The solution to this problem would have been the isolation of the semiochemical source on the body surface in such a fashion as to permit its facultative use. The numerous potential strategies which could effect this all have the common characteristic of sequestering the germinal portion of the integumentary gland "below" the body surface. Such is readily accomplished by simple, progressive, gradualistic invagination. Sequestering the *germinal* component implies that daughter cells can mature in a different environment than typical surface keratinocytes. "Secretion(s)" can now be voided to the external environment in various ways. The organism can actively apply the appropriate body segment to a chosen substrate (*mechanical expression*), or contract myoepithelial cells surrounding the germinal and duct portions of the gland (*physiological expression*). Within the operational confines of these definitions, we can explore the *form* of the secretion.

The "simplest" type of secreted material would result from desquamated, loosely-joined plugs of keratinous materials: such characterize the preanal organs of lizards (Chiu and Maderson, 1975; Chiu et al., 1975; Cole, 1966; Maderson, 1968a, 1970, 1972c, 1985; Maderson and Chiu, 1970, 1981; Menchel and Maderson, 1975), some chelonian glands (Waagen, 1972; Winokur and Legler, 1975) and the plantar glands of certain mustelid carnivores (Buskirk et al., this symposium). None involve major changes in the fundamental differentiative pathways of the keratinocytes. Although such epidermal specializations warrant further study of nearly every type -- all we have are data on their morphology and some inferences from behavior -- these secretions have to be physically expressed.

Sequestering maturing cells within an invagination permits extreme modulation of basic keratinocyte metabolism. Such glands may exude their "oily" or "watery" secretions in every possible way from "weeping" to "squirting," but what is of extreme interest is their phylogenetic distribution. Physiologically controlled glands are found in modern amphibians (Madison, 1977), the Desert Tortoise *Gopherus* (see refs. in Winokur and Legler, 1975), various chelonian musk glands (Waagen, 1972), glands in marine turtles (Solomon, 1984), a few non-serpentiform lepidosaurs (Burkholder and Tanner, 1974), and crocodilians (Wright and Moffat, 1985). They also include the avian uropygial gland (Jacob and Ziswiler, 1982) and the extensive variety of glands all over the skin of mammals.

This brief review of tetrapod integumentary glands permits two conclusions. First, it supports a previous generalization -- one should be cautious in trying to establish homologies in vertebrate integumentary specializations (Maderson, 1972a). Second, the data imply an enormous diversity of semiochemicals, but we are appallingly ignorant of their chemical nature, especially in reptiles.

THE INTEGUMENT AS A SOURCE OF SEMIOCHEMICALS WHERE "GLANDULAR" SEPCIALIZATIONS ARE LACKING

The last sentence of the preceding section pertains to two nagging problems. First, in lizards, two types of gland have been described. The "generation gland" (Maderson, 1967, 1968a, b, 1970, 1972c, 1985; Maderson and Chiu, 1970, 1981; Maderson et al., 1977, 1979) produces its secretion in association with periodic skin-shedding. The "preanal organ" (see refs. in preceding section) shows cellular activities quite independent of the unique, pan-body, synchronized proliferative and differentiative activities which characterize the lepidosaurian epidermis. Our knowledge of the hormonal mechanisms controlling the activity of these structures is reviewed elsewhere (Maderson, 1984). The secretions of generation glands lie on the ventral body surface where they must be abraded during normal locomotion (Maderson et al., 1977). Logically, all species which have such structures should shed their skins quite frequently to expose new secretory material: however, many species in many families inhabiting many different environments, shed infrequently (Alexander and Maderson, 1972). Second, beginning with certain studies by Noble and his colleagues, there is nearly 50 year's worth of data testifying that snakes (see review, Halpern and Kubie, 1984) and lizards (Cooper and Vitt, 1984a, b; Duvall, 1982) do interact by touching the conspecifics' skin with the tongue and then placing the tongue tip close to, if not in, the buccal opening of the Jacobson's organ. The problem is that few snakes have integumentary glands away from the cloaca, which does not seem to be dramatically involved in these behavioral repertoires. Studies of lizards do not indicate specific attention being paid to typical gland-bearing body regions, although there is some evidence that such regions are facultatively applied to substrate surfaces.

Two further complications of the situation with lizards warrant comment. The Gekkonidae (1) have a multitude of skin glands and a well-developed Jacobson's organ, (2) have acute vision, perhaps reflecting their primarily nocturnal habit (they are neither conspicuously colored nor do they have elaborate postural displays), and (3) their acute hearing is matched with conspicuous vocal tendencies (Gans and Maderson, 1973). Thus at least one of the 19 accepted lacertilian families seems to have pursued a "multimodality sensory trip" throughout its evolution! By contrast, the Iguanidae, the other large family for which a considerable body of behavioral data is available, have, in their North American representatives (e.g., *Dipsosaurus*, *Uma*, *Uta*), conspicuous visual displays and exceptionally well-developed preanal organs which appear to be active during the breeding season (Cole, 1966). The numerous South American genera, for which fewer behavioral data are available, do not seem to show such elaborate posturing as their North American cousins, nor are preanal organs so widespread but, these are some of the lizards with "generation glands" which shed infrequently!

The integument of all extant amniotes, and probably that of their ancestors, is/was to some degree, an uncontrolled semiochemical source. This is because the organ has its own flora and fauna which produce semiochemicals as metabolic byproducts. While this phenomenon has been most widely-studied in mammals (Albone, 1984), the role of the pelage in the housing of such a flora and fauna suggests that birds should be studied in this context. The hinge regions of the scaled integuments of both extinct and extant reptiles, and their anthracosaurian ancestors (Maderson, 1972a), could similarly house microorganisms. Furthermore, the characteristic Oberhäutchen of the superficial surface of the squamate β-layer is another protoadapted site (Smith et al., 1984) for the retention of biodegradable molecules (see below). Whether such putative semiochemicals in extant reptiles and birds have demonstrable biological roles awaits investigation.

Even in the absence of microorganisms, epidermal materials lost from the body could be an uncontrollable source of semiochemicals (see below). What of the intact epidermis, i.e., that population of living and dead materials whose primary biological role is mechanical and physiological protection? All modern amniotes show behaviors where the oro-nasal region of one individual is applied to the integument of a conspecific. If we interpret such behaviors as "seeking" a semiochemical signal, the question arises, does "the seeker" get the signal from the extreme outer surface of the epidermis or from deeper cell populations, perhaps including the dermal tissues?

When corneous materials are lost from the body surface there remain remnants of intercellular cement. Ultrastructural studies of squamate skin-shedding (see refs. in Maderson, 1985), clearly reveal the association of these materials with the Oberhäutchen (Smith et al., 1984). While such could be semiochemicals, their efficacy would diminish rapidly due to abrasion during environmental contact. However, between closely overlapping scales e.g., in snakes, one could predict a longer functional half-life.

With regard to the deeper epidermal tissues, it must be realized that no known corneous tissue is 100% waterproof (Lillywhite and Maderson, 1982). Thus, in principle, many molecule(s) within or beneath such tissues could "seep" through to the skin surface, but several problems exist. First, amniote corneous tissues contain intra- and extracellular lipids which affect the waterproofing. Thus lipidaceous semiochemicals could seep through only if the waterproofing lipids themselves function in this context, i.e., they would assume a secondary biological role. However, if this were the case, the efficacy of the primary barrier role would steadily diminish necessitating regular replacement of the dead tissues. Second, if the semiochemical is sequestered within mature corneous tissues, its release would be passive. In either or both situations, as in the case of putative surface semiochemicals, the "strength" of the signal would wax and wane. The corollary of such models is that either signal recognition would need to be "all or nothing" and/or epidermal replacement would have to be sufficiently frequent to keep the signal above a minimum strength. Selection could produce replacement patterns of appropriate frequences as long as the water economy of the organism were above a minimum strength. Selection could produce replacement patterns of appropriate frequences as long as the water economy of the organism were accommodated (Maderson, 1984).

The work being done by Dr. Duvall and his colleagues on crotalid denning, warrants comment on skin development. An ultrastructural study of integumentary embryogenesis in *Lacerta muralis* revealed features not seen previously in adult material (Dhouailly and Maderson, 1984). The authors argued that if future research confirms these features to be unique to the embryo, reflecting biochemical differences in primary epidermal histogenesis, different biological roles may characterize neonate skin-shedding. These suggestions are verified by data in Duvall et al. (1985) and Graves et al. (this volume). The first, embryonic epidermal generation has inferior properties with regard to inhibiting cutaneous water loss. The first post-natal shed occurs approximately 10 days after birth. The new epidermis has improved properties as a barrier against cutaneous water loss confirming Dhouailly and Maderson's (1984) predictions. However, of even greater interest are Graves and Duvall's (1983b) data implying that shed skins from neonates contain lipid semiochemicals which facilitate their finding their way to winter denning sites (Duvall et al., 1985 and Graves et al., this symposium). Primary, embryonic epidermal differentiation in all amniotes could be a mechanism for establishing semiochemical communication within a population. The data from Dhouailly and Maderson

(1984) and Duvall et al. (1985) imply possible semiochemical differences between neonate and adult epidermal tissues. Coopersmith and Leon (1984) have shown that mother rats produce specific chemical signals in their post-parturative period which facilitates their recognition by their offspring; study of the reverse possibility i.e., the neonates producing recognition factors could yield interesting results.

Verification of possible correlations between "shedding behavior" (sensu lato) and intraspecific semiochemical communcation in a variety of behavioral contexts, necessitates consideration of the vicissitudes of the organism's water economy in its natural environment. Such correlative data are beginning to become available and data like those discussed above show the intricate subtleties of the interactions between diverse biological roles.

Can the skin act as a route to the external environment for semiochemicals not of integumentary origin? I am prepared to admit responsibility for drawing to the attention of Drs. Garstka and Crews certain data on reptile skin which led to those authors' (1981) suggestion regarding the mode of release of semiochemicals in *Thamnophis sirtalis parietalis*. My own studies on comparative integumentary morphology continue to provide more and more examples of "bizarre" modifications. Rosenberg and Russell (1980) described a series of caudal glands in the gekkonid *Diplodactylus*. Material is "squirted" from these glands when the animal is stressed, an action apparently effected by myoepithelial cells. The secretion has been characterized as a mixture of polypeptides which seem to be species-specific (Rosenberg et al., 1984).

To any who doubt the plausibility of Garstka and Crews' (1981) interpretation of the mode of delivery of a semiochemical, I offer these comments. First, there is no need to assume that the mechanism is either unique and/or exclusive of any other delivery system. Second, the study of semiochemical communication in non-mammals still being in its infancy, it is the more apparent that few subjects demand more interdisciplinary consideration. As a skin biologist I emphasize that a preoccupation with traditional definitions of glandular structures, with inferences as to their mode of action, can be extremely misleading -- even when mammals are being studied (Buskirk et al., this symposium).

One aspect of the Garstka and Crews' (1981) model has implications for alternative modes of delivery of semiochemicals. That a squamate reptile can and does "damage its own skin" without serious consequences to its water balance is supported by the fact that I have observed countless "minor traumas" in various stages of repair in field collected material (Maderson, unpublished data). The *Thamnophis* model offers the only plausible explanation for the evolution of the nucho-dorsal glands in other colubrids (Madison, 1977). Several extensive French studies of lepidosaurian skin morphology refer to "pockets" of vacuolated cells in hinge regions (see refs. in Madison, 1977). These cells probably contain lipids (Duvall, et al., 1985; Graves and Duvall, 1983b; Graves et al., this volume). There are therefore no less than three, quite different, potential sites for lipid deposition and/or, passage through, the squamate epidermis: (1) those sequestered in the "waterproofing mesos layer," (see refs. in Maderson, 1984, 1985), (2) those "squeezed out" of the dermal tissues (Garstka and Crews, 1981), and (3) the pockets of cells in the hinge regions. The latter are of great interest. The proliferative kinetics of the squamate epidermis are still poorly understood (Maderson, 1985). New living cells could be produced in the hinge region even during the "resting phase" of the cycle. Their intracellular contents could vary according to the physiological status of the organism. If lipids were produced, they could be extruded through minute breaks in the thin corneous tissues. Further histochemical studies of the skin in appropriately fixed fresh material could reveal numerous possible transformation series in gland evolution from a "rupture" system, to intra-epidermal sequestration, to a full-fledged, physiologically-controlled gland. The fundamental issue needing resolution is the question of the evolution of an appropriate delivery system for a semiochemical signal.

Much of the foregoing discussion could provide an explanation for the puzzling presence of generation glands in lizard species which shed infrequently. The syncytial β-layer of the squamate integument must impede the passage of semiochemicals to some degree. However, in those species where generation glands occur, the histology suggests some modification of the β-layer. One explanation for a "holocrine secretory specialization" in species which shed infrequently could be as follows. The surface modifications are not themselves the secretions, but rather they act as "wicks" facilitating the release of semiochemicals which have seeped through from the deeper cells or tissues (Duvall, 1981, 1982; Smith et al., 1984). This suggestion does not preclude the possibility that some generation glands do indeed produce holocrine secretions. What is important is that there are obviously many different strategies which have been exploited in different taxa. In the same way that microscopic study revealed two fundamentally different forms of generation gland (Maderson, 1972c), further biochemical and/or behavioral investigations might reveal that there are several different modes of action for specializations which have similar histological structure.

CONCLUDING COMMENTS

In the 15 years which have elapsed since Quay (1972) reviewed integumentary gland evolution in vertebrates, many new data have emerged. My own contribution has been refutation of the long-standing assertion that reptiles lack integumentary glands. Emboldened by a considerable body of data demonstrating that semiochemical signalling is important in the lives of lower tetrapods, I have attempted to show that there is a myriad of strategies which can and have been employed by Nature in using the integument as a direct or indirect source for such signals. I emphasize the lack of appropriate data for drawing even the most modest conclusions pertaining to the "whole story." I can but hope that my observations and speculations regarding the evolution of semiochemical sources will both help others to interpret existing data concerning the signals and the biological roles thereof, and encourage younger workers to enter this fascinating field of zoology.

ACKNOWLEDGMENTS

I would like to thank Drs. D. Crews and D. Duvall for their comments on a first draft of this manuscript. The original work on which this review is based has been supported by various CUNY intramural and NIH grants over the past 18 years. Brooklyn College provided funds to facilitate my attendance at this symposium.

REFERENCES

Albone, E. S., 1984, "Mammalian Semiochemistry: the Investigation of Chemical Signals Between Mammals," J. Wiley and Sons, New York.

Alexander, S. A., and Maderson, P. F. A., 1972, Further observations on holocrine epidermal specializations in iguanid lizards, Am. Zool., 12:731.

Bateson, G., 1966, Problems in cetacean and other mammalian communication, in: "Whales, Dolphins and Porpoises," K. S. Norris, ed., University of California Press, Berkeley.

Billett, F., Gans, C., and Maderson, P. F. A., 1985, Why study reptilian development?, in: "Biology of the Reptilia: Development," volume 14, C. Gans, F. Billett, and P. F. A. Maderson, eds., J. Wiley and Sons, Inc., New York.

Bock, W. J., and von Wahlert, G., 1965, Adaptation and the form - function complex, Evolution, 19: 269.

Burkholder, R. L., and Tanner, W. M., 1974, A new gland in Sceloporus graciosus males (Sauria, Iguanidae), Herpetologica, 30:368.

Carroll, R. L., 1982, Early evolution of reptiles, Ann. Rev. Ecol. Syst., 13:87.

Chiu, K. W., and Maderson, P. F. A., 1975, The microscopic anatomy of epidermal glands in two species of gekkonine lizards, with some observations on testicular activity, J. Morph., 147:23.

Chiu, K. W., Maderson, P. F. A., Alexander, S. A., and Wong, K. L., 1975, Sex steroids and epidermal glands in two species of gekkonine lizards, J. Morph., 147:9.

Cole, C. J., 1966, Femoral glands of the lizard Crotaphytus collaris, J. Morph., 118:119.

Cooper, W. E., and Vitt, L. J., 1984a, Detection of conspecific odors by the female broad-headed skink (Eumeces laticeps), J. Exp. Zool., 229:49.

Cooper, W. E., and Vitt, L. J., 1984b, Conspecific odor detection by the male broad-headed skink (Eumeces laticeps): Effects of sex and site of odor source and of male reproductive condition, J. Exp. Zool., 230:199.

Coopersmith, R., and Leon, M., 1984, Enhanced neural response to familiar olfactory cues, Science, 225:849.

Crews, D., and Greenberg, N., 1981, Function and causation of social signals in lizards, Am. Zool., 21:273.

Dhouailly, D., and Maderson, P. F. A., 1984, Ultrastructural observations on the embryonic development of the integument of Lacerta muralis (Lacertilia, Reptilia), J. Morph., 179:203.

Duvall, D., 1981, Western fence lizard (Sceloporus occidentalis) chemical signals. II. A replication with naturally breeding adults and a test of the Cowles and Phelan hypothesis of rattlesnake olfaction, J. Exp. Zool., 218:351.

Duvall, D., 1982, Western fence lizard (Sceloporus occidentalis) chemical signals. III. An experimental ethogram of conspecific body licking, J. Exp. Zool., 221:23.

Duvall, D., King, M. B., and Graves, B. M., 1983, Fossil and comparative evidence for possible chemical signaling in the mammal-like reptiles, in: "Chemical Signals in Vertebrates, III," D. Muller Schwarze, and R. M. Silverstein, eds., Plenum, New York.

Duvall, D., King, M. B., and Gutzwiller, K. J., 1985, Behavioral ecology and ethology of the prairie rattlesnake, Nat. Geog. Res., 1:80.

Edwards, D., 1980, Early land floras, in: "The Terrestrial Environment and the Origin of Land Vertebrates," A. L. Panchen, ed., Academic Press, London and New York.

Fergusson, B., Bradshaw, S. D., and Cannon, J. R., 1985, Hormonal control of femoral gland in the lizard Amphibolurus ornatus, Gen. Comp. Endocrinol., 57:371.

Flaxman, B. A., and Maderson, P. F. A., 1976, Growth and differentiation, J. Invest. Dermatol., 67:8.

Freeman, S. K., 1978, Pheromones (olfactory communication), J. Soc. Cosmet. Chem., 29:47.

Gans, C., 1979, Momentarily excessive construction as the basis for protoadaptation, Evolution, 33:227.

Gans, C., and Maderson, P. F. A., 1973, Sound producing mechanisms in recent reptiles: Review and comment, Am. Zool., 13:1195.
Gartska, W. R., and Crews, D., 1981, Female sex pheromone in the skin and circulation of a garter snake, Science, 214:681.
Grassman, M. A., Owens, D. W., McVey, J. P., and Marquez, E. M., 1984, Olfactory based orientation in artificially imprinted sea turtles, Science, 224:83.
Graves, B. M., and Duvall, D., 1983a, A role for aggregation pheromones in the evolution of mammal-like reptile lactation, Am. Nat., 122:835.
Graves, B. M., and Duvall, D., 1983b, Occurrence and function of prairie rattlesnake mouth gaping in a non-feeding context, J. Exp. Zool., 227:471.
Greenberg, N., ed., 1977, "Social Behavior in Reptiles," Am. Zool., 17:151.
Gunderson, H. L., 1976, "Mammalogy," McGraw Hill, New York.
Hadley, N. F., 1985, "The Adaptive Role of Lipids in Biological Systems," J. Wiley and Sons, New York.
Halpern, M., and Kubie, J. L., 1984, The role of the vomeronasal system in species-typical behavior, Trends Neurosci., 7:472.
Hara, T. J., ed., 1982, "Chemoreception in Fishes," Elsevier Scientific Publishing Co., Amsterdam and New York.
Jacob, J., and Ziswiler, V., 1982, The uropygial gland, in: "Avian Biology VI," D. S. Farner, J. R. King and K. C. Parkes, eds., Academic Press, London and New York.
Kurosumi, K., Shibasaki, S., and Ito, T., 1984, Cytology of the secretion in mammalian sweat glands, Int. Rev. Cytol., 87:253.
Lancet, D., 1984, Molecular view of olfactory reception, Trends Neurosci., 7:35.
Lillywhite, H. B., and Maderson, P. F. A., 1982, Skin structure and permeability, in: "Biology of the Reptilia: Physiology C.," vol. 12, C. Gans and H. Pough, eds., Academic Press, New York.
Little, C., 1983, "The Colonisation of Land: Origins and Adaptations of Terrestrial Animals," Cambridge University Press, London and New York.
Maderson, P. F. A., 1967, The histology of the escutcheon scales of Gonatodes (Gekkonidae), with a comment on the squamate sloughing cycle, Copeia, 1967:743.
Maderson, P. F. A., 1968a, The epidermal glands of Lygodactylus (Gekkonidae, Lacertilia), Breviora, #228:1.
Maderson, P. F. A., 1968b, On the presence of "escutcheon scales" in the Eublepharine gekkonid (Coleonyx), Herpetologica, 24:99.
Maderson, P. F. A., 1970, Lizard glands and lizard hands: Models for evolutionary study, Form. Funct., 3:179.
Maderson, P. F. A., 1972a, When? Why? and How? Some speculations on the evolution of the vertebrate integument, Am. Zool., 12:159.
Maderson, P. F. A., 1972b, On how an archosaurian scale might have given rise to an avian feather, Am. Nat., 176:424.
Maderson, P. F. A., 1972c, The structure and evolution of holocrine epidermal glands in sphaerodactyline and eublepharine gekkonid lizards, Copeia, 1972:559.
Maderson, P. F. A., 1975, Embryonic tissue interactions as the basis for morphological change in evolution, Am. Zool., 15:315.
Maderson, P. F. A., 1981, Developmental-functional comments on punctuation and gradualism, Am. Zool., 21:1009.
Maderson, P. F. A., 1983, An evolutionary overview of epithelial-mesenchymal interactions, in: "Epithelial-Mesenchymal Interactions in Development," R. H. Sawyer, and J. F. Fallon, eds., Praeger Scientific, New York.
Maderson, P. F. A., 1984, The squamate epidermis: New light has been shed, Symp. Zool. Soc. Lond., 52:111.
Maderson, P. F. A., 1985, Some developmental problems of the reptilian integument, in: "Biology of the Reptilia: Development," Volume 14, C. Gans, F. Billett, and P. F. A. Maderson, eds., J. Wiley and Sons, New York.

Maderson, P. F. A., Baranowitz, S. A., and Azulay, D., 1984, The phylogeny of tetrapod keratins, Am. Zool., 24:63A.

Maderson, P. F. A., and Chiu, K. W., 1970, Epidermal glands in gekkonid lizards: Phylogeny and evolution, Herpetologica, 26:233.

Maderson, P. F. A., and Chiu, K. W., 1981, The effects of androgens on the β-glands of the tokay Gekko gecko: Modification of an hypothesis, J. Morph., 167:109.

Maderson, P. F. A., and Chiu, K. W., 1985, Endocrine relationships of secondary sexual derivatives in tetrapods, in: "Proc. IXth Int. Symp. Comp. Endocrinol., December 1981," B. Lofts and W. N. Holmes, eds., University of Hong Kong Press.

Maderson, P. F. A, Chiu, K. W., Weiner, B. E., and Guber, H. A., 1977, The evolutionary origin of gekkonine epidermal glands, Am. Zool., 17:896.

Maderson, P. F. A., Guber, H. A. and Weiner, B. E., 1979, Hormone-tissue interactions in tokay epidermal glands, Am. Zool., 19:979.

Madison, D. M., 1977, Chemical communication in amphibians and reptiles, in: "Chemical Signals," D. Muller-Schwarze, and M. M. Mozell, eds., Plenum, New York.

Menchel, S., and Maderson, P. F. A., 1975, The post-natal development of holocrine epidermal specializations in gekkonid lizards, J. Morph., 147:23.

Mittal, A. J., and Banerjee, T. K., 1980, Keratinization versus mucus secretion in fish epidermis, Symp. Linn. Soc., London, 9:1.

Olson, E. C., 1975, Permo-Carboniferous paleoecology and morphotypic series, Am. Zool., 15:371.

Price, A. H., and LaPointe, J. L., 1981, Structure-functional aspects of the scent glands in (Lampropeltis getulus splendida), Copeia, 1981:138.

Quay, W. B., 1972, Integument and the environment: Glandular composition, function and evolution, Am. Zool., 12:95.

Rosenberg, H.I., and Russell, A. P., 1980, Structural and functional aspects of tail squirting: A unique defense mechanism of Diplodactylus (Reptilia, Gekkonidae), Can. J. Zool., 58:865.

Rosenberg, H. I., Russell, A. P., and Kapoor, M., 1984, Preliminary characterization of the defense secretion of Diplodactylus (Reptilia, Gekkonidae), Copeia, 1984:1025.

Schmalhausen, I. I., 1968, "The Origin of Terrestrial Vertebrates," Academic Press, New York and London.

Solomon, S. E., 1984, The characterisation and distribution of cells lining the axillary gland of the adult green turtle (Chelonia mydas L.), J. Anat., 138:267.

Smith, H. M., Duvall, D., Graves, B. M., Jones, R. E., and Chiszar, D., 1984, The function of squamate epidermatoglyphics, Bull. Phil. Herpetol Soc., 30:3.

Waagen, G. N., 1972, "Musk Glands in Recent Turtles," Unpublished Master of Science Thesis, University of Utah.

Winokur, R.M., and Legler, J. W., 1975, Chelonian mental glands, J. Morph., 147:275.

Wright, D. E., and Moffat, L. A., 1985, Morphology and ultrastructure of the chin and cloacal glands of juvenile Crocodylus porosus (Reptilia, Crocidilia), in: "Biology of Australian Frogs and Reptiles," R. Shine, G. Grigg, and M. Ehrman, eds., in press.

Yamada, K. M., ed., 1983, "Cell Interaction and Development: Molecular Mechanisms," J. Wiley and Sons, New York.

Yapp, W. B., 1970, "The Life and Organization of Birds," American-Elsevier Publishing Co., New York.

MAMMALIAN SEMIOCHEMISTRY: ISSUES AND FUTURES, WITH SOME EXAMPLES FROM A STUDY OF CHEMICAL SIGNALLING IN CATTLE

Eric S. Albone[1], Nicholas B. Blazquez[2], Jane French[2], Susan E. Long[2] and Graham C. Perry[2]

[1]School of Chemistry, University of Bristol, Cantock's Close, Bristol, BS8 1TS, UK and [2]Department of Animal Husbandry, University of Bristol, Langford, Bristol, BS8 7DU UK

INTRODUCTION

The science of semiochemistry (derived from the Greek "semeion," sign or signal) deals with the nature of those substances, or semiochemicals, by which organism interacts with organism in their shared environment. That chemical interactions occur between mammals has long been known and numerous behavioural and physiological effects have been investigated (Doty, 1976; Stoddart, 1980a, 1980b; Vandenbergh, 1983; Brown and Macdonald, 1985), while International Symposia such as this testify to an increasing awareness of the importance of chemical signals in the lives of many mammals. Yet in spite of the centrality of chemistry to a full understanding of the subject and its applications, the number of studies in which chemistry has been employed is remarkably few. The future of this subject clearly depends on a much greater emphasis on the integration of chemical studies with the work of biologists.

Mammalian semiochemistry (Albone, 1984) differs from the analogous study of insects, in which chemistry has played a prominent role, in that mammalian systems are characterized by a greater complexity. Unlike insect chemoreceptors, mammalian olfactory chemoreceptors are not tuned to specific compounds but detect a wide range of substances and odors (which may be associated with quite complex mixtures of compounds) that will be differentiated at a higher level in the nervous system (Shirley, 1984).

Similarly, definitive behavioural assays are much more difficult to design for mammals because of the effects of conditioning and habituation, which are associated with the use of test animals of high cognitive ability. Further, the mammal does not generally respond to a chemical stimulus in isolation but integrates the stimulus with other non-chemical cues and with factors depending on its physiological state and its past experience. Commonly, mammals will respond to a large range of novel stimuli whatever their natural biological significance, will rapidly learn to respond to test situations in certain ways, for example, show habituation and will exhibit very considerable variations in response from individual to individual.

In those cases where a long-term physiological semiochemical effect (primer pheromone effect) can be monitored, such problems do not arise. Such an effect (synchronization of estrus) has been assigned to estrous bovine cervical mucus (Izard, 1983).

Although it is preferable to embark on a semiochemical study using a simple, firm biological assay which, by repeated application, can guide the chemist through repeated fractionations of a semiochemical substrate towards biologically active components which may be present even as minor components of highly complex mixtures, such a "response-guided strategy" may only be applicable to a limited range of chemically mediated interactions in mammals. Since behavioural responses are complex, simple rapid assays are not always available. Even when they are, difficulties arise if biological activity is associated with complex mixtures of components, for all activity may be lost even on preliminary fractionation.

A valuable approach is to use the instantaneous physiological response of a mammal to individual components separated from a semiochemical substrate to indicate which may be worthy of detailed behavioural study. Instantaneous heart rate changes have been monitored telemetrically in rabbits constrained to monitor continuously the effluent from a gas chromatographic column on which rabbit scent components are being separated (Goodrich et al., 1981). It is quite possible, however, that components which do not elicit such a response could have an important semiochemical role.

A complementary approach to such response-guided strategies is to examine as far as possible the entire "chemical image" which one mammal presents to another, paying particular attention to those features which vary with important biological parameters, for these contain information which the species could exploit semiochemically. An emphasis on the patterns of occurrence of compounds associated with a semiochemical substrate would also accord with the view of Beauchamp et al. (1976), that mammalian chemical signals are more likely to be encoded in the patterns of occurrence of compounds rather than in individual biologically active substances.

Studies in mammalian semiochemistry are intrinsically interdisciplinary for although chemistry brings with it important insights and powerful analytical techniques, it depends for its success on the sensitive guidance of biologists. The significance of semiochemical components requires assessment by the physiologist, ethologist and ecologist while an understanding of the semiochemical sources, their anatomical and histological structures, their hormone dependence and the possible role of commensal bacteria (Albone et al., 1977) suggest aspects of the chemical image which merit more detailed study.

In this paper we refer to current work of an interdisciplinary group at Bristol in the reproductive semiochemistry of cattle, and in particular on the chemical signals associated with estrus. Any findings which facilitate estrus detection in cattle would be of economic importance (Bailie, 1982) for every estrus missed, delays insemination for three weeks with the result that annual milk yields and beef production are reduced.

ANIMAL BEHAVIOUR

Observations of cattle behaviour have shown that olfactory and gustatory investigations by bulls and other cows towards a cow's anogenital region increases at proestrus and estrus. This effect may be detected before behavioural estrus (Williamson et al., 1972; Reinhart, 1983) and Stevens (1983) showed a significant effect four days before estrus. This accords with the peaking or repeated flehmen reactions of bulls towards cows at estrus (Hradecky et al., 1983). That the cow's chemical image changes is evidenced by the fact that laboratory rats have been conditioned to discriminate between estrous and anestrous cow urine with 65% success (Ladewig and Hart, 1981) while dogs have been trained to discriminate proestrous and estrous vaginal swabs with 80% success (Kiddy et al., 1978). That this identity is not confined to a single body fluid is shown by the findings of Kiddy et al. (1984), whose dogs were able to discriminate with success rates of from 86% to 99% between bovine estrous and non-estrous odors associated with swabs of the vagina and vestibule, with fluids from deep within the vagina, with voided urine, with catheterized bladder urine, and with milk and with blood plasma.

Behavioural tests with cattle to define the semiochemical source have been directed toward the discrimination of estrous odors in urine, vaginal mucus and faeces. Conflicting results have been reported, however, reflecting inconsistencies of the various testing programs employed. Studies at Bristol (Stevens, 1983) and elsewhere have shown that in simple testing situations, where biological samples are presented in isolation, problems of rapid habituation to the test situation are particularly severe and may lead to differing conclusions. Thus, Donovan (1967) suggested that faeces contained the estrus signal while Paleologou (1977) concluded that it was associated with cervico-vaginal mucus. While Sambraus and Waring (1975) observed that estrous urine was more attractive to the bull than diestrous urine, Stevens (1983) found that bulls investigated all samples of cow urine regardless of their reproductive state, but only those bulls with extensive herd mating experience discriminated between estrous and diestrous samples.

Some problems associated with the presentation of samples in isolation are overcome when samples are applied to non-estrous carrier cows. Here the test animal receives a greater range of olfactory and other stimuli but, although this sustains the interest of the bull, it also makes it less easy to distinguish which stimulus or combination of stimuli evokes the response, so that results are again conflicting. We found for example, that the degree of olfactory discrimination shown by bulls between estrous and non-estrous urine and vaginal mucus samples applied to carrier cows was very variable and depended considerably on the bull's individual temperament and previous sexual experience.

Studies reported by Nishimura et al. (1984) have shown that the identity of the carrier cow is also important. They found that vaginal mucus taken at estrus and applied to the same cow at diestrus induced mounting by other members of a herd of cows, while this did not happen if the estrous vaginal mucus was applied to a different diestrous cow.

A further problem with most test regimes is the limited number of behavioural responses which can normally be measured and compared with the wide behavioural repertoire exhibited in the natural environment. In our present work at Bristol we are making a detailed inventory of the behaviour patterns exhibited by a small herd of cattle at pasture to determine in detail the change in behaviour pattern which accompanies the approach of estrus. This information can then be used to assess the semiochemical significance of various biological materials.

Such is the chemical complexity of biological materials that it is generally not possible to obtain a total chemical description of any such material (Zlatkis et al., 1979). From the outset biologically informed selections must be made of which compounds or groups of compounds are to be examined. Thus, interest in the pig pheromone C_{19}-Δ16 steroids is such that these odorous compounds have been sought, using radioimmunoassay techniques, as trace components in the highly complex mixture which comprises human axillary sweat (Claus and Alsing, 1976; Bird and Gower, 1980).

Since odorants must be volatile, most attention to date has been given to those semiochemical compounds which have in common sufficient volatility to be amenable to the methods of headspace analysis (Zlatkis and Shanfield, 1979). Here volatiles emitted from an odor source are taken up in a stream of inert gas and trapped, over a period of time, on an inert absorbent material such as Tenax. The volatiles are subsequently desorbed at high temperature directly on a gas chromatographic column and analysed. Figure 1 shows a typical gas chromatogram of headspace volatiles obtained from estrous cow urine using a 50m x 0.32mm i.d. fused silica capillary column of CP-Wax-57CB (Chrompack) operated isothermally at 60° for 5 min. and subsequently programmed to 190° at 4°/min.

Valuable as the method is, it is limited in that important odorants, such as the pig pheromone steroids, lack sufficient volatility to register in such a profile. In addition, the odor thresholds of different volatile compounds vary over a very wide range so that trace components may dominate the perceived odor (Buttery et al., 1969). Also, the relative proportions of the components registered is heavily dependent on the experimental procedure used (Jennings and Filsoof, 1977) so that interest surrounds variations in profiles obtained under identical analytical conditions.

Headspace analysis of this type has been used in the semiochemical study of rabbit scent glands and faecal pellets (Goodrich, 1983) and of hamster vaginal secretion components (Singer et al., 1976). In the absence of a behavioural assay, a possible significance could be inferred for the observed change in the chemical image of male red fox urine during the breeding season associated with the peaking of the volatile organosulfur compound, 3-methylbutyl methyl sulfide (Bailey et al., 1980) while a similar suggestion attended the discovery of the novel testosterone-dependent volatiles, 2-*sec*-Butyldihydrothiazole and dehydro-*exo*-brevicomin in the headspace of mouse urine. Subsequently it has been found that while inactive individually, together these substances elicit aggression from other male mice similar to that produced by intact adult male mouse urine (Novotny et al., 1984; Novotny et al., 1985). Similarly, estrogen-dependent volatiles, principally pentyl acetate, *z*-2-penten-1-yl acetate, *p*-toluidine and heptan-2-one, have been identified in female mouse urine (Schwende et al., 1984b) while statistical analysis has revealed differences, principally in octanal, phenylacetone, *o*-toluidine and phenol composition, associated with semiochemically distinguishable strain differences in the closely similar headspace volatile profiles of urines from congenic female mice (Schwende et al., 1984a).

A recent study has also documented variations in female-beagle-urine headspace volatiles across the estrous cycle (Schultz et al., 1985).

Visual inspection of our cow urine headspace volatile profiles monitored over the estrous cycle (the stage of the cycle being assessed

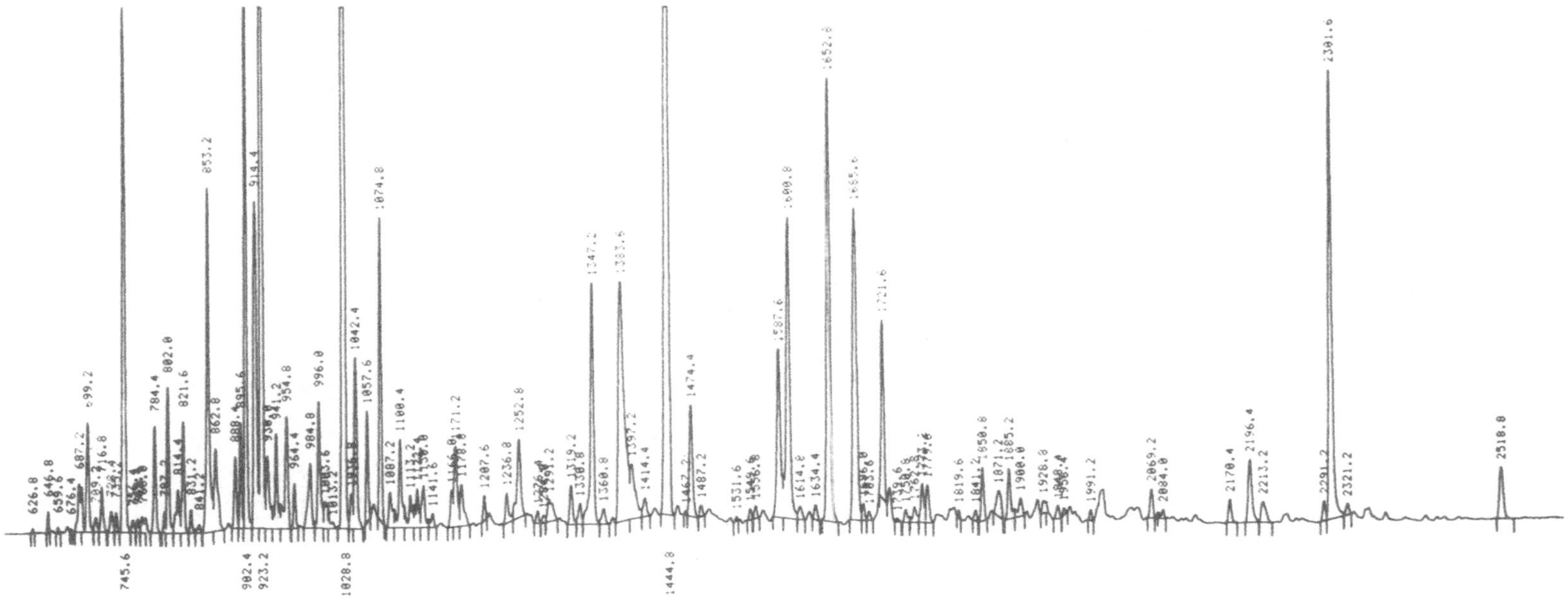

Figure 1. Typical capillary gas chromatogram of headspace volatiles obtained from the urine of a cow showing behavioural estrus. The numbers are retention times in seconds. Conditions are given in the text.

behaviourally and by estrogen, progesterone and luteinizing hormone radioimmunoassays of blood samples) showed that while no major change accompanied estrus, and while the profiles of individual cows were individually distinctive, these profiles were highly dependent on diet (silage/concentrates or grass or hay/concentrates).

However, the application of statistical techniques to urine headspace volatile profiles obtained from two cows over two complete cycles discriminated reproducibly between profiles obtained at different stages of the estrous cycle (Morgans, 1984). Data from 49 normalized peak areas from 83 chromatograms on canonical variates analysis (MacFie et al., 1978; MacFie and Gutteridge, 1982) produced a plot showing good separation of samples according to the day of the cycle on which they were collected, and good clustering of samples on those days for both animals for both cycles. That the clustering was dependent on the pattern of the majority of the 49 components was revealed by the loss of sample separation and clustering when the number of peaks was reduced from 49 to 25 by stepwise discriminant analysis.

Components tentatively identified in cow headspace volatiles on the basis of mass spectral data included aldehydes (to C_9 and benzaldehyde), ketones (to C_{10}), volatile carboxylic acids, phenols, sulfides and hydrocarbons.

Although most attention has been directed to the investigation of volatiles, it is becoming increasingly apparent that involatile components are semiochemically important (Albone, 1984). Not only may they act as precursors for volatiles and also modulate the release of volatiles from a semiochemical source, but they may act semiochemically in their own right in cases where direct contact is made with the source. In this context Nishimura et al. (1984) found that the mounting semiochemical associated with bovine estrous vaginal mucus was of low volatility and was heat stable. A semiochemical component of hamster estrous vaginal secretion which induces mounting has also been found to be associated with macromolecules (Singer et al., 1980).

ANATOMY AND HISTOLOGY

While behavioural and chemical programs are in progress, a further complementary approach to the study of any semiochemical system involves an examination of the semiochemical source and of factors which modulate that source's activity in ways which could be semiochemically significant (Adams, 1980).

The anogenital region in the cow comprises a complex semiochemical source with inputs from faeces, bladder urine, secretions from the upper reproductive tract, vagina and vestibular glands, and from the enlarged sweat and sebaceous glands of the vulval skin. On passing to the exterior all of these secretions may be mixed with the others and all may be acted upon by the changing resident microbial populations.

One input which, by virtue of its structure, location and hormonal dependence, could be of particular interest, is that of the major vestibular (Bartholin's) gland. These mucus secreting glands, which in the cow comprise a pair of yellow structures, each weighing up to 16g and approximately 3cm x 2cm x 2cm in size, are situated either side of the vestibule of the vagina. Each empties on to the lateral wall of the vestibule some 6cm from the vulva. The structure and location of the glands suggest a pulsatile mode of secretion, for secretion accumulating in the large ducts could be expelled on contraction of the surrounding

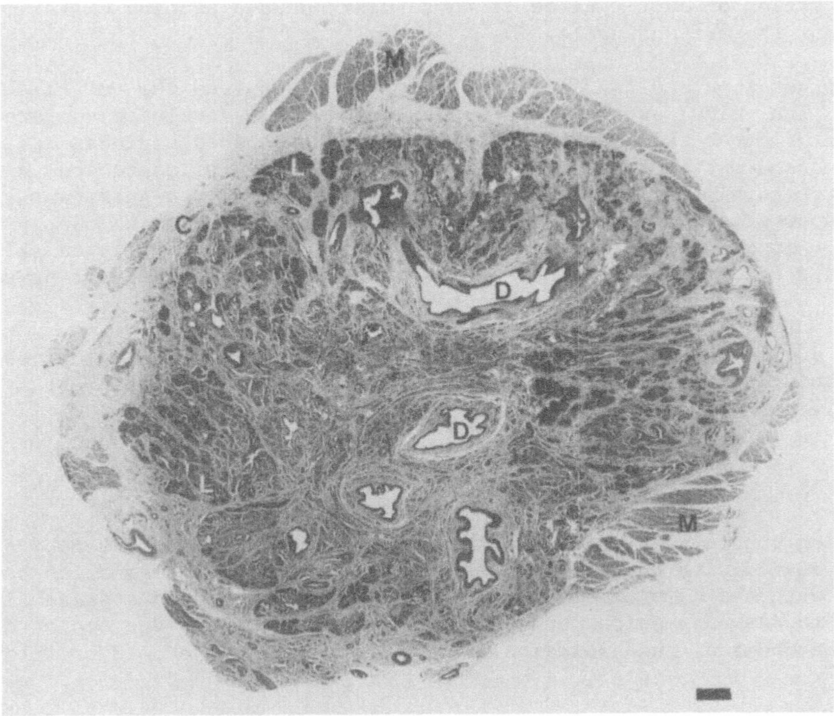

Fig. 2 Major vestibular (Bartholin's) gland of a cow. Section stained with alcian blue - hematoxylin - eosin. C, connective tissue; D, excretory duct; M. bulboglandularis muscle fibers; L, lobule. The bar represents 1 mm.

vestibular and urethralis muscles during urination. Similarly, contraction of the smooth muscle of the stroma or of fibers of the bulboglandularis muscle might also be expected to cause expulsion of secretion.

Histological studies (Friess, 1972; Reutner and Morgan, 1948; Cribb and Flood, personal communication) have shown the gland to be composed of branched, mucus-secreting tubules organized into lobules surrounded by connective tissue. The glands are extremely sensitive to estrogen and at estrus secretory activity is maximal (Hammond, 1927). This is thought to be a result of increased mucus production (Reutner and Morgan, 1984). There is also a marked ductular hyperplasia (Friess, 1972). Such is the extent of these morphological changes that these glands are used as a sensitive indicator of whether veal calves have been administered growth-promoting estrogens (Kroes, 1971). Figure 2 shows a section through a major vestibular gland at estrus revealing the lobules and very large excretory ducts. At the edge of the gland the connective tissue condenses to form a quasi-capsule in which the fibers of the bulboglandularis muscle are embedded.

Our histochemical studies have shown that the tubule cells are filled to varying degrees with neutral sialated and sulfated mucins, while the duct cells tend to contain only neutral mucins in their apices. Routine lipid stains have failed to indicate the presence of lipid in any of the cells. We are presently quantifying the different types of mucins and the ratios of duct-type cells to tubular mucus-producing cells throughout the estrous cycle.

Another input to the anogenital region arises from the vulval skin glands. The vulval sweat glands are extremely well developed compared to those of other skin regions. Similarly, the sebaceous glands are well developed and multilobulated. They empty into large pilosebaceous canals. These structural specialisations may be semiochemically significant since modified sweat and sebaceous glands have been associated with odor production (Adams, 1980). However, it should be noted that enlarged skin glands are a common feature of mucocutaneous junctions and their presence does not necessarily imply a semiochemical function.

In due course we plan to subject the secretions of the major vestibular glands and the vulval skin glands to chemical analysis and behavioral assay.

CONCLUSION

Major problems which attend the future development of the subject are associated with the need to mesh the complexities of mammalian behaviour with the complexities of chemical composition. An approach in which both response-guided and chemical image strategies are employed to unite the chemical and biological inputs seems to be the most profitable.

ACKNOWLEDGMENTS

We thank Dr. D. Mottram for valuable discussion. Studies at Bristol have been funded by the Wellcome Trust, Agricultural and Food Research Council, the Milk Marketing Board and the Royal Society.

REFERENCES

Adams, M. G., 1980, Odour producing organs of mammals, Symp. Zool. Soc. Lond., 45:57.

Albone, E. S., 1984, "Mammalian Semiochemistry," Wiley, Chichester and New York.

Albone, E. S., Gosden, P. E., and Ware, G. C., 1977, Bacteria as a source of chemical signals in mammals, in: "Chemical Signals in Vertebrates," D. Muller-Schwarze, and M. M. Mozell, eds., Plenum, New York.

Bailey, S., Bunyan, P. J., and Page, J. M. J., 1980, Variation in the levels of some components of the volatile fraction of urine from captive red fox (Vulpes vulpes) and its relationship to the state of the animal, in: "Chemical Signals: Vertebrates and Aquatic Invertebrates," D. Müller-Schwarze and R. M. Silverstein, eds., Plenum, New York.

Bailie, J. H., 1982, Management and economic effects of different levels of oestrus detection in the dairy herd, Vet. Rec., 110:218.

Beauchamp, G. K., Doty, R. L., Moulton, D. G., and Mugford, R. A., 1976, The pheromone concept in mammalian chemical communication: A brief critique, in: "Mammalian Olfaction, Reproductive Processes and Behavior," R. L. Doty, ed., Academic Press, New York.

Bird, S., and Gower, D. B., 1980, Measurement of 5α-androst-16-en-3-one in human axillary secretions by radioimmunoassay, J. Endocrinol., 85:8P.

Brown, R. E., and Macdonald, D. W., eds., 1985, "Mammalian Social Odours," Oxford University Press, Oxford.

Buttery, R. G., Seifert, R. M., Guadagni, D. G., and Ling, L. C., 1969, Characterization of some volatile constituents of bell peppers, J. Agric. Food Chem., 17:1322.

Claus, R., and Alsing, W., 1976, Occurrence of 5α-androst-16-en-one, a boar pheromone, in man and its relationship to testosterone, J. Endocrinol., 68:483.

Donovan, C. A., 1967, Some clinical observations on sexual attraction and deterrence in dogs and cattle, Vet. Med./Small Animal Clinician, 62:1047.

Doty, R. L., ed., 1976, "Mammalian Olfaction, Reproductive Processes and Behavior," Academic Press, New York.

Friess, A. E., 1972, Histochemische Untersuchungen un den Glandulae vestibulares maiores (Bartholini) des Rindes, Acta Histochem., 44:62.

Goodrich, B. S., 1983, Studies on the chemical composition of secretions from skin glands of the rabbit, Oryctolagus coniculus, in: "Chemical Signals in Vertebrates 3," D. Muller-Schwarze, and R. M. Silverstein, eds., Plenum, New York.

Goodrich, B. S., Hesterman, E. R., Shaw, K. S., and Mykytowycz, R., 1981, Identification of some volatile compounds in the odor of fecal pellets of the rabbit, Oryctolagus cuniculus, J. Chem. Ecol., 7:817.

Hammond, J., 1927, "The Physiology of Reproduction in the Cow," Cambridge University Press, Cambridge.

Hradecky, P., Sis, R. F., and Klemm, W. R., 1983, Distribution of flehmen reactions of the bull throughout the bovine estrous cycle, Theriogenology, 20:197.

Izard, M. K., 1983, Pheromones and reproduction in domestic animals, in: "Pheromones and Reproduction in Mammals," J. G. Vandenbergh, ed., Academic Press, New York.

Jennings, W. G., and Filsoof, M., 1977, Comparison of sample preparation techniques for gas chromatographic analysis, J. Agric. Food Chem., 25:440.

Kiddy, C. A., Mitchell, D. S., Bolt, D. J., and Hawk, H. W., 1978, Detection of estrus-related odors in cows by trained dogs, Biol. Reprod., 19:389.

Kiddy, C. A., Mitchell, D. S., and Hawk, H. W., 1984, Estrus-related odors in body fluids of dairy cows, J. Dairy Sci., 67:388.

Kroes, R., 1971, Public health aspects in oestrogen administration of veal calves, and histological tests to prove oestrogen administration, Tijdschrift voor Diergeneeskunde, 96:1196.

Kroes, R., Ruitenberg, E. J., and Berkvens, J. M., 1970, Histological changes in the genital tract of the female calf after administration of diethylstilbestrol and hexestrol, Zbl. Vet. Med., 17A:440.

Ladewig, J., and Hart, B. L., 1981, Demonstration of estrus-related odors in cow urine by operant conditioning of rats, Biol. Reprod., 24:1165.

Lloyd, D. H., Dick, W. D. B., and Jenkinson, D. McE., 1979, Location of the microflora in the skin of cattle, Br. Vet. J., 135:519.

MacFie, H. J. H., Gutteridge, C. S., and Norris, J. R., 1978, Use of canonical variate analysis in the differentiation of bacteria by pyrolysis gas-liquid chromatography, J. Gen. Microbiol., 104:67.

MacFie, H. J. H., and Gutteridge, C. S., 1982, Comparative studies on some methods for handling quantitative data after generation by analytical pyrolysis, J. Anal. Appl. Pyrol., 4:175.

Morgans, T., 1984, "Chemical Identification of The Active Components in Cattle and Sheep Oestrous Secretions," Ph.D. Thesis, University of Bristol, Bristol, UK.

Nishimura, K., Okano, T., Utsumi, K., and Yuhara, M., 1984, Partial separation of mounting-inducing pheromones from vaginal mucus of

Holstein heifer, Proc. 10th Int. Congr. Anim. Reprod. Artificial Insemination (University of Illinois, Urbana-Champaign), 3:291.

Novotny, M., Schwende, F. J., Wiesler, D., Jorgenson, J. W., and Carmack, M., 1984, Identification of a testosterone-dependent unique volatile constituent of male mouse urine; 7-exo-ethyl-5-methyl-6, 8-dioxabicyclo [3.2.1]-3-octene, Experientia, 40:217.

Novotny, M., Scott, H., Jemiolo, B., and Alberts, J., 1985, Synthetic pheromones that promote inter-male aggression in mice, Proc. Nat. Acad. Sci., (USA), 82:2059.

Paleologou, A. M., 1977, Detecting oestrus in cows by a method based on bovine sex pheromones, Vet. Rec., 100:319.

Reinhart, V., 1983, Flehmen, mounting and copulation among members of a semi-wild cattle herd, Anim. Behav., 32:641.

Reutner, T. F., and Morgan, B. B., 1948, A study of the bovine vestibular gland, Anat. Rec., 101:193.

Sambraus, H. H., and Waring, C. H., 1975, Effect of urine from oestrous cows on libido in bulls, Z. Säugetierkunde, 40:49.

Schultz, T. H., Kruse, S. McK., and Flath, R. L., 1985, Some volatile constituents of female dog urine, J. Chem. Ecol., 11:169.

Schwende, F. J., Jorgenson, J. W., and Novotny, M., 1984a, Possible chemical basis for histocompatability-related mating preference in mice, J. Chem. Ecol. 10:1603.

Schwende, F. J., Wiesler, D., and Novotny, M., 1984b, Volatile compounds associated with estrus in mouse urine: Potential pheromones. Experientia, 40:213.

Shirley, S. G., 1984, Mammalian chemoreception, in: "Mammalian Semiochemistry," E. S. Albone, ed., Wiley, Chichester and New York.

Singer, A. G., Agosta, W. C., O'Connell, R. J., Pfaffmann, C., Bowen, D. V., and Field, F. H., 1976, Dimethyl disulphide: An attractant pheromone in hamster vaginal secretion, Science, 191:948.

Singer, A. G., Macrides, F., and Agosta, W. C., 1980, Chemical studies of hamster reproductive pheromones, in: "Chemical Signals: Vertebrates and Aquatic Invertebrates," D. Muller-Schwarze, and R. M. Silverstein eds., Plenum, New York.

Stevens, K., 1983, "The Role of Olfaction During Oestrus Detection in Sheep and Cattle," Ph.D. Thesis, University of Bristol, Bristol, UK.

Stoddart, D. M., ed., 1980a, "Olfaction in Mammals," (Symp. Zool. Soc., Lond., 45) Academic Press, London and New York.

Stoddart, D. M., 1980b, "The Ecology of Vertebrate Olfaction," Chapman and Hall, London and New York.

Vandenbergh, J. C., ed., 1983, "Pheromones and Reproduction in Mammals," Academic Press, New York.

Williamson, N. B., Morris, R. S., Blood, D. C., Cannon, C. W., and Wright, P. J., 1972, A study of oestrous behaviour and oestrous detection methods in a large commercial dairy herd, Vet. Rec., 91:58.

Zlatkis, A., Lee, K. Y., Poole, C. F., and Holzer, G., 1979, Capillary gas chromatographic profile analysis of volatile compounds in sera of normal and virus-infected patients, J. Chromatog., 163:125.

Zlatkis, A., and Shanfield, H., 1979, Concentration techniques for volatile samples, in: "Practical Mass Spectrometry," B. S. Middleditch, ed., Plenum, New York.

MOTOR PATTERNS DEDICATED TO SENSORY FUNCTIONS

David Chiszar

Department of Psychology, Box 345
University of Colorado
Boulder, Colorado 80309

INTRODUCTION

This paper started out to be a review of our recent work on snake chemoreception (Chiszar et al., 1983a, b, 1984, 1985; O'Connell et al., 1985, Radcliffe et al., 1983, 1984, 1985), but it has metamorphosed into a theoretical presentation which I hope will be of general interest to chemical ecologists and sensory physiologists. The main idea will be brought into focus by describing a survey of experiments reported in Muller-Schwarze and Silverstein (1980).

ORIENTING AND INVESTIGATORY BEHAVIORS AS DEPENDENT VARIABLES IN CHEMICAL ECOLOGY EXPERIMENTS

Most experiments reported in Muller-Schwarze and Silverstein (1980) presented putative chemical cues and control stimuli to animals and measured behavioral or physiological dependent variables to assess the effects of the chemicals under consideration. I partitioned the experiments into two categories: (1) studies that measured orienting or investigatory behaviors, and (2) studies that measured other dependent variables. Typically these latter studies measured action patterns such as escaping, marking, settling on hosts, copulating, or fighting, but some concentrated on physiological processes. Next I calculated the percentage of experiments from each category that reported significant effects of the chemicals under examination. My results are shown in Table 1 where it can be seen that only 44% of the experiments in category 1 reported significant effects whereas 99% of the experiments in category 2 did so (Chi square = 37.23, df = 1, $p < .01$; see also Beauchamp et al., 1983).

Of course, there are many reasons why an experiment can fail to reject the null hypothesis, including the possibility that the putative chemical signal was not, in fact, biologically effective. Nevertheless, I urge you to consider the idea that some aspects of orienting and investigatory behavior are <u>elicited</u> motor patterns (Sokolov, 1963; Voronin et al., 1965). Moreover, these behaviors may be elicited by events associated with the presentation of chemicals rather than by the chemicals themselves. Internal and autonomous coordinating processes then govern the subsequent flow of behavioral components, especially during relatively brief test sessions conducted in environments containing no other relevant stimulation.

Table 1. Survey of Experiments Described in Muller-Schwarze and Silverstein (1980)

	Percent of Experiments Reporting	
	Insignificant Effects	Significant Effects
Category I: Experiments measuring orienting and/or investigatory behaviors	56	44
Category II: Experiments measuring all other dependent variables	1	99

Because control and experimental conditions can be expected to elicit these initial aspects of orientation and investigation, it is not surprising that the two conditions often fail to differ significantly when these behaviors are the dependent variables.

It will be instructive to review two studies that did obtain significant effects of chemical cues on orienting or investigatory behavior (i.e., Beauchamp et al., 1980; and Singer et al., 1980).

Beauchamp et al. (1980) showed that male guinea pigs spent more time sniffing, licking and headbobbing when female urine was the stimulus than when water was presented. The Monell group has established that guinea pig urine contains information regarding genus, species, sex, individual identity, and dietary history. Perhaps the extraction of all of this information requires extensive sampling, and perhaps this sampling is encouraged by a hedonic effect of urine detected by the vomeronasal system. This fascinating notion suggests that vomeronasal stimulation can reinforce investigatory behavior, causing it to occur more frequently and to gather more data than might otherwise be the case (Beauchamp et al., 1980, 1983; Kubie & Halpern, 1979). Hence, there may be an additional effect superimposed upon the elicitory or signal properties of the experimental condition, creating a positive feedback system for the male and making his investigatory behavior rise above the baseline rate defined by the control condition.

Singer et al. (1980) studied the effects of hamster vaginal discharge on the behavior of males. These animals investigated the unscented forequarters of experimental and control surrogates, and the hindquarters of control surrogates were also investigated to about the same extent as either type of forequarter. These facts indicate that some degree of investigation is released by any surrogate. However, the hindquarters of experimental surrogates (scented with fresh vaginal discharge) received much more attention than any of the above-described areas, reflecting the additional sniffing and licking called forth by the attractant pheromone (dimethyl disulfide). My suggestion is that investigatory behavior directed to control surrogates contains elicited components that are cued by visual, tactile, or chemical stimuli. The presence of the attractant pheromone brings about an increased orientation towards the hindquarters where the mounting pheromone can be detected. That is, the attractant pheromone may be potentiating investigation through a positive feedback system analogous to the one postulated by the Monell group, thereby causing the detection of the mounting pheromone (Singer et al., 1980, p. 368).

In these studies, measures of investigatory behavior revealed significant differences between experimental and control chemicals. But, the control scores were never zero, and at least some of the investigatory behaviors seen in the control conditions had properties consistent with the idea that they were elicited (and internally coordinated) responses. Both studies suggest that reinforcing or hedonic effects can be superimposed upon the low baseline levels of these motor patterns (e.g., sniffing, headbobbing), driving their frequencies higher in the experimental than in the control conditions. Investigatory motor patterns are not restricted to the experimental conditions; rather, they simply occur more frequently in them because of positive feedback effects. As a secondary point, I offer the guess that such effects are uncommon and that orienting responses are typically organized to do their data-gathering jobs efficiently and without the need for hedonically-reinforced supplementation. In some ways the position here advocated is consistent with that taken by Bronson (1968) in his use of the term "signaling pheromone." In general, I suggest that the capacity to detect pheromones and other stimuli depends not only upon sensory receptors but also upon readily elicited motor processes dedicated to them (e.g., sniffing, licking, flicking, etc.). The "polyvalent" nature of these investigatory behaviors in many animals is proposed as the reason why they occur in control as well as in experimental conditions and, hence, as the reason why such behaviors frequently do not distinguish between these conditions.

STRIKE-INDUCED CHEMOSENSORY SEARCHING (SICS)

This motor pattern consists of an increase in the rate of tongue flicking by rattlesnakes following the envenomation of rodent prey (Chiszar et al., 1977, 1983a, b, c). Rattlesnakes typically release adult rodents after the strike, remaining in actual contact with the prey only about 0.3 sec (Kardong, 1985). The prey is allowed to wander freely while venom takes effect, and a rodent can get several hundred cm from the site of an attack before becoming immobile (Brock, 1980; Estep et al., 1981; Hayes & Galusha, 1984). SICS contributes to the snake's ability to locate and follow trails left by envenomated rodents (Chiszar et al., 1983a; Dullemeijer, 1961; Golan et al., 1982; see also Burghardt & Pruitt, 1975; Duvall et al., 1980).

The rate of tongue flicking seen after a predatory strike does <u>not</u> differ depending upon the presence vs. absence of chemical cues in the post-strike environment (Chiszar et al., 1977; Golan et al., 1982; Scudder et al., 1983). Also, SICS will continue for about 2 1/2 hr following an envenomating strike if no prey carcass is available (Chiszar et al., 1982). Long-term zoo captives do not differ from wild-caught rattlesnakes in these respects, indicating that SICS is not degraded as a consequence of standard husbandry techniques used in most zoos (i.e., feeding dead rodents to rattlesnakes; O'Connell et al., 1985; Chiszar et al., 1985).

These data have led us to think of SICS as a fixed action pattern (FAP) that is released by the act of striking prey and that has a relatively stereotyped topography and duration (Chiszar et al., 1977, 1983a, b, 1985; Golan et al., 1982). Although the rate of tongue flicking does not change upon contacting a trail, chemical cues arising from the trail exert taxic effects on the directionality of tongue flicking. The immediate consequence is that tongue flicks are aimed at the trail; the long-range consequence is that the snake is drawn along the trail to the mouse carcass at its termination. Therefore, we can think of SICS as an elaborate motor pattern, dedicated to the vomeronasal receptors (Halpern & Kubie, 1983).

THE CRAIG-LORENZ-TINBERGEN CONCEPTUALIZATION OF CONSUMMATORY AND APPETITIVE BEHAVIORS

An important part of my argument is that some aspects of orientation and investigation are innate and elicited rather than learned and emitted. The innate components are not under voluntary control and are not as variable in intensity, frequency, or duration as the emitted components. The ethological tradition has, however, lumped all forms of searching and exploration into the category of appetitive behavior:

> ". . . Contrary to the consummatory action, appetitive behavior is not characterized by a stereotyped motor pattern, but rather by (1) its variability and plasticity and (2) its purposiveness. . . The consummatory act is relatively simple, at its most complex, it is a chain of reactions, each of which may be a simultaneous combination of a taxis and a fixed pattern. But appetitive behavior is a true purposive activity, offering all the problems of plasticity, adaptiveness, and of complex integration that baffle the scientist in his study of behavior as a whole. Appetitive behavior is a conglomerate of many elements of very different order, of reflexes, of simple patterns like locomotion, of conditioned reactions, of 'insight' behavior, and so on." (Tinbergen, 1951, pp. 105-106)

Although this quotation acknowledges the existence of appetitive reflexes, the emphasis is clearly upon the more complex types of behavior that exhibit voluntary control, variability, and learning. Indeed, the history of ethology since the early publications of Craig (1918), Lorenz (1937) and Tinbergen (1951) has continued to focus upon these complex forms of appetitive behavior, leading eventually to the emergence of optimal foraging theory (Kamil & Sargent, 1981). This work has taught us much about the organization of these behaviors and about their sensitivity to energetic and/or risk contingencies; however, the simpler and reflexive types of appetitive behavior have been all but forgotten by European and American ethologists.

THE RUSSIAN REFLEXOLOGISTS

Lorenz (1981) recognized that some appetitive behaviors are FAPs, endogenously released and terminated upon detection of the key stimulus for the consummatory act that is the objective of the search. But, he did not develop the theoretical implications of this idea. Instead we must turn to the Russian physiologists who have spent 50 years studying orienting reflexes (Sokolov, 1963; Voronin et al., 1965, see also Berlyne, 1960).

The Russians have focused upon the stimuli that elicit orienting reactions, the CNS structures subserving them, the physiological concomitants of orienting behaviors, and the contribution of such behaviors to the process of Pavlovian conditioning. The literature is voluminous, and it contains both comparative and ontogenetic dimensions. Hence, this work leaves no doubt about the widespread existence of innate and topographically-fixed orienting reflexes.

There is a terminological problem in this literature; namely, the Russians have never recognized the importance of distinguishing between reflexes and FAPs. In fact, the latter term does not occur very often in the literature dealing with orienting reflexes. However, certain behaviors studied by Russian physiologists are undoubtedly appetitive FAPs (Anohkin, 1965; Gerd, 1965; see also Lorenz, 1981).

Nevertheless, the Russians have provided a rich description of motor patterns that surely facilitate the operation of sensory receptors. They have shown that some of these are readily extinguishable whereas others are practically inextinguishable (this corresponds to an operational definition of the difference between FAPs and reflexes given by Lorenz, 1981). More importantly, the Russians have established that orienting reactions are elicited by multiple stimuli. Thus, a behavior such as sniffing can be released by auditory, visual, and tactile cues as well as by chemical cues. This, of course, makes great sense when we consider the function of orienting reactions, but this is a major difference between appetitive reactions and consummatory behaviors (which are released by relatively specific sign stimuli).

DEDICATED MOTOR PATTERNS

Because some aspects of orientation and investigation are certainly learned whereas others are innate, and because some of the innate forms are reflexes whereas others are FAPs, it is useful to avoid a terminology that is committal with respect to mechanism. Consequently, I suggest that we simply recognize the existence of motor patterns that are dedicated to sensory functions. For short, I offer the term "dedicated motor patterns" (DMPs). In this essay, I have been primarily interested in innate DMPs, both reflexes and FAPs, because they provide a potential explanation for the relatively frequent failure of investigatory behavior to distinguish between experimental and control conditions in chemical ecology experiments. Now I would like to point to other implications of DMPs. For example, some DMPs should be conceptualized as adaptations in their own right, their contribution to fitness coming from the service they provide to sensory receptors. Furthermore, it can be hypothesized that the existence of DMPs reduces selective pressure upon receptors. In the extreme case, receptor tissues may stop evolving after DMPs begin to serve them. Consequently, we may discover advanced or specialized animals making use of primitive receptor systems because DMPs have evolved to enhance the effectiveness of those systems (Wells et al., 1971).

As the Russian work suggests, a DMP can be released by many stimuli, including cues from modalities other than the one(s) served by the DMP. Therefore, DMPs like sniffing or tongue flicking can occur even when no chemical cues are available. This means that a DMP can tell us what sensory system an animal is attempting to use but not if the system is actually detecting stimulation. It is frequently assumed, sometimes incorrectly, that a high rate of tongue flicking implies vomeronasal utilization of chemical information or that the vomeronasal system is guiding investigation (Chiszar et al., 1977, 1983a, c). Similar assumptions are sometimes made about the utilization of visual information by human infants based on the fact that eye movements are released by certain visual stimulus configurations (Fagan, 1984; Fantz, 1961). Although some of these assumptions are tenable, it must be remembered that certain components of occular orienting behavior occur in congenetically blind persons (Zemstova et al., 1965). Therefore, it seems appropriate to consider tongue flicking and (at least some) eye movements to be polyvalent DMPs and to recognize that statements about the utilization of the sensory systems to which DMPs are dedicated requires electrophysiological or other confirmation. In other words, DMPs are only an indirect indication of sensation and cue utilization.

This brings up the critical point that we need a systematic analysis of DMPs. We must have topographical descriptions, phyletic comparisons, and we must know more about the eliciting stimuli. Of equal importance, we must develop a quantitative understanding of the contribution of DMPs to sensation and to fitness. The Russians have made a start, but their work has not gone far enough. It is hoped that by calling attention to these specialized behaviors, DMPs will now receive as much experimental and theoretical effort as the tissues they serve.

Finally, two additional points must be made. First, because some DMPs are topographically fixed, and because it may be advantageous for individuals to attend to DMPs occurring in conspecifics, we can expect that such behaviors can acquire a secondary role as ritualized intraspecific signals. Muller-Schwarze (1980) provided an excellent example in the alert reaction of black-tailed deer. Table 2 in his paper leaves no doubt that this behavior enters into stochastic relationships with other behaviors, and that the alert reaction therefore constitutes a source of information. Second, behavioral sequences often begin with elicited DMPs and progress through learned ones, eventually terminating with a consummatory pattern. Learned aspects of searching behavior are likely to reveal idiosyncracies of personal history, yet these behaviors will probably flow smoothly from the initial DMPs to the final consummatory actions, creating the impression of a unitary behavior pattern. Consequently, the job of fractionating appetitive behavior sequences into their component DMPs is going to be as formidable a task as that of fractionating the organic compounds of interest to chemical ecologists. Nevertheless, this is what must be done if we are to understand these important behaviors from the perspective of evolution, ecology, ontogeny and sensory physiology.

ACKNOWLEDGMENTS

I thank the M. M. Schmidt Foundation for its financial support, and I thank the various colleagues with whom I have discussed these ideas: J. W. Chiszar, D. Duvall, C. W. Radcliffe, and K. M. Scudder. Finally, I thank J. Albersheim and M. L. Brown for typing the manuscript.

REFERENCES

Anokhin, P. K., 1965, The role of the orienting-exploratory reaction in the formation of the conditioned reflex, in: "Orienting Reflex and Exploratory Behavior," L. G. Voronin, A. N., Leontiev, A. R. Luria, E. N. Sokolov, and O. S. Vinogradova, eds., Amer. Inst. Biol. Sci., Washington, D. C.

Beauchamp, G. K., Martin, I. G., Wellington, J. L., and Wysocki, C. J., 1983, The accessory olfactory system: Role in maintenance of chemo-investigating behavior, in: "Chemical Signals in Vertebrates," D. Muller-Schwarze, and R. M. Silverstein, eds., Plenum Press, New York.

Beauchamp, G. K., Wellington, J. L., Wysocki, C. J., Brand, J. C., Kubie, J. L., and Smith, A. B., 1980, Chemical communication in the guinea pig: Urinary components of low volatility and their access to the vomeronasal organ, in: "Chemical Signals: Vertebrates and Aquatic Invertebrates," D. Muller-Schwarze, and R. M. Silverstein, eds., Plenum Press, New York.

Berlyne, D. E., 1960, "Conflict, Arousal, and Curiosity," McGraw-Hill, New York.

Brock, O. G., 1980, Predatory Behavior of Eastern Diamondback Rattlesnakes (Crotalus adamanteus): Field Enclosure and Y-maze Laboratory Studies, Emphasizing Prey Trailing Behavior, unpublished doctoral dissertation, Florida State University.

Bronson, F. H., 1968, Pheromonal influences on mammalian reproduction, in: "Reproduction and Sexual Behavior," M. Diamond, ed., Indiana Univ. Press, Bloomington.

Burghardt, G. M., and Pruitt, G. H., 1975, The role of the tongue and senses in feeding of naive and experienced garter snakes, Physiol. Behav., 14:185.

Chiszar, D., O'Connell, B., Greenlee, R., Demeter, B., Walsh, T., Chiszar, J., Moran, K., and Smith, H. M., 1985, Duration of strike-induced chemosensory searching in long-term captive rattlesnakes at National Zoo, Audubon Zoo, and San Diego Zoo, Zoo Biol., in press.

Chiszar, D., Radcliffe, C. W., O'Connell, B., and Smith, H. M., 1982, Analysis of the behavioral sequence emitted by rattlesnakes during feeding episodes. II. Duration of strike-induced chemosensory searching in rattlesnakes (Crotalus viridis, C. enyo), Behav. Neural. Biol., 34:261.

Chiszar, D., Radcliffe, C. W., and Scudder, K. M., 1977, Analysis of the behavioral sequence emitted by rattlesnakes during feeding episodes. I. Striking and chemosensory searching, Behav. Biol., 21:418.

Chiszar, D., Radcliffe, C. W., Scudder, K. M., and Duvall, D., 1983a, Strike-induced chemosensory searching by rattlesnakes: The role of envenomation-related chemical cues in the post-strike environment, in: "Chemical Signals in Vertebrates," D. Muller-Schwarze, and R. M. Silverstein, eds., Plenum Press, New York.

Chiszar, D., Stimac, K., and Boyer, T., 1983b, Effect of mouse odors on visually-induced and strike-induced chemosensory searching in prairie rattlesnakes (Crotalus viridis), Chem. Senses, 7:301.

Chiszar, D., Stimac, K., Poole, T., Miller, T., Radcliffe, C. W., and Smith, H. M., 1983c, Strike-induced chemosensory searching in cobras (Naja naja kaouthia, N. mossambica pallida), Zeitschr. Tierpsychol., 63:51.

Chiszar, D., Walsh, T., Demeter, B., Davenport, M., Chiszar, J., and Smith, H. M., 1984, No social induction of chemosensory searching in rattlesnakes at National Zoo. Paper presented at meetings of the Animal Behavior Society, Eastern Washington University, Cheney, WA.

Craig, W., 1918, Appetites and aversions as constituents of instincts, Biol. Bull., 34:91.

Dullemeijer, P., 1961, Some remarks on the feeding behavior of rattlesnakes, Kon. Nederland. Acad. Wetensch., Series C, 64:383.

Duvall, D., Scudder, K. M., and Chiszar, D., 1980, Rattlesnake predatory behaviour: Mediation of prey discrimination and release of swallowing by cues arising from envenomated mice, Anim. Behav., 28:674.

Estep, K., Poole, T., Radcliffe, C. W., O'Connell, B., and Chiszar, D., 1981, Distance traveled by mice (Mus musculus) after envenomation by prairie rattlesnakes (Crotalus viridis), Bull. Psychon. Sci., 18:108.

Fagan, J. F., 1984, The intelligent infant: Theoretical implications, Intelligence, 8:1.

Fantz, R. L., 1961, The origin of form perception, Sci. Amer., 204:66.

Gerd, M. A., 1965, The orienting and exploratory reflexes and their role in the process of animal training, in: "Orienting Reflex and Exploratory Behavior," L. D. Voronin, A. N. Leontiev, A. R. Luria, E. N. Sokolov, and O. S. Vinogradova, eds., Amer. Inst. Biol. Sci., Washington, D. C.

Golan, L., Radcliffe, C. W., Miller, T., O'Connell, B., and Chiszar, D., 1982, Trailing behavior in prairie rattlesnakes (Crotalus viridis), J. Herpetol., 16:287.

Halpern, M., and Kubie, J. L., 1983, Snake tongue flicking behavior: Clues to vomeronasal system functions, in: "Chemical Signals in Vertebrates," D. Muller-Schwarze, and R. M. Silverstein, eds., Plenum Press, New York.

Hayes, W. R., and Galusha, J. C., 1984, Effects of rattlesnake (Crotalus viridis oreganus) envenomation upon mobility of male wild and laboratory mice (Mus musculus), Bull. MD Herpetol. Soc., 20:135.

Kamil, A. C., and Sargent, T. D., Eds., 1981, "Foraging Behavior: Ecological, Ethological, and Psychological Approaches," Garland STPM Press, New York.

Kardong, K. V., 1985, The strike behavior of the rattlesnake, Crotalus viridis oreganus, in preparation.
Kubie, J. L., and Halpern, M., 1979, Chemical senses involved in garter snake prey trailing, J. Comp. Physiol. Psychol., 93:648.
Landis, C., and Hunt, W. A., 1939, "The Startle Pattern," Farrar and Rinehart, New York.
Lorenz, K., 1937, Über die Bildung des Instinktbegriffs, Die Naturwissensch, 25:289, 307, 324.
Lorenz, K. Z., 1981, "The Foundations of Ethology," Springer-Verlag, New York.
Muller-Schwarze, D., 1980, Chemical signals in alarm behavior of deer, in: "Chemical Signals: Vertebrates and Aquatic Invertebrates," D. Muller-Schwarze, and R. M. Silverstein, eds., Plenum Press, New York.
Muller-Schwarze, D., and Silverstein, R. M., eds., 1980, "Chemical Signals: Vertebrates and Aquatic Invertebrates," Plenum Press, New York.
O'Connell, B., Greenlee, R., Bacon, J., and Chiszar, D., 1982, Strike-induced chemosensory searching in old world vipers and new world pit vipers at San Diego Zoo, Zoo Biol., 1:287.
O'Connell, B., Greenlee, R., Bacon, J., Smith, H. M., and Chiszar, D., 1985, Strike-induced chemosensory searching in elapid snakes (cobras, taipans, tiger snakes and death adders) at San Diego Zoo, Psychol. Rec., in press.
Radcliffe, C. W., Estep, K., Boyer, J., and Chiszar, D., 1985, Stimulus control of predatory behaviour in red spitting cobras (Naja mossambica pallida) and prairie rattlesnakes (Crotalus v. viridis), Anim. Behav., in press.
Radcliffe, C. W., Poole, T., Feiler, F., Warnock, N., Byers, T., Radcliffe, A., and Chiszar, D., 1983, Immobilization of mice following envenomation by cobras (Naja mossambica pallida), Bull. Psychon. Soc., 21:243.
Radcliffe, C. W., Stimac, K., Smith, H. M., and Chiszar, D., 1983, Effects of prey size on post-strike behavior of juvenile red spitting cobras (Naja mossambica pallida), Trans. Kans. Acad. Sci., 87:59.
Scudder, K. M., Chiszar, D., and Smith, H. M., 1983, The effect of environmental odors on strike-induced chemosensory searching by rattlesnakes, Copeia, 1983:519.
Singer, A. G., Macrides, F., and Agosta, W. C., 1980, Chemical studies of hamster reproductive pheromones, in: "Chemical Signals: Vertebrates and Aquatic Invertebrates," D. Muller-Schwarze, and R. M. Silverstein, eds., Plenum Press, New York.
Sokolov, E. N., 1963, Higher nervous functions: The orienting reflex, Ann. Rev. Physiol., 25:545.
Tinbergen, N., 1951, "The Study of Instinct," Oxford University Press, London.
Voronin, L. G., Leontiev, A. N., Luria, A. R., Sokolov, E. N., and Vinogradova, O. S., eds., 1965, "Orienting Reflex and Exploratory Behavior," Amer. Inst. Biol. Sci., Washington, D. C.
Wells, K. E., Smith, H. M., and Spaur, R. C., 1971, Correlation of ophidian sensory modalities with gross brain proportion, J. Herpetol., 5:200.
Zemstova, M. I., Novikova, L. A., Paramonova, N. P., and Sokolov, E. N., 1965, Investigation of the orienting reflex to sound stimuli in blind persons, in: "Orienting Reflex and Exploratory Behavior," L. G. Voronin, A. N. Leontiev, A. R. Luria, E. N. Sokolov, and O. S. Vinogradova, eds., Amer. Inst. Biol. Sci., Washington, D. C.

CHEMICAL CONTROL OF FEEDING IN HERBIVOROUS AND CARNIVOROUS FISH

Michael A. Adams and Peter B. Johnsen

Monell Chemical Senses Center
University of Pennsylvania
3500 Market Street, Philadelphia, PA 19104

It is now generally recognized that chemical control of feeding is an important modulator of physiological well-being in most animal species (Lindstedt, 1971; Harborne, 1977; Kare and Maller, 1977; Croll, 1983; Simpson and Bernays, 1983). Feeding is an elective process with the animal selecting specific food items from a variety of nutrient sources within the environment. This selection is mediated by two separate factors: chemical constituents of the food item and the animal's own chemosensitivity to those constituents. Fish may detect food by a variety of sensory mechanisms, in particular vision and olfaction, but taste generally plays the key role in determining whether a potential food item is swallowed or rejected (Adron and Mackie, 1978). The taste systems of carnivorous fish appear to be well-tuned for the detection and ingestion of edible animal tissue (e.g., amino acids; Mackie, 1982) and there is evidence that the receptor response spectrum is highly species-specific (Goh and Tamura, 1980). This "sensory tuning" or possession of chemoreceptors which have particular sensitivities may form the basis of feeding regulation and feeding niche separation in fishes. Due to the chemical similarity of all prey items in the carnivore diet, the gustatory system need not have the ability to respond to a variety of chemicals differing widely in structure; most of the information concerning the suitability of a food item can be conveyed by a few amino acids. Thus, an experimental examination of sensory regulation in diet selection is difficult when studying carnivorous fish alone.

We believed that a comparative study with herbivorous fish would be more fruitful. With a diet very different from carnivores and restricted to plant matter, the chemosensitivities of the taste systems of these fish might be expected to be different from carnivores and perhaps more complex in their response to potential food items. Substances present in the diet of an animal may act as feeding stimulants, feeding enhancers (that is, they induce the animal to eat more than it normally would), or as feeding deterrents. The presence or absence of these compounds in the diet will determine whether a food item is eaten or rejected and, to some extent, the quantity consumed. For example, the amino acids of plant tissue would be sufficient stimulation to induce an herbivorous fish to feed, but a secondary sensitivity may operate to prevent the fish from ingesting harmful quantities of toxic plants. Sun and Fenical (1979) have presented an example of this stimulation-deterrence system. When compounds isolated from *Rhipocephalus phoenix* were incorporated into food pellets, the pellets were rejected by the herbivorous fish Eupomacentrus leucostictus, demonstrating the presence of a feeding deterrent. We believe that herbivorous fish present an interesting and unique model system for studying regulation of feeding. By studying the chemosensitivities of herbivorous fish to compounds isolated from plants, and comparing these with the chemosensitivities of carnivores (which would not normally encounter these compounds) we are examining the sensory mechanisms of the regulation of feeding.

The principal hypothesis we are examining in our continuing study is that the gustatory sensitivities of herbivores and carnivores correspond to important chemical constituents of appropriate food items. We are using as our model herbivore a member of the genus Tilapia (Tilapia zillii). We are searching for substances having the ability to enhance, stimulate, or deter feeding in these fish, as measured by behavioral bioassay, from plants upon which the fish feed in their natural setting; and as the isolation procedures yield pure compounds, we are structurally characterizing them. By determining the nature of compounds which regulate the feeding behavior of herbivores and comparing chemosensitivities of herbivores and carnivores, we are elucidating the chemical and sensory mechanisms of regulation of feeding in fishes.

An additional and very practical reason for studying herbivorous fish stems from their widespread use in aquaculture as a human food resource. It is generally acknowledged that animals low on the food chain (i.e., herbivores) can be produced more economically than animals higher on the food chain (i.e., carnivores). By having an understanding of the regulatory mechanisms of food intake in herbivorous fish, it should be possible to manipulate the feeding behavior of these fish through the use of chemical stimulants or feeding enhancers in a manner that would increase food consumption and growth, or that would induce consumption of inexpensive, nutritious plant material that the fish might not normally eat because of reasons of taste.

Our interest in Tilapia stems from the fact that these fish have been harvested since antiquity as a source of food for man and, in the last 200 years, have become one of the most widely cultivated fish species. This fact is reflected in statistics from Israel, which show that Tilapia now account for more than 25% of the fish harvested for human consumption in that country (Sarig and Arieli, 1980).

STUDIES WITH CARNIVOROUS FISH

There is a dearth of information on feeding in herbivorous fish. To date, published investigations into the chemical regulation of feeding in fishes have involved only 24 species. All of the fish studied are carnivores; with the exception of the previously mentioned study involving plant-derived toxins, no work has been conducted with herbivorous fish. Through the studies on carnivorous fish, general techniques for chemical fractionation procedures and testing bioassays have been developed. Application of this well-developed methodology, with some modification, has allowed us to make rapid progress in understanding feeding preferences in T. zillii.

Carr (1982) has succinctly stated the principal requirements for a quantitative study of the chemical regulation of feeding:

"In studies of chemically stimulated feeding behavior, it is imperative to employ quantitative bioassay procedures together with quantitative chemical analyses if an effort is being made to account for the stimulatory capacity of a tissue extract or a biological fluid. Quantitative bioassay procedures permit the investigator to compare directly the stimulatory capacity of a total extract (or biological fluid) with that of an extract fraction or a synthetic preparation. Likewise, quantitative chemical analyses of the components in an extract or biological fluid enable the investigator to relate the effective concentration of a stimulant(s) to the actual concentration at which this stimulant(s) occurs in a preparation of biological origin."

Some general properties of feeding stimulants for fish, derived from animal tissue, have been described by Carr (1982). They are: 1) low molecular weight (< 1000), 2) non-volatile, 3) nitrogenous, and 4) amphoteric. These generalizations apply to stimulatory compounds like betaine, amino acids and inosine, which have been identified in experimental feeding studies using behavioral bioassays. Efficacy of these stimulants have been compared to stimulatory efficacies of whole-animal extracts and the results have commonly been expressed as percent effectiveness. Table 1 summarizes the findings of investigations of feeding stimulants for 22 of the carnivorous species studied to date.

All regulation of energy consumption begins with oral chemoreceptors acting as sentinels for the gastrointestinal system. Atema (1971) provided evidence that alerting, locating, and ingesting behaviors are regulated by these chemoreceptors. The electrophysiological response sensitivities of these receptors do not alone indicate the behavioral efficacy of feeding stimulants, which must be verified by behavioral experiments. However, Goh and Tamura (1980) report close agreement of gustatory electrical responses with behavioral feeding stimulant effectiveness in red sea bream (*Chrysophrys major*). In testing amino acids for stimulatory effects on the lip taste receptors of the puffer fish (*Fugu pardalis*), Hidaka et al. (1978) found five stimulatory (alanine, glutamine, glycine, proline, serine) and three ineffective (arginine, monosodium glutamate, taurine) amino acids. Amino acids which were non-stimulatory in the electrophysiological preparation also failed to elicit feeding responses in separate behavioral experiments, whereas stimulatory amino acids did elicit feeding in the behavioral experiments.

Analysis of the amino acid constituents of the short-necked clam *Tapes japonica* (Konosu et al., 1965) demonstrated the presence of 18 amino acids. Kiyohara et al. (1975), recording from lip receptors of the puffer tested these compounds at concentrations found in clam tissue (5.3×10^{-2} *M* to 1.6×10^{-4} *M*) and found only four (glycine, alanine, proline, aspartic acid) to be stimulatory. When concentrations were increased to 0.01 *M*, electrophysiological responses were obtained only with the same four amino acids. These four amino acids were the second, third, eighth, and tenth most abundant in clam tissue. Gustatory recordings from the Hawaiian goatfish, *Parupeneus porphyreus*, (Holland, 1978) showed that amino acids most abundant in the tissues of the Hawaiian crab *Podophthalmus vigil* (proline, glycine, arginine, glutamic acid, alanine) are all stimulatory, whereas serine, histidine, tyrosine, and cysteine, which when combined constitute only 8% of the total amino acid composition in the tissue, were ineffective as gustatory stimulants.

Goh and Tamura (1980) have suggested, on the basis of electrophysiological recordings, that the amino acid response spectrum is highly species-specific. It has been suggested by Holland (1978) that "sensory

Table 1. Feeding stimulants for several carnivorous fish species. All amino acids are L unless otherwise noted. Bet = betaine, Tau = taurine.

Species	Stimulatory Compounds	Reference
Anguilla japonica (eel)	18 amino acids stimulatory, ineffective individually	Hashimoto et al.,1968
Bathystoma rimator	amino acids ineffective	Haynes et al., 1967
Chrysophrys major (red sea bream)	1/2 (Ala + Bet)+1/2 (Gly + Bet) > Ala> Val > Gly =Ser> Pro >Arg	Goh and Tamura, 1980
Clupea harengus (herring, larvae)	Gly, Pro	Dempsey, 1978
Cyprinus carpio (carp)	17 amino acids: Lys + His + Arg + Asp + Thr + Ser + Glu + Pro + Gly + Ala + Cys + Val + Met + Ile + Leu + Tyr + Phe	Tsushima and Ina, 1978
Fugu pardalis (puffer)	Bet + Ala + Gly + Pro + Ser	Hidaka et al., 1978; Ohsugi et al., 1978
Fundulus heteroclitus (mummichog)	Ala, Gly, δ-aminobutyric acid, His, released biting behavior at 10^{-4} M individually	Sutterlin, 1975
Gadus morhua (cod)	Gly + Ala + Thr + Glu + Leu + Ser	Pawson, 1977
Ictalurus natalis (bullhead)	Cys· HCl	Bardach et al., 1967
Ictalurus puntactus (channel catfish)	Cys makes trout diet more palatable	Little, 1977
Lagodon rhomboides (pinfish)	Bet + 20 amino acids gives 20% activity of shrimp extract, Bet = 1% activity	Carr and Chaney, 1976

Limanda limanda (dab)	synergistic effects: (Pro + Gly)> (Tau + Ala + Arg) > (Tau + Pro + Gly + Ala + Arg)	Mackie, 1982
Menidia menidia (silverside)	Ala, β-Ala, Thr, Met	Sutterlin, 1975
Myxine glutinoso (hagfish)	24 single amino acids non-attractive	Sutterlin, 1975
Orthopristis chrysopterus (pigfish)	Bet + 20 amino acids = shrimp extract, Bet alone = 40% activity	Carr, 1976
Pleuronectes platessa (plaice)	Tau, Pro, Gly, Ala, Arg synergistic effects: (Tau + Pro + Gly + Ala + Arg)>(Pro + Gly)>(Tau + Ala + Arg)	Mackie, 1982
Petromyzon marinus (sea lamprey)	isoleucine methyl ester	Kleerekoper cited in Hara, 1975
Prionotus carolinus (sea robin)	Phe + Tyr + Asp	Bardach and Case, 1965
Pseudopleuronectes americanus (white flounder)	Gly>Ala, Met, Asp, Cys, Glu + Leu	Sutterlin, 1975
Salmo gairdneri (rainbow trout)	Tyr, Phe + Lys or His stimulatory Gly inactive Pro, Tau + Ala + Arg deterrents L-forms stimulant, D-forms repellent	Adron and Mackie, 1978
Scophthalmus maximum (turbot)	inosine	Mackie and Adron, 1978
Solea solea (dover sole)	fish of $\leq$ 2.5 gm wt.: Gly = Ala >Lys> Phe >Glu for fish of $\geq$ 50 gm wt., Gly + Bet	Mackie et al., 1980

Table 2. Relative stimulatory effectiveness of 12 amino acids for 6 species of carnivorous fish. (+) = tested-gave response; (0) = tested-no response; (-) = not tested; numbers indicate that effectiveness was ranked with 1 the most stimulatory.

Chemicals	Catfish[1]	Eel[2]	Puffer[3]	Carp[4]	Goatfish[5]	Minnow[6]
Cys	5	0	0	0	0	2
Met	8	–	0	0	–	8
Phe	–	–	0	0	–	5
Glu	10	0	–	3	+	–
Leu	9	–	0	0	–	7
Ala	1	3	2	2	+	3
Ser	3	5	5	5	0	–
Asp	–	0	4	6	–	–
Thr	6	–	0	0	–	4
Pro	7	4	3	1	+	1
Arg	2	1	0	0	+	–
Gly	4	2	1	4	+	6

[1]Cited from Caprio, 1975; [2]Cited from Yoshii et al., 1979; [3]Cited from Kiyohara et al., 1975; [4]Cited from Marui et al., 1983; [5]Cited from Holland, 1978; [6]Cited from Kiyoara et al., 1981.

tuning" (i.e., possessing a "correct" set of receptors which would allow optimum detection of prey), corresponding to specific chemical components of prey tissue, may be important in the regulation of diet; thus considerable effort has been expended over the years to characterize the chemical sensitivities of chemoreceptors in many species of fish (Caprio, 1975, 1977, 1978; Funakoshi et al., 1981; Goh and Tamura, 1980; Hidaka et al., 1976, 1978; Holland, 1978; Kaku et al., 1980; Kiyohara et al., 1975, 1981; Marui et al., 1983; Yoshii et al., 1979).

Six species of fish have been studied in detail and the relative stimulatory effectiveness of some amino acids have been ranked in Table 2.

Among the various fish species there are obvious differences in sensitivities to amino acids. It is interesting to note that all studies to date have been confined exclusively to carnivorous species. The carnivorous fish all have diets which are reasonably similar in chemical composition. The herbivorous species in our study has a diet completely unlike any previously studied species. By taking a comparative approach, we hope to examine the concept of "sensory tuning" and the way in which selection of diet is regulated by sensory control.

The biological effects that we are searching for are of three types (Figure 1). The first is feeding stimulation. This is the ability of a substance to induce a fish to feed in the absence of any "whole" food stimulus. The second effect is that of feeding enhancement. The effect is manifested by an increase in consumption of an enhancer-modified food item over the amount of the same food item consumed without added enhancer. Finally, a feeding deterrent effect is marked by the ability of a substance to inhibit a fish from feeding upon a food that is normally found to be completely palatable. We measure these effects by a behavioral bioassay, which is described in detail below.

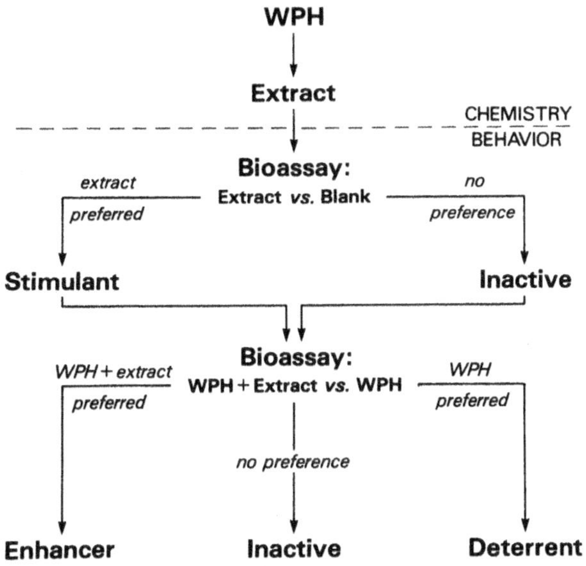

Figure 1. Flow chart depicting the process of determining feeding stimulants, enhancers, and deterrents using the agar block bioassay technique (WPH is whole plant homogenate).

EXPERIMENTAL PROCEDURES

Romaine lettuce (a plant readily eaten by *T. zillii*) was selected for use as a reference standard in our bioassays. It was purchased from commercial suppliers locally and immediately weighed and homogenized in a Waring blender prior to further processing. This homogenate, after centrifuging to remove suspended solids, is used directly as an assay standard. In similar fashion, any food plant can be so processed to give a whole plant homogenate for use as an assay standard.

Using our extraction procedure (a modification of the procedure described by Uhazy and co-workers; Uhazy et al., 1978) (Johnsen and Adams, 1985a) we were able to separate an aqueous homogenate of Romaine lettuce into several different fractions, based on substance type (see Figure 2). This procedure is summarized as follows: The aqueous plant homogenate (usually a 2:1 water-plant suspension, v/v) was centrifuged to remove solids and the resulting supernatant liquid was filtered to provide further clarification of the solution. Proteins were precipitated by addition of trichloroacetic acid (TCA) followed by centrifugation to pellet the proteins. The resulting supernatant liquid was decanted and this protein-depleted solution was extracted with diethyl ether to remove lipids; the lipids were recovered for bioassay by evaporating the ether, dissolving the residue in chloroform, and partitioning the solution against water. Evaporation of the chloroform solution gave the lipid fraction. The aqueous residue from ether extraction was lyophyllized and then chromatographed on a strong cation exchange resin column. Elution with 0.01 *N* hydrochloric acid gave a fraction containing organic acids and sugars. A second elution with 4 *N* NH_4OH produced an amino acid-containing fraction. The fraction containing organic acids and sugars was lyophyllized and rechromatographed on a DEAE-cellulose column. Elution with water gave a sugars fraction; subsequently organic acids were eluted from the column by washing with 4 *N* formic acid. The biological activity of all fractions was determined by bioassay (see

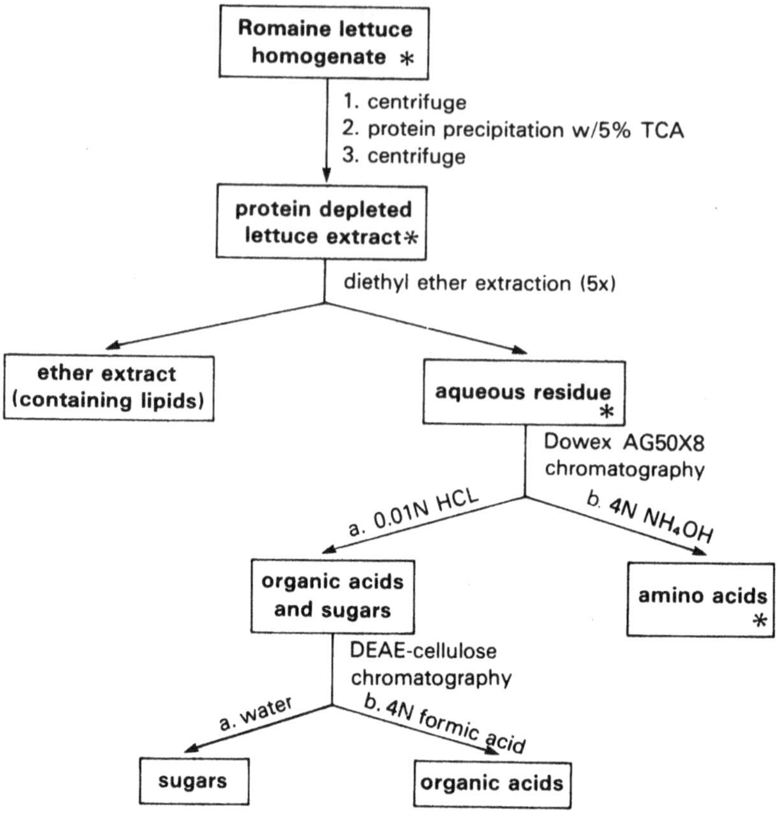

Figure 2. Fractionation scheme for Romaine lettuce. An asterisk denotes a fraction with biological activity.

below) as they were produced; thus, feeding activity was reliably tracked through the fractionation scheme.

BIOASSAY PROCEDURE

The standard method of our bioassay is a two-choice preference test in which compounds are presented to fish in a disk-shaped agar matrix (Johnsen and Adams, 1985b). The agar itself elicits minimal feeding activity and what little consumption there is appears to be exploratory; the fish take a few small bites when investigating the presented sample, but eat no more than that. Bioassay samples are prepared in the following way. Two percent Noble agar (w/v) is brought to a boil in deionized water and allowed to dissolve. The heat is then removed and the solution is allowed to cool. The appropriate quantity of a test sample is then added at any temperature above the solidification point (ca. 40°C) of the agar. The sample-containing agar solution is cast in petri dishes in which are suspended 2.5 cm plastic rods (computer printout binding posts). After the agar cools and solidifies, a circular cutter is used to punch out disks of agar which have the plastic rod protruding from them. The plastic rod is attached to a mounting bracket and this assembly is placed in the test tank (Figure 3). The mounting brackets hold both an experimental and a reference disk. Each disk is weighed before and again after a two hour test period during which

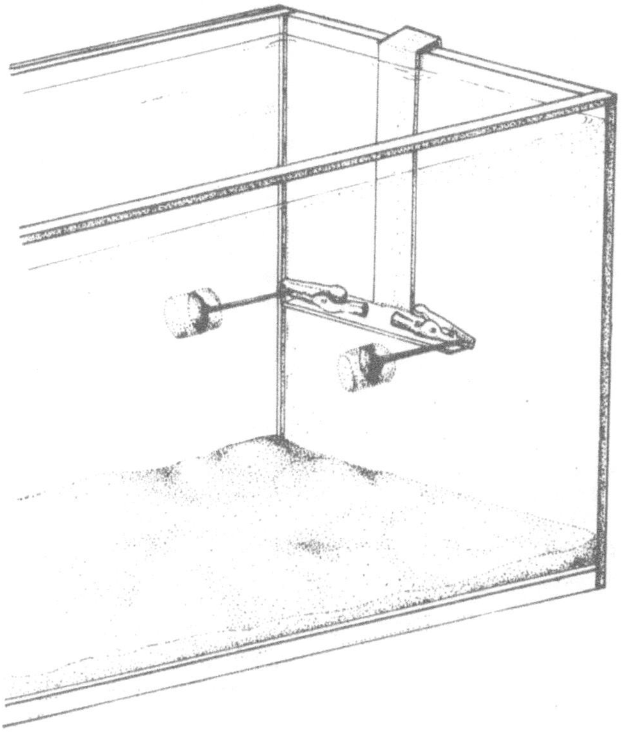

Figure 3. Bioassay apparatus. Two agar disks are shown, attached to a mounting bracket and suspended in an aquarium.

the disks are placed in a tank containing a single fish. The weight differences of the disks are used to calculate consumption and preference. A set of control disks is also placed into water for the two hour test period to permit water gain by the test disks to be measured. The agar disks typically increase in weight by one to two percent during the two hour test. This hydration factor is then used to calculate a corrected initial disk weight.

The agar disk bioassay has several advantages over previously used methods. Actual consumption is measured rather than the intensity of feeding activity or arousal. The agar itself is relatively non-stimulatory, permitting baseline consumptions to be easily determined. Noble agar is commercially available in purified form and, because test samples are presented in a solid matrix, only small quantities are required. Table 3 presents sample data collected from *Tilapia zillii* using our procedure. The results of Test 1 in which plain agar (the reference) was paired with agar containing 0.01 *M* glutamic acid (an experimental stimulus), demonstrate that the fish prefer the sample containing glutamic acid, indicating that sufficient material is released from the matrix during the two hour test to elicit feeding. Test 2, 0.01 *M* glutamic acid *vs*. 0.01 *M* glutamic acid, indicates that the fish exhibited no position preference in the two-choice test; both disks are sampled equally. Fish are able to detect and respond to differences in the concentration of amino acid released from the agar matrix, as shown in Test 3, 0.1 *M* glutamic acid *vs*. 0.01 *M* glutamic acid. Test 4, 0.01 *M* alanine *vs*. agar alone, shows that another amino

Table 3. Results of selected bioassays using the agar disk bioassay technique.

Test No.	Experimental	Reference	No. of trials	mean % preference for expt'l sample $\bar{x}$ ± (S.D.)
1.	0.01 M glutamic acid	blank	14	76 ± (13)*
2.	0.01 M glutamic acid	0.01 M glutamic acid	15	54 ± (15)
3.	0.1 M glutamic acid	0.01 M glutamic acid	16	84 ± (19)*
4.	0.01 M alanine	blank	11	65 ± (13)*
5.	0.01 M glutamic acid	0.01 M alanine	14	72 ± (17)*

* = $p < 0.01$, Student's t test.

acid is an adequate feeding stimulus while Test 5, 0.01 M glutamic acid vs. 0.01 M alanine, demonstrates that glutamic acid is preferred over alanine. These results demonstrate the type of tests and results which may be produced by using the agar disk bioassay in feeding studies.

The agar disks allow small amounts of test material to diffuse slowly into the water over a relatively long period of time. Thus, the time for the blocks to lose their flavor has been determined to be significantly longer than the test time. Because only small amounts of test material are added to the agar, the test is very conservative of sample. This greatly reduces the scale of the preparative chemistry needed to extract and purify compounds which we wish to assay. Another advantage of this method is that actual consumption may be measured directly. Previous studies have suffered from the necessity of using a more indirect index of "interest" based on observations of feeding behaviors (Hidaka et al., 1978; Carr et al., 1976). Preference in our tests is calculated as mean percent preference, and a Student's t-test is used to compare between the two means.

RESULTS

General

Using the extraction procedures detailed previously, we have separated an aqueous homogenate of Romaine lettuce into several different fractions based on substance type (see Figure 2). Romaine lettuce was selected as our standard plant reference material because it is inexpensive, it is readily eaten by *T. zillii*, is available throughout the year, and good analytical data exists on its amino acid and organic acid content. Each fraction was assayed for feeding activity and fractions which stimulated feeding in our test fish are marked with an asterisk in Figure 2. Romaine lettuce homogenate served as the standard of activity to which all other extracts were related, at least qualitatively. We found that if we removed proteins from the homogenate by precipitating them with TCA, the fish fed on the protein-depleted extract to the same degree that they fed on whole lettuce. Diethyl ether extraction of the protein-depleted lettuce extract produced two subfractions: the ether extract and the aqueous residue. The lipid-containing ether extract had no feeding activity; the feeding activity resided entirely in the aqueous residue. When this residue was chromatographed on a column of strong cation exchange resin, it was possible by washing with 0.01 N hydrochloric acid to recover a fraction containing mainly organic acids and

Table 4. Feeding assay results for *Tilapia zillii*. All amino acids are L. Samples are made with 2% agar, dyed green (to mask color effects). Percent lettuce refers to the amount of whole lettuce homogenate added to the agar matrix. Lettuce fractions are 10% solutions. Pref stands for preference.

Experimental	Reference	n	% Pref ± (S.D.)	p < 0.01?*
Lettuce homogenate:				
blank	blank	10	46 ± (22)	n
1% lettuce	blank	12	73 ± (10)	y
10% lettuce	blank	13	83 ± (16)	y
10% lettuce	1% lettuce	8	71 ± (8)	y
Lettuce fractions:				
protein depleted extract	blank	18	83 ± (11)	y
lipids fraction	blank	18	67 ± (27)	n
aqueous residue (w/o lipids)	blank	15	84 ± (12)	y
total amino acids	blank	18	94 ± (6)	y
organic acids and sugars	blank	18	57 ± (20)	n
total organic acids	blank	10	56 ± (17)	n
total sugars	blank	11	48 ± (19)	n
Amino acids:				
0.01 M glutamic acid	blank	14	76 ± (13)	y
0.01 M aspartic acid	blank	13	72 ± (19)	y
0.01 M lysine	blank	10	75 ± (16)	y
0.01 M serine	blank	17	70 ± (17)	y
0.01 M alanine	blank	11	65 ± (13)	y
0.01 M betaine	blank	10	59 ± (13)	n
0.01 M cysteine	blank	14	51 ± (15)	n
0.01 M glutamine	blank	13	60 ± (18)	n
0.01 M proline	blank	14	54 ± (26)	n
0.01 M asparagine	blank	8	69 ± (19)	n
0.01 M taurine	blank	8	44 ± (10)	n
0.01 M valine	blank	18	56 ± (15)	n
0.01 M glycine	blank	14	50 ± (25)	n
Sugars:				
0.01 M glucose	blank	14	50 ± (17)	n
0.01 M fructose	blank	9	43 ± (22)	n
Organic acids:				
0.01 M citric acid	blank	15	86 ± (9)	y
0.001 M citric acid	blank	13	53 ± (24)	n
0.01 M malic acid	blank	27	62 ± (24)	n

* = $p < 0.01$, Student's t test.

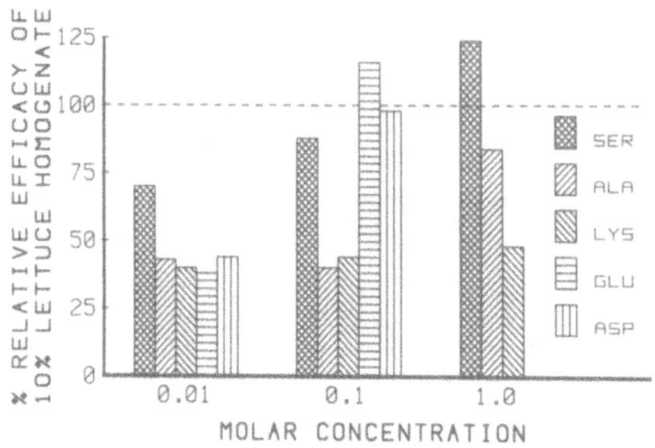

Figure 4. Relative feeding efficacies of five stimulatory amino acids for T. zillii. Whole Romaine lettuce extract is the standard of comparison. Glutamic acid and aspartic acid were not tested at 1.0 M because their feeding efficacies were close to or over 100% when tested at 0.1 M.

sugars. Washing the column with 4 N ammonium hydroxide solution eluted the amino acids. Again, feeding assays were done on each of the fractions and these indicated that feeding activity resided in the amino acid fraction. Results of all these tests are displayed in Table 4.

Stimulatory Substances

An extensive investigation of the stimulatory efficacy of amino acids was carried out to determine the feeding responsiveness of T. zillii to these substances. In general, carnivorous fish prefer amino acids such as alanine, glutamine, glycine, proline, arginine, taurine and betaine (see Table 1). Tilapia in our tests prefer glutamic and aspartic acids more than any of the other amino acids tested. Interestingly, glutamic acid is the most abundant amino acid in Romaine lettuce (Uhazy et al., 1978).

The stimulatory efficacies of 13 amino acids found in significant quantities in Romaine lettuce (Uhazy et al., 1978) or squid extract (Mackie, 1982) were determined at 0.01 M; only glutamic acid, aspartic acid, lysine, alanine, and serine elicited feeding responses (Table 4). The relative efficacies of these five compounds were compared with 10% whole lettuce homogenate (Figure 4). When tested at 0.01 M, all five amino acids produced an average efficacy of 45% (ala = 34; asp = 44; glu = 38; lys = 40; ser = 70). At 0.1 M the efficacy of aspartic acid and serine rose to 98% and 88% respectively, while glutamic acid elicited more feeding than the 10% whole lettuce homogenate, resulting in an efficacy value of 116%. The ten-fold increase in concentration of alanine and of lysine only increased the efficacies slightly, to 40% and 44% respectively. An alanine concentration of 1.0 M produced a relative efficacy of 84%, and serine reached 124%. The efficacy of lysine remained quite low; at the highest concentration tested it was only 48% as effective as whole lettuce homogenate. In addition to measuring the relative efficacy of these five stimulatory amino acids, pair-wise preference tests were conducted to rank them for feeding preference (Table 5). Glutamic acid was preferred over all others, with aspartic acid, lysine, serine and alanine not differentiated by the fish.

Table 5. Pair-wise comparisons of amino acids, tested at 0.01 M.

Experimental	Reference	Number of trials	Mean % pref. ± (S.D.)	Experimental preferred?*
glu	asp	30	62 ± (23)	y
glu	ala	14	72 ± (17)	y
glu	lys	19	77 ± (19)	y
glu	ser	17	73 ± (21)	y
asp	lys	25	57 ± (27)	n
asp	ala	19	47 ± (23)	n
asp	ser	31	53 ± (18)	n
lys	ala	14	55 ± (28)	n
lys	ser	20	45 ± (24)	n
ala	ser	30	48 ± (13)	n

* = $p < 0.01$, Student's t test.

The organic acids and sugars fractions showed no activity of itself, but we chromatographed this material to separate the organic acids from the sugars so that they could be bioassayed separately. To effect this separation, the fraction was chromatographed on a column of DEAE-cellulose. Elution with water gave the sugars, and elution with 4N formic acid released the organic acids. Bioassay of these subfractions revealed no stimulatory activity. Analysis of the organic acid content of Romaine lettuce showed that two acids, malic and citric, were present in fairly large quantities. Accordingly, we decided to test these substances individually in the feeding assay. We found that malic acid had no effect on feeding but citric acid evoked a feeding response (i.e., acts as a stimulant). We performed additional experiments to determine the threshold of sensitivity of *Tilapia* to citric acid and found that the threshold lies between 0.001 M and 0.01 M for plain agar containing only citric acid (Table 4).

DISCUSSION

The principal objective of this study was to determine the chemical nature of feeding stimuli found in plants readily eaten by the herbivorous fish *Tilapia zillii*. By chemically fractionating a whole homogenate of Romaine lettuce, we were able to isolate fractions containing proteins, lipids, sugars, organic acids, and amino acids. These fractions were tested in our bioassay for stimulatory efficacy, and only the amino acid fraction elicited a feeding response (Table 1). The free amino acids found in Romaine lettuce (Uhazy et al., 1978) were then assayed at a concentration of 0.01 M and five were found to be stimulatory (glutamic acid, aspartic acid, serine, lysine, and alanine). These amino acids were then assayed for relative preference (glutamic acid > aspartic acid = serine = lysine = alanine). The relative efficacies of these five amino acids, with respect to whole plant homogenate, were determined and were ranked as glutamic acid > aspartic acid > serine > alanine > lysine (Figure 4). It is interesting to note that this herbivore can be stimulated to feed by single amino acids. In contrast to our results with *Tilapia*, some investigators (Hashimoto et al., 1968; Konosu et al., 1968; Adron and Mackie, 1978) working with carnivorous fish report that, whereas mixtures of amino acids are effective feeding stimuli, the individual components of the mixtures, when presented singly, are often lacking in stimulatory efficacy.

As with the herbivore *Tilapia zillii*, amino acids appear to play a principal role in stimulating feeding in carnivorous fish (for reviews see Carr, 1982; Mackie, 1982). However, the specific amino acids which stimulate feeding appear to be quite different for the two groups.

Amino acids which stimulate feeding in the sea bream (*Chrysophrys major*) are alanine, betaine, glycine, and valine (Ina and Matsui, 1980). The Dover sole (*Solea solea*) is stimulated to feed by betaine, glycine, and alanine (Mackie et al., 1980) while taurine, proline, glycine, alanine and arginine are stimulatory for the plaice (*Pleuronectes platessa*) (Mackie, 1982). The puffer (*Fugu pardalis*) is stimulated to feed by betaine, glycine, proline, and serine (Hidaka et al., 1978). Examination of the amino acid composition of squid (Mackie, 1982) reveals that the amino acids found to be most stimulatory for these four carnivores are also among the most abundant. In order of abundance they are: proline, betaine, glycine, taurine, alanine, and arginine. Comparison of stimulatory amino acids for the herbivore *T. zillii* with their order of abundance in lettuce tissue (Uhazy et al., 1978) shows a similar concordance. Glutamic acid, aspartic acid, serine, lysine, and alanine were found to be stimulatory while their order of relative abundance is first, second, fourth, sixth and thirteenth respectively. Compounds which were found to be among the most stimulatory for carnivores (proline, valine, taurine, glycine, and betaine) failed to elicit feeding responses in the herbivore, demonstrating the very different response sensitivities of the two feeding types. Like *Tilapia*, the herbivorous marine gastropod, *Aplysia californica*, exhibits feeding responses to a limited spectrum of amino acids, seven of twenty tested, and glutamic and aspartic acids were found to be the most stimulatory (Jahan-Paywar, 1975).

CONCLUSIONS AND FUTURE PROSPECTS

Our results with *T. zillii* suggest that the regulation of feeding in this species is profoundly affected by the chemical makeup of the plants upon which the fish feed. These chemicals form the basis of acceptance or rejection of food items in much the same way that food taste characteristics control ingestion in terrestrial vertebrates. Evidence from the study of carnivorous fish indicates that these fish make feeding decisions based on chemical information released by their prey. The fact that some mechanism of sensory tuning functions in a manner that restricts herbivores to feeding upon plants and constrains carnivorous fish to animal-based foods possibly serves to maintain feeding niche separation of these two types of fish.

We are currently examining our extracts for activities as feeding enhancers or deterrents. Our results so far indicate that enhancers and deterrents are powerful contributors to the chemical regulation of feeding in *Tilapia*. Further work in this area will be directed to an investigation of the plants upon which *T. zillii* normally feed. We expect that the results of these studies will contribute to the knowledge of mechanisms of sensory function and regulation in fish, as well as to provide information useful to the aquaculturist interested in raising herbivorous fish as a source of high quality protein for human consumption.

REFERENCES

Adron, J. W., and Mackie, A. M., 1978, Studies on the chemical nature of feeding stimulants for rainbow trout, Salmo gairdneri Richardson, J. Fish Biol., 12:303.

Atema, J., 1971, Structures and functions of the sense of taste in the catfish (Ictalurus natalis), Brain Behav. Evol., 4:273.

Bardach, J. E., and Case, J., 1965, Sensory capabilities of the modified fins of squirrel hake (Urophycis chuss) and searobins (Prionotus carolinus and P. evolans), Copeia, 1965:194.

Bardach, J. E., Todd, J. H., and Crickmer, R., 1967, Orientation by taste in fish of the genus Ictalurus, Science, 155:1276.

Caprio, J., 1975, High sensitivity of catfish taste receptors to amino acids, Comp. Biochem. Physiol., 52A:247.

Caprio, J., 1977, Electrophysiological distinctions between the taste and smell of amino acids in catfish, Nature, 266:850.

Caprio, J., 1978, Olfaction and taste in the channel catfish: An electrophysiological study of the responses to amino acids and derivatives, J. Comp. Physiol., 123:357.

Carr, W. E. S., 1976, Chemoreception and feeding behavior in the pigfish, Orthopristis chrysopterus: Characterization and identification of stimulatory substances in a shrimp extract, Comp. Biochem. Physiol., 55A:153.

Carr, W. E. S., and Chaney, T. B., 1976, Chemical stimulation of feeding behavior in the pinfish, Lagodon rhomboides: Characterization and identification of stimulatory substances extracted from shrimp, Comp. Biochem. Physiol., 54A:437.

Carr, W. E. S., Gondeck, A. R., and Delanoy, R. L., 1976, Chemical stimulation of feeding behavior in the pinfish, Lagodon rhomboides: A new approach to an old problem, Comp. Biochem. Physiol., 54A:161.

Carr, W., 1982, Chemical stimulation of feeding behavior, in: "Chemoreception in Fishes," T. J. Hara, ed., Elsevier, New York.

Croll, R. P., 1983, Gastropod chemoreception, Biol. Rev., 58:293.

Dempsey, C. H., 1978, Chemical stimuli as a factor in feeding and intraspecific behavior of herring larvae, J. Mar. Biol. Assoc., UK., 58:739.

Funakoshi, M., Kawakita, K., and Marui, T., 1981, Taste responses in the facial nerve of the carp, Cyprinus carpio (L.), Japan J. Physiol., 31:381.

Goh, Y., and Tamura, T., 1980, Olfactory and gustatory responses to amino acids in two marine teleosts - red sea bream and mullet, Comp. Biochem. Physiol., 66C:217.

Hara, T. J., 1975, Olfaction in fish, Prog. Neurobiol., 58:739.

Harborne, J. B., 1977, "Introduction to Ecological Biochemistry," Academic Press, New York.

Hashimoto, Y., Konosu, S., Fusetani, N., and Nose, T., 1968, Attractants for eels in the extracts of short-necked clam - I. Survey of constituents eliciting feeding behavior by the omission test, Bull. Japan Soc. Sci. Fish., 34:78.

Haynes, L. T., Sangster, A. W., Steven, D. M., and Thomas, S., 1967, Chemical factors inducing exploratory feeding behavior in fish - E. F. B. - inducing properties of marine invertebrates, Comp. Biochem. Physiol., 20:755.

Hidaka, I., Nyu, N., and Kiyohara, S., 1976, Gustatory response in the puffer - IV. Effects of mixtures of amino acids and betaine, Bull. Fac. Fish., Mie Univ., 3:17.

Hidaka, I., Ohsugi, T., and Kubomatsu, T., 1978, Taste receptor stimulation and feeding behaviour in the puffer, Fugu paradalis, I. Effect of single chemicals, Chem. Senses Flavour, 3:341.

Holland, K., 1978, Chemosensory orientation to food by a Hawaiian goatfish (Parupeneus porphyreus, Mullidae), J. Chem. Ecol., 4:173.

Ina, K., and Matsui, H., 1980, Survey of feeding stimulants for the sea bream (Chrysophrys major) in the marine worm Perinereis vancaurica tetradentata, J. Agric. Chem. Soc. Jpn., 54:7.
Jahan-Pawar, B., 1975, Chemoreception in gastropods, in: "Olfaction and Taste V" D. A. Denton, and J. P. Coghlan, eds., Academic Press, New York.
Johnsen, P. B., and Adams, M. A., 1985, Chemical feeding stimulants for the herbivorous fish, Tilapia zillii, Comp. Biochem. Physiol., in press.
Johnsen, P. B., and Adams, M. A., 1985b, A solid matrix bioassay for determining chemical feeding stimulants, Prog. Fish-Cult., in press.
Kaku, T., Tsumagari, M., Kiyohara, S., and Yamashita, S., 1980, Gustatory responses in the minnow, Pseudorasbora parva, Physiol. Behav., 25:99.
Kare, M. R., and Maller, O., 1977, "The Chemical Senses and Nutrition," Academic Press, New York.
Kiyohara, S., Hidaka, I., and Tamura, T., 1975, Gustatory responses in the puffer - II. Single fiber analyses, Bull. Jpn. Soc. Sci. Fish., 41:383.
Kiyohara, S., Yamashita, S., and Harada, S., 1981, High sensitivity of minnow gustatory receptors to amino acids, Physiol. Behav., 26:1103.
Konosu, S., Fujimoto, K., Takashima, Y., Matsushita, T., and Hashimoto, Y., 1965, Constituents of the extracts and amino acid composition of the protein of short-necked clam, Bull. Jpn. Soc. Sci. Fish., 31:680.
Konosu, S., Fusetani, N., Nose, T., and Hashimoto, Y., 1968, Attractants for eels in the extracts of short-necked clam. II. Survey of constituents eliciting feeding behavior by fractionation of the extracts, Bull. Jpn. Soc. Sci. Fish., 34:84.
Lindstedt, K. J., 1971, Chemical control of feeding behavior, Comp. Biochem. Physiol., 39A:553.
Little, E. E., 1977, Conditioned aversion to amino acid flavors in the catfish, Ictalurus punctatus, Physiol. Behav., 19:743.
Mackie, A. M., and Adron, J. W., 1978, Identification of inosine and inosine-5'-mono-phosphate as the gustatory feeding stimulants for the turbot, Scophthalmus maximus, Comp. Biochem. Physiol., 60A:79.
Mackie, A. M., Adron, J. W., and Grant, P. T., 1980, Chemical nature of feeding stimulants for the juvenile Dover sole, Solea solea (L.), J. Fish Biol., 16:701.
Mackie, A. M., 1982, Identificaiton of the gustatory feeding stimulants, in: "Chemoreception in Fishes," T. J. Hara, ed., Elsevier, New York.
Marui, T., Harada, S., and Kasahara, Y., 1983, Gustatory specificity for amino acids in the facial taste system of the carp, Cyprinus carpio (L.), J. Comp. Physiol., 153:299.
Mazid, M. A., Tanaka, Y., Katayama, T., Simpson, K., and Chichester, C., 1978, Metabolism of amino acids in aquatic animals - III. Indispensable amino acids for Tilapia zillii, Bull. Jpn. Soc. Sci. Fish., 44:739.
Ohsugi, T., Hidaka, I., and Ikeda, M., 1978, Taste receptor stimulation and feeding behaviour in the puffer, Fugu pardalis. II. Effects produced by mixtures of constituents of clam extracts, Chem. Senses Flavour, 3:355.
Pawson, M. G., 1977, Analysis of natural chemical attractant for whiting Merlangius merlangius (L.) and cod Gadus morrhua (L.) using a behavioral bioassay, Comp. Biochem. Physiol., 56A:129.
Sarig, S., and Arieli, Y., 1980, Growth capacity of Tilapia in intensive culture, Bamidgeh, 32:57.
Simpson, S. J., and Bernays, E. A., 1983, The regulation of feeding: Locusts and blowflies are not so different from mammals, Appetite, 4:313.
Sun, H. H., and Fenical, W., 1979, Rhipocephalin and rhipocephenal; toxic feeding deterrents from the tropical marine alga Rhipocephalus phoenix, Tetrahedron Lett., 685.

Sutterlin, A. M., 1975, Chemical attraction of some marine fish in their natural habitat, J. Fish. Res. Board Can., 32:729.

Tsushima, J., and Ina, K., 1978, Survey of feeding stimulants for carps, *Cyprinus carpio* (Studies on the feeding stimulants for fishes Part III), J. Agric. Chem. Soc. Jpn., 52:225. (In Japanese, with an English abstract.)

Uhazy, L. S., Tanaka, R. D., and MacInnis, A. J., 1978, *Schistosoma mansoni*: Identification of chemicals that attract or trap its snail vector, *Biomphalaria glabrata*, Science, 201:924.

Yoshii, K., Kamo, N., Kurihara, K., and Kobatake, Y., 1979, Gustatory responses of eel palatine receptors to amino acids and carboxylic acids, J. Gen. Physiol., 74:301.

CHARACTERISTICS OF EARTHWORM WASHINGS DETECTED BY THE VOMERONASAL SYSTEM OF SNAKES

Mimi Halpern, Nancy Schulman and *Donald M. Kirschenbaum

Departments of Anatomy and Cell Biology and *Biochemistry
Downstate Medical Center
Brooklyn, New York 11203

Garter snakes of the genus *Thamnophis* respond in a species-typical manner, by rapidly tongue-flicking and attacking objects coated with aqueous washes of their preferred prey (Burghardt, 1966, 1967a, 1967b, 1968, 1969; Campbell, 1970; Wilde, 1938). The snakes are highly selective in this response, responding with greatest vigor to washes of prey that form part of their natural diet. Although prey wash preference can be altered (Arnold, 1978; Burghardt et al., 1973; Fuchs and Burghardt, 1971), newborn garter snakes with no previous neonatal feeding experience demonstrate a strong preference for the washes of prey that form the diet of adult members of their species or subspecies living in similar geographic areas and ecological niches (Arnold, 1977, 1980, 1981; Burghardt, 1970b; Gove and Burghardt, 1975).

The vomeronasal system has been strongly implicated in detection of the chemical prey washes that snakes respond to by either increased tongue flicks, attack or trailing (Burghardt and Hess, 1968; Burghardt and Pruitt, 1975; Chiszar et al., 1981; Halpern and Frumin, 1979; Kahmann, 1932; 1934; Kubie and Halpern, 1979; Naulleau, 1966; Wilde, 1938). As part of our continuing efforts to understand the vomeronasal system and contrast it to the olfactory system, we have attempted to isolate and characterize the active components in the chemoattractant portion of the wash of one prey of snakes, the common earthworm, *Lumbricus terrestris*.

Early work by Sheffield et al. (1968), using the Wilde-Burghardt (Burghardt, 1966; Wilde, 1938) swab technique, demonstrated that the active component in earthworm wash was a non-volatile, water soluble, remarkably stable macromolecule, resistant to heating, lyophilization and pH manipulation.

GENERAL PROCEDURE

We pursued these studies, initially replicating many of the results of Sheffield et al. (1968), but utilizing a somewhat different bioassay (Reformata et al., 1983). In our studies snakes are individually housed in plastic fish tanks (13 x 18 x 15 cm). One hour prior to a trial, the contents of the cage are removed. Test samples are presented on a discrimination tray that holds two petri dishes. One dish is coated with test sample, the other with a control (distilled water, dH_2O, or saline) sample.

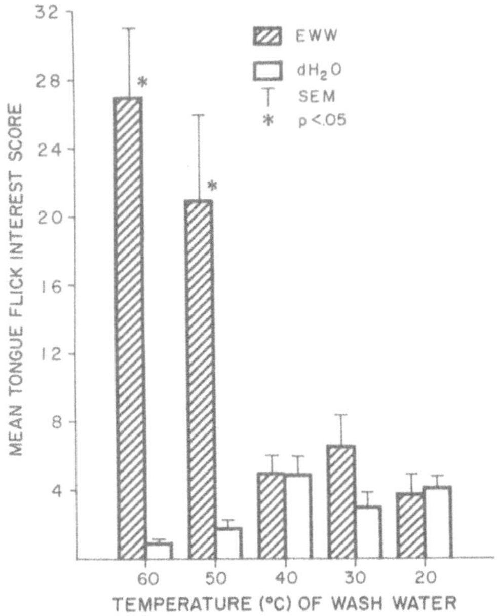

Fig. 1. Mean tongue flick interest score of snakes presented with earthworm wash prepared at 60, 50, 40, 30 and 20°C. Asterisk indicates that mean tongue flick interest score to worm wash was significantly different from mean tongue flick interest score to dH_2O at the .05 level of confidence using a paired T test. SEM = standard error of the mean.

The side of the discrimination tray on which the test sample is placed is determined by a random numbers table.

The discrimination tray is lowered into the snake's cage, and a trial begins when the snake approaches the tray. A trial is two minutes in duration; tongue flicks toward and time spent at each petri dish are recorded. A tongue flick interest score is calculated for each sample on each trial (tongue flicks x time at dish/total time at both dishes). An n x n Latin square design is used to insure that each animal in the study is tested with each sample and that the order of testing of samples is balanced throughout the experiment.

Earthworm wash (EWW) is normally prepared by placing 6 gms of clean earthworms in 20 cc of 60° dH_2O for one or two minutes (our earlier studies used one minute, our current studies use two minutes). This is the standard (1X) concentration of EWW. The temperature at which EWW is prepared is critical to the production of the chemoattractant (Halpern et al., 1984; Sheffield et al., 1968). At temperatures lower than 40°C, no discriminable chemoattractant is produced (Fig. 1).

Characteristics of the Chemoattractant

Our early studies demonstrated that garter snakes reliably spend more time and tongue-flick more frequently at a dish containing EWW than a dish containing dH_2O (Reformato et al., 1983). Garter snakes respond differentially to washes of worms that form part of their normal diet (Lumbricus),

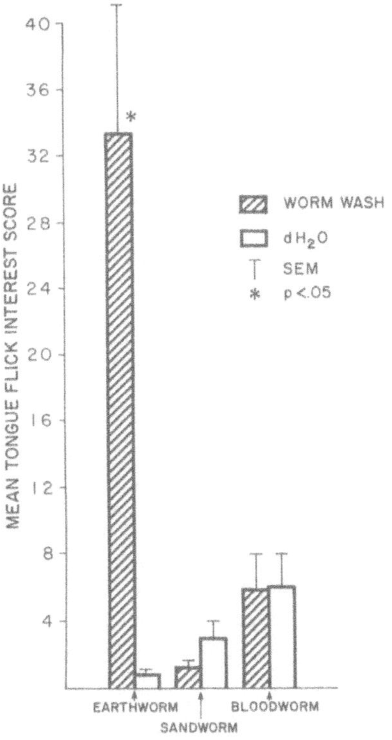

Fig. 2. Mean tongue flick interest score of snakes presented with washes of earthworm, sandworm and bloodworm. See Fig. 1 for other abbreviations.

but not worms they would not normally eat (Nereis virens, the sandworm, or Glycera debranciata, the blood worm) (Fig. 2).

The discriminability of the EWW is directly related to its concentration; successive 1:3 dilutions of EWW with dH_2O result in increasingly poor discriminability. Dilutions greater than 1/243X are not discriminated by the snakes (Reformato et al., 1983).

The differential response of snakes to EWW is dependent on a functional vomeronasal system since snakes with their vomeronasal ducts sutured closed fail to respond differentially to EWW compared to a control substance (Reformato et al., 1983).

The chemoattractant material in EWW absorbs at 280nm and 230 nm, contains carbohydrate, is ninhydrin-positive, gives positive Bradford and Lowry test for protein and is precipitated by 90% - 100% saturated ammonium sulfate (Halpern et al., 1984). Earthworm wash retains its chemoattractant properties after dialysis through 50,000 molecular weight (MW) cut off tubing, and lyophilization. However, repeated lyophilization, freeze-drying and thawing, do diminish and eventually destroy the chemoattractant in EWW (Table 1).

Earthworm wash - dialyzed, using 3500 MW cut-off tubing, lyophilized, reconstituted in 0.15M saline and placed on a Sephadex G-75, AcA 44 or AcA 34 column and eluted using 0.15M saline as eluent - produces an elution profile similar to that illustrated in Fig. 3. These experiments (Halpern et al., 1984) consistently result in a chromatographic pattern consisting

Table 1. Response of Garter Snakes to Earthworm Wash Subjected to Repeated Shell Freezing and Repeated Lyophilization.

	Mean TFIS	±SEM		Mean TFIS	±SEM	
EWW	33.93	5.43	dH_2O	0.02	0.02	a
EWW_D	14.97	3.37	dH_2O	0.36	0.19	a
EWW_{S1}	11.21	2.63	dH_2O	0.34	0.19	a
EWW_{S2}	12.17	2.67	dH_2O	0.74	0.33	a
EWW_{S3}	10.62	3.20	dH_2O	0.54	0.12	a
EWW_{L1}	10.67	2.10	dH_2O	0.46	0.09	a
EWW_{L2}	11.39	1.57	dH_2O	0.62	0.20	a
EWW_{L3}	6.41	1.38	dH_2O	1.87	0.53	

Abbreviations: EWW = earthworm wash; D = dialyzed (MW cut-off 3500); S_1 - S_3 = shell frozen 1 to 3 times; L_1 - L_3 = shell frozen and lyophilized 1 to 3 times; TFIS = tongue flick interest score; SEM = standard error of the mean; a = probability of difference between mean TFIS for EWW and $dH_2O < .05$.

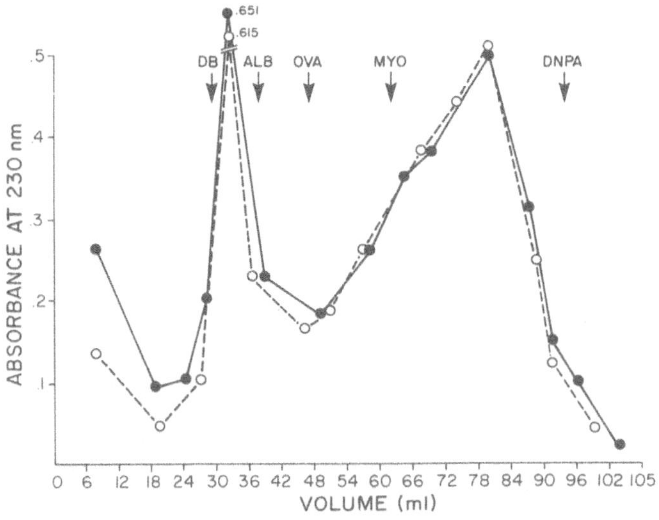

Fig. 3. Elution profile of solution of lyophilized earthworm wash separated on a 1.6 x 42.5 cm column of AcA 44 using 0.15 M NaCl as eluent. Two ml fractions were collected. •-•-• 17.6 mg EWW in 2 ml 0.15 M NaCl; 0-0-0 21.3 mg EWW in 2 ml 0.15 M NaCl. DB = Blue Dextran; ALB = bovine serum albumin; OVA = ovalbumin; MYO = myoglobin; DNPA = DNP-alanine.

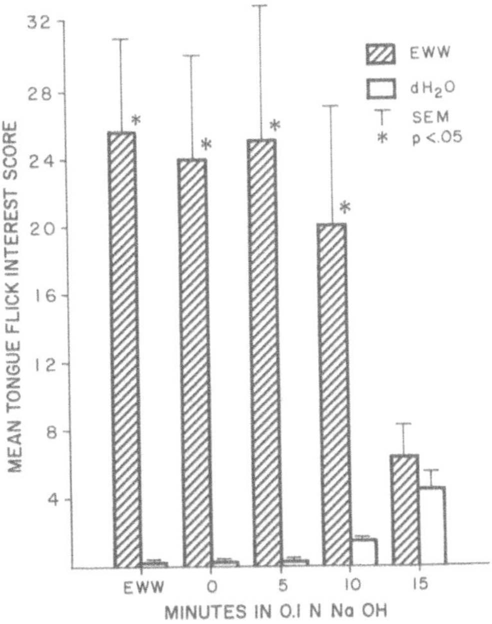

Fig. 4. Mean tongue flick interest score of snakes presented with earthworm wash (EWW) and earthworm wash heated in a boiling water bath in 0.1 N NaOH for 0, 5, 10 or 15 minutes, cooled and neutralized to pH 5.8. See Fig. 1 for other abbreviations.

of two peaks, a high MW peak, F2, where the chemoattractant activity resides, and a low MW peak, F4, devoid of chemoattractant activity. Between these 2 peaks minor amounts of material have been detected by 280 nm and 230 nm absorption. When EWW is dialysed through 50K Da dialysis tubing the low MW peak, F4, is no longer detected and chemoattractant activity is retained in the F2 peak. F2 appears in front of the albumin marker location and just behind the Blue Dextran (DB) marker location. This places the MW of the F2 peak well above the MW of albumin which is 65K Da. From preliminary HPLC analysis the MW appears to be as large as 400K Da. Collagen derived from earthworm cuticle is known to be a large, high MW molecule and we believe a fragment of it to be the chemoattractant (see below).

Carbohydrate determination mimics the pattern found by 280 nm and 230 nm absorption measurement. Thus F2 and F4 peaks are found by this method (Kirschenbaum et al., 1985).

pH and Temperature Effects

The chemoattractant property of EWW is remarkably resistant to pH and temperature variation (Halpern et al., 1984; Reformato et al., 1983; Sheffield et al., 1968). Snakes continue to respond to EWW, which has a normal pH of 5.8, when highly acidified (pH2) or highly alkalinized (pH11) (Reformato et al., 1983). Neutral EWW continues to retain its chemoattractant properties after boiling for 2 hours, but loses it after 4 hours (Halpern et al., 1984). EWW in 0.1 N HCl, heated in a boiling water bath for more than 15 minutes, loses its chemoattractant properties (Kirschenbaum et al., 1985), but retains an F2 peak for as long as 3 hours of heating. EWW in 0.1 N NaOH is more labile; when heated in a boiling water bath chemoattractant activity is lost after 10 minutes (Fig. 4), as is the F2 peak as

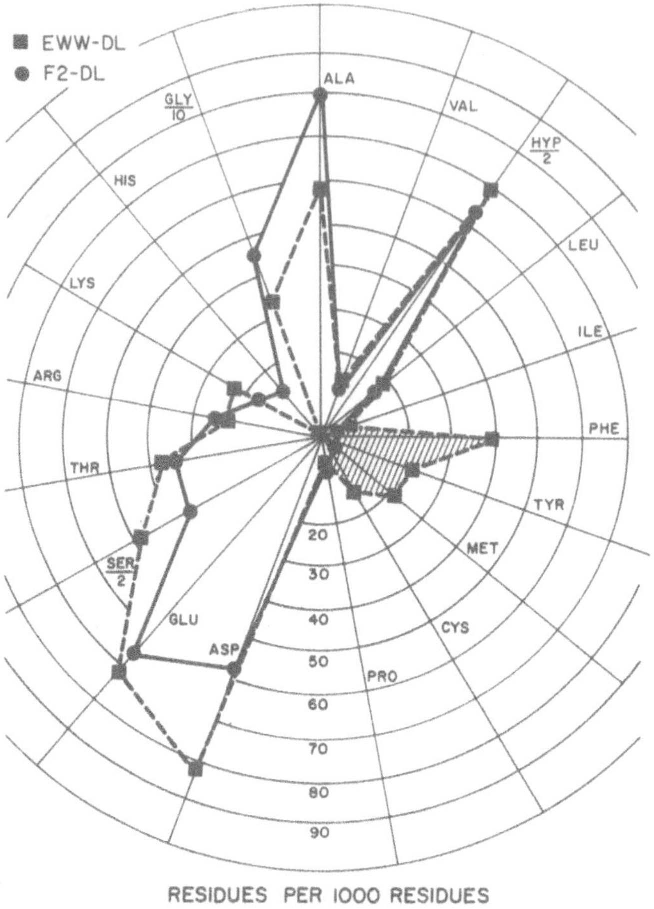

Fig. 5. Amino acid contents, as residues per thousand, of dialyzed, lyophilized earthworm wash (EWW-DL) and the dialyzed, lyophilized high molecular weight peak fraction (F2) of EWW.

demonstrated by the loss of 280 nm absorption, lack of carbohydrate and lack of ninhydrin positive material (Kirschenbaum et al., 1985).

There are two kinds of temperature effects to be seen when working with EWW chemoattractant. One kind can be detected in initial stages of EWW preparation. Examination of washes prepared at different temperatures, 20° - 60°, reveal that only those made at 50° and 60° contain chemoattractant and give an F2 and F4 peak. In addition, lyophilized 60° preparations result in fine sheets of material resembling tissue paper. Preparations made at 20° and 40° and lyophilized result in meager powdery preparations not at all comparable to the 60° preparation. It is of interest that garter snakes respond to EWW chemoattractant _in vivo_ at ambient temperatures while it is necessary to place worms in 50° H_2O to extract the chemoattractant. We discuss this apparent inconsistency below.

The second temperature effect takes place when the EWW is simultaneously submitted to pH and heating perturbations. The changes effected can be detected chemically and behaviorally. In 0.1 N NaOH at 37° chemoattractant activity can be detected after 30 minutes of incubation but not at 60 minutes. Examination of the elution profiles from gel exclusion chromatography at 0, 30 and 60 minutes reveals at each time an F2 and an F4 peak by UV absorption and carbohydrate determination. This is quite different from what occurs in 0.1 N NaOH at boiling water temperature. Here chemoattractant activity is lost after 10 minutes as is the F2 peak. The F4 peak persists and in fact increases. The results of these experiments led us to look for an explanation for this temperature-pH lability. Among several possibilities (Kirschenbaum et al., 1985), O-glycosidically linked carbohydrates, which are susceptible to β-elimination of carbohydrate, appeared to be most relevant.

Amino Acid Analysis

Fig. 5 represents the amino acid analyses obtained for dialyzed, lyophilized EWW and the F2 prepared from it. This method for displaying amino acid analyses results in a pattern called a "rose" (Robinson et al., 1975). Each amino acid is assigned a radius and each radius is marked at every 20 residues per thousand. When the results of each analysis are plotted on this circular grid two patterns are obtained which are fairly similar except for small amounts of sulfur-containing amino acids, methionine and cystine, and aromatic acids, phenylalanine and tyrosine, which are present in the wash but not in the F2 fraction obtained from the wash by gel exclusion chromatography. The important relationships to observe here, which are distinctive for oligochaete cuticle collagen, are the large content of hydroxyproline, the large content of serine plus threonine, the very large content of glycine, approximately 1/3 of the number of residues, and the very small amount of proline. Not shown on the diagram is the absence of hydroxylysine, and amino acid prevalent in vertebrate collagens (Kirschenbaum et al., 1986).

Earthworm Cuticle Collagen: A Chemoattractant?

The amino acid profile of earthworm chemoattractant so closely corresponds to the amino acid profiles reported for earthworm cuticle collagen (EWC) (Spiro and Bhoyroo, 1980) and earthworm cuticle gelatin (EWG) (Watson, 1958) that we proceeded to determine if EWC had structural and behavioral properties similar to earthworm chemoattractant.

Earthworm cuticle collagen was purified by the methods of Murray et al. (1982) and Josse and Harrington (1964). Lyophilized EWC placed on an AcA 44 column produced an elution profile (Fig. 6) very similar to the elution profile of EWW (Fig. 3). Large F2 peaks and large F4 peaks were observed. Behavioral testing showed that the F2 peak contained chemoattractant activity (Fig. 7) as did the EWC prior separation on the column. To determine if earthworms that had been decuticlized could be used to produce chemoattractant, we removed the cuticles of 50 worms by placing the worms in anhydrous ether for 15 minutes and then stripping the cuticles from the worms (Watson, 1958). The decuticlized worms were placed in 60°C H_2O for 2 minutes, discarded, and the wash water centrifuged. The supernatant was tested for chemoattractant activity as described above and found to have no chemoattractant activity.

Our usual procedure for producing chemoattractant is to bathe worms in 60°C dH_2O for 1 or 2 minutes. Earthworm cuticle has a melting point of 22°-23°. We reasoned that repeated washes in 60°C dH_2O should remove all chemoattractant if we were melting off the cuticle, and this was the source of the chemoattractant. Worms were placed in a 60°C dH_2O bath for one minute, removed and the same worms placed into a second 60°C dH_2O bath for one minute, removed and the same worms placed in a third 60°C H_2O bath. The worms were discarded and each water bath centrifuged. The supernatants were tested for chemoattractant activity as described above. Earthworm

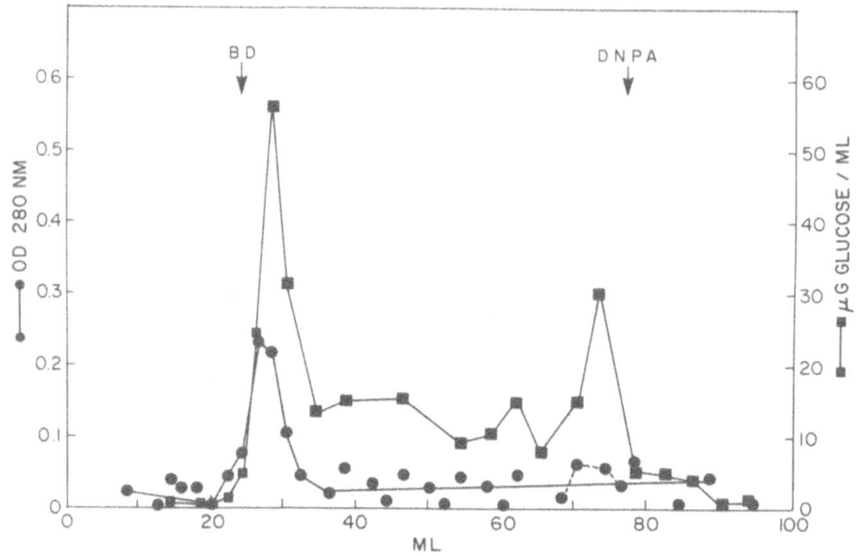

Fig. 6. Elution profile of solution of 15 mg lyophilized earthworm cuticle in 2 ml of 0.15 M NaCl separated on a 1.6 x 42.5 cm column of AcA 44. Two ml fractions were collected. ●-●-●, 280 nm absorption; ■-■-■, carbohydrate (glucose) determination. Abbreviations as in Fig. 3.

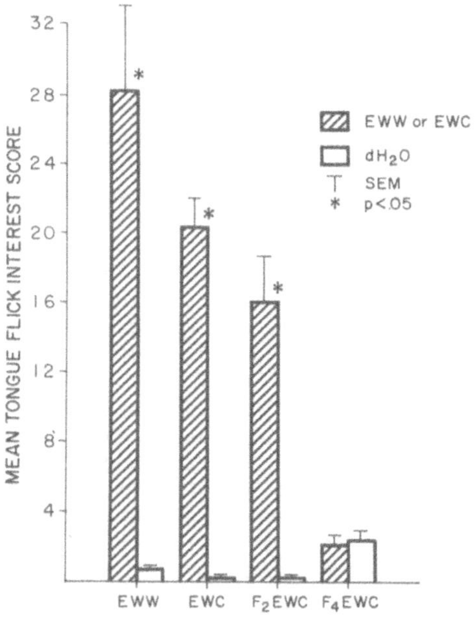

Fig. 7. Mean tongue flick interest score of snakes presented with earthworm wash (EWW), solutions of earthworm cuticle collagen (EWC), the high molecular weight peak fractions of earthworm cuticle collagen (F2 or F_2EWC), and the low molecular weight peak fractions of earthworm cuticle collagen (F4 or F_4EWC).

wash from worms that had not been used previously had chemoattractant activity; water wash of worms that had been subjected to one previous 60° bath also had chemoattractant activity; however, water wash of worms that had been previously subjected to two 60°C baths did not have chemoattractant activity. These results suggest that the two one minute baths removed all of the surface substances that normally provide the source of chemoattractant (Kirschenbaum et al., 1986).

As suggested above, it is puzzling that to produce EWW, earthworms must be bathed in 60° H_2O whereas the melting point of earthworm cuticle collagen is 22-23°. This apparent inconsistency may relate to the difference between the denaturation temperature (T_D) and shrinkage temperature (T_S) of EWC (Kimura, 1971). The denaturation temperature is determined on a solution of collagen by observing the optical rotation or the reduced viscosity of the solution as a function of temperature. The midpoint of the transition is the denaturation temperature. In contrast, the shrinkage temperature is determined by observing the changes in length of the piece of collagen, a fiber, or even the cuticle in water at different temperatures. The midpoint of the transition is the shrinkage temperature. Thus one technique starts with a solution of collagen (T_D) while the other starts with an insoluble mass (T_S) which remains so during the determination. When a variety of annelid collagens were examined (Kimura, 1971), T_S was usually some 15° higher than T_D.

This finding of the relationship between T_S and T_D may be pertinent to the preparation of EWW. When the earthworms are put into water at 60° the controlling factor might well be the shrinkage temperature of the cuticle. As the heating continues, the cuticle solubilizes and the solution of EWG has a much less viscous consistency. The MW of the material in solution drops to 600K from 1.9×10^6 (Josse and Harrington, 1964) upon thermal treatment. Whether this represents a disassociation into subunits is not known for EWG, but for *Nereis* it has been demonstrated that its collagen consists of two kinds of subunits with a MW of 5×10^5 (Kimura and Tanzer, 1977).

Current knowledge available to us about earthworm cuticle collagen/gelatin does not help us explain how the snake, at ambient temperatures, is able to tongue flick extremely small particles of earthworm cuticle collagen, abraded as the earthworm moves through the ground, and transfer them to its vomeronasal organ where receptors for earthworm cuticle collagen/gelatin presumably exist that function at ambient temperatures.

Perhaps an enzyme system in the tongue or the vomeronasal organ of the snake processes the chemoattractant to make it detectable by the snake. During this extremely short period, between pick-up by the flicking tongue and delivery to the vomeronasal organ, a labile bond may be cleaved making active chemoattractant. This enzymatically mediated conversion may be equivalent to the extraction of EWW at 50° or 60° to make it a chemoattractant suggesting that there may be a thermolabile bond. Rather than considering only bond cleavage of some kind, a conformational change, such as collagen to gelatin conversion, may also be considered. Cuticle collagen is known to denature in solution at 22°-24° (Josse and Harrington, 1964; Maser and Rice, 1962, 1963) and renature very slowly. A conformational change of this type is known to affect structure. The product of the denaturation of EWC is EWG, known to be more susceptible to enzymatic attack.

While much of our work has been done with chemoattractant, we have also found some things that are not chemoattractants but are part of earthworm cuticle collagen. The tripeptide, glycylhydroxyprolylserine, which has been isolated from earthworm cuticle collagen (Goldstein and

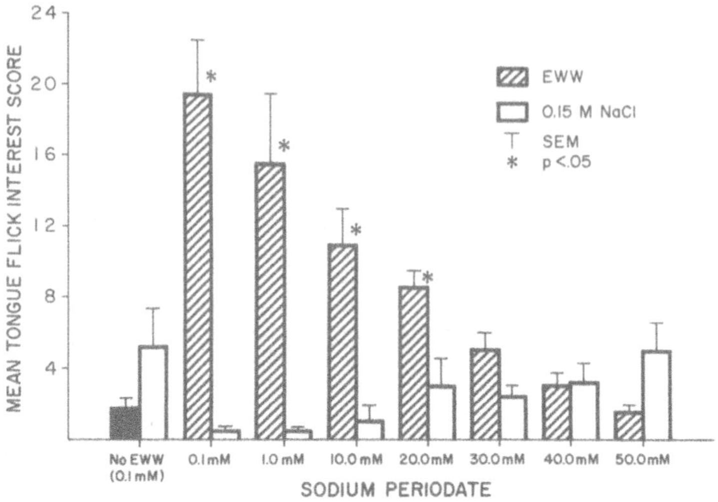

Fig. 8. Mean tongue flick interest score of snakes presented with control sample (0.1 mM sodium periodate with no earthworm wash) or earthworm wash treated with 0.1, 1.0, 10.0, 20.0, 30.0, 40.0, and 50.0, mM sodium periodate for 1 h at room temperature. See Fig. 1 for other abbreviations.

Adams, 1970) and galactose, a major carbohydrate of earthworm cuticle collagen are inactive as chemoattractants for garter snakes.

Periodate Effects

Our studies to this point strongly suggested, based on the relative heat stability, protein and carbohydrate content of the chemoattractant, that we were dealing with a glycoprotein and more specifically an earthworm cuticle/gelatin-like material. We have based this latter suggestion upon the similarity of amino acid analyses, the large galactose content and its ability to undergo a β-elimination. Thus we became interested in determining the relationship between carbohydrate and chemoattractant activity.

Periodate, IO_4^-, is known to cleave cis-diols under very mild conditions in aqueous solution (Montgomery, 1972). It has been so used to determine the structure of glycoproteins and the relationship of carbohydrate structure to function in glycoproteins. In some very preliminary work, we have studied the effect of periodate at several concentrations and at two different times on the chemoattractant activity of EWW.

From control studies, we learned that periodate and ethylene glycol, separately and in mixture, behave neither as attractants nor as aversive agents for garter snakes. Treatment of EWW with different concentrations of periodate for 15 minutes at room temperature (RT) resulted in loss of chemoattractant activity at 2.8 mM periodate while at 2.1 mM periodate activity could still be detected. When the reaction time was increased to 1 hour at RT, activity was retained up to 0.83 mM periodate but lost at 1.25 mM periodate (Fig. 8). At 8.3 mM periodate at RT, inactivation was instantaneous. (The periodate concentrations shown in Fig. 8 are the concentrations used prior to dilution with EWW. In the text, final periodate concentrations are used. The equivalent values are, initial concentrations: 0.1, 1.0, 10.0, 20.0, 30.0, 40.0, 50.0 mM; final concentrations: 0.007, 0.04, 0.42, 0.83, 1.25, 1.67, 2.1 mM, respectively.)

The loss of chemoattractant activity by reaction with periodate leads to the inevitable conclusion that the carbohydrate residues of EWW are directly related to chemoattractant activity. The galactose residues of EWC are found as mono-, di- and trisaccharides, mainly as the di- and trisaccharides, O-glycosidically linked to serine and threonine residues of the protein chain (Muir and Lee, 1969, 1970). We believe that the loss of activity due to alkali treatment is due to β-elimination of carbohydrate from the serine and threonine residues (Pigman et al., 1974). Periodate is not considered as a protein modifying reagent although certain free amino acids, not in peptide linkage, may be modified by treatment with periodate (Maekawa and Kushibe, 1954). The amino acids attacked when in peptide linkage, cystine, methionine, tyrosine and tryptophane (Maekawa and Kushibe, 1954) are either absent or present in low concentration in EWW. In addition the rate of attack is much slower on amino acid residues than on carbohydrate residues.

To summarize: treatment of EWW with periodate at room temperature results in the loss of chemoattractant activity. A possible explanation for the loss of this activity is the destruction of some or all of the carbohydrate residues needed to interact with the vomeronasal receptors of the snake. These receptors may recognize the galactose residues of the active site of the chemoattractant and respond. Modification of the chemoattractant results in loss of the ability of the vomeronasal receptors to recognize the active site. A question of major import to which we are directing our efforts is the determination of the minimum change in the structure of the chemoattractant which results in the loss of chemoattractant activity. We also plan on isolating the di- and trisaccharides from EWC and from the chemoattractant to determine the effects on chemoattractant activity of these substances and of the protein portion devoid of carbohydrate.

Annelid Cuticle Collagens Compared

Washes of earthworm (*Lumbricus*), sand worm (*Nereis*), and blood worm (*Glycera*) were tested as chemoattractants for the garter snake. The results are shown in Fig. 2. Only the wash of *Lumbricus* functioned as a chemoattractant. Examination of washes of sand worm and blood worm by gel exclusion chromatography showed that both washes had a high MW peak with a MW in excess of 65K Da and a low MW peak, less than 16K Da. The high MW peak and low MW peak both contained carbohydrate. Thus, although both worms provided chemical data similar to that found for an earthworm wash, neither wash contained chemoattractant for the garter snake. Is there any chemical structure in the material in the washes to which this differential response of the snake can possibly be allocated? No definitive answer can be provided at present. Nevertheless, it would be profitable to compare what is known of the chemical constituents and physical characteristics of the *Lumbricus* and *Nereis* surface collagens from which, in the case of *Lumbricus*, a chemoattractant may be obtained.

Cuticle collagens from *Nereis* and *Lumbricus* are very large molecules, ca 1.7×10^6 Da for *Nereis* and 1.9×10^6 Da for *Lumbricus* (Josse and Harrington, 1964; Kimura and Tanzer, 1977). Table 2 presents the results of amino acid analyses and carbohydrate analyses of the cuticle collagens of the two annelids, one an oligochaete and the other a polychaete. Included in this table are quantitative analyses and relationships between amino acids whose presence constitute the hallmarks of collagen: glycine, proline, hydroxyproline, and hydroxylysine; the hallmark amino acids of invertebrate collagens: the imino acids, hydroxyproline and proline, the hydroxyamino acids, serine and threonine; the amino acids that define the charge characteristics of a protein: the acidic amino acids, glutamic acid and aspartic acid; the basic amino acids lysine, hydroxylysine, arginine and histidine; and

Table 2. The Amino Acid Analyses and Carbohydrate Analyses of the Cuticle Collagen of the Oligochaete *Lumbricus* and the Polychaete *Nereis* (adapted from Spiro and Bhoyroo, 1980).

AMINO ACID[a]	Sand Worm *Nereis virens*	Earthworm *Lumbricus terrestris*
4 Hydroxyproline (HYP)	165	179
Proline (PRO)	56	7
HYP/PRO	2.95	25.6
Threonine (THR)	43	50
Serine (SER)	69	87
THR/SER	0.62	0.5
Glutamic Acid (GLU)	134	84
Aspartic Acid (ASP)	13	60
GLU + ASP	147	144
GLU/ASP	10.3	1.4
Glycine	319	318
Cystine + Methionine	1	2.1
Tyrosine	1	1.5
Hydroxylysine (HYL)	0	0
Lysine (LYS)	1.2	18
Arginine (ARG)	79	25
Histidine (HIS)	0.3	1
HYL + LYS + ARG + HIS	80.5	44
LYS + ARG + HIS/GLU + ASP	0.55	0.31
CARBOHYDRATE		
Galactose	9.1	57
Glucuronic Acid	2.0	0
Mannose	2.2	1.7

[a]Residues per thousand residues of amino acid.

the amino acids usually present in very small amounts or totally lacking in all types of collagen: the sulfur containing amino acids, cystine and methionine, and the aromatic amino acid responsible for the 280 nm absorption: tyrosine. Tryptophan is usually absent from both kinds of collagens. Both cuticles are qualitatively similar. Quantitatively they are quite different except for the content of glycine present, to about one-third of the total amino acid content. Both imino acids, hydroxyproline and proline, are present but in *Lumbricus* collagen the ratio of one to the other is 10-fold greater than in *Nereis* collagen due to the much lower proline content in *Nereis* collagen. The large content of 4-hydroxyproline is preserved in both invertebrate collagens. In contrast, vertebrate collagens usually have a proline content somewhat larger than the hydroxyproline content (Kimura et al., 1969). The hydroxyamino acids, threonine and serine, are present in approximately the same amounts. The acidic amino acids, glutamic acid and aspartic acid, are present in the same amount, but their proportions are quite different. There is approximately ten-fold as much glutamic acid as aspartic acid in *Nereis* collagen.

If we examine the ratio of basic to acidic amino acids, HLY + LYS + ARG + HIS/ GLU + ASP, we can learn something about the charge characteristics of the protein. There are almost twice as many of the basic amino acids in *Nereis* collagen as in *Lumbricus* collagen. Most of this is contributed by massive amounts of the strongly basic amino acid arginine. In *Lumbricus* collagen the amide group is present as 115 residues per thousand residues (Maser and Rice, 1962) which are found as glutaminyl and asparaginyl residues. This could make an acidic protein, where the ratio of basic to acidic amino acids is only 0.31, become a basic protein if all the amide groups are assigned to glutamic and aspartic acids. The ratio of basic to acidic amino acid residues would change drastically to 1.52, a five-fold difference. No information is available for the *Nereis* collagen amide content.

Vertebrate collagens have hydroxylysine residues to which the carbohydrate residues are attached. Invertebrate collagens usually lack hydroxylysine and use the hydroxy groups of serine and threonine to attach their carbohydrate residues. Thus, in addition to other criteria mentioned above for the identification of invertebrate collagen, the almost total absence of hydroxylysine and the presence of large amounts of serine and threonine can be added.

If we now turn our attention to the carbohydrates found in the two collagens (Spiro and Bhoyroo, 1980) two results are outstanding: the 6-fold excess of galactose of *Lumbricus* collagen over that of *Nereis* collagen and the presence of two residues of glucuronic acid in *Nereis* collagen and its absence in *Lumbricus* collagen. The glucuronic acid is present as the disaccharide 6-O-α-D-glucuronosyl-O-D-mannose. The small content of galactose is distributed in the following proportion: monosaccharide:disaccharide:trisaccharide, 1:2:2. The galactose residues of *Lumbricus* collagen are found mainly as di- and trisaccharide derivatives O-glycosidically attached to serine and threonine. The residues of galactose are present in the following proportion: monosaccharide:disaccharide:trisaccharide, 1:17:5. The major portion of the galactose is found as the disaccharide 2-O-α-D-galactosyl-pyranosyl-D-galactose (Muir and Lee, 1969; Muir and Lee, 1970).

From this short description of what is known concerning the chemistry of *Nereis* and *Lumbricus* collagens and the knowledge that alkali treatment destroys the ability of *Lumbricus* collagen to act as a chemoattractant for garter snakes we can conjecture that the chemoattractant could be a small fragment of annelid oligochaete collagen having O-glycosidically linked carbohydrates. In addition, the carbohydrates should be neutral, as galactose, and not acidic, as glucuronic acid. Until further data are available, very little more can be added to these requirements.

The combined physical, chemical and behavioral evidence presented above strongly suggests the notion that at least a component of earthworm chemoattractant is a part of the earthworm cuticle collagen molecule.

ACKNOWLEDGMENTS

The research described in this chapter was generously supported by NIH grant NS 11713. We thank Tina Salerno for typing the manuscript.

REFERENCES

Arnold, S. J., 1977, Polymorphism and geographic variation in the feeding behavior of the garter snake *Thamnophis elegans*, *Science*, 197:676.

Arnold, S. J., 1978, Some effects of early experience on feeding responses in the common garter snake, Thamnophis sirtalis, Anim. Behav., 26:455.
Arnold, S. J., 1980, The microevolution of feeding behavior, in: "Foraging Behavior: Ecological, Ethological and Psychological Perspectives," A. Kamil, and T. Sargent, eds., Garland Press, New York.
Arnold, S. J., 1981, Behavioral variation in natural populations. I. Phenotypic, genetic and environmental correlations between chemoreceptive responses to prey in the garter snake, Thamnophis elegans, Evolution, 35:489.
Burghardt, G. M., 1966, Stimulus control of the prey attack response of naive garter snakes, Psychon. Sci., 4:37.
Burghardt, G. M., 1967a, Chemical perception in newborn snakes, Psychol. Today, 1:50.
Burghardt, G. M., 1967b, Chemical-cue preferences of inexperienced snakes: Comparative aspects, Science, 157:718.
Burghardt, G. M., 1968, Chemical preference studies on newborn snakes of three sympatric species of Natrix, Copeia, 732.
Burghardt, G. M., 1969, Comparative prey-attack studies in newborn snakes of the genus Thamnophis, Behaviour, 33:77.
Burghardt, G. M., 1970a, Chemical perception in reptiles, in: "Communication by Chemical Signals," J. W. Johnston, D. G. Moulton, and A. Turk, eds., Appleton-Century-Crofts, New York.
Burghardt, G. M., 1970b, Intraspecific geographical variation in chemical food cue preferences of newborn garter snakes (Thamnophis sirtalis), Behaviour, 36:246.
Burghardt, G. M., and Hess, E. H., 1968, Factors influencing the chemical release of prey attack in newborn snakes, J. Comp. Physiol. Psychol., 66:289.
Burghardt, G. M., and Pruitt, C. H., 1975, Role of the tongue and senses in feeding of naive and experienced garter snakes, Physiol. Behav., 14:185.
Burghardt, G. M., Wilcoxon, H. C., and Czaplicki, J. A., 1973, Conditioning in garter snakes. Aversion to palatable prey induced by delayed illness, Anim. Learn. Behav., 1:317.
Campbell, H. W., 1970, Prey selection in naive, Elaphe obsoleta (Squamata: Serpentes) - A reappraisal, Psychon. Sci., 21:300.
Chiszar, D., Taylor, S. V., Radcliffe, C. W., Smith, H. M., and O'Connell, B., 1981, Effect of chemical and visual stimuli on chemosensory searching by garter snakes and rattlesnakes, J. Herpetol., 15:415.
Fuchs, J. L., and Burghardt, G. M., 1971, Effects of early feeding experience on the responses of garter snakes to food chemicals, Learn. Motiv., 2:271.
Goldstein, A., and Adams, E., 1970, Glycylhydroxyprolyl sequences in earthworm cuticle collagen: glycylhydroxyprolylserine, J. Biol. Chem., 245:5478.
Gove, D., and Burghardt, G. M., 1975, Responses of ecologically dissimilar populations of the water snake Natrix s. sipedon to chemical cues from prey, J. Chem. Ecol., 1:25.
Halpern, M., and Frumin, N., 1979, Roles of the vomeronasal and olfactory systems in prey attack and feeding in adult garter snakes, Physiol. Behav., 22:1183.
Halpern, M., Schulman, N., Scribani, L. and Kirschenbaum, D. M., 1984. Characterization of vomeronasally-mediated response-eliciting components of earthworm-wash II, Pharmacol. Biochem. Behav., 21:655.
Josse, J., and Harrington, W. G., 1964, Role of pyrolidine residues in the structure and stabilization of collagen, J. Mol. Biol., 9:269.
Kahmann, H., 1932, Sinnesphysiologische Studien an Reptilien - I. Experimentalle Untersuchungen über das Jacobsonsche Qrgan der Eidechsen und Schlangen, Zool. Jb. Abt. Allg. Zool. Physiol., 51:173.
Kahmann, H., 1934, Zur Chemorezeption der Schlangen (ein Nachtrag.) Zool Anz., 107:249.

Kimura, S., 1971, Marine invertebrate collagens. 3. Characterization of cuticle collagens in annelids, Bull. Jpn. Soc. Sci. Fish., 37:419.
Kimura, S., Nagaoko, Y., and Kubota, M., 1969, Studies on marine invertebrate collagens - I. Some collagens from Crustaceans and Molluscs, Bull. Jpn. Sci. Fish., 35:743.
Kimura, S., and Tanzer, M. L., 1977, Nereis cuticle collagen: Isolation and characterization of two distinct subunits, Bioch., 16:2554.
Kubie, J. L., and Halpern, M., 1979, Chemical senses involved in garter snake prey trailing, J. Comp. Physiol. Psychol., 93:648.
Kirschenbaum, D. M., Schulman, N., and Halpern, M., 1986, Earthworms produce a collagen-like substance detected by the garter snake vomeronasal system, Proc. Nat. Acad. Sci., in press.
Kirschenbaum, D. H., Schulman, N., Yao, P., and Halpern, M., 1985, Chemoattractant for the garter snake: Characterization of vomeronasally-mediated response-eliciting components of earthworm wash III, Comp. Biochem. Physiol., in press.
Maekawa, K., and Kushibe, M., 1954, Les effets du hyperiodate sur le spectre d'absorption du tryptophane, de la tyrosine, et de la proteine, Scientific Repts. of the Makuyama Agric. College, 12:33.
Maser, M. D., and Rice, R. V., 1962, Biophysical and biochemical properties of earthworm-cuticle collagen, Biochem. Biophys. Acta, 63:255.
Maser, M. D., and Rice, R. V., 1963, The denaturation and renaturation of earthworm cuticle collagen, Biochem. Biophys. Acta, 74:283.
Montgomery, R., 1972, Periodation of glycoproteins and glycopeptides, in: "Glycoproteins, Their Composition, Structure and Function," Part B, Gottschalk, A., ed., Elsevier Publishing Co., New York.
Muir, L., and Lee, Y. C., 1969, Structures of the D-galactose oligosaccharides from earthworm cuticle collagen, J. Biol. Chem., 244:2343.
Muir, L., and Lee, Y. C., 1970, Glycopeptides from earthworm cuticle collagen, J. Biol. Chem., 245:502.
Murray, L. W., Waite, J. H., Tanzer, M. L., and Hauschka, P. V., 1982, Preparation and characterization of invertebrate collagens, Methods Enzymol., 82:65.
Naulleau, G., 1966, La Biologie et le Comportement Prédateur de Vipera aspis au Laboratoire et dans la Nature, These. P. Fanlac, Paris as reported by Burghardt, 1970a.
Pigman, W., Moschera, J., Downs, F., and Wakabayashi, R., 1974, Mechanisms and conditions for beta-elimination of ologosaccharides from glycoproteins, in: "Méthodologie de la Structure et du Métabolisme des Glycoconjugates (Glyproteins et Glycolipides)," Symposium International sur les Glycoconjugates. 2nd, Editions du Centre Natural de la Recherche Scientifique, Paris, vol. 1.
Reformato, L. S., Kirschenbaum, D. M., and Halpern, M., 1983, Preliminary characterization of response-eliciting components of earthworm extract, Pharmac. Biochem. Behav., 18:247.
Robinson, C., Lowe, N. R., and Weatherall, J. A., 1975, Amino acid composition, distribution and origin of tuft protein in human and bovine dental enamel, Arch. Oral Biol., 20:29.
Sheffield, L. P., Law, J. H., and Burghardt, G. M., 1968, On the nature of chemical food sign stimuli for newborn snakes, Comm. Behav. Biol., 2:7.
Spiro, R. G., and Bhoyroo, V. D., 1980, Studies on the carbohydrate of collagens. Characterization of a glucuronic acid-mannose disaccharide unit from Nereis cuticle collagen, J. Biol. Chem., 255:5347.
Watson, M. R., 1958, The chemical composition of earthworm cuticle, Biochem. J., 68:416.
Wilde, W. S., 1938, The role of Jacobson's organ in the feeding reaction of the common garter snake, Thamnophis sirtalis sirtalis (Linn.), J. Exp. Zool., 77:445.

CHEMICAL AND BIOLOGICAL INVESTIGATIONS OF FEMALE MOUSE PHEROMONES

Bozena Jemiolo, Franca Andreolini, and Milos Novotny

Department of Chemistry
Indiana University
Bloomington, IN

The importance of chemical communication in the reproductive behavior of rodents is well documented. Female mice or rats can modulate the ovarian cycles of other females in their social group through pheromones (Bronson, 1971; McClintock, 1983). The urine of singly caged estrous females (Drickamer, 1982) and of pregnant or lactating female mice (Drickamer and Hoover, 1979; Hoover and Drickamer, 1979) has the capacity to accelerate puberty in young females and to promote estrous in adults. Odors from diestrous female rats shorten the estrous cycle, while odors from proestrous females cause prolongation. In addition, female rats that are in the acyclic state of persistent estrous are capable of inducing a similar state in other females (McClintock and Adler, 1978).

However, female mice kept in groups of more than four animals and isolated from contact with a male produce a urinary chemosignal which has a contrary effect and acts to retard sexual maturation in other females, although here the timing and duration of the urine exposure is critical (Drickamer, 1977, 1983). The delay chemosignal(s) is contained in the excreted urine from grouped, but not singly caged females, and is present in the bladder urine of all females regardless of age or caging density (McIntosh and Drickamer, 1977). The presence of the delay chemosignal in urine is influenced by adrenalectomy (Drickamer and McIntosh, 1980) but not ovariectomy (Drickamer et al., 1978). Singly caged females treated daily with urine from grouped females will produce the delay-puberty chemosignal in their own urine (Drickamer, 1982). There is substantial evidence that female rodents produce multiple female-specific chemosignals with different mechanisms of hormonal regulation, different degrees of volatility, as well as different sites of chemical production and release (Nyby, 1983).

Based on the recent study and the suggestions that the puberty-delaying activity is lost after exposure of urine to air at room temperature (Coppola and Vandenbergh, 1985) and that the puberty-inhibition affect may be associated with the adrenal function (Drickamer and McIntosh, 1980), we decided to study quantitative chemical differences between the urinary excretion patterns of intact and adrenalectomized grouped females. The urinary volatile substances were subsequently identified and tested for biological activity. The samples were analyzed using the headspace technique described previously by our laboratory (Novotny et al., 1974).

The volatiles were sparged from 1 ml urine samples at room temperature with purified helium gas and adsorbed into a precolumn packed with TENAX-GC (Schwende et al., 1984). The samples were subsequently desorbed in the heated injection port of a gas chromatograph, equipped with a flame ionization detector, and retrapped into a cooled section of a glass capillary. The analytical column was 60 m x 0.25mm, i.d., glass capillary column coated statically with UCON HB-2000 and deactivated through addition of benzyltriphenylphosphonium chloride (Franken, 1976). A subsequent temperature program of the column results in complex chromatograms from typical samples of the mouse urine (Schwende et al., 1985). While quantitative differences are obtained through peak integration routines, identity of the individual profile's constituent is established through a combined gas chromatograph/mass spectrometer. Structures are further verified through authentic chemicals.

Comparison of numerous urinary volatile profiles obtained from the two sample types revealed that the urines of adrenalectomized animals tend to be more "dilute." However, consistent quantitative differences are readily seen between these two sample types. The chromatographic peaks 1-6 are generally depressed following adrenalectomy (Fig. Compounds 1-6 belong to three structurally distinct classes (Table 1): acetate esters (A); ketones (K); and pyrasine (P).

The compounds listed in Table 1 were added, individually or in groups, to the urine of adrenalectomized female mice in order to determine their biological activity (Novotny et al., 1985). Our bioassay procedure and urine collection were similar to those described by Drickamer and coworkers (1978). As seen from Table 2, the mice treated with pyrazine alone and the acetates, or both types of compounds together, spiked into the urine from adrenalectomized, grouped females reached puberty significantly later than young females treated with ketones plus urine from adrenalectomized females, or water (control). The urine of adrenalectomized animals with all six compounds resulted in a full recovery of biological activity.

Urinary volatiles from female mice in different reproductive status were examined next. Our preliminary data indicate that levels of the three ketones are dependent on the caging and reproductive conditions (Table 3

Table 1. Structural Information on the Compounds Decreased by Adrenalectomy.

Peak Designation	Structure	Mass-spectral Information m/e (relative intensity)
1	2-heptanone (K)	114(9), 99(7), 85(7), 71(26), 58 (80), 43(100)
2	n-pentyl acetate (A)	73(13), 70(31), 61(25), 55(24), 43(100)
3	cis-2-penten-1-yl-acetate (A)	128(3), 113(3), 99(6), 86(30), 85 (13), 68(46), 67(45), 57(23), 53 (20), 43(100)
4	trans-4-hepten-2-one (K)	112(4), 97(6), 94(8), 69(9), 43 (100), 41(26), 39(18)
5	trans-5-hepten-2) one (K)	112(9), 97(8), 94(12), 69(13), 43(100), 41(29), 39(15),
6	2,5-dimethyl pyrazine (P)	108(100), 81(12), 42(55), 40(12), 39(22)

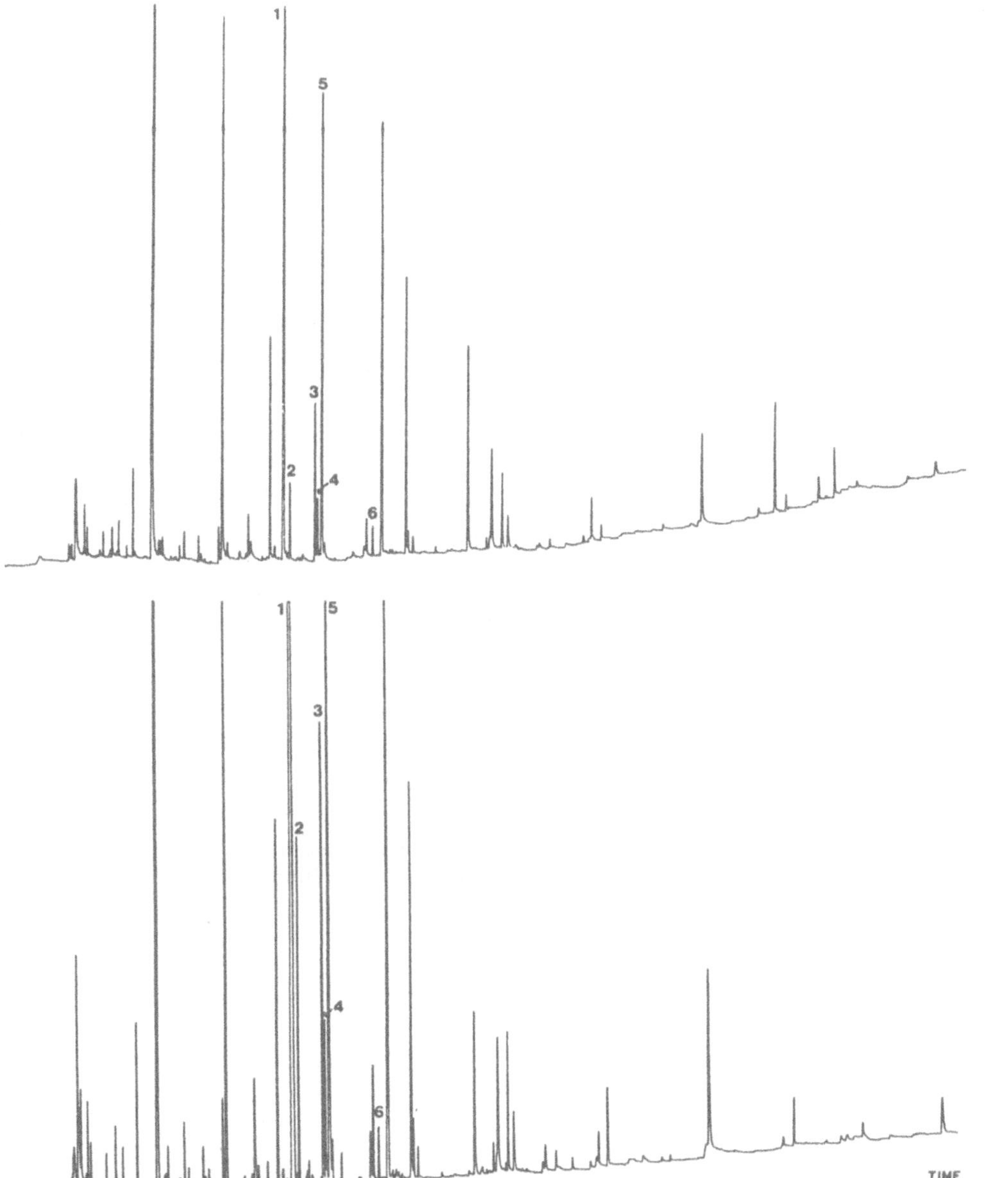

Fig. 1. Gas chromatograms of urine volatiles from adrenalectomized (upper) or intact (lower), ICR/Alb female mice.

Table 2. Average Day of the First Vaginal Estrous (±S.E.M.) for Young Females Painted Daily on the External Nares with Excreted Urine from Adrenalectomized Grouped Females, Mixed with Synthetic Compounds, or Water (Control). N = 30 Mice/Treatment.

Treatment		Average Day to Reach First Estrous
Water (control)		29.8 (0.3)
(1,4,5)	Three ketones	29.8 (0.3)
(2,3)	Two acetate esters	31.3 (0.3)
(2,3,6)	Two acetate esters + pyrazine	31.5 (0.3)
(6)	Pyrazine	32.2 (0.3)
(1-6)	All six compounds	33.5 (0.3)

Those means not connected by the same vertical line are significantly different at the 0.01 level [Duncan's new multiple range test; $F\ (5,175) = 21.08$, $p < 0.001$]. Urine alone was not tested.

and Fig. 2). The largest peak areas were observed in the urine of females in estrous and metestrous stage, in singly housed animals and in pregnant premiparous females. Conversely, large decreases are observed following adrenalectomy, or during lactation. However, the acetate esters appear abundant in the urine of group caged females, singly caged in metestrous stage and pregnant primiparous. The pyrazine was found in very large peak areas only in urine of group caged and pregnant females.

The results of these investigations support the conclusion that the six profile components, whose concentration is dependent on the adrenal glands, play an important role in puberty delay of young female mice. However, the changes in their concentration in the urines collected from females in the different stages of estrus cycle, or during lactation,

Table 3. Peak Areas (arbitrary units) for the Compounds Found in Female Urine, in Dependence on Caging or Reproductive Conditions.

Peak Designation	1(K)	2(A)	3(A)	4(K)	5(K)	6(P)
Urine Source:						
Intact, group caged	54.4	4.5	7.3	3.5	18.1	1.2
Adrenalectomized, group caged	19.1	1.9	2.3	1.5	9.2	0.5
Intact, singly caged						
proestrus	34.8	1.1	1.9	3.6	9.1	0.2
estrous	73.3	2.8	5.2	5.3	21.8	0.4
metestrous	76.9	5.6	5.3	4.7	22.8	0.5
diestrous	43.3	2.4	2.9	2.8	13.9	0.4
Lactating[a]	17.9	0.4	0.2	0.4	1.3	0.2
Pregnant (primiparous)[b]	113.9	15.1	13.3	6.0	26.1	2.3

[a]Collection from 1st to 20th day of lactation.
[b]Collection from 4th to 7th day of pregnancy.

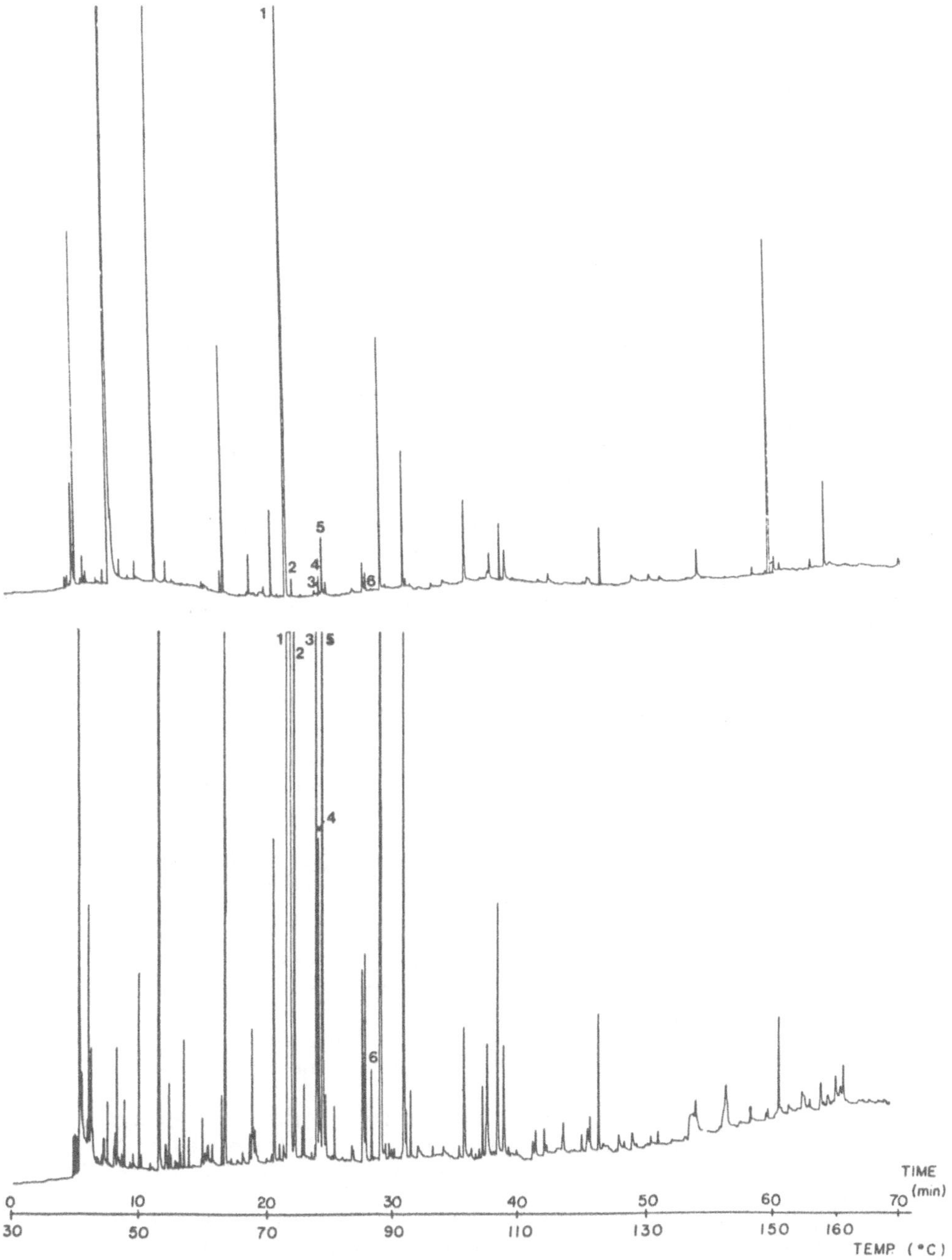

Fig. 2. Gas chromatograms of urine volatiles from lactating (upper) or pregnant primiparous (lower), ICR/Alb female mice.

indicate that the ovaries could partially be involved in controlling the appearance of inhibiting pheromones.

The urine of lactating and pregnant females has the capacity to lengthen the period of estrus in other adult females (Hoover and Drickamer, 1979). However, the time of puberty of young females was not affected by the urines from the early and late stages of pregnancy (Cowley and Wise, 1972). Our results seem to indicate that the levels of the six compounds are very low in samples from lactating females, but very high in the urine of pregnant primiparous (4-7 day) females. Further biological effects of these compounds are under investigation.

REFERENCES

Bronson, F. H., 1971, Rodent pheromones, Biol. Reprod., 4:344.

Coppola, D. M., and Vandenbergh, J. G., 1985, Puberty delay pheromone in female mice: Density, duration of grouping and age of urine stimulus, J. Reprod. Fert., 73:517.

Cowley, J. J., and Wise, D. R., 1972, Some effects of mouse urine on neonatal growth and reproduction, Anim. Behav., 20:499.

Drickamer, L. C., 1977, Delay of sexual maturation in female house mice by exposure to grouped females or urine from grouped females, J. Reprod. Fert., 51:77.

Drickamer, L. C., 1982, delay and acceleration of puberty in female mice by urinary chemosignals from other females, Dev. Psychobiol., 15:433.

Drickamer, L. C., 1983, Effect of period of grouping of donors and duration of stimulus exposure on delay of puberty in female mice by a urinary chemosignal from grouped females, J. Reprod. Fert., 69:723.

Drickamer, L. C., and Hoover, J. E., 1979, Effect of urine from pregnant and lactating female house mice on sexual maturation of juvenile females, Dev. Psychobiol., 12:545.

Drickamer, L. C., and McIntosh, T. K., 1980, Effect of adrenalectomy on the presence of a maturation-delaying pheromone in the urine of female mice, Horm. Behav., 14:146.

Drickamer, L. C., McIntosh, T. K., and Rose, E. A., 1978, Effect of ovariectomy on the presence of maturation-delaying pheromone in the urine of female mice, Horm. Behav., 11:131.

Franken, J. J., Rutten, G. A. F. M., and Rijks, J. A., 1976, Preparation of glass capillary columns coated with polar phase for high-temperature gas chromatography, I. Study of roughening of the glass surface, J. Chromatogr., 126:132.

Hoover, J. E., and Drickamer, L. C., 1979, Effects of urine from pregnant and lactating female house mice on estrus cycle of adult females, J. Reprod. Fert., 55:237.

McClintock, M. K., 1983, Pheromonal regulation of the ovarian cycle: Enhancement, suppression and synchrony, in: "Pheromones and Reproduction in Mammals," J. G. Vandenbergh, ed., Academic Press, New York.

McClintock, M. K., and Adler, N. T., 1978, Induction of persistent estrus by airborne chemical communication among female rats, Horm. Behav., 11:414.

McIntosh, T. K., and Drickamer, L. C., 1977, Excreted urine, bladder urine and the delay of sexual maturation in female house mice, Anim. Behav., 25:999.

Novotny, M., Lee, M. L., and Bartle, K. D., 1974, Some analytical aspects of the chromatographic headspace concentration method using a porous polymer, Chromatographia, 7:333.

Novotny, M., Jemiolo, B., Harvey, S., Wiesler, D., and Marchlewska-Koj, A., 1985, Adrenal-mediated endogenous metabolites inhibit puberty in female mice, submitted for publication.

Nyby, J., 1983, Volatile and nonvolatile chemosignals of female rodents: Difference in hormonal regulation, in: "Chemical Signals in Vertebrates," D. Muller-Schwarze, and R. M. Silverstein, eds., Plenum Press, New York.

Schwende, F. J., Wiesler, D., and Novotny, M., 1984, Volatile compounds associated with estrus in mouse urine: Potential pheromones, Experientia, 40:213.

Schwende, F. J., Wiesler, D., Jorgenson, J. W., Carmack, M., and Novotny, M., 1985, Urinary volatile constituents of the house mouse, *Mus musculus*, and their endocrine dependency, J. Chem. Ecol., in press.

BEHAVIORAL FUNCTION AND CHEMISTRY OF VOLATILES FROM FECES AND ANAL SECRETIONS OF HOUSE MICE, MUS MUSCULUS - A PRELIMINARY REPORT

B. S. Goodrich[1], S. Gambale[2], P. R. Pennycuik[3], and T. D. Redhead[2]

Division of Wildlife and Rangelands Research, CSIRO
[1]C/-Chemistry Department, Macquarie University, North Ryde, N. S. W., Australia 2113, [2]P.O. Box 84, Lyneham A. C. T., Austraila 2602, [3]P. O. Box 184, North Ryde N. S. W., Australia 2113

INTRODUCTION

The occurrence in the anal region of odoriferous gland secretions which, either on their own or in combination with feces, function for behavioral purposes has been reported in a number of mammalian species. In the wild rabbit, Oryctolagus cuniculus, the territorial function of the secretion from anal glands and feces has been studied particularly intensively (Mykytowycz, 1970), and ethological enquiries have been backed by chemical investigation (Goodrich et al., 1978, 1981a, 1981b).

In the house mouse, Mus musculus, a great deal of attention has been paid in the past to the behavioral properties of its urine (Albone, 1984, and references quoted therein) and to a much lesser extent on other sources of odor such as the coagulating glands (Jones and Nowell, 1973), and the preputial glands (Bronson and Caroom, 1971). Even less information is available on the function of the odor of feces with which the secretion from glands located in the anal region is often associated.

It has been emphasised that urine plays an important role in the formation and maintenance of territories in mouse populations, and in male to male aggression (Bronson, 1971, 1979; Desjardin et al., 1973; Maruniak et al., 1974). Earlier, Eibl-Eibesfeldt (1950) had suggested that urine as well as feces are used by mice to form trails. Welch (1953) reported that mice establish "posts" composed of fecal pellets, dust and debris and urine along paths within their home range.

Plagues of house mice cause serious economic losses in rural areas of southern and eastern Australia. A recent study of the ecology of mice in cereal farms in New South Wales, suggested that changes in spacing behavior are involved in the formation of plagues (Redhead, 1982). Examination of this hypothesis necessitates specific studies of the territorial behavior of the species.

As a first step, we attempted to reproduce the urine patterns reported for adjacently-caged mice (Bronson, 1979). It was noticed that deposition of feces was more frequent than that of urine. It was also observed that

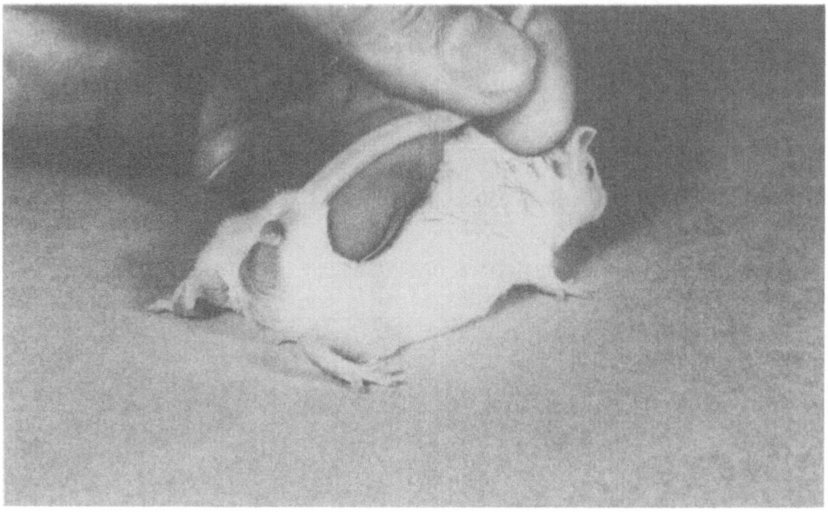

Fig. 1. Adult male mouse, Mus musculus (BALB/c) excreting clear liquid bubble from the anus.

the mice excreted a clear bubble of liquid from the anus (Fig. 1). This liquid was subsequently smeared on the floor as the mouse dragged its hindquarters.

Fecal pellets, in the course of their excretion from the lower bowel to the exterior, are coated with mucus from the goblet cells embedded in the epithelium of the tubular glands lining the colon and rectum (Hummel et al., 1966). As feces pass through the anus they also become coated with a secretion from the anal glands. These glands are modified sebaceous glands.

This paper presents preliminary observations on the behavioral function and chemistry of the odor of feces and secretion from the anus.

BEHAVIORAL STUDIES

Both laboratory and wild mice were used in the following studies. Males were used in these initial investigations because the propensity for aggression between them is thought to be the basis of formation and maintenance of territories in populations of house mice (Bronson, 1979).

Observations on the Frequency of Marking with Feces and with Secretion from Anus

Twenty adult male BALB/c mice were used in this study. A sheet of plexiglass (160 x 60 cm) was covered first with polythene film (0.1 mm) and then a layer of clean newsprint. The area of the newsprint was partitioned into five compartments (approx. 30 x 60 cm) using glass slabs (15 cm high). Each mouse was placed for 10 minutes into the compartments sequentially. Then the polythene and newsprint were replaced, the glass slabs were washed, dried and replaced, and the next mouse was tested. Frequency of urination, defecation, smearing of secretion by dragging of hindquarters and ambulation (as a measure of exploratory activity) were recorded.

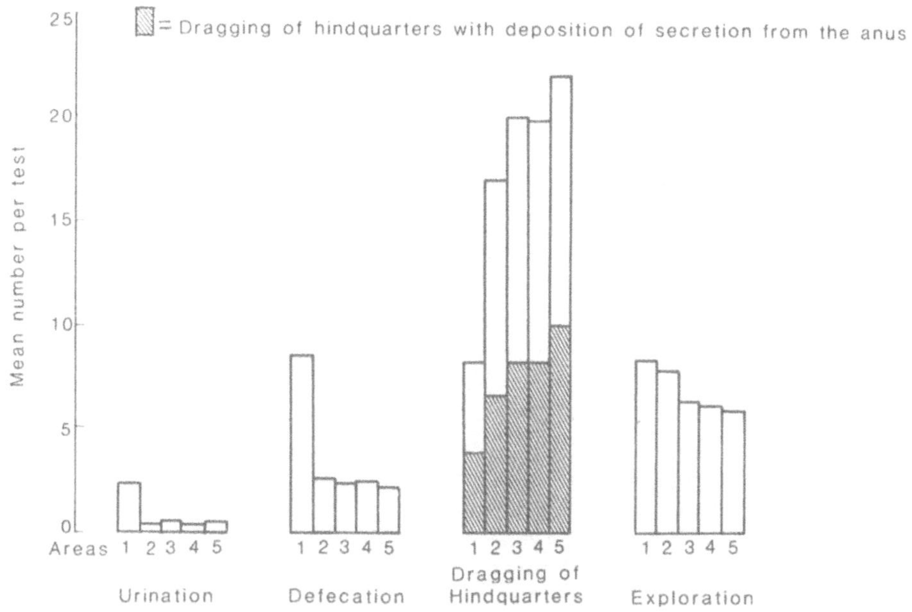

Fig. 2. Marking and exploratory (ambulatory) activities of 20 adult male mice allowed to occupy, sequentially, five novel areas (each for 10 min).

The results (Fig. 2) indicate that the number of urinations and fecal pellets deposited decreased from the first compartment to the second and then remained constant. In contrast, the number of draggings of hindquarters, and signs of deposition of smearings of secretion from the anus increased from the first compartment to the second, and then steadily to the fifth. The exploratory activities of the mice remained at a constant level throughout the test.

These results suggest that mice, when repeatedly introduced into a new space, will continue to attempt to deposit feces. The dragging of their hindquarters and smearing of the anal secretion may suggest that this secretion is for the purpose of marking novel areas.

Effect of the Proximity of Another Mouse on the Distribution of Feces

In this study 24 adult male mice which had been caught in the field were tested in pairs. Before testing, the animals were caged singly and kept at 20°C in the laboratory for a minimum of four weeks.

Two plexiglass test cages (30 x 30 x 30 cm) with plexiglass lids and metal grid floors were placed next to each other above a sheet of blotting paper. The lower half of the adjacent walls were painted black to prevent visual contact between the animals (Fig. 3). This setup enabled collection of excreta and circulation of air between the cages. A mouse was introduced into each of the cages at 1630 hr and left in total darkness until 0930 hr the next day. Food and water were available *ad lib* and each mouse had available to it a small plastic tube for shelter. At the conclusion of the test, the blotting paper under each cage was subdivided into three equal areas and the location of fecal pellets and urine spots, which were detected with the aid of ultraviolet light, were recorded.

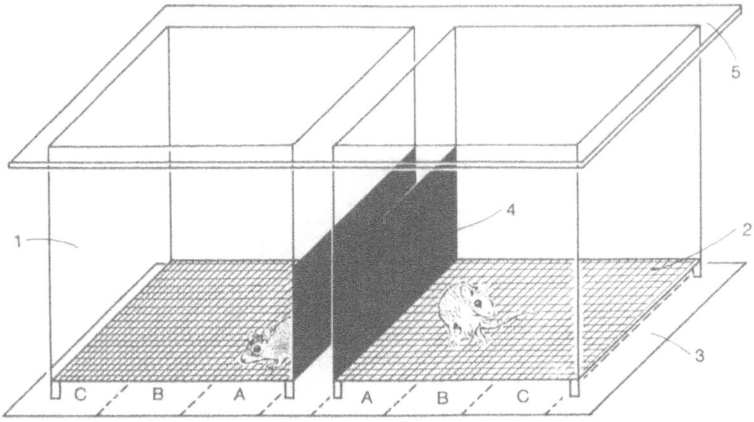

Fig. 3. Cages for testing the non-visual effect of the proximity of adult male mice on the distribution of excreta. 1 = plexiglass cages; 2 = metal grid floor; 3 = blotting paper under each enclosure subdivided into three equal areaa; 4 = opaque walls of cages; 5 = gladd lid.

The counts of fecal pellets and urine spots in each third of the two cages during 12 tests are shown in Table 1. The most useful comparison is that between areas A and C, which each contain two corners. Urine spots were slightly more numerous in area A than in area C and there were very few in area B. This suggests that mice prefer to urinate in areas containing corners. For feces, however, there was no such effect. Similar number of pellets were found in areas B and C. The number of feces in area A however, was more than twice that found in either areas B or C.

The results described above indicate that the close proximity of another mouse strongly stimulated the deposition of fecal pellets.

Confidence-giving Effect of Odor of Own Feces and of Secretion from Anus

Studies on wild rabbits have shown that when a male individual is confronted with another male in a neutral environment, where feces from one of them are already present, there is a high probability that the donor of the feces will dominate the other one (Mykytowycz et al., 1976). To determine whether a similar effect would operate with mice, laboratory and wild strains were tested using feces and secretion from the anus

Table 1. Number of fecal pellets and urine spots deposited by 24 adult male mice in three areas under two adjoining cages.

	Area A			Area B			Area C		
	Number	Mean	% of total	Number	Mean	% of total	Number	Mean	% of total
Fecal pellets	1488	62.0	55	615	25.6	23	586	24.4	22
Urine spots	109	8.2	55	4	0.2	2	85	3.5	43

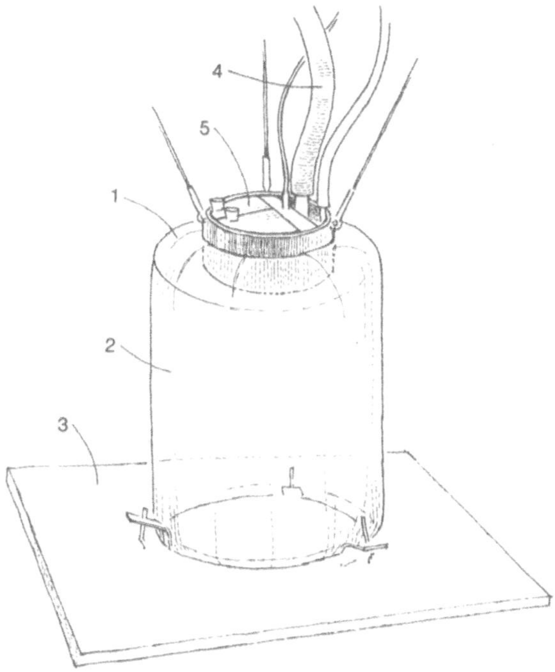

Fig. 4. Diagram of the neutral chamber used for testing the behavioral reaction of mice to the presence of odor of their own feces or secretion from the anus. 1 = polythene film lining; 2 = plexiglass cylinder; 3 = baseborad; 4 = air inlet tube; 5 = separate compartments in the chamber head for holding mice prior to indroduction into chamber.

separately. The neutral chamber (50 cm dia. x 90 cm high) used in the studies (Fig. 4) was similar to that used with (Hesterman et al., 1981).

Fecal pellets, uncontaminated with urine, were collected from individual animals housed for five consecutive nights in cages with wire-mesh floors. The feces were then stored at -18°C until required. Secretion from the anus deposited on aluminum foil by individual mice was transferred to glass vials and also stored at -18°C.

For each study, adult male mice trapped in the field were used. They were caged individually at 20°C in a reversed 12:12 hourly light cycle. To prevent mice from becoming acquainted with one another, they were divided into two groups and kept in different rooms. For comparison, the same number of laboratory strain mice (BALB/c), kept under conditions similar to the wild strain, were used. All animals were given food and water *ad lib*. Pairs of mice from different rooms, and matched by weight, were selected for each test. Feces (0.5 g) or secretion from the anus (50 mg) from one of them selected randomly (donor mouse) were placed on filter papers on the floor of the test chamber. Tests were conducted in darkness under an infrared light. The tail of the donor mouse was marked with a strip of reflective tape. Both contestants were placed in two separate compartments in the chamber head and were introduced onto the floor simultaneously by remote-control. Mice in the chamber were viewed through a video camera fitted with a Philips-brand image-intensifier, and

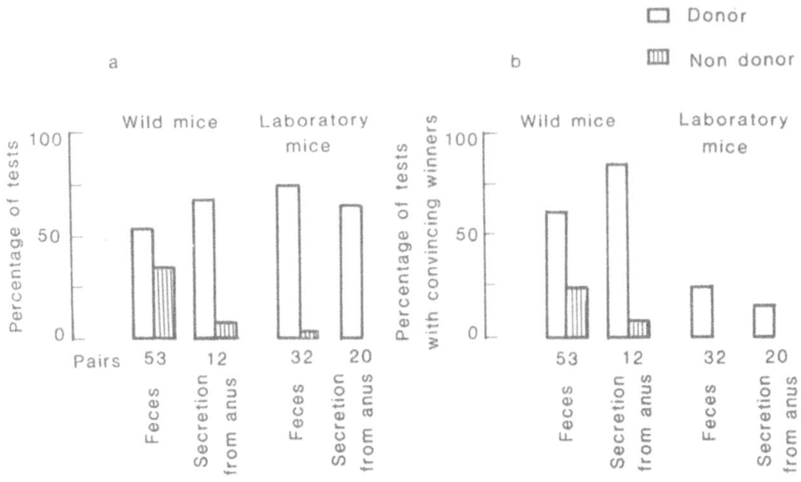

Fig. 5. Degree of confidence as demonstrated by (a) initiation of contact, and (b) dominance within pairs of adult male mice from wild and laboratory strains introduced into a neutral chamber containing the odor of one of them (donor).

their behavior recorded for 10 minutes starting from the time when first contact was established between them. The behavior was viewed by two observers on a monitor screen remote from the test area. By observing time spent by each mouse on exploration of the chamber, sniffing, grooming, biting, chasing, and nudging of the other mouse, the two observers independently decided which mouse was dominant. In addition, the time elapsed between release and first contact was also recorded, since early approach of an individual indicates its higher degree of confidence.

One week after all the pairs had been tested, a new series of tests was carried out with the donor and non-donor roles of the contestants reversed.

The results, shown in Fig. 5, indicate that mice of both laboratory and wild strains, when in the presence of their own odors, be it in the form of feces or secretion from the anus, tend to dominate their opponents.

CHEMICAL STUDIES

In order to understand better how chemical signals are formed, it is essential to study the composition of odors. Chemical investigations were carried out on pooled samples obtained from laboratory strain mice (BALB/c).

Feces were collected as required and used immediately. Glands were dissected out and stored at -18°C. Prior to use, the glands (100 mg) were macerated in sodium chloride solution (0.5 ml; 0.1% w/v) at 5°C in a micro glass homogenizer. The secretion from the anus was mixed with an equal volume of water immediately upon collection and stored at -18°C until required. The secretion, which is a clear, colorless, viscous liquid with a slightly pungent odor, gives positive tests for proteins and carbohydrates. It is freely miscible with water when fresh, but forms an insoluble, translucent mass on standing in air. When placed on filter

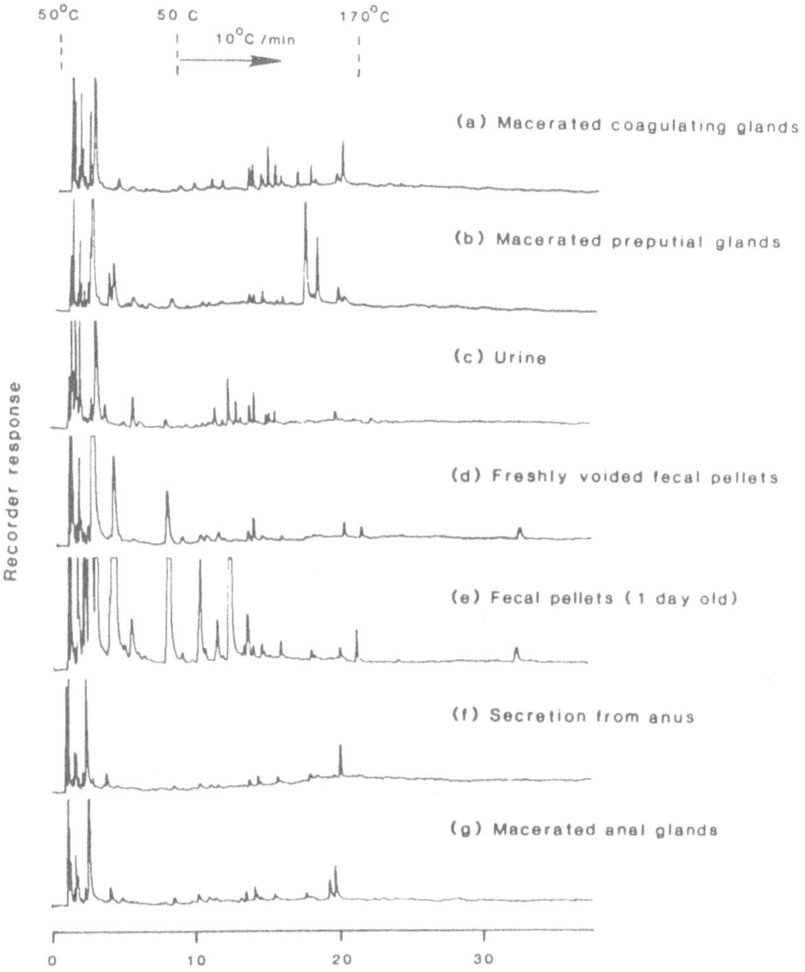

Fig. 6. Gas chromatograms on bonded phase (BP 20), vitreous silica capillary column of headspace volatiles above odoriferous secretions from adult male mice, Mus musculus.

paper, in long-wave ultraviolet light the dry secretion shows a pale yellow fluorescence, which can be distinguished from that produced by urine.

Gas Chromatography and Gas Chromatograph - Mass Spectrometry

The volatile constituents emanating from fecal pellets, macerated anal glands, and secretion from the anus (headspace volatiles) were collected on Chromosorb 105 and introduced through a modified chromatographic inlet (Goodrich et al., 1978, 1981a) onto a polar, bonded phase (BP 20), vitreous silica capillary column (25 m; 0.2 mm ID). For the purpose of comparison, headspace volatiles from coagulating and preputial glands and from urine were similarly fractionated.

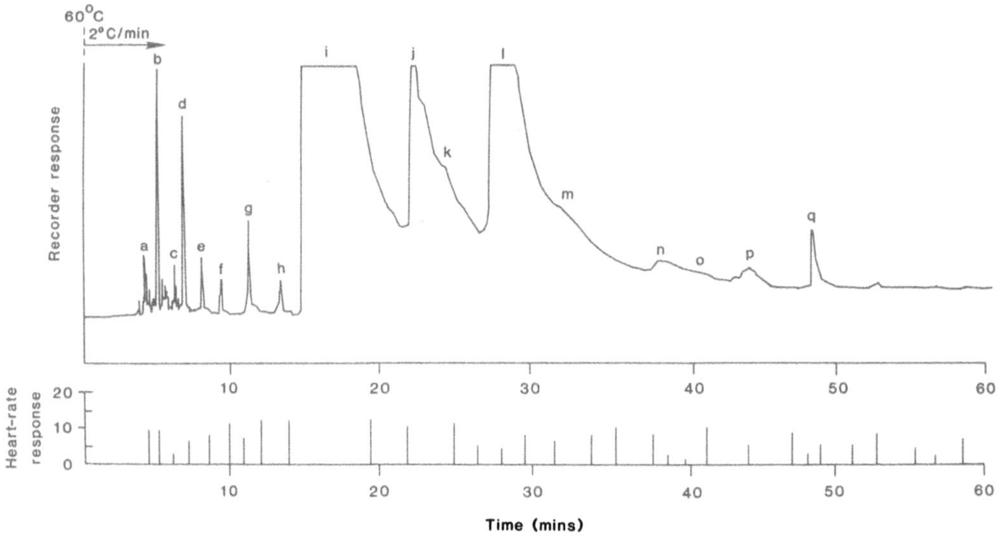

Fig. 7. Gas chromatogram on Carbowax 20M glass SCOT column of headspace volatiles from fecal pellets of adult male mice, Mus musculus, with indication of the number of mice showing heart-rate responses to the separated components. Identity of major constituents by gas chromatography-mass spectrometry: (a) n-propyl ethanoate; (b) n-propanol; (c) n-propyl n-butanoate; (d) n-butanol; (e) heptan-2-one; (f) 3-methylbutan-1-ol; (g) n-pentanol; (h) n-hexanol; (i) ethanoic acid; (j) propanoic acid; (k) 2-methylpropanoic acid; (l) n-butanoic acid; (m) 3-methylbutanoic acid; (n) 2-phenylethanol; (o) n-pentanoic acid; (p) phenol; (q) indole.

The headspace volatiles were collected by passing oxygen-free nitrogen (40 ml/min) for 15 min over freshly voided and 1 day old fecal pellets (0.5 g), macerated anal, coagulating, and preputial glands (0.1 ml), and urine (0.05 ml). The secretion from the anus (10 mg) was first diluted with sodium chloride solution (0.2 ml; 0.1% w/v) and the headspace volatiles collected as above.

Examination of the odor of fecal pellets by gas chromatography-mass spectrometry was carried out on the headspace volatiles of feces (0.5 g) collected as above for 16 hr. Identities of the major constituents of the odor are shown in Fig. 7.

The gas chromatograms of all samples (Fig. 6) show variations in the composition of volatiles present. The odor of anal glands, secretion from the anus, and feces appear to contain a few components which are in common, and which may be absent in the headspace volatiles from coagulating, and preputial glands, or urine.

Freshly voided feces, when allowed to stand in air for 24 hours, show volatile profiles in which there are changes in the size of some peaks as well as the appearance of additional volatile components, of which the volatile fatty acids are the major constituents (see Fig. 7).

Measurement of Cardiac Response

In order to determine which of the volatile constituents of fecal odor may be behaviorally important, a pair of adult male mice were positioned to sniff the effluents from a glass SCOT column coated with a polar phase (Carbowax 20M), and changes in their heart-rates in response to the eluted components were recorded using the technique employed during studies on rabbits (Hesterman et al., 1981; Goodrich et al., 1978, 1981a). In the present investigation, no distinction has been made between accelerative and decelerative changes in heart-rate. A total of 18 different mice was used. Headspace volatiles were collected from pooled samples of fecal pellets (0.5 g) for 16 hr as described above.

Changes in cardiac responses were obtained throughout the fractionation of the fecal odor (see Fig. 7). However, mice showed a greater number of responses to some components than to others. In addition, the animals consistently responded at the front, and on the tail, of some large chromatographic peaks, e.g., ethanoic acid.

DISCUSSION

Behavioral involvement of feces and of the secretions from glands located in the anal region in mice have largely been overlooked in the past. However, the results which are reported here support the few earlier suggestions (Eibl-Eibesfeldt, 1950; Welch, 1953) that fecal pellets (and/or the secretion from the anus which animals deposit by dragging of their hindquarters) are important to this species for communicating messages related to territoriality. They also indicate that, in the presence of its own odor from the above sources, an individual acquires a higher territorial confidence as found previously for rabbits.

Different levels of defecation by laboratory rodents have been interpreted previously as a measure of the emotional state of individuals (Archer, 1973). If, as our results suggest, feces are used for territorial signaling, the relationship between the two functions (fear and territorial marking) is in need of further examination.

Gas chromatographic comparison of headspace volatiles collected from above feces, the secretion from the anus and macerated anal glands (see Fig. 6), shows the presence of several common components which do not appear to be present in the odor of urine or macerated coagulating and preputial glands. This could be expected in view of the likelihood that different messages are embodied in the odor from different sources. Variations in the chromatographic profiles of the odors as a consequence of the action of enzymes or microorganisms on high molecular weight substrates such as proteins and carbohydrates must also be considered. Thus volatile constituents found in 1 day old fecal pellets varied considerably from those of pellets which were freshly voided. The behavioral message contained within these samples is also likely to be different.

In the present study, like earlier ones on rabbits (Hesterman et al., 1981; Goodrich et al., 1978, 1981a), heart-rate responses of mice have been used merely to indicate the presence of volatiles which could be detected olfactorily, and to which animals selectively pay greater attention. The assumption is made that compounds which produce cardiac responses are more likely to be components of complex odors which the animals use for olfactory communication, rather than those to which mice are inattentive or olfactorily insensitive.

As has previously been reported for rabbits (Goodrich et al., 1978), mice also responded consistently at the front, and on the tail, of some large chromatographic peaks (see Fig. 7). Since odor quality is often dependent on the concentration of eluting volatiles, presentation of odor components in concentrations which vary greatly from those which mice would perceive under natural conditions could result in false behavioral responses in the animals.

Gas chromatography-mass spectrometry of the volatile constituents of fecal pellets revealed the presence of several classes of organic compounds. Some major components to which mice reacted by way of change in cardiac response, such as volatile fatty acids (see Fig. 7), occur also in the fecal pellets of the rabbit, O. cuniculus (Goodrich et al., 1981) and in secretions of skin glands of a number of mammals including, in the anal gland of the red fox, Vulpes vulpes (Albone and Fox, 1971) and the anal sac of the coyote, Canis latrans (Preti et al., 1976). Several constituents have yet to be identified and work is in progress to determine their structures. The results of this and other related studies point to the complexity of odors which mammals use for communication with conspecifics.

ACKNOWLEDGMENTS

The authors wish to thank their colleagues Miss L. Krick, Mr. W. Price and Mr. C. Reece for their technical assistance, and Dr. R. Mykytowycz and Mr. E. R. Hesterman for their helpful comments in the preparation of this manuscript.

REFERENCES

Albone, E. S., 1984, "Mammalian Semiochemistry," John Wiley and Sons Limited, New York.

Albone, E. S., and Fox, M. W., 1971, Anal gland secretion of the red fox, Nature, 233:569.

Archer, J., 1973, Tests for emotionality in rats and mice, Anim. Behav., 21:205.

Bronson, F. H., 1971, Rodent pheromones, Biol. Reprod., 1:344.

Bronson, F. H., 1979, The reproductive ecology of the house mouse, Q. Rev. Biol., 54:265.

Bronson, F. H., and Caroom, D., 1971, Preputial gland of the male mouse: Attractant function, J. Reprod. Fert., 25:279.

Desjardins, C., Maruniak, J. A., and Bronson, F. H., 1973, Social rank in the house mouse: Differentiation revealed by ultraviolet visualisation of urinary marking patterns, Science, 182:939.

Eibl-Eibesfeldt, I., 1950, Beiträge zur Biologie der Haus- und der Ährenmaus nebst einigen Beobachtungen an anderen Nagern, Z. Tierpsychol., 7:558.

Goodrich, B. S., Hesterman, E. R., Murray, K. E., Mykytowycz, R., Stanley, G., and Sugowdz, G., 1978, Identification of behaviorally significant volatile compounds in the anal gland of the rabbit, Oryctolagus cuniculus, J. Chem. Ecol., 4:581.

Goodrich, B. S., Hesterman, E. R., Shaw, K. S., and Mykytowycz, R., 1981a, Identification of some volatile compounds in the odor of fecal pellets of the rabbit, Oryctolagus cuniculus, J. Chem. Ecol., 7:813.

Goodrich, B. S., Hesterman, E. R., and Mykytowycz, R., 1981b, Effect of volatiles collected above fecal pellets on behavior of the rabbit, Oryctolagus cuniculus, tested in an experimental chamber. II. Gas chromatographic fractionation of trapped volatiles, J. Chem. Ecol., 7:947.

Hesterman, E. R., Goodrich, B. S., and Mykytowycz, R., 1981, Effect of volatiles collected above fecal pellets on behavior of the rabbit, *Oryctolagus cuniculus*, tested in an experimental chamber. I. Total volatiles and some chemically prepared fractions, *J. Chem. Ecol.*, 7:795.

Hummel, K. P., Richardson, F. L., and Fekete, E., 1966, Anatomy, *in*: "Biology of the Laboratory Mouse," E. L. Green, ed., McGraw-Hill, New York.

Jones, R. B., and Nowell, N. W., 1973, The coagulating glands as a source of aversive and aggression-inhibiting pheromone(s) in the male albino mouse, *Physiol. Behav.*, 11:455.

Maruniak, J. A., Owen, K., Bronson, F. H., and Desjardins, C., 1974, Urinary marking in male house mice: Responses to novel environmental and social stimuli, *Physiol. Behav.*, 12:1035.

Mykytowycz, R., 1968, Territorial marking in rabbits, *Sci. Amer.*, 218:116.

Mykytowycz, R., 1970, The role of skin glands in mammalian communication, *in*: "Advances in Chemoreception, Vol. I, Communication by Chemical Signals," J. W. Johnston, D. G. Moulton, and A. Turk, eds., Appleton-Century-Crofts, New York.

Mykytowycz, R., Hesterman, E. R., Gambale, S. and Dudzinski, M. L., 1976, A comparison of the effectiveness of the odors of rabbits, *Oryctolagus cuniculus*, in enhancing territorial confidence, *J. Chem. Ecol.*, 2:13.

Preti, G., Muetterties, E. L., Furman, J. M., Kennelly, J. J. and Johns, B. E., 1976, Volatile constituents of dog (*Canis familiaris*) and coyote (*Canis latrans*) anal sacs, *J. Chem. Ecol.*, 2:177.

Redhead, T. D., 1982, Reproduction, growth, and population dynamics of house mice in irrigated and non-irrigated cereal farms in New South Wales, Ph.D. Thesis, Australian National University.

Welch, J. F., 1953, Formmation of urinating "posts" by house mice (*Mus*) held under restricted conditions, *J. Mammal.*, 34:502.

THE EVOLUTION OF CHEMICAL ALARM SIGNALS IN FISHES

R. Jan F. Smith

Biology Department
University of Saskatchewan
Saskatoon, Sask., Canada, S7N 0W0

GENERAL DESCRIPTION OF CHEMICAL ALARM SYSTEMS IN FISHES

A wide variety of living things, ranging from trees (Baldwin and Schultz, 1983) to mammals (MacKay-Sim and Laing, 1981) respond to chemical stimuli emitted by injured conspecifics. The known chemical alarm signals in fish belong to this general type of response. Mechanical injury to the "sender" releases chemicals that are detected by the "receiver". The receiver then responds by performing defensive behavior that protects it from predators.

Two analogous chemical alarm systems have been described in fishes and there is evidence that other comparable systems exist. I will deal primarily with the best known of these systems, the alarm substance or Schreckstoff system found in the ostariophysan fishes (e.g. minnows, suckers, catfish), first described by von Frisch (1938) and recently reviewed by Pfeiffer (1977, 1982) and Smith (1977, 1982a). The terms "alarm substance" and "alarm substance cells" (ASCs) will be used only for the ostariophysan system. The second established alarm system is found in darters of the Family Percidae (Smith, 1979, 1982b). The darter system will be specifically designated when it is described.

In the ostariophysan system the alarm substance is contained in specialized cells, the ASCs, that lie in the epidermis outside the protective scales (Fig. 1A). The alarm substance may be released by relatively minor damage. The receivers detect the alarm substance using olfaction (von Frisch, 1941; Kasumyan and Pashenko, 1982; Pfeiffer et al., 1984) and perform a species specific fright reaction that is assumed to be defensive behavior against predators, in the sense of Edmunds (1974). Other conspecifics may respond visually to the fright reaction and spread the response quickly through a group of fish. The darters also show defensive behavior in response to chemicals from damaged skin and they possess specialized sacciform cells in the epidermis (Fig. 1B), but the relationship between the alarm pheromone and the cells is not proven (Smith, 1982b).

Animal alarm signals pose an evolutionary problem (Williams, 1966; Charnov and Krebs, 1975; Maynard-Smith, 1976; Harvey and Greenwood, 1978; Smith, 1977, 1982a; Weldon, 1983). What benefit does the sender gain by warning conspecifics? It used to be presumed that the benefit to the group or species would account for the evolution of alarm signals but this

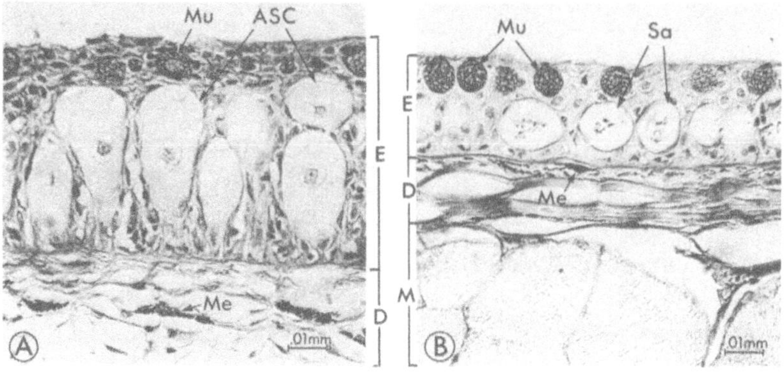

Fig. 1. The skin of two groups of fish that show alarm responses to skin extract. (A) The skin of an ostariophysan fish, the bluehead chub (Nocomis leptocephalus), illustrating the alarm substance cells (ASCs) in the epidermis; (b) The skin of a darter, the tesselated darter (Etheostoma olmstedi), illustrating the sacciform cells (Sa) that may contain darter alarm pheromone.
Legend: ASC=alarm substance cell, D=dermis, E=epidermis, M-=muscle, Me=melanophores, Mu=mucus cells, Sa=sacciform cells.

mechanism now seems unlikely. We must seek explanations based on individual selection that account for increased risk or energy expenditure apparently for the benefit of other individuals. The problem seems particularly difficult in the case of the ostariophysan alarm system in which the sender develops specialized morphological structures, the ASCs, that have no known function other than alarm signalling.

Evolution will act on each of the various attributes of the system. These include the response of the receivers to specific components of the chemical mix released, the form and duration of the defensive response and the stages in the life cycle at which the animal can send the signal or respond to it.

In order to discuss the evolution of the alarm system I will divide it into two main segments: the evolution of the sender's portion of the system and the evolution of the receiver's responses.

SELECTION ACTING ON THE SENDER

Why should ostariophysan fishes differentiate and maintain specialized cells that may constitute over 30% of their epidermis when these cells have no known function other than warning other individuals? Explanations for this situation fall into two main categories. What I will term "alarm theories" are based on some benefit of the alarm signal redounding to improve the fitness of the sender. "Non-alarm theories" are based on some primary function for the ASCs and their contents that benefits the possessor directly without reference to the alarm response of other individuals.

Alarm Theories

Costs. The benefits of sending the signal must exceed the costs for the alarm signal function to drive the evolution of the ASCs. In the case of the fish alarm systems the cost will be the production and storage of the alarm pheromone. Injury or death during the release process would presumably occur anyway and are not costs of the signal system. There is no evidence that the senders sacrifice themselves in any way. Although predators might be attracted by the alarm pheromone there is no evidence at present to suggest that it is more attractive than any other chemical component of injured skin.

Direct benefits. Some animal alarm signals may benefit a sender directly, if the sender survives. For example Trivers (1971) has suggested that decreasing a predator's success rate in a region will make it less likely to hunt in that region or that failure to capture the sender's type of prey will mean that the predator is less likely to form a search image for that prey type. In each case the predator is less likely to hunt the sender in the future. It has been shown experimentally that predatory fish will preferentially hunt prey that are easy to catch over those that have more effective evasive behaviour (Vinyard, 1980). Charnov and Krebs (1975) have suggested that the sender of an alarm signal might benefit if the predator was distracted by the sudden movements of warned prey animals. Advantages to the alarm sender based on reciprocal altruism (Trivers, 1971) or on the advantages of retaining school members or territorial neighbours (Smith, 1985) have also been postulated. Each of these hypotheses depends on the sender surviving the encounter in order to benefit from reduced predation.

The fish alarm system seems poorly designed to take advantage of such post-signal benefits. The only known mechanism of signal release is mechanical damage to the skin. In both ostariophysans (von Frisch 1941) and darters (Smith 1979), threatened but uninjured fish do not release chemical alarm signals. The remaining question then is how often would these fish escape after releasing an effective amount of alarm pheromone? On the face of it such escape seems unlikely, and there have been virtually no data available on the escape of ostariophysan prey fish after injury in nature, until recently.

Smith and Lemly (1985) report one situation in which 16% of the individuals in a concentration of fathead minnows, *Pimephales promelas*, had been injured by piscivorous birds. The fish possessed ASCs, as would be expected in non-breeding fathead minnows. The degree of damage in the fish examined would have been sufficient to release alarm substance since scale loss had occurred and the ASCs lie in the epidermis, outside the scales. Some of the injured fish were kept in aquaria without medical treatment. Many still remain healthy almost a year after their capture and several have spawned successfully and produced viable eggs (unpublished data). This suggests that post-signal benefits cannot be ignored in the evolution of fish chemical alarm systems. There is a real possibility of fish surviving a damaging injury and living to benefit from the results of sending the alarm signal.

Indirect benefits. Sending the alarm might be adaptive if it improved survival of receivers that shared genes with the sender. For such selection to be effective, the fish must be in groups that include fairly close kin. Strong selection would occur if the sender's offspring were protected by the alarm signal. Williams (1966), in a discussion of alarm signals in general, has argued that only the warning of offspring or mates would be beneficial and that retention of alarm signals outside the season of parental care is not adaptive. Similarly, Staton (1978), discussing crocodilian distress calls, argued that the distress calls of adults were non-functional vestiges of the calls juveniles use to draw adult assistance.

The attributes of the ostariophysan alarm system make it unlikely that it functions to warn offspring or to draw parental assistance. Schools of fish will seldom contain both parents and offspring because fish growth is indeterminate and fish school with individuals of similar size, in response to the hydrodynamic relationships between size and swimming speed. The majority of ostariophysan fish do not provide parental care to free-swimming young, and, in the six species in which development age of the fright reaction has been examined, the young do not show the fright reaction until they are about one to two months old (Schutz, 1956; Pfeiffer, 1963a,b, 1966a; Waldman, 1982). This is well beyond the period when they might be close to the parents, particularly since none of these species provide parental care to free-swimming young. Unfortunately the development of the alarm system has not been examined in catfish, the one ostariophysan group where parental care continues after hatching.

Siblings or other relatives of approximately the same age and size may benefit from the warning signal. It is not known to what extent natural fish schools contain related individuals. There are factors that will tend to obscure such relationships. First, schooling fish tend to be similar in appearance. Predators cull out distinctive individuals (Ohguchi, 1981). Individual fish are often indistinguishable to human observers and relationships between them are hard to discern. Second, for many fish, schooling with any conspecific will be an advantage over solitary life. This means that unrelated individuals will school together and we are left with trying to detect a difference in preference for schooling with kin over non-kin or a preferential association with kin in a mixed school. It is also difficult to trace individuals through the period from fertilization to schooling in order to check kin relationships in wild populations.

Despite this unpromising situation, there is some evidence that fish schools are more than random groups of unrelated fish. First there is the argument of proximity. When fish hatch, the closest individuals in space and in stage of development are likely to be the siblings spawned in the same mating. The ontogeny of schooling has not been studied in most ostariophysan fishes, but it seems likely that when schooling starts siblings will be available as school mates. On a broader scale, if fish remain resident in an area or return to a home spawning area then a local population of related individuals may occur. Eberhard (1975) has argued that low-cost, high-benefit acts might be favoured even at low levels of kinship. The darters, for example, are not typically schooling fish but are resident in discrete locations to the point that endemic species often evolve in small geographic areas (Page, 1983; Kuehne and Barbour, 1983).

Examination of gene frequencies using enzyme electrophoresis indicates that schooling is associated with "genotype structuring" in common shiners, *Notropis cornutus*, cyprinid fishes with alarm substance (Ferguson and Noakes, 1981). This suggests that some natural schools may contain closely related individuals. Within schools, some individuals tend to stay in one region in the school e.g. sockeye salmon fingerlings, Oncorhynchus nerka (Healey and Prieston, 1973) and mackerel, *Scomber scombrus* (Pitcher et al., 1982). If such behavior occurs in fish with alarm signals, then kin may stay preferentially in the same part of the school and benefit from each other's signals.

The whole question of kinship benefits arising from fish alarm signals remains speculative. On the other hand evidence for kin recognition is accumulating from other vertebrates. Anurans, for example, are capable of kin recognition (Waldman and Adler, 1979; Waldman, 1981; Waldman, 1982; Blaustein et al., 1984) and some tadpoles have a chemical alarm system very similar to that found in fish (Richmond, 1947; Eibl-Eibesfeldt, 1949; Hrbacek, 1950; Kulzer, 1954).

Non-alarm Theories

The evolution of alarm substance cells might not be driven exclusively by benefits arising from the alarm response of school mates. Instead, the so called "alarm substance" and the specialized cells might serve some other function that benefits the "sender" independently of the response of other individuals. The characteristics of the cells and their contents may give some clues about their possible functions.

Alarm substance cells. Ostariophysan ASCs are epidermal cells belonging to the general type referred to as club cells or clavate cells (Mittal et al., 1981; Whitear and Mittal, 1983). These cells usually lack pores to the surface and are generally found in the middle layer of the epidermis. As epidermal structures, they are not in close contact with the circulatory system. These conditions would seem to rule out functions associated with continuous secretion of a substance onto the body surface or functions associated with endocrine secretion into the blood stream. The internal structure of the cells suggests a secretory rather than a mechanical support role although there is a spiral arrangement of filaments in the outer cytoplasm (Whitear and Mittal, 1983) and the possibility of providing support by means of hydrostatic pressure cannot be ruled out. Morphologically the cells seem best suited for release of their contents during mechanical damage to the skin.

Club cells, so similar to ostariophysan ASCs that they can be considered homologous, occur in the polypteriform fish Calamoichthys calabricus (Whitear, 1981). The Order Polypteriformes has not been tested for the presence of a fright reaction (Pfeifer, 1977). Other, generally similar, club cells occur in eels, Anguilla anguilla (Leonard and Summers, 1976) that have been tested, with negative results (Schutz, 1956), for the fright reaction. The term "club cell" has been rather broadly applied to teleost epidermal secretory cells, and it is sometimes difficult to determine the degree of resemblance among the various cell types.

ASCs also occur in some groups of ostariophysan fishes that lack the fright reaction. These special cases include the blind cave fishes Anoptichthys jordani (Astyanax mexicanus) and A. antrobius (Pfeiffer, 1963b, 1967, 1977) and Caecobarbus geertsi (Thines and Legrain, 1973) and the pirhanas and silver dollars (Pfeiffer, 1963b, 1977). Skin extract from these fishes will elicit a fright reaction in other species. Other specialized ostariophysans have neither the reaction nor the cell type. These include the Gymnotidae, Loricarridae, Aspredininae and several species in the Suborder Characoidei (Pfeiffer, 1974, 1977).

The occurrence of ASCs in species that lack the fright reaction indicates either that the cells are vestigial in those species or that there is some other function favouring their retention. Fragile secretory cells in the epidermis might contain substances that deterred predators or parasites or that combat infection (Smith, 1977). The ASCs of goldfish, Carassius auratus, may become confluent during skin injury providing a solid barrier of secretion (Whitear and Mittal, 1983). In fish that are known to have toxic or distasteful compounds in the skin such as the coral gobie, Gobiodon quinquestrigatus (Hashimoto et al., 1974), the toxins are often sequestered in epidermal secretory cells that resemble the ASCs in their general size and location although they may differ in specific staining properties or in such details as the location of the nucleus. This suggests that location and structure of ASCs are appropriate for defensive secretions as well as for alarm pheromone.

There are some indications of a direct link between crinotoxicity and ASCs. Some ostariophysan catfishes, particularly the madtoms, Noturus sp. (Reed, 1907, 1924), and the marine catfishes such as Plotosus lineatus (Pawlowsky, 1913, 1914) and Arius felis (Halstead et al., 1953) have very effective poison glands associated with their dorsal and pectoral spines.

Similar, but less toxic, glandular areas are associated with the spines of other Siluroid fishes. As with all known fish venom glands, the poison is released when epidermal secretory cells are mechanically disrupted (Cameron and Endean, 1972). The cells of the poison glands are morphologically similar to the ASCs. Reed (1907) felt there was a continuous series of stages between the clavate cells (ASCs) of the epidermis and the venom cells of the poison glands in the stonecat, Noturus flavus.

Some toxicity is associated with the unspecialized skin of catfish. Birkhead (1967, 1972) found that skin extracts from the caudal fin of two North American catfishes (Ictalurus melas and I. natalis) was toxic to mosquitofish (Gambusia affinis) although less toxic than poison gland extract, which contains a much higher concentration of clavate cells than the caudal skin. Calton and Burnett (1975) found chemical fractions in the caudal fins of white catfish (Ictalurus catus) to be lethally toxic to mice although less toxic than venom gland extracts or fin extracts from perch (Perca flavescens) and bluegill (Lepomis macrochirus), fish that do not possess ASCs.

Since Williams (1964) proposed that the alarm substance acts as a predator deterrent, there has been one direct test of his hypothesis. Bernstein and Smith (1983) treated fathead minnows (Pimephales promelas) with androgen, a process that eliminates ASCs from the skin (Smith, 1973). They then observed the feeding behavior of rainbow trout presented with androgen-treated minnows without ASCs and untreated control minnows with ASCs. The trout showed no significant preference for either type of minnow even though the same test procedure did detect a preference for liver over commercial food pellets. The results indicate that the presence of ASCs in the skin does not act as a deterrent to rainbow trout. This is consistent with the widespread use of minnows, including fathead minnows, as bait fish by sports fishermen.

There is also evidence against the predator deterrence hypothesis from two other sources. Blind cave characins (Anoptichthys jordani) and cyprinids (Caecobarbus geertsi) respond with feeding behaviour to skin extracts that elicit fright reactions in other ostariophysans (Thines and LeGrain, 1973). Similarly, Waldman (1982) found that juvenile zebra danios (Brachydanio rerio) did not respond to skin extract until they were 42 days old, then between 42 and 47 days they were attracted to the point of alarm substance release and showed no darting or fleeing behaviour. By 52 days they had developed a normal fright reaction. The responses of both the cave fishes and the juvenile danios indicate that the skin extract may be inherently attractive rather than distasteful or deterring.

Alarm substance. The chemical nature of the alarm pheromone might give some indication of what other functions it could perform that might benefit the sender. Argentini (1976) suggests that the alarm pheromone in the European minnow (Phoxinus phoxinus) is hypoxanthine-3(N)-oxide although a few other heterocyclic compounds such as isoxanthopterin, 2,6-diamino-4-oxodihydropteridine and 6-acetonylisoxanthopterin elicited fright reactions in the giant danio, (Danio malabricus) (Pfeiffer, 1978). These compounds do have some biological activity. Isoxanthopterin, for example, interferes with nucleic acid synthesis (Lagowski and Forrest, 1967; Harris and Forrest, 1967; Kokolis, 1975), interferes with tail regeneration in the newt (Triturus vulgaris) (Kokolis et al., 1972b) and inhibits tumor development (Petit et al., 1972; Kokolis et al., 1972a). These effects are similar to those of some antibiotics and suggest an antipathogenic function for alarm substance (Smith, 1982). Hypoxanthine-3-oxide and the related 3-hydroxyxanthine are potent oncogens (Sugiura et al., 1970; McDonald et al., 1973; Lam and Parham, 1982) which may also be consistent with deterrent or antipathogenic functions.

Other published statements about the chemical nature of alarm pheromone are not sufficiently precise to provide information on potential non-alarm functions. These include the statement by Ahsan and Prasad (1982) that the alarm pheromone is a polypeptide or protein molecule and the suggestion by Tucker and Suzuki (1972) that the alarm pheromone may be a mixture of peptides, amino acids and derivatives. Similarly Lebedeva et al. (1975) isolated active fractions from minnow skin (Phoxinus phoxinus) but did not identify specific compounds although they found the molecular weight of their active components to be over 950, well above the molecular weight of hypoxanthine-3-oxide. Kasumyan and Lebedeva (1977, 1979) similarly do not provide sufficient detail about the fractions isolated to allow conclusions about other functions for alarm substance.

Both Reed et al. (1972) and Ruddy and Baeder (1973) elicited fright reactions with histamine but histamine is present in all vertebrates while the alarm pheromone is specific to ostariophysans. Histamine "Occurs widely in nature as a result of putrefactive processes" (Stecher et al., p. 533, 1969). So the response reported by Reed et al. (1972) and Ruddy and Baeder (1973) may be an adaptive avoidance response to odours of putrefaction rather than a specific response to conspecific alarm pheromone. Neither group of researchers tested the response of non-ostariophysan fish to histamine.

There have been no studies on the chemical composition of the alarm pheromone in the darters or the cyprinodont species that may have alarm pheromone systems.

Ostariophysan fish that do not have ASCs. There are two situations in which ostariophysan fishes do not possess ASCs. Some taxonomic subgroups within the superorder lack ASCs throughout life (Pfeiffer, 1977). In some other species, males or both sexes lose the ASCs during the breeding season and regain them after breeding is over (Smith, 1976).

The ostariophysans that lack ASCs altogether are a mixture of specialized groups and are probably examples of secondary loss of ASCs rather than of a primary condition preceding ASC development (Pfeiffer, 1974). Chilodus punctatus is a "headstander", a characin that assumes a head-down body orientation when it is at rest. It is a common aquarium fish with peaceful habits. There is no obvious reason why it should differ from most other characins by lacking both ASCs and the fright reaction but its biology in the field does not seem to have been studied. The Nannostominae or "pencil fishes", are small, secretive characins that often have the habit of resting in a "head-up" posture. Again there is no obvious relationship between their known biology and their lack of a chemical alarm system. The other three cases are the heavily armoured loricarid catfishes, the electric gymnotiformes and Bunocephalus bicolor, a cryptic catfish that burys itself during the day (Sterba, 1962). These last three groups are well armoured, extremely cryptic or specialized for electric communication, leading Pfeiffer (1977) to emphasize their difference from the usual, less specialized, ostariophysan way of life. Even if these fish are specialized, it is hard to see why they would lose their ASCs when other fishes retain them. Other armoured catfish such as the Callichthyidae retain ASCs as do the extremely cryptic Chaca chaca and the strongly electric Malapterurus electricus. Perhaps we will have to wait until these fish have been studied in the field before we understand why they lost their ASCs.

Seasonal ASC loss was first reported in breeding male fathead minnows (Pimephales promelas), a species in which the males rub vigorously on a spawning surface during the breeding season (Smith, 1973; Smith and Murphy, 1974; McMillan and Smith, 1974). This suggested the hypothesis that ASC loss was an adaptation to abrasive spawning, preventing "false alarms" caused by skin damage associated with spawning (Smith, 1973, 1976a). Fathead minnow ASC loss remains the most intensively studied case. ASCs dissappear from the skin in response to gonadal androgens (Smith, 1973) and can be eliminated by treatment with 17α-methyltestosterone (Smith, 1973, 1974). Females do not lose their ASCs during normal reproductive cycles but do lose them if treated with 17α-methyltestosterone (Smith, 1974, 1978). Males and females both retain their response to alarm substance throughout the breeding season (Smith, 1976b).

It now appears that there are at least four different responses to reproduction shown by ostariophysan ASCs (Table 1). Some species retain their ASCs at normal levels in both sexes throughout the breeding cycle. Others, like the fathead, show seasonal loss of ASCs during the breeding season in males only and, where tested, seasonal loss is controlled by androgen and female ASCs are androgen sensitive. A third group shows ASC loss in both sexes during breeding and, where tested, neither male nor female ASCs are responsive to androgens (J.D. Smith, 1982). Species in the fourth category show reduction in male ASCs relative to female ASCs but retain substantial numbers of ASCs in both sexes. In one species, the zebra danio (Brachydanio rerio), the reduction in males is under androgen control (Smith, unpublished). In the other species, the common sucker (Catostomus commersoni) breeding males have fewer ASCs than adult females (Smith, 1976a) but juveniles are similar in ASC counts and their ASCs are unresponsive to 17α-methyltestosterone (J.D. Smith, 1982).

The abrasion hypothesis seems consistent with reproduction in the species in which ASC loss occurs only in breeding males (Table 1). The one possible exception, the red shiner (Notropis lutrensis) has very plastic breeding habits (Pflieger, 1975) and may sometimes spawn on submerged logs in an abrasive manner similar to Notropis whipplei. The species with ASC loss in both sexes present a less convincing picture. The breeding male stoneroller (Campostoma anomalum) is a nest builder and is covered with sharp tubercles. There are reports of female stonerollers with scale loss and eye injuries after spawning (O.R. Smith, 1935). Pearl dace (Semotilus margarita) spawning has not been exhaustively described, but the descriptions of Langlois (1929) do not suggest unusual abrasion. O.R. Smith (1935), however, describes patches of replaced scales as typical of pearl dace, including females. The spawning behaviour of the finescale dace (Chrosomus neogaeus) (Stasiak, 1978) apparently resembles that of the lake chub (Couesius plumbeus) (Brown et al., 1970) suggesting that two species with very similar behaviour differ in the seasonal pattern of their ASC numbers.

The adaptive significance of androgen controlled reduction of ASCs to about half the female level in zebra danios (Brachydanio rerio) is unclear. There is a correspondence between ASC counts and the threshold for the fright reaction. Gandolfi et al. (1968) reported that male zebra danios had a lower threshold of response to alarm pheromone than females. If there is a tendency for danios to form monosexual schools in nature, then this may be a case of matching threshold to emission rate, maintaining an optimum active space by keeping the Q/K ratio constant.

The two species in which ASC loss occurs in both sexes but is not controlled by androgen, the pearl dace and the finescale dace, both develop some androgen-induced secondary sexual characteristics, including tubercles and breeding colours, in the autumn and retain them through the winter prior to early spring spawning (J.D. Smith, 1982). In contrast, they retain their ASCs until the time of actual spawning then lose them briefly during the spawning period. So, for the winter season there is a dissociation of the normal attributes of breeding fish. Colour and tubercles are present but ASCs are absent. This may account for non-androgen control of ASC loss. It also suggests that losing the ASCs is less tolerable than gaining colour

Table 1. Species of ostariophysan fishes examined for seasonal loss or retention of ASCs with information on endocrine control where available, and an assessment of the degree of abrasion during reproductive behavior. + indicates some evidence of abrasive behavior, - indicates that reproduction is not known to be abrasive.

Species	Abrasion	ASC loss	Control	Reference*
Pimephales promelas	+	male	androgen	1,2
Pimephales notatus	+	male	?	3
Pimephales vigilax	+	male	?	3
Nocomis biguttatus	+	male	?	3
Nocomis leptocephalus	+	male	?	4
Notropis cornutus	+	male	androgen	3,5
Notropis venustus	+?	male	?	6
Notropis whipplei	+	male	?	6
Notropis lutrensis	-?	male	?	4
Semotilus atromaculatus	+	male	androgen	3,5
Semotilus margarita	-?	both	not androgen	3,5
Chrosomus neogaeus	-	both	not androgen	5,6
Campostoma anomalum	+	both	?	3
Brachydanio rerio	-	male reduced	androgen	7
Catostomus commersoni	-	male reduced	not androgen?	3,5
Couesius plumbeus	-	neither	?	3
Rhinichthyes cataractae	-	neither	?	3
Notropis chiliticus	?	neither	?	4
Clinostomus funduloides	?	neither	?	4

*1. Smith, 1973; 2. Smith, 1974; 3. Smith, 1976a; 4. Smith & Lemly, in prep.; 5. J.D. Smith, 1982; 6. Smith & Smith, 1983; 7. Smith, in prep.

and tubercles several months prior to breeding.

If the abrasion hypothesis accounts for some of the cases of ASC loss then this is an area in which the response of conspecifics to the alarm substance may have influenced the evolution of ASCs. The disadvantage of alarming conspecifics during mating would presumably have exceeded any advantage, including direct benefits, derived from retaining ASCs.

SELECTION ACTING ON THE RECEIVER

It seems reasonable that an animal might benefit by responding to stimuli associated with injury to a conspecific, and such responses are very widespread in nature. This effect may occur purely through selection on the receiver with no cooperation being required of the sender. The receiver, however, will be subjected to selection pressures acting on the specificity of the response, on the conditions under which the response is given and on the form of the response.

Stimulus Specificity

Fish restrict their defensive response to a limited number of chemicals out of the large number of molecules that are released from damaged tissue. In the ostariophysans the response is restricted to some component or components of the ASCs (Pfeiffer, 1974; Smith, 1973), possibly hypoxanthine-3(N)-oxide (Argentini, 1976; Pfeiffer, 1982). In the darters, skin extract is much more effective in releasing the defensive response than extracts of other tissues (Smith, 1979).

Several advantages might accrue from restriction of the response to a limited range of stimuli. In carnivorous fish, confusion between chemicals normally released from the receiver's prey and chemicals from damaged conspecifics would be reduced. Even in fish that were not carnivorous, false alarms would occur if the fish responded to chemicals from prey animals that did not share the same predators (Atema and Stenzler, 1977; Stenzler and Atema, 1977; Smith, 1977). Conspecifics of the same size will be the best indicators of danger but other similar species may also share the same predators, that is they may be members of the same prey guild.

Studies of cross-reaction between fish species have emphasized taxonomic relationship rather than shared membership in a prey guild. Schutz (1956) for example, found that there was decreasing response with more distant taxonomic relationship. In darters, cross reactions occur among the three genera in the group although reactions to hetero-specific skin extracts are sometimes weaker than reactions to conspecific extracts (Smith, 1979, 1982b).

Pfeiffer and Lemke (1973) and Pfeiffer (1978) have described strong cross reactions between giant danio, *Danio malabricus* and the European minnow, *Phoxinus phoxinus*, but this is in contrast to the taxonomic variation in response found by Schutz (1956), Aoki and Kuroki (1975) and Ahsan and Prasad (1984). It is possible that there is a single alarm-releasing compound but that the response is tempered by other compounds indicating the degree of relationship between the sender and receiver. If secondary chemicals from heterospecific animals, for example, inhibited the response, then "pure" alarm substance would give a strong response in all species, but skin extract from an unrelated fish would give a reduced response. This idea has not been examined experimentally in fish, but a similar hypothesis was tested by Snyder (1967) on snails. Heterospecific extracts did not inhibit the snail fright reaction.

A particular compound or set of compounds may have the best set of properties for transmission and detection of the alarm message. For example they may stimulate the receptors most strongly or be most easily distinguished from background levels of chemicals in the natural environment or have the appropriate life span for a warning signal. If this is the case, the response of the receiver might evolve toward exclusive reliance on the "alarm pheromone".

Nature of the Fright Reaction

There is no direct experimental evidence to show that the fright reaction actually improves the survival of the receiver. There are, however, a number of correlations between the fright reactions shown by various fish when they detect alarm substance and their reactions to other indicators of predation. The most general type of reaction in schooling fish is for the school to aggregate and move away from the area of the stimulus. This type of response was first recorded in the European minnow by von Frisch (1938) and has since been observed in many species (Pfeiffer, 1977).

This schooling reaction has been quantified in common shiners, *Notropis cornutus* (Heczko and Seghers, 1981), and zebra danios, *Brachydanio rerio* (Waldman 1982). Both species showed a statistically significant tendency to clump together and the danios also moved to the bottom of the tank. The fish in these studies were not fed at the time of pheromone introduction so the response is not confused by the conditioned feeding responses that are present when the traditional method is used of training the fish to come for food then presenting the alarm substance.

Other types of fright reaction occur among fishes (Pfeiffer, 1963b). Some bottom-dwelling species sink to the bottom and remain motionless; other species that normally live near the surface flee to the water surface and may jump out of the water (Schutz, 1956). In general the fright reaction is appropriate to the biology of the species.

There is one reported case in which the fright reaction changes during the development of the fish, as the fish changes its food habits and other behaviour. Creek chub, Semotilus atromaculatus, undergo a major shift in their diet, activity cycles, and other behaviour between the first and second year of life (Newsome, 1975, Magnan and Fitzgerald, 1984). As juveniles, their fright reaction is typical of small schooling cyprinids, darting away and hiding. The adults, in contrast, sink quietly to the bottom and remain motionless when they detect alarm substance (Newsome, 1975). This change has not yet been directly tied to differences in predation on the two age groups.

Predator specific fright reactions occur in the common shiner. They respond to alarm substance by seeking cover in vegetation if a kite, simulating an piscivorous bird, is flown over their tank, but they do not enter cover if a predatory fish is present in the tank (Heczko, 1980; Heczko and Seghers, 1981).

The darters, with an alarm system analogous to the ostariophysan system, show only minor variations in fright reaction among the five species that have been extensively tested. Iowa darters, Etheostoma exile, johnny darters, E. nigrum, gulf darters, E. swaini, naked sand darters, Ammocrypta beani, and blackbanded darters, Percina nigrofasciata all become immobile, "freeze", when they detect skin extract (Smith, 1979, 1982b). Blackbanded darters and gulf darters also adopt more cryptic, background matching colouration after freezing, and both species may precede the freeze with one or two rapid darts or with a slow movement into cover.

The naked sand darter is the least responsive of the species tested, but it normally spends much of its time buried in the sand where it is probably hard for predators to find. Interestingly, both the sand darter and the johnny darter freeze on top of the sand rather than burying in response to the skin extract although both species are capable of burying themselves rapidly. Immobility, freezing, is an effective form of primary defense against predators. Primary defensive mechanisms reduce the chance that a predator will detect prey or recognize prey (Edmunds, 1974). The relatively conspicuous burying movements may call attention to prey while remaining still will allow it to go undetected.

Fish That Do Not Show the Fright Reaction

Within the taxonomic groups that typically possess alarm pheromones there are several situations in which the fright reaction is absent, either permanently or temporarily. Permanent loss of the fright reaction occurs in several taxonomic groups either with retention of ASCs as in the blind cave fishes (Pfeiffer, 1974; Thines and Legrain, 1973) or with concurrent loss of ASCs as in the groups were listed earlier in this review. Temporary absence of the fright reaction also occurs. For example, all juvenile fish tested show a delay of about one to two months between hatching and the development of the fright reaction (Schutz, 1956; Pfeiffer, 1963a, b, 1966; Waldman, 1982). The significance of this delay is not immediately obvious. One suggestion is that the fright reaction becomes effective as schooling improves, but some non-schooling species do show fright reactions as adults. At the other end of the developmental sequence, there are recurring reports that the fright reaction is reduced in very large, and hence older, fish (e.g. Ahsan and Prasad, 1982). Larger individuals may be subjected to less predation and selection for the fright reaction may be relaxed.

Temporary suppression or loss of the fright reaction may also occur in response to specific contingencies. For example, hiding from predators is probably incompatible with feeding, courting mates and other activities. A fish must balance its responses so that it stays uneaten, while eating and mating sufficiently often to propogate its genotype. The specific conflict between feeding and fright reaction has not been quantified in ostariophysans, although the traditional test procedure of training the fish to come for food then presenting the alarm pheromone does involve a conflict. The original descriptions of the fright reaction used by von Frisch (1941), Pfeiffer (1974) and others include mention of the fish darting back to the feeding area and snatching particles of food, indicating some tendency for feeding stimuli to override predator defense.

Iowa darters show a clear change in the priority of their fright reaction when they are deprived of food for 24 to 48 hours (Smith, 1981). After a one to two day fast they respond to a mixture of skin extract and feeding stimuli (water that had contained food organisms) with a feeding response. After 48 hours of abundant food,the same darters respond to the same stimulus with immobility, a defensive reaction. Similar alterations in the priority of feeding and fright reactions have been observed in the response of snails to alarm pheromones (e.g. Stenzler and Atema, 1977).

Another case of temporary reduction of the fright reaction occurs in harlequin fish, Rasbora heteromorpha. They do not show as much school cohesion in response to alarm substance when they are tested in the dark during the night phase of their activity cycle as when they are tested, in light or dark, during the day phase of the cycle (Thines and Vandenbussche, 1966). Cohesion may be more effective against typical diurnal predators than against nocturnal predators.

Researchers working with alarm pheromones have sometimes noted that fish become habituated to unreinforced exposure to alarm substance (e.g. von Frisch, 1941) but this effect does not seem to have been extensively studied. In contrast, Waldman (1982) found no effect of prior exposure in zebra danios and Gandolfi (1968) found a sensitization effect from prior exposure, also in zebra danios.

FINAL COMMENT

In examining the chemical alarm systems of fishes, it is important to try to see the system as it would function in the wild. The greatest need, at the moment, is more study in the field of predation and defence and the specific roles played by skin secretions, whether as alarm signals, deterrents or protective substances. The elements of these systems can vary among species, among the members of a species or even within the life span of an individual. The ostariophysan fishes are a diverse group composed of about 5000 to 6000 species, the darters include another 150 species. Many marine groups have not yet been examined for chemical alarm systems. Accounting for the evolution of these alarm systems and for their variations remains a fascinating challenge.

ACKNOWLEDGEMENTS

I thank Dr. A.D. Lemly and Mrs. M.J. Smith for their helpful comments on the manuscript and Dennis Dyck for preparing the figure. The preparation of this manuscript was supported in part by a grant from the Natural Sciences and Engineering Research Council of Canada.

REFERENCES

Ahsan, S. N., and Prasad, M., 1982, Occurrence of fright reaction in Indian fishes, Biol. Bull. India, 4:41-47.

Aoki, I., and Kuroki, T., 1975, Alarm reaction of three Japanese cyprinid fishes, Tribolodon hakonensis, Gnathopogon elongatus elongatus, Rhodeus occelatus occelatus, Bull. Jpn. Soc. Sci. Fish., 41:507-513.

Argentini, M., 1976, Isolierung des Schreckstoffes aus der Haut der Elritze Phoxinus phoxinus (L), Univ. of Zurich, Zurich, Ph.D. Thesis.

Atema, J., and Stenzler, D., 1977, Alarm substance of the marine mud snail, Nassarius obsoletus: biological characterization and possible evolution, J. Chem. Ecol., 3:173-187.

Baldwin, I. T., and Schultz, J. C., 1983, Rapid changes in tree leaf chemistry induced by damage: evidence for communication between plants, Science, 221:277-279.

Bernstein, J. W., and Smith, R. J. F., 1983, Alarm substance cells in fathead minnows do not affect the feeding preference of rainbow trout, Env. Biol. Fish., 9: 307-311.

Birkhead, W. S., 1967, The comparative toxicity of stings of the ictalurid catfish genera Ictalurus and Schilbeodes, Comp. Biochem. Physiol., 22:101-111.

Birkhead, W. S., 1972, Toxicity of sting of arriid and ictalurid catfishes. Copeia, 1972:790.

Blaustein, A. R., O'Hara, R. K., and Olson, D. H., 1984, Kin preference is present after metamorphosis in Rana cascadae frogs, Anim. Behav., 32:445-450.

Brown, J. H., Hammer, U. T., and Koshinsky, G. D., 1970, Breeding biology of the lake chub, Couesius plumbeus, at Lac la Ronge, Saskatchewan, J. Fish. Res. Board Can., 27:1005-1015.

Calton, G. J., and Burnett, J. W., 1975, Catfish (Ictalurus catus) fin venom, Toxicon, 13:399-403.

Cameron, A. M., and Endean, R., 1972, The venom glands of teleost fishes, Toxicon, 10:301-303.

Charnov, E. L. and Krebs J. R. 1975, The evolution of alarm calls: altruism or manipulation?, Amer. Nat., 109:107-112.

Eberhard, M. J. W., 1975, The evolution of social behavior by kin selection, Quart. Rev. Biol., 50:1-34.

Edmunds, M., 1974, "Defence in animals," Longman, N.Y.

Eibl-Eibesfeldt, I., 1949, Über das Vorkommen von Schreckstoffen bei Erdkrötenquappen, Experientia, 5:236.

Ferguson, M. M., and Noakes, D. L. G., 1981, Social grouping and genetic variation in common shiners, Notropis cornutus (Pisces, Cyprinidae), Env. Biol. Fish., 6:357-360.

Frisch, K. von, 1938, Zur Psychologie des Fisch-Schwarmes, Naturwissenschaften, 26:601-606.

Frisch, K. von, 1941, Über einen Schreckstoff der Fischhaut und seine biologische Bedeutung, Z. vergl. Physiol., 29:46-145.

Gandolfi, G., 1968, Reazione di paura nel ciprinide Brachydanio rerio: variaioni della reativita indotte sperimentalmente, Arch. Zool. Ital., 53:245-259.

Gandolfi, G., Mainardi, D., and Rossi, A. C., 1968, The fright reaction of zebra fish. Atti della Societa Italiana di Scienze Naturali e del Museo Civico di Storia Naturale di Milano, 107:74-88.

Halstead, B. W., Kuninobu, L. S., and Hebard, H. G., 1953, Catfish stings and venom apparatus of the Mexican catfish Galeichthys felis (Linnaeus), Trans. Am. microsc. Soc., 72:297.

Harris, S. E. and Forrest, H. S., 1967, Inhibition by certain pteridines of ribosomal RNA and DNA synthesis in developing Oncopeltus eggs, Proc. Natl. Acad. Sci. U.S.A., 58:89-94.

Harvey, P. H., Greenwood, P. J., 1978, Antipredator defense strategies: some evolutionary problems, in: "Behavioural Ecology, an Evolutionary Approach," J. R. Krebs, and N. B. Davies, eds., Sinauer Associates, Sunderland, MA.

Hashimoto, Y., Shiomi, K., and Aida, K., 1974, Occurrence of a skin toxin in coral-gobies Gobiodon spp., Toxicon, 12:523-528.

Healey, M. C., and Prieston, R., 1973, The interrelationships among individuals in a fish school, Fish. Res. Bd. Canada Tech. Rept., #389.

Heczko, E., 1980, The effect of Schreckstoff on the behaviour of the common shiner, Notropis cornutus Cyprinidae, M.Sc. thesis University of Western Ontario.

Heczko, E., and Seghers, B. H., 1981, Effects of alarm substance on schooling in the common shiner (Notropis cornutus, Cyprinidae), Env. Biol. Fish., 6:25-29.

Hrabcek, I., 1950, On the flight reaction of the tadpoles of the common toad caused by chemical substances, Experientia, 6:100-101.

Kasumyan, A. O., and Lebedeva, N. E., 1977, On the chemical nature of the alarm substance from skin of the European minnow (Phoxinus phoxinus), Biol. Nauke, 1:37-41.

Kasumyan, A. O., and Lebedeva, N. E., 1979, Novye svedeniya o prirode feromona trevogi karpovykh ryb, Vopr. Ikhtiol., 19:890-895.

Kasumyan, A. O., and Paschenko, N. I., 1982, The role of olfaction in the defense reaction of grass carp, Ctenopharyngodon idella (Cypinidae), to alarm pheromone, J. Ichthyol., 22:122-126.

Kokolis, N., 1975, The influence of folic acid and isoxanthopterin on renal nucleic acids and protein synthesis, Folia Bioch. et Biol. Graeca, 12:11-17.

Kokolis, N., Mylonas, N., and Ziegler, I., 1972a, Pteridine and riboflavin patterns during tail regeneration in Triturus species and the effects of chloramphenicol, isoxanthopterin and reserpine, Z. Naturforsch., 27b:285-291.

Kokolis, N., Mylonas, N., and Ziegler, I., 1972b, Pteridine and riboflavin in tumor tissue and the effect of chloramphenicol and isoxanthopterin, Z. Naturforsch., 27b:292-295.

Kuehne, R. E., and Barbour, R. W., 1983, "The American Darters," University Press of Kentucky, Lexington.

Kulzer, E., 1954, Untersuchungen über die Schreckreaktion bei Erdkrötenquappen (Bufo bufo L.), Z. vergl. Physiol., 36:443-463.

Lagowski, J. M., and Forrest, H. S., 1967, Interaction between isoxanthopterin and DNA, Proc. Natl. Acad. Sci. U.S.A., 58:1541-1547.

Lam, F. L., and Parham, J. C., 1982, Photochemistry of purine 3-oxides in hydroxylic solvents, Tetrahedron, 38:2371-2376.

Langlois, T. H., 1929, Breeding habits of the northern dace, Ecology, 10:161-163.

Lebedeva, N. Ye., Malyukina, G. A., and Kasumyan, A. O., 1975, The natural repellent in the skin of cyprinids/Phoxinus phoxinus gel chromatography physicochemical properties, Journal of Ichthyology, 15:472-480.

Leonard, J. B., and Summers, R. G., 1976, The ultrastructure of the integument of the American eel, Anguilla rostrata, Cell Tiss. Res., 171:1-30.

Mackay-Sim, A., and Laing, D. G., 1981, Rats' responses to blood and body odors of stressed and non-stressed conspecifics, Physiol. Behav., 27:503-510.

Magnan, P., and Fitzgerald, G. J., 1984, Ontogenetic changes in diet activity, food habits and spatial distribution of juvenile and adult creek chub, Semotilus atromaculatus, Env. Biol. Fish., 11:301-307.

Maynard Smith, J., 1976, Group Selection, Quart. Rev. Biol., 51:277-283.

McDonald, J. J., Stohrer, G., and Brown, G. B., 1973, Oncogenic purine N-oxide derivatives as substrates for sulfotransferase, Cancer Res., 33:3319-3323.

McMillan, V. E., and R. J. F. Smith, 1974, Agonistic and reproductive behaviour of the fathead minnow (Pimephales promelas Rafinesque), Z. Tierpsychol., 34:25-58.

Mittall, A. J., Whitear, M., and Bullock, A. M., 1981, Sacciform cells in the skin of teleost fish, Z. mikrosk.-anat. Forsch., Leipzig, 95:559-585.

Newsome, G. E., 1975, A study of prey preference and selection by creek chub, Semotilus atromaculatus, in the Mink River, Manitoba, Ph.D. Thesis, University of Manitoba.

Ohguchi, O., 1981, Prey density and selection against oddity by three-spined sticklebacks, Z. Tierpsychol. Suppl., 23:1-79.

Page, L. M., 1983, "Handbook of Darters", T.F.H. Publications, Neptune City, N.J.

Pawlowsky, E. N., 1913, Sur la structure des glandes à venin de certains poissons et en particulier de celles de Plotosus, C.r. Seance. Soc. Biol., 74:1033.

Pawlowsky, E. N., 1914, Über den Bau der Giftdrüsen bei Plotosus und anderen Fischen, Zool. Jb., 38:427.

Petit, G. R., Houghton, L. E., Rogers, N. H., Coomes, R. M., Berger, D. F., Reucroft, P. R., Day, J. F., Hartwell, J. L., and Wood, H. B. Jr., 1972, Butterfly wing antineoplastic agents, Experientia, 28:381-382.

Pfeiffer, W., 1963a, The fright reaction in North American Fish, Can. J. Zool., 41:69-77.

Pfeiffer, W., 1963b, Alarm substances, Experientia, 19:113-123.

Pfeiffer, W., 1966, Die Schreckreaktion der Fische und Kaulquappen, Naturwissenschaften, 53:565-570.

Pfeiffer, W., 1967, Schreckreaktion und Schreckstoffzellen bei Ostariophysi und Gonorhynchiformes, Z. vergl. Physiol., 56:380-396.

Pfeiffer, W., 1974, Pheromones in fish and amphibia, In: "Pheromones", Frontiers of Biology, M.C. Birch, ed., Vol. 32, North-Holland Publishing Co., Amsterdam.

Pfeiffer, W., 1977, The distribution of fright reaction and alarm substance cells in fishes, Copeia, 1977:653-665.

Pfeiffer, W., 1978, Heterocyclic compounds as releasers of the fright reaction in the giant danio Danio malabricus (Jordon) (Cyprinidae, Ostariophysi, Pisces), J. Chem. Ecol., 4:665-673.

Pfeiffer, W., 1982, Chemical signals in communication, in: "Chemoreception in Fishes," T. J. Hara, ed., Elsevier, Amsterdam.

Pfeiffer, W., and Lemke, J., 1973, Untersuchungen zur Isolierung und Identifizierung des Schreckstoffes aus der Haut der Elritze, Phoxinus phoxinus (L.) (Cyprinidae, Ostariophysi, Pisces), J. Comp. Physiol., 82:407-410.

Pfeiffer, W., Mangold-Wernado, U., and Neusteuer, P., 1984, Identification of the nerve bundle in the tractus olfactorius of the tench, Tinca tinca L., which conducts the nervous excitation elicited by the alarm substance, Experientia, 40:219-220.

Pflieger, W. L., 1975, "The Fishes of Missouri," Missouri Dept. of Conservation, Columbia, Mo.

Pitcher, T. J., Wyche, C. J., and Magurran, A. E., 1982, Evidence for position preferences in schooling mackerel, Anim. Behav., 30:932-934.

Reed, H. D., 1907, The poison glands of Noturus and Shilbeodes, Amer. Natur., 41:553-566.

Reed, H.D., 1924, The morphology of dermal glands in nematognathous fishes, Z. Morph. Anthrop., 24:227.

Reed, J. R., Wieland, W., and Kimbrough, T. D., 1972, A study on the biochemistry of alarm substances in fish, Proc. of the 26th Ann. Conference, S.E. Assoc. of Game and Fish Commissioners, Knoxville.

Richmond, N. D., 1947, Life history of *Scaphiopus holbrookii hollbrookii* (Harlan, Part 1: Larval development and behavior, Ecology, 28:53-67.

Ruddy, L. L., and Baeder, D. H., 1973, Effect of antihistamines on fish behaviour induced by pheromone and histamine, Fed. Proc., Fed. Am. Soc. Exp. Biol., 32:818.

Schutz, F., 1956, Vergleichende Untersuchungen über die Schreckreaktion bei Fischen und deren Verbreitung, Z. Vergl. Physiol., 38:84-135.

Smith, J. D., 1982, The control of alarm substance cells in some minnows, M.Sc. Thesis, Biology Dept., University of Saskatchewan, Saskatoon.

Smith, O. R., 1935, The spawning habits of the stoneroller minnow (*Campostoma anomalum* Rafinesque), Trans. Am. Fish. Soc., 65:148-151.

Smith, R. J. F., 1973, Testosterone eliminates alarm substance in male fathead minnows, Can. J. Zool., 51:875-876.

Smith, R. J. F., 1974, Effects of 17 α-methyltestosterone on the dorsal pad and tubercles of fathead minnows (*Pimephales promelas* Rafinesque), Can. J. Zool., 52:1031-1038.

Smith, R. J. F., 1976a, Seasonal loss of alarm substance in North American cyprinid fishes and its relation to abrasive spawning behavior, Can. J. Zool., 54:1172-1182.

Smith, R. J. F., 1976b, Male fathead minnows (*Pimephales promelas* Rafinesque) retain their fright reaction to alarm substance during the breeding season, Can. J. Zool., 54:2230-2231.

Smith, R. J. F., 1977, Chemical communication as adaptation: alarm substance of fish, in: "Chemical Signals in Vertebrates," D. Muller-Schwarze, and M. M. Mozell. eds., Plenum Publ. Co., N.Y.

Smith, R. J. F., 1978, Seasonal changes on the histology of the gonads and dorsal skin of the fathead minnows *Pimephales promelas*, Can. J. Zool., 56:2103-2109.

Smith, R. J. F., 1979, Alarm reaction of Iowa and johnny darters (*Etheostoma*, Percidae, Pisces) to chemicals from injured conspecifics, Can. J. Zool., 57:1278-1282.

Smith, R.J.F., 1981, Effect of food deprivation on the reaction of Iowa darters (*Etheostoma exile*) to skin extract, Can. J. Zool., 59:558-560.

Smith, R. J. F., 1982a, The adaptive significance of the alarm substance - fright reaction system, in: "Chemoreception in Fishes," T. J. Hara ed., Elsevier, Amsterdam.

Smith, R. J. F., 1982b, Reaction of *Percina nigrofasciata*, *Ammocrypta beani*, and *Etheostoma swaini* (Percidae, Pisces) to conspecific and intergeneric skin extracts, Can. J. Zool., 60:1067-1072.

Smith, R. J. F., 1985, Evolution of Alarm Signals: Role of Benefits Derived From Retaining Group Members or Territorial Neighbors, (unpublished ms).

Smith, R. J. F., and Lemly, A. D., 1985, Survival of fathead minnows after injury by predators and its' possible role in the evolution of alarm signals, Env. Biol. Fishes, (in press).

Smith, R. J. F., and Murphy, B. D., 1984, Functional morphology of the dorsal pad in fathead minnows (*Pimephales promelas*), Trans. Amer. Fish. Soc., 103:65-72.

Smith, R. J. F., and Smith, J. D., 1983, Seasonal loss of alarm substance cells in *Chrosomus neogaeus*, *Notropis venustus* and *N. whipplei*, Copeia, 1983:822-826.

Snyder, N. F. R., 1967, An alarm reaction of aquatic gastropods to intraspecific extract, Cornell Univ. Agr. Sta. Mem., #403.

Stasiak, R. H., 1978, Reproduction, age and growth of the finescale dace, *Chrosomus neogaeus*, in Manitoba, Trans. Am. Fish. Soc., 107:720-723.

Staton, M.A., 1978, 'Distress Calls' of crocodilians - whom do they benefit? Am. Nat., 112:327-332.

Stecher, P. G., Windholz, M., Leahy, D. S., Bolton, D. M., and Eaton, L. G., eds.,1969, "The Merck Index, eighth edition," Merck Co., Rahway, N.J.

Stenzler, D., and J. Atema, 1977, Alarm response of the marine mud snail, Nassarius obsoletus: specificity and behavioral priority, J. Chem. Ecol., 3:159-171.

Sterba, G., 1962, "Freshwater Fishes of the World," Pet Library Ltd., N.Y.

Sugiura, K., Teller, M. N., Parham, J. C., and Brown, G. B., 1970, A comparison of the oncogenicities of 3-hydroxyxanthine, guanine 3-N-oxide, and some related compounds, Cancer Res., 30:184-188.

Thines, G., and Legrain, J. M., 1973, Effects of alarm substance on the behavior of the cavefish Anoptichthys jordani and Caecobarbus geertsi, Ann. Speleol., 28: 291-297.

Thines, G., and Vandenbussche, E., 1966, The effects of alarm substance on the schooling behaviour of Rasbora heteromorpha Duncker in day and night conditions, Anim. Behav., 14: 296.

Trivers, R. L., 1971, The evolution of reciprocal altruism, Quart. Rev. Biol., 46: 35-57.

Tucker, D. and Suzuki, N., 1972, Olfactory responses to Schreckstoff of catfish, in: "Olfaction and Taste" 4. Proc. 4th Int. Symp. held in Starnberg, Germany, August 2-4, 1971; D. Schneider, ed., (Wissenschaftl. Verlagsges., Stuttgart) pp. 121-127.

Vinyard, G. L., 1980, Differential prey vulnerability and predator selectivity: effects of evasive prey on bluegill (Lepomis macrochirus) and pumpkinseed (L. gibbosus) predation, Can. J. Fish. Aquat. Sci., 37:2294-2299.

Waldman, B., 1981, Sibling recognition in toad tadpoles: the role of experience, Z. Tierpsychol., 56:341-358.

Waldman, B., 1982, Quantitative and developmental analysis of the alarm reaction in the zebra danio, Brachydanio rerio, Copeia, 1982:1-9.

Waldman, B., and Adler, K., 1980, Toad tadpoles associate preferentially with siblings, Nature, 282:611-613.

Weldon, P. J., 1983, The evolution of alarm pheromones, in: "Chemical Signals in Vertebrates 3," D. Muller-Schwarze, and R. M. Silverstein, eds.,Plenum, N.Y.

Whitear, M., 1981, Secretion in the epidermis of polypteriform fish, Z. Mikrosk. anat. Forsch., Leipzig, 95:531-543.

Whitear, M., and Mittal, A. K., 1983, Fine structure of the club cells in the skin of ostariophysan fish, Z. mikrosk, -anat. Forsch., Leipzig, 97:141-157.

Williams, G.C., 1964, Measurement of consociation among fishes and comments on the evolution of schooling, Publ. Mus. Mich. State University Biol. Ser., 2:349-384.

Williams, G.C., 1966, "Adaptation and Natural Selection," Princeton University Press, Princeton, N.J.

FISH REPRODUCTIVE PHEROMONES

N.E. Stacey[1], A.L. Kyle[2], and N.R. Liley[3]

[1]Department of Zoology, University of Alberta
Edmonton, Alberta, Canada T6G 2E9
[2]Department of Anatomy, University of Calgary
Calgary, Alberta, Canada T2N 1N4
[3]Department of Zoology, University of British Columbia
Vancouver, British Columbia, Canada V6T 2A9

INTRODUCTION

Chemical communication in fish is known or suspected to be involved in a wide range of functions including general social attraction, species and individual recognition, parent-young interactions, alarm or fright reactions, migration, and a variety of reproductive responses including stimulation of spawning readiness, sexual attraction, and elicitation of courtship behavior (Colombo et al., 1982; Liley, 1982; Liley and Stacey, 1983). We will deal exclusively with reproductive aspects of chemical communication in fish (i.e. chemical communication between individuals which directly alters reproductive development or the likelihood of sexual behavior) and will not be concerned with non-reproductive social interactions. Emphasis will be placed on developments which have occurred since the publication of earlier reviews, with particular attention being given to the possible chemical nature of fish reproductive pheromones, the information which these chemical signals might transmit, and the mode of action by which they are thought to induce the observed reproductive responses. As little is known of many of these aspects of chemical communication in fish, it is hoped that the present article will serve as a focus and stimulus for further research in this area.

Numerous observations and experiments show that chemicals released by one fish can alter the behavior or physiology of another. However, as discussed previously by Liley (1982), it is not clear in the majority of examples from fish whether these chemicals should be considered pheromones - discrete chemical signals which have evolved as a component of the species' communication system - or whether they may simply be metabolic products eliciting specific adaptive responses. Animals may be genetically programmed to respond adaptively to a variety of abiotic physical and chemical stimuli which would not be considered communication signals. What distinguishes such chemical responsiveness from chemical communication is that the latter exhibits evidence of evolutionary development of the signalling mechanism such as the presence of specialized secretory structures, specialized chemical products, or behaviors associated with pheromone release. With few exceptions, however, there is little evidence in fish of such a development and specialization in the mechanisms generating chemical signals. Thus, while there are many demonstrations that chemicals released from one fish transmit information which evokes specific reproductive responses in a second, in the majority of cases these 'communication' systems appear to have resulted from the evolution of responsiveness to a preexisting, fortuitous, and apparently unspecialized stimulus source. In this respect, chemical signalling in fish appears to contrast with pheromone communication in animals relying on airborne pheromones. For example, in most terrestrial insect and mammal species in which pheromone communication has been identified, it is possible to recognize distinct glandular organs capable of the synthesis of highly specialized volatile products. The apparent difference between fish and terrestrial forms is almost certainly a reflection of the difference in the media in which they live. To be transmitted any distance through air, a chemical must be volatile; in an aquatic medium, a chemical must be soluble. Whereas the range of potential airborne pheromones is restricted by the decrease in volatility and increase in cost of synthesis correlated with an increase in molecular weight, the range of soluble biochemical products which may be perceived by another individual, and thus might potentially serve as chemical signals, is immense (Wilson, 1975). Furthermore, the disadvantages associated with chemical communication - slow transmission due to dependence on diffusion or currents, and slow fade-out - are almost certainly more severe for soluble waterborne chemicals than for airborne odors (Wilson, 1975). Perhaps as a consequence of this last feature, waterborne signals are restricted to the transmission of relatively simple, tonic messages - arousal and orientation - and because of this limitation chemical signalling in fish has remained evolutionarily unspecialized. A low level of evolutionary specialization combined with the diversity of soluble chemicals potentially available to serve as pheromones has permitted fish to be opportunistic and flexible in incorporating fortuitous chemical stimuli into their communication repertoires.

A major reason for caution in interpreting much of the evidence for the existence of reproductive pheromones in fish concerns the nature of the experimental procedures used to demonstrate chemical communication. The problems here are of two basic types and involve the determination of both the nature of the response and the specificity of the response. Where the response is simply an increase in activity or an orientation, it is often not clear whether the reaction should be considered to be a sexual response rather than a more general social attraction or excitation. Often, the sexual nature of the observed behavior may be established by demonstrating that the response is affected in a meaningful way by the reproductive condition of either the pheromone donor or the pheromone recipient. The second problem, that of demonstrating response specificity, arises when experiments do not include sufficient controls (alternate stimulus types) for the stimulus being tested. The demonstration that water from one fish attracts another of opposite sex cannot be taken as evidence for the presence of a sexual pheromone unless control treatments demonstrate that the same response is not elicited by odors from other sources such as same-sex or heterospecific fish, or food. Although a number of studies have used experimental designs which clarify the nature and specificity of the response, the majority have not.

RESPONSES TO REPRODUCTIVE PHEROMONES

As is the case with pheromonally induced responses in other animal groups, pheromones involved in reproductive processes of fish have been categorized either as 'releasers' - which trigger a more or less immediate response in the pheromone recipient - or 'primers' - which induce a response only after a delay of hours or days. Although these terms are useful in indicating the speed with which pheromone-stimulated sexual responses can occur, it must be kept in mind that, in the absence of evidence to the contrary, the distinction between releasing and priming pheromones is probably a reflection of differences in response times of the processes they stimulate, rather than an indication of differences in the nature or initial mode of action of the pheromones themselves.

Male Pheromones

Exposure of females to chemical stimuli from conspecific males induces two basic types of response suggested to serve a reproductive function: attraction to the male or a source of water bearing the male odor, and stimulation of ovarian growth and ovulation. Attraction of females has been demonstrated in a variety of species including rainbow trout, *Salmo gairdneri* (Newcombe and Hartman, 1973), a blenny, *Blennius pavo* (Laumen et al., 1974), zebrafish, *Brachydanio rerio* (Bloom and Perlmutter, 1977), several Ictalurid species (Kendle, 1970; Rubec, 1979), five Belontiid species (Lee and Ingersoll, 1979), and the black goby, *Gobius jozo* (Colombo et al., 1980, 1982). Stimulation of ovarian growth and ovulation have been observed in *B. rerio* (Chen and Martinich, 1975) and a cichlid, *Pterophyllum scalare* (Chien, 1973).

Although it is clear that females are attracted to chemicals released by males, it is not always clear whether such a response represents a general social attraction among conspecifics, or whether it is restricted to a sexual context. One indication that the response is related to reproduction would be the demonstration that it is restricted to the time when the female is sexually active. In oviparous species which do not form pair bonds prior to spawning, female sexual behaviors normally are performed only during the brief period between ovulation and completion of oviposition. Thus, the finding that in several Belontiid species attraction to 'male water' occurred only in ovulated females, or increased when females had ovulated, strongly suggests that female attraction is indeed a sexual response indicative of spawning readiness (Lee and Ingersoll, 1979). Comparable differences in attraction of ovulated and unovulated *G. jozo* to a water source containing etiocholanolone glucuronide (a proposed male pheromone in this species), also suggest that this female response occurred in a sexual context. Further evidence in this case is that some ovulated *G. jozo* oviposited during the test period (Colombo et al., 1980). However, in cases where females were attracted to water from both conspecific males and females (*B. rerio*; Bloom and Perlmutter, 1977: *S. gairdneri*; Newcombe and Hartman, 1973), and where responses of only unovulated (*B. rerio*) or ovulated (*S. gairdneri*) fish were examined, it is possible that the females were simply exhibiting a general social response to conspecifics. Laumen et al. (1974) have shown that water from male *B. pavo* attracts female conspecifics even though they are not ovulated (authors' interpretation); however, additional information from this study (see CONTROL OF PHEROMONE PRODUCTION) indicates that a male sexual pheromone is produced in this species.

Not surprisingly, the most convincing evidence for male reproductive pheromones comes from species with reproductive strategies in which the adaptive significance of chemical communication from male to female is obvious. In Belontiids, for example, male territoriality and parental behavior (which prevent the male from searching for receptive females), in conjunction with turbid waters of the natural habitat (which reduce the effectiveness of visual communication), have almost certainly been strong selective pressures favoring the evolution of a chemical communication system for attracting receptive females to the male's bubble nest.

Female Pheromones

In males of a number of teleost species, chemical stimuli from conspecific females have been shown to attract males or to cause an increase in male reproductive activities (Liley, 1982). In contrast to the situation in females, it is clear that in many cases the responses of males to female products occur in a sexual context. For example, in experiments with oviparous species in which males are exposed to water from both ovulated and unovulated conspecifics, it has generally been found that water from ovulated fish is more effective in attracting males (goldfish, Carassius auratus; Partridge et al., 1976: five species of Belontiids; Lee and Ingersoll, 1979: ayu, Plecoglossus altivelis; Honda, 1979: S. gairdneri; Honda, 1980a: loach, Misgurnus anguillicaudatus; Honda, 1980b: rose bitterling, Rhodeus ocellatus, and slender bitterling, Acheilognathus lanceolatus; Honda, 1982a: yamame salmon, Oncorhynchus masou, and amago salmon, O. rhodurus; Honda, 1982b) and in stimulating male reproductive behaviors (frillfin goby, Bathygobius soporator; Tavolga, 1956: Sarotherodon mossambicus; Kim and Liley, unpublished results). Males of the cichlid, Haplochromis burtoni, display more behaviors normally associated with sexual interaction following the introduction of water from 'gravid' females than after exposure to water from males or from sexually 'neutral' females (Crapon de Caprona, 1980); unfortunately, it is not clear whether the gravid females were ovulated. Thus, these studies indicate that in oviparous species release of female sexual pheromone(s) is restricted to, or increases during, the periovulatory period, presumably in conjunction with the onset of female sexual behaviors.

In bitterlings, R. ocellatus and A. lanceolatus (Honda, 1982a), and salmonids, O. masou and O. rhodurus (Honda, 1982b), males in a Y-maze preferred water from ovulated conspecifics over water from ovulated heterospecifics, indicating species specificity of female pheromone. However, only in the bitterling study (Honda, 1982a) was it determined that males did not discriminate between waters from unovulated conspecifics and heterospecifics, an important control in demonstrating that preference for ovulated conspecifics is due to pheromone produced at ovulation.

There also is good evidence that, in a number of ovoviviparous Poeciliid fishes, females release pheromone(s) which serve not only to attract conspecific males but also to increase male sexual behaviors (guppy, Poecilia reticulata; Amouriq, 1964, 1965; Gandolfi, 1969; Crow and Liley, 1979; Meyer and Liley, 1982; Vinogradova and Manteyfel, 1982: Poecilia chica; Brett and Grosse, 1982). In the guppy, these responses are seen only in males and are species and sex specific (Gandolfi, 1969; Crow and Liley, 1979; Vinogradova and Manteyfel, 1982). There is also evidence of microgeographic variation in the female guppy pheromone. Snyder (1979), working with two geographically isolated populations from Trinidad, demonstrated that males of both populations showed a greater sexual response to ovariectomized (pheromonally inert) females when in the presence of chemical cues derived from females of their own population.

A further indication that the responses of male Poeciliids to conspecific female odors are sexually significant is that the effectiveness of the female in inducing a response in males is dependent on her reproductive condition. Female Poeciliids undergo regular cycles of ovarian development, oocytes being fertilized within the ovarian follicles, and 'ovulation' occurring several weeks later, in conjunction with development of the next crop of oocytes and just prior to parturition. Although male guppies evidently respond sexually to chemical cues from females at all stages of the ovarian cycle (Gandolfi, 1969; Crow and Liley, 1979; Vinogradova and Manteyfel, 1982), there is good evidence that in the guppy (Crow and Liley, 1979) and other Poeciliids (P. sphenops, P. melanogaster, Xiphophorus helleri; Parzefall, 1973: P. chica; Brett and Grosse, 1982) the nonvirgin female is most stimulating in the immediate postpartum period, the time of maximum female sexual receptivity in the guppy (Liley, 1966). In the guppy, Liley and Wishlow (1974) have shown that virgin females are continually receptive, while in P. chica water from virgin females is as stimulating for males as is water from immediately postpartum females (Brett and Grosse, 1982). Thus in the guppy, and perhaps in other Poeciliid fishes as well, the function of female sexual pheromone(s) appears to be to attract males at a time when female behavior and ovarian development are most appropriate for successfull fertilization. As will be discussed below, there is good evidence in Poeciliids for endocrine involvement in this synchronization of ovarian development, female behavior, and pheromone production.

SOURCES AND CHEMICAL NATURE OF FISH REPRODUCTIVE PHEROMONES

Despite the considerable observational and experimental evidence for the existence of reproductive pheromones in fish, less is known concerning the sites of pheromone production, while only in a few cases have the chemical identities of fish reproductive pheromones been proposed. This is unfortunate, for a better understanding of the origins and chemical natures of fish reproductive pheromones will not only allow a much greater sophistication in the experimental analysis of pheromone function, but can also be expected to provide insight into the evolutionary development of chemical communication systems in fish.

Sites of Pheromone Production

In the majority of species examined, reproductive pheromones of both male and female fish appear to be of gonadal origin, although skin, secondary sex structures, and urine also have been implicated (Colombo et al., 1982; Liley, 1982). Unfortunately, the evidence that a pheromone is of gonadal origin is frequently equivocal, due to the manner in which samples of 'gonadal' pheromone have been collected. Ideally, test preparations should be obtained from gonads carefully dissected from donor fish. However, the usual procedure is to 'strip' ovulated and spermiated donors, by applying pressure to the body walls, to obtain gametes and genital (or coelomic) cavity fluids. Thus, preparations collected by stripping are likely to be contaminated to some extent by skin mucous, urine, feces, and bile, all of which have been suggested as sources of pheromone.

In three species, a pondsmelt, *Hypomesus olidus* (Okada et al., 1978), Pacific herring, *Clupea harengus pallasi* (Stacey and Hourston, 1982), and *G. jozo* (Colombo et al., 1980, 1982), there is evidence that male reproductive pheromones are produced in the genital tract. In herring, a suspension of dissected testes from spermiated males rapidly induced normal spawning behavior of both sexes, while courtship behavior of male pondsmelt in heterosexual groups was stimulated by the aqueous supernatant of stripped milt. In the case of *G. jozo*, it has been proposed that the male pheromone is etiocholanolone glucuronide, the major 5β-reduced androgen produced by specialized mesorchial glands. Although it is clear that ovulated *G. jozo* are attracted to this steroid, we are not aware of any published data showing that females are attracted to male odors or to testis preparations.

Only in *B. pavo* is there good evidence for an extra-gonadal source of male reproductive pheromone (Laumen et al., 1974). Males of this species possess specialized anal fin appendages which enlarge at maturity. Water from mature males with well developed appendages attracts female conspecifics, whereas water from females or immature males does not; significantly, females are not attracted to water from males which have had the appendages removed. The caudal glands found in males of the glandulocaudine fishes (e.g. *Corynopoma riisei*) also have been suggested to function in pheromone production (Nelson, 1964; Atkins and Fink, 1979). Although the secretory cells become smaller when males are kept in isolation, and enlarge when courtship activity is resumed, there is no direct experimental evidence that these male secondary sex structures release a reproductive pheromone.

In female fish, considerable evidence points to the ovary as a source of reproductive pheromones. In his classic study of the frillfin goby, B. soporator, Tavolga (1956) found that the male's courtship response to water from ovulated females could also be induced by fluid from dissected ovulated ovaries and by swabs of the urogenital area, whereas ovarian fluids from unovulated fish and non-ovarian fluids (urine, blood) from ovulated fish were, at best, marginally effective. These results, indicating a brief peri-ovulatory period of ovarian pheromone production, have since been extended to a variety of species.

In H. olidus, ovarian fluid, obtained either by stripping or dissecting ovulated fish, stimulates male courtship behavior (Okada et al., 1978). Similar effects have been demonstrated in other Salmonoids. In P. altivelis, fluid from stripped ovulated eggs and homogenates of recently ovulated ovaries caused an increase in ventilatory rates of males, a response also seen when males were exposed to water in which ovulated fish had been held (Honda, 1979). In S. gairdneri, males in choice tests preferred fluid from stripped ovulated eggs and homogenates of recently ovulated ovaries over homogenates of previously rinsed ovulated eggs; only the first two preparations were effective in stimulating males to court an unovulated female (Honda, 1980a). These results are in agreement with the earlier finding (Emanuel and Dodson, 1978) that ovarian fluid from stripped ovulated eggs stimulates positive rheotropic responses in male rainbow trout. Studies of Cyprinoid species also indicate that female pheromone is present in the ovaries and ovarian cavity of ovulated females. In Y-maze tests, male M. anguillicaudatus (Honda, 1980b) and bitterlings, R. ocellatus and A. lanceolatus (Honda, 1982a), each of which preferred the odor of ovulated females over that of unovulated females, also preferred stripped ovarian fluid and homogenate of ovulated ovaries over homogenate of previously washed ovulated eggs. Similar results have been obtained in goldfish (Partridge et al., 1976) and B. rerio (van den Hurk and Lambert, 1983).

As in oviparous teleosts, the ovary appears to be the source of female sex pheromone in the ovoviviparous guppy. In a choice test, males showed a preference for water which had held females in the first few days of their gestation cycle over water holding midcycle (pregnant) or ovariectomized females (Crow and Liley, 1979). Furthermore, male courtship of an ovariectomized female was stimulated within minutes by water from intact females, but not by water from ovariectomized females (Crow and Liley, 1979). While these results implicate the ovary in pheromone production, Meyer and Liley (1982) were unable to show that ovarian extracts stimulate male courtship. This result contradicts that of Amouriq (1964) who found that ovarian extracts stimulated locomotory activity in male guppies. However, as discussed by Crow and Liley (1979), the considerable latency of the response to the extract (several hours), and the fact that a number of unrelated compounds exerted similar effects, raise questions as to the functional significance of the male responses observed by Amouriq. Although other experiments involving hypophysectomy and estrogen treatment (Meyer and Liley, 1982) also provide evidence that the guppy ovary plays some role in pheromone production (see CONTROL OF PHEROMONE PRODUCTION), they fail to confirm that the ovary is the source of the pheromone.

While much evidence indicates that the gonads are sources of reproductive pheromones in fish, virtually nothing is known of the control of pheromone production and release. As pheromonal activity has been demonstrated in the genital cavity fluids of many species, the most likely route to the exterior is the gonopore. The pheromone may leak passively through the gonopore, and would certainly be released with the gametes at spawning. In the guppy, the period when the female is most attractive to the male - for several days after parturition (Crow and Liley, 1979) - coincides with the time when the proximal end of the oviduct is open to the ovarian lumen (Stolk, 1950).

One aspect of the sexual interactions of many teleost species (e.g. goldfish) is the smelling or mouthing of the partner's cloacal area, presumably to assess the partner's reproductive condition by sampling pheromone. Although such behavior would be expected if pheromone is released from the gonopore, it would also be expected if pheromone is released in the urine. Indeed, Colombo et al. (1982) have found that male courtship behaviors in G. jozo are more effectively stimulated by urine from ovulated females than by fluid from dissected ovulated ovaries, while females tested in a Y-maze are attracted by male urine. If gonadal and urinary pheromones are the same in these cases, these findings would indicate that pheromones produced in the gonad may be released in the urine.

The Chemical Identity of Reproductive Pheromones

Attempts to characterize reproductive pheromones by chemical fractionation of fish gonads have led to a variety of proposals regarding the chemical nature of the active fractions: an ether soluble ovarian substance which attracts male goldfish (Partridge et al., 1976); a proteinaceous substance from ovaries and/or testes of H. olidus which stimulates male courtship (Okada et al., 1978); a water-ether soluble basic ovarian substance which stimulates male ventilatory rate in P. altivelis (Honda, 1979); two water-ether soluble basic ovarian substances which attract and stimulate courtship of male S. gairdneri (Honda, 1980a); ovarian steroid glucuronides which stimulate attraction and courtship of male B. rerio (van den Hurk et al., 1982; van den Hurk and Lambert, 1983). Only in two species, however, is there good evidence for the chemical identity of reproductive pheromones.

In B. rerio, attraction of the male to the ovulated female evidently is mediated by steroid glucuronides. Water soluble ovarian extracts (lipids and free steroids removed) prepared either from ovulated ovaries (van den Hurk and Lambert, 1983) or from recently spent ovaries incubated with testosterone (van den Hurk et al., 1982) attract male B. rerio, whereas females show no response. Male B. rerio also are attracted to a subfraction of the aqueous ovarian extract containing steroid glucuronides, and to synthetic glucuronides of estradiol and testosterone (van den Hurk and Lambert, 1983).

Colombo et al. (1980, 1982) have also proposed that steroid glucuronides function as sex pheromones in three other teleost species. Male G. jozo possess a well developed mesorchial gland, rich in Leydig cells, which appears to be specialized for the production of large amounts of conjugated 5β-reduced androgens. Using a still-water testing system in which females were required to locate a nest-site which released pheromone, Colombo et al. (1980) found that etiocholanolone glucuronide, the major conjugated steroid produced by the male, attracts ovulated females and induces oviposition, whereas preovulatory and spent females show virtually no response. While these results indicate that etiocholanolone glucuronide functions as a male pheromone in a species in which receptive females must detect and approach territorial males, there also is evidence that this steroid is a female pheromone in species where males are attracted to ovulated females. When etiocholanolone glucuronide was added to one end of a semipartitioned aquarium, male goldfish spent significantly more time in the end containing steroid, whereas females showed no preference; males were also attracted to ovarian fluid, but not to charcoal-treated ovarian fluid, testosterone, estradiol, or its 3-glucuronide (Colombo et al., 1982). Using a similar testing procedure, Colombo et al. (1982) found that male guppies were attracted to etiocholanolone glucuronide (but not to its epimer, androsterone glucuronide), whereas females showed no response.

To date, metabolites of androgens and estrogens are the only steroids implicated as sex pheromones in fish. However, at the time of ovulation and spawning, two metabolites of progesterone, 17α-hydroxyprogesterone (17P) and 17α,20β-dihydroxy-4-pregnen-3-one (17,20P), have been shown to increase in the blood of both males and females of a number of teleost species (Fostier et al., 1983; Scott et al., 1984). The periovulatory increase of progestins in females is not surprising, as 17,20P has been shown to be a potent inducer of oocyte final maturation in many teleosts (Goetz, 1983). Although the function of similar increases in males is not well understood, progestins may act directly on the testes to increase milt volume. Injection of 17P and 17,20P increases milt volume in goldfish, an effect which is not blocked by olfactory tract section or acute hypophysectomy (Stacey, unpublished results). In addition, recent findings (Stacey, unpublished results) indicate that progestins may act as pheromones to increase milt volume.

Goldfish were stripped of milt, placed in aquaria containing steroids or control vehicle, and stripped again the following day. At a water concentration of 1.5 ng/ml, 17P and 17,20P stimulated dramatic increases in milt volume, whereas progesterone was only minimally effective and other steroids (pregnenolone, estradiol, testosterone, 11-ketotestosterone) had no effect. Surprisingly, neither etiocholanolone glucuronide nor glucuronides of estradiol and testosterone, steroids suggested to function as behavioral pheromones in goldfish (Colombo et al., 1982) and B. rerio (van den Hurk and Lambert, 1983), had any effect on milt volume. At a water concentration of 30 pg/ml (approximately 10^{-10} M), 17,20P is still effective in increasing milt volume, although 17P is not. A pheromonal action is indicated by the fact that olfactory tract section blocks the milt response to 17,20P exposure.

An important question concerning a possible pheromonal action of progestins in stimulating milt increase in goldfish concerns the source of the pheromone. Progestins may be released by the ovulated female to attract and stimulate males. Alternatively, progestins may be released with milt at spawning, and thus may serve to synchronize spawning readiness of males in a local area. In this regard, Ueda et al. (1984) have shown that, when incubated with 17P, sperm of S. gairdneri synthesize large amounts of 17,20P, suggesting that 17,20P may normally reach high concentrations in the milt. Whether progestins are released in the free or conjugated form is not known. However, van den Hurk and Lambert (personal communication) have found that water soluble extracts of B. rerio testes, which induce ovulation in females, also contain glucuronides of 17,20P and other steroids.

CONTROL OF PHEROMONE PRODUCTION

Generally, sexual activity in fish is restricted to a brief period in the adult lifespan. This restriction is most obvious in females, where sexual activity in oviparous species occurs for only a short period (hours or days) after ovulation. As the primary function of most reproductive pheromones in fish is evidently to attract and stimulate sexual arousal in potential conspecific partners, it is to be expected that pheromone production should be synchronized with transient episodes of sexual competence, and that these processes might share common regulatory mechanisms.

Although there is evidence that male pheromones in three teleost species (H. olidus; Okada et al., 1978: C. harengus; Stacey and Hourston, 1982: G. jozo; Colombo et al., 1982) are of testicular origin, there is no evidence that pheromone output increases during the breeding season. However, in the case of G. jozo, where the pheromone is believed to be a testicular steroid, it is reasonable to speculate that pheromone production is regulated by gonadotropin. Indeed, plasma levels of testosterone glucuronide peak during the spawning period in this species (Bonnin, 1979). In addition, *in vitro* synthesis of androgen glucuronides is stimulated by gonadotropin in male tilapia, Sarotherodon mossambicus (Kime and Hyder, 1983), while plasma levels of androgen glucuronides are maximal during spermiation in brown trout, Salmo trutta (Kime and Manning, 1982). Whether these conjugated steroids function as male pheromones in tilapia and trout is not known.

Only in B. pavo, where male pheromone is produced in the anal fin appendages (Laumen et al., 1974), has evidence been presented that pheromone production in males is related to sexual maturation. Water from 'immature' males (gonadal condition unknown) with poorly developed anal fin appendages did not attract females, whereas treatment of these males with mammalian luteinizing hormone (LH) induced a transient increase in pheromone production which disappeared with further LH treatment. These increases and decreases in attractiveness of male water corresponded with a transient increase in size of the posterior anal appendage. As methyltestosterone treatment did not stimulate either pheromone production or anal appendage size in the immature males, Laumen et al. (1974) suggest that gonadotropin, rather than steroids, regulates pheromone production. However, because these workers did not investigate the actions of other steroids (e.g. 11-ketotestosterone, known to be a potent inducer of male secondary sex characters in fish; Fostier et al., 1983), and administered steroid to fish in which previous LH treatment had led to cessation of pheromone production, these conclusions should be regarded with caution.

Unlike the situation in males, there is considerable evidence that production of female reproductive pheromones is restricted to, or maximal during, the period shortly after ovulation. In M. anguillicaudatus, for example, pheromone is released for only three to four hours after ovulation (Honda, 1980b). Research to date indicates that ovulation in fish is preceded by an increase in plasma gonadotropin which affects synthesis of ovarian steroids and prostaglandins (Goetz, 1983). Some recent evidence suggests that these periovulatory ovarian changes are responsible for stimulating pheromone production at the time of ovulation and sexual responsiveness.

In B. rerio, where synthesis of the presumed female pheromones (steroid glucuronides) is retricted to the immediate postovulatory period (van den Hurk and Lambert, 1983), it seems likely that pheromone production is ultimately regulated by the preovulatory increase in blood gonadotropin. Synthesis of testosterone glucuronide has also been shown to increase shortly after ovulation in the African catfish, Clarias lazera (Lambert and van den Hurk, 1982), although pheromonal studies in this species have not been reported. In female guppies, attractiveness to males is maximal shortly after parturition (Crow and Liley, 1979), coincident with maximum steroidogenesis (Lambert, 1970). Attractiveness is lost if donor females are hypophysectomized or ovariectomized, but can be restored by treating hypophysectomized females with either gonadotropin or estradiol (Meyer and Liley, 1982). As estrogen does not restore attractiveness in ovariectomized donor females (Meyer and Liley, 1982), it is suggested that estrogen acts on the ovary to restore attractiveness, perhaps by being metabolized to some conjugated form.

Estradiol has also been implicated in the production of a female pheromone in goldfish. Males actively courted hypophysectomized males which had been fed a diet containing estradiol, but failed to respond to hypophysectomized males which had received testosterone or control diets (Yamazaki and Watanabe, 1979). Both the regressed gonadal condition of the hypophysectomized stimulus fish, and conspicuous histological changes in the kidneys of the estrogen-treated fish, led Yamazaki and Watanabe to suggest that estrogen acts on the kidneys to stimulate pheromone release. As plasma estrogen levels in goldfish are elevated for a considerable period prior to ovulation (Kagawa et al., 1983), such a mechanism may allow male goldfish to identify female conspecifics, rather than to determine whether they have ovulated.

Prostaglandins (PGs) perform important reproductive functions in oviparous fish (Stacey and Goetz, 1982). Ovarian PGs not only act within the ovary to induce ovulation (Goetz, 1983), but also appear to be carried in the circulation to the brain to stimulate female sexual behavior (Liley and Stacey, 1983). More recent findings indicate that, in addition to synchronizing sexual behavior with ovulation, PG may play a role in pheromone production. Male goldfish were stimulated to court an unovulated female following addition of water from either an ovulated female or an unovulated female injected with prostaglandin F2 alpha, but did not respond to water from an unovulated female injected with saline (Stacey, in press; cf. Sorensen et al., this volume). Similar results have been obtained in the fathead minnow, Pimephales promelas (Cole and Smith, in prep.). How PG might stimulate pheromone release is not known. However, as courtship of male goldfish and fathead minnows is not induced when prostaglandin F2 alpha is placed directly in the tank water, it would appear that injected PG either stimulates release of a pheromone, or itself functions as a pheromone after being metabolized to an active form.

MECHANISMS OF ACTION AND CONTROL OF RESPONSIVENESS

Between pheromonal input and behavioral or physiological output, there exists a 'black box' of neural mechanisms that mediate pheromone action. At the present time, potential pheromone-mediating neural mechanisms are best identified by their sensitivity to compounds that may be chemically related to pheromones, or by their involvement in behavioral or physiological processes that are influenced by pheromones.

Involvement of the Olfactory System

Olfaction is thought to be the dominant chemosensory modality by which reproductive pheromones are detected in fish, although there is evidence that taste mechanisms are involved in some species (Liley, 1982). Fish made anosmic by occlusion of the nares, cautery of the nasal epithelium, or section of the olfactory tracts showed reduced sexual behavior or ability to discriminate pheromone-bearing water (Tavolga, 1956; Partridge et al., 1976; Yamazaki and Watanabe, 1979; Honda, 1979, 1980a, 1980b, 1982a, 1982b; Stacey and Kyle, 1983). The following discussion, therefore, will focus on the olfactory system and its role in mediating the observed effects of reproductive pheromones in fish, beginning at the level of the sensory receptor and ending with third order neurons in the brain.

In fish, two structural classes of olfactory receptors are generally recognized: those with ciliated and those with microvillar dendritic processes (Yamamoto, 1982). Based on extracellular recordings from the olfactory epithelium and bulb, two functional receptor classes have been suggested: those maximally sensitive to amino acids and those maximally sensitive to bile steroids (Thommesen, 1982; Doving et al., 1980). In the char, Salmo alpinus, a correlation between the distribution of the ciliated receptors over the olfactory epithelium and the distribution of extracellular olfactory receptor potentials elicited by bile steroids led Thommesen (1983) to suggest that the ciliated receptor cell is responsible for sensitivity to bile steroid-like substances. It is tempting to speculate that the ciliated cells may be the mediators of responsiveness to other types of conjugated steroids, such as those thought to be sexual pheromones in some fish.

Within the olfactory bulb, synapses are made between the axons of the primary olfactory receptors and the dendrites of the secondary olfactory neurons, the mitral cells. These mitral cells then project to the forebrain via the olfactory tracts, where they innervate various telencephalic and diencephalic targets (Oka et al., 1982). In fish, there is evidence for an anatomical and functional separation of the olfactory bulb into lateral and medial subdivisions. Mitral cell activity in the medial and lateral parts of the bulb is regulated by distinct medial and lateral populations of inhibitory interneurons, the granule cells (Satou et al., 1983), and mitral cell axons extend to the brain via separate medial (MOT) and lateral (LOT) bundles of the olfactory tract (Satou et al., 1983). Bile steroids evoked extracellular potentials from the medial part of the bulb, while amino acids stimulated potentials in the lateral portion (Doving et al., 1980; Thommesen, 1982). Taken together, these studies suggest that the medial and lateral subdivisions of the bulb and its tract operate relatively independently and may subserve different chemosensory functions (Satou et al., 1983). As will be described below, there is good evidence in some fishes that the medial subdivision is concerned with sexual behavior and, presumably, pheromone detection and processing.

Stimulation of a sub-bundlet of the medial olfactory tract (MOT) in cod, Gadus morhua, resulted in a head-up body posture and rapid contraction of the body muscles similar to behaviors performed during spawning (Doving and Selset, 1980); in contrast, stimulation of the LOT evoked searching and feeding activity. While it is not known whether pheromones play a role in cod reproduction, the results suggest that some olfactory stimulus transmitted via the MOT is involved. In male goldfish, there is evidence that courtship behavior (Partridge et al., 1976) and milt volume increase (Stacey, unpublished results) are stimulated by pheromones. Significantly, MOT lesion severely impaired male spawning behavior (Stacey and Kyle, 1983) while MOT stimulation caused sperm release (Demski and Dulka, 1984). LOT lesion and stimulation experiments in goldfish gave no evidence that this tract plays a significant role in reproduction.

As well as containing mitral cell axons from the medial olfactory bulb, the MOT also carries axons of the terminal nerve (TN) (Demski and Northcutt, 1983; Springer, 1983; Stell et al., 1984), raising the possibility that the TN could be responsible for the observed effects of MOT lesion and stimulation. In teleosts, the TN ganglion is embedded within the olfactory system (either in the olfactory nerve in species with pedunculate bulbs such as goldfish, or located between the bulbs and telencephalon in species with sessile bulbs such as platyfish, Xiphophorus maculatus) and projects to the forebrain and retina (Munz et al., 1982; Demski and Northcutt, 1983; Halpern-Sebold and Schreibman, 1983; Springer, 1983; Stell et al., 1984). Although there is some, largely correlative, evidence that the TN may have a role in reproduction (reviewed by Demski and Northcutt, 1983), there is yet no evidence that the TN is directly involved in pheromone reception or even that TN dendrites contact the free surface of the olfactory epithelium. Demski and Northcutt (1983) did find, however, that application of horseradish peroxidase to the nasal epithelium resulted in transport of this enzyme to the TN ganglion and its axon terminals in the telencephalon, suggesting proximity of TN dendrites and olfactory epithelium. Indeed, immunocytochemical visualization of the TN of goldfish revealed processes extending from the ganglion cells along the branches of the olfactory nerves to the base of the olfactory lamellae, but no processes were observed in the sensory epithelium itself (Kyle, unpublished results). Clearly, a role for the TN in pheromone detection has yet to be established.

Involvement of the Olfactory Terminal Fields

Upon entering the forebrain, the medial olfactory tract subdivides and innervates terminal fields in the contralateral olfactory bulb, telencephalon, preoptic area, and hypothalamus. Theoretically, some of these terminal fields can be considered as the central nervous effectors of the behavioral and physiological responses elicited by sexual pheromones. Viewed in this way, the distinction between 'releaser' pheromones (that trigger a more or less immediate response in the recipient) and 'primer' pheromones (that induce a delayed response which may be mediated by altered endocrine or physiological parameters) likely results from differences in the terminal fields that are activated. Unfortunately, there are very few species for which sufficient information is available to allow speculation on the identity of terminal fields with releasing and/or priming functions. An exception is the goldfish, where both specific MOT terminal fields and some of the brain areas controlling reproductive behavior and endocrinology have been identified.

Since spawning behavior and production and release of milt in male goldfish are influenced by the medial but not by the lateral olfactory tract, any terminal fields that are innervated by the medial tract, but not by the lateral tract, are likely candidates for mediating the effects of pheromones. Two such areas occur in the goldfish: the 'medial terminal field' of the ventromedial telencephalon and the anterior preoptic area (Ichikawa and Ueda, 1978; von Bartheld et al., 1984), both of which are implicated in the control of reproductive events.

Lesions placed in the supracommissural portion of the ventral telencephalon (area ventralis telencephali pars supracommissuralis and the posterior area ventralis telencephali pars ventralis; Vs-pVv) severely disrupted male spawning behavior (Kyle and Peter, 1982; Kyle et al., 1982; Koyama et al., 1984). The authors suggested that this effect was due, in part, to destruction of the medial terminal field and thus pheromone perception, although responses to pheromone alone were not examined. Lesioned fish were not different from controls in either relative gonad weight or serum gonadotropin concentration (Kyle et al., 1982), suggesting that the ventral telencephalon acts as a 'releaser' of behavior, rather than priming endocrine events.

The anterior preoptic area, or nucleus preopticus periventricularis (NPP), is clearly a complex and multifunctional region in the goldfish brain, having both behavioral and endocrine roles. It appears to be part of the neural pathway involved in sperm release (Demski and Hornby, 1982) and is involved in the performance of male coursthip behavior and spawning (Koyama et al., 1984). The NPP also is implicated in the control of gonadotropin release (Peter, 1983) and thus may affect the levels of gonadal steroids and steroid-dependent events, such as milt production and sexual motivation.

Control of Responsiveness to Pheromones

As previously discussed, the production of pheromones tends to be synchronized with periods of sexual competence. As well, the ability to perceive and/or respond appropriately to pheromones may be restricted to the time of mating readiness in some species. Such mechanisms would ensure that both the sender and recipient of the pheromonal message are in a state appropriate for reproduction.

In male goldfish (Partridge et al., 1976) and Pacific herring (Stacey and Hourston, 1982), responses to sexual pheromones were seen in mature (spermiated) males but not in immature individuals. Similarly, only ovulated females of herring (Stacey and Hourston, 1982), *G. jozo* (Colombo et al., 1982), and several Belontiids (Lee and Ingersoll, 1979) responded to natural and synthetic pheromones, whereas unovulated females were unresponsive.

There are three general mechanisms by which responsiveness to pheromones could be coordinated with reproductive maturity: 1) reproductive hormones might directly affect primary receptor structure or function, 2) olfactory bulb activity could be modulated by efferent, hormone-sensitive fibres from the brain, and 3) the occurrence of appropriate behaviors could be influenced by hormone-dependent changes in sexual motivation.

Several studies suggest that sexual hormones can alter the structure, and hence the sensitivity, of the olfactory epithelium of fishes. Yamazaki and Watanabe (1979) observed that treatment with methyltestosterone increased the thickness and cellular development of the epithelium of immature goldfish and reversed some of the regressive effects of hypophysectomy in mature animals. In platyfish (Schreibman et al., 1984) and the European eel, Anguilla anguilla (Pankhurst and Lythgoe, 1983), marked morphological changes in the nasal epithelium occurred at the time of sexual maturation, suggesting that epithelial structure is under the control of reproductive hormones.

Sex steroid treatment of goldfish not only changes the structure of the olfactory epithelium, but also alters the spontaneous and evoked electrical activity of the olfactory bulbs (Oshima and Gorbman, 1968; Hara, 1967). Based on the observation that section of the medial olfactory tract also altered bulbar electrical activity, Hara and Gorbman (1967) suggested that some of the effects of sex steroids on the bulb could be mediated by efferent fibres from the forebrain. Subsequently, it was found that the ventral telencephalon (medial terminal field; Vs-pVv area) of the goldfish has both afferent and efferent connections with the olfactory bulbs (Oka et al., 1982; von Bartheld, 1984) and concentrates sex steroid hormones (Kim et al., 1978).

When a courtship response is used as an indicator of pheromone sensitivity, the failure of immature animals to respond may be a reflection of a reduced state of sexual motivation rather than an inability to detect pheromonal odors. In a wide variety of male and female teleosts, sufficient levels of circulating sex steroids are required before sexual behavior is expressed (Liley and Stacey, 1983). The sex steroid concentrating areas of the ventral telencephalon and preoptic area are thought to mediate this effect by controlling behavioral output, a function distinct from their possible role in the detection of pheromones per se (see discussions in Kyle and Peter, 1982; Kyle et al., 1982; Demski and Hornby, 1982).

REFERENCES

Amouriq, L., 1964, L'activite et le phenomene social chez Lebistes reticulatus (Poeciliidae: Cyprinodontiformes), C. R. Hebd. Seances Acad. Sci., 259:2701.

Amouriq, L., 1965, Origine de la substance dymogene emise par Lebistes reticulatus femelle (Poisson: Poeciliidae: Cyprinodontiformes), C. R. Hebd. Seances Acad. Sci., 260:2334.

Atkins, D. L., and Fink, W. L., 1979, Morphology and histochemistry of the caudal gland of Corynopoma riisei Gill, J. Fish Biol., 14:465.

Bloom, H. D., and Perlmutter, A., 1977, A sexual aggregating pheromone system in the zebrafish, Brachydanio rerio (Hamilton-Buchanan), J. Exp. Zool., 199:215.

Bonnin, J. P., 1979, Variations saisonnières de la testosterone plasmatique chez un poisson téléostéen, Gobius niger, C. R. Hebd. Seances Acad. Sci., 288:627.

Brett, B. L. H., and Grosse, D. J., 1982, A reproductive pheromone in the Mexican Poeciliid fish Poecilia chica, Copeia, 219.

Chen, L. C., and Martinich, R. L., 1975, Pheromonal stimulation and metabolite inhibition of ovulation in the zebrafish, Brachydanio rerio, Fish. Bull., 73:889.
Chien, A. K., 1973, Reproductive behavior of the angelfish Pterophyllum scalare (Pisces: Cichlidae). II. Influence of male stimuli upon the spawning rate of females, Anim. Behav., 21:457.
Cole, K. S., and Smith, R. J. F., Prostaglandin induced release of a female sex pheromone in the fathead minnow, Pimephales promelas, (in prep.).
Colombo, L., Marconato, A., Belvedere, P. C., and Frisco, C., 1980. Endocrinology of teleost reproduction: a testicular steroid pheromone in the black goby, Gobius jozo L., Boll. Zool., 47:355.
Colombo, L., Belvedere, P. C., Marconato, A., and Bentivegna, F., 1982, Pheromones in teleost fish, in: "Proc. Int'l. Symp. Reprod. Physiol. Fish," C. J. J. Richter and H. J. Th. Goos, eds., Pudoc, Wageningen.
Crapon, de Caprona, M. D., 1980, Olfactory communication in a cichlid fish, Haplochromis burtoni, Z. Tierpsychol., 52:113.
Crow, R. T., and Liley, N. R., 1979, A sexual pheromone in the guppy, Poecilia reticulata (Peters), Can. J. Zool., 57:184.
Demski, L. S., and Hornby, P. J., 1982, Hormonal control of fish reproductive behavior: brain-gonadal steroid interactions, Can. J. Fish. Aquatic Sci., 39:36.
Demski, L. S., and Northcutt, R. G., 1983, The terminal nerve: a new chemosensory system in vertebrates?, Science, 220:435.
Demski, L. S., and Dulka, J. G., 1984, Functional-anatomical studies of sperm release evoked by electrical brain stimulation of the olfactory tract in goldfish, Brain Res., 291:247.
Doving, K. B., and Selset, R., 1980, Behavior patterns in cod released by electrical stimulation of olfactory tract bundlets, Science, 207:559.
Doving, K. B., Selset, R., and Thommesen, G., 1980, Olfactory sensitivity to bile acids in salmonid fishes, Acta Physiol. Scand., 108:123.
Emanuel, M. E., and Dodson, J. J., 1979, Modification of the rheotropic behavior of male rainbow trout (Salmo gairdneri) by ovarian fluid, J. Fish. Res. Board Can., 36:63.
Fostier, A., Jalabert, B., Billard, R., Breton, B., and Zohar, Y., 1983, The gonadal steroids, in: "Fish Physiology," Vol. IXA, "Endocrine Tissues and Hormones," W. S. Hoar, D. J. Randall, and E. M. Donaldson, eds., Academic Press, New York.
Gandolfi, G., 1969, A chemical sex attractant in the guppy Poecilia reticulata Peters (Pisces: Poeciliidae), Monit. Zool. Ital., 3:89.
Goetz, F. W., 1983, Hormonal control of oocyte final maturation and ovulation in fishes, in: "Fish Physiology," Vol. IXB, "Behavior and Fertility Control," W. S. Hoar, D. J. Randall, and E. M. Donaldson, eds., Academic Press, New York.
Halpern-Sebold, L. R., and Schreibman, M. P., 1983, Ontogeny of centers containing luteinizing hormone releasing hormone in the brain of platyfish (Xiphophorus maculatus) as determined by immunocytochemistry, Cell Tissue Res., 229:75.
Hara, T. J., 1967, Electrophysiological studies on the olfactory system of the goldfish, Carassius auratus L. III. Effects of sex steroids on olfactory activity, Comp. Biochem. Physiol., 22:209.
Hara, T. J., and Gorbman, A., 1967, Electrophysiological studies of the olfactory system of the goldfish, Carassius auratus L. I. Modification of the electrical activity of the olfactory bulb by other central nervous system structures, Comp. Biochem. Physiol., 22:185.
Honda, H., 1979, Female sex pheromone of the ayu, Plecoglossus altivelis, involved in courtship behavior, Bull. Jap. Soc. Sci. Fish., 45:1375.
Honda, H., 1980a, Female sex pheromone of rainbow trout, Salmo gairdneri, involved in courtship behavior, Bull. Jap. Soc. Sci. Fish., 46:11-9.
Honda, H., 1980b, Female sex behavior of the loach, Misgurnus anguillicaudatus, involved in courtship behavior, Bull. Jap. Soc. Sci. Fish., 46:1223.

Honda, H., 1982a, On the female sex pheromones and courtship behavior in the bitterlings Rhodeus ocellatus ocellatus and Acheilognathus lanceolatus, Bull. Japan. Soc. Sci. Fish., 48:43.
Honda, H., 1982b, On the female sex pheromones and courtship behavior in the salmonids, Oncorhynchus masou and O. rhodurus, Bull. Japan. Soc. Sci. Fish., 48:47.
Ichikawa, M., and Ueda, K., 1979, Electron microscopic study of the terminal regions of the olfactory tract fibres in the goldfish telencephalic hemisphere, J. Fac. Sci. Univ. Tokyo, Sect. 4, 14:127.
Kagawa, H., Young, G., and Nagahama, Y., 1983, Changes in plasma steroid hormone levels during gonadal maturation in female goldfish, Carassius auratus, Bull. Japan. Soc. Sci. Fish., 49:1783.
Kendle, E. R., 1970, Sexual discrimination by olfaction in the black bullhead, Ictalurus melas, State Nebr. Game Parks Comm. Job Prog. Rep., 21.
Kim, Y. S., Stumpf, W. E., and Sar, M., 1978, Topography of estrogen target cells in the forebrain of goldfish, Carassius auratus, J. Comp. Neurol., 182:611.
Kime, D. E., and Manning, N. J., 1982, Seasonal patterns of free and conjugated androgens in the brown trout Salmo trutta, Gen. Comp. Endocrinol., 48:222.
Kime, D. E., and Hyder, M., 1983, The effect of temperature and gonadotropin on testicular steroidogenesis in Sarotherodon (Tilapia) mossambicus in vitro, Gen. Comp. Endocrinol., 50:105.
Koyama, Y., Satou, M., Oka, Y., and Ueda, K., 1984, Involvement of the telencephalic hemispheres and the preoptic area in sexual behavior of the male goldfish, Carassius auratus: a brain-lesion study, Behav. Neural Biol., 40:70.
Kyle, A. L., and Peter, R. E., 1982, Effects of forebrain lesions on spawning behavior in the male goldfish, Physiol. Behav., 28:1103.
Kyle, A. L., Stacey, N. E., and Peter, R. E., 1982, Ventral telencephalic lesions: effects on bisexual behavior, activity and olfaction in the male goldfish, Behav. Neural Biol., 36:229.
Lambert, J. G. D., 1970, The ovary of the guppy Poecilia reticulata: the granulosa cells as sites of steroid biosynthesis, Gen. Comp. Endocrinol., 15:464.
Lambert, J. G. D., and van den Hurk, R., 1982, Steroidogenesis in the ovaries of the African catfish, Clarias lazera, before and after an HCG induced ovulation, in: "Proc. Int'l. Symp. Reprod. Physiol. Fish," C. J. J. Richter and H. J. Th. Goos, eds., Pudoc, Wageningen.
Laumen, J., Pern, U., and Blum, V., 1974, Investigations on the function and hormonal regulation of the anal appendices in Blennius pavo (Risso), J. Exp. Zool., 190:47.
Lee, C. T., and Ingersoll, D. W., 1979, Social chemosignals in five Belontiidae (Pisces) species, J. Comp. Physiol. Psychol., 93:117.
Liley, N. R., 1966, Ethological isolating mechanisms in four sympatric species of Poeciliid fishes, Behaviour, Suppl. 13:1.
Liley, N. R., 1982, Chemical communication in fish, Can. J. Fish. Aquatic Sci., 39:22
Liley, N. R., and Wishlow, W. P., 1974, The interaction of endocrine and experiential factors in the regulation of sexual behavior in the female guppy, Poecilia reticulata, Behaviour, 148:185.
Liley, N. R., and Stacey, N. E., 1983, Hormones, pheromones and reproductive behavior in fish, in: "Fish Physiology," Vol. IXB, "Behavior and Fertility Control," W. S. Hoar, D. J. Randall, and E. M. Donaldson, eds., Academic Press, New York.
Meyer, J. H., and Liley, N. R., 1982, Hormonal control of pheromone production in the guppy (Poecilia reticulata Peters), Can. J. Zool., 60:1505.
Munz, H., Claas, B., Stumpf, W. E., and Jennes, L., 1982, Centrifugal innervation of the retina by luteinizing hormone releasing hormone (LHRH)-immunoreactive telencephalic neurons in teleostean fishes, Cell Tissue Res., 222:313.

Nelson, K., 1964, Behavior and morphology in the glandulocaudine fishes (Ostariophysi, Characidae), Univ. Calif., Berkeley, Publ. Zool., 75:55.
Newcombe, C., and Hartman, G., 1973, Some chemical signals in the spawning behavior of rainbow trout (Salmo gairdneri), J. Fish. Res. Board Can., 30:995.
Oka, Y., Ichikawa, M., and Ueda, K., 1982, Synaptic organization of the olfactory bulb and central projections of the olfactory tract, in: "Chemoreception in Fishes," T. J. Hara, ed., Elsevier, Amsterdam.
Okada, H., Sakai, D. K., and Sugiwaka, K., 1978, Chemical stimulus on the reproductive behavior of the pondsmelt, Sci. Rep. Hokkaido Fish Hatchery, 33:89.
Oshima, K., and Gorbman, A., 1968, Modification by sex hormones of the spontaneous and evoked unitary activity in the olfactory bulb of goldfish, J. Endocrinol., 40:409.
Pankhurst, N. W., and Lythgoe, J. N., 1983, Changes in vision and olfaction during sexual maturation in the European eel Anguilla anguilla (L.), J. Fish Biol., 23:229.
Partridge, B. L., Liley, N. R., and Stacey, N. E., 1976, The role of pheromones in the sexual behavior of the goldfish, Anim. Behav., 24:291.
Parzefall, J., 1973, Attraction and sexual cycle of Poeciliids, in: "Genetics and Mutagenesis of Fish," J. H. Schroeder, ed., Springer-Verlag, Berlin.
Peter, R. E., 1983, The brain and neurohormones in teleost reproduction, in: "Fish Physiology," Vol. IXA, "Endocrine Tissues and Hormones," W. S. Hoar, D. J. Randall, and E. M. Donaldson, eds., Academic Press, New York.
Rubec, P. J., 1979, Effect of Pheromones on Behavior of Ictalurid Catfish, Ph. D. Thesis, Texas A & M University, College Station, Texas.
Satou, M., Fujita, I., Ichikawa, M., Yamaguchi, K., and Ueda, K., 1983, Field potential and intracellular potential studies of the olfactory bulb in the carp: evidence for a functional separation of the olfactory bulb into lateral and medial subdivisions, J. Comp. Physiol., 152:319.
Schreibman, M. P., Margolis-Kazan, H., Halpern-Sebold, L., O'Neill, P., and Silverman, R. C., 1984, Structural and functional links between the olfactory and reproductive systems: puberty-related changes in olfactory epithelium, Brain Res., 302:180.
Scott, A. P., MacKenzie, D. S., and Stacey, N. E., 1984, Endocrine changes during natural spawning in the white sucker, Catostomus commersoni. II. Steroid hormones, Gen. Comp. Endocrinol., 56:349.
Snyder, B. E., 1979, A Comparison of Responses to the Sexual Pheromones Produced by Two Allopatric Races of Guppy (Poecilia reticulata Peters), B. Sc. Thesis, University of British Columbia, Vancouver, British Columbia.
Springer, A. D., 1983, Centrifugal innervation of goldfish retina from ganglion cells of the nervus terminalis, J. Comp. Neurol., 214:404.
Stacey, N. E., Prostaglandins and female sexual behavior in goldfish and other externally fertilizing vertebrates, in: "Proc. Ninth Int'l. Symp. Comp. Endocrinol.," B. Lofts, ed., Univ. Hong Kong Press, Hong Kong, (in press).
Stacey, N. E., and Goetz, F. W., 1982, Roles of prostaglandins in fish reproduction, Can. J. Fish. Aquatic Sci., 39:92.
Stacey, N. E., and Hourston, A. S., 1982, Spawning and feeding behavior of captive Pacific herring, Clupea harengus pallasi, Can. J. Fish. Aquatic Sci., 39:489.
Stacey, N. E., and Kyle, A. L., 1983, Effects of olfactory tract lesions on sexual and feeding behavior in the goldfish, Physiol. Behav., 30:621.
Stell, W. K., Walker, S. E., Chohan, K. S., and Ball, A. K., 1984, The goldfish nervus terminalis: a luteinizing hormone-releasing hormone and molluscan cardioexcitatory peptide immunoreactive olfactoretinal pathway, Proc. Natl. Acad. Sci., 81:940.

Stolk, A., 1950, The ovarian occlusion apparatus in the viviparous cyprinodonts Lebistes reticulatus (Peters) and Xiphophorus helleri (Heckel), Proc. K. Ned. Akad. Wet., 53:526.

Tavolga, W. N., 1956, Visual, chemical and sound stimuli as cues in the sex discriminatory behavior of the gobiid fish Bathygobius soporator, Zoologica (N.Y.), 41:49.

Thommesen, G., 1982, Specificity and distribution of receptor cells in the olfactory mucosa of char (Salmo alpinus L.), Acta Physiol. Scand., 115:47.

Thommesen, G., 1983, Morphology, distribution, and specificity of olfactory receptor cells in salmonid fishes, Acta Physiol. Scand., 117:241.

Ueda, H., Kambegawa, A., and Nagahama, Y., 1984, In vitro 11-ketotestosterone and 17α,20β-dihydroxy-4-pregnen-3-one production by testicular fragments and isolated sperm of rainbow trout, Salmo gairdneri, J. Exp. Zool., 231:435.

van den Hurk, R., Hart, L. A.'t, Lambert, J. G. D., and van Oordt, P. G. W. J., 1982, On the regulation of sexual behavior of male zebrafish, Brachydanio rerio, Gen. Comp. Endocrinol., 46:403.

van den Hurk, R., and Lambert, J. G. D., 1983, Ovarian steroid glucuronides function as sex pheromones for male zebrafish, Brachydanio rerio, Can. J. Zool., 61:2381.

Vinogradova, V. M., and Manteyfel, Yu. B., 1982, The role of chemical signals and olfaction in the sexual behavior of the guppy, Poecilia reticulata (Cyprinodontiformes, Poeciliidae), J. Ichthyol., 22:107.

von Bartheld, C. S., Meyer, D. L., Fiebig, E., and Ebbesson, S. O. E., 1984, Central connections of the olfactory bulb in the goldfish, Carassius auratus, Cell Tissue Res., 238:475.

Wilson, E. O., 1975, "Sociobiology: The New Synthesis," Harvard University Press, Cambridge, MA.

Yamamoto, M., 1982, Comparative morphology of the peripheral olfactory organ in teleosts, in: "Chemoreception in Fishes," T. J. Hara, ed., Elsevier, Amsterdam.

Yamazaki, F., and Watanabe, K., 1979, The role of steroid hormones in sex recognition during spawning behavior in the goldfish, Carassius auratus L., Proc. Indian Natl. Sci. Acad. Part B, 45:505.

CHEMOSENSORY ORIENTATION MECHANISMS OF FISH

Peter B. Johnsen

Monell Chemical Senses Center
3500 Market Street
Philadelphia, PA 19104

INTRODUCTION

In the aquatic environment, solubility rather than volatility is the limiting factor for chemical dispersion, and thus, compounds of wide variety become potential information messengers. Because most bodies of water are in motion, these stimuli and their information can be carried over considerable distances, and animals such as fish, which are very sensitive to chemical stimulation, may exploit this information for many important functions in their life histories. Chemical signals are known to play significant roles in reproduction, homing, feeding, schooling, defense and parent/young interactions of fish.

Whether these chemical cues are feeding stimuli, inter- and intra-specific signals or stimuli which give identity to specific locations, all can function to guide the fish from its present position to the source of stimulation. Oriented movements may cover distances varying from a few centimeters to thousands of kilometers. Kleerekoper (1982) has tabulated numerous studies which indicate the use of chemical signals for oriented movements by a wide variety of fish species.

This paper will be concerned with studies examining the sensory and behavioral mechanisms by which fish make use of chemosensory information to facilitate oriented movements within the environment. This process is dependent on three separate factors: 1) the physical distribution of the stimuli, 2) the chemosensitivity of the fish and 3) the particular behavioral responses of the animal.

The factors which influence the dispersal of chemical stimuli and the resultant spatial distribution patterns have been discussed in considerable detail by Bossert and Wilson (1963) and will not be covered here. The chemosensitivities of fish have also been studied extensively, and both electrophysiological and behavioral studies reveal that these animals have extreme sensitivities to many classes of compounds and thus are capable of detecting greatly diluted stimuli. This means that the studied species are able to infer the presence of a stimulation source at great distances. It is the fish's specific behavioral responses to chemical stimuli which will be considered in this paper.

Unlike other environmental cues, such as light and sound, chemical signals have no directive component and vary only in intensity. This lack of directivity therefore imposes several restrictions on possible mechanisms by which fish might orient to a stimulus source. For a fish to discern the direction from which chemical signals emanate, it must make use of additional information to obtain a directional component to the signal.

There are a number of possible mechanisms by which a fish might orient within the environment. It could respond to chemical gradients by klinotaxis and discern the axis of a chemical gradient by measuring concentration differences through repeated samplings in time or space. This, however, has never been demonstrated for fish. A few fish species have morphological adaptations which would permit them to orient by tropotaxis and compare concentrations between paired receptors at the same time, thereby perceiving the axis of the gradient across their body. There is limited evidence that fish can use tropotaxis. Chemical signals may also serve as sign-stimulus releasers. Fish, after detecting the stimulus, could orient to another directive cue such as water current. A fourth mechanism involves the release of some stereotypic swimming pattern which has a specific probability of resulting in the animal's net displacement in the direction of the stimulus source. I will provide experimental evidence that the oriented movements of many fishes are, in fact, the results of various combinations of these general categories of orientation mechanisms.

SALMONID HOMING: A PROTOTYPICAL EXAMPLE OF CHEMOSENSORY ORIENTATION

Clearly, one of the most spectacular examples of chemosensory orientation is the homing migration of salmon. Because our recent work has indicated that migrating salmon may make use of a variety of orientation mechanisms, they will serve as our model animal for chemosensory orientation.

Background

Members of the genera *Oncorhynchus* and *Salmo* are known to undergo spawning migrations of up to 4000 km and negotiate complex river channels making hundreds of choices at river junctions to return to specific natal sites. The first published account of the salmon's return to the homestream seems to be by Konrad Gesner in 1558. In his first edition of *The Compleat Angler*, Isaak Walton (1653) described experiments of Gesner who tied ribbons through the tails of young Atlantic salmon (*Salmo salar*). These fish were reported to be recaptured at a later date in the same weirs as their tagging, thereby indicating homing.

However, the concept that salmon return to their natal stream was not universally accepted until the early 1940's (Harden-Jones, 1968). Skepticism was due in a large part to the lack of a clearly understood mechanism which could possibly guide salmon thousands of kilometers from the ocean to a specific spawning tributary (Rich, 1937). The early twentieth century saw many studies involving the marking of young salmon and the subsequent recapture of returning adults. These studies, reviewed by Scheer (1939), overwhelmingly confirmed the parent stream concept of salmon homing.

With the acceptance of the salmon's faithful return to the stream of their birth, an hypothesis concerning the mechanisms of finding the homestream was required. The mechanism must indicate to the fish that it should move upstream, thus providing a directive component, and it must also have a segregating component to permit recognition of different homestreams. Two proposed hypotheses, one involving temperature (Ward,

1921a, 1921b, 1939) and the other CO_2 tension gradients (Powers, 1939, 1940, Powers and Clark, 1943) failed to provide a mechanism for the salmon's return to different streams. A third hypothesis, based on olfaction, does meet both the directive and segregational criteria.

The olfactory hypothesis has existed in various forms for over 100 years (e.g., Buckland, 1880) but A. D. Hasler (1954) gave modern expression to the concept and proposed methods for testing it in the laboratory and field. The hypothesis (Hasler, 1954) states:

1. Each stream has a characteristic odor.
2. Salmon can differentiate among them.
3. Young salmon become imprinted on these odors.
4. Adult salmon use this information to relocate the homestream.

Evidence supporting this hypothesis has been gathered along several lines of research. Hasler and Wisby (1951) demonstrated by classical conditioning of avoidance and reward that streams do have characteristic odors and that fish can differentiate between them. In numerous field studies, it has been shown that fish rendered experimentally anosmic are less successful at homing than intact fish (Wisby and Hasler, 1954; others reviewed by Stasko, 1971), while other types of sensory impairment (e.g., blinding) have little to no effect on homing success.

Studies involving transfer of young pre-smolt fish from the site where they were born to different sites have been conducted by many investigators (reviewed by Ricker, 1972; Hasler et al., 1978). These fish, after residing a short time in the new stream and smolting, migrate out into the ocean. As adults, the fish return to the stream in which they were released rather than the stream in which they were born. Transfers of fish after smolting resulted in adult fish straying to many sites rather than returning to a particular location (Peck, 1970). These investigations show that salmon learn the cues to identify the homestream during the smolt stage. Studies with coho salmon (Oncorhynchus kisutch) (Jensen and Duncan, 1971) and Atlantic salmon (Carlin, 1968) indicate that exposure of only a few days to a site at the smolt stage is sufficient to insure successful homing and thus support the concept of imprinting.

To conduct the critical test of the hypothesis, experiments using artificial imprinting have been performed by Hasler and his associates (Scholz et al., 1976; Hasler et al., 1978). In these experiments, fish were exposed to synthetic chemical odors in fish hatcheries at the time of smolting and later were decoyed as adults to streams scented with the same odors. This work confirmed the concept of olfactory imprinting.

While these experiments demonstrate the importance of imprinted chemical cues, the actual mechanisms by which oriented movements are accomplished were not specified. More recent experimental work has revealed many of the mechanistic details of the process.

The homing migration of salmon has been divided traditionally into two phases: the oceanic and stream, and are thought to involve different cues and mechanisms (Hasler and Scholz, 1983). Some investigators add a third, intermediate phase between the oceanic and stream migrations involving movement through the near shore regions. The cues and mechanisms involved in this phase may represent a combination of those used in the other two phases (Nordeng, 1977). I will, however, present data that indicate that open water movements in the near shore regions are more like those in the stream than those in the ocean. While both are believed to involve chemosensory orientation, they will be discussed separately.

Instream Orientation Mechanisms

The most important feature of a river or stream is that it represents a "bounded" environment where its edges and bottom are accessible to the animal. An additional critical aspect is that the water is typically in motion and the direction of movement can be perceived by the fish using the boundaries of the environment as a reference system. Therefore, rheotropic responses may be used for orientation.

In a dendritic river system, the patterns of odor distribution are complex but exist in two general forms which have particular importance in the homeward migration of salmon. Downstream from a source, odors are distributed fully across the width of the stream. They may not be present in uniform concentrations owing to discontinuous mixing, but are, nevertheless, present at suprathreshold concentrations. A second pattern of distribution exists at the confluence of two streams. The stream carrying the odor of interest is scented from bank to bank, while the other tributary does not carry any of this scent. Where these two channels unite, there exists for a distance downstream, two discrete plumes, one with odor and one without. As these plumes of water move downstream they become mixed. This pattern of distribution and mixing is repeated throughout the river system as the water flows to the ocean.

In experiments to study the movements of coho salmon under each of these two conditions, imprinting odor distributions were established in two experimental channels which simulated these two configurations. The movements of individual fish which had been imprinted to these and other synthetic compounds were recorded in relation to controlled distributions of the stimuli (Johnsen and Hasler, 1980). When imprinting odors were distributed across the full width of the stream, fish imprinted to that compound exhibited straight, positive rheotaxis. If the homestream (imprinting) compound was absent, fish exhibited negative rheotaxis and actively moved downstream until they again detected the home scent. Thus, both the directive and segregative requirements of a homing hypothesis are met. The segregation of fish from different home sites is based on differential rheotactic responses in the presence or absence of the imprinted homestream stimulus.

To simulate the conditions found at the confluence of two streams, an imprinting chemical was released into the stream so that its distribution was confined to one half of the stream channel, thus creating adjacent scented and unscented water masses. We observed that fish confined their movements to the side of the stream which contained the imprinting odor (Johnsen and Hasler, 1980). Detailed analysis of the fish's position with respect to the odor within the stream revealed that the fish made use of the interface between the scented and unscented masses of water to a greater extent than the odor corridor itself (Johnsen, 1978; see examples published in Hasler and Scholz, 1983). By zig-zagging along the edge of the odor plume, fish made sure and steady progress up the trail.

These observations of fish movements in relation to controlled distributions of homestream odor allowed the development of a behavioral control model which describes the movements of homing salmonids as they make their way from the sea to their homestream in a dendritic river system. The model, previously described (Johnsen, 1982, 1984), makes use of two feedback loops which represent swimming behaviors dependent on the spatial patterns of odor distribution encountered along the way (Figure 1).

The two major patterns of upstream movement include straight positive rheotaxis in water scented from bank to bank and zig-zag swimming along the interface of the scented plume at the confluence of two tributaries.

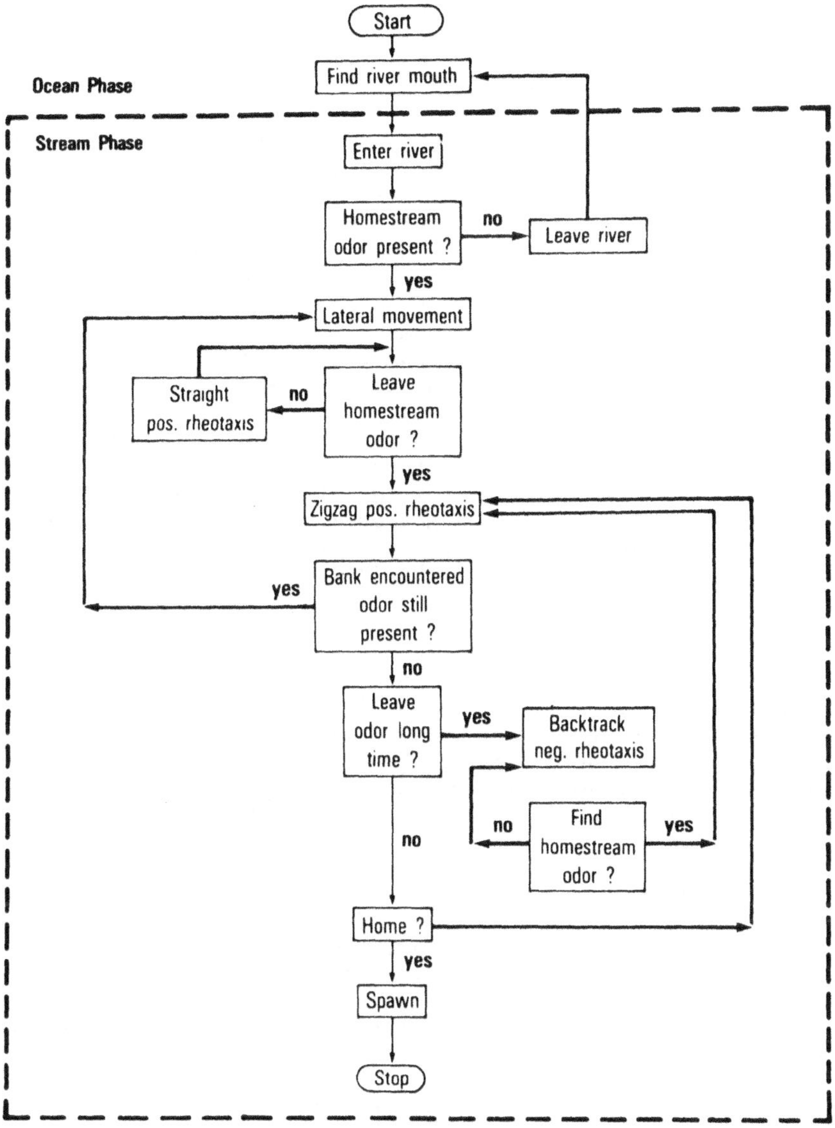

Fig. 1. Proposed behavioral control model for upstream movement and homestream selection in migrating salmonids (from Johnsen, 1982).

An additional feature includes negative rheotaxis in the absence of the homestream odor. I believe that these simple behaviors, positive and negative rheotaxis and zig-zagging at the confluence of two streams, are sufficient to guide fish to the homestream.

While differences in the specific swimming patterns are dependent on the spatial distribution of the stimuli, all behaviors within the river are rheotropic responses. For a fish to discern the direction of the current flow, it must have some fixed reference feature (Arnold, 1974). In the river, this is accomplished by maintaining visual and tactile contact with the bottom. However, fish migrating in the open ocean and in the near-shore regions are not in contact with the physical boundaries of the environment and, therefore, do not have this reference system. To move in an oriented manner, the migrating fish must use other mechanisms.

Near Shore Open-Water Orientation

When a fish in open water ascends in the water column so that it no longer maintains physical contact with the structural edges of its environment, it loses this reference system. After breaking contact with the bottom, it becomes difficult for the fish to monitor its displacement by water currents. In a sense, the surrounding environment is drifting with the fish so there are no reference features available to determine current direction. Sign stimuli provided by chemical cues are then of little value. While there are horizontal chemical gradients, they are much too weak to be perceived by tropotactic means, though the fish might theoretically be able to compare concentrations in time as it moves through the environment. But even for this mechanism there must be some additional cue so that the fish can maintain or alter its direction between samplings (Westerberg, 1982). Again, the requirement of a suitable reference system is presented. A novel solution to this problem was suggested by Westerberg (1982) who proposed that the fish might be able to use information found in the vertical distributions of the chemical stimuli along with the local current structure rather than in the horizontal concentration distribution or absolute current directions.

In the near-shore environment, there is considerable vertical stratification owing to thermal and salinity induced density differences. In effect, bands or layers of water with different chemical composition are present in the water column. Across the gradient layer between adjacent water masses there will be a current shear, the direction of which is determined by the relative motion of the two layers. By moving in and out of this boundary layer the fish then has a reference system to monitor its direction of movement.

Using a depth sensitive sonic transmitter, Westerberg (1982) demonstrated that migrating Atlantic salmon made extensive vertical movements and often swam for extended periods at selected depths. These depths were often associated with strong thermoclines indicating the interface between two water masses. Examination of the precise movements of the fish when swimming at a particular depth showed that they swam in vertical zig-zag movements which had a very similar appearance to the horizontal zig-zags reported by Johnsen (1982, 1984). A single experiment, in which the olfactory nerves of a salmon were cut, indicated that the fish did not respond to the stratified hydrographic features as did intact fish in the same study (Westerberg, 1982). This suggested that olfaction may be important in recognizing and responding to the stratified features.

To test this hypothesis, we initiated a tracking study in which intact and anosmic fish were followed during their homing migration through a Norwegian fjord (Døving et al., 1985). In addition, electrophysiological recordings from the olfactory system were made in response to water samples taken from different layers in the fjord. Two basic hypotheses were to be tested: (1) fish with intact olfactory systems would respond to hydrographic stratification while anosmic fish would fail to do so, and (2) electrophysiological evidence would indicate that the different layers could be detected and discriminated on the basis of olfactory cues.

In the tracking studies we observed that fish with intact olfactory systems made large amplitude excursions and showed preferences for particular water layers by swimming at those depths for extended periods (Figure 2). When selecting a particular depth, the salmon were, in fact, making small amplitude vertical movements up and down. The anosmic fish

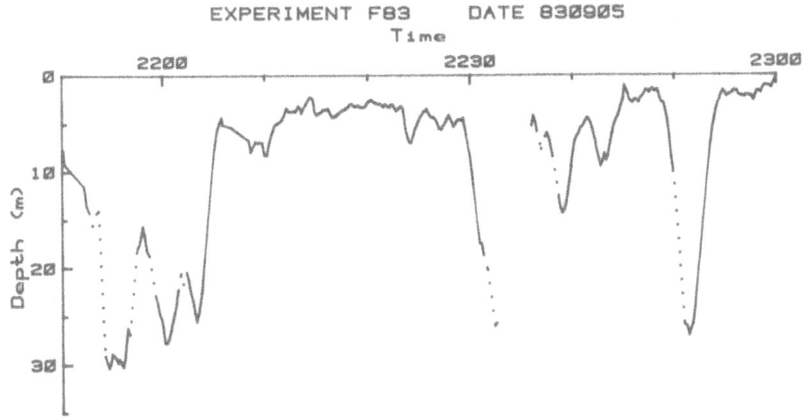

Fig. 2. One hour swimming depth record for salmon F. The continuous line shows successive 10 second averages, the dotted portions indicate periods of low signal-to-noise ratio and data drop-out (from Døving et al., 1985).

made large, continuous movements up and down in the water column and typically descended much deeper than the intact fish, often penetrating well into the lower basin of the fjord. These anosmic fish were also observed to follow bottom contours. The trauma caused by the surgical incisions to make fish anosmic did not prevent the fish from active swimming, and a fish with unilateral sectioning of the olfactory nerve successfully returned to the river of release.

During these tracking studies we collected water samples from different stratified layers within the water column. These samples were then used to stimulate the olfactory epithelium of hatchery-reared fish while we measured electrical activity from single olfactory bulb neurons. Ninety per cent of the responding neurons, presumably mitral cells, showed differential responses to the water samples (Figure 3) demonstrating the capacity of the olfactory system to discriminate among stratified water layers found in the ocean.

We believe this work indicates that migrating salmonids are capable of using olfaction to discriminate between layers of water which are found in stratified, near-shore regions. These olfactory discontinuities may serve to aid the fish in locating a region where it can find suitable reference features in the forms of layer edges and/or the sheer currents found in the boundary region. At this time we have no evidence that the fish are selecting any particular odor layers (i.e., the homestream) or are simply responding to the olfactory differences between layers. However, by locating a "meaningful" olfactory layer and determining its direction of movement by monitoring the sheer currents across its body, a fish could orient in an appropriate direction and eventually locate the source of stimulation.

Olfaction is, then, believed to be important in both the near-shore environs as well as the riverine stages of the homing migration of salmon. While the orientation mechanism within the river has been demonstrated to be released rheotropic responses dependent on the distribution of imprinted stimuli, near-shore movements are less clearly understood. Our recent experiments, however, indicate that responses to chemical stimuli and water currents may be important here as well.

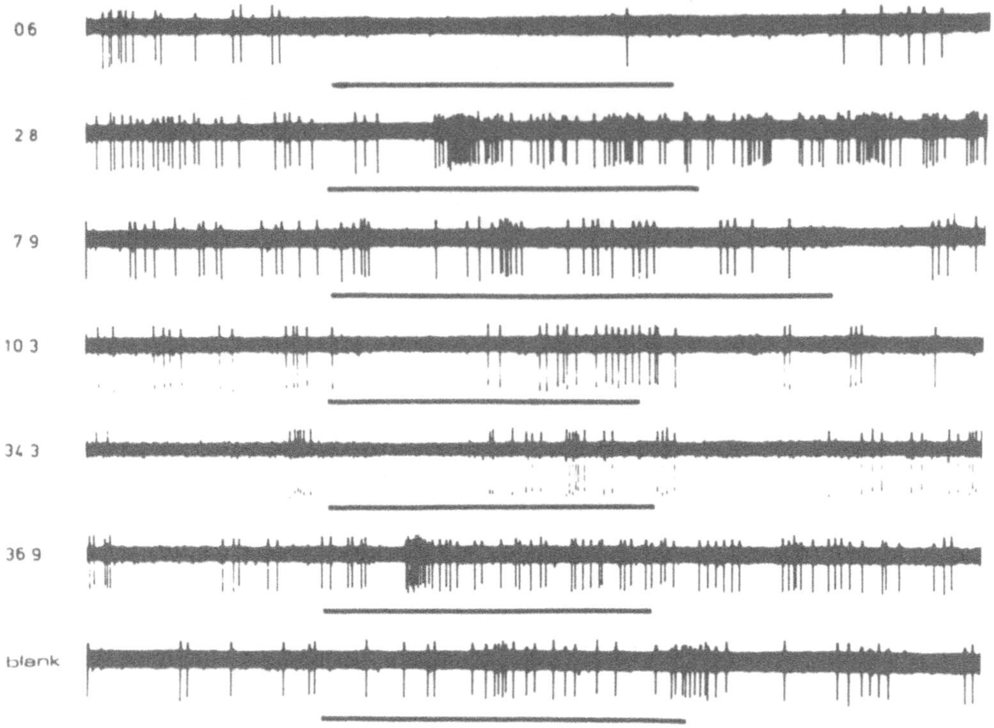

Fig. 3. The response of a single olfactory bulb neuron of an Atlantic salmon to blank and five water samples. The numbers to the left of each trace indicate the depth in the fjord where samples were collected. Stimulation period is indicated by the horizontal line under each trace. Total trace length equals 20 seconds (from Døving et al., 1985).

ALTERNATIVE ORIENTATION MECHANISMS

Tropotaxis

While rheotropic responses are presumably important in guiding salmon to the source of chemical stimulation, the possibility of tropotactic responses is very small. As has been observed previously, the slope of the gradient in the horizontal plane is very slight relative to the physical separation of the incurrent nares of the fish. In fact, only two clear examples of tropotaxis have been demonstrated for fish. In both cases the fish in question, brown bullhead (*Ictalurus nebulosus*) and the bonnethead shark (*Sphyrna tiburo*) possess specialized chemoreceptive structures.

Johnsen and Teeter (1982) demonstrated that bullheads could be conditioned to respond to chemical concentration differences presented across their maxillary barbels. These structures contain taste receptive fields innervated by the facial (VII) nerve. Fish responded to concentration differences as small as 10%, but the minimum discernible difference in concentration depended on the concentration of the stimulus. At higher concentrations, larger differences between the left and right barbels were required for the fish to show an appropriate behavioral response. As the absolute concentration of the stimulus was reduced, the ability of the animal to perceive smaller concentration differences improved. In similar experiments, Johnsen and Teeter (1985) demonstrated that naive bonnethead sharks would respond to concentration differences presented to the two incurrent nares. Response capabilities at different stimulus concentrations were not determined for the sharks.

While it has been demonstrated that gustatory and olfactory systems having specialized structures which increase the physical separation of the receptors are capable of tropotaxis, actual use of this orientation mechanism has never been demonstrated and may be reserved for special circumstances. In particular, fish may make use of this mechanism to perceive the very distinct interface between scented and non-scented water mass. Orientation up long and uniform gradients is probably quite rare due to shallow gradients and, more probably, the limited existence of uniform gradients; in most cases stimulus mixing is quite discontinuous. If an animal were to respond to chemical gradients by always turning in the direction of the highest concentration measured across its paired receptors, it is quite possible for the fish to become "trapped" in an eddy containing relatively higher amounts of stimulus than the surrounding region.

Released Behaviors

Both horizontal and vertical zig-zag movements have been observed for salmon swimming at the interface of two discrete water masses (Johnsen, 1978; Westerberg, 1982; Døving et al., 1985). The regulation of zig-zag behavior has been discussed in detail previously (Johnsen 1982, 1984) and is believed to be an internally gated behavior released by the detection of rapidly changing concentrations.

This behavior may represent a general behavioral "solution" to a common problem and is similar to the orientation of insects to sex attractants and food odors. Considerable evidence for zig-zagging behavior patterns in the presence of a pheromone exists for a wide variety of insects (Kennedy, 1977). Anemotaxis, or orientation to wind, provides the directive factor in homing of insects to the pheromone source, while the presence or absence of the pheromone and its relative concentrations regulate specific flight patterns. At the boundary of scented and unscented air, the insects show a behavior termed "reversing anemomenotaxis" in which the animal flies at an angle to the wind (Kellogg et al., 1962; Traynier, 1968; Farkas and Shorey, 1972). At short, irregular intervals, it switches from one side of the wind line to the other, resulting in zig-zags moving upwind. It has been demonstrated in wind tunnels that when the insects move from the presence of the odor into unscented air, an immediate extra-anemomenotactic reversal occurs, bringing the insect back into the odorous stream (Kramer, 1975). In this way, control of the frequency of reversals by the presence or absence of the pheromone results in an upwind orientation along the edge of the pheromone plume, directing the insect to the source of the pheromone in a manner which appears similar to the behavior of salmon seeking the source of the imprinting odor.

Zig-zagging has been observed, but not well described, in other fish including the nurse shark (Ginglymostoma cirratum) (Hodgson and Mathewson, 1971), cod (Gadus morhua) (Pawson, 1977), and rainbow trout (Salmo gairdneri) (Emanuel and Dodson, 1979). These fish displayed to-and-fro movements of the head and swam in and out of the presumed edge of the odor trail as they localized the odor source from downstream.

In an effort to simulate the interface between scented and unscented water masses, rapidly changing concentrations were presented to swimming bonnethead sharks. These tethered animals were fitted with head mounts

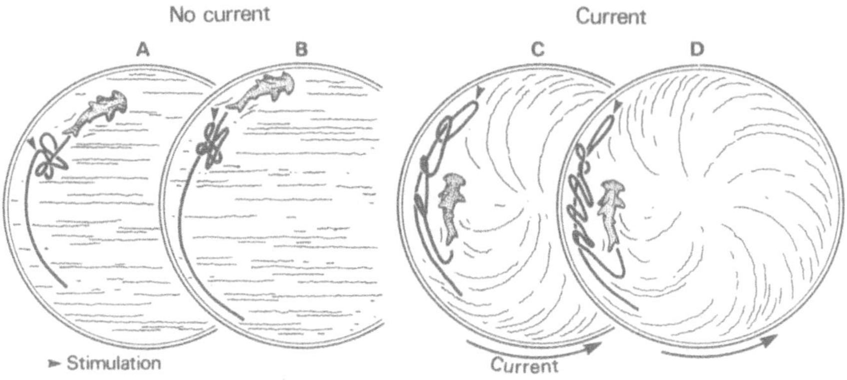

Fig. 4. Swimming paths of bonnethead sharks recorded by an overhead video camera. A - Response to 10 ml sample of crab homogenate introduced in still water; B - Response to 0.5 ml sample delivered via headmount; C - Response to 10 ml sample delivered into 15 cm/sec water current; D - Response to 0.5 ml sample delivered via headmount in presence of 15 cm/sec water current (from Johnsen and Teeter, 1985).

so that controlled deliveries of stimuli could be injected into the fish's nares via plastic tubes (Johnsen and Teeter, 1985). We had hoped to stimulate zig-zag behaviors but the shark's swimming behavior of circling the perimeter of the tank prevented this. We were however, able to conduct other experiments which gave interesting results.

When the swimming sharks encountered large volume (10 ml) samples of stimuli applied directly into the water, they demonstrated the usual circling and turning behavior which has been described for sharks for over seventy years (Parker, 1914). The previously presumed mechanism by which a shark remains in the vicinity of the stimulus involves the animal reversing its swimming direction when it swims beyond the water containing suprathreshold quantities of stimuli. It also may be a result of increased turning rates due to chemical stimulation (Kleerekoper, 1982). This can occur in both stagnant and moving water. However, our headmount allowed for a more precise delivery and distribuiton of stimuli. We could present samples of small volume and short duration to the sharks. After the stimulus sample passed through the nasal structure, its concentration was diluted to below threshold, yet the fish showed the same pattern of circling and turning at the stimulation site as with the large volume stimulation. Even more strikingly, sharks stimulated in the presence of a current circled in the projected site which was moving down current even though the stimulus had diluted to below threshold levels (Figure 4). This behavior indicated that the turning behavior exhibited by these sharks is an internally generated one released by a chemical signal. In this way, chemical stimuli may serve only to modulate the released stereotypic behaviors which result in the animal's displacement in the direction of the stimulus source. Again, Kleereloper et al. (1970), have described specific "handedness" in turning behaviors. This may be part of the internally generated swimming behaviors.

The aquatic environment is ideally suited for animals sensitive to chemical signals. By the nature of the transmission medium (water), a wide variety of compounds become suitable messengers. In addition, the general motion of most bodies of water leads to widespread dispersal of chemical signals. The one limiting feature of orientation by chemical signals is that no directional information is conveyed by the stimulus itself. An animal can, however, obtain two bits of information from the stimulus, the identity (or quality) of the stimulus and its intensity. Directional information must be ascertained from other features in the environment.

We have demonstrated that if a fish perceives the presence or absence of a signal of interest and responds appropriately to the directional information given by water currents, it may locate the source of stimulation. The limitation of this method of orientation is that there must be a reference system for the fish to establish its position and relative motion. Most typically, a fish moving in a river or along the bottom may use visual and tactile cues to maintain contact with the physical environment which functions as the required reference system. This is the proposed chemosensory orientation mechanism utilized by fish moving in rivers and other physically "bounded" environments.

When a fish in open water ascends in the water column so that it can no longer maintain physical contact with the structural edges of its environment, it loses this reference system. Westerberg (1982) suggested that fish might possibly detect the vertical concentration gradients of stimuli found in horizontal layers and then respond to sheer currents resulting from the relative motion of the two adjacent water layers. In this way, the animal finds a necessary reference system. Salmon have demonstrated the ability to discriminate by olfaction differences among water layers and exhibit specific swimming behaviors dependent on olfactory information.

Thus, both riverine and open water orientation by salmon appear to involve rheotropic responses dependent on the distribution of chemical stimuli and zig-zag swimming behaviors. Future work on understanding the mechanisms of chemosensory orientation must address the basis of zig-zag behaviors. I have hypothesized that when a fish traveling within a scented watermass encounters a rapidly changing stimulus concentration, thus indicating an interface, the behavior of swimming at an angle to the current, rather than against the current, is released. This movement allows the fish to move in the axis of greatest probable concentration change. After establishing the presence of the interface, the fish starts the zig-zag swimming behavior. To determine the factors which are responsible for each turn, one can examine the stimulus intensity at each of the turn points during the behavior. One turn occurs at a relatively low concentration and the other at a relatively higher one. Two different strategies may be employed by the fish to control the turning. Under one scheme, with the animal turning back from the unscented water and swimming at an angle to the current, the fish enters the scented plume (below threshold concentration) and continues until it encounters a threshold concentration. It then turns and swims on the opposite tack traveling through the plume until the concentration goes below threshold, thus initiating another turn. A second scheme involves the animal turning at some critical concentration difference with respect to the lower and higher concentrations. This is different from the first strategy in that the animal can perceive the presence of the stimulus, but does not turn until some specific concentration is reached. Computer simulation models of zig-zag behavior generate the same path using both schemes indicating that further experimental work is required to resolve the actual mechanism.

In both fishes and insects, the amplitude of zig-zags seems to be related to the slope of the gradient across the edge of the plume. In cases where the gradient is quite shallow, the animals show larger amplitudes in their zigs and zags. As the gradient steepens, the turning rate increases, and the amplitude decreases. This is consistent with both schemes and again suggests the need for further experimentation.

Effects of adaptation, or the decrement of sensitivity after prolonged exposure to a stimulus, would greatly influence the behavioral responses of the fish. Both when the fish is searching for some absolute difference in concentration between the lower and upper concentrations, and when the fish is turning when the concentration reaches threshold, turns would become less frequent due to the depressed sensitivities of the fish. The fish, as it enters the plume, is not at threshold concentration, yet the olfactory system is exposed to stimuli and the process of sensory adaptation is underway. As a result, the threshold sensitivity of the animal slides upward. The longer the animal remains in the subthreshold concentration, the higher the required concentration for threshold stimulation. The same thing would occur for an animal seeking some concentration difference. Because the apparent concentration is different than the actual, the animal would move to higher and lower concentrations before initiating a turn.

Cardiac-conditioning experiments have been conducted to determine the consequences of prolonged exposure to chemical stimuli. There is a proportional increase in the concentration of stimulus required for detection as the background concentration rises (Johnstone, 1980). More studies on the effects of prolonged subthreshold-stimulus exposure are required.

Another important aspect of the zig-zag behavior is the detection of the edge of the plume and the absence of stimuli. We know from peripheral nerve recordings that the response of olfactory nerves does not totally adapt after prolonged exposure. Thus, the animal is presumably able to sense the off-set of a stimulus and would be capable of detecting movement out of the stimulus. However, there are other explanations of the animal's ability to detect the edge of the plume. These may include re-stimulation after rinsing out by the subthreshold water and a re-stimulation by suprathreshold water found at the discontinuously mixed interface. Experiments on the fish's ability to detect decreasing concentrations will be required to determine the actual mechanism of the release of zig-zag behavior.

All chemosensory-mediated orientation by fish is not, however, accomplished by rheotropic responses. The ability to orient in a steep chemical gradient using tropotaxis has been demonstrated for two species of fish, one using the gustatory system and the other olfaction. In both cases the fish possess specialized receptive structures which serve to increase the physical separation of the chemoreceptors. The actual use of tropotaxis in the natural environment has not been demonstrated, nor is it known under what conditions this orientation mechanism might be utilized. Work on establishing the conditions under which these behaviors might operate and their functions are needed.

An interesting mechanism involving the release of stereospecific swimming behaviors has been proposed recently (Johnsen, 1984; Johnsen and Teeter, 1985). It has been demonstrated that sharks could be induced in the laboratory to exhibit persistent swimming behaviors with very short term chemical stimulations. These could be released in both still and moving water and are interpreted to be internally generated swimming patterns which eventually result in the displacement of the fish in the direction of the stimulus source. We are currently attempting to perform similar experiments with free swimming sharks in the open ocean by utilizing a radio-controlled stimulation device.

REFERENCES

Arnold, G. P., 1974, Rheotropism in fishes, Biol. Rev., 49:515.

Bossert, W. H., and Wilson, E. O., 1963, The analysis of olfactory communication among animals, J. Theor. Biol., 5:443.

Buckland, F., 1880, "Natural History of British Fishes," Unwin, London.

Carlin, B., 1968, Salmon conservation, tagging experiments and migration of salmon in Sweden, Lect. Ser. Atl. Salm. Assoc., Montreal.

Døving, K. B., Westerberg, H., and Johnsen, P. B., 1985, Role of olfaction in the behavioral and neuronal responses of Atlantic salmon, Salmo salar, to hydrographic stratification, Can. Fish. Aquat. Sci., in press.

Emanuel, M. E., and Dodson, J. J., 1979, Modification of the rheotropic behavior of male rainbow trout (Salmo gairdneri) by ovarian fluid, J. Fish. Res. Board Can., 36:63.

Farkas, S. R., and Shorey, H. H., 1972, Chemical trail-following by flying insects. A mechanism for orientation to a distant odor source, Science, 178:67.

Harden-Jones, F. R., 1968, "Fish Migration," Edward Arnold Ltd., London.

Hasler, A. D., and Wisby, W. J., 1951, Discrimination of stream odors by fishes and relation to parent stream behavior, Am. Nat., 85:223.

Hasler, A. D., 1954, Odor perception and orientation in fishes, J. Fish. Res. Board Can., 11:107.

Hasler, A. D., Scholz, A. T., and Horrall, R. M., 1978, Olfactory imprinting and homing in salmon, Am. Sci., 66:347.

Hasler, A. D., and Scholz, A. T., 1983, "Olfactory Imprinting and Homing in Salmon," Springer-Verlag, New York.

Hodgson, E. S., and Mathewson, R. F., 1971, Chemosensory orientation in sharks, in: "Orientation: Sensory Basis," H. E. Adler, ed., Ann. NY Acad. Sci., 188:175.

Jensen, A. L., and Duncan, R. N., 1971, Homing of transplanted coho salmon, Prog. Fish Cult., 33:216.

Johnsen, P. B., 1978, Behavioral Mechanisms of Upstream Migration and Homestream Selection in Coho Salmon (Oncorhynchus kisutch), Ph.D. Thesis, Univ. Wisc., Madison.

Johnsen, P. B., and Hasler, A. D., 1980, The use of chemical cues in the upstream migration of coho salmon, Oncorhynchus kisutch Walbaum, J. Fish. Biol., 17:67.

Johnsen, P. B., and Teeter, J. H., 1982, Spatial gradient detection of chemical cues by catfish, J. Comp. Physiol., 140:95.

Johnsen, P. B., 1982, A behavioral control model for homestream selection in salmonids, in: "Proceedings of the Salmon and Trout Migratory Behaviour Symposium," E. L. Brannon, and E. O. Salo, eds., School of Fisheries, Univ. Washington, Seattle.

Johnsen, P. B., 1984, Establishing the physiological and behavioral determinants of chemosensory orientation, in: "Mechanisms of Migration in Fishes," J. D. McCleave, G. P. Arnold, J. J. Dodson, and W. H. Neill, eds., Plenum Publishing Corp., New York.

Johnstone, A. D. F., 1980, The detection of dissolved amino acids by the Atlantic cod, Gadus morhua L., J. Fish. Biol., 17:219.

Kellogg, F. E., Frizel, D. E., and Wright, R. H., 1962, The olfactory guidance of flying insects, IV, Drosophila, J. Can. Entomol., 94:884.

Kennedy, J. S., 1977, Olfactory responses to distinct plants and other odor sources, in: "Chemical Control of Insect Behavior, Theory and Application," H. H. Shorey, and J. J. McKelvey, Jr., eds., Wiley-Interscience, New York.

Kleerekoper, H., Timms, A. M., Westlake, G. F., Davy, F. B., Malar, T., and Anderson, V. H., 1970, An analysis of locomotor behaviour of goldfish (Carassius auratus), Anim. Behav., 18:317.

Kleerekoper, H., 1982, The role of olfaction in the orientation of fishes, in: "Chemoreception in Fishes," T. J. Hara, ed., Elsevier, New York.

Kramer, E., 1975, Orientation of the male silkmoth to the sex attractant Bombykol, in: "Olfaction and Taste 5," D. Denton, and J. D. Coghlan eds., Academic Press, New York.

Nordeng, H., 1977, A pheromone hypothesis for homeward migration in anadromous salmonids, Oikos, 28:155.

Parker, G. H., 1914, The directive influence of the sense of smell in the dogfish, Fish. Bull., 33:63.

Pawson, M. G., 1977, The responses of cod (Gadus morhua) to chemical attractants in moving water, J. Cons. Int. Explor. Mer., 37:316.

Peck, J. N., 1970, Straying and reproduction of coho salmon, Oncorhynchus kisutch, planted in a Lake Superior tributary, Trans. Am. Fish. Soc., 99:591.

Powers, E. B., 1939, Chemical factors affecting the migratory movements of the Pacific salmon, Publ. Am. Assn. Adv. Sci., 8:72.

Powers, E. B., and Clark, R. T., 1943, Further evidence on chemical factors affecting the migratory movements of fishes, especially the salmon, Ecology, 24:109.

Rich, W. H., 1937, Homing of Pacific salmon, Science, 85:477.

Ricker, W. E., 1972, Hereditary and environmental factors affecting certain salmonid populations, in: "The Stock Concept in Pacific Salmon," R. C. Simon, and P. A. Larkin, eds., MacMillan Lectures in Fisheries, Univ. British Columbia, Vancouver.

Scheer, B. T., 1939, Homing instinct in salmon, Quart. Rev. Biol., 14:408.

Scholz, A. T., Horrall, R. M., Cooper, J. C., and Hasler, A. D., 1976, Imprinting to chemical cues: The basis for homestream selection in salmon, Science, 192:1247.

Stasko, A. B., 1971, A review of field studies on fish orientation, Ann. NY Acad. Sci., 188:12.

Traynier, R. M. M., 1968, Sex attraction in the Mediterranean flour Moth Anagaster kuhniella: Location of the female by the male, Can. Entomol., 100:5.

Walton, I., 1653,"The Compleat Angler or the Contemplative Man's Recreation," London.

Ward, H. B., 1921a, Some features in the migration of the sockeye salmon and their practical significance, Trans. Am. Fish. Soc., 50:387.

Ward, H. B., 1921b, Some of the factors controlling the migration and spawning of the Alaska red salmon, Ecology, 2:235.

Ward, H. B., 1939, Factors controlling salmon migrations, Publ. Am. Assn. Adv. Sci., 8:60.

Westerberg, H., 1982, Ultrasonic tracking of Atlantic salmon. II. Swimming depth and temperature stratification, Inst. of Fresh. Res., Drottningholm Rep., 60:102.

Westerberg, H., 1984, The orientation of fish and the vertical stratification of fine and microstructure scales, in: "Mechanisms of Migration in Fishes," J. D. McCleave, W. H. Neill, J. J. Dodson, and G. P. Arnold, eds., Plenum Publ. Corp., New York.

Wisby, W. J., and Hasler, A. D., 1954, Effect of olfactory occlusion on migrating silver salmon, Oncorhynchus kisutch, J. Fish. Res. Board Can., 11:472.

RELEASE OF SPAWNING PHEROMONE(S) BY NATURALLY-OVULATED AND PROSTAGLANDIN-INJECTED, NONOVULATED FEMALE GOLDFISH

P. W. Sorensen, N. E. Stacey, and P. Naidu

Department of Zoology
University of Alberta
Edmonton, Alberta, Canada T6G 2E9

INTRODUCTION

Although several studies have demonstrated the importance of smell in goldfish spawning behavior, they have not substantiated the release of a discrete odor signal, or pheromone, by recently ovulated females. Partridge et al. (1976) found male goldfish tested in a Y-maze preferred water in which ovulated females had been held. However, a preference was also exhibited toward food odor. When placed with ovulated and nonovulated females intact males interacted selectively with ovulated females, but olfactory-ablated males exhibited the same preference although to a lesser degree. Similarly, Stacey and Kyle (1983) found that while olfactory tract sectioning reduces male courtship behavior it does not eliminate it. Goldfish are attracted to conspecific odor (Le Martret and Saglio, 1982); the possibility remains that ovulated females merely release increased quantities of conspecific odor and that visual and tactile signals play preeminent roles in spawning. To study the effects of odor alone, this experiment monitored the behavior of solitary males in an open-field maze into which control, food, nonovulated female, and ovulated female odors were introduced. To determine the social relevance of behaviors in the maze, male response to the odors was also observed in an aquarium containing nonovulated mature females with whom the males were familiar.

In goldfish and several other oviparous teleost fish, prostaglandin (PG) acts within the ovary to induce ovulation and appears to be carried simultaneously to the brain to evoke female spawning behavior (Goetz, 1983; Stacey and Goetz, 1982; Stacey et al., this volume). Injections of PG quickly evoke seemingly normal female spawning behavior in both male and female goldfish. Male goldfish spawn normally with PG-injected females and appear incapable of distinguishing between them and ovulated females (Stacey, 1981). Because female oviparous fish, including goldfish, are believed to release sex pheromones at the time of ovulation, PG may simultaneously mediate female spawning behavior and ovulatory pheromone(s) release, thereby maximizing the chances of successful spawning. To test this hypothesis, the response of male goldfish to the odor of PG-injected females was assayed and compared with male response to the odor of ovulated fish. PG alone was also tested to see if it functions as a pheromone.

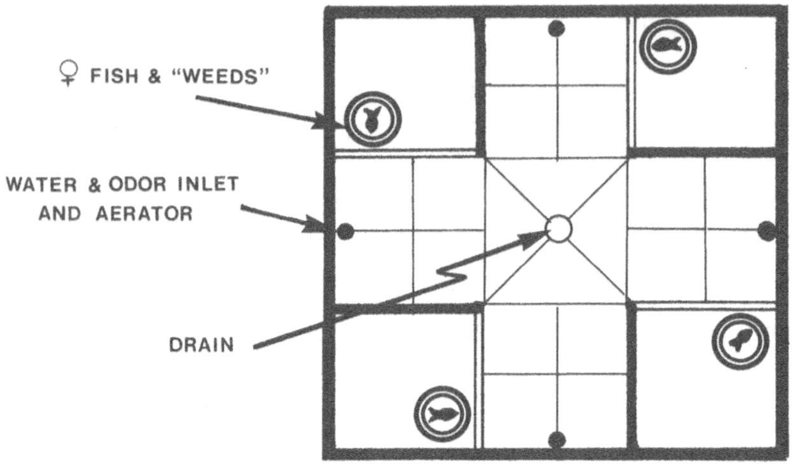

Fig. 1. The 1.2 m square open-field maze. Clear walls are not darkened.

METHODS

Male goldfish obtained from Grassyfork Fisheries Co., Martinsville, IN, were kept in running 15°C dechlorinated water on a simulated natural (Edmonton) photoperiod and fed ad lib with Nutrafin (Hagen Co., Montreal). Two weeks prior to testing, they were placed in tanks which were gradually warmed to 20°C, the normal spawning temperature (Stacey and Kyle, 1983). The photoperiod was also shifted to 16L:8D with lights-on at 11 AM. Assays were conducted between 0900-1600h. This insured that fish were assayed within a few hours of their dawn, when they normally spawn (Stacey, 1981). Fish were not fed for 24 hours prior to testing. Experiments were conducted May-June, the normal spawning season, and only healthy males with expressable milt and tubercles on both their opercula and pectoral fins were used. Males had an average head-fork length of 9.3 cm and gonadosomatic index (GSI) of 3.

Male goldfish were tested in a 1.2 m square open-field maze with 4 cross-arms, a depth of 20 cm, and a volume of 160 l (Fig. 1). Dechlorinated 20°C blank water flowed into each arm at 2 l min^{-1} (gentle flow) and out a central drain. An odor inlet merged with each blank inlet tube and aerators beside their joint outlets masked extraneous noise. Dye tests revealed an approximately linear flow with wisps of dye reaching the center in a minute and a solid cloud present in 5 min. Dye entered "non-odor" arms after 10 min and drifted to the base of "non-odor" arms after 15 min, although a concentration gradient was still evident.

The maze floor was divided into 20 equal areas (4 20-cm squares and a central triangle comprised each arm) to measure swimming activity. White gravel covered the bottom providing a substrate to sample for food. Maze walls were blue except for a clear plexiglass side wall in each arm. A clear cylinder containing a blinded (preventing interaction) female goldfish and synthetic weeds (in which goldfish spawn) behind each plexiglass wall acted as a visual spawning cue. The maze was enshrouded in black plastic and lit by overhead fluorescent lights. Fish were viewed through a side peephole and via an overhead video camera.

Single male fish were placed in the maze and allowed to familiarize themselves. If they were not slowly swimming throughout the maze and sampling for food after 30 min they were not tested. Roughly 80% of the fish met these criteria. Assays were divided into two parts: a 15 min pre-test in the absence of odor and a 15 min experimental period in the presence of odor. Four behavioral parameters were recorded on a Rustrak event recorder: movements between areas (across lines) as an estimate of activity, time spent in each arm, time spent attempting to reach/interact with females, and frequency of sampling the bottom for food. In the pre-test, behavior was observed until the fish had entered each arm. If a fish took more than 5 min to do this it was not used, but 90% of all fish met this criterion, and then monitored for 15 additional min. The 15 min experimental period commenced when the fish swam into a new arm. Odor was then released into this arm and the one adjacent to it. After 15 min, the fish was removed to a test aquarium (described later) and the maze was drained, rinsed, and refilled. Each odor was assayed with 10 fish and fish were only tested once. By using equal sample sizes the magnitude of observed responses could be compared.

Odors were introduced at 100 ml min^{-1} into each arm (3 l per test). Dechlorinated 20°C water was used as the blank. Food odor was created by placing 10 ml of supernatant from a 1% Nutrafin suspension into 3 l of blank water; it was equivalent to adding 1 g of food to the 280 l of water employed during an experimental period. Ovulated-female water was created by placing a spontaneously ovulated, unspawned female (ovulation was induced by placing the female in warm water with weeds overnight; Stacey et al., 1979) into 7 l of aerated blank water for 1 h early on the morning of ovulation. PG-injected female odor was created by injecting intramuscularly 3 vitellogenic females (average GSI of 6) with 0.4 ug PGF Tris salt (Sigma) per g body weight in a saline buffer and placing them in 20 l of aerated blank water for 1 h. This dosage induced female spawning behavior and is similar to that used by Stacey and Kyle (1983). The odor of saline-injected females was created by injecting 3 vitellogenic females (average GSI of 6) with saline buffer and placing them in 20 l of aerated blank water for 1 h. PG odor was formulated by adding 3 ug $PGF_{2\alpha}$ (in 6 ul of buffer) to 3 l of blank water; this represented approximately twice the amount of PG which could have been released from the injected females.

In the analysis, the pre-test time was the most recent 15 min period which began with the fish entering the arm into which the odor was later introduced. This allowed a direct comparison of the pre-test and experimental periods as it removed any bias individual fish may have had for particular areas, even though the maze was, on average, unbiased ($P < 0.10$). Pre-test values were compared between experiments using the Kruskal-Wallis test. The Wilcoxon matched pairs signed ranks test (Daniel, 1978) was used to compare the number of lines crossed per minute and feeds per minute, percent time in the odor arms (or those arms which became odor arms) and percent time trying to interact with females for pre-test and experimental periods for each odor. Two-tailed probability values were calculated and both medians and ranges plotted.

After testing, male fish were placed in 70 l glass aquaria containing synthetic weeds, 2 vitellogenic nonovulated females, and flowing 20°C water. By familiarizing males with females, the spontaneous chasing which characterizes sexually-active males was minimized. The next morning, fish were observed for a 10 min pre-test and a 10 min experimental period in which the odor tested the previous day was injected through a blank water inlet tube at 35 ml min^{-1}. The resulting odor concentration was about half that tested in the maze. Fish were not tested if they chased continuously in the pre-test period. This eliminated a quarter of all fish. Percent time chasing females (moderate-to-fast swimming while pursuing a female) and the frequency of nudging a female were recorded. These behaviors characterize sexual arousal and were examined by Partridge et al. (1976) and Stacey and Kyle (1983). Data were analyzed in the same manner as those from the maze.

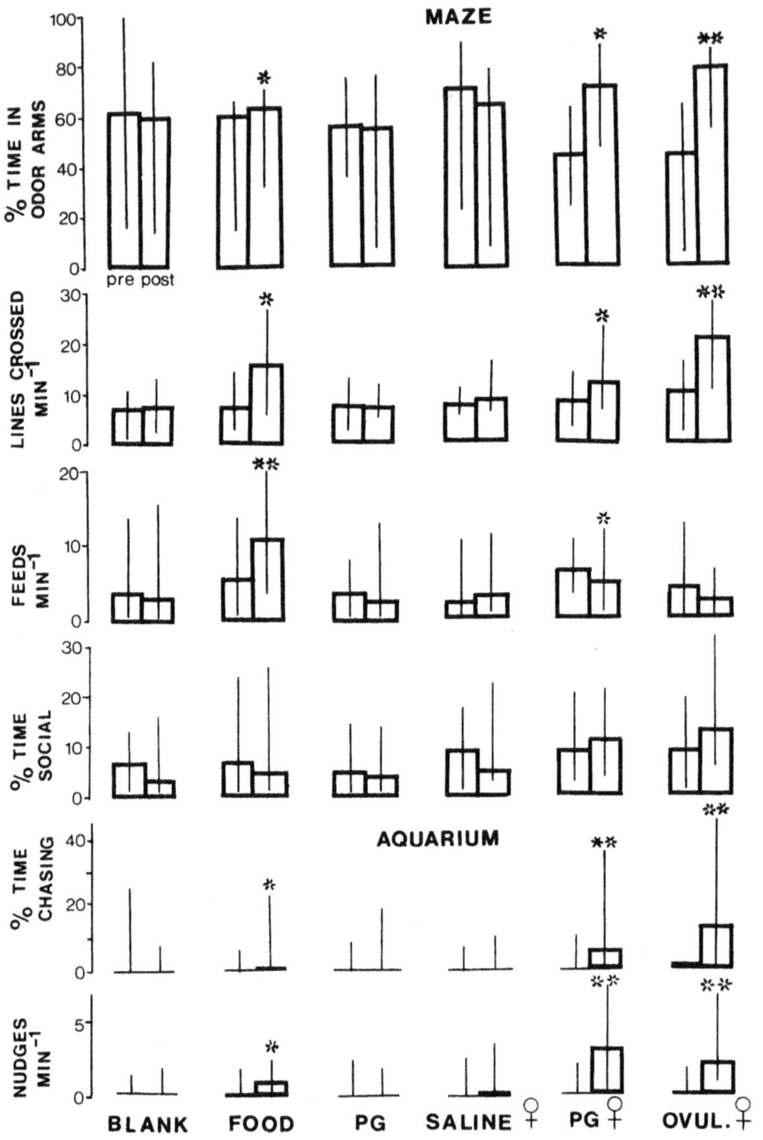

Fig. 2. Behavioral responses of solitary male goldfish to odors in an open-field maze and in an aquarium with females. Histograms represent median values and bars represent ranges. Results of pre-tests made in the absence of the odor are presented alongside experimental periods when odor was present. Differences significant at $P \leq 0.05$ are designated by "*" while "**" designates differences significant at $P \leq 0.01$.

RESULTS

Results are summarized in Figure 2. Pre-test values of the behavioral parameters did not differ ($P < 0.10$) between experiments. Except for time spent trying to interact with females in maze ("% time social"), which decreased slightly ($P = 0.07$), none of the behavioral parameters were signficantly different ($P < 0.10$) after the introduction of blank water. Neither the odor of PG nor odor of saline-injected females influenced male behavior. Food odor was moderately attractive and caused an increase in activity ($P \leq 0.05$), a dramatic increase in feeding ($P \leq 0.01$), and had no apparent influence on social activity in the maze. In the aquaria, food odor evoked small increases in chasing and nudging ($P \leq 0.05$).

The pattern of behavioral changes evoked by the odor of PG-injected females and ovulated females was similar in all instances and differed from that induced by food odor. Both odors were attractive and caused an increase in activity ($P \leq 0.05$), with ovulated-female odor evoking the strongest responses. PG-female and ovulated-female odors caused increases in social activity in the maze which, although nonsignificant, contrasted with the decreases associated with the control and all other odors. Similarly, in the aquaria, both odors evoked dramatic increases in chasing and nudging ($P \leq 0.01$). These odors also caused reductions in feeding activity although only that induced by PG-female odor was significant ($P \leq 0.05$). In both instances, this decrease in feeding activity was most evident in the first 5 min after odor introduction (data not shown).

DISCUSSION

Female goldfish appear to release a discrete chemical signal, or pheromone, coincident with or soon after ovulating as suggested by Partridge et al. (1976). The odor of recently ovulated female goldfish evokes specific and dramatic changes in the behavior of mature male goldfish, changes which differ markedly from those evoked by odors of nonovulated females and food. Responses to ovulatory odor include strong attraction, heightened locomotor activity, increased interest in/contact with nonovulated females that previously aroused little interest, and a decrease in feeding attempts. Male response to nonovulated females in the presence of ovulated female odor is similar to, although less intense than, that exhibited toward ovulated females (Partridge et al., 1976; Stacey, 1981). Preliminary experiments indicate that immature (nonspermiated) males do not respond to the odor of ovulated females; ovulatory odor appears to have a specific behavioral message. It is clear that this pheromone "advertises" the proximity of a female in spawning condition and that males are able to locate regions of higher concentration in relatively still water; however, it is unclear whether the pheromone is used to locate and identify individual females. Vision and touch almost certainly play important supplemental roles in goldfish spawning behavior.

In this study, the odor of PG-injected nonovulated females induced changes in male behavior indistinguishable from those induced by the odor of ovulated females. PG injection induces normal spawning behavior in nonovulated females, and the present finding that PG also stimulates pheromone release may explain why male goldfish find ovulated and PG-treated nonovulated females equally attractive as spawning partners (Stacey, 1981).

Virtually nothing is known of how PG stimulates pheromone release or the nature of the pheromone released. However, as the odor of PG-injected males does not appear to induce the same responses as PG-female odor (Sorensen et al., unpublished results), it appears that PG may act on the ovary to stimulate pheromone production/release. Because PG does not affect male behavior when added directly to tank water, injected PG must either be metabolized to an active form or stimulate the release of a second substance(s). Two ovarian steroids have been proposed to function as sex pheromones in goldfish: 17α, 20β-dihydroxy-4-pregnene-3-one may act via the olfactory system to increase milt levels (Stacey et al., this volume), and etiocholanolone glucuronide has been shown to attract males in still water (Colombo et al., 1982). Neither of these compounds had any effect on male behavior when added directly to the open-field maze and aquaria, however (Sorensen et al., unpublished data).

In several oviparous teleosts, including goldfish, prostaglandins play a role in ovulation and increase in the blood during the post ovulatory period (Goetz, 1983). The results of earlier work suggest that, at least in goldfish, the postovulatory increase in PG acts on the brain to trigger female sexual behavior when ovulated eggs are ready for release (Liley and Stacey, 1983). Results of this study suggest that by stimulating release of a sex pheromone, PG functions to synchronize female sexual attractiveness with female sexual behavior. The idea of such a coordinating role for PG is appealing in its simplicity, but it remains to be determined whether the pheromone released by ovulated females is chemically the same as that released in response to PG treatment.

ACKNOWLEDGMENTS

The Alberta Heritage Foundation for Medical Research generously funded this study while supporting the senior author.

REFERENCES

Colombo, L., Belvedere, P. C., Marconato, A., and Frisco, C., 1982, Pheromones in teleost fish, in: "Proc. Intl. Symp. Reprod. Physiol. Fish," C. J. J. Richter, and H. J. Th. Goos, eds., Pudoc, Wageningen.

Daniel, W. W., 1978, "Applied Nonparametric Statistics," Houghton Mifflin Co., Boston.

Goetz, F. W., 1983, Hormonal control of oocyte final maturation and ovulation in fishes, in: "Fish Physiology, Vol. IXB, Behavior and Fertility Control," W. S. Hoar, D. J. Randall, and E. M. Donaldson, eds., Academic Press, New York.

Le Martret, M.-A., and Saglio, P., 1982, Communication intraspecifique d'origine phéromonale chez le carassin dore immature, Carassius auratus L., Biol. of Behav., 7:41.

Liley, N. R., and Stacey, N. E., 1983, Hormones, pheromones and reproductive behavior in fish, in: "Fish Physiology, Vol. IXB, Behavior and Fertility Control," W. S. Hoar, D. J. Randall, and E. M. Donaldson, eds., Academic Press, New York.

Partridge, B. L., Liley, N. R., and Stacey, N. E., 1976, The role of pheromones in the sexual behavior of the goldfish, Anim. Behav., 24:291.

Stacey, N. E., 1981, Hormonal regulation of female reproductive behavior in fish, Am. Zool., 21:305.

Stacey, N. E., Cook. A. F., and Peter, R. E., 1979, Ovulatory surge of gonadotropin in the goldfish, Carassius auratus, Gen. Comp. Endocrinol., 37:246.

Stacey, N. E., and Goetz, F. W., 1982, Roles of prostaglandins in fish reproduction, Can. J. Fish. Aquatic Sci., 39:92.

Stacey, N. E., and Kyle, A. L., 1983, Effects of olfactory tract lesions on sexual and feeding behavior in goldfish, Physiol. Behav., 30:621.

THE STRUCTURAL AND FUNCTIONAL RELATIONSHIPS BETWEEN OLFACTORY AND REPRODUCTIVE SYSTEMS FROM BIRTH TO OLD AGE IN FISH

M. P. Schreibman, H. Margolis-Nunno and L. Halpern-Sebold

Department of Biology
Brooklyn College
Brooklyn, NY 11210

INTRODUCTION

The existence of a structural and functional axis between the olfactory and reproductive systems has long been suspected and is now becoming better understood because of increased interest and study. As a result, exciting new information is now available which clearly demonstrates that changes in olfactory epithelium can be directly related to naturally occurring and experimentally induced events in the neuroendocrine system of vertebrates (e.g., menstrual and estrous cycles, castration, and sex steroid administration). The frequency of studies in these areas decreases, however, as one descends the phylogenetic scale. In addition, there is a patent paucity of temporal analyses of the olfactory system during the postnatal development and senescence of the reproductive system for vertebrates in general.

In this chapter we will trace the development of the olfactory epithelium and associated structures from birth through old age in sibling platyfish (Xiphophorus maculatus) that are genetically determined to reach sexual maturity at different ages. These observations will be evaluated in the light of events occurring in the neuroendocrine axis that are responsible for regulating reproductive system maturation and operation. Special attention will be devoted to the nucleus olfactoretinalis (presumably homologous to the nervus terminalis, or terminal nerve; see Wedekind, 1978; Crapon de Caprona and Fritzch, 1983; Demski and Northcutt, 1983; Demski, 1984), a unique LHRH-containing nucleus in the brain.

The Platyfish (Xiphophorus maculatus; Poeciliidae) as an Experimental Model

The platyfish, a small freshwater teleost, has been studied intensively for more than 60 years (Kallman, 1975). Sufficient information has been amassed on its neuroendocrine system and on the hereditary mechanisms that control its age at sexual maturation, to make the platyfish invaluable for the study of the ontogeny, functioning and aging of the brain-pituitary-gonadal (BPG) axis.

Variation in the time of onset of sexual maturation of male and female platyfish is due to genetic factors (Kallman and Schreibman, 1973; Kallman, et al., 1973; Kallman, 1975; Schreibman and Kallman, 1977; Schreibman and Kallman, 1978). A sex-linked gene, P, controls the age at which the BPG axis becomes activated and sexual maturation ensues. Five P alleles (P^1 through P^5 have been identified in natural populations and laboratory stocks (Kallman and Borkoski, 1978) and depending upon genotype, puberty occurs at a specific predictable age between 8 and 104 weeks. There is no overlap in maturation time when different genotypes are raised under identical conditions (Kallman and Schreibman, 1973; Kallman et al., 1973; Schreibman and Kallman, 1977). The various P factors are carried on the sex chromosomes (X and Y) and are closely linked to a number of genes for body pigmentation that serve as their phenotypic markers.

The life span of platyfish is 2 to 3 years; however, fish raised in isolation and under optimum conditions have lived almost twice as long (Kallman, unpublished; see Schreibman et al., 1983). The structure and function of the BPG axis from birth to senescence has been under intensive investigation in our laboratory. The platyfish presents us with a unique model system with which to study the effect of the interaction of genetic and environmental factors on the development and senescence of integral parts of the nervous and endocrine systems and on the phenomenon of aging in general. By using known genetic stocks, some with large degrees of homozygosity, we can eliminate the variability of animal populations that plagues the analysis of so many aging studies.

Methods of Analysis

Essentially, the experimental design involves the examination of early (P^1P^2) and late (P^5P^2) maturing males, and early (P^1P^1) or P^5P^1), and occasionally late (P^5P^5), maturing females at specific periods between birth and 30 months of age, the average life span of platyfish (see Schreibman et al., 1983 and Margolis-Kazan and Schreibman, 1984). All fish studied are of known genotype and are derived from stocks maintained in the Genetics Laboratory at the New York Aquarium or from their descendants reared in our laboratory at Brooklyn College.

Histology was carried out on Bouin's fixed, paraffin embedded materials as previously detailed (Schreibman, 1964). Representative sections of the brain and pituitary were stained with Masson's trichrome procedure, periodic acid-Schiff (PAS), or Nissl stain. Immunocytochemical studies utilized the peroxidase anti-peroxidase (PAP) method of Sternberger et al. (1970) and were performed as previously described (Margolis-Kazan and Schreibman, 1981) using antisera generated against luteinizing hormone releasing hormone (LHRH), serotonin (5HT), tyrosine hydroxylase (TH) and the beta subunit of carp gonadotropin (GTH). In all cases standard immunocytochemical control methods were utilized to determine the specificity of the immunoreactive response (see Margolis-Kazan and Schreibman, 1981, and Halpern-Sebold and Schreibman, 1983 for details). All comparisons of levels of immunoreactivity were made for material processed at the same time using identical methods. Cell and nuclear areas were determined using procedures detailed by Leatherland (1970) and Schwanzel-Fukuda et al. (1981) or by using a computer-based image analyzer (Zeiss "Videoplan"). All statistical analyses were carried out according to Zivin and Bartko (1976) and significant differences between groups were selected at P equals less than 0.05.

Nasal epithelium was prepared for scanning electron microscopic (SEM) study according to methods previously reported (Schreibmen et al., 1984). Gross morphology of the gonads was ascertained at autopsy. Histochemical methods for analyses of enzyme activity in the gonads were performed as previously described (Schreibman et al., 1982a). Histological analyses of gonads and the olfactory epithelium were carried out on material stained with hematoxylin and eosin or Masson's trichrome procedures.

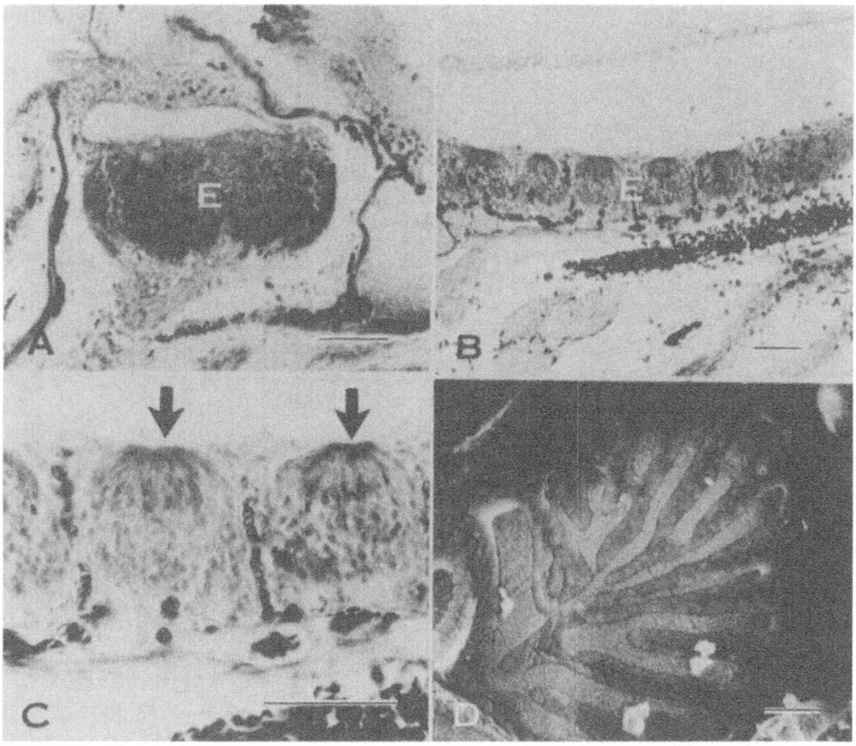

Fig. 1. A: Cross-section through the olfactory epithelium (E) of a one day old platyfish. Note the arrangement of the epithelium as a single thickened plate with the olfactory nerve extending from its base. B: Cross-section of sexually mature olfactory epithelium (E) extending from one side of the photograph to the other. Compare with Figure 1A. C: Magnified area of Figure 1B. Note the separation of neuroepithelial cells (arrows) by vascularized connective tissue septa. Goblet cells are evident on the surface between neuroepithelial cells. D: Scanning electron micrograph of the nasal epithelium from a mature fish. Fan-like radiations of sensory epithelium are separated by darker non-sensory tissue. Bar in each figure represents 50 micra. (From Schreibman et al., 1984.)

Metamorphosis of the anal fin into a gonopodium (the intromittent organ) is dependent upon increasing levels of androgens (Grobstein, 1948). Thus, determination of the stage of anal fin transformation by gross inspection is invaluable for ascertaining the state of gonadal maturation (Kallman and Schreibman, 1973; Schreibman et al., 1982a). All male platyfish, regardless of genotype, enter into stage 1 of anal fin transformation at 5 weeks of age. Stage 2 commences with the initiation of sexual maturation (early maturers, 10-11 weeks; late, 25-26 weeks). At the final stage, 6, fish are fully mature and have functional testes (early, 18 weeks; late, 52 weeks). Females follow a timetable for maturation that is comparable to males of similar P allele constitution (Schreibman and Kallman, 1977).

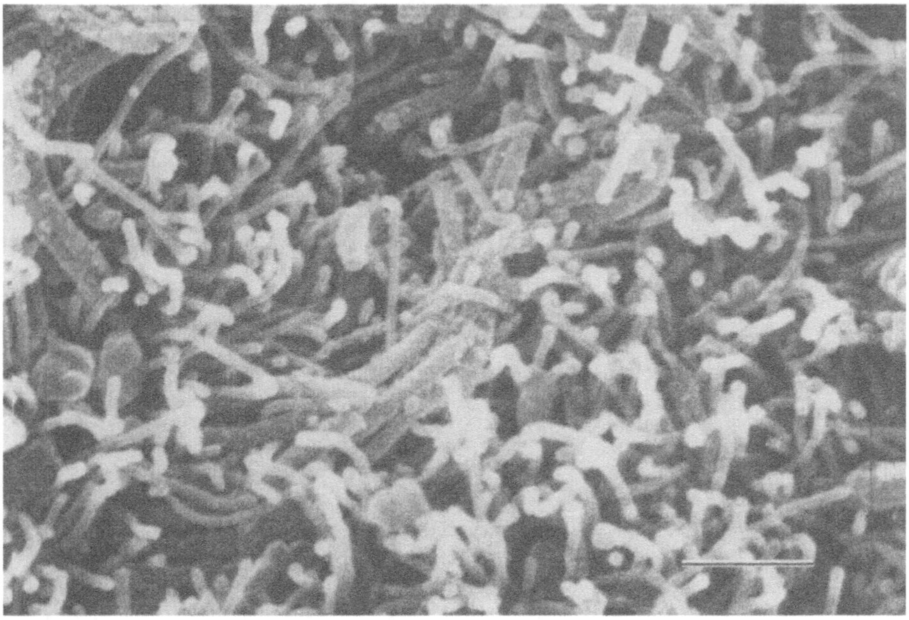

Fig. 2. Scanning electron micrograph of sensory nasal epithelium from a 6 month old platyfish. Note the variety of cilia and projections from the apices of cells. See text for description. Bar, one micron.

BIRTH TO SEXUAL MATURITY (PUBERTY)

The Olfactory System

Much of the knowledge on nasal structure in fishes has been compiled by Hara (1975), and more recently by Kleerekoper (1982) and Yamamoto (1982). In platyfish, the olfactory epithelium is located on the medial wall of each nasal cavity close to the incurrent pore which is located near the mouth (the excurrent pores are near the eyes). At birth, the olfactory epithelium in all platyfish consists of single thickened plates in each nostril and contains both neuroepithelial and non-neuroepithelial cells with few or no goblet cells on their surface (Fig. 1A). Fascicles of olfactory nerve combine close to the base of each plate and extend posteriorly as single units to the olfactory bulb which lies immediately ventral to the telencephalon (i.e., non-pedunculated type). A single blood vessel is located in proximity to the olfactory epithelium and stratified squamous epithelium covers the remaining surface of each cavity. The only changes in this histological description that occur during prepubertal stages are an increase in the size, without apparent change in cellular arrangement, of each neuroepithelial plate, and a thickening of the stratified squamous epithelium in the nasal cavities (Schreibman et al., 1984).

In sexually mature fish, the neuroepithelium, as viewed in histological section, is separated into smaller segments by highly vascularized connective tissue septa which extend from the basal surface, and a large population of mucous-secreting goblet cells is apparent on the luminal surface (Figs. 1B, 1C). The surface of the olfactory epithelial plate (approximately 0.3mm^2 in area) of adult platyfish, as viewed with the scanning electron microscope, is characterized by peninsular-like strips of sensory epithelia radiating out in a fan shape from the anterior end of the

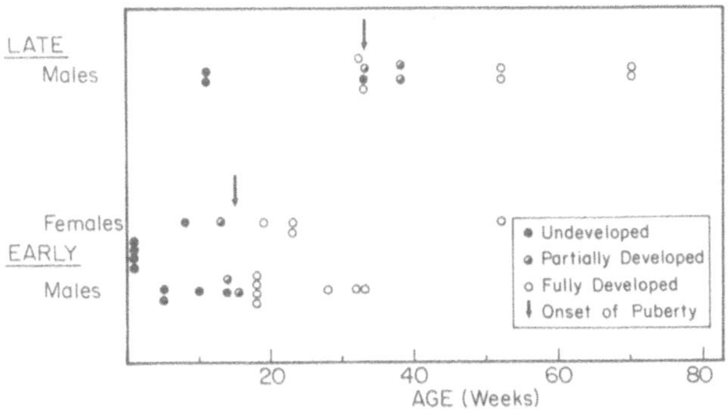

Fig. 3. Correlation of changes in the olfactory epithelium with sexual maturation and age. Transformation of nasal epithelium (circles) in early and late maturing platyfish is directly related to stage of maturation (arrows) and not chronological age.

plate (Fig. 1D). The surface of the sensory epithelium has a "shag-like" appearance generated by numerous cilia and other projections from the apices of cells. The cilia vary in length and shape from ones which are long, slender and "flexible" in appearance to broader ones that occur in clumps, are more "rigid" in appearance, and lie closer to the surface (Fig. 2). Frequently one observes several cilia projecting from a single "knob-like" structure which lies close to the surface of the epithelium (Silverman et al., 1983 and unpublished with C. Forest). The nonsensory epithelium, which separates the neuroepithelium, is composed of a distinctive irregular pattern of microridges and grooves which give it the appearance of a fingerprint. Light, spherical, slightly elevated patches are randomly distributed throughout, however, they appear to increase in number near the periphery of the plate.

The profound structural changes which characterize the transition from prepubertal to adult neuroepithelia in platyfish require several weeks to occur. This process is related to stage of sexual development (beginning at a time approximating the onset of sexual maturation) rather than chronological age, in males and females of both genotypes (Fig. 3).

Several reports for other teleosts have demonstrated that there is a difference between juvenile and adult forms in the spatial relationship of sensory and non-sensory nasal epithelia (Holl, 1965; Zeiske, 1973, 1974; Zeiske et al., 1976, 1979; Bruecker et al., 1979; Theisen et al., 1980). These differences were generally correlated with the differing locomotory habits of juveniles and adults and not with the changes occurring in the reproductive system. In eels, hormonally induced sexual precocity was shown to lead to a marked degeneration of olfactory sensory epithelia and to a reduction of its mucous cells (Pankhurst and Lythgoe, 1983). These changes, which have been associated with alterations in migratory patterns of eels related to reproduction, and probably not to the administration of the hormone, are contrary to those seen in platyfish at the time of puberty. In rainbow trout (Salmo gairdneri), it has been suggested that age-dependent increases in the size of lamellae in the olfactory epithelium are associated with physiological changes during sexual maturation, however details to support this contention are not available (Halama, 1982). Studies in a closely related species, Atlantic salmon (Salmo salar), showed that fish that had been rendered anosmic by cauterization of the olfactory mucosa went on to complete sexual development (Stabell and Refstie, 1980).

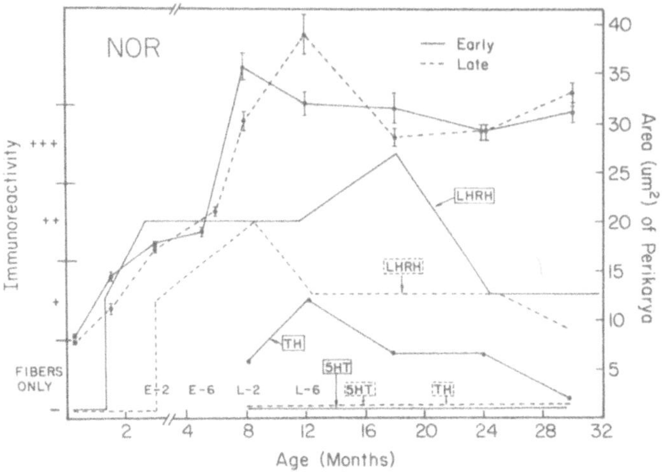

Fig. 4. Graphic representation of changes in cell area (mean ± S.E.; right ordinate) and relative intensity (- to +++) of immunoreactivity (left ordinate) in the nucleus olfactoretinalis (NOR) of male platyfish from birth to 30 months of age. See text for description. Ages at gonopodial stages 2 and 6 are indicated for early (E) and late (L) maturing genotypes; LHRH, luteinizing hormone releasing hormone; 5HT, serotonin; TH, tyrosine hydroxylase.

The Nucleus Olfactoretinalis (NOR; Nervus Terminalis)

Using antisera generated against synthetic LHRH, we have localized immunoreactive (ir-) material in three brain nuclei: the NOR, the nucleus preopticus periventricularis (NPP), and the nucleus lateralis tuberis pars posterioris (NLT) (Schreibman et al., 1979; Schreibman et al., 1982b). We believe that the NOR plays an important role in orchestrating the translation of the genetically-encoded information (P allele genotype) into the activity that is essential for the development and functioning of the reproductive system, and in transmitting or translating environmental cues, such as odor, to centers in the brain concerned with reproductive phenomena. This statement is based on observations of the anatomical position of the NOR, and of its physiological activity in male and female platyfish of early and late maturing genotype (Schreibman et al., 1982b, 1984; Halpern-Sebold and Schreibman, 1983).

In platyfish, the olfactory bulb is immediately ventral to the telencephalon; the NOR is located on the basal surface of the brain at the boundary between the ventral telencephalon and the olfactory bulb. Münz and his associates (1981, 1982) have used horseradish peroxidase (HRP) to demonstrate that the NOR sends neuronal projections to both the olfactory and optic nerves. We have demonstrated further, that ir-LHRH containing nerve processes extend between the NOR and the olfactory bulb, optic nerve, pineal region (habenular nucleus), NPP, NLT and pituitary gland in sexually mature platyfish (Schreibman et al., 1982b, 1982c, 1984).

The NOR is the first region in the brain to contain demonstrable ir-LHRH which appears at 5 weeks of age in early, and 9-11 weeks in late, maturing genotypes. Ir-LHRH then appears in the NPP and NLT at times coinciding with the initiation of sexual maturation or gonopodial stage 2 (10-11 weeks,

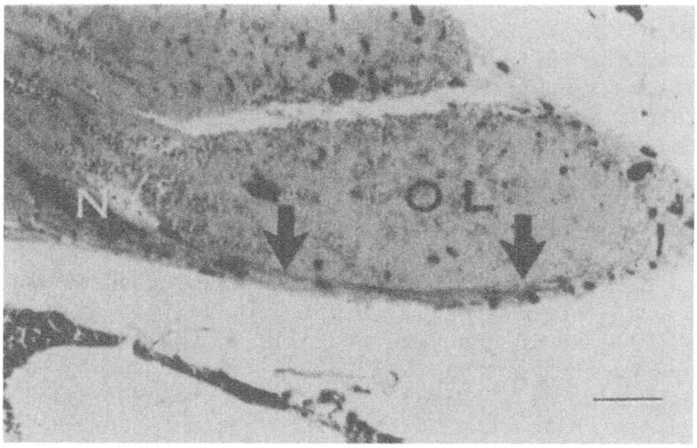

Fig. 5. Sagittal section showing ir-LHRH tracts (arrows) extending from the NOR (N) along the ventral margin of the olfactory bulb (OL). Section has been counterstained with Masson's trichrome. Anterior is the right. Bar, 50 micra. (From Schreibman et al., 1984.)

early; 25-26 weeks, late). Anatomically, the pattern of development in the brain is sequential, proceeding from anterior to posterior, and it is essentially similar in both genotypes. We have termed this process the "cascade effect" because of the apparent continuum in the pattern of development of the NOR, NPP, NLT, and pituitary gland (Schreibman et al., 1982b; Halpern-Sebold and Schreibman, 1983).

The cellular and nuclear areas of NOR perikarya in early and late maturers increase from one to five (gonopodial stage 1) weeks (Fig. 4). Ir-NOR perikarya are at their maximum number when they first appear, and this number which is similar in both genotypes, remains constant into adulthood. There are, however, significant increases in the dimensions of NOR perikarya in both genotypes from stage 1 to stage 2, a period that is characterized by the appearance of ir-fibers between the NOR, and the NPP and NLT. These observations suggest increased cellular activity, a concept supported by cytological observations of increased basophilia, paler nuclei and more prominent nucleoli at stage 2 (Halpern-Sebold, 1984; Schreibman et al., 1985). In early maturers, NOR perikaryal dimensions remain similar between stages 2 and 6 and rise to maximum values at 8 months of age. In late maturers, the area of NOR perikarya increases after stage 2 and attains its maximum value when fish reach stage 6 (12 months) (Fig. 4).

Ir-LHRH first appears in the neuronal tracts between the NOR and olfactory bulb at about 10 weeks of age in both genotypes, suggesting that this may be an age-dependent phenomenon. However, the ir-response of these tracts increases in intensity at times that coincide with the initiation of puberty (gonopodial stage 2) and with the changes in the olfactory epithelium described above (Schreibman et al., 1984). Ir-LHRH containing processes extend anteriorly from the NOR along the ventral border of the olfactory bulb and continue up to, but not beyond, the junction of the olfactory bulb and olfactory nerve (Fig. 5). For both genotypes, the greatest staining intensity is found when fish are sexually mature.

The BPG Axis

We have demonstrated that the initial appearance of ir-LHRH containing brain nuclei follows a specific sequence whose timing is related to stage of sexual maturation and not chronological age. The development of these centers precedes, and is essential for, the ontogeny of functional gonadotropes in the caudal pars distalis (CPD). In turn, these functional gonadotropes are necessary for completion of gonad maturation (Kallman and Schreibman, 1973; Schreibman and Kallman, 1977; Schreibman et al., 1985).

In the pituitary gland of neonates, both ir-GTH and ir-LHRH can be localized in the PAS-positive cells (GTH^P; c.f. Schreibman, 1986) of the pars intermedia (PI) and in clusters of cells in the lateral pockets of the CPD; gonadotropes of the ventral (v) CPD are represented by a few scattered ir-cells (Schreibman et al., 1982d). The gonadotropes of the PI are numerous at 1 week of age while those of the vCPD begin to appear just prior to the onset of sexual maturation and then increase markedly in number, size and activity. There is a direct correlation between the total number of ir-LHRH containing neurons in the three brain areas and the number of ir-GTH containing CPD and GTH^P cells in the pituitary of both early and late maturers from one week to adulthood, and early maturers have a greater number of ir-CPD cells than late maturers (Halpern-Sebold, 1984; Halpern-Sebold et al., 1983, 1984a).

The testis in neonatal platyfish consists of 2 lobes composed of small clusters (cysts) of spermatogonia surrounding a central stroma. Enzymes necessary for sex steroid formation (3 beta-hydroxysteroid dehydrogenase, 3BHSD and glucose-6-phosphate dehydrogenase, G6PD) have been identified histochemically in Leydig cells located in the central stroma and among the spermatogonia in the two lateral regions of the testes. As testicular development continues, spermatogonia proliferate and efferent ducts appear (Schreibman et al., 1982a). All immature males which are still in stage 1 of gonopodial development have testes that have progressed in spermatogenesis up to, but not including, the primary spermatocyte stage (Kallman and Schreibman, 1973; Schreibman and Kallman, 1977). With the initiation of the process of sexual maturation, cells containing steroid synthesizing enzymes become more numerous and display greater activity at the periphery of the testes and in the stroma around efferent ducts.

In female platyfish, oogenesis proceeds only up to the yolk deposition stage in the absence of functional gonadotropes in the vCPD (Schreibman and Kallman, 1977). In 3 day old females, 3BHSD and G6PD activities are found in the stroma of gonads comprised of previtellogenic oogonia. With subsequent ovarian development, these cells apparently increase in number and activity (Schreibman et al., 1982a).

MATURITY TO SENESCENCE

Olfactory System

Our study of aging changes in the structure of the olfactory system is in its preliminary phase and as a result the extent of our knowledge is limited. Generally, the changes in the olfactory epithelium observed at the scanning electron microscope level show considerable variability, and tissue from aging platyfish has ranged from nearly normal to abnormal in appearance. However, with increasing age there is usually a breakdown in the structural uniformity and arrangement of both sensory and nonsensory areas that comprise the olfactory epithelial plate (Silverman et al., 1983 and unpublished observations). In two and three year old platyfish, significant alterations are represented by a marked decrease in the density of

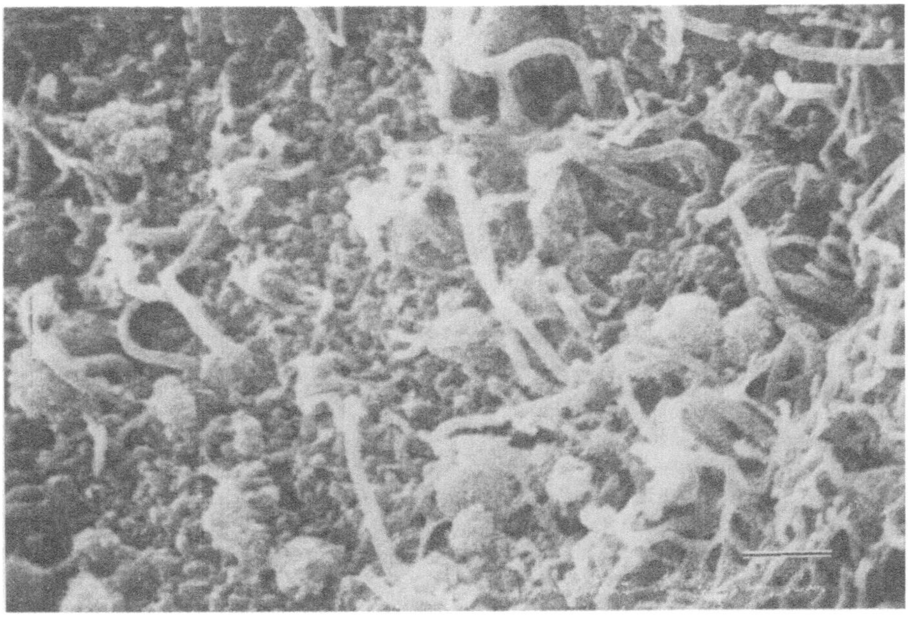

Fig. 6. Scanning electron micrograph of sensory nasal epithelium of a 2 year old platyfish. Compare with Figure 2 and see text. Bar, one micron.

projections from the neuroepithelium resulting in a loss of the dense "shag-like" pattern typical of younger fish; however, both thin and thick cilial elements and olfactory "knobs" that are characteristic of younger fish are seen (Fig. 6). There is also an increase in the intertwining of cilial elements in old fish, and in many cases white spherical masses of unknown function or etiology appear on the neuroepithelial surface. Nonsensory epithelium from aging platyfish shows a flattening of the microridges and frequent interruptions of their generally continuous pattern. Scanning and transmission electron micrographs of the olfactory epithelium from the closely related swordtail, Xiphophorus helleri, have been published (Zeiske et al., 1976). However, these are of limited value since the age, sex and origin of the subjects examined were not reported.

At the histological level, the neuroepithelium of aging platyfish appears to be recessed below the nonsensory epithelium. The typical uniform pattern that is seen in cross section of the apical end of sensory epithelium from young fish, is replaced by an irregular one which probably reflects the changes in cilial density and structure observed at the SEM level. We are currently determining if age-related changes occur in the number or type of cells that compose the olfactory epithelium in order to better understand the relationship between function and the structural changes that occur with increasing age. It is known that several morphological classes of receptor and epithelial cells exist in fish, that receptor cell turnover probably persists through life, and that the variation in cell type and their distribution may be as diverse as the number of species of teleosts and their habitats (Hara, 1971; Zeiske et al., 1976, 1979; Thommesen, 1983; Muller and Marc, 1984).

The NOR

In early maturing males, the average area of NOR cell bodies reaches a maximum value at 8 months and then decreases gradually between 8 and 24 months, while the nuclear area decreases between 8 and 30 months. Late maturing males differ from early maturers in several parameters related to nerve cell dimensions (Fig. 4). In addition, cytoplasmic basophilia at 8 months is somewhat darker in late (not yet mature) than in early maturers (sexually mature); however, as in early maturers, nuclei are consistently round and pale staining, and nucleoli are prominent at all ages (see Schreibman et al., 1985).

In early maturing males, the intensity of ir-LHRH in NOR perikarya and fibers is similar at 8 and 12 months (Fig. 4). At 18 months, there is a significant increase in the immunoreactive response, which later drops to low levels at 24 and 30 months. By contrast, in early maturing females, immunoreactivity is consistently light at all ages between 8 and 30 months (Halpern-Sebold et al., 1984b; Schreibman et al., 1985). In late maturing males, ir-LHRH is most intense at 8 months; however, late maturers at this age are still sexuallly immature. Ir-LHRH intensity then decreases to a constant level between 12 and 24 months and then declines slightly at 30 months. Thus, the increase in ir-LHRH intensity seen in the NOR of early maturers at 18 months does not occur in late maturing males (Fig. 4). The intensity of the ir-response to anti-LHRH in the neuronal processes of the olfactory bulb, or the distribution of these processes, does not change with age in the groups we examined.

Ir-tyrosine hydroxylase (TH; Joh et al., 1973), an enzyme essential for synthesis of catecholamines, has been localized in perikarya of the NOR, but not in the NPP or NLT, in early maturing male and female platyfish (Halpern-Sebold et al., 1985). Processes containing ir-TH extend dorsocaudally and ventrocaudally from NOR perikarya, but not anteriorly into the olfactory bulb. In early maturers, ir-TH can be seen in NOR perikarya only at 12 months, although ir-material is present in NOR processes at all ages from 8 to 24 months (Fig. 4 and unpublished observations). Ir-TH is not present in the NOR of late maturing males at any age.

Although ir-serotonin (5HT; Immunoclear, Inc.) has been localized in other forebrain nuclei (Margolis-Kazan et al., 1984, 1985; Margolis-Nunno et al., 1985) we have not identified immunoreactivity to anti-5HT in the NOR of platyfish of either genotype at all ages examined (Fig. 4 and unpublished observations).

The BPG Axis

There are distinct changes in the distribution and relative quantity of neuropeptides, neurotransmitters and pituitary and gonadal hormones, as well as in the structure of the cells that manufacture, transport or are target organs for them during the life span of platyfish. These events, many of which are sexually dimorphic and genotype dependent, have been recently summarized (Schreibman et al., 1985). With increasing age (8 to 30 months) there are dramatic fluctuations in the distribution of ir-LHRH and ir-5HT in the brain, with major changes occurring at 18 and/or 24 months. These aging shifts follow, or occur concomitantly with, distinct changes in the morphometry, cytology and/or immunocytochemistry (LHRH, GTH, 5HT) of pituitary gonadotropes. These age-related alterations in the brain and pituitary gland could conceivably be translated into the "degenerative" transformations which are noted in the gonads, especially in the testes of early maturers at 18 months (e.g., the distortion of the highly ordered structure of spermatozeugmata, the infiltration of tubular elements by connective tissue, and the decrease in metabolic and steroidogenic enzymes). However, we have demonstrated that in the oldest animals examined (up to 54 months), testes and ovaries retain the capacity for gametogenesis and the pituitary distribution of ir-GTH and ir-LHRH is the same as in younger fish (Schreibman et al., 1983).

The interdigitating cyclical events which occur during platyfish development and senescence suggest that there is an exchange of information among the components of the BPG axis. The olfactory system needs to be examined in the light of these changes in order to determine the specific sequence of events, and cause and effect relationships between olfactory and neuroendocrine systems, especially at 18 and 24 months when marked changes occur in the brain, pituitary and gonads. These parameters need to be investigated in both males and females since our observations suggest that patterns of aging of the brain, pituitary and gonads are not the same in both sexes (Margolis-Kazan and Schreibman, 1984; Halpern-Sebold et al., 1984b; Schreibman et al., 1985).

EXPERIMENTAL MANIPULATIONS

We have begun to explore the nature of olfactory-brain-pituitary-gonad (OBPG) axis function by performing experimental manipulations which, although in their initial phases, further support our contention that a fundamental structural and functional association exists among the components of the OBPG axis.

Lesioning of the NOR

The NOR of sexually mature male and female platyfish was bilaterally destroyed with heat. Three to five weeks following surgery, testes and ovaries appeared markedly reduced in size, and no yolky oocytes were seen in the majority of females.

Histological analysis of the nasal epithelium of NOR-destroyed (confirmed by histological examination) fish shows extreme convolution and a dramatic reduction in the number of apical projections, as well as an increase in pyknotic nuclei and a decrease in vascular elements (unpublished observations with F. Caracheo). In histological preparations, testes displayed active spermatogenesis, and Leydig and Sertoli cells surrounding developing spermatogenic cysts were greatly enlarged. Spermatozeugmata (packets of mature sperm) were reduced in number relative to other stages and they were located in large numbers in the efferent ducts. Ovaries lacked yolky eggs and the most advanced stages of oogenesis (where yolk deposition was only just beginning) were found discharged into ovarian ducts rather than positioned in the stroma as was observed in control fish. Thecal cells were generally hypertrophied.

The above histological picture suggests that following the destruction of the NOR, gametes are actively produced and released without storage, or in the case of females, without completion of vitellogenesis. These observations in platyfish appear to be related to those in goldfish where the release of sperm followed the electrical stimulation of the medial olfactory tracts (Demski et al., 1982; Demski and Dulka, 1984) which contain nervus terminalis processes (Demski, 1984). In goldfish, the medial olfactory tracts must be intact in order for normal courtship behavior to occur (Stacey and Kyle, 1983), and Demski and Northcutt (1983) have suggested that they are essential for the male response to female goldfish pheromones. In goldfish, terminal nerve fibers terminate in the supracommissural nuclei; bilateral lesioning of the area ventralis telencephali pars supracommissuralis or the area ventralis telencephali pars ventralis, but not the NPP (the NOR was not examined), reduces the number of spawning males (Kyle and Peter, 1982). Similarly, in male platyfish, mating behavior is affected when their forebrains are lesioned (Kamrin and Aronson, 1954). Demski and Hornby (1982) suggest that in goldfish the preoptic area controls sperm release during spawning while the supracommissural telencephalon is involved in sex arousal. The system for sperm release in goldfish, as well as in other teleosts, appears to be sympathetic, with neural pathways extending between the preoptic area to the rostral spinal cord and into the muscle layers of the sperm ducts (Demski and Hornby, 1982; Dulka and Demski, 1982; Demski, 1982, 1984; Demski and Sloan, 1985). Our observations in platyfish suggest that a neuronal pathway may extend from the NOR and perhaps from the olfactory epithelium itself (see below). Octapeptides must also be considered in elucidating the mechanisms that regulate sperm release. This is suggested by our localization of arginine vasotocin, neurophysin and oxytocin in the NPO of platyfish (Schreibman and Halpern, 1980 and unpublished observations), an area which is also rich in neuronal tracts from the NOR (Halpern-Sebold and Schreibman, 1983), and the demonstration that neurohypophysial hormones can stimulate the contraction of smooth muscle in the gonads of teleosts and thus induce spawning responses (Heller, 1972; c.f. Liley and Stacey, 1983). Whether the observations of gamete release in platyfish are strictly neuronal (perhaps with LHRH and/or neurohypophysial factors acting as a neurotransmitter), or whether other hormonal (tropic) components also operate, raises exciting questions for further study.

Destruction of the Nasal Epithelium

Sexually mature male and female platyfish were rendered anosmic by bilateral surgical removal, or cautery, of the olfactory epithelium. Surgical procedures were confirmed by histological analysis. In all fish examined, at least 3 weeks following the experimental procedure, the immunoreactive response to anti-LHRH was markedly intensified in NOR cell bodies and in their dorsal and posterior projections (unpublished with F. Caracheo). In addition, in fish where the nasal epithelium was removed by surgery, and not in those where it was destroyed by cautery, enhanced ir-LHRH staining was also localized in tracts on the ventral region of the olfactory bulb between the NOR and the inception of the olfactory nerve. At autopsy, gonads of these fish appeared either normal or reduced in size. Gonads and pituitary glands have not yet been examined histologically.

Our preliminary observations of male and female platyfish following NOR lesioning or olfactory epithelium removal support our hypothesis that a basic neuroendocrine system of communication is present between the NOR and the nose. Our results further suggest that information can proceed in either direction between these two structures to affect the development, functioning and senescence of the reproductive system. This work is currently being expanded and subjected to additional scrutiny.

SUMMARY AND SUGGESTIONS FOR FUTURE RESEARCH

Our study, which correlates changes in the olfactory system with events occurring in the BPG axis during the life span of platyfish, demonstrates that an important primary axis of neuroendocrine interaction exists between the olfactory and reproductive systems and that the NOR (nervus terminalis) serves as a crucial relay center. This investigation of the olfactory-brain-pituitary-gonad (OBPG) axis from birth to senescence in male and female platyfish genetically constructed to reach sexual maturity at different ages, indicates clearly that profound structural (and thus presumably functional) modifications occur in the olfactory epithelium in relation to specific stages of sexual maturation that are independent of chronological age. In addition, the structure of the olfactory epithelium undergoes what appears to be degenerative modifications at times when there are age-related alterations taking place in the structure and function of the reproductive system. Considerably more work is required to determine whether any correlations exist between these observations.

We have clearly established that the NOR is an especially important component of the OBPG axis in immature, mature, and senescent platyfish. The evidence strongly suggests that the NOR initiates physiological events in the BPG axis that are essential for puberty to occur and that it probably continues to play a decisive role in reproductive physiology during adulthood (c.f. Schreibman et al., 1985). In addition, by virtue of its anatomical position and structural features, the NOR may be capable of sorting out, analyzing and translating information received by the olfactory system (and other receptors of environmental cues) into reproductive system activity.

In short, we believe that the NOR can serve as the structural and functional link between the olfactory and reproductive systems. It is obvious, however, that we need to learn more of the specific mechanisms of interaction between the NOR and the olfactory system. Among the more basic questions that require clarification are:

1. Where, and equally important, how, are the events that culminate in sexual maturation initiated in the OBPG axis?

2. What is the specific role of the olfactory system in the process of puberty, and what is the nature of its interaction with the brain, pituitary gland and gonads?

3. What is the direction of "information" transfer between the olfactory and neuroendocrine systems, especially with regard to the LHRH-containing brain nuclei?

4. What are the specific cytological transformations that occur in the olfactory epithelium during aging? Is there a division of labor of the cell types that comprise the neuroepithelium so that some function solely in reproductive system-related phenomena?

5. What happens in the platyfish olfactory system when the dramatic age-related fluctuations in neuroendocrine activity occur at 18 and 24 months of age?

6. How is the olfactory system related to the "clock" for sexual maturation (and perhaps reproductive aging) that resides within the genotype, and how is it involved with the yet to be determined site and mechanism of genetic control of development and aging of the BPG axis?

The system we are describing is not restricted to platyfish; it has significance in all classes of vertebrates. The NOR is presumably homologous to the nervus terminalis, a cranial nerve that has been conserved throughout vertebrate evolution. They are both similar in their ontongeny, morphology and cytology, including immunocytochemistry (Larsell, 1918; Pearson, 1941; Schwanzel-Fukuda and Silverman, 1980; Schwanzel-Fukuda et al., 1981; Crapon de Caprona and Fritzch, 1983; Demski and Northcutt, 1983; Fernald and Finger, 1984; Stell et al., 1984). Although the terminal nerve was described in mammals as early as 1913 (Huber and Guild) it is only in recent years that it has become prominent in studies related to sexual maturation, reproductive behavior, olfaction, vision, and comparative neurophysiology and neuroanatomy (Schwanzel-Fukuda et al., 1981, 1985; Münz et al., 1981, 1982; Silverman et al., 1982; Stell et al., 1982, 1984; Demski and Northcutt, 1983; Springer, 1983; Stacey and Kyle, 1983; Witkin and Silverman, 1983; Demski and Dulka, 1984; Ekstrom, 1984; Fernald and Finger, 1984; Schreibman et al., 1984). Studies in rats suggest that the ir-LHRH containing terminal nerve, which is also linked to the olfactory system, may influence the development of the brain-pituitary-gonad axis (Schwanzel-Fukuda et al., 1985). Thus, striking similarities exist in the OBPG axis in fish and mammals.

The observations reported in this chapter provide support for, and facilitate the interpretation of, the wealth of data amassed from studies on the olfactory system begun almost 250 years ago (cf. Kleerekoper, 1982) which have dealt with the description of the structure of the olfactory epithelium, the neuroanatomical links between the nose and the brain, the importance of scent and pheromone detection in animal behavior, and with experimentally induced changes in the olfactory system. Much of the work we have presented is in its early phases of study. Thus, this chapter is not meant to be definitive, but rather we hope that by presenting new concepts, approaches and questions, we will serve to instigate study (perhaps with a different direction, scope or emphasis) on the structural and functional interrelationships of the olfactory and reproductive systems in all classes of vertebrates.

ACKNOWLEDGMENTS

The research of our laboratory has been supported by the National Institute on Aging (AGO 1938), the National Science Foundation (PCM 77-15981) and the City University of New York (PSC-CUNY).

REFERENCES

Breucker, H., Zeiske, E., and Melinkat, R., 1979, Development of the olfactory organ in the rainbow fish, Nematocentris maccullochi (Atheriniformes, Melantaeniidae), Cell Tis. Res., 200:53.
Crapon de Caprona, M. D., and Fritzch, B., 1983, The development of the retinopetal nucleus olfactoretinalis of two cichlid fish as revealed by horseradish peroxidase, Dev. Brain Res., 11:281.
Demski, L. S., 1982, Electrical stimulation of the brain, in: "Fish Neurobiology and Behavior," R. G. Northcutt, and R. E. Davis, eds., Univ. of Michigan Press, Ann Arbor, Michigan.
Demski, L. S., 1984, The evolution of neuroanatomical substrates of reproductive behavior: Sex steroid and LHRH-specific pathways including the terminal nerve, Am. Zool., 24:809.
Demski, L. S., and Dulka, J. G., 1984, Functional-anatomical studies on sperm release evoked by electrical stimulation of the olfactory tract in goldfish, Brain Res., 291:241.
Demski, L. S., and Hornby, P. J., 1982, Hormonal control of fish reproductive behavior: brain-gonadal steroid interactions, Can. J. Fish. Aquat. Sci., 39:36.
Demski, L. S., and Northcutt, R. G., 1983, The terminal nerve: A new chemosensory system in vertebrates? Science, 220:435.
Demski, L. S., and Sloan, H. E., 1985, A direct magnocellular-preopticospinal pathway in goldfish: Implications for control of sexual behavior, Neurosci. Letters, in press.
Demski, L. S. Dulka, J. G., and Northcutt, R. G., 1982, Chemosensory control of spawning mechanisms in goldfish, Neuroscience, Abstract, 8:611.
Dulka, J. G., and Demski, L. S., 1982, Sperm duct contractions correlated with sperm release evoked by brain stimulation in goldfish, Am. Zool., 22:887.
Ekstrom, P., 1984, Central neural connections of the pineal organ and retina in the teleost, Gasterosteus aculeatus L., J. Comp. Neurol., 226:321.
Fernald, R. D., and Finger, T. E., 1984, Catecholaminergic neurons of the locus coeruleus project to the ganglion cells of the nervus terminalis (NT) in goldfish, Neuroscience, Abstract, 10:50.
Grobstein, D., 1948, Endocrine and developmental studies of gonopod differentiation in certain poeciliid fishes. I. The structure and development of the gonopod in Platypoecilus maculatus, Univ. Calif. Publ. Zool., 47:1.

Halama, L., 1982, Morphometry of the olfactory organs of the rainbow trout, Salmo gairdneri, So. Acta. Biol. Cracov. Ser. Zool. 105:112.
Halpern-Sebold, L. R., 1984, The ontogeny and functional significance of luteinizing hormone releasing hormone (LHRH) containing centers of the brain of the freshwater teleost, Xiphophorus maculatus, Unpublished Ph.D. Dissertation, City University of New York, New York.
Halpern-Sebold, L. R., and Schreibman, M. P., 1983, Ontogeny of centers containing luteinizing hormone-releasing hormone in the brain of platyfish Xiphophorus maculatus as determined by immunocytochemistry, Cell Tis. Res., 229:75.
Halpern-Sebold, L., Schreibman, M. P., Margolis-Kazan, H., and Goos, H. J. Th., 1983, A relationship between the number of luteinizing hormone releasing hormone (LHRH) containing neurons and gonadotrops in platyfish, Am. Zool., Abstract, 23:882.
Halpern-Sebold, L., Schreibman, M. P., and Margolis-Kazan, H., 1984a, Ontogeny of luteinizing hormone releasing hormone (ir-LHRH) containing centers in the brain and their relationship to sexual maturation, Neuroscience, Abstract, 10:456.
Halpern-Sebold, L. R., Schreibman, M. P., and Margolis-Kazan, H., 1984b, Sexually dimorphic age-related changes in the distribution of immunoreactive LHRH, Am. Zool., Abstract, 24:71A.
Halpern-Sebold, L. R., Margolis-Kazan, H., Schreibman, M. P., and Joh, T., 1985, Immunoreactive tyrosine hydroxylase in the brain and pituitary gland of the platyfish, Proc. Soc. Exper. Biol. Med., 178:486.
Hara, T. J., 1971, Chemoreception, in: "Fish Physiology," W. S. Hoar, and D. J. Randall, eds., Academic Press, New York.
Hara, T. J., 1975, Olfaction in fish, Prog. Neurobiol., 5:271.
Heller, H., 1972, The effect of neurohypophysial hormones on the reproductive tract of lower vertebrates, Gen. Comp. Endocrinol., Suppl., 3:703.
Holl, A., 1965, Vergleichende morphologische und histologische Untersuchungen am Geruchsorgan der Knochenfische, Z. Morph. Oekol. Tiere, 54:707.
Huber, G. E., and Guild, S. R., 1913, Observations on the peripheral distribution of the nervus terminalis in mammals, Anat. Rec., 7:253.
Joh, T. H., Gregham, G., and Reis, D. J. C., 1973, Immunochemical demonstration of increased tyrosine hydroxylase protein in sympathetic ganglia and adrenal medulla elicited by reserpine, Proc. Nat. Acad. Sci., 70:2767.
Kallman, K. D., 1975, The Platyfish, Xiphophorus maculatus, in: "Handbook of Genetics," R. C. King, ed., Plenum Publishing Corporation, New York.
Kallman, K. D., and Borkoski, V., 1978, A sex-linked gene controlling the onset of sexual maturation in female and male platyfish, Xiphophorus maculatus, fecundity in females and adult size in males, Genetics, 89:79.
Kallman, K. D., and Schreibman, M. P., 1973, A sex-linked gene controlling gonadotrope differentiation and its significance in determining the age of sexual maturation and size of the platyfish, Xiphophorus maculatus, Gen. Comp. Endocrinol., 21:287.
Kallman, K. D., Schreibman, M. P., and Borkoski, V., 1973, Genetic control of gonadotrope differentiation in the platyfish, Xiphophorus maculatus (Poeciliidae), Science, 58:678.
Kamrin, R. P., and Aronson, L. R., 1954, The effect of forebrain lesions on mating behavior in the male platyfish, Xiphophorus maculatus, Zoologica, 39:133.
Kleerekoper, H., 1982, Research on olfaction in fishes: Historical aspects, in: "Chemoreception in Fishes," T. J. Hara, ed., Elsevier, New York and Amsterdam.
Kyle, A. L., and Peter, R. E., 1982, Effects of forebrain lesions on spawning behaviour in the male goldfish, Physiol. Behav., 28:1103.

Larsell, O., 1918, Studies on the nervus terminalis: Mammals, J. Comp. Neurol., 30:423.
Leatherland, J. F., 1970, Histological investigation of pituitary homotransplants in the marine form Trachurus of the three-spine stickleback, Gasterosteus aculeatus L., Z. Zellforsch., 104:337.
Liley, N. R., and Stacey, N. E., 1983, Hormones, pheromones, and reproductive behavior in fish, in: "Fish Physiology," W. S. Hoar, D. J. Randall, and E. M. Donaldson, eds., Academic Press, New York.
Margolis-Kazan, H., and Schreibman, M. P., 1981, Cross-reactivity between human and fish pituitary hormones as demonstrated by immunocytochemistry, Cell Tis. Res., 221:257.
Margolis-Kazan, H., and Schreibman, M. P., 1984, Sexually dimorphic, age-related changes in pituitary gonadotrope distribution, Mech. Ageing Devel., 24:325.
Margolis-Kazan, H., Halpern-Sebold, L. R., and Schreibman, M. P., 1984, Aging changes in the distribution of immunoreactive serotonin (ir-5HT) in platyfish, Am. Zool., Abstract, 24:70A.
Margolis-Kazan, H., Halpern-Sebold, L. R., and Schreibman, M. P., 1985, Immunocytochemical localization of serotonin in the brain and pituitary gland of the platyfish, Xiphophorus maculatus, Cell Tis. Res., in press.
Margolis-Nunno, H., Halpern-Sebold, L., and Schreibman, M. P., 1985, Immunocytochemical changes in serotonin in the forebrain and pituitary of aging fish, Neurobiol. Aging, in press.
Muller, J. F., and Marc, R. E., 1984, Three distinct morphological classes of receptors in fish olfactory organs, J. Comp. Neurol., 222:482.
Münz, H., Stumpf, W. E., and Jennes, L., 1981, LHRH systems in the brain of platyfish, Brain Res., 221:1.
Münz, H., Claas, B., Stumpf, W. E., and Jennes, L., 1982, Centrifugal innervation of the retina by luteinizing hormone releasing hormone (LHRH)-immunoreactive telencephalic neurons in teleostean fishes, Cell Tis. Res. 222:313.
Pankhurst, N. W., and Lythgoe, J. N., 1983, Changes in vision and olfaction during sexual maturation in the european eel, Anguilla anguilla L., J. Fish Biol., 23:229.
Pearson, A. A., 1941, The development of the nervus terminalis in man, J. Comp. Neurol., 75:39.
Schreibman, M. P., 1964, Studies on the pituitary gland of Xiphophorus maculatus, the platyfish, Zoologica, 49:217.
Schreibman, M. P., 1986, Pituitary gland morphology, in: "Vertebrate Endocrinology: Fundamentals and Biomedical Implications," P. K. T. Pang, and M. P. Schreibman, eds., Academic Press, New York, in press.
Schreibman, M. P., and Halpern, L. R., 1980, The demonstration of neurophysin and arginine vasotocin by immunocytochemical methods in the brain and pituitary gland of the platyfish, Xiphophorus maculatus, Gen. Comp. Endocrinol., 40:1.
Schreibman, M. P., and Kallman, K. D., 1977, The genetic control of the pituitary-gonadal axis in the platyfish, Xiphophorus maculatus, J. Exp. Zool., 200:277.
Schreibman, M. P., and Kallman, K. D., 1978, The genetic control of sexual maturation in the teleost, Xiphophorus maculatus (Poeciliidae); a review, Ann. Biol. Anim. Bioch. Biophys., 18:957.
Schreibman, M. P., Halpern, L. R., Goos, H. J. Th., and Margolis-Kazan, H., 1979, Identification of luteinizing hormone-releasing hormone (LHRH) in the brain and pituitary gland of a fish by immunocytochemistry, J. Exp. Zool., 210:153.
Schreibman, M. P., Berkowitz, E. J., and Hurk, R. van den, 1982a, The histology and histochemistry of the platyfish Xiphophorus maculatus testis and ovary from birth to sexual maturity, Cell Tis. Res., 224:81.
Schreibman, M. P., Margolis-Kazan, H., Halpern-Sebold, L., and Goos, H. J. Th., 1982b, The functional significance of nucleus olfactoretinalis in the platyfish Xiphophorus maculatus, in: "Proceedings of the

International Symposium on Reproductive Physiology of Fishes," C. J. J. Richter, and H. J. Th. Goos, eds., Pudoc, Wacheningen.
Schreibman, M. P., Margolis-Kazan, H., Halpern-Sebold, L., O'Neill, P. A., and Caracheo, F., 1982c, An LHRH containing center in the brain, connecting olfactory and reproductive systems, Am. Zool., 22:856.
Schreibman, M. P., Margolis-Kazan, H., and Halpern-Sebold, L. R., 1982d, Immunoreactive gonadotropin and luteinizing hormone-releasing hormone in the pituitary gland of neonatal fish, Gen. Comp. Endocrinol., 47:385.
Schreibman, M. P., Margolis-Kazan, H., Bloom, J. L., and Kallman, K. D., 1983, Continued reproductive potential in aging platyfish as demonstrated by the persistence of gonadotropin, luteinizing hormone-releasing hormone and spermatogenesis, Mech. Ageing Devel., 22:105.
Schreibman, M. P., Margolis-Kazan, H., Halpern-Sebold, L., O'Neill, P. A., and Silverman, R. C., 1984, Structural and functional links between olfactory and reproductive systems: puberty-related changes in olfactory epithelium, Brain Res., 302:180.
Schreibman, M. P., Margolis-Nunno, H., and Halpern-Sebold, L., 1985, Aging of the neuroendocrine system, in: "Hormones and Reproduction in Fishes, Amphibians and Reptiles," D. O. Norris, and R. E. Jones, eds., Plenum Press, New York.
Schwanzel-Fukuda, M., and Silverman, A. J., 1980, The nervus terminalis of the guinea pig: a new luteinizing hormone-releasing hormone (LHRH) neuronal system, J. Comp. Neurol., 191:213.
Schwanzel-Fukuda, M., Robinson, J. A, and Silverman, A. J., 1981, The fetal development of the luteinizing hormone-releasing hormone (LHRH) neuronal systems of the guinea pig brain, Brain Res. Bull., 7:293.
Schwanzel-Fukuda, M., Morrell, J. I., and Pfaff, D. W., 1985, Ontogenesis of neurons producing luteinizing hormone-releasing hormone (LHRH) in the nervus terminalis of the rat, J. Comp. Neurol., in press.
Silverman, A. J., Antunes, J. L., Abrams, G. M., Nilaver, G., Thau, R., Robinson, J. A., Ferin, M., and Krey, L. C., 1982, The luteinizing hormone-releasing hormone pathways in rhesus, Macaca mulatta and pig-tailed, Maccaca nemestrina monkeys, J. Comp. Neurol., 211:309.
Silverman, R. C., Schreibman, M. P., and Forest, C. L., 1983, A scanning electron microscope study of the olfactory epithelium in mature and aging platyfish, Xiphophorus maculatus, Am. Zool., 23:926.
Springer, A. D., 1983, Centrifugal innervation of goldfish retina from ganglion cells of the nervus terminalis, J. Comp. Neurol., 214:404.
Stabell, O. B., and Refstie, T., 1980, A note on the significance of the olfactory sense upon sexual maturation in atlantic salmon, Salmo salar, Aquaculture, 21:165.
Stacey, N. E., and Kyle, A. L., 1983, Effects of olfactory tract lesions on sexual and feeding behavior in the goldfish, Physiol. Behav., 30:621.
Stell, W. K., Chohan, K. S., Lam, D. M. K., and Kozlowski, G. P., 1982, Luteinizing hormone-releasing hormone (LHRH)-immunoreactive fibers in goldfish retina, Invest. Opthal. Vis. Sci., 22:278.
Stell, W. K., Walker, S. E., Chohan, K. S., and Ball, A. K., 1984, The goldfish nervus terminalis: an LHRH- and FMRFamide-immunoreactive olfactoretinal pathway, Proc. Nat. Acad. Sci., 81:940.
Sternberger, L. A., Hardy, P. H., Cuculis, J. J., and Meyer, H. G., 1970, The unlabeled antibody enzyme method of immunohistochemistry: preparation and properties of soluble antigen-antibody complex (horseradish peroxidase) and its use in identification of spirochetes, J. Histochem. Cytochem., 18:315.
Theisen, B., Breucker, H., Zeiske, E., and Melinkat, R., 1980, Structure and development of the olfactory organ in the garfish, Belone belone L. (Teleostei, Atheriniformes), Acta Zool., 61:161.
Thommesen, G., 1983, Morphology, distribution, and specificity of olfactory receptor cells in salmonid fishes, Acta Physiol. Scand., 117:241.
Wedekind, H. P., 1978, Riesenzellen im bulbus olfactorius von Teleostiern, Zoomorphologie, 91:217.

Witkin, J. W., and Silverman, A. J., 1983, Luteinizing hormone-releasing hormone in rat olfactory systems, J. Comp. Neurol., 218:426.
Yamamoto, M., 1982, Comparative morphology of the peripheral organs in teleosts, in: "Chemoreception in Fishes," T. J. Hara, ed., Elsevier, New York and Amsterdam.
Zeiske, E., 1973, Morphologische Untersuchungen am Geruchsorgan von Zahnkarpzen (Pisces, Cyprinodontoidea), Z. Morph. Tiere, 74:1.
Zeiske, E., 1974, Morphologische und morphometrische Untersuchungen am Geruchsorgan oviparer Zahnkarpfen (Pisces), Z. Morph. Tiere, 77:19.
Zeiske, E., Melinkat, R., Breucker, H., and Kux, J., 1976, Ultrastructural studies on the epithelia of the olfactory organ of cyprinodonts (Teleostei, Cyprinodontoidea), Cell Tis. Res., 172:245.
Zeiske, E., Breucker, H., and Melinkat, R., 1979, Gross morphology and fine structure of the olfactory organ of rainbow fish (Atheriniformes, Melantotaeniidae), Acta Zool., 60:173.
Zivin, J. A., and Bartko, J. J., 1976, Statistics for disinterested scientists, Life Sciences, 18:15.

THE EVOLUTION OF SALAMANDER COURTSHIP PHEROMONES

Lynne D. Houck

The University of Chicago
Department of Biology
940 East 57th Street
Chicago, IL 60637 USA

INTRODUCTION

Among vertebrates, the importance of chemical signals and chemosensory perception is unquestioned, especially for mammals (Birch, 1974b; Doty, 1976; Muller-Schwarze and Mozell, 1977). Chemosensory perception also is well documented for many reptiles (Burghardt, 1970), fish (Bardach and Todd, 1970; Liley, 1982; Pfeiffer, 1974; Stacey, 1983), and even some birds (Grubb, 1974; Stager, 1967; Walcott and Lednor, 1983). It is less widely appreciated that the perception and production of chemical information represents a critical mode of communication for amphibians (Madison, 1977), salamanders in particular. Salamanders produce chemical secretions from a glandular skin, from specialized glands located on the head, or from discrete glands within the cloaca (e.g., Brodie, 1968; Noble, 1931; Sever, 1976). For many salamanders, skin secretions include noxious poisons that are effective at repelling potential predators (Labanick and Brandon, 1981; Madison, 1977). Within species, salamanders use chemical perception as one of several significant cues when migrating long distances to breeding sites (Dolmen, 1982; Grant et al., 1968; Kleeberger and Werner, 1982; Twitty, 1959, 1961) or when returning to "home" areas from which they have been displaced (Holomuzki, 1982; Madison 1969, 1972; Madison and Shoop, 1970). Forester (1977, 1979) has shown that females of at least one plethodontid species (Desmognathus ochrophaeus) returned from months of foraging in the forest litter to the precise streamside site where they had oviposited in a previous year. Whether or not chemosensory information is the primary cue in breeding migrations or long-distance homing, chemical perception is highly critical. Its significance can be demonstrated at the level of individual interactions, especially during competitive interactions of terrestrial species (Hairston, 1980, 1983; Jaeger, this volume; Jaeger et al., 1982; Jaeger et al., 1985; Keen, 1982; Keen and Sharp, 1984; Tristram, 1977; Wrobel et al., 1980). In addition, well documented observations reveal that odors and pheromones are used directly for species identification, gender recognition, and for distinguishing among individual conspecifics (Arnold, 1976; Dawley, 1984a, 1984b; Jaeger and Gergits, 1979; Madison, 1975; Simon and Madison, 1984).

Chemical communication in salamanders is especially important in reproduction. Males distinguish between potential mates and potential rivals by their odors. The perception of male odors incites aggressive or defensive reactions from other males, and female odors act as sex attractants and sexual stimuli for males. One particular sub-group of reproductive pheromones are those delivered by the male to the female during courtship. Courtship pheromones are defined as chemical signals that are transmitted only between potential sexual partners and only during courtship (Arnold and Houck, 1982; also see Butler, 1970). This definition excludes pheromones such as sex attractants because these attractants are produced and function long before sexual partners are courting. Courtship pheromones also are distinct from priming pheromones; primers usually work on a longer time scale (many hours to weeks, not just during courtship) and primers function to alter a female's basic reproductive physiology, not merely her sexual receptivity (e.g., Marchlewska-Koj, 1982; Whitten and Bronson, 1970). Male courtship pheromones most likely function to increase the probability that the female will become receptive to the male's subsequent insemination attempts. Courtship pheromones sometimes are called aphrodisiacs, but this is a misleading term that should be eliminated from scientific literature (Birch, 1974a). The word aphrodisiac implies the automatic outcome of female sexual receptivity as a response to the application of the male pheromone. Instead, the response to a courtship pheromone--certainly for vertebrates--is better viewed as behavior that is subsequently more probable, but by no means a fixed response.

The phylogenetic occurrence of courtship pheromones is very uneven (Arnold and Houck, 1982). Examples of courtship pheromones, like examples of sex attractants, are especially widespread among insect species (see Birch, 1974a; Butler, 1970). Courtship pheromones are much less common within vertebrate groups. Despite the overwhelming importance of chemical communication among mammals, for instance, no examples of true courtship pheromones are known for this group. A close parallel exists for swine, however, and serves to distinguish mate attractants from courtship pheromones. In direct response to the odor of a sexually active boar, a sow will assume a mating stance (the "standing reaction") whereby the female braces her feet and stands still in response to pressure (presumably that of the male) on her back (Signoret, 1976). Although the odor of the boar affects female receptivity, this odor technically is not a courtship pheromone since it is produced at all times by a sexually active male, not just when the male and female already have come together as potential mates. In fact, the boar's odor alone (and not the presence of the male) will elicit the standing response in estrous sows (Signoret and du Buisson, 1961). No examples of true courtship pheromones are known for birds either, but this is not surprising since interactions in this group are mediated primarily by visual and auditory stimuli. Reptiles and anurans also lack examples of courtship pheromones, but some fish appear to produce these specialized reproductive chemicals. Among species of glandulocaudine fish (Characidae), there is great diversity in male glands and scales located at the base of the tail (Eigenmann and Myers, 1929; Nelson, 1964). During courtship, the male probably releases pheromones from these glands prior to spawning. Courtship pheromones are not widely characteristic of courtship in fish, however. Among vertebrates, salamanders are unique in that courtship pheromones are known for the majority of species for which information concerning courtship behavior is available. Evidence for salamander courtship pheromones is summarized below. But the potential significance of courtship pheromones deserves special attention here. Why is it worth distinguishing courtship pheromones from other reproductive pheromones?

The evolution of all chemical signals could be affected by two kinds of selection, natural selection and sexual selection. In the broadest sense, the term natural selection intuitively includes all factors that affect an individual's reproductive success. But Darwin (1859, 1871) made a distinction between sexual selection and natural selection. Sexual selection acts only through reproductive variance that results from differences in mating success. A male's mating success is defined as the number of mates that produce offspring sired by that male. (Ideally, estimates of male mating success variance are based on the lifetime number of mates for each male, to control for differences in longevity.) The focus is on male rather than female mating success since reproductive variance due to differences in numbers of mates typically is much greater for males than for females; a male usually can increase its individual reproductive success much more than a female can simply by having more mates (Bateman, 1948). The two agents that affect male mating success are mate choice (female choice of male mates) and male-male competition for mates. In contrast, the agents of natural selection are any factors that cause differences in survivorship, fertility, age at first reproduction, etc., but not differences in mating success. This view of natural selection, in the narrow sense, is used throughout this paper. An easily-visualized trait that is affected both by sexual selection and by natural selection is the colorful breeding plumage of many male birds: imagine the exotic birds of paradise. Such plumage represents a conspicuous male display that is favored by sexual selection to the extent that this display influences female mate choice. Male plumage also is affected by natural selection, however, if this plumage renders the male more vulnerable to predators.

The example of breeding plumage in male birds helps illustrate two important points concerning the evolution of traits affected by sexual selection:

(1) Sexual selection may favor the development of a trait in a direction (and to an extreme) not favored by natural selection. One can imagine that the peacock's tail, while effective at stimulating female mate choice (sexual selection), may also lead to increased costs of predation and survivorship for the male (natural selection). Sexual selection therefore can operate in the opposite direction from natural selection. The nature of evolutionary change depends on the relative importance, and the direction, of natural selection and sexual selection acting on a particular phenotype (e.g., the size of a peacock's tail or the amount of courtship pheromone produced).

(2) Traits, that are affected primarily by sexual selection can evolve to an extreme not favored by natural selection acting alone. Recent theoretical work (Lande 1980, 1981, 1982; Kirkpatrick, 1982) shows that the expected change in a phenotypic trait over time (i.e., the equilibrium value of this trait) is strikingly different if the trait is affected primarily by natural selection or primarily by sexual selection. Under sexual selection, the values of phenotypic traits at equilibrium will depend largely on the initial conditions that affected those traits. Closely related species may show an exceptionally wide variety of phenotypes for sexually selected traits. This variety would be the predicted result of sexual selection magnifying chance differences among initial populations, with relatively minor constraints imposed by natural selection. If a trait were affected purely by sexual selection, the direction of change within any population would be unrelated to environmental conditions and largely the result of historical events. In comparison, expectations for changes in traits affected primarily by natural selection are in marked contrast to the magnitude and varied directions of phenotypic change predicted under sexual selection. If natural selection prevails, there will be a unique equilibrium value of the trait under selection (Lande, 1980). Differences among related species therefore may be less pronounced for traits affected solely by natural selection. The potential differences in evolutionary outcome generated by differences in the relative forces of sexual selection and

natural selection are not intuitively obvious. It is the strength of evolutionary models that these models reveal expectations that could not be achieved by intuition alone.

My goals in this paper are to review the role of courtship pheromone delivery in salamander reproduction, and to consider this behavior in terms of sexual selection. I first summarize breeding patterns of aquatic and terrestrial salamanders [since the timing and mode of delivery of these pheromones can be very different in an aquatic versus a terrestrial medium].

SALAMANDER REPRODUCTIVE PATTERNS

Reproduction in salamanders is typified by two generalized patterns, aquatic and terrestrial (Salthe, 1967). Aquatic-breeding salamanders spend most of the year on land and then migrate to a stream or temporary pond where mating and egg laying occur. Eggs develop into aquatic larvae that subsequently metamorphose and leave the water. After several years, metamorphs periodically return to the water as breeding adults (e.g., see Bell, 1977; Gill, 1985). This general aquatic pattern is common to most species in the Families Ambystomatidae and Salamandridae. The other major reproductive pattern in salamanders is a terrestrial one. Mating occurs on land and eggs are deposited in moist (but not aquatic) sites. Larval development occurs within the egg, and a hatchling emerges in the same form as the adult. There is no dependence on free-standing water. This terrestrial pattern is the most common one within the Plethodontidae.

Fertilization is internal in most aquatic and terrestrial salamander species. The male lacks an intromittent organ, however, and sperm are transferred by means of a spermatophore. The two-part spermatophore consists of a gelatinous base that supports and elevates a mass of sperm. Usually, the male will not deposit a spermatophore except in the presence of a receptive female. Courtship studies of salamanders, therefore, can use the presence of a spermatophore (or a spermatophore base) as a reliable indicator of both male and female sexual receptivity.

In terrestrial plethodontids and other species that have a many-month breeding season, the female has a well-developed sperm-storage organ, the spermatheca, in which viable spermatozoa can be maintained for periods of weeks or months, perhaps years (Boisseau and Joly, 1975). In most cases, ova are not fertilized until just before eggs are laid (but see Joly, 1966). The ability to store sperm permits a decoupling of insemination and ovulation; the female can oviposit fertilized eggs in a small, secluded refugium without requiring the presence of the male.

Breeding in salamanders can be divided into three stages: mate location, courtship, and sperm transfer. The focus of this paper is the courtship stage during which the male applies tactile, visual and chemical stimuli to the female. These stimuli are signals in the manipulative sense of Krebs and Dawkins (1984); the result of the male's pheromone delivery presumably is to elevate the female's level of sexual receptivity. If the female's receptivity rises above some threshold level, the female gives behavioral cues that normally provoke the male to deposit a spermatophore. Although the male may continue applying courtship pheromones for a short while after the female first indicates active courtship interest, the pair shifts relatively quickly to the sperm transfer stage.

There are three levels of evidence for salamander courtship pheromones. (A) Observations of pheromone delivery during courtship: direct observations of courtships during which the male delivers secretions to the female from specialized, sexually dimorphic glands. (B) The morphological presence of putative courtship glands and accessory structures: in species where courtship has not been observed, adult males have glands that are homologous to those of other, closely related species where males have been observed to use these glands only during courtship [in some plethodontid species, premaxillary or mandibular teeth act as accessory structures during pheromone delivery (discussed below). Typically, both glands and teeth are well developed only in males that are producing mature spermatozoa]. (C) Experimental manipulation of pheromone delivery: delivery of courtship pheromones can be prevented or enhanced during courtship, and measures of receptivity compared between females that have and have not received pheromones.

Evidence from these three levels--courtship observations, morphology and experimental courtship manipulations--is summarized below. Most information is available for salamanders in three families (Ambystomatidae, Salamandridae, and Plethodontidae) which account for the majority of salamander species living today. The information below is not intended to duplicate detailed reviews of salamander reproduction and courtship behavior (e.g., Arnold, 1972, 1976, 1977; Houck, 1977, 1982; Joly, 1966; Organ, 1961; Salthe, 1967). Instead, I focus on specific examples that highlight the diversity of modes with which males deliver courtship pheromones to females.

(A) Observations of Pheromone Delivery During Courtship

Different salamander species employ different modes of delivering male courtship pheromones to the female. These modes are described below in terms of two major characteristics: the occurrence and nature of male physical contact with the female, and whether the courtship pheromones are delivered indirectly (via a water current) or are applied directly to the female's body. For completeness, I also mention species in which courtship pheromones have not been implicated in male-female interactions.

(1) No obvious delivery; little contact with female. Some aquatic-breeding ambystomatids are characterized by courtships in which the male makes relatively little physical contact with the female. Courtship in *Ambystoma maculatum*, *A. annulatum*, *A. opacum*, and *A. texanum* is rapid, involving brief and intermittent contact between the male and female (Arnold, 1972, 1976; Noble and Brady, 1933; Spotila and Beumer, 1970). The male contacts the female with his snout, nudging her and crawling under her, but there is no obvious delivery of courtship pheromones during this contact. The male spends much time nearby (but apart from) the female in search of suitable spermatophore deposition sites. Other species in which courtship pheromones are not implicated include the plethodontids (*Bolitoglossa macrinii* (Papenfuss et al., 1983) and *Ensatina eschscholtzii* (Stebbins, 1954)), and salamandrids in the genera *Salamandra* and *Euproctus* (reviewed by Arnold, 1972).

(2) Indirect delivery; little contact with female. In some aquatic species, a male courts a female by relying on water currents to transport his pheromones to her. This describes courtship in *Triturus* and *Cynops* (Salamandridae). Excellent analyses by Halliday (1974, 1975, 1977) of the courtship of the smooth newt, *Triturus vulgaris*, show that the male moves his tail in a fan-like fashion to create a water current from the male's clocal glands towards the female's nares (see Halliday, 1975, Fig. 4).

Males possess enlarged cloacal glands only during the breeding season. Recent evidence from dissections showed that the abdominal gland (found only in males) is greatly enlarged in breeding males (Verrell and Halliday, pers. comm.). This gland also has been strongly implicated as a source of courtship pheromones in T. cristatus (Cedrini and Fasolo, 1971; Giacoma et al., 1979; Malacarne et al., 1984).

(3) Indirect delivery; increased contact with the female. Courtship in another aquatic species, Ambystoma tigrinum, represents a mode of courtship pheromone delivery in which the male makes much more physical contact with the female, although the female never is clasped or physically restrained (Arnold, 1976). If rival males threaten to interfere, a courting A. tigrinum male may transport the female by nudging her midsection and by using his tail to propel them both a short (but critical) distance away from rivals (Arnold, 1976, Fig. 9). During physical contact, courtship pheromones from the male's cloaca presumably stimulate the female to behave in a manner that provokes spermatophore deposition. Female A. tigrinum also orient to the male's cloaca during the sperm transfer phase of courtship (Arnold, 1976; Arnold and Houck, 1982, Fig. 4). This kind of courtship also is found in Ambystoma dumerili (Brandon, 1970) and A. mexicanum (Arnold, 1972; Gasco, 1881).

(4) Direct delivery; physical contact but female not restrained. In most plethodontid salamanders, courtship is terrestrial and the male does not clasp the female or restrain her in any way. The male, nevertheless, makes much physical contact with the female, crawling over and under her, or rubbing her snout with his head. In particular, the male frequently delivers secretions from specialized cephalic glands directly to the female. In Plethodon jordani, the male has an enlarged mental (submandibular) gland that he slaps over the female's nares (Arnold, 1977, Fig. 8). In a related species, P. yonahlossee, the male also slaps his mental gland over the female's snout (Arnold, 1972). Neither P. jordani nor P. yonahlossee males use their teeth to aid in secretion delivery. In other species, such as Desmognathus ochrophaeus, the male has sexually dimorphic, elongated premaxillary teeth (Dunn, 1926; Arnold and Houck, 1982, Fig. 7) in addition to a mental gland. The male uses these teeth to abrade the female's skin in the same area that is rubbed with the mental gland secretions (Arnold, 1977, Fig. 7; Arnold and Houck, 1982, Fig. 8). Presumably, the use of teeth facilitates the "injection" of courtship pheromones into the female's superficial circulatory system. Other species in which the male uses specialized teeth in conjunction with mental gland secretions include: Bolitoglossa subpalmata (Arnold, 1972), Eurycea bislineata (Arnold, 1977), Desmognathus monticola and D. fuscus (Organ, 1961).

(5) Direct delivery; female is physically restrained. (a) No male courtship glands known. Males of an ambystomatid, Ambystoma laterale make much physical contact with a female during an aquatic courtship (Arnold, 1976; Storez, 1969 ; Uzzell, 1969; pers. obs.). The male will clasp the female with his forelimbs and use his tail to move the pair through the water, especially when rival males are present. While clasping the female, the male rubs his chin over the female's snout, probably applying glandular secretions to her nares. The male also moves his cloaca over the female's head during courtship. During the sperm transfer stage, the female orients to and nudges the male's cloacal papillae. This type of courtship also is found in Ambystoma gracile (Licht, 1969), A. jeffersonianum (Uzzell, 1964, 1969), and A. macrodactylum (Anderson, 1961).

(b) Male has specialized courtship gland(s). Male newts in the genera Taricha and Notophthalmus will restrain an initially unresponsive female during courtship by clasping her. During clasping, pheromones from specialized cephalic glands are directly applied to the female's nares (Arnold, 1972; Davis and Twitty, 1964; Verrell, 1982a). Direct courtship pheromone delivery has the advantage of being more accurate than delivery via water currents. Also, restraining the female offers the male a chance to deliver pheromones to a female that otherwise might flee. Male Taricha clasp the female with both their fore and hind limbs. While clasping, the male slides forward periodically to rub his submandibular glands across the female's nares (Arnold, 1977, Fig. 6). In a different version of courtship pheromone application during clasping, a male N. viridescens uses only his hind limbs to clasp the female's forebody; the male then regularly applies secretions from his genial (cheek) glands to the female's snout, and periodically rubs her snout with his forelimbs (Arnold, 1977, Fig. 5). Courtship pheromone delivery in a terrestrial salamander that involves physical restraint of the female is known in only one plethodontid species, Desmognathus wrighti (Houck, 1980, in prep.). A courting male clasps the female in his jaws, and holds on to her even when she struggles vigorously (pers. obs.). The male has sexually dimorphic, barb-shaped teeth which pierce the female in close proximity to the external apertures of the male's mental gland. Glandular secretions presumably are introduced into the female's circulatory system in this manner.

(B) Morphological Presence of Courtship Glands and Accessory Structures

Courtship behavior either has not been observed or is incompletely known for many salamander species in which an adult male has a mental gland. These species include many neotropical plethodontids (Houck, 1977): Bolitoglossa engelhardti, B. flavimembris, B. franklini, B. morio, B. occidentalis, B. resplendens, B. rostrata, Dendrotriton bromeliacia, Pseudoeurycea brunnata, P. goebeli, and P. rex. One neotropical genus, Thorius, is composed of diminutive species which secondarily have lost much adult cranial ossification and dentition (Wake, 1966; Hanken, 1984). Despite these losses, most adult males retain at least one prominent premaxillary tooth (Hanken, 1984). The retention of this tooth very likely indicates its importance during courtship as an aid to courtship pheromone delivery. The presence of a mental gland in the adult male also has been noted for North American plethodontids (Brodie, 1968), including Plethodon hoffmani (Angle, 1969), P. neomexicanus (Williams, 1972), and P. vehiculum (Peacock and Nussbaum, 1973). Courtship observations in all of these species probably would reveal that secretions from these glands are used as courtship pheromones.

(C) Experimental Manipulation of Pheromone Delivery

The first experimental evidence that courtship pheromones affect female receptivity was reported by Rogoff (1927) for the red-spotted newt Notophthalmus viridescens. Rogoff reported courtship results from three separate experiment: (i) when male genial glands were blocked preventing secretions from reaching the female, (ii) when the female's nares were blocked, preventing her perception of the male's genial gland secretions, and (iii) when the amount of male genial gland secretions was enhanced (by injecting the male with pilocarpine hydrochloride and by holding the female's nares next to the male's glands). Females subjected to enhanced pheromone delivery (expt. iii) tended to orient towards and follow the male (signs of female receptivity) more than did a control group of females. When delivery of glandular secretions was diminshed (expts. i and ii), a female tended not to respond to the male. These results support the contention that male courtship pheromones influence female receptivity in N. viridescens. Rogoff published only a brief abstract of his work, however, and the experimental investigation of courtship pheromones in N. viridescens has been continued only recently. Paul Watson (pers. comm.) has devised a new technique to occlude genial glands in N. viridescens. Using a two-step process, Watson applied a surgical adhesive to males, covering the genial gland apertures in an experimental courtship group and covering a nearby portion of skin in a control courtship group. Using these two groups of males, Watson observed courtships in which a male first clasped an initially unresponsive female and then rubbed his genial glands across her nares. Watson's assay of female receptivity was whether the female remained passive (relatively inactive; not attempting independent movement) when she was clasped by the male or whether the female struggled (attempted to free herself from the male's hold). In 12 courtships with experimental males (no genial gland secretions), 7 of the 12 females struggled; in 7 courtships with control males, none of the females struggled. Struggling was never observed in 25 other courtships in which males had not been experimentally treated. These results support the hypothesis that genial gland secretions promote female receptivity.

Similar evidence is available for the crested newt, Triturus cristatus. By occluding the female's nares with petroleum jelly, Malacarne et al. (1984) found that these females would be less likely to respond to male courtship than would control females. The removal of the plugs subsequently restored a female's courtship receptivity.

The only other experimental attempt to evaluate the effects of male courtship pheromones in salamanders is work conducted with the terrestrial plethodontid, Desmognathus ochrophaeus (Houck and Arnold, in prep.). Courtship interactions for this species (described above) involve pheromone delivery from the male's mental gland, and the probable use of premaxillary teeth to introduce this pheromone into the female's circulatory system. In their experiment, Houck and Arnold had three experimental male groups: (1) Each male had its mental gland surgically removed, (2) Each male had its premaxillary teeth removed, and (3) Each male had both its mental gland and its premaxillary teeth removed. There were control (sham-operated) male groups, one for each of the experimental groups. Each control and each experimental male was given a courtship opportunity with a gravid female. Pairs were placed together from dusk until approximately 0800 hr the following morning. The occurrence of courtship was inferred by the presence of a spermatophore (or just the spermatophore base) on the substrate. The female's cloaca was checked for the presence of a sperm mass to determine whether she was inseminated. Surprisingly, the number of courtships and inseminations for males in the experimental groups did not differ significantly from those in the control groups. The experimental protocol used in this study, however, permitted courtship encounters to last throughout the night. Data concerning the absolute time until spermatophore deposition were not recorded, and consequently we do not know whether spermatophore deposition occurs later in those pairs where the male lacks a courtship gland.

DISCUSSION

Current studies of male courtship pheromones focus on the contribution of these pheromones to individual male mating success. This emphasis on individual reproductive success is a distinct shift from earlier assumptions that courtship pheromones served primarily for species identification (e.g., Butler, 1970, p. 43). This earlier emphasis prevailed at a time when the species concept was receiving much attention; mechanisms that prevented interspecific matings were of special interest. Courtship pheromones, however, only function after potential conspecific mates have located one another. Courtship pheromones affect a female's receptivity towards a particular conspecific male, and consequently affect the mating success of that male relative to the success of other conspecifics.

Courtship pheromones most likely promote individual male reproductive success, but the delivery of these pheromones is not an absolute guarantee of female receptivity. Observations of courtships in a variety of salamander species show that many courtships do not lead to spermatophore deposition or insemination, even though a male assiduously delivers pheromones during these courtships (Arnold, 1976; Halliday, 1977; Malacarne et al., 1984; Verrell, 1982a). Nevertheless, persistent stimulation of the female, including pheromone delivery, apparently increases the male's probability of inseminating the female. Verrell (1982a) found that the number of male courtship actions in N. viridescens was positively correlated with the number of spermatophores deposited. For the crested newt, Triturus cristatus, Malacarne and Curtassa (1983) also found that time spent in courtship was correlated with spermatophore deposition and insemination success. For the smooth newt, T. vulgaris, Teyssedre and Halliday (in press) conducted a model study of the relative timing of male courtship behaviors, including pheromone delivery. A courtship display bout in T. vulgaris consists of a series of male displays at short (∿ 4 seconds) intervals. Courtships in which display bouts were longer were more likely to lead to spermatophore deposition (Ibid). Courtships culminating in spermatophore deposition, therefore, were characterized not only by the performance of particular male actions, but also by the relative timing of these actions. The importance of timing as opposed to pheromone delivery per se also is suggested by the study of courtship in the plethodontid Desmognathus ochrophaeus (discussed above). Thus, although courtship phermone delivery doesn't guarantee mating success, females receiving these pheromones probably will indicate receptivity sooner than would females courted by males without pheromones.

Whether a female actually mates with a given male is, of course, a function of many factors in addition to the presence or absence of male pheromone delivery. Multiple sensory modes of communication certainly affect reproductive receptivity in mammals (e.g., Whitten and Bronson, 1970). Major sensory stimuli important to salamander reproduction can include visual stimuli (e.g., Himstedt, 1979; Malacarne and Curtassa, 1983; Verrell, 1982b), tactile stimuli, and olfactory stimuli other than courtship pheromones (Margolis, 1980; Roth, 1976). Moreover, the effects of any sensory signals on any animal can depend strongly on daily and seasonal changes in motivational states (Halliday, 1977; Halliday and Sweatman, 1976). Female receptivity in D. ochrophaeus, for example, is correlated with the time of year (highest in spring; Houck et al., 1985a), but also with the amount of time since the female was last inseminated (pers. obs.). In general, courtship efforts also may be affected by male sexual experience (Dewsbury, 1981; Farr, 1980; Payne, 1978) or by the novelty of the sexual partner (Dewsbury, 1981). Although many factors can affect reproductive success, courtship pheromones are important because they might increase the probability of female receptivity. Even a slight increase in receptivity could make the difference between male insemination success and the loss of a reproductive opportunity.

For salamanders, the potential magnitude of courtship pheromone effects undoubtedly varies among species depending on the length of the breeding season and the length of the average male-female courtship encounter. Courtship pheromones are less likely to be significant in species such as Ambystoma maculatum where the breeding season is short (usually measured in days), courtships are extremely rapid, and the male spends relatively little time in contact with the female, never clasping her (Arnold, 1976). At the opposite extreme, terrestrial plethodontids such as D. ochrophaeus have a courtship season that lasts many months. Complete courtships have not been observed in the field for this species, but encounters staged in the laboratory can last for at least 20-60 minutes, sometimes much longer (pers. obs.). Terrestrial plethodontids also are good candidates for studies of courtship pheromones. The male produces courtship pheromones from specialized glands on the head and pheromone delivery can be observed easily. Moreover, delivery from cephalic glands can be manipulated (enhanced or prevented) more readily than in aquatic species that rely upon glands secreting into the cloaca. Some aquatic-breeding salamanders (especially newts such as Notophthalmus or Taricha spp.), also are promising candidates for pheromone studies. These species have a many week courtship season, they also have cephalic courtship glands, and they use easily observable physical restraint (clasping) of the female during pheromone delivery. Another advantage of studying aquatic salamanders is that courtships in some species can be observed in the field. Recently, courtships of N. viridescens have been studied in natural ponds in Virginia by Adrianne Massey (pers. comm.). Massey's observations of over 70 courtships included five in which the female was inseminated, as well as 20 in which sexual interference occurred (rival male disrupts courtship of original male).

Evidence for the production and delivery of male pheromones during courtship is consistent with the hypothesis that these pheromones affect male mating success. Until further data are obtained, however, other possible hypotheses also should be entertained, and some are discussed by Arnold and Houck (1982). These include potential contributions of courtship pheromones to female fertility or female survivorship. Some male insects, for example, transfer nutrients to the female during courtship: male salivary secretions are ingested by female scorpionflies during copulation (Thornhill, 1981). Salamander courtship pheromones are not ingested, however, and the common mode of delivery to the female's nares make it unlikely that these pheromones play a nutritional role. Another possibility is that male pheromones act as primers (sensu Wilson and Bossert, 1963). These could facilitate the maturation of ova or induce ovulation (hence fertilization) of ovarian eggs. In support of this view are observations that a female salamander rarely oviposits spontaneously if she has not been courted and inseminated. But some newts (N. viridescens) are inseminated in courtships where the male completely bypasses courtship pheromone delivery (Arnold, 1972; Verrell, 1982a). Also, most plethodontids can court and be inseminated many months prior to ovulation. These observations suggest that courtship pheromones do not act as primers, at least not as this action is known for other vertebrates (Marchlewska-Koj, 1982). Another possible function of male pheromone delivery in salamanders might be to make the female repellent to rival males. The courting male most often delivers his pheromones directly to the female's nares, however, rather than applying the secretions to her body in a manner that would allow subsequent detection by another male. Courtship pheromones might change the female's behavior towards subsequent males, but immediate changes in female behavior usually result in receptivity that is increased, not decreased. Although alternative hypotheses should be explored, the evidence currently available for salamanders best supports the hypothesis that a primary function of courtship pheromones is to affect female receptivity, hence male reproductive success.

Sexual selection acts on differences in mating success among males, but only when the measure of mating success includes mates that actually bear progeny sired by the inseminating male. In many animal species, insemination success is not always correlated with fertilization success. Females of many salamander species mate multiple times (Halliday and Verrell, 1984; Labanick, 1983; Tilley and Hausman, 1976). Spermatozoa are stored in the spermatheca until they are needed for fertilization. Many insect species also store sperm from multiple inseminations. Depending on the particular insect species, paternity of clutches can be attributed to the first sire alone, some combination of all or most sires, or only the last sire (Parker, 1970; Gwynne, 1984). The only evidence for sperm competition currently available for salamanders indicates that all or most of a female's mates sire at least some of her progeny (Houck et al., 1985b; Houck and Schwenk, 1984). Conclusions from these data are only tentative, but it appears as though the female has a "reservoir" of spermatozoa within her spermatheca. The probability of a given male's sperm being utilized may be proportional to its amount relative to other sperm in the female's spermatheca. If indeed this is the general case for salamanders, then male fertilization success still would be positively correlated with mating success, even when a male inseminates the same female more than once. Multiple insemination and sperm storage in salamanders therefore does not appear to affect the crucial relationship between male mating success and fertilization success.

There are two levels at which the evolution of courtship pheromones can be considered. At the broader level, courtship pheromones are promising candidates for testing theoretical predictions of how sexually selected traits evolve (Lande, 1980, 1981, 1982; Kirkpatrick, 1982). If courtship pheromones primarily affect mating success, rather than survivorship or fertility, then these pheromones are strongly subjected to sexual selection. In this sense courtship pheromones are better candidates for the study of sexual selection than are other characters (such as the sexually dimorphic plumage of many birds) that probably are also affected by natural selection to a more substantial degree. Species groups that are well characterized would provide an ideal data base for comparing levels of variation between traits subjected to sexual selection and traits affected primarily by natural selection. Unfortunately, it will be many years before there are sufficient data concerning vertebrate courtship pheromones to test the predictions based on sexual selection theory.

Meanwhile, these data will come from studies that are significant in their own right, studies that concentrate on courtship pheromone evolution at the population level. Several questions pertinent to this level of analysis are:

(1) Is there significant variation among males in mating success?
(2) To what extent do courtship pheromones account for among-male variance in mating success?
(3) What is the nature of individual variation in male courtship pheromones, in pheromone production rates, and in pheromone delivery?
(4) By what mechanism(s) do courtship pheromones affect female behavior?

The first question was addressed by a recent laboratory study conducted with *D. ochrophaeus* by Houck et al. (1985a). They used an experimental design that permitted the measurement of among-male variation in mating success. This variation was highly significant, even though it was a minimum estimate since potentially contributing factors (such as male-male competition and mate encounter rate) were held constant. The importance of significant among-male variance in mating success is that it offers an opportunity for sexual selection to act on the phenotypic characters (including behaviors) that affect mating success (Crow, 1958; Wade and Arnold, 1980). Houck et al. (1985a) measured the opportunity for sexual selection acting on male mating success variance in *D. ochrophaeus* using equations derived by Arnold and Wade (1984a, 1984b). They also were able to calculate the maximum amount that the mean of a phenotypic character could shift if sexual selection acted directionally on that character. For *D. ochrophaeus*, the maximum extent of change in a male phenotypic character affecting mating success is 25% of one standard deviation. This represents a relatively large opportunity for sexual selection to act.

The next step is to examine the relationship between mating success variance and courtship pheromones. A comparison of this variance might be made for a group of males in which pheromone delivery first was prevented and then restored for each individual. One potential problem in interpreting results from courtship experiments is that uncontrollable differences among female mates may contribute to differences in male mating success scores. If multiple courtships are experimentally feasible for each individual, however, special statistical designs that assign particular combinations of male-female pairs can be used to factor out variance due to possible differences in initial receptivity among females (Houck et al., 1985a). Simple scores of mating success (e.g., insemination) can be used to estimate male variance, but measurements of female behavioral responses during courtship also would be desirable.

The investigation of courtship pheromone effects would be enhanced considerably if the chemical nature of particular pheromones were known. Glandular secretions could be identified using techniques well known to those studying insect pheromones (e.g., Birch, 1974b; Payne et al., 1985). Pheromones probably vary significantly among individuals as well as among species (e.g., Johnston, 1982; Muller-Schwarze, 1974; Perrigo and Bronson, 1983). Investigating pheromones in a South American primate (Saguinus fuscicollis), Smith et al. (1985) report exciting possibilities of using computer programs to identify individual odor patterns; gender and subspecies of unknown donors can be correctly assigned using computer analysis of high resolution gas chromatographic patterns.

Critical investigations are just beginning, but analyses of salamander courtship pheromones potentially can offer compelling insights at the population level into sources of individual variation in reproductive success. At another level, these pheromones are candidates for examining predictions of sexual selection theory. At either level, the taxonomic variety and evolutionary persistence of courtship pheromone delivery in salamanders is a testimony to the reproductive significance of these chemical signals.

ACKNOWLEDGMENTS

I thank Tim Halliday, Adrianne Massey, Paul Verrell and Paul Watson for permission to include data from their studies of salamander courtship. Stevan Arnold made constructive comments on the manuscript. Supported by NSF BSR-17517.

REFERENCES

Anderson, J. D., 1961, The courtship behavior of Ambystoma macrodactylum croceum, Copeia, 1961:132.
Angle, J. P., 1969, The reproductive cycle of the northern ravine salamander, Plethodon richmondi richmondi, in the Valley and Ridge Province of Pennsylvania and Maryland, J. Wash. Acad. Sci., 59:192.
Arnold, S. J., 1972, The evolution of courtship behavior in salamanders. Ph.D. dissertation, University of Michigan, Ann Arbor.
Arnold, S. J., 1976, Sexual behavior, sexual interference and sexual defense in the salamanders Ambystoma maculatum, Ambystoma tigrinum and Plethodon jordani. Z. Tierpsychol., 42:247.
Arnold, S. J., 1977, The evolution of courtship behavior in New World salamanders with some comments on Old World salamandrids, in: "The Reproductive Biology of Amphibians," D. H. Taylor, and S. I. Guttman, eds., Plenum Press, New York.

Arnold, S. J., and Houck, L. D., 1982, Courtship pheromones: Evolution by natural and sexual selection, in: "Biochemical Aspects of Evolutionary Biology," M. H. Nitecki, ed., Univ. of Chicago Press, Chicago.
Arnold, S. J., and Wade, M. J., 1984a, On the measurement of natural and sexual selection: Theory, Evolution, 38:709.
Arnold, S. J., and Wade, M. J., 1984b, On the measurement of natural and sexual selection: Applications, Evolution, 38:720.
Bardach, J. E., and Todd, J. H., 1970, Chemical communication in fish, in: "Communication by Chemical Signals," J. W. Johnston, Jr., D. G. Moulton, and A. Turk, eds., Appleton-Century-Crofts, New York.
Batemen, A. J., 1948, Intra-sexual selection in Drosophila, Heredity, 2:349.
Bell, G., 1977, The life of the smooth newt (Triturus vulgaris) after metamorphosis, Ecol. Monogr., 47:279.
Belvedere, P. C., Colombo, L., Mezzalira, F., and Malacarne, G., 1984, Steroid biosynthesis by the abdominal gland and the testis of the crested newt, Triturus cristatus carnifex Laur., Boll. Zool., 51, Suppl. 9.
Birch, M. D., 1974a, Aphrodisiac pheromones in insects, in: "Pheromones," M. D. Birch, ed., North-Holland Publ. Co., Amsterdam.
Birch, M. D., ed., 1974b, "Pheromones," North-Holland Publ. Co., Amsterdam.
Blair, W. F., ed., 1961, "Vertebrate Speciation," Univ. Texas Press, Austin.
Boisseau, C., and Joly, J., 1975, Transport and survival of spermatozoa in female Amphibia, in: "The Biology of Spermatozoa," E. S. E. Hafez, and C. G. Thibault, eds., S. Karger AG, Basel, Switzerland.
Brandon, R. A., 1970, Courtship, spermatophores and eggs of the Mexican Achoque, Ambystoma (Bathysiredon) dumerili (Duges), Zool. J. Linn. Soc., 49:247.
Brodie, E. D., Jr., 1968, Observations on the mental hedonic gland-clusters of western salamanders of the genus Plethodon, Herpetologica, 24:248.
Burghardt, G. M., 1970, Chemical perception in reptiles, in: "Communication by Chemical Signals," J. W. Johnston, Jr., D. G. Moulton, and A. Turk, eds., Appleton-Century-Crofts, New York.
Butler, C. G., 1970, Chemical communication in insects: Behavioral and ecologic aspects, in: "Communication by Chemical Signals," J. W. Johnston, Jr., D. G. Moulton, and A. Turk, eds., Appleton-Century-Crofts, New York.
Cedrini, L., and Fasolo, A., 1971, Olfactory attractants in sex recognition of the crested newt. An electrophysiological research, Monit. Zool. Ital. (N.S.), 5:223.
Crow, J. F., 1958, Some possibilities for measuring selection intensities in man, Human Biol., 30:1.
Darwin, C., 1859, "The Origin of Species by Means of Natural Selection or the Preservation of Favoured Races in the Struggle for Life," John Murray, London.
Darwin, C., 1871, "The Descent of Man and Selection in Relation to Sex," John Murray, London.
Davis, W. C., and Twitty, V. C., 1964, Courtship behavior and reproductive isolation in the species of Taricha (Amphibia, Caudata), Copeia, 1964:601.
Dawley, E. M., 1984a, Recognition of individual, sex and species odours by salamanders of the Plethodon glutinosus - P. jordani complex, Anim. Behav., 32:353.
Dawley, E. M., 1984b, Identification of sex through odors by male red-spotted newts, Notophthalmus viridescens, Herpetologica, 40:101.
Dewsbury, D. A., 1981, Effects of novelty on copulatory behavior: The Coolidge effect and related phenomena, Psych. Bull., 89:464.
Dodd, C. K., Jr., Johnson, J. A. and Brodie, E. D., Jr., 1974, Noxious skin secretions of an eastern small Plethodon, P. nettingi hubrichti, J. Herpetol., 8:89.

Doty, R. L., ed., 1976, "Mammalian Olfaction, Reproductive Processes, and Behavior," Academic Press, New York.
Dunn, E. R., 1926, "The Salamanders of the Family Plethodontidae," Smith College, Mass.
Eigenmann, C. H., and Myers, G. S., 1929, The American Characidae, Part V, Mem. Mus. Comp. Zool. Harvard, XLII:429.
Farr, J. A., 1980, The effects of sexual experience and female receptivity on courtship-rape decisions in male guppies, Poecilia reticulata (Pisces: Poeciliidae), Anim. Behav., 28:1195.
Forester, D. C., 1977, Comments on the female reproductive cycle and philopatry by Desmognathus ochrophaeus (Amphibia, Urodela, Plethodontidae), J. Herpetol., 11:311.
Forester, D. C., 1979, Homing to the nest by female mountain dusky salamanders (Desmognathus ochrophaeus) with comments on the sensory modalities essential to clutch recognition, Herpetologica, 35:330.
Gasco, F., 1881, Les amours des axolotls, Bull. Soc. Zool. France, 6:151.
Giacoma, C., Malacarne, G., Andreoletti, G. C., and Vellano, C., 1979, The significance of the abdominal gland in the reproductive behavior of the crested newt: a preliminary study, Mon. Zool. Ital. (N.S.), 13:204.
Gill, D. E., 1985, Interpreting breeding patterns from census data: a solution to the Husting dilemma, Ecology, 66:344.
Grant, D., Anderson, O. and Twitty, V., 1968, Homing orientation by olfaction in newts (Taricha rivularis), Science, 160:1354.
Grubb, T. C., 1974, Olfactory navigation to the nesting burrow in Leach's Petrel (Oceanodroma leucorrhoa), Anim. Behav. 22:192.
Gwynne, D. T., 1984, Male mating effort, confidence of paternity, and insect sperm competition, in: "Sperm Competition and the Evolution of Animal Mating Systems," R. L. Smith, ed., Academic Press, New York.
Hairston, N. G., 1980, Species packing in the salamander genus Desmognathus: What are the interspecific interactions involved?, Am. Nat., 115:354.
Hairston, N. G., 1983, Alpha selection in competing salamanders: Experimental verification of an a priori hypothesis, Am. Nat., 122:105.
Halliday, T. R., 1974, Sexual behaviour of the smooth newt, Triturus vulgaris (Urodela, Salamandridae), J. Herpetol., 8:277.
Halliday, T. R., 1975, An observational and experimental study of sexual behaviour in the smooth newt, Triturus vulgaris (Amphibia: Salamandridae), Anim. Behav., 23:291.
Halliday, T. R., 1977, The effects of experimental manipulation of breathing behaviour on the sexual behaviour of the smooth newt, Triturus vulgaris, Anim. Behav., 25:39.
Halliday, T. R., and Sweatman, H. P. A., 1976, To breathe or not to breathe: The newt's problem, Anim. Behav., 24:551.
Halliday, T. R., and Verrell, P. A., 1984, Sperm competition in amphibians, in: "Sperm Competition and the Evolution of Animal Mating Systems," R. L. Smith, ed., Academic Press, N.Y.
Hanken, J., 1984, Miniaturization and its effect on cranial morphology in plethodontid salamanders, genus Thorius (Amphibia, Plethodontidae). I. Osteological variation, Biol. Jour. Linn. Soc., 23:55.
Himstedt, W., 1979, The significance of colour signals in partner recognition of the newt Triturus alpestris, Copeia, 1979:40.
Holomuzki, J. R., 1982, Homing behavior of Desmognathus ochrophaeus along a stream, J. Herpetol., 16:307.
Houck, L. D., 1977, Life history patterns and reproductive biology of neotropical salamanders, in: "The Reproductive Biology of Amphibians," D. H. Taylor, and S. I. Guttman, eds., Plenum Press, New York.
Houck, L. D., 1980, Courtship behavior in the plethodontid salamander, Desmognathus wrighti, Amer. Zool., 20:825.
Houck, L. D., 1982, Male tail loss and courtship success in the plethodontid salamander Desmognathus ochrophaeus, J. Herpetol., 16:335.

Houck, L. D., Arnold, S. J., and Thisted, R., 1985a, A statistical analysis of sexual selection in the plethodontid salamander Desmognathus ochrophaeus, Evolution, 39:370.
Houck, L. D., and Schwenk, K., 1984, The potential for long-term sperm competition in a plethodontid salamander, Herpetologica, 40:410.
Houck, L. D., Tilley, S. G., and Arnold, S. J., 1985b, Sperm competition in a plethodontid salamander: Preliminary results, J. Herpetol., 19:420.
Jaeger, R. G., and Gergits, W. F., 1979, Intra- and interspecific communication in salamanders through chemical signals on the substrate, Anim. Behav., 27:150.
Jaeger, R. G., Goy, J., Tarver, M., and Marquez, C., 1985, Salamander territoriality: Pheromonal markers as advertisement by males, Anim. Behav., in press.
Jaeger, R. G., Kalvarsky, D., and Shimizu, N., 1982, Territorial behavior of the red-backed salamander: Expulsion of intruders, Anim. Behav., 30:490.
Johnston, R. E., 1982, Mechanisms of individual discrimination in hamsters, in: "Chemical Signals in Vertebrates III,", D. Muller-Schwarze and R. M. Silverstein, eds., Plenum Press, New York.
Joly, J., 1966, Sur l'éthologie sexuelle de Salamandra salamandra L., Z. Tierpsychol., 23:8.
Keen, W. H., 1982, Habitat selection and interspecific competition in two species of plethodontid salamanders, Ecology, 63:94.
Keen, W. H., and Sharp, S., 1984, Responses of a plethodontid salamander to heterospecific and conspecific intruders, Anim. Behav., 32:58.
Kirkpatrick, M., 1982, Sexual selection and the evolution of female choice, Evolution, 36:1.
Kleeberger, S. R., and Werner, J. K., 1982, Home range and homing behavior of Plethodon cinereus in Northern Michigan, Copeia, 1982:409.
Krebs, J. R., and Dawkins, R., 1984, Animal signals: Mind-reading and manipulation, in: "Behavioural Ecology: An Evolutionary Approach," Second Ed., J. R. Krebs, and N. B. Davies, eds., Sinauer, Sunderland, MA.
Kumpf, D. F., 1934, The courtship of Ambystoma tigrinum, Copeia, 1934:7.
Labanick, G. M., 1983, Inheritance of the red-leg and red-cheek traits in the salamander Desmognathus ochrophaeus, Herpetologica, 30:114.
Labanick, G. M., and Brandon, R. A., 1981, An experimental study of Batesian mimicry between the salamanders Plethodon jordani and Desmognathus ochrophaeus, J. Herpetol., 15:275.
Lande, R., 1980, Sexual dimorphism, sexual selection, and adaptation in polygenic characters, Evolution, 34:292.
Lande, R., 1981, Models of speciation by sexual selection on polygenic traits, Proc. Natl. Acad. Sci. USA, 78:3721.
Lande, R., 1982, A quantitative genetic theory of life history evolution, Ecology, 62:607.
Lanza, B., 1949, Il corpo ghiandolare mento niero dei Plethodontidae (Amphibia, Caudata), Monitore Zool. Ital., 67:15.
Licht, L. E., 1969, Observations on the courtship behavior of Ambystoma gracile, Herpetologica, 25:49.
Liley, N. R., 1982, Chemical communication in fish, Can. J. Fish. Aquat. Sci., 39:22.
Losey, G. S., 1969, Sexual pheromone in some fishes of the genus Hypsoblennius, Gill., Science, 163:181.
Madison, D. M., 1969, Homing behaviour of the red-cheeked salamander, Plethodon jordani, Anim. Behav., 17:25.
Madison, D. M., 1972, A mechanism of homing orientation in salamanders involving chemical cues, in: "NASA Spec. Publ. 262," S. R. Galler, K. Schmidt-Koenig, G. L. Jacobs, and R. B. Bellville, eds., NASA.
Madison, D. M., 1975, Intraspecific odor preferences between salamanders of the same sex: Dependence on season and proximity of residence, Can. J. Zool., 53:1356.

Madison, D. M., 1977, Chemical communication in amphibians and reptiles, in: "Chemical Signals in Vertebrates," D. Muller-Schwarze, and M. M. Mozell, eds., Plenum Press, New York.

Madison, D. M. and Shoop, C. R., 1970, Homing behavior, orientation, and home range of salamanders tagged with Tantalum-182, Science, 168:1484.

Malacarne, G., Bottoni, L., Massa, R., and Vellano, C., 1984. The abdominal gland of the crested newt: a possible source of courtship pheromones. Preliminary ethological and biochemical data, Monit. Zool. Ital., (N.S.), 18:33.

Malacarne, G., and Curtassa, R., 1983, Sexual selection in the crested newt, Anim. Behav., 31:1256.

Marchlewska-Koj, A., 1982, Priming pheromones in mice, in: "Chemical Signals in Vertebrates III," D. Muller-Schwarze, and R. M. Silverstein, eds., Plenum Press, New York.

Margolis, S. E., 1980, Ethological criteria of chemical sensitivity of the aquatic forms of newts, Zoolog. Zhurnal., 59:1683.

Moore, R. E., 1965, Olfactory discrimination as an isolation mechanism between Peromyscus maniculatus and Peromyscus polionotus, Amer. Midl. Natur., 73:85.

Muller-Schwarze, D., 1974, Olfactory recognition of species, groups, individuals and physiological states among mammals, in: "Pheromones," M. D. Birch, ed., North-Holland Publ. Co., Amsterdam.

Muller-Schwarze, D., 1977, Complex mammalian behavior and pheromone bioassay in the field, in "Chemical Signals in Vertebrates," D. Muller-Schwarze, and M. M. Mozell, eds., Plenum Press, New York.

Nelson, K., 1964, Behavior and morphology in the Glandulocaudine fishes (Ostariophysi, Characidae), Univ. Calif Publ. Zool., 75:59.

Noble, G. K., 1927, The plethodontid salamanders; some aspects of their evolution, Am. Mus. Novit., 249:1.

Noble, G. K., 1929, The relations of courtship to secondary sex characters of the two-line salamander, Eurycea bislineata (Green), Am. Mus. Novit., 362:1.

Noble, G. K., 1931, "The Biology of the Amphibia," Dover, New York.

Noble, G. K., and Brady, M. K., 1931, The courtship of plethodontid salamanders, Copeia, 1930:52.

Noble, G. K., and Brady, M. K., 1933, Observations on the life history of the marbled salamander, Ambystoma opacum Grabenhorst, Zoologica, N.Y., 11:89.

O'Connell, R. J., 1977, From insect to mammal: Complications of the bioassay, in: "Chemical Signals in Vertebrates," D. Muller-Schwarze, and M. Mozell, eds., Plenum Press, New York.

Organ, J. A., 1961, Studies on the local distribution, life history, and population dynamics of the salamander genus Desmognathus in Virginia, Ecol. Monogr., 31:189.

Papenfuss, T. J., Wake, D. B., and Adler, K., 1983, Salamanders of the genus Bolitoglossa from the Sierra Madre del Sur of Southern Mexico, J. Herpetol., 17:295.

Parker, G. A., 1970, Sperm competition and its evolutionary consequences in the insects, Biol. Rev., 45:525.

Payne, A. P., 1978, Attractant properties of the Harderian gland and its products on male golden hamsters of differing sexual experience, Proc. Soc. Endocr., 8-30/9-2.

Peacock, R. L., and Nussbaum, R. A., 1973, Reproductive biology and population structure of the western red-backed salamander, Plethodon vehiculum (Cooper), J. Herpetol., 7:215.

Payne, T., Cardé, R., and Boech, J., eds., 1985. "Mechanisms of Perception and Orientation to Insect Olfactory Signals," Oxford Univ. Press, Oxford.

Perrigo, G., and Bronson, F. H., 1983, Communication disparities between genetically-diverging populations of deermice, in: "Chemical Signals in Vertebrates III," D. Muller-Schwarze, and R. M. Silverstein, eds., Plenum, New York.

Pfeiffer, W., 1974, Pheromones in fish and amphibia, in: "Pheromones," D. Birch, ed., North-Holland Publ. Co., Amsterdam.
Pool, T. B., and Dent, J. N., 1977, The ultrastructure and the hormonal control of product synthesis in the hedonic glands of the red-spotted newt Notophthalmus viridescens, J. Exp. Zool., 201:177.
Rogoff, J. L., 1927, The hedonic glands of Triturus viridescens: a structural and morphological study, Anat. Rec., 34:132.
Roth, G., 1976, Experimental analysis of prey catching behavior of Hydromantes italicus Dunn (Amphibia, Plethodontidae), J. Comp. Physiol., 109:47.
Salthe, S. N., 1967, Courtship patterns and phylogeny of the urodeles, Copeia, 1967:100.
Sever, D. M., 1975, Morphological and seasonal variation of the mental hedonic glands of the dwarf salamander, Eurycea quadridigitata (Holbrook), Herpetologica, 31:241.
Sever, D. M., 1976, Morphology of the mental hedonic gland clusters of plethodontid salamanders (Amphibia, Urodela, Plethodontidae), J. Herpetol., 10:227.
Signoret, J. P., 1976, Chemical communication and reproduction in domestic mammals, in: "Mammalian Olfaction, Reproductive Processes, and Behavior," R. L. Doty, ed., Academic Press, New York.
Signoret, J. P., and du Mesnil du Buisson, F., 1961, Etude de comportement de la Truie en oestrus, in: "4th Congres Inter. de Reprod. Animale et Insemination Artificielle."
Simon, G. S., and Madison, D. M., 1984, Individual recognition in salamanders: Cloacal odours, Anim. Behav., 32:1017.
Smith, A. B., III, Belcher, A. M., Epple, G., Jurs, P., and Lavine, B., 1985, Computerized pattern recognition: A new technique for the analysis of chemical communication, Science, 228:175.
Spotila, J. R., and Beumer, R. J., 1970, The breeding habits of the ringed salamander, Ambystoma annulatum (Cope), in northeastern Arkansas, Amer. Midl. Nat., 84:77.
Stacey, N., 1983, Hormones and pheromones in fish sexual behavior, BioScience, 9:552.
Steger, K. E., 1967, Avian olfaction, Amer. Zool., 7:415.
Stebbins, R. C., 1954, Natural history of the salamanders of the plethodontid genus Ensatina, Univ. Calif. Publ. Zool., 54:47.
Storez, R. A., 1969, Observations on the courtship of Ambystoma laterale, J. Herpetol., 3:87.
Teyssedre, C., and Halliday, T., Cumulative effect of male's displays in the sexual behaviour of the Smooth Newt Triturus vulgaris (Urodela, Salamandridae), Z. Tierpsychol., in press.
Thornhill, R., 1981, Panorpa (Mecoptera: Panorpidae) scorpionflies: Systems for understanding resource-defense polygyny and alternative male reproductive efforts, Ann. Rev. Ecol. Syst., 12:355.
Tilley, S. G., and Hausman, J. S., 1976, Allozymic variation and occurrence of multiple inseminations in populations of the salamander Desmognathus ochrophaeus, Copeia, 1976:734.
Tristram, D. A., 1977, Intraspecific olfactory communication in the terrestrial salamander Plethodon cinereus, Copeia, 1977:597.
Twitty, V. C., 1959, Migration and speciation in newts, Science, 130:1735.
Twitty, V. C., 1961, Experiments on homing behavior and speciation in Taricha, in: "Vertebrate Speciation," W. F. Blair, ed., Univ. Texas Press, Austin.
Uzzell, T., 1964, Relations of the diploid and triploid species of the Ambystoma jeffersonianum complex (Amphibia, Caudata), Copeia, 1964:257.
Uzzell, T., 1969, Notes on spermatophore production by salamanders of the Ambystoma jeffersonianum complex, Copeia, 1969:602.
Verrell, P., 1982a, The sexual behaviour of the red-spotted newt, Notophthalmus viridescens (Amphibia: Urodela: Salamandridae), Anim. Behav., 30:1224.

Verrell, P., 1982b, Newts can be colorful too, BioScience, 32:244.
Wade, M. J., and Arnold, S. J., 1980, The intensity of sexual selection in relation to male sexual behaviour, female choice, and sperm precedence, Anim. Behav., 28:446.
Wake, D. B., 1966, Comparative osteology and evolution of the lungless salamanders, family Plethodontidae. Mem. S. Calif. Acad. Sci., 4:1.
Walcott, C. and Lednor, A. J., 1983, Bird navigation, in: "Perspectives in Ornithology," A. H. Brush, and G. A. Clark, Jr., eds., Cambridge Univ. Press, Cambridge.
Whitten, W. K., and Bronson, F. H., 1970, The role of pheromones in mammalian reproduction, in: "Communication by Chemical Signals," J. W. Johnston, Jr., D. G. Moulton, and A. Turk, eds., Appleton-Century-Crofts, New York.
Williams, S., 1972, Reproduction and ecology of the Jemez Mountains salamander, Plethodon neomexicanus, M. S. Thesis, Univ. New Mexico.
Wilson, E. O., and Bossert, W. H., 1963, Chemical communication among animals, Rec. Prog. Horm. Res., 19:673.
Wrobel, D. J., Gergits, W. F., and Jaeger, R. G., 1980, An experimental study of interference competition among terrestrial salamanders. Ecology, 61:1034.

PHEROMONAL MARKERS AS TERRITORIAL ADVERTISEMENT BY TERRESTRIAL SALAMANDERS

Robert G. Jaeger

Department of Biology
University of Southwestern Louisiana
Lafayette, Louisiana 70504, U.S.A.

The existence of territoriality in salamanders was firmly established only about seven or eight years ago. Before that time, there was some evidence that individuals of several species engaged in intraspecific conflicts and that non-random distributions of individuals occurred. However, so little was known about the behavior of salamanders (other than courtship and homing) that it was impossible to piece together a comprehensive picture of spatial interactions. Today there is still less information about territorial behavior of salamanders than there is about almost any other group of vertebrates, and less than for many groups of invertebrates. The evolution, function, mechanisms, and ontogeny of territoriality are only vaguely understood. What knowledge does exist is confined to the salamander family Plethodontidae (the lungless salamanders), and within this family, only seven or eight species have been critically studied. These species are distributed among the three genera _Aneides_, _Desmognathus_, and _Plethodon_. The tendency thus far has been for researchers to study a few species in detail, rather than to examine superficially numerous species.

Studies of pheromonal advertisement during territorial defense are, essentially, just beginning. For that reason, in this paper I shall attempt to review what little information is available about territorial pheromones and, additionally, to put this information in the context of the natural history of selected species.

NATURAL HISTORY

The red-backed salamander, _Plethodon cinereus_, has been the most intensively studied species in terms of territorial behavior and consequent pheromonal communication. Therefore, I shall briefly describe important aspects of its life history. It occurs in forested areas throughout much of the eastern United States (Highton, 1972). Individuals are never found in standing or running water, but instead they inhabit the leaf-litter environment on the forest floor and subsurface tunnels in the soil. There is no larval stage, and eggs hatch directly into an immature stage that looks very much like the adult, except for smaller size. Sexual maturity is reached at the end of the second growing season (Sayler, 1966), at which point males court every year (autumn and spring) and females court every second year. The female lays a small clutch of seven to nine large eggs (Fraser, 1980) in late spring and provides parental care for the eggs for about a month. Adults reach a maximum size of about 52 mm snout-vent length (Fraser, 1980) and probably attain a maximum life span of about 18 years.

Note that these life history tactics are typical of a so-called K-selected species (Pianka, 1970): extended age to maturity, iteroparity with a small clutch size, parental (maternal, in this case) care, and a relatively long life. If one agrees with the principles of r- and K-selection theory, then one might suspect that *Plethodon cinereus* inhabits a highly competitive environment. This, indeed, is true.

INTRASPECIFIC COMPETITION AND SPACING

During periods of rainfall, individuals of *Plethodon cinereus* forage in the leaf-litter, ingesting large numbers of invertebrates (Jaeger, 1972). During rainless periods, though, salamanders are forced out of the drying leaf-litter (by the danger of desiccation), and they occupy moisture conserving patches under logs and rocks or they go underground; at these times, the stomachs of most salamanders are nearly devoid of prey (Jaeger, 1972). Thus, there are bottleneck periods when competition for prey becomes intense (Jaeger, 1980a). However, some members of the population are successful foragers (gaining far more calories than needed to sustain metabolism and movement) during both wet and dry periods while other members are unsuccessful foragers (negative energy budgets) during both periods (Jaeger, 1980a). This differential foraging success suggests that some individuals may gain access to rich patches of prey while preventing other foragers from doing so. This would be expected if territoriality controls access to prey.

Territoriality should also induce spacing among individuals, or at least among adults. In field studies, this has been shown to occur for *Plethodon cinereus* (Jaeger, 1979), in which interindividual spacing (males and females) increases from spring to summer, and in a larger congener, *P. glutinosus* (Wells, 1980), in which males tend not to associate with one another.

The overall conclusion from field studies, then, is that the periodicity of rainfall provides a situation in which prey are available to *Plethodon cinereus* at certain (wet) times and less available at other (dry) times. This periodicity in food availability causes bottleneck points when competition for food is intense, resulting in spacing among individuals and differential access to patches of prey. A similar situation occurs between species.

INTERSPECIFIC COMPETITION AND PARAPATRY

When different *Plethodon* species of similar body sizes are found together, they often maintain parapatric distributions, with little geographic overlap where the species meet (Highton, 1972). The cause of such parapatry has been worked out for a few pairs of species. *Plethodon cinereus* and *P. shenandoah*, in Shenandoah National Park, Virginia, are in strong competition for prey, with competition again being severe during rainless periods when salamanders are confined under rocks, logs, and underground (Jaeger, 1970, 1971a, 1971b, 1972, 1974). In this case, *Plethodon cinereus* excludes *P. shenandoah* from habitats of deep soil, forcing the latter species into small refuges of talus slopes. Due to its inability to compete successfully with *Plethodon cinereus* in areas of deep soil and due to frequent perturbations of moisture in the dry talus slopes, Plethodon shenandoah is presently undergoing extinction (Jaeger, 1980b). The competitive superiority of Plethodon cinereus in this case is due to interspecific territoriality, as was shown by a series of field and laboratory experiments by Gergits (1982). The mechanisms of the territorial superiority of Plethodon cinereus are not yet clear, but they apparently are governed by pheromonal communication (Jaeger and Gergits, 1979, to be discussed below) and agonistic behavioral interactions (Gergits, 1982).

In the Great Smoky Mountains and Balsam Mountains of North Carolina, Plethodon jordani inhabits higher elevations and Plethodon glutinosus occurs at lower elevations. Where the two species meet in the Great Smoky Mountains, geographic overlap is narrow while overlap is much greater in the Balsam Mountains. In a series of brilliant field experiments (which are perhaps the best test of competition theory in all of community ecology), Hairston (1973, 1980a, 1980b, 1981, 1983) demonstrated that competition between the two species is intense where the overlap is small and interspecific competition is reduced where overlap is extensive. In a subsequent study, Nishikawa (1985) demonstrated (by mapping generally nonoverlapping "home ranges" of individuals) that territoriality exists within each species, and it is likely (based on observations of interspecific aggression) that the two species defend territories against each other.

Fraser (1976a, 1976b) examined two dissimilarly sized, sympatric species of salamanders in the mountains of Virginia: Plethodon cinereus and P. hoffmani. He found no evidence of direct competition for food, and as an alternative, he proposed that interspecific competition occurs for space. This was the first substantial indication that spacing behavior via territoriality might occur within or between species of salamanders.

TERRITORIALITY

I prefer the definition of territoriality as given by Brown and Orians (1970): "a defended area." There are four essential characteristics by which territorial behavior can be detected (Gergits, 1982): the animal must (1) have a fixed site (or sites) that (2) is advertised (as by pheromones) and (3) is defended, and from which there (4) is a high probability of expelling an intruding competitor.

All of these characteristics have been studied with respect to interspecific territoriality between Plethodon cinereus and P. shenandoah (Gergits, 1982) and for intraspecific territoriality within Plethodon cinereus (Jaeger, 1981, 1984; Jaeger et al., 1982). Details of the behavioral interactions of the latter species will be given in another paper, because they are outside the scope of the present review. Here, I shall give only a brief synopsis.

During and shortly after rainfalls, individuals of Plethodon cinereus forage in the leaf-litter of the forest floor, and food is generally not a limiting resource at that time (Jaeger, 1978, 1980a). As the leaf-litter dries, the population apparently divides into two groups: those that have established territories return to or remain in them, under rocks and logs (Jaeger, 1980c); those that lack territories (apparently primarily juveniles) go underground into a maze of tunnels provided by the burrowing of other animals (Taub, 1961). This species is not well adapted for subsurface foraging, so those underground obtain little or no food during periods between rains (Fraser, 1976b). However, rocks and logs do harbor a small but generally consistent population of invertebrates (Jaeger, unpublished data), and these form the defendable patches (area) that both males and females utilize as feeding territories (Jaeger et al., 1982). Whether these territories are also used during periods of breeding is not yet clear.

Keen (1982) found that microhabitat utilization and activity of Desmognathus fuscus was dependent in part upon the presence or absence of Desmognathus monticola. He concluded that interference competition might be the cause of such distributional changes. Keen and Sharp (1984) later provided convincing evidence that males and females of Desmognathus monticola maintain intra- and interspecific territories. In their experiments, territorial residents were able to expel intruding conspecifics and even more so, intruding Desmognathus fuscus; aggressive interactions were common, with larger individuals being generally dominant to smaller ones, both within and between species.

Males of Aneides aeneus also apparently defend territories (Cupp, 1980). In laboratory experiments, territorial residents were successfully able to defend areas against intruders, and the same was true when the roles of individuals as intruders and residents were reversed. Aggressive interactions were common.

Species of the genus Plethodon inhabit forest floors, never enter water, and have no larval stage. Species of Desmognathus inhabit the areas in and around streams and have an aquatic larval stage. Aneides aeneus inhabits rock crevices and also has no aquatic larval phase in the life cycle. Because territorial behavior has been demonstrated in all three genera, this may indicate that territoriality has evolved broadly among the many genera of the family Plethodontidae.

For Plethodon cinereus, a territorial resident has three lines of defense against intruders. First, pheromones are placed on the substrate, and these advertise the presence of a defended area (Jaeger and Gergits, 1979; Jaeger et al., 1986). Second, if an intrusion does occur, residents and intruders will exchange aggressive and submissive visual displays (Jaeger, 1984). Third, if the intruder does not depart during the visual display phase, the resident will often attack and expel the intruder; this is done by biting the chemosensory organs (the nasolabial grooves) or the fat storage organ (the tail) of the intruder (Jaeger, 1981).

At this point, I shall discuss the first in this series of defensive tactics: the territorial pheromones.

PHEROMONAL ADVERTISEMENT OF TERRITORIES

Nothing is known about the biochemistry of the pheromones used in territorial advertisement. This is unavoidably true partly because the study of territoriality in salamanders is so new, partly because the existence of pheromones has been inferred only from behavioral responses of individuals under experimental conditions, and partly because none of the glands producing the suspected pheromones have been conclusively identified. In short, there is so little background information about territorial pheromonal communication in salamanders that sophisticated studies probably will not be conducted for several years yet.

In this section, then, I shall review the behavioral evidence for the existence of territorial pheromones, suggest some possible receptors involved, and mention the glands suspected of producing pheromones.

Behavior of the Receiver

Virtually all tests of territorial pheromones have involved the response of a receiver salamander in the presence or absence of presumed pheromones from a sender. Four basic experimental designs have been used.

(1) A salamander is placed on a substrate half of which has been "marked" by another salamander and half of which has been "marked" by the test animal (or left as a blank control). If the test salamander can distinguish between the two halves such that it spends statistically significantly more time on one half than the other, then the test salamander must be receiving some information about the presence or absence of another animal's secretions. The presence of pheromones is inferred from such behavior (Jaeger and Gergits, 1979).

(2) One salamander is allowed to "mark" a chamber, a second salamander is allowed to "mark" a second chamber, and a third chamber is left untouched. One of the salamanders is then picked up and sequentially placed alone in each chamber. If certain behavior patterns (such as feeding rate, sampling with the nasolabial cirri, etc.) vary significantly among the chambers, then presence or absence of pheromones is inferred (Jaeger et al., 1981, 1983).

(3) A salamander is allowed to walk down a Y-tube, with air flowing from both arms of the Y. The air streams, before reaching the test animal, pass over two constrained salamanders, or over one salamander paired with a blank control, or over differently "marked" substrates. If the test animal makes a significant choice for one air source over the other, pheromonal differences are inferred (Madison, 1975).

(4) A test salamander is given a choice between substrates (or burrows) containing different sources of fecal pellets (or cloacal washes). The differences might be between a test animal's own pellet versus a pellet of another animal, the test animal's pellet versus a surrogate (control) pellet, or a conspecific salamander's pellet versus a surrogate. If the test animal makes significant choices between substrates (or burrows) on the basis of type of pellet, the presence or absence of pheromones is inferred (Jaeger et al., 1986; Simon and Madison, 1984).

Madison (1975) conducted the first critical experiment using the Y-maze approach with *Plethodon jordani*. He found that before the breeding season, both males and females were attracted more to members of their own sex who had been neighbors in the forest than to members of the same sex who had not been forest neighbors. However, during the breeding season, neither males nor females demonstrated an intrasexual odor preference. This experiment was conducted before much was known about salamander territoriality, and so Madison (1975) concluded that the choices were based on kinship factors. Another possible explanation may be that individuals of *Plethodon jordani* were exhibiting "dear enemy recognition," in which territorial neighbors are less aggressive (and more submissive) toward one another than are non-neighbors. This phenomenon is well known in territorial species of birds (Fisher, 1954) and has been shown to occur in *Plethodon cinereus* (Jaeger, 1981). In either case, Madison (1975) did demonstrate that individuals of *Plethodon jordani* can differentiate the identities of other individuals and that this recognition must be based on chemical cues produced by the individuals themselves: i.e., pheromones.

Using experimental procedure (2) above, Tristram (1977) found that indivdiuals of *Plethodon cinereus* tapped their nasolabial cirri (chemo-receptors) at a significantly faster rate to a substrate marked with their own secretions than to a conspecific marked substrate or to an unmarked substrate. This suggested that an individual could distinguish its own marked area from that of another salamander.

Extending Tristram's (1977) research, McGavin (1978) used experimental procedure (2) above to determine whether "dear enemy recognition" occurs in *Plethodon cinereus*. She allowed certain individuals to become experienced with the substrate markings of certain other individuals while remaining ignorant of the markings of another set of individuals. When tested for recognition between the two types of markings, the salamanders tapped their nasolabial cirri at a significantly faster rate to substrates marked with familiar odors than to those marked with unfamiliar ones. As with Madison's (1975) study, this demonstrated that salamanders of this species also can distinguish among pheromones of conspecifics: i.e., individual recognition is possible.

Jaeger and Gergits (1979) used experimental procedure (1) above to determine whether pheromones could communicate information interspecifically as well as intraspecifically, and between sexes as well as within sexes. Within *Plethodon cinereus*, both adult males and females avoided substrates previously marked by males but neither avoided (nor were attracted to) substrates previously marked by females. These experiment were conducted outside the breeding season and with non-gravid females to avoid the influence of courtship pheromones. The results indicate that males place pheromones on the substrate but females do not; or if females do produce pheromones, they are ignored. In retrospect, these results are surprising, because adult males and females are now known to be territorial (Jaeger et al., 1982). Thus, it would seem that males can advertise their territories by chemical cues as well as by visual displays (Jaeger, 1984) while females may be able to utilize visual displays only (Jaeger, 1984). Current research in my laboratory is attempting to clarify this apparent sex difference of territorial pheromones.

Jaeger and Gergits (1979) also examined interspecific responses of *Plethodon cinereus* and *P. shenandoah*, two species known to be in intense competition in forest habitats with the former species being dominant to the latter. They found that males and females of *Plethodon shenandoah* avoided substrates marked by males and females of *Plethodon cinereus*; also, males of *Plethodon cinereus* avoided substrates marked by both sexes of *Plethodon shenandoah*. Female *Plethodon cinereus* avoided substrates marked by congeneric females but not those marked by congeneric males. Overall, then, congeners avoided each others marked areas in nearly every case. These results are important, because they demonstrate interspecific communication via pheromones. Such communication would be necessary for the evolution of interspecific territoriality, which was demonstrated by Gergits (1982). Note that males of both *Plethodon cinereus* and *P. shenandoah* avoided substrates marked by congeneric females, while males of the former species did not respond to substrates of conspecific females. This implies that females of both species must be producing a pheromone that is detectable between species, if not within a species. This confusing situation requires further experimentation.

Using experimental procedure (2) above, Jaeger et al. (1981) allowed individuals of *Plethodon cinereus* to forage in their own marked territories, in territories previously marked by conspecifics, and in an unmarked control area. Results showed that foraging efficiency (capture rates, net caloric gain) decreased in the second and third conditions compared to the first. In their own territories, salamanders foraged as would be predicted from optimal foraging theory (Pyke, 1984). In territories of conspecifics, salamanders forwent foraging and devoted more time to submissive posturing (Jaeger, 1984), as if to allay an attack from the territorial resident (who was not even present). In unmarked chambers, salamanders reduced foraging and devoted more time to "marking" (with the cloaca) behavior, as if establishing territories of their own. This marking behavior was significantly faster for males than for females. This experiment again shows that territorial pheromones do exist in Plethodon cinereus and that such chemical cues tend to elicit submissive behavior from intruders. Thus, these pheromones alone may be an effective deterrent to territorial intrusions.

This research was extended by Jaeger et al. (1983) using experimental procedure (2), to incorporate "dear enemy recognition". Salamanders were tested, for foraging efficiency as above, on an unmarked substrate, on a substrate previously marked by pheromones with which the test animal was familiar (through prior experience), and on a substrate previously marked by pheromones with which the test animal was unfamiliar (no prior experience). The salamanders were unaffected in foraging behavior by the presence of familiar pheromones, but the presence of unfamiliar pheromones caused a reduction in prey capture rates and net caloric gain. As in some previous studies (e.g., Madison, 1975), this experiment demonstrated that salamanders can differentiate among other individuals on the basis of pheromones (individual recognition) and that they are not particularly threatened by the pheromones familiar to them via prior experience. This would be expected of "dear enemy recognition" in territorial situations.

Jaeger (1981) looked at the interactive agonistic behavior of Plethodon cinereus that were either familiar or unfamiliar with each others' pheromones. Using a variation of experimental procedure (2), I allowed pairs of salamanders to sample each others' marked substrates without ever coming into physical or visual contact with each other. When placed together in neutral (unmarked) chambers, the paired salamanders exhibited few aggressive and many submissive displays (Jaeger, 1984) toward each other. When the same salamanders were paired with animals with which they had had no prior pheromonal experience, aggressive displays (and biting) significantly increased and submissive displays significantly decreased. This experiment provides the best evidence yet that pheromones are not only placed in territories, but that they also control the level of aggression among salamanders.

Two attempts have been made to identify the source of the pheromones used in territorial communication by salamanders. The first was by Simon and Madison (1984), using experimental procedure (4) and Plethodon cinereus. They provided each male salamander with a choice of two substrates, each marked with either fecal pellets or cloacal washes. On one side, the pellets (or washes) were the test animal's own; on the other side, the pellets (or washes) were from a conspecific. None of the salamanders had prior experience with each other. For both pellets and washes, salamanders significantly chose their own marked substrates over those of conspecifics. This was the first indication that fecal pellets are used to place pheromones in territories. In addition, the data from cloacal washes suggest that pheromones are produced by glands in or around the cloaca, and are not a by-product of the chemical composition of the feces per se.

Jaeger et al. (1986), again using experimental procedure (4) and Plethodon cinereus, asked whether male salamanders could distinguish between burrows based on the source of fecal pellets placed at their entrances. Each salamander was tested four times: with own versus conspecific pellets, own versus surrogate (wadded paper) pellets, conspecific versus surrogate pellets, and a control of two surrogate pellets. In the first case, salamanders spent significantly more time in the burrows with their own pellets and spent significantly more time submissively displaying to the burrows with the conspecifics' pellets. In the second case, significantly more time was spent in "own burrows" than in "surrogate burrows," and there was negligible submission to either type of burrow. In the third case, salamanders spent significantly more time in "surrogate burrows" than in "conspecific burrows," and there was significantly more submissive displaying to the latter than to the former. The fourth case indicated no position bias in the experiment. These data demonstrate, then, that male salamanders of this species are not only attracted to areas marked with their own pheromones, but are repulsed (and threatened) by areas marked by pheromones of other males. Territorial advertisement is inferred from these data. Parallel tests with females are now being performed.

In summary, virtually all of the information on territorial pheromones (through behavioral inferences of their existence) comes from single species, Plethodon cinereus. Evidence to date indicates that at least males (and perhaps females) place chemical substances on the substrate, that these substances are contained in fecal pellets and are produced in or near the cloaca (though other glandular sources are not ruled out), that these substances tend to deter intruders from entering a marked (territorial) area, and that the aggressive and submissive interactions of territorial residents and intruders are governed in part by these substances (and in part by visual displays). Thus, pheromones, whatever they are, are used by Plethodon cinereus (and perhaps other species of salamanders) to advertise territorial possession.

Sensory Receptors

Apparently the nares are the principal centers for receiving pheromonal information. Olfactory receptors in the nares are able to discriminate among a variety of odors in salamanders of the family Ambystomatidae (Kauer, 1974; Kauer and Moulton, 1974; Kauer and Shepherd, 1975, 1977). However, members of this family are not known to be territorial (Martin et al., in press).

Much less neurophysiological research has been performed with species of Plethodontidae, perhaps because of the small size of many species (such as Plethodon cinereus). All species of plethodontid salamanders possess a complex chemosensory apparatus attached to the nares (Fig. 1). A thin capillary tube (nasolabial groove) descends from each naris to the upper lip and terminates in a nasolabial cirrus. In Plethodon cinereus, the two cirri are alternately tapped to the substrate (Fig. 1B) as the salamander walks (Fig. 1A), apparently monitoring chemical cues. Brown (1968), using Ensatina, found that chemicals pass up the nasolabial grooves, to the external nares, into the nasal chambers, over the chemoreceptors of the Jacobson's organs, and down through the internal nares into the pharynx. Consequently, plethodontid salamanders have an efficient mechanism for sampling chemicals on the substrate, such as pheromones placed there in fecal pellets or by other cloacal exudates.

Both Tristram (1977) and McGavin (1978), using Plethodon cinereus, observed changes in the rate of tapping of the nasolabial cirri to the substrate when pheromonal cues were changed (own versus conspecific markings, familiar versus unfamiliar conspecific markings, respectively). This implies that the cirri are used for monitoring pheromones. Jaeger (1984) found that the rate of tapping cirri to fecal pellets increased ten-fold for a territorial resident after an intruder entered the territory. Jaeger et al. (1986) also found a significant increase in rate with which males tapped cirri to fecal pellets of conspecific males versus control surrogate pellets.

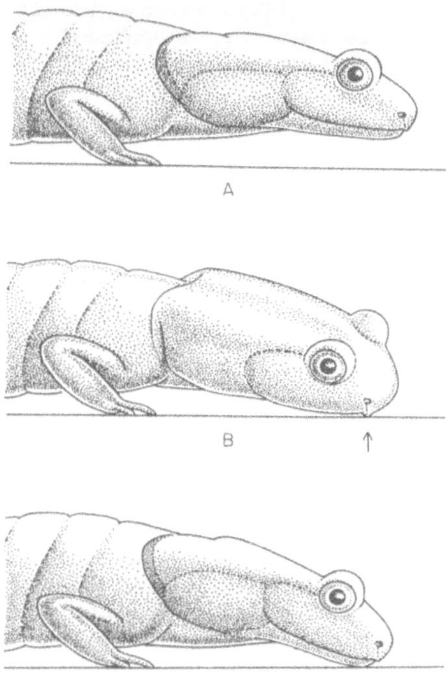

Fig. 1. Three behavior patterns of Plethodon cinereus: (A) normal walking posture, (B) tapping the nasolabial cirri to the substrate, and (C) "chin scraping," which may be involved in pheromonal marking.

Given these data, there is considerable evidence that the nasolabial cirri, and ultimately the chemoreceptors in the nares, are the principal monitors of territorial pheromones.

Glands Involved in Production of Pheromones

Since males of *Plethodon cinereus* place pheromones into fecal pellets (Jaeger et al., 1986; Simon and Madison, 1984), the cloaca is the logical source of those pheromones. Of the many glands in the cloaca, the vent gland is the most likely to be involved in the production of pheromones utilized in territorial advertisement. This gland is in the epidermal area of the posterior region of the cloacal chamber (Sever, 1978a). It is found in all species of Plethodontidae thus far examined (Sever, 1983), and it is the only cloacal gland in males that remains active outside the courtship season (Sever, 1978a): i.e., when territories are being maintained.

The data from Jaeger and Gergits (1979) suggest that female *Plethodon cinereus* do not produce territorial pheromones, because both conspecific males and females are unresponsive to substrates upon which females have resided. Females lack the vent gland, and indeed they lack all cloacal glands except for the spermatheca (Sever, 1978b). If females do produce cloacal pheromones, they must be derived from the spermatheca. This is a possibility, because female Plethodon cinereus do mark substrates that are recognizable to males and females of Plethodon shenandoah (Jaeger and Gergits, 1979).

However, both males and females of Plethodon cinereus have numerous granular glands on the ventral surface of the tail, just posterior to the cloacal orifice (Sever, pers. comm.). Sever has suggested that these granular glands might produce chemical substances that could be smeared onto a recently deposited fecal pellet as the tail moves over it. Also, both males and females frequently touch the cloacal (and tail) area to substrates while walking, which appears to be a marking behavior (Jaeger et al., 1981). Obviously, much more research is needed on the function of cloacal and surrounding glands before their significance in pheromonal communication can be understood.

There is another possible source of territorial pheromones in Plethodon cinereus: the submandibular region. Both Tristram (1977) and Jaeger et al. (1981) noted that both males and females frequently tap the chin to the substrate (Fig. 1C). Chin tapping increases significantly when a salamander leaves its own territory and enters either a previously unmarked area or one previously marked by a conspecific (Jaeger et al., 1981). This is a very puzzling behavior, because while males have a mental gland on the chin, such glands are unknown for females of any species (Sever, 1975). Whether the submandibular region produces pheromones used in territorial advertisement is still a mystery.

OTHER USES OF PHEROMONES

There are at least two functions of salamander pheromones other than in territorial advertisement: homing and courtship. A salamander that is displaced from its home range (or territory) often will quickly return. This has been shown for Plethodon jordani (Madison, 1969; Madison and Shoop, 1970) and for Plethodon cinereus (Kleeberger and Werner, 1982; Samuelsen, 1977). Madison (1969, 1972) showed that homing is based on olfactory responses, and he suggested that chemicals produced by conspecifics living around the displaced animal's home area may be important as site specific odor cues. Thus, individual recognition, through pheromones, of neighbors might help a salamander find its way home.

There is a large literature on courtship behavior of salamanders, but only a few papers will be cited here. Dawley (1984a) showed that, with few exceptions, males and females of Plethodon jordani and P. glutinosus can recognize not only members of their own sex via airborne odors, but they can recognize conspecific members of the opposite sex. Dawley 1984b) also found that male newts, Notophthalmus viridescens, can orient toward water-borne odors of females. These studies indicate that pheromones may provide a pre-mating reproductive isolating mechanism and may allow males to track females during courtship seasons. During courtship itself, males of many species produce pheromones that affect the female's behavior; pheromonal production by males in turn is affected by the female's behavior (Arnold, 1976; Arnold and Houck, 1982).

The purpose of this scant review of homing and courtship is to emphasize that salamanders produce a variety of pheromones, probably from a variety of glands, and having a diversity of functions. However, there is not a single case where the composition and origin of a pheromone is well known, a factor that severely inhibits the study of pheromonal communication during territorial defense by salamanders.

REFERENCES

Arnold, S. J., 1976, Sexual behavior, sexual interference and sexual defense in the salamanders Ambystoma maculatum, Ambystoma triginum and Plethodon jordani, Z. Tierpsychol., 42:247.

Arnold, S. J., and Houck, L. D., 1982, Courtship pheromones: Evolution by natural and sexual selection, in: "Biochemical Aspects of Evolutionary Biology," M. H. Nitecki, ed., University of Chicago Press, Chicago.

Brown, C., 1968, Additional observations on the function of the nasolabial grooves of plethodontid salamanders, Copeia, 1968:728.

Brown, J. L., and Orians, G. H., 1970, Spacing in mobile animals, Ann. Rev. Ecol. Syst., 1:239.

Cupp, P. V., 1980, Territoriality in the green salalander, Aneides aeneus, Copeia, 1980:463.

Dawley, E. M., 1984a, Recognition of individual, sex and species odours by salamanders of the Plethodon glutinosus-P. jordani complex, Anim. Behav., 32:353.

Dawley, E. M., 1984b, Identification of sex through odors by male red-spotted newts, Notophthalmus viridescens, Herpetologica, 40:101.

Fisher, J., 1954, Evolution and bird sociality, in: "Evolution as a Process," J. Huxley, A. C. Hardy, and E. B. Ford, eds., Allen and Unwin, London.

Fraser, D. F., 1976a, Coexistence of salamanders of the genus Plethodon: A variation of the Santa Rosalia theme, Ecology, 57:238.

Fraser, D. F., 1976b, Empirical evaluation of the hypothesis of food competition in salamanders of the genus Plethodon, Ecology 57:459.

Fraser, D. F., 1980, On the environmental control of oocyte maturation in a plethodontid salamander, Oecologia, 46:302.

Gergits, W. F., 1982, Interference Competition and Territoriality Between the Terrestrial Salamanders Plethodon cinereus and Plethodon shenandoah, M. S. Thesis, State University of New York at Albany, Albany.

Hairston, N. G., 1973, Ecology, selection and systematics, Breviora, 414:1.

Hairston, N. G., 1980a, Evolution under interspecific competition: Field experiments on terrestrial salamanders, Evolution, 34:409.

Hairston, N. G., 1980b, The experimental test of an analysis of field distributions: Competiton in terrestrial salamanders, Ecology, 61:817.

Hairston, N. G., 1981, An experimental test of a guild: Salamander competition, Ecology, 62:65.

Hairston, N. G., 1983, Alpha selection in competing slamanders: Experimental verification of an a priori hypothesis, Am. Nat., 122:105.

Highton, R., 1972, Distributional interactions among eastern North American salamanders of the genus Plethodon, Virginia Polytechnic Inst. Res. Div. Monogr., 4:139.

Jaeger, R. G., 1970, Potential extinction through competition between two species of terrestrial salamanders, Evolution, 24:632.

Jaeger, R. G., 1971a, Moisture as a factor influencing the distributions of two species of terrestrial salamanders, Oecologia 6:191.

Jaeger, R. G., 1971b, Competitive exclusion as a factor influencing the distributions of two species of terrestrial salamanders, Ecology, 52:632.

Jaeger, R. G., 1972, Food as a limited resource in competition between two species of terrestrial salamanders, Ecology, 53:535.

Jaeger, R. G., 1974, Competitive exclusion: Comments on survival and extinction of species, BioScience, 24:33.

Jaeger, R. G., 1978, Plant climbing by salamanders: Periodic availability of plant-dwelling prey, Copeia, 1978:686.

Jaeger, R. G., 1979, Seasonal spatial distributions of the terrestrial salamander Plethodon cinereus, Herpetologica, 35:90.
Jaeger, R. G., 1980a, Fluctuations in prey availability and food limitation for a terrestrial salamander, Oecologia, 44:335.
Jaeger, R. G., 1980b, Density-dependent and density-independent causes of extinction of a salamander population, Evolution, 34:617.
Jaeger, R. G., 1980c, Microhabitats of a terrestrial forest salamander, Copeia, 1980:265.
Jaeger, R. G., 1981, Dear enemy recognition and the costs of aggression between salamanders, Am. Nat., 117:962.
Jaeger, R. G., 1984, Agonistic behavior of the red-backed salamander, Copeia, 1984:309.
Jaeger, R. G., and Gergits, W. F., 1979, Intra- and interspecific communication in salamanders through chemical signals on the substrate, Anim. Behav., 27:150.
Jaeger, R. G., Goy, J. M., Tarver, M., and Marquez, C. E., 1986, Salamander territoriality: Phermonal markers as advertisement by males, Anim. Behav., 34: in press.
Jaeger, R. G., Joseph, R. G., and Barnard, D. E., 1981, Foraging tactics of a terrestrial salamander: Sustained yield in territories, Anim. Behav., 29:1100.
Jaeger, R. G., Kalvarsky, D., and Shimizu, N., 1982, Territorial behaviour of the red-backed salamander: Expulsion of intruders, Anim. Behav., 30:490.
Jaeger, R. G., Nishikawa, K. C. B., and Barnard, D. E., 1983, Foraging tactics of a terrestrial salamander: Costs of territorial defence, Anim. Behav., 31:191.
Kauer, J. S., 1974, Response patterns of amphibian olfactory bulb neurones to odour stimulation, J. Physiol., 243:695.
Kauer, J. S., and Moulton, D. G., 1974, Responses of olfactory bulb neurones to odour stimulation of small nasal areas in the salamander, J. Physiol., 243:717.
Kauer, J. S., and Shepherd, G. M., 1975, Olfactory stimulation with controlled and monitored step pulses of odor, Brain Res., 85:108.
Kauer, J. S., and Shepherd, G. M., 1977, Analysis of the onset phase of olfactory bulb unit responses to odour pulses in the salamander, J. Physiol., 272:495.
Keen, W. H., 1982, Habitat selection and interspecific competition in two species of plethodontid salamanders, Ecology, 63:94.
Keen, W. H., and Sharp, S., 1984, Responses of a plethodontid salamander to conspecific and congeneric intruders, Anim. Behav., 32:58.
Kleeberger, S. R., and Werner, J. K., 1982, Home range and homing behavior of Plethodon cinereus in northern Michigan, Copeia, 1982:409.
Madison, D. M., 1969, Homing behaviour of the red-cheeked salamander, Plethodon jordani, Anim. Bahav., 17:25.
Madison, D. M., 1972, Homing orientation in salamanders: A mechanism involving chemical cues, in: "Animal Orientation and Navigation," S. R. Galler, K. Schmidt-Koenig, G. J. Jacobs, and R. E. Belleville, eds., National Aeronautics and Space Administration, Washington, D.C.
Madison, D. M., 1975, Intraspecific odor preferences between salamanders of the same sex: Dependence on season and proximity of residence, Can. J. Zool., 53:1356.
Madison, D. M., and Shoop, C. R., 1970, Homing behavior, orientation, and home range of salamanders tagged with Tantalum-182, Science, 168:1484.
Martin, D. L., Jaeger, R. G., and Labat, C. P., Territoriality in an Ambystoma salamander? Support for the null hypothesis, Copeia, in press.
McGavin, M., 1978, Recognition of conspecific odors by the salamander Plethodon cinereus, Copeia, 1978:356.

Nishikawa, K. C. B., 1985, The Ecology and Evolution of Aggressive Behavior in Two Species of Terrestrial Salamanders, Ph.D. Dissertation, University of North Carolina, Chapel Hill.

Pianka, E. R., 1970, On r- and K-selection, Am. Nat., 104:592.

Pyke, G. H., 1984, Optimal foraging theory: A critical review, Ann. Rev. Ecol. Syst., 15:523.

Samuelsen, P., 1977, The Home Range and Homing Behavior of the Red-backed Salamander, Plethodon cinereus, M. S. Thesis, University of Rhode Island, Kingston.

Sayler, A., 1966, The reproductive ecology of the red-backed salamander, Plethodon cinereus, in Maryland, Copeia, 1966:183.

Sever, D. M., 1975, Morphology and seasonal variation of the mental hedonic glands of the dwarf salamander, Eurycea quadridigitata (Holbrook), Herpetologica, 31:241.

Sever, D. M., 1978a, Male cloacal glands of Plethodon cinereus and Plethodon dorsalis (Amphibia: Plethodontidae), Herpetologica, 34:1.

Sever, D. M., 1978b, Female cloacal anatomy of Plethodon cinereus and Plethodon dorsalis (Amphibia: Urodela, Plethodontidae), J. Herpetol., 12:397.

Sever, D. M., 1983, Cloacal anatomy of male salamanders in the plethodontid subfamily Desmognathinae, Herpetologica, 39:16.

Simon, G. S., and Madison, D. M., 1984, Individual recognition in salamanders: Cloacal odours, Anim. Behav., 32:1017.

Taub, F. B., 1961, The distribution of the red-backed salamander, Plethodon c. cinereus, within the soil, Ecology, 42:681.

Tristram, D. A., 1977, Intraspecific olfactory communication in the terrestrial salamander Plethodon cinereus, Copeia, 1977:597.

Wells, K. D., 1980, Spatial associations among individuals in a population of slimy salamanders (Plethodon glutinosus), Herpetologica, 36:271.

THE RECOGNITION AND USE OF CHEMICAL SIGNALS BY A NESTING SALAMANDER

Don C. Forester

Department of Biological Sciences
Towson State University
Towson, Maryland 21204 U.S.A.

LIFE HISTORY

The mountain dusky salamander (Desmognathus ochrophaeus Cope) is a small plethodontid (snout-vent length = 30-40 mm), indigenous to the Appalachian highlands of eastern North America. At low elevations, the species inhabits moist microhabitats associated with stream margins and seepages; at high elevations, the species exhibits pronounced terrestrialism, ranging far up wooded slopes (Hairston, 1949). Males and females reach sexual maturity at an age of three to five years (Tilley, 1977, 1980), and males court females during the fall and spring (Houck, 1982). Oviposition occurs annually (Forester, 1977) between mid-June and mid-July (Tilley, 1977; Forester, 1979a). Nesting sites are located beneath natural debris along stream margins and in seepages (Bishop, 1943). Following egg deposition, the female parent remains with the eggs until they hatch (46-69 days, Tilley, 1972; Forester, 1974). During this period, the female contributes to the survival of the eggs by: (1) reducing predation (as a result of aggressive nest defense; Forester, 1978, 1979a, 1983); (2) reducing pathogenic infestation (Forester, 1979a); (3) providing mechanical stimulation resulting in increased oxygenation and reduced yolk layering (Forester, 1979a); (4) reducing the rate of clutch desiccation (Forester, 1984).

There is a significant volume of evidence (both direct and indirect) to support the importance of chemical cues to female mountain dusky salamanders during the prolonged nesting season. In this paper, I shall review work pertaining to the role of chemical signals during philopatry and homing to the nest following experimental displacement. In addition, I shall examine the phenomenon of egg recognition by focusing on the following questions: (1) Does an egg mass possess an individually unique chemical signature? (2) Does an egg mass possess a species-specific chemical signature? (3) Do chemical cues enable females to determine the chronological age of an egg mass? (4) What is the relationship between egg recognition and practice of oophagy (egg eating) by the attending parent?

PHILOPATRY AND HOMING TO THE NEST

Female mountain dusky salamanders are reported to nest in the same stream segment in successive years (Forester, 1977). During the summer of 1972, I located 119 brooding females along the margin of a small montane stream. The females were collected, photographed and returned to their nest site. Although only seven were recaptured the following year, of these, all were nesting within 5 m of their previous nesting site. While it is adaptive for females to practice philopatry (i.e., to return to the site of previous reproductive success), there is no direct evidence to indicate that chemical cues are involved in the process. However, olfaction is known to be important during location of reproductive sites by the western, red-bellied newt, Taricha rivularis, (Twitty, 1961; Grant et al., 1968), and the eastern red-spotted newt, Notophthalmus viridescens, (Hershey and Forester, 1981). In addition, a related congener, Desmognathus fuscus, has been shown to experience impaired homing ability when rendered anosmic (Barthalmus and Bellis, 1972).

The ability of female plethodontids to relocate their nests following experimental displacement is well documented (Aneides aeneus: Gordon, 1961; Woods, 1968; Snyder, 1971; Plethodon cinereus: Bachmann, 1964; Desmognathus auriculatus: Rose, 1966; D. ochrophaeus: Forester, 1981). The mechanism by which the nest is relocated is unclear, but likely involves olfactory cues. Rose (1966) observed that homing success decreased with distance and that females displaced with their eggs did not attempt to return to the nest site. He concluded that a chemical bond existed between the female and her eggs. Bachmann (1964) observed that displaced female red-backed salamanders would vigorously nudge and bite at their eggs upon relocation. She speculated that the females were trying to rub something on or off the eggs."

When I displaced 107 attending females 2 m from their nests, 78 (73%) homed within 24 hours (Forester, 1981). Nesting sites were often in close proximity to one another, and on one occasion, seven females were found nesting on a single, moss-covered rock (25 x 25 cm). When marked and released, 5 of 7 returned to the rock and rejoined their own eggs despite the presence of unattended clutches only a few cm away.

The apparent ability of females to return to their nest following displacement suggested that, not only could they locate an egg mass, but they could also distinguish their own from those of conspecifics. When I displaced females and switched their eggs from one nesting cavity to another, they returned to their eggs and not the nest cavity (Forester, 1974), suggesting a unique chemical signature associated with the eggs themselves.

INDIVIDUAL RECOGNITION

During the brooding period, female salamanders reduce their foraging and become generally less active, remaining in the nest cavity more than 80% of the time (Bachmann, 1964; Forester, 1981). When they leave their eggs, either voluntarily (to forage) or involuntarily (to escape predators or as a result of seasonal flooding), they are confronted with the problems of nest location and egg recognition.

In an early attempt to document individual egg recognition (Forester, 1979b), I placed females in a circular arena containing their own eggs and those of two conspecfic females. The egg masses were positioned in dishes spaced equidistantly about the perimeter of the arena (Fig. 1). Females were observed brooding their own eggs after one hour in 32 of 51

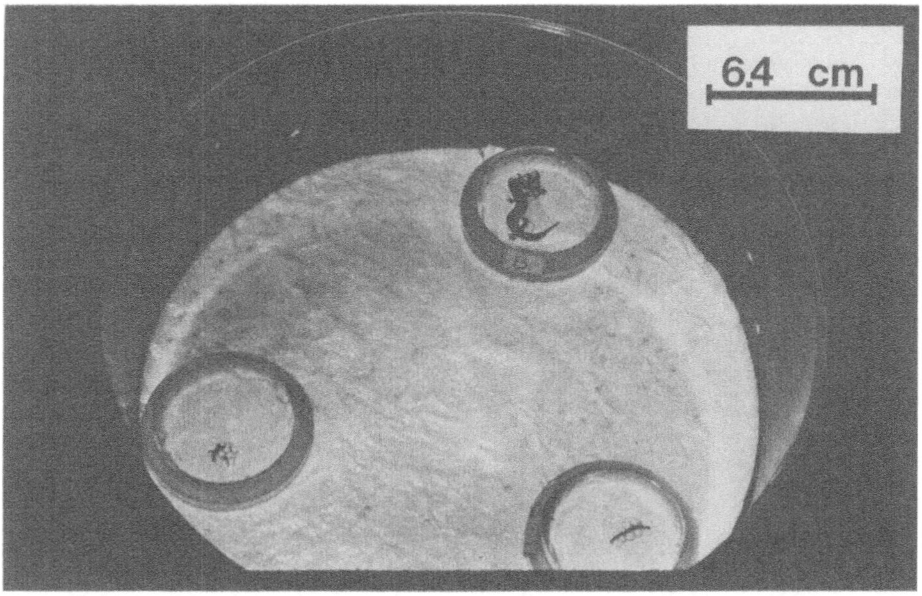

Fig. 1. Egg discrimination apparatus used in three choice experiments designed to test a female's ability to recognize her own eggs in the presence of other unattended clutches.

trials ($X^2 = 19.88$, df = 2, $P < 0.005$). Since the experiments were conducted in total darkness, I hypothesized that clutch recognition involved airborne chemoreception rather than vision. Airborne chemical cues are known to be important to plethodontid salamanders attempting to relocate their home areas following experimental displacement (Madison, 1969; Madison and Shoop, 1970), and for social recognition (Dawley, 1984). Nevertheless, I was troubled by the fact that, during numerous trials, the unselected egg masses had been displaced from their respective dishes, indicating contact by the female during the one hour test period. Furthermore, I had anticipated greater clutch recognition accuracy and had assumed that females would move directly to their own eggs. Was it possible that egg recognition relied upon contact chemoreception facilitated by trial and error searching? In collaboration with two undergraduate students, I designed an apparatus which would enable me to accurately determine a female's ability to recognize her own clutch as well as her tendency to remain with her eggs during an extended test period.

For convenience, I elected to examine egg recognition and clutch fidelity in two congeneric species -- Desmognathus ochrophaeus (n = 22) and D. fuscus (n = 20). These species were selected for investigation because nesting females are readily available throughout the summer. The females of each species were individually placed in confinement cylinders of one of five identical discrimination chambers (Fig. 2). Following a 30 min acclimation period (during which the female could become familiar with the olfactory environment of the test chamber), the cylinder was removed and each test female allowed to move at will within the chamber for 24 h. A simulated nest cavity had been constructed at each end of the chamber. During a given trial, one nest cavity contained eggs of the test female, and the other contained a conspecific egg mass of similar developmental age. A pair of infrared light beams connected to separate channels

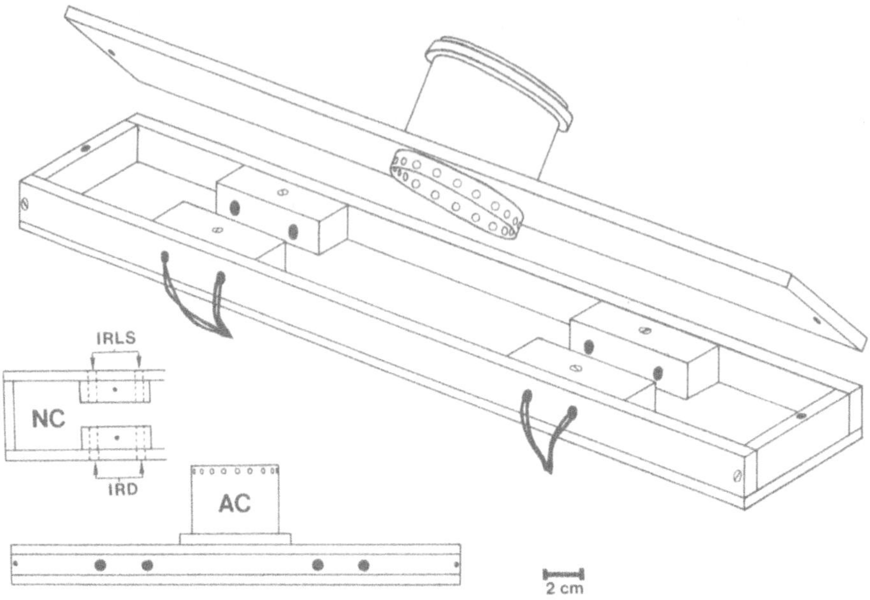

Fig. 2. Egg discrimination apparatus. NC = nest chamber, AC = acclimation cylinder, IRLS = infrared light source, IRD = infrared detector. Scale refers to the two dimensional diagrams only. In the two dimensional diagram (lower left) the acclimation cylinder has been inverted and placed as a cover for the opening in the lid. (From Forester et al., 1983).

of an Esterline-Angus event recorder made it possible to monitor the movement of females into and out of each nest cavity. The nest cavity holding the eggs of the test female was alternated during successive trials, and females were tested only once.

Four aspects of female behavior were considered as indicative of egg recognition: (1) whether or not a female made initial contact with her own clutch, (2) the percentage of the test period spent with an egg mass (brooding time), (3) the percentage of the brooding time spent with her own eggs, (4) the tendency of a female to be in contact with her own eggs after 24 h. For the purpose of statistical analysis, females which spent significantly more than 50% of the brooding time in the presence of their own eggs were said to exhibit clutch fidelity.

The results of these tests are presented in Table 1. Females of both species exhibited clutch fidelity and a significant number were in the presence of their own eggs after 24 h. The only difference between species involves the initial response of females toward the two egg masses. The fact that *D. fuscus* females did not exhibit an initial preference for their own eggs, coupled with the fact that the initial preference of *D. ochrophaeus* was just significant, leads me to hypothesize that individual clutch recognition involves perception of airborne chemical cues of low volatility and/or contact chemoreception relying upon trial and error searching. Females were most active during the first six h following their release from the cylinder, and regardless of their initial choice, most females contacted both egg masses during the 24 hour test period.

Table 1. Data summary from experiments designed to test for the presence of a unique, individual chemical marker among salamander eggs. OCH = _ochrophaeus_, FUS = _fuscus_, CS = conspecific.

		Initial Response		After 24 H		$\bar{X}$ Brooding Time		Clutch Fidelity	
Species	n	own	CS	own	CS	Total	own	own	CS
OCH	22	16	6	15	4	90%	77%	16	6
		$X^2 = 4.55$ $P < 0.05$		$X^2 = 6.37$ $P < 0.025$				$X^2 = 4.55$ $P < 0.05$	
FUS	20	9	11	15	5	88%	70%	15	5
		$X^2 = 0.2$ $P > 0.05$		$X^2 = 5.0$ $P < 0.05$				$X^2 = 5.0$ $P < 0.05$	

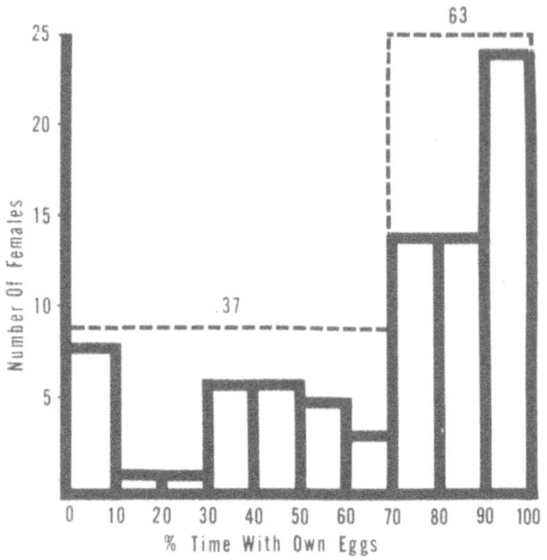

Fig. 3. Individual variation in the percent time spent brooding. The dotted lines emphasize behavioral variability. Numbers over the dotted lines represent the proportion of the females comprising each group. N = 82. (From Forester et al., 1983).

Factors Influencing The Persistence of Egg Recognition

Parental investment theory (Trivers, 1972) predicts that a female should maintain or improve her egg recognition and maternal drive throughout the brooding period. Experiments were conducted to determine the effect of temporal isolation on the ability of females to recognize their eggs, as well as to determine if egg recognition varies over the course of the brooding season (Forester et al., 1983).

Eighty-five attending females were divided into two groups based on the approximate age of the eggs they were brooding (egg staging based on Tilley, 1972). Females comprising one group had been attending eggs for less than 21 days (herein referred to as "early" eggs), while females comprising the second group had been attending for at least 44 days (herein referred to as "late" eggs). The eggs of this latter group were within 9 days of hatching.

Each "age group" was further divided into five subgroups. Prior to egg discrimination experiments, the females from each subgroup were isolated from their eggs for 0, 24, 48, 72, or 96 h, respectively. During periods of isolation exceeding 24 h, each clutch was rinsed in aged tap water and rolled about the dish to disrupt the growth of fungi. The females of each isolation class were tested in the experimental apparatus described above.

Egg recognition and clutch fidelity did not change from the beginning to the end of the brooding season. Collectively, the females of each age group spent significantly more than 50% of their brooding time with their own eggs (early--$\bar{X}$ = 69%, X^2 = 14.44, df = 1, $P < 0.001$; late--$\bar{X}$ = 62%, X^2 = 5.65; late--H = 7.557, $P < 0.025$). Single factor analysis of variance by ranks revealed no difference in the % time spent with "own eggs" between isolation classes within either age group (early--H = 6.609,

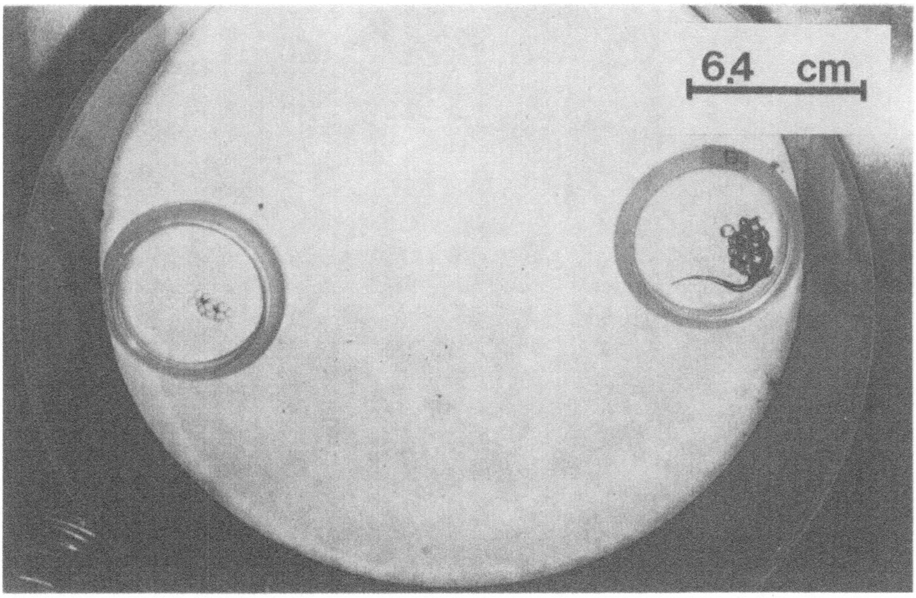

Fig. 4. Egg discrimination apparatus used in two choice experiments designed to test a female's preference for conspecific eggs based on their developmental age.

P > 0.05; late--H = 7.557, P > 0.05). Nor was there a significant difference between age groups (U = 534, df = 82, P > 0.05). Nevertheless, females exhibited pronounced individual variation (Fig. 3). It is unknown what effect isolation greater than 96 h might have on female behavior, but this is likely a moot point. In the field, unattended, developmentally young eggs experienced 80% fungal contamination within 6 days; and all unattended clutches, regardless of their age, are subject to predation (Forester, 1974, 1979a).

Nest Odor Experiments

The return of a displaced female to her own eggs is predicated upon her ability to relocate her nesting site. In nature, this process may involve airborne chemical cues (Madison, 1979; Madison and Shoop, 1970), but it is unlikely that such cues are important to clutch recognition (see the results of the egg recognition experiments). Clutch recognition appears to require trial and error searching facilitated by contact chemoreception of substrate-borne cues. The significance of substrate-borne chemical cues to other plethodontid salamanders has been demonstrated (Shlotzhauer, 1982; Jaeger and Gergits, 1979; Tristram, 1977). Potentially, secretions produced by the attending female represent one of the most important classes of these substrate odors. Such odors were absent in all of the experiments described earlier.

In a series of experiments designed to test the importance of a female's body secretions as a behavioral signature facilitating nest site recognition following displacement, 39 attendants were divided into three experimental groups (Forester et al., 1983).

Females of the first group (N = 15) were placed in the discrimination chambers described in Figure 2. The boxes each contained two conspecific

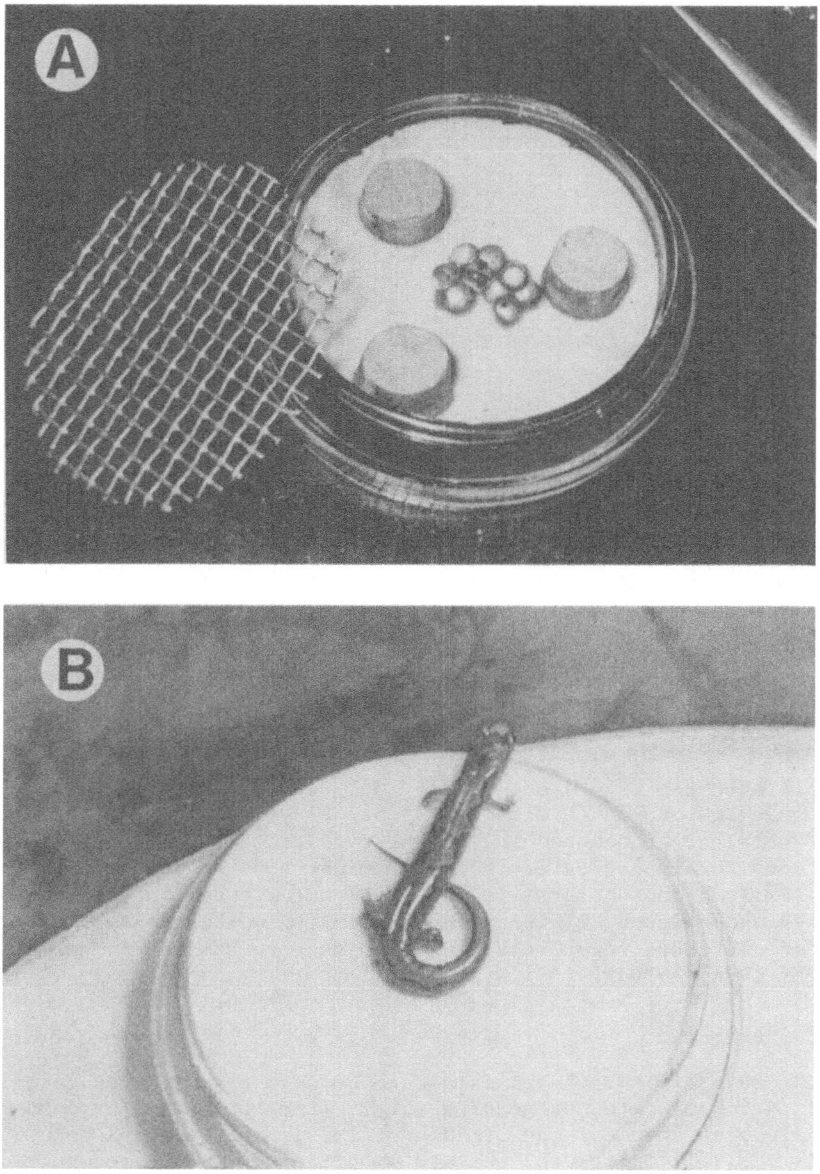

Fig. 5. Sensory modality experiment. (A) Construction of the false-bottom dish. (B) Female resting atop the grid over the advanced egg mass.

egg masses corresponding in developmental age to that of the test female (the significance of matching developmental ages will be discussed below). After a 30 min acclimation period, a test female was allowed to move at will within the box for 24 h.

The second group (N = 12) was exposed to the same set of experimental parameters as the first, with the exception that one of the conspecific egg masses rested upon a filter paper disc which had been previously marked by the test female. Filter papers were marked by removing the test female from her eggs and isolating her upon a moistened disc (diam. = 4.25 cm) for 96 h prior to the experimental period. During a given test, one egg cavity not only contained the odors of a conspecific clutch, but also those of the test female herself.

Females of the third group (N = 12) were allowed to select between their own eggs and a conspecific egg mass of similar developmental age, resting upon a marked disc of filter paper. Test females were thus confronted with their own, familiar egg odors in one chamber and unfamilar egg odors coupled with familiar body odors in the other chamber.

The bottom of the control chamber and egg dishes were lined with two layers of moistened filter paper. At the conclusion of each test, the filter paper was removed, the inside of the box wiped dry, and clean filter paper installed.

During the tests involving two conspecific egg masses, females spent an average of 93% of their time in the presence of eggs, but did not exhibit a preference for one egg mass over the other ($X^2 = 3.028$, df = 1, $P > 0.05$). In the related two choice experiments, females did not exhibit a preference for conspecific eggs resting on a marked disc of filter paper ($X^2 = 0.910$, df = 1, $P > 0.05$), but did select their own eggs over conspecific eggs resting on a marked filter paper disc ($X^2 = 7.182$, df = 1, $P < 0.01$).

These data suggest that the buildup of female secretions in the nest cavity is of little value during nest location and the clutch recognition process. It is not possible, however, to rule out the importance of other non-egg, olfactory cues, not to discount the possibility that the marking procedure was inadequate.

SPECIES RECOGNITION

In the previous experiments, I provided circumstantial evidence to support individual clutch recognition by chemical means. This led to a logical question: do egg masses also possess a unique species-specific marker?

To examine this question, I collected and tested brooding females of two species: (1) *Desmognathus ochrophaeus* (n = 18); (2) *D. fuscus* (n = 15). During a given trial, a female was placed in one of the five discrimination chambers (previously described). One nest cavity contained the eggs of a conspecific, while the other cavity those of a congener. After 30 min females were released from the acclimation cylinder and allowed to move at will for 24 h. Females which spent significantly more than 50% of their brooding time in the presence of a conspecific clutch were considered to exhibit species fidelity.

The results of these species recognition experiments are summarized in Table 2. The females of both species exhibited a strong tendency to brood eggs, but there was no evidence of species fidelity in either experimental group. As a consequence, I reject the hypothesis that desmognathine eggs exhibit a species-specific chemical marker. The absence of a species marker does little to decrease a female's inclusive fitness. Even broadly sympatric species usually exhibit spatial and/or temporal nest site isolation. In the rare situations where two species are known to exist side by side within the same seepage (e.g., *D. ochrophaeus* and *D. santeetlah*, personal observation), individual clutch fidelity would simultaneously promote species fidelity.

Table 2. Data summary from experiments designed to test for the presence of a species-specific chemical marker among salamander eggs. OCH = *ochrophaeus*, FUS = *fuscus*, CS = conspecific, CG = congener.

		Initial Response		After 24 H		$\bar{X}$ Brooding Time		Species Fidelity	
Test Species	N	CS	CG	CS	CG	Total	CS	CS	CG
OCH	18	11	7	8	9			10	8
		$X^2 = 0.89$		$X^2 = 0.06$		82.6%	57%	$X^2 = 0.222$	
		$P > 0.05$		$P > 0.05$				$P > 0.05$	
FUS	15	11	3	9	4			8	7
		$X^2 = 4.57$		$X^2 = 1.92$		86%	61%	$X^2 = 0.67$	
		$P < 0.05$		$P > 0.05$				$P > 0.75$	

DETERMINATION OF CHRONOLOGICAL AGE

Chemical signals emanating from the eggs may provide an attendant with information pertaining to their chronological age. In this section, I review behavioral experiments which support this hypothesis. I shall also speculate on the adaptive significance of such signals.

During three choice tests designed to document clutch recognition, a significant number of females (32 of 51) selected their own eggs (Forester, 1979b). In 12 of the instances when a female attended eggs other than her own, the clutch selected was chronologically older and, hence, more developed than hers. Prior to this, no attempt had been made to match the developmental age of the three egg masses. This observation led me to question whether females deprived of their eggs exhibit a preference for conspecific clutches based on the developmental stage of the eggs. To test this question, an advanced egg mass and a recently deposited egg mass were placed on opposite sides of a circular test chamber (Fig. 4). A female not previously associated with either clutch was released in the center and allowed access to the egg masses. Prior to being used in the egg discrimination trials, the natural clutch of each female was aged chronologically and categorized as: (1) early (gastrula or younger), (2) intermediate (Neurula to hindlimb bud), (3) advanced (dorsal pigmentation or older). Of the 36 females which responded by attending an egg mass (respondents), 12 had been removed from early clutches, 13 from intermediate clutches, and 11 from advanced clutches.

Twenty-nine of the respondents selected and attended the advanced clutch ($X^2 = 13.44$, df = 1, $P < 0.005$). The experiment was conducted in total darkness; thus, I hypothesized that olfactory cues are essential during the clutch selection process. Only 5 of 12 females separated from early clutches selected the more advanced egg mass during the trials, while all 24 females isolated from intermediate and advanced clutches selected the more developed eggs. I interpret this to mean that the chemical cues emanating from advanced clutches have a priming effect on the behavior of the attendant.

To test my olfactory hypothesis, I constructed false bottoms in each of two glass dishes (diam. = 64mm) by placing a circular piece of hardware cloth (6.4mm mesh) on top of three cork discs (6mm thick) resting on the bottom of each dish (Fig. 5a). The wire screens were covered with a disc of moistened filter paper. The role of olfaction was tested by concealing an advanced clutch under one screen and a freshly deposited clutch under the other. The concealed eggs were not visible through the side of the dish. A positive response was recorded if, after 30 min, the female was on top or along the edge of the dish containing the advanced eggs.

Nineteen of 34 females responded by crawling on top of a dish (Fig. 5b). However, in the apparent absence of visual and/or tactile reinforcement, they seldom remained there for more than a few minutes. It was not possible, therefore, to discern a preference without a modification of the experimental procedure.

By placing a freshly deposited clutch on each of the screens, a responding female was exposed to the tactile and visual stimulation of a freshly oviposited clutch resting on top of each dish, while olfactory stimulation resulted from the interaction of chemical cues emanating from the exposed and concealed clutches of each dish.

Females responded in 24 of 69 trials. Of these respondents, 21 selected and attended the clutch above the advanced eggs ($X^2 = 10.67$, df = 1, $P < 0.005$). These data suggest that olfactory cues are essential to the clutch recognition process, but they are most effective in the presence of visual and/or tactile reinforcement. Baerends (1959) stated that, in birds, incubation can be shown to continue effectively only when "the proper feedback stimuli are released by tactile- and thermoreceptors of the brood patches; otherwise, the appetitive activities of incubation set in, often mixed with displacement activities (preening, etc.)." This sensory-feedback hypothesis was first applied to salamanders by Bachmann (1964) and is supported by my results.

Bachmann (1964) suggested that the developmental stage of an egg mass may affect the strength of the bond held by the female for her eggs. My data support this hypothesis and implicate chemical cues in the recognition process. As the time spent brooding an egg mass increases, so does the period of fasting (Forester, 1983). A strengthening of the female-clutch bond with time may be necessary to prevent or greatly reduce foraging behavior.

CHEMICAL CUES AND OOPHAGY

Unattended egg masses (in both the field and the laboratory) are prone to fatal fungal infestation within a few days. In contrast, fungal infestation of attended clutches is rare (Stebbins, 1954; Bachmann, 1964; Snyder, 1971; Tilley, 1972; Forester, 1979). Highton and Savage (1961) believed that fungal growth appeared only on eggs that had already died and that living embryos avoided contamination by secreting an antipathogenic substance. This hypothesis was challenged by Vial and Preib (1967) and Snyder (1971) who observed fungal mycelia growing on "viable and vigorous eggs." Tilley (1972) observed heavy infestations of fungal hyphae growing on the jelly embryos of viable D. ochrophaeus but suggested they could "almost always be traced to one or two dead eggs." He concluded that fungal infestations seldom, if ever, were the initial cause of egg mortality.

Snyder (1971) observed egg eating (oophagy) by female green salamanders (Aneides aeneus) and suggested that this behavior might serve an antipathogenic function. Oophagy is part of the normal behavioral repertoire of D. ochrophaeus, enabling the females to respond to eggs as a food source and to remove dead eggs from their clutches. Tilley (1972) hypothesized that removal of dead eggs might decrease fungal contamination by elimination of a potential growth medium.

I conducted experiments to test this hypothesis and to determine how females recognize dead and diseased eggs (Forester, 1979a). My experiments addressed the question: do females respond differentially to living and dead eggs added to their egg mass?

Each of fourteen attendants was presented (on consecutive days) with: (1) the living egg of another female, (2) the dead egg of another female. In addition, 10 females were presented with an egg from their own clutch which had been removed and killed. Eggs were killed by immersion in boiling water for three minutes.

Females responded differentially to living and dead eggs added to their clutches. All dead eggs, regardless of their origin, were consumed within 12 h. On the other hand, only 2 of 14 living conspecific eggs were consumed during a 12 h period.

Each of the experiments described above were conducted in total darkness to eliminate reliance upon visual cues. As a consequence, I concluded that females identify dead eggs by chemical means. This ability to recognize dead eggs is highly adaptive since it enables the female to remove a substrate which is favorable for fungal growth. When a dead, contaminated egg was added to the clutch of 15 females whose mouths were sewn closed to prevent oophagy, all succumbed to fungal infestation within 8 days. Similarly treated control females consumed the dead egg and their clutches developed normally (Forester, 1979a).

SUMMARY AND SUGGESTIONS FOR FUTURE RESEARCH

The experimental results reviewed in this paper provide strong circumstantial evidence for the importance of chemical cues during the nesting period of lungless salamanders. These cues appear to facilitate egg recognition and clutch fidelity while promoting a strong maternal bond between the female and her eggs. Nevertheless, clutch recognition is not absolute (see Fig. 3). Only 63% of my test females spent more than 70% of their brooding time with their own eggs. Although statistically significant, this value is much lower than I had originally anticipated. It must be remembered, however, that female salamanders normally oviposit in a secluded nest cavity and remain with their clutch until it hatches. As a consequence, they are seldom forced to choose between their own eggs and those of another female. It is quite possible, therefore, that selection has not been intense enough to promote uniform chemical recognition in all members of the species.

Even females which exhibit a strong attraction to their own eggs do not show a tendency to contact them first during two choice experiments. However, females do exhibit a remarkable ability to home to their nest site following experimental displacement. These observations lead me to hypothesize that displaced females use airborne, olfactory cues to locate the general vicinity of the nest cavity, but that individual clutch recognition requires contact chemoreception. If this scenario is correct, clutch discrimination would involve trial and error search similar to that previously observed for _Plethodon cinereus_ (Bachmann, 1964). Confirmation of my hypothesis requires a series of bioassay experiments. Efforts are presently underway to isolate one or more of the necessary chemical markers using the technique of gas chromatography.

ACKNOWLEDGMENTS

This work would not have been possible without continued financial support from the Faculty Research Committee of Towson State University and grants-in-aid from the Highlands Biological Station, Highlands, NC. A special thanks goes out to the many graduate and undergraduate students who provided assistance and companionship in the field and laboratory during the past decade. Finally, I thank the late B. S. Martof who instilled in me the value of scientific contribution, however modest.

REFERENCES

Bachmann, M. D., 1964, "Maternal Behavior of The Red-backed Salamander, _Plethodon cinereus_," Ph.D. dissertation, University of Michigan, Ann Arbor, Michigan.

Baerends, G. P., 1959, The ethological analysis of incubation behaviour, _Ibis_, 101:357.

Barthalmus, G. L., and Bellis, E. D., 1972, Home range, homing and the homing mechanism of the salamander, _Desmognathus fuscus_, _Copeia_, 1972:632.

Bishop, S. C., 1943, "The Handbook of Salamanders," Cornell University Press, Ithaca, NY.
Dawley, E. M., 1984, Recognition of individual, sex and species odours by salamanders of the Plethodon glutinosus-P. jordani complex, Anim. Behav., 32:353.
Forester, D. C., 1974, Parental care in Desmognathus ochrophaeus Cope (Urodela: Plethodontidae): A behavioral study. Ph.D. dissertation, North Carolina State University, Raleigh, NC.
Forester, D. C., 1977, Comments on the female reproductive cycle and philopatry in Desmognathus ochrophaeus (Amphibia, Urodela, Plethodontidae), J. Herpetol., 11:311.
Forester, D. C., 1978, Laboratory encounters between attending Desmognathus ochrophaeus (Amphibia, Urodela, plethodontidae) females and potential predators, J. Herpetol., 12:537.
Forester, D. C., 1979a, The adaptiveness of parental care in Desmognathus ochrophaeus Cope (Urodela: Pelthodontidae), Copeia 1979:332.
Forester, D. C., 1979b, Homing to the nest by female mountain dusky salamanders (Desmognathus ochrophaeus) with comments on the sensory modalities essential to clutch recognition, Herpetologica, 35:330.
Forester, D. C., 1981, Parental care in the salamander Desmognathus ochrophaeus: Female activity pattern and trophic behavior, J. Herpetol., 15:29.
Forester, D. C., 1983, Duration of the brooding period in the mountain dusky salamander (Desmognathus ochrophaeus) and its influence on aggression toward conspecifics, Copeia 1983:1098.
Forester, D. C., 1984, Brooding behavior by the mountain dusky salamander: Can the female's presence reduce clutch desiccation? Herpetologica, 40:105.
Forester, D. C., Harrison, K. and McCall, L., 1983, The effects of isolation, the duration of brooding, and non-egg olfactory cues on clutch recognition by the salamander Desmognathus ochrophaeus, J. Herpetol., 17:308.
Gordon, R. E., 1961, The movement of displaced green salamanders, Ecology, 42:200.
Grant, D., Anderson, O. and Twitty, V., 1968, Homing orientation by olfaction in newts (Taricha rivularis), Science, 160:135.
Hershey, J. L., and Forester, D. C., 1980, Sensory orientation in Notophthalmus viridescens (Amphibia: Salamandridae), Can. J. Zool., 58:266.
Highton, R., and Savage, T., 1961, Functions of the brooding behavior in the female red-backed salamander, Plethodon cinereus, Copeia, 1961:95.
Houck, L. D., 1982, Male tail loss and courtship success in the plethodontid salamander Desmognathus ochrophaeus, J. Herpetol., 16:335.
Jaeger, R. G., and Gergits, W. F., 1979, Intra- and interspecific communication in salamanders through chemical signals on the substrate, Anim. Behav., 17:27.
Madison, D. M., 1969, Homing behavior of the red-cheeked salamander, Plethodon jordani, Anim. Behav., 17:27.
Madison, D. M., and Shoop, C. R., 1970, Homing behavior, orientation, and home range of salamanders tagged with tantalum-182, Science, 168:1484.
Rose, F. L., 1966, Homing to nests by the salamander Desmognathus auriculatus, Copeia 1966:251.
Shlotzhauer, T. L., 1982, Interspecific response to substrate markings by plethodontid salamanders, Bios, 53:244.

Snyder, D. H., 1971, The function of brooding behavior in the plethodontid salamander, _Aneides aeneus_: A field study, Ph.D. dissertation, University of Notre Dame, South Bend, Indiana.
Stebbins, R. C., 1954, Natural history of the salamanders of the plethodontid genus _Ensatina_, _Univ. Calif. Publ. Zool._, 54:47.
Tilley, S. G., 1972, Aspects of parental care and embryonic development in the salamander _Desmognathus ochrophaeus_, _Copeia_, 1972:532.
Tilley, S. G., 1977, Studies of life histories and reproduction in North American plethodontid salamanders, _in_: "The Reproductive Biology of Amphibians," D. H. Taylor, and S. I. Guttman, eds., Plenum Press, New York.
Tilley, S. G., 1980, Life histories and comparative demography of two salamander populations, _Copeia_, 1980:806.
Tristram, D. A., 1977, Intraspecific olfactory communication in the terrestrial salamander _Plethodon cinereus_, _Copeia_, 1977:597.
Trivers, R. L., 1972, Parental investment and sexual selection, _in_: "Sexual Selection and the Descent of Man," B. Campbell, ed., Aldine Pub. Co., Chicago.
Twitty, V. C., 1961, Experiments on homing behavior and speciation in _Taricha_, _in_: "Vertebrate Speciation," W. F. Blair, ed., University of Texas Press, Austin.
Vial, J. L., and Preib, F. B., 1967, Antibiotic assay of dermal secretions from the salamander, _Plethodon cinereus_, _Proc. Mo. Acad. Sci._, 1:37.
Woods, J. E., 1968, The ecology and natural history of Mississippi populations of _Aneides aeneus_ and associated salamanders, Ph.D. dissertation, University of Southern Mississippi, Hattiesburg, Mississippi.

EVOLUTION OF CHEMICAL SIGNALS AS A PREMATING ISOLATING MECHANISM IN A COMPLEX OF TERRESTRIAL SALAMANDERS

Ellen M. Dawley

Section of Ecology and Systematics
Cornell University
Ithaca, New York 14853

INTRODUCTION

Salamanders in the family Plethodontidae are an important component of forest ecosystems in North America (Burton and Likens, 1975). Species of *Plethodon* are completely terrestrial amphibians; all aspects of their life history, including courtship and egg deposition, occur on land. Both sexes are believed to be solitary, spending most of their time in cool, moist retreat sites and are active on the surface only when atmospheric conditions are suitably cool and moist (Feder, 1983). My research on chemical communication with pairs of closely related species within this genus has shown that salamanders can identify sex and species through odors. Because *Plethodon* species are nocturnal, and because sexes and sympatric species are nearly indistinguishable morphologically, it is likely that chemoreception is the most important sensory mode. I suggest that chemical communication functions as a premating isolating mechanism between sympatrically occurring sibling species.

GENERAL METHODS AND RESULTS

Male and female salamanders were allowed to choose between volatile airborne odors in a Y-maze. The maze consisted of a 2.5 cm diameter glass Y-tube with a 30.5 cm main body and two 15 cm side arms (see Fig. 3 in Dawley, 1984a). Each side arm was connected to a sample jar containing a living salamander. Air flowed past these odor source salamanders into each side arm for at least 1 hour before trials began (flow rates to each side arm were measured and equalized; see Dawley, 1984a for more detail). The Y-tube was enclosed in a light-proof box supplied with an observation hole. A test salamander was introduced in a glass approach tube, facing away from the main body of the Y-tube. A salamander was given 20 minutes to turn around and enter the main body of the Y-tube, and 30 min to walk down the tube and into the side arm of its choice. If it failed to do so, the trial was terminated and repeated the next day.

For the sex discrimination tests one sample jar contained a male and the other contained a female. *Plethodon glutinosus*, *P. jordani*, and *P. aureolus* were used; 72 males and 36 females among these species were tested. Both males and females significantly preferred the odors of females over males (Dawley, 1984a). All tests were done during the nonbreeding season. Jaeger and Gergits (1979) found that male and female P. cinereus preferred their own marked substrates over those marked by males, but showed no preference between their own and female marked substrates. They concluded that individual salamanders were avoiding male marked substrates because of pheromonal markers that function in territorial defense. My experiments suggest that the preference for female odors does not necessarily reflect an avoidance of male odors. In additional tests, male P. jordani preferred female odors over no salamander odor (an empty sample jar) but randomly chose between male odor and no salamander odor (Dawley, 1984a). Female responses varied according to their reproductive condition. Female P. jordani preferred conspecific male odors over no salamander odor during the breeding season while comparable tests done with P. aureolus and P. glutinosus during the nonbreeding season showed that these females preferred no salamander odor over male odors or randomly chose between them (Dawley, 1984b).

In species discrimination tests male salamanders were given a choice between a conspecific female in one sample jar and a sympatric heterospecific female in the other jar. Similarly, females were given a choice between conspecific and sympatric heterospecific male odors. All animals were collected during the spring and fall courtship seasons to ensure that most salamanders were reproductively active. Testing was begun immediately following collection, but it was inevitable that some tests were performed several weeks after, when salamanders were no longer in reproductive condition. Female responses were unavoidably influenced by this factor.

In every pair of nonhybridizing species tested (P. teyahalee and P. aureolus, P. teyahalee and P. jordani from two different localities, and P. glutinosus and P. kentucki) males preferred conspecific female odors over heterospecific female odors (Dawley, 1984a, b). Clearly, males of these species pairs can discriminate between conspecific and heterospecific females using olfactory signals alone. This suggests that olfaction is an important premating isolating mechanism in these salamanders. Electrophoretic evidence demonstrates the lack of hybridization between these sympatric species (Highton and Henry, 1970; Highton, 1983; Highton and MacGregor, 1983) even though their habitat preferences overlap extensively (personal observations), and I observed courtship in all five species. Courtship sequences of all five species were indistinguishable (personal observations) even though courtship is stereotyped and complex. When males were given an opportunity to court both conspecific and heterospecific females, they preferred to court conspecific females (Dawley, 1984b). This preference appeared to be determined before courtship began because males always initiated courtship first with conspecific females (Dawley, 1984b).

Females tested during the breeding season (P. teyahalee and P. jordani, P. glutinosus and P. kentucki) preferred conspecific male odors over heterospecific odors, but females tested when out of breeding condition preferred heterospecific male odors or randomly chose between the two odors (Dawley, 1984a, b). It may be that females are only attracted to conspecific male odors when they are reproductively active. This conclusion is supported by the results of the sex choice tests (see above) in which females preferred other female odors over male odors during the summer nonbreeding season. It is also supported by additional tests in which females were given the choice between no salamander odor or conspecific or heterospecific male odor (see above). Females of those species tested during the breeding season preferred conspecific male odor over no salamander odor. Females of species tested after the breeding season preferred no salamander odor over conspecific male odor. They also preferred no salamander odor over heterospecific male odors or randomly chose between those two odors. These tests suggest that females are avoiding conspecific male odors during the nonbreeding season.

Species of Plethodon most likely evolved in geographic isolation (Highton and Henry, 1970). In this respect they appear to conform to Templeton's (1980) adaptive divergence mode of speciation, in which some extrinsic barrier to gene flow splits a population into two subpopulations. As the two subpopulations become more and more genetically different, Templeton contends that reproductive isolating barriers arise as a by-product of this divergence. However, it is unclear whether the species-specific differences in chemical cues that I observed are a result of direct selection to reduce hybridization in sympatry (after secondary contact), or a pleiotropic consequence of divergence. I tested P. teyahalee and P. jordani from a hybridizing and a nonhybridizing locality for the ability to distinguish between conspecific and heterospecific odors. If chemoreception does function as a premating isolating mechanism, then nonhybridizing pairs should be able to distinguish between species' odors while hybridizing pairs may not be able to do so. Whether or not hybridizing species have the ability to distinguish between species' odors may shed some light on how this reproductive isolating mechanism may have evolved.

In extensive areas of the Great Smoky Mountains the ranges of P. teyahalee and P. jordani overlap with little or no hybridization, while in other overlap zones there is extensive hybridization (Highton and Henry, 1970). Within both species there is little genetic variation between populations (Highton, personal communication), suggesting either that there still is significant gene flow among populations of the same species or that the different populations shared a recent common ancestor. Males of both species from the nonhybridizing locality preferred conspecific female odors over sympatric heterospecific female odors. Only male P. teyahalee from the hybridizing locality preferred conspecific female odors, while P. jordani from that locality randomly chose between conspecific and heterospecific female odors (Dawley, in prep.). Only female P. teyahalee from the nonhybridizing locality showed a significant response when given a choice between conspecific and heterospecific male odors, preferring heterospecific odors. All other females (P. jordani from the nonhybridizing locality and both species from the hybridizing locality) randomly chose between the two male odors (Dawley, in prep.).

Because females were tested at the end of their courtship season and may not have been in reproductive condition, additional tests were performed in order to allow females to choose between male odors and no salamander odor. In these tests, females who can discriminate between species should prefer no salamander odor over male odor in either the conspecific or heterospecific male choice tests. Females of both species from the nonhybridizing locality preferred no salamander odor over conspecific male odors, suggesting that they are avoiding male conspecifics, and can discriminate between species' odors. However, females of both species from the hybridizing locality randomly chose between male odor and no salamander odor (Dawley, in prep.). This suggests that these females probably cannot discriminate between species' odors.

Comparisons between hybridizing and nonhybridizing species provide the strongest evidence that the ability to distinguish between species' chemical cues is an important premating isolating mechanism. Pairs of salamander species able to distinguish conspecifics by chemical cues do not hybridize, while at the hybridizing locality male P. jordani and females of both species cannot recognize conspecifics and are probably making "mistakes" leading to hybridization. Although males initiate courtship and might be expected to take the more active role in reproductive isolation, females have many opportunities to reject courting males, and mate recognition is probably important in both sexes.

These data also suggest that selection has not yet acted upon male P. jordani and females of both species at the hybridizing locality to reduce hybridization through a premating isolating mechanism. In contrast, selection may have played an important role in the evolution of premating isolation at the nonhybridizing locality. It is possible that premating isolating barriers had started to evolve in allopatry and were completed in sympatry because of the action of selection (Mayr, 1969; Ayala, 1982). The key feature of this example is that natural selection probably played a major role in the evolution of chemoreception as a premating isolating mechanism for the salamanders at one locality. At the hybridizing locality reproductive isolation is not complete, and it is unclear whether in time selection will favor the development of reproductive isolation, or whether the two gene pools will merge.

ACKNOWLEDGEMENTS

Richard Highton made this study possible by supplying locality information that was the result of years of work on his part. I also thank Kent Wells and Robert Dawley for their help and comments on all aspects of my work.

REFERENCES

Ayala, F. J., 1982, Gradualism versus punctuation in speciation: reproductive isolation, morphology, genetics, in: "Mechanisms of Speciation," C. Barigozzi, ed., Alan R. Liss Inc., New York.

Burton, T. M., and Likens, G. E., 1975, Energy flow and nutrient cycling in salamander populations in the Hubbard Brook Experimental Forest, New Hampshire, Ecology, 56:1068.

Dawley, E. M., 1984a, Recognition of individual, sex and species odours by salamanders of the Plethodon glutinosus-P. jordani complex, Anim. Behav., 32:353.

Dawley, E. M., 1984b, Chemical Communication and Premating Isolating Mechanisms in the Plethodon glutinosus-P. jordani Species Complex, Ph.D. thesis, Univ. Connecticut.

Feder, M. E., 1983, Integrating the ecology and physiology of plethodontid salamanders, Herpetologica, 39:291.

Highton, R., 1983, A new species of woodland salamander of the Plethodon glutinosus group from the southern Appalachian Mountains, Brimleyana, 9:1.

Highton, R., and Henry, S. A., 1970, Evolutionary interactions between species of North American salamanders of the genus Plethodon, in: "Evolutionary Biology," vol. 4, T. Dobzhansky, M. K. Hecht, and W. C. Steere, eds., Appleton-Century-Crofts, New York.

Highton, R., and MacGregor, J. R., 1983, Plethodon kentucki Mittleman: a valid species of Cumberland Plateau woodland salamander, Herpetologica, 39:189.

Jaeger, R. G., and Gergits, W. F., 1979, Intra- and interspecific communication in salamanders through chemical signals on the substrate, Anim. Behav., 27:150.

Mayr, E., 1969, "Populations, Species and Evolution," Belknap, Cambridge.

Templeton, A. R., 1980, Modes of speciation and inferences based on genetic distances, Evolution, 43:719.

CHEMICAL ECOLOGY OF KIN RECOGNITION IN ANURAN AMPHIBIANS

Bruce Waldman

Department of Organismic and Evolutionary Biology
Harvard University
The Biological Laboratories
Cambridge, Massachusetts 02138

The study of social behavior involves the examination of how individuals interact in groups, the consequences of these interactions on fitness, and the mechanisms mediating these interactions. As data linking social structure with ecological and evolutionary factors now accumulate, studies on the mechanisms underlying complex social behaviors are gaining new impetus. One of the basic assumptions of sociobiological theory is that organisms can discriminate behaviorally between their genetic relatives and non-relatives. As proposed by Hamilton (1964), cooperative behaviors that are of negligible or even negative personal benefit can be selected if they aid kin. Moreover, some advantage may accrue to individuals that avoid breeding with close kin, or those that "optimally outbreed" (Bateson, 1983; Shields, 1983). When populations are highly structured, kin discriminations can be accomplished effectively by simple rules: familiar individuals or those encountered in predictable social circumstances can be treated as kin with some reliability. Where populations are more fluid, however, specialized kin recognition mechanisms may be favored by natural selection.

Recognition of kin (other than parents or offspring) has been studied experimentally only during the past few years. Initially, investigators sought simply to document that organisms can recognize relatives, and that kin discrimination abilities are separable from familiarity effects (e.g., individual recognition of known conspecifics, which may bear no particular resemblance to one another). The next step is to characterize the mechanisms underlying such abilities. We can consider the components of a kin recognition system as we would consider those of any communication system. Thus we can analyze the signals emitted, or the traits expressed, by the sender. Singly or together, these components of the organism's "label" encode its kinship identity. We can examine how these labels are propagated through the environment, and the form in which they are detected by the receiver. Finally, we can study how the receiver evaluates these cues to assess its genetic relatedness to the sender. By comparing perceived labels with those stored in its own "template," the receiver can identify kin through a process of "phenotype matching" (Alexander, 1979; Waldman, 1981; Holmes and Sherman, 1982).

Because in most cases kinship cues have not yet been well characterized, at present, kin recognition systems can be only tentatively considered to effect communication. Simple morphological characters, such as body shape or color patterns, may convey information about kinship identity; yet they often fail to satisfy properties generally ascribed to signals in communication systems. These properties include nonconstancy (i.e., signals are discrete, well-defined, and of limited duration) and specialization (structural or behavioral adaptations exist for the production and detection of signals) (Green and Marler, 1979; Halliday, 1983). In some organisms, specialized organs may have evolved, and signals may be produced under limited circumstances or at particular times. In others, general morphological characters may continuously broadcast kinship identity. Regardless of our precise definition of what constitutes a signal, by experimentally examining kin recognition systems within this framework, we should be able to generate specific testable hypotheses concerning the physiological mechanisms underlying kin recognition abilities, and the modes in which such abilities will be expressed under various ecological conditions.

Chemical cues may signal kinship identity in a variety of vertebrates. Some fishes recognize their siblings by chemical cues (e.g., coho salmon, Oncorhynchus kisutch, Quinn and Busack, 1985). Many mammals discriminate their kin based on odors (e.g., spiny mice, Acomys cahirinus, Porter et al., 1978; rats, Rattus norvegicus, Wills et al., 1983; ground squirrels, Spermophilus tridecemlineatus, Holmes, 1984). But other sensory modalities may also be important in vertebrate kin recognition systems. For example, primates may rely to some extent on visual cues (e.g., pigtail macaques, Macaca nemestrina, Wu et al., 1980; Fredrickson and Sackett, 1984). Birds can recognize kin based on visual cues (Japanese quail, Coturnix coturnix japonica, Bateson, 1982) or vocal cues (bank swallows, Riparia riparia, Beecher and Beecher, 1983). Evidence that genetic relatives are recognized in the absence of prior experience of them (Wu et al., 1980; Bateson, 1982; Davis, 1982; Grau, 1982; Holmes and Sherman, 1982; Kareem and Barnard, 1982; Hayashi and Kimura, 1983; Hepper, 1983) is suggestive of a genetic component to the production and detection of kinship cues, but further tests are needed.

Among vertebrates, mechanisms underlying kin recognition abilities of anuran amphibians (frogs and toads) have been perhaps most extensively studied. Anuran larvae can be easily manipulated for developmental or sensory experiments. Their recognition abilities are thus much more amenable to analysis than are those of birds or mammals. In addition, they can be obtained and reared in large numbers and show easily measurable kin association responses in a variety of laboratory and field testing situations. Here I review recent experimental studies on the sensory mechanisms mediating kin recognition abilities of anuran amphibians. I then discuss how these recognition abilities develop and the possible roles of genetic and environmental factors in the specification of association preferences during ontogeny. Finally, I suggest some aspects of the chemical ecology of anuran larvae that may have direct ramifications on the fitness of individuals in kin groups, and present data indicating a possible relationship between the kinship composition of groups and growth rates.

SENSORY MODE OF RECOGNITION

American toad (Bufo americanus) tadpoles associate preferentially with their siblings when groups are placed in laboratory pools (Waldman and Adler, 1979; Waldman, 1981). Laboratory-reared larvae, when released in natural ponds, form schools consisting predominantly of siblings (Waldman, 1982b). For many years, anuran larvae were thought to aggregate in response to extrinsic environmental cues (Brattstrom, 1962; Bragg, 1965) or because they were visually attracted to their conspecifics (Wassersug and Hessler, 1971; Wassersug, 1973). Proximate factors mediating the formation and maintenance of sibling aggregations appear, however, to be different. Work in several

Table 1. Summary of studies on kin recognition abilities of anuran amphibians and their sensory basis.

Species (Investigator)	Sensory cues Chemical	Sensory cues Visual	Discrimination abilities *UF*S/*UF*NS	*F*S/*F*NS	*F*S/*UF*NS	*UF*S/*F*NS
B. americanus (Waldman)	+		✓	¤[a]	✓	✓
B. americanus (Dawson)	+	-			✓	
B. boreas (O'Hara and Blaustein)				¤	✓	
R. cascadae (Blaustein and O'Hara)	+	×	✓	¤	✓	
R. sylvatica (Waldman)				✓	✓	

UF unfamiliar; *F* familiar
S sib; NS non-sib
\+ chemical cues are sufficient to elicit response
\- visual cues, when available, make chemical cues ineffective
× visual cues have no effect
✓ discrimination has been demonstrated
¤ lack of behavioral discrimination has been demonstrated
[a] laboratory test results; in field tests using different protocols, tadpoles discriminate between familiar sibs and familiar non-sibs

laboratories suggests that anuran kin recognition abilities are based largely on the production and detection of chemical cues (Table 1).

Dawson (1982) reared B. americanus tadpoles in sib groups, and then recorded the time spent by individuals on either side of a porcelain tray, near two compartments enclosed by visually opaque cheesecloth partitions. When five of the test subject's siblings were placed in one of the compartments and five non-siblings were placed in the other, subjects showed a significant tendency to swim toward their siblings. Subjects given a choice between siblings in one compartment and a second empty compartment also oriented toward their siblings, but individuals presented with non-siblings in one compartment and an empty compartment made no choice. These results suggest that B. americanus tadpoles may be attracted to some waterborne cue released by their siblings. Visual cues are not necessary; indeed, when Dawson supplemented waterborne cues with visual cues (by replacing the cheesecloth partitions with ones made of aluminum screening), subjects demonstrated no preference for siblings over non-siblings. Water previously conditioned by sib groups elicited a similar, albeit weaker, sib preference response when applied to both sides of the testing apparatus at repeated intervals. Subjects showed no preference, however, for water conditioned by siblings over blank (plain dechlorinated) water. Dawson's results are difficult to evaluate as he fails to present numerical data. Nonetheless, his experiments complement those I was simultaneously (and independently) conducting.

In my experiments (Waldman, 1985a), individual B. americanus larvae were placed in a glass Y-maze in which they were simultaneously presented with water flowing from two stimulus containers. When one container held 25 of the test subject's siblings and the other 25 non-siblings, individuals oriented significantly toward the water flowing from their siblings (Fig. 1). Neither stimulus group was visible to test subjects, so as in Dawson's (1982) experiments, discrimination was clearly based on non-visual cues. To eliminate the possibility that kinship identity was being communicated through

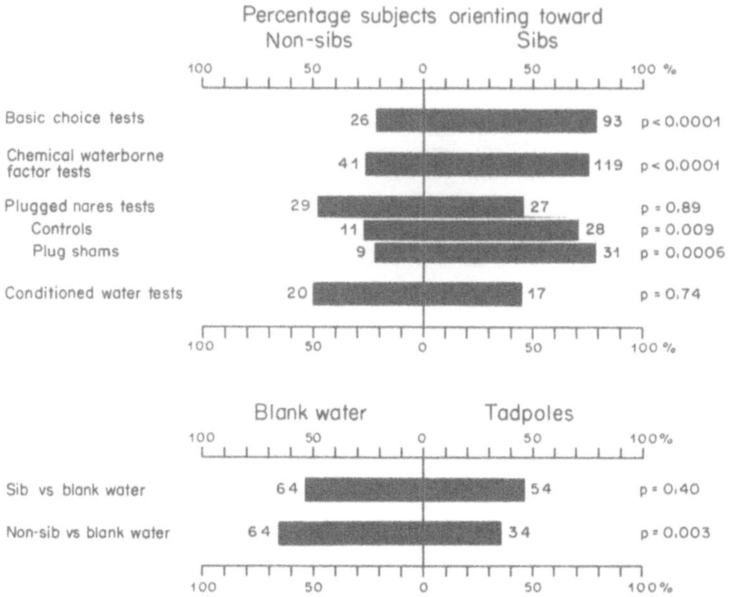

Fig. 1. Percentage of B. americanus tadpoles in a Y-maze choice apparatus preferring sib and non-sib cues. A subject was considered to prefer a stimulus if it spent more than half its response time in the corresponding stimulus arm. In basic choice tests, water was fed directly from stimulus containers holding sib groups to the arms of the Y-maze. Testing procedures were the same in chemical waterborne factor tests, but water was first fed from the stimulus containers through intermediary reservoirs. Experimental subjects in plugged nares tests had their nares occluded with a gelatinous paste immediately before testing, and most were retested after the plugs had been removed (controls). The paste was applied outside the nasal cavities of plug shams. In conditioned water tests, sib groups were allowed to swim in stimulus containers; after they were removed, the containers were stored for periods of 24-30 h prior to testing. In sib versus blank water and non-sib versus blank water tests, subjects were simultaneously presented with blank water through one stimulus arm and water flowing from a container holding either their siblings or non-siblings through the other arm. Numbers of individuals oriented toward each stimulus are shown beside bars. Differences between these data were evaluated by the binomial test. Two-tailed probabilities are indicated. See text and Waldman (1985a) for further details.

vocal, vibratory, or electrical signals, stimulus water was allowed to drip into intermediary reservoirs before entering the testing device. Tadpoles nonetheless oriented toward their siblings in preference to non-siblings. They ceased to discriminate between these stimuli when their nares were blocked with a gelatinous paste, but later oriented toward siblings when the plugs became dislodged. Subjects given a choice between water flowing from a container holding their siblings and blank water demonstrated no preference, whereas those presented with water flowing from a non-sibling container and blank water avoided non-siblings. Time spent oriented toward the stimulus groups was less, however, in these tests than in the others. In contrast to Dawson (1982), my results suggest that B. americanus tadpoles tend to avoid non-siblings rather than approach siblings. The experiments are not directly comparable, however; vocal or vibratory communication was possible in Dawson's experiment but not in mine. The formation of sibling schools under natural conditions probably results from visually mediated mutual attraction among conspecifics (without regard to genetic relationship) coupled with a negative, klinokinetic response to chemical factors released by non-siblings (Waldman, 1985a). In Dawson's testing apparatus, the visual component of the association response may have been stronger than the chemosensory component, leading to a lack of discrimination when visual cues were also available. Possibly the small size of his stimulus groups precluded the transmission of adequate concentrations of chemical cues.

Cues used by larval Cascades frogs (Rana cascadae) in recognizing kin also have come under investigation. Blaustein and O'Hara (1982a) tested individual tadpoles in a rectangular tank in which different stimulus groups (sib groups) were placed behind partitions on each side. As in Dawson's (1982) experiments, individuals were released in the center and their movement pattern recorded. When the partitions consisted of plastic mesh screens, allowing the diffusion of chemical cues, subjects spent significantly more time near their siblings than near non-siblings, regardless of whether visual cues were also available. When the partitions consisted of clear glass panes, allowing visual but no chemical communication, subjects showed no preference for either stimulus group. Hence, unlike Dawson's findings on B. americanus larvae, visual cues appear not to affect the sibling association response of R. cascadae larvae. Although Blaustein and O'Hara (1982a) did not control for other waterborne cues in their experimental design, they monitored noise levels in their testing tank and failed to detect any sound production. The constraints on sound propagation in small enclosed tanks make accurate measurement of underwater sound quite difficult, however, so further investigation seems warranted. In interpreting their results, Blaustein and O'Hara (1983) suggest that R. cascadae tadpoles demonstrate a positive preference for kin rather than an avoidance of non-kin, but they seem not to have explicitly tested this point.

CHARACTERIZATION OF CHEMICAL CUES

Little is known about the actual waterborne factors used by anuran larvae in communicating kinship identity. In all the experiments just discussed, stimulus animals were removed from their rearing tanks and placed into fresh water for tests. Hence, subjects are not simply responding to cues from their rearing containers. Amphibian larvae may release some metabolite or other waste product into their environment that incidentally serves to communicate kinship identity to conspecifics. Alternatively, pheromones might be produced by specialized organs for this purpose. In either case, such signals might consist of a mixture of components. These mixtures ("medleys") would serve well as kinship signals if the number of components shared between individuals correlated with their genetic relatedness. Kinship identity might also be encoded in the relative proportions of the mixture components. Signal diversity sufficient to uniquely identify very many sibships could be generated in these ways, and the mixtures would be particularly effective at close range (Wilson, 1968, 1970).

For B. americanus larvae, we have some information on how the signal decays in time, in space, and with heat treatment. Water in which sib groups have been housed for at least 24 h fails to elicit a discrimination response if it is held for periods of 24-30 h after the tadpoles are removed (Fig. 1; Waldman, 1985a). Dawson (1982) noted a decreased responsiveness with time when testing subjects' responses to water conditioned by sib and non-sib groups, suggesting that kinship cues diffuse rapidly. Alternatively, this may indicate rapid sensory accommodation to the recognition labels. Boiling of sib-conditioned water followed by reconstitution of the residue in aqueous solution eliminates the effectiveness of the stimulus in attracting siblings (Dawson, 1982). These results are all consistent with the hypothesis that the signals are relatively labile, have fast fade-out times, and are effective over limited active space (Waldman, 1985a). In natural conditions, B. americanus larvae typically swim with rapid undulatory tail movements (see Beiswenger, 1975). Although this behavior may serve primarily to stir up food, the swimming patterns may also facilitate signal transmission by directing chemical secretions toward conspecifics.

Recent investigations on population-specific chemical substances in fishes may provide a model for future work on anurans. Salmon (Salmo alpinus and S. salar) respond to odorants released by conspecifics, orienting toward those released by their siblings or other members of their population (Selset and Døving, 1980; Stabell, 1982). Substantial interpopulation variability is found in these substances (Stabell et al., 1982), which may consist of bile salts or other solutes carried in bile salt micelles (possibly metabolic by-products of the intestinal microfauna) (Døving et al., 1980; Selset, 1980). Among amphibians, bile salt mixtures are very complex, showing substantial variability in constituent proportions among populations and perhaps among sibships (Haslewood, 1978, personal communication). Other steroids and their derivatives are also present in the skin of both adult and larval Bufo (e.g., bufodienolides; Flier et al., 1980) and are biologically important. Some of these substances may function as toxins for protection from predators, but they might also serve as alarm pheromones to alert conspecifics (Kulzer, 1954). Additional work is needed to determine whether these compounds vary in structure among sibships and under what if any conditions they can be actively secreted by larvae.

CHEMICAL CUES AND THE ONTOGENY OF KIN RECOGNITION

Genetic relatedness and familiarity have been confounded in many empirical studies of recognition abilities (see Table 1). When extrapolating to natural conditions, such experiments are not necessarily flawed, however, because in structured populations familiar individuals are often genetic relatives. Nonetheless, with amphibians one can disentangle the effects of innate and experiential factors using some relatively simple procedures. For example, individuals can be reared in social isolation with no experience of conspecifics. B. americanus embryos reared from eggs in individual petri dishes subsequently discriminate between their siblings and non-siblings, associating more closely with their siblings in a laboratory pool (Waldman, 1981). R. cascadae tadpoles, similarly reared in social isolation, also discriminate between siblings and non-siblings, orienting toward their siblings (Blaustein and O'Hara, 1981). Rather than implying that social experience has no effect on the ontogeny of kin recognition abilities, these results suggest that such abilities can develop in the absence of exposure to conspecifics.

Even when reared in social isolation, individuals might learn something about their own phenotype and then use these self-perceived traits as a basis for comparing themselves with conspecifics. Kin obviously would have a higher probability of sharing genetically determined traits than would non-kin. Kinship assessments might also be based on other traits acquired in common by family members. These labels might result from exposure to localized environmental cues, mixing of individual traits (e.g., odors), or traits induced through various maternal effects. B. americanus tadpoles reared in social isolation can discriminate between full siblings and paternal half siblings, but evidently not between full siblings and maternal half siblings (Waldman, 1981). R. cascadae tadpoles reared in groups preferentially associate with full siblings over maternal half siblings, and with maternal half siblings over paternal half siblings (Blaustein and O'Hara, 1982b). These results suggest that some maternal factor (cytoplasmic or extracellular) may be important in the ontogeny of kin recognition abilities. One intriguing possibility is that the chemicals secreted by the maternal parent into the jelly surrounding her eggs may be important in inducing histocompatibility responses (Waldman, 1981), and that these in turn may influence the expression of behavioral kin recognition responses (for recent reviews, see Boyse et al., 1982; Yamazaki et al., 1983; Schwende et al., 1984; Lenington and Egid, 1985).

While the development of kin recognition abilities is not dependent on social experience in either B. americanus or R. cascadae tadpoles, under natural conditions larvae of both species invariably develop in social groups comprised of one or more sibships. Indeed, in both these species, as well as the others that have been studied, social interactions during development do affect the ontogeny of kin discrimination responses. Moreover, such effects vary among species. This can be illustrated by results of parallel experiments I have done on two species, the American toad, B. americanus, and the wood frog, Rana sylvatica. Experiments on both species were conducted using largely similar rearing protocols and identical testing procedures.

Larvae were reared from a very early embryonic stage (before gastrulation) either with their siblings only (in 10 l or 75 l tanks) or exposed to both siblings and non-siblings (each sib group was housed in a fiberglass-mesh basket which was placed, together with seven other baskets, into a 425 l tank). As water was actively pumped among all compartments in mixed groups, larvae were continuously exposed to waterborne cues emanating from their siblings and from non-siblings. After three to eight weeks, 25 members of each of two sibships were differentially marked with vital stains, and were then released together in a laboratory test pool. Spatial distributions were recorded, and from these, nearest-neighbor distances were computed. The results demonstrate that both B. americanus and R. sylvatica tadpoles, when reared with just their siblings, associate preferentially with these familiar siblings over unfamiliar non-siblings. When reared in mixed groups, however, B. americanus tadpoles fail to show a significant preference for their siblings, whereas R. sylvatica tadpoles associate preferentially with theirs (Fig. 2). Thus, social experience appears to affect the ontogeny of kin recognition in B. americanus to a greater extent than in R. sylvatica (Waldman, 1981, 1984).

Other aspects of kin association behavior differ between the two species, as well. Whether reared in sib groups or mixed groups, nearest-neighbor distances between B. americanus larvae (siblings or non-siblings) appear smaller than those between R. sylvatica larvae (Fig. 2). To obtain a baseline measure for comparison, nearest-neighbor distances between siblings and between non-siblings can be computed, assuming that individuals move randomly in the test pool. Under typical test conditions, siblings would be expected on average to be 126 mm apart and non-siblings would be 122 mm apart*.

*The discrepancy between these measures is a consequence of excluding an individual's distance to itself when determining its nearest neighbors. Standardized distance scores can be devised to equate sib and non-sib expected measures; however, use of these numerical transformations necessitates a series of statistical and biological assumptions that complicate the interpretation of results. Because the expected difference is relatively small and represents a conservative error, I have chosen to analyze direct measurements uniformly in my studies. A slight tendency for tadpoles to be closer to non-siblings than siblings is thus expected, and is evidenced in results of many control tests.

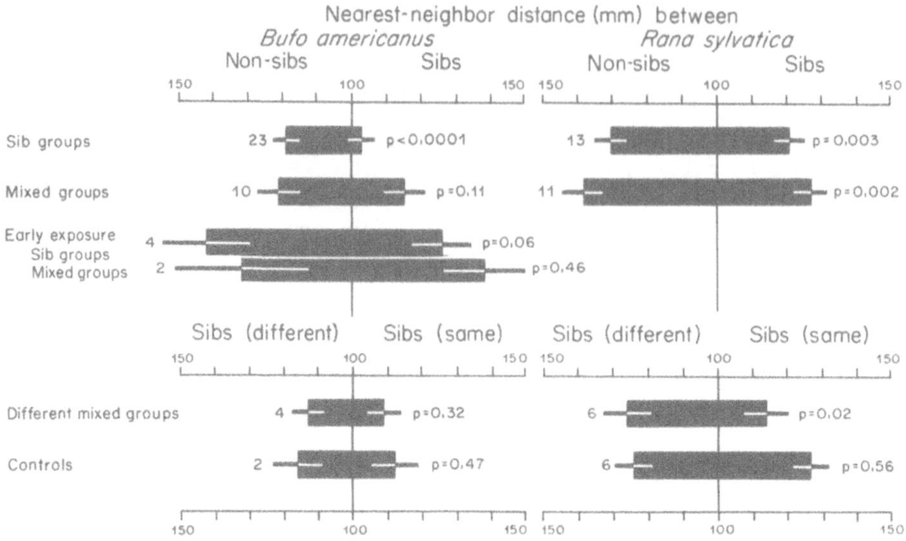

Fig. 2. Mean nearest-neighbor distances between siblings and between non-siblings in a laboratory test pool. Tadpoles were reared under various ontogenetic regimes, and then 25 members of each of two sibships were differentially marked and released in a 115 x 115 cm test pool. Spatial distributions of groups were recorded twice daily over four or five day periods. To determine whether tadpoles discriminated between siblings and non-siblings (above), larvae were reared either with siblings only (sib groups), with siblings and non-siblings (mixed groups), or (for B. americanus) under both regimes in succession with early exposure (during the first 18 days) either to sib groups or mixed groups. To determine whether recognition labels were transferred among individuals during ontogeny (below), siblings were split between different mixed groups and were subsequently tested using the same procedures for their tendencies to discriminate between familiar siblings and unfamiliar siblings. In control tests, sibships were split between different baskets in the same mixed group tank and were similarly tested for differential association tendencies. The number of test groups included in each experiment is indicated to the left of each bar. Error bars denote ± 2 SEM. Differences between sib and non-sib nearest-neighbor distances were statistically evaluated by repeated measures analyses of variance (see Waldman, 1984, 1985b, for further details).

In fact, B. americanus larvae are substantially closer to their siblings than this expected distance, but are approximately randomly distributed with respect to non-siblings. In contrast, R. sylvatica larvae appear to be randomly distributed with respect to their siblings, but are farther apart from non-siblings than would be expected if they moved randomly. In natural conditions, B. americanus tadpoles are usually found in dense schools, whereas R. sylvatica tadpoles are rarely found in schools but are more often observed to be dispersed or regularly spaced throughout a pond (see Waldman, 1984, and references therein). Measurements of spatial distributions in laboratory test pools thus fit nicely with observations on natural populations. (N.B.: In the laboratory pool tests described here, as in natural conditions, tadpoles can interact by visual, chemical, acoustic, or other cues. As I described previously, different effects can be observed when testing paradigms limit interactions to the chemosensory mode.)

Nearest-neighbor analyses further reveal that B. americanus tadpoles reared in social isolation behave differently from their socially reared conspecifics. Although B. americanus tadpoles reared under either social regime subsequently swim closer to their siblings than to non-siblings, spatial distributions of B. americanus tadpoles reared in isolation resemble those of socially reared R. sylvatica larvae. Isolate-reared individuals, interacting with conspecifics for the first time, distribute themselves approximately randomly in the test pool with respect to their unfamiliar siblings (120 $\pm$ 3 mm; $\bar{x} \pm$ SEM, 7 tests) but are considerably farther from non-siblings (138 $\pm$ 4 mm) than would be expected if they moved randomly. This suggests that when they first interact with conspecifics, tadpoles rely to a large extent on chemical cues rather than visual or mechanical stimuli. Like socially reared B. americanus tadpoles tested for their chemosensory responses, isolate-reared individuals appear to move away from non-siblings rather than approach siblings. Exposure to conspecifics, though not necessary for the development of sibling recognition abilities per se, thus may play an important role in the development of social behaviors such as schooling (cf. Shaw, 1961; Williams, 1976).

The ontogeny of kin recognition in B. americanus tadpoles is affected by waterborne cues, quite likely pheromones, metabolites, or other chemicals released by the developing larvae. Chemical interactions during an early developmental period, before or soon after hatching, are particularly important in determining the specificity of the kin recognition system. Larvae that have been reared just with their siblings for 18 days or more appear to develop specific sib recognition abilities. Even after being transferred to mixed groups in which they are reared with non-siblings for several weeks, these tadpoles associate preferentially with their siblings. In the reciprocal experiment, individuals reared first in mixed groups for 18 days and later transferred to tanks holding only their siblings subsequently fail to discriminate between siblings and non-siblings (Fig. 2). The lack of behavioral discrimination is particularly striking because many of these individuals have cumulatively spent much more time in sib groups than in mixed groups, and their most recent experience has been in sib groups. B. americanus larvae may effectively "learn" traits of their siblings sometime during the first 18 days of life. If different sibships come into contact during this period, traits of non-siblings as well as those of siblings (and the individual's own) may be incorporated into a recognition template. After it is formed during this early sensitive period, the template appears to be resistant to further environmental effects. In principle, a simple developmental process could give rise to template formation. For example, Rana temporaria tadpoles exposed to various organic compounds during early development later express altered responses to these specific solutes (Kiseleva, 1984). Recognition specificity might be determined as tadpoles habituate to odors of conspecifics when they are first exposed to them early in development. B. americanus tadpoles appear to move away from non-siblings rather than approach siblings. Hence, responses such as swimming movements that were selectively elicited by individuals that smelled differently from the learned odor would constitute behavioral kin discriminations. While early social experience is undoubtedly important in the ontogeny of kin recognition, perception of one's own phenotype may also be involved. When B. americanus tadpoles are given a choice between familiar non-siblings and unfamilar siblings in a Y-maze apparatus, subjects orient toward their unfamiliar siblings (Waldman, 1986). These results suggest that tadpoles may retain an ability to make kinship assessments in some circumstances based just on their own phenotypic characters rather than on all those stored in a template (Waldman, 1986).

The development of kin recognition abilities in R. sylvatica appears also to be affected by chemical cues, but in a different manner. Unlike B. americanus, R. sylvatica tadpoles reared in mixed groups can discriminate between the siblings and non-siblings with which they have been reared. This discrimination could not possibly be based on a template that nonselectively incorporated traits of familiar conspecifics, and indicates either that tadpoles use their own traits as a basis for comparison or that the template has a genetic component. The labels expressed by R. sylvatica larvae are, however, influenced by waterborne factors released by conspecifics during development. When a single sibship of R. sylvatica is split between two mixed groups consisting largely of different sibships, the tadpoles preferentially associate with their familiar siblings over unfamiliar siblings (Waldman, 1984). This suggests that recognition labels of siblings reared in different social environments diverge from one another. B. americanus tadpoles, similarly reared and tested, do not behaviorally discriminate between familiar siblings and unfamiliar siblings (Fig. 2). A variety of additional experiments confirm that while the recognition template of B. americanus larvae is influenced by the social environment, the expression of their recognition labels is not similarly affected (Waldman, 1985b). If recognition labels of R. sylvatica tadpoles become more similar as they interact, their recognition templates may also be adjusted to correspond with these labels. Moreover, this social convergence must be incomplete because tadpoles are able to discriminate between familiar siblings and familiar non-siblings.

Ontogenetic differences between species correlate with differences in their early larval ecology. B. americanus larvae develop in proximity to their siblings; clutches consist of eggs encased in continuous gelatinous strings, and embryos and hatching larvae are thus directly surrounded by their siblings. Because pairs usually do not all oviposit at a common site, under natural conditions larvae are likely to learn cues of their siblings during early development. In contrast, R. sylvatica egg masses are almost always deposited in large communal clumps in which egg masses of different pairs directly abut one another (see Waldman, 1982a). When emerging from an egg mass, R. sylvatica larvae are likely to intermingle immediately both with siblings and non-siblings. While it may make good sense for B. americanus larvae to learn traits of their siblings soon after hatching, especially if siblings do not all share identical recognition labels, this process of template specification would be inappropriate for R. sylvatica larvae.

The ontogeny of kin recognition in R. cascadae differs both from B. americanus and R. sylvatica. In discriminating siblings from nonsiblings, R. cascadae tadpoles apparently use recognition templates that are genetically determined or ontogenetically fixed prior to hatching. For instance, R. cascadae tadpoles reared in containers in which they are exposed only to non-siblings can discriminate between unfamiliar siblings and unfamiliar non-siblings (Blaustein and O'Hara, 1983). Thus it has been suggested that they may use a "genetic recognition system" (Blaustein and O'Hara, 1981). Nonetheless, the labels by which kinship identity is communicated are strongly influenced by the larval social environment. Indeed, labels of conspecifics reared together appear to converge fully, regardless of the genetic relatedness of the individuals. R. cascadae tadpoles reared in mixed groups of siblings and non-siblings subsequently fail to discriminate among these familiar individuals, even though they associate preferentially with unfamiliar siblings over unfamiliar non-siblings (O'Hara and Blaustein, 1981). The spawning behavior of R. cascadae is similar to that of R. sylvatica. Pairs deposit their egg masses in large, communal clumps usually consisting of between 30 and 70 egg masses (Sype, 1975). Larvae are therefore likely to interact both with siblings and non-siblings during development, and all individuals may develop a common recognition label. Whether R. cascadae tadpoles demonstrate sibling association tendencies under natural conditions is thus still problematic. It is possible, however, that kin recognition abilities expressed by larval anurans reflect epigenetic processes that have selective benefit only after metamorphosis (Waldman, 1983). Labels that are transferred in an aquatic medium might not be similarly transferred on land. Indeed, recent evidence suggests that newly metamorphosed *R. cascadae* appear to distinguish between familiar siblings and unfamiliar non-siblings, hopping toward their familiar siblings in a testing device (Blaustein et al., 1984).

Bufo boreas tadpoles also can discriminate between their siblings and non-siblings, preferentially associating with their familiar siblings over unfamiliar non-siblings (O'Hara and Blaustein, 1982). As in *B. americanus*, recognition specificity is influenced by the social experience of the larvae. Tadpoles reared both with siblings and non-siblings subsequently fail to discriminate between these familiar groups (O'Hara and Blaustein, 1982). Unlike *B. americanus*, however, no evidence has been found to suggest that template specification occurs during an early sensitive period in *B. boreas*. Tadpoles exposed to non-siblings, even during late developmental stages, seem to lose their ability to discriminate between familiar siblings and familiar non-siblings (O'Hara and Blaustein, 1982). Interspecific differences may indeed exist or these differing results may instead simply reflect the different testing procedures used.

FITNESS EFFECTS OF CHEMICAL CUES

For many years, investigators have been aware that waterborne factors released by growing anuran larvae play some role in regulating the growth rates of smaller conspecifics (Richards, 1958, 1962; Rose, 1960; West, 1960; Rose and Rose, 1961; Akin, 1966; Licht, 1967; Heusser, 1972). Until recently however, kinship was not considered to be a factor influencing the effects of such growth regulators. Consequently, the genetic relatedness of groups tested in most previous studies is unknown. To investigate whether regulating substances have specific kin-directed effects, Shvarts and Pyastolova (1970a) conducted a series of tests on *Rana arvalis* tadpoles. Their results suggest indeed that larger tadpoles release waterborne factors that inhibit the growth of smaller conspecifics, and that siblings are affected more strongly by such factors than are non-siblings. Essentially the same results are obtained whether groups are actually allowed to socially interact or are exposed only to water conditioned by other groups. More individuals that are reared in mixed groups (with non-siblings or exposed to their cues) metamorphose than those that are reared in pure groups (with siblings or exposed to their cues). In addition, body lengths of individuals reared with their siblings are significantly smaller than those of individuals reared in mixed groups (Shvarts and Pyastolova, 1970a). While the growth rates of smaller individuals are inhibited, those of larger individuals appear to be accelerated. The differential inhibitory effects observed may be accompanied by specific acceleratory effects: smaller individuals appear to increase the growth rates of their larger siblings more than those of comparably sized non-siblings, though this difference was not statistically evaluated (Shvarts and Pyastolova, 1970b; cf. Rose and Rose, 1961; Steinwascher, 1979). These effects may also be mediated by waterborne factors released by the smaller individuals.

While Shvarts and Pyastolova (1970a, 1970b) believe that differential growth regulation has been favored by group selection to maintain genetic variability within populations, their results can also be interpreted from the perspective of a kin selection model. Many frogs breed in transient habitats to avoid potential predation or to reduce interspecific competition. Under these conditions, resources may be limiting or ephemeral (for example, ponds may dry before individuals metamorphose). Those individuals that initially grow most quickly have the highest chance of success. If a slowly growing individual is unlikely to metamorphose, it still may increase its inclusive fitness (Hamilton, 1964) by ceasing to compete with its larger siblings or by actually accelerating their growth (see discussion in Waldman, 1982b). Indeed, my own observations suggest that once B. americanus larvae are shunted into a slow growth track early in development, they can be reared under good conditions (uncrowded, ad lib food) for periods of six months or more and never metamorphose. In more competitive situations, these runts may be cannibalized by their siblings. Similar to Shvarts and Pyastolova (1970b), I observed that when B. americanus tadpoles are kept under conditions of high population density and limited food, the largest individuals are the only ones to survive, and they sometimes metamorphose much more rapidly than in natural conditions. Through their effects on kin, growth regulators effectively may adjust clutch size to local environmental conditions, which are unpredictable from year to year.

To test these ideas, I have been conducting experiments on how chemical factors released by B. americanus larvae influence the growth of kin and non-kin. In initial experiments, 10 sibships (2000 subjects) were each randomly divided into three test groups: one group was exposed to cues released by siblings, another was exposed to cues released by non-siblings, and a third (control) group was exposed to fresh water. Individuals were all singly reared from the same early embryonic stage (before neural plate formation) in frosted petri dishes so that they were deprived of any direct social interactions. Their water was regularly changed either with water from a 75 l tank housing 150 of their siblings, water from a tank housing 150 members of another sib-ship, or fresh dechlorinated water. Source water was filtered through a fine nylon sieve (mesh size, 0.1 mm) immediately before it was introduced into subjects' dishes. Individuals were all fed equal portions of chopped spinach daily. After periods of 30-40 days, the body mass, length, and developmental stage of individuals in each treatment group were measured and statistically compared.

As in Shvarts and Pyastolova's (1970a) experiments, growth rates of individuals exposed to water conditioned by siblings and those exposed to water conditioned by non-siblings significantly differed. In some tests, however, individuals exposed to their siblings grew more slowly while in others individuals exposed to non-siblings grew more slowly. Effects were not necessarily reciprocal, as some sibships appeared to inhibit the growth rate of members of a second sibship while not themselves being differentially inhibited by that sibship. Although most experimental groups grew more slowly than the controls, in some tests, individuals exposed to waterborne cues released by non-siblings actually grew more quickly than did their siblings exposed to fresh water. These data are intriguing but nonetheless puzzling in that it is difficult to account for the diversity of effects observed. Moreover, though source tanks consisted of populations of equal density that were treated identically, statistical analyses reveal that sibship effects were confounded by tank effects. Water from some tanks inhibited both siblings and non-siblings more than water from others. This might reflect variation among sibships in the effects of the regulating factors they release, or may simply represent a random error resulting from the experimental design.

I subsequently repeated these experiments with six additional sibships. Procedures were basically the same, but tadpoles in experimental treatments were exposed to cues released by siblings or non-siblings reared singly in petri dishes, instead of water conditioned by socially reared sib groups. Day-old water obtained from all control individuals in each test series was collected, filtered, and mixed with fresh dechlorinated water (in 1:1 or 2:3 mixtures). It was then immediately introduced into the dishes of the

Table 2. Mean measures of growth at 30 days of B. americanus tadpoles reared in social isolation with exposure to sib cues, non-sib cues, or blank (dechlorinated tap water).

	N	Length (mm)	Mass (mg)	Stage (Gosner, 1960)
Series A				
Sib	55	* 168	* 430.7	**** 31.2
Non-sib	62	179	516.0	33.8
Blank	42	198 ****	655.1 ****	34.8 ****
Treatment		$F_{2,151}$=25.66 ****	$F_{2,151}$=25.37 ****	$F_{2,153}$=26.91 ****
Clutch		$F_{1,151}$= 0.63	$F_{1,151}$= 2.23	$F_{1,153}$= 0.53
Interaction		$F_{2,151}$= 3.74 *	$F_{2,151}$= 3.87 *	$F_{2,153}$= 5.66 **
Series B				
Sib	20	* 188	611.9	34.3
Non-sib	46	179	591.7	35.4
Blank	12	201 ****	680.3 *	35.8 *
Treatment		$F_{2,72}$=11.40 ****	$F_{2,72}$= 2.80	$F_{2,72}$= 2.86
Clutch		$F_{1,72}$= 1.35	$F_{1,72}$= 0.00	$F_{1,72}$= 3.63
Interaction		$F_{2,72}$= 4.47 *	$F_{2,72}$= 5.47 **	$F_{2,72}$= 6.20 **
Series C				
Sib	71	**** 186	**** 711.3	*** 34.3
Non-sib	73	172	535.2	33.1
Blank	40	220 ****	1047.6 ****	37.0 ****
Treatment		$F_{2,177}$=75.50 ****	$F_{2,178}$=77.71 ****	$F_{2,178}$=47.24 ****
Clutch		$F_{1,177}$= 3.90 *	$F_{1,178}$= 3.07	$F_{1,178}$= 0.18
Interaction		$F_{2,177}$= 5.20 **	$F_{2,178}$=10.75 ****	$F_{2,178}$= 9.10 ***

Effects of treatment (sib, non-sib, or blank water), clutch (sibship), and their interaction were evaluated by analyses of variance (General Linear Models Procedure, SAS Release 82.4). Treatment effects were further analyzed by orthogonal contrasts (sib vs non-sib, experimental vs control).
* $P < 0.05$; ** $P < 0.01$; *** $P < 0.001$; **** $P < 0.0001$

experimental groups. Tadpoles were measured twice, at 30 days and at 52 days. As in the previous tests, growth inhibitory effects appeared to differ significantly among treatment groups. Experimental subjects in all cases grew more slowly than controls. In the first test series (A), however, individuals exposed to sib-conditioned water grew more slowly than those exposed to non-sib cues, whereas in the other two test series (B, C), those exposed to sib cues grew more rapidly than those exposed to non-sib cues (Table 2). Later in development, these initial effects were still largely evident. Although series A individuals exposed to sib and non-sib cues no longer significantly differed in body length or mass, individuals reared in sib-conditioned water were retarded in development as compared to those reared in water conditioned by non-siblings (Table 3). Space limitations preclude more detailed data presentation here, but as in the previous tests, effects were not necessarily reciprocal. Some sibships appeared to have affected growth of members of another sibship though they were not affected in the same manner by factors released by that sibship (see Treatment x Clutch interaction terms, Tables 2 and 3). These results, unlike those of the previous experiments, reflect substantial mortality among test groups, but deaths were not concentrated in particular treatment groups (controls were affected just as experimentals). Individuals all appeared healthy and bacterial cultures failed to reveal pathogenic organisms. Mortality was attributed to excessively warm temperatures in the laboratory and consequent oxygen stress on individuals reared in petri dishes.

My findings support the hypothesis that tadpoles release waterborne substances into their surroundings that differentially affect the growth of conspecifics according to their degree of genetic relatedness. While a

Table 3. Mean measures of growth at 52 days of B. americanus tadpoles reared in social isolation with exposure to sib cues, or blank (dechlorinated tap) water.

	N	Length (mm)	Mass (mg)	Stage (Gosner, 1960)
Series A				
Sib	47	206	751.8	** 33.3
Non-sib	34	203 ****	696.7 ****	35.4 ****
Blank	37	227	945.7	36.5
Treatment		$F_{2,112}$= 8.36 ***	$F_{2,105}$=11.91 ****	$F_{2,112}$=20.93 ****
Clutch		$F_{1,112}$= 0.14	$F_{1,105}$= 2.23	$F_{1,112}$= 0.07
Interaction		$F_{2,112}$= 1.13	$F_{2,105}$= 6.75 **	$F_{2,112}$= 2.12
Series B				
Sib	19	219	1003.3	36.6
Non-sib	33	210	917.2	36.5 *
Blank	8	222	959.1	37.9
Treatment		$F_{2,54}$= 2.27	$F_{2,54}$= 1.32	$F_{2,54}$= 3.73 *
Clutch		$F_{1,54}$= 1.56	$F_{1,54}$= 2.56	$F_{1,54}$= 3.44
Interaction		$F_{2,54}$= 3.21 *	$F_{2,54}$= 5.78 **	$F_{2,54}$= 2.07
Series C				
Sib	42	** 206	*** 963.5	**** 36.4
Non-sib	43	197 ****	794.4 ****	34.8 ****
Blank	25	239	1442.1	39.9
Treatment		$F_{2,103}$=39.90 ****	$F_{2,100}$=47.04 ****	$F_{2,104}$=58.92 ****
Clutch		$F_{1,103}$= 5.27 *	$F_{1,100}$= 8.19 **	$F_{1,104}$= 0.00
Interaction		$F_{2,103}$= 6.31 **	$F_{2,100}$= 9.71 ****	$F_{2,104}$= 3.23 *

For explanation of statistical comparisons, see Table 2.

runt may increase its inclusive fitness by selectively responding to growth inhibitory factors released by its siblings, larger individuals would do best to reduce intraspecific competition by directing the effects of such inhibitors toward non-kin. Whether either or both of these mechanisms are operative, and how environmental and social conditions influence their effects, remain entirely a matter of speculation. Indeed, the possibility that these effects are epiphenomenal cannot be discounted at this time. Although I believe that evidence supporting a kin-directed component to growth-regulating factors is strong, the observed effects are complex and additional experiments will be needed before we begin to understand the dynamics of growth regulation in anuran larvae. Further investigations may also reveal the relationship, if any, between the chemical factors mediating sib-specific growth regulation and those serving as cues for behavioral kin identifications.

Tadpoles that recognize and school with their siblings may accrue greater benefits than indiscriminate schoolers, as any cooperative behaviors they express will benefit their siblings. Schooling B. americanus tadpoles may thus increase their inclusive fitness by deterring predators through their distastefulness, by releasing alarm pheromones to alert other members of the school of a predator's presence, or by responding to growth regulators released by their siblings (discussed in Waldman, 1982b). When individuals form sibling aggregations, the benefits of these traits may be directed principally to siblings by behavioral means even if the signals mediating the responses lack sib-specificity. For example, any growth-regulating substances released should quickly become diluted as they diffuse away from the school, so that members of other schools will experience them at much lower concentrations, if at all. Alarm pheromones may similarly have been favored by natural selection because their effects are behaviorally directed toward siblings. When the epidermis of a Bufo bufo tadpole is injured, a a chemical substance is released into the water which elicits a fleeing reaction among conspecifics that come in contact with it (Eibl-Eibesfeldt, 1949; Hrbacek, 1950). Larvae of many other bufonids respond in a similar manner (Pfeiffer, 1966). Like analogous alarm pheromones in ostariophysan fishes (reviewed by Smith, 1977, 1982) the substance apparently cannot be actively secreted (see Kulzer, 1954; Pfeiffer, 1974; Watanabe, 1971). As the signal is only released when an individual's skin is broken (presumably in natural conditions by a predator), the signaler may not directly gain from its effects, but the benefit to nearby conspecifics is probably substantial. Whether alarm substances, like those regulating growth, have sib-specific components has not yet been examined.

SUMMARY

Larvae of many anuran amphibians recognize their siblings. Kin recognition abilities may facilitate the formation of sibling schools or may influence other social interactions. Kinship apparently is perceived through chemical cues ("labels"). The ontogeny of kin recognition varies among species. Chemical cues released by conspecifics during development may be important in the specification of a kin recognition "template." In some species, kinship labels are modified by the social environment, whereas in others they are not. Individuals recognize as kin those conspecifics whose labels match some aspect of their template ("phenotype matching"). Environmental influences on templates and labels may be generally greater when larvae are predictably aggregated in sib groups during early development. Although the cues communicating kinship identity have yet to be identified, they appear to have limited active space and to decay rapidly. Growth-regulating substances released by larvae into their environment can have differential effects on kin and non-kin. The relationship of these substances to behavioral kin recognition mechanisms and their influence on larval population dynamics are presently unclear but merit further study.

ACKNOWLEDGMENTS

My research has been supported by NSF grant BNS-7909119, the Theodore Roosevelt Memorial Fund (American Museum of Natural History), the Gaige Fund (American Society of Ichthyologists and Herpetologists), the George D. Harris Foundation, Sigma Xi, and Harvard University. Computer resources were provided by the University of Cambridge Computing Service. I thank D. Perlman and L. Rosen for reading the manuscript.

REFERENCES

Akin, G. C., 1966, Self-inhibition of growth in Rana pipiens tadpoles, Physiol. Zool., 39:341.

Alexander, R. D., 1979, "Darwinism and Human Affairs," University of Washington Press, Seattle.

Bateson, P., 1982, Preferences for cousins in Japanese quail, Nature, 295:236.

Bateson, P., 1983, Optimal outbreeding, in: "Mate Choice," P. Bateson, ed., Cambridge University Press, Cambridge.

Beecher, I. M., and Beecher, M. D., 1983, Sibling recognition in bank swallows (Riparia riparia), Z. Tierpsychol., 62:145.

Beiswenger, R. E., 1975, Structure and function in aggregations of tadpoles of the American toad, Bufo americanus, Herpetologica, 31:222.

Blaustein, A. R., and O'Hara, R. K., 1981, Genetic control for sibling recognition? Nature, 290:246.

Blaustein, A. R., and O'Hara, R. K., 1982a, Kin recognition cues in Rana cascadae tadpoles, Behav. Neural Biol., 36:77.
Blaustein, A. R., and O'Hara, R. K., 1982b, Kin recognition in Rana cascadae tadpoles: Maternal and paternal effects, Anim. Behav., 30:1151.
Blaustein, A. R., and O'Hara, R. K., 1983, Kin recognition in Rana cascadae tadpoles: Effects of rearing with nonsiblings and varying the strength of the stimulus cues, Behav. Neural Biol., 39:259.
Blaustein, A. R., O'Hara, R. K., and Olson, D. H., 1984, Kin preference behaviour is present after metamorphosis in Rana cascadae frogs, Anim. Behav., 32:445.
Boyse, E. A., Beauchamp, G. K., Yamazaki, K., Bard, J., and Thomas, L., 1982, Chemosensory communication. A new aspect of the major histocompatibility complex and other genes in the mouse, Oncodev. Biol., Med., 4:101.
Bragg, A. N., 1965, "Gnomes of the Night. The Spadefoot Toads," University of Pennsylvania Press, Philadelphia.
Brattstrom, B. H., 1962, Thermal control of aggregation behavior in tadpoles, Herpetologica, 18:38.
Davis, L. S., 1982, Sibling recognition in Richardson's ground squirrels (Spermophilus richardsonii), Behav. Ecol. Sociobiol., 11:65.
Dawson, J. T., 1982, "Kin Recognition and Schooling in the American Toad (Bufo americanus)," Ph.D. thesis, State University of New York, Albany.
Døving, K. B., Selset, R., and Thommesen, G., 1980, Olfactory sensitivity to bile acids in salmonid fishes, Acta Physiol. Scand., 108:123.
Eibl-Eibesfeldt, I., 1949, Über das Vorkommen von Schreckstoffen bei Erdkrötenquappen, Experientia, 5:236.
Flier, J., Edwards, M. W., Daly, J. W., and Myers, C. W., 1980, Widespread occurrence in frogs and toads of skin compounds interacting with the oubain site of Na^+ and K^+ATPase, Science, 208:503.
Fredrickson, W. T., and Sackett, G. P., 1984, Kin preferences in primates (Macaca nemestrina): Relatedness or familiarity? J. Comp. Psychol., 98:29.
Grau, H. J., 1982, Kin recognition in white-footed deermice (Peromyscus leucopus), Anim. Behav., 30:497.
Green, S., and Marler, P., 1979, The analysis of animal communication, in: "Handbook of Behavioral Neurobiology," Vol. 3, P. Marler, and J. G. Vandenbergh, eds., Plenum Press, New York.
Halliday, T., 1983, Information and communication, in: "Animal Behaviour," Vol. 2, T. R. Halliday, and P. J. B. Slater, eds., W. H. Freeman, New York.
Hamilton, W. D., 1964, The genetical evolution of social behaviour. I, II, J. Theor. Biol., 7:1.
Haslewood, G. A. D., 1978, "The Biological Importance of Bile Salts," North-Holland, Amsterdam.
Hayashi, S., and Kimura, T., 1983, Degree of kinship as a factor regulating preferences among conspecifics in mice, Anim. Behav., 31:81.
Hepper, P. G., 1983, Sibling recognition in the rat, Anim. Behav., 31:1177.
Heusser, H., 1972, Intra- und interspezifische Crowding-Effekte bei Kaulquappen der Kreuzkröte, Bufo calamita Laur., Oecologia, 10:93.
Holmes, W. G., 1984, Sibling recognition in thirteen-lined ground squirrels: Effects of genetic relatedness, rearing association, and olfaction, Behav. Ecol. Sociobiol., 14:225.
Holmes, W. G., and Sherman, P. W., 1982, The ontogeny of kin recognition in two species of ground squirrels, Am. Zool., 22:491.
Hrbacek, J., 1950, On the flight reaction of tadpoles of the common toad caused by chemical substances, Experientia, 6:100.
Kareem, A. M., and Barnard, C. J., 1982, The importance of kinship and familiarity in social interactions between mice, Anim. Behav., 30:594.
Kiseleva, E. I., 1984, Responses of behavioural reactions in the Rana temporaria tadpoles to chemical stimuli under the conditions of pretreatment, Zool Zh., 63:1046.

Kulzer, E., 1954, Untersuchungen über die Schreckreaktion der Erdkrötenkaulquappen (Bufo bufo L.), Z. vergl. Physiol., 36:443.
Lenington, S., and Egid, K., 1985, Female discrimination of male odors correlated with male genotypes at the T-locus: A response to T-locus or H-2-locus variability? Behav. Genet., 15:53.
Licht, L. E., 1967, Growth inhibition in crowded tadpoles: Intraspecific and interspecific effects, Ecology, 48:736.
O'Hara, R. K., and Blaustein, A. R., 1981, An investigation of sibling recognition in Rana cascadae tadpoles, Anim. Behav., 29:1121.
O'Hara, R. K., and Blaustein, A. R., 1982, Kin preference behavior in Bufo boreas tadpoles, Behav. Ecol. Sociobiol., 11:43.
Pfeiffer, W., 1966, Die Verbreitung der Schreckreaktion bei Kaulquappen und die Herkunft des Schreckstoffes, Z. vergl. Physiol., 52:79.
Pfeiffer, W., 1974, Pheromones in fish and amphibia, in: "Pheromones," M. C. Birch, ed., North-Holland, Amsterdam.
Porter, R. H., Wyrick, M., and Pankey, J., 1978, Sibling recognition in spiny mice (Acomys cahirinus), Behav. Ecol. Sociobiol., 3:61.
Quinn, T. P., and Busack, C. A., 1985, Chemosensory recognition of siblings in juvenile coho salmon (Oncorhynchus kisutch), Anim. Behav., 33:51.
Richards, C. M., 1958, The inhibition of growth in crowded Rana pipiens tadpoles, Physiol. Zool., 31:138.
Richards, C. M., 1962, The control of tadpole growth by alga-like cells, Physiol. Zool., 35:285.
Rose, S. M., 1960, A feedback mechanism of growth control in tadpoles, Ecology, 41:188.
Rose, S. M., and Rose, F. C., 1961, Growth-controlling exudates of tadpoles, Symp. Soc. Exp. Biol., 15:207.
Schwende, F. J., Jorgenson, J. W., and Novotny, M., 1984, Possible chemical basis for histocompatibility-related mating preference in mice, J. Chem. Ecol., 10:1603.
Selset, R., 1980, Chemical methods for fractionation of odorants produced by char smolts and tentative suggestions for pheromone origins, Acta Physiol. Scand., 108:97.
Selset, R., and Døving, K. B., 1980, Behaviour of mature anadromous char (Salmo alpinus L.) towards odorants produced by smolts of their own population, Acta Physiol. Scand., 108:113.
Shaw, E., 1961, The development of schooling in fishes. II, Physiol. Zool. 34:263.
Shields, W. M., 1983, Optimal inbreeding and the evolution of philopatry, in: "The Ecology of Animal Movement," I. R. Swingland, and P. J. Greenwood, eds., Clarendon Press, Oxford.
Shvarts, S. S., and Pyastolova, O. A., 1970a, Regulators of growth and development of amphibian larvae. I. Specificity of effects, Sov. J. Ecol., 1:58.
Shvarts, S. S., and Pyastolova, O. A., 1970b, Regulators of growth and development of amphibian larvae. I. Diversity of effects, Sov. J. Ecol., 1:122.
Smith, R. J. F., 1977, Chemical communication as adaptation: Alarm substance of fish, in: "Chemical Signals in Vertebrates," D. Muller-Schwarze, and M. M. Mozell, eds., Plenum Press, New York.
Smith, R. J. F., 1982, The adaptive significance of the alarm substance-fright reaction system, in: "Chemoreception in Fishes," T. J. Hara, ed., Elsevier, Amsterdam.
Stabell, O. B., 1982, Detection of natural odorants by Atlantic salmon parr using positive rheotaxis olfactometry, in: "Salmon and Trout Migratory Behavior Symposium," E. L. Brannon, and E. O. Salo, eds., School of Fisheries, University of Washington, Seattle.
Stabell, O. B., Selset, R., and Sletten, K., 1982, A comparative chemical study on population-specific odorants from Atlantic salmon, J. Chem. Ecol., 8:201.

Steinwascher, K., 1979, Host-parasite interaction as a potential population-regulating mechanism, Ecology, 60:884.
Sype, W. E., 1975, "Breeding Habits, Embryonic Thermal Requirements and Embryonic and Larval Development of the Cascade Frog, Rana cascadae Slater," Ph.D. thesis, Oregon State University, Corvallis.
Waldman, B., 1981, Sibling recognition in toad tadpoles: The role of experience, Z. Tierpsychol., 56:341.
Waldman, B., 1982a, Adaptive significance of communal oviposition in wood frogs (Rana sylvatica), Behav. Ecol. Sociobiol., 10:169.
Waldman, B., 1982b, Sibling association among schooling toad tadpoles: Field evidence and implications, Anim. Behav., 30:700.
Waldman, B., 1983, "Kin Recognition and Sibling Association in Anuran Amphibian Larvae," Ph.D. thesis, Cornell University, Ithaca, New York.
Waldman, B., 1984, Kin recognition and sibling association among wood frog (Rana sylvatica) tadpoles, Behav. Ecol. Sociobiol., 14:171.
Waldman, B., 1985a, Olfactory basis of kin recognition in toad tadpoles, J. Comp. Physiol. A., 156:565.
Waldman, B., 1985b, Sibling recognition in toad tadpoles: Are kinship labels transferred among individuals? Z. Tierpsychol., 68:41.
Waldman, B., 1986, Preference for unfamiliar siblings over familiar non-siblings in American toad (Bufo americanus) tadpoles, Anim. Behav., 34:48.
Waldman, B., and Adler, K., 1979, Toad tadpoles associate preferentially with siblings, Nature, 282:611.
Wassersug, R. J., 1973, Aspects of social behavior in anuran larvae, in: "Evolutionary Biology of the Anurans," J. L. Vial, ed., University of Missouri Press, Columbia.
Wassersug, R., and Hessler, C. M., 1971, Tadpole behaviour: Aggregation in larval Xenopus laevis, Anim. Behav., 19:386.
Watanabe, Y., 1971, Fright substance in tadpoles, Biol. Sci., Tokyo, 22:169.
West, L. B., 1960, The nature of growth inhibitory material from crowded Rana pipiens tadpoles, Physiol. Zool., 33:232.
Williams, M., 1976, Rearing environments and their effects on schooling of fishes, Pubbl. Staz. Zool. Napoli, 40:238.
Wills, G. D., Wesley, A. L., Sisemore, D. A., Anderson, H. N., and Banks, L. M., 1983, Discrimination by olfactory cues in albino rats reflecting familiarity and relatedness among conspecifics, Behav. Neural. Biol., 38:139.
Wilson, E. O., 1968, Chemical systems, in: "Animal Communication," T. A. Sebeok, ed., Indiana University Press, Bloomington.
Wilson, E. O., 1970, Chemical communication within animal species, in: "Chemical Ecology," E. Sondheimer, and J. B. Simeone, eds., Academic Press, New York.
Wu, H. M. H., Holmes, W. G., Medina, S. R., and Sackett, G. P., 1980, Kin preference in infant Macaca nemestrina, Nature, 285:225.
Yamazaki, K., Beauchamp, G. K., Wysocki, C. J., Bard, J., Thomas, L., and Boyse, E. A., 1983, Recognition of H-2 types in relation to the blocking of pregnancy in mice, Science, 221:186.

PHEROMONES AND REPRODUCTION IN GARTER SNAKES

William R. Garstka
Dept. Biol. Sci.
University of Alabama
Huntsville, AL 35899

David Crews
Inst. Reprod. Biol.
University of Texas
Austin, TX 78712

Darwin (1871) distinguished between sexual and natural selection, viewing them as separate, often apposing processes. Evolutionary biologists have shown how those behavioral traits associated with perpetuation of the individual's genome, such as courtship displays, are under different selective pressures than are those associated with survival of the individual, such as thermoregulation. In the first instance, sexual selection may lead to conspicuous or extremely rapid evolution of a trait, whereas in the latter instance, natural selection will give rise to less dramatic and/or rapid changes (Arnold and Houck, 1982; Kirkpatrick, 1982; Lande, 1980, 1981).

In his analysis of sexual selection, Darwin (1871) concentrated on the problem of female choice; that is, he dealt with the conditions under which females should choose among males. Other authors have followed his lead in concentrating on the communicative value of the male in reflecting some aspect(s) of his fitness to potential mates (Arnold, 1983; Emlen and Oring, 1977; Halliday, 1978, 1983; Orians, 1969; Zahavi, 1975). Until recently, the female has received scant attention; there has largely been a tacit assumption that females will be found attractive irrespective of their relative fitness, for any female capable of bearing young is worth mating. However, at least in garter snakes, it is clear that males choose to court certain females and not to court others, or at least to court them less intensively. Stated simply, large, unmated, reproductively intact females are preferentially courted and small, ovariectomized, or mated females are usually not. There are strong adaptive reasons for this. In garter snakes, female body size is positively correlated with fecundity; the larger a female is, the more offspring she can produce (Aldridge, 1979; Fitch, 1965; Ford and Ball, 1977; Gregory, 1977) and the more likely she is to reproduce in a given year (Gregory, 1977). Further, in mating, male snakes deposit secretions of their renal sex segment, a modified portion of the nephron possibly homologous in function to the mammalian prostate gland/seminal vesicle complex (Bishop, 1959; Cuellar et al., 1972; Fox, 1979; Prasad and Reddy, 1972; Regaud and Policard, 1903), in the female's cloaca (Devine, 1975, 1977a; Fitch, 1965; Ross and Crews, 1977, 1978). In garter snakes, this secretion coagulates in the female's cloaca and may form a physical barrier to further mating (Devine, 1975, 1977a; Whittier and Crews, 1985).

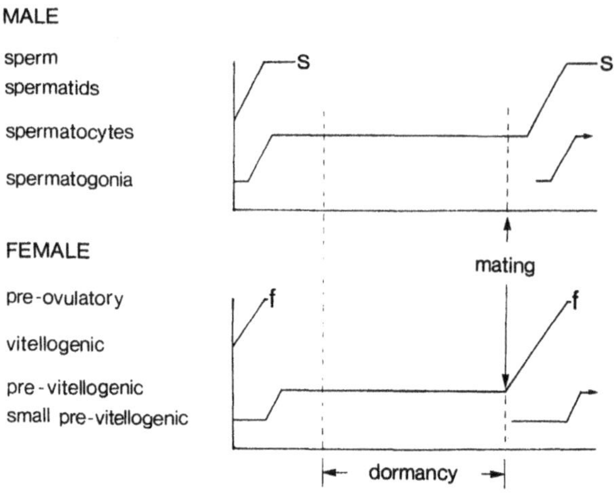

Fig. 1. The annual cycle of gametogenesis in Thamnophis sirtalis in Canada. Mature spermatozoa (S) and ovarian follicles (F) are shed into the accessory sex structures as indicated at s and f. From Garstka et al. (1982), with permission.

The problem we will address in this paper is how the status or suitability of the female as a mate is communicated to the male garter snake. We will also discuss how our findings contribute to an understanding of the evolution of chemical communication in vertebrates.

NATURAL HISTORY OF GARTER SNAKES AND THEIR APPLICABILITY TO STUDIES OF CHEMICAL COMMUNICATION AND REPRODUCTION

Garter snakes, genus Thamnophis, are New World colubrid snakes of the subfamily Natricinae, and are most closely related to Nerodia and the other seven New World natricine genera (Rossman and Eberle, 1977). There are four natural groups and approximately 19 species within the genus (Ruthven, 1908; Smith 1942). Garter snakes may be among the most successful North American terrestrial vertebrates. They are certainly the most abundant and widespread North American snakes, occurring from the lowland tropical forests of Central America to the subarctic of Canada. Further, many features of the reproductive biology of these snakes are found among all species so far studied and may thus be said to be evolutionarily conservative. Therefore, generalizations based on studies with these snakes may be applicable to other equally diverse vertebrate groups.

Perhaps the most striking feature of the reproduction of these snakes is that both sexes of all species of garter snake so far investigated exhibit dissociated (=postnuptial; see Crews, 1984) gametogenesis; gametic maturation (spermiogenesis and folliculogenesis) occurs at a time not directly preceding nor simultaneous with breeding (Cieslak, 1945; Fox, 1952, 1954; Garstka and Crews, 1982; Gregory, 1977; Hawley and Aleksiuk, 1976). In most seasonally breeding vertebrates, including other snakes, spermiogenesis and follicular maturation coincide with seasonal breeding (see Bermant and Davidson, 1974; Crews, 1984; Farner and Follett, 1979 for general reviews). Further, the endocrine changes necessary for the expression of sexual behavior, specifically the production and release of gonadal steroids, are attendant with gonadal recrudescence and thus breeding occurs only when the gametes are mature in both sexes. Causation of breeding in species with dissociated gametogenesis is at best little understood; gonadal steroids seem not to be involved in the control of sexual behavior in adults (at least in male garter snakes; see Camazine et al., 1980; Crews 1983; Crews et al., 1984; Friedman and Crews, 1984a, 1984b; Garstka et al., 1982).

Courtship behavior of males of the various *Thamnophis* species is very similar and is highly stereotyped (Aleksiuk and Gregory, 1974; Ball, 1979; Camazine et al., 1980; Carpenter, 1977; Crews, 1976; Fitch, 1965; Garstka et al., 1982; Hawley and Aleksiuk, 1975; List, 1950; Munro, 1948; Noble, 1937; Pisani, 1976). After initial visual contact is established, the male investigates the female by tongue flicking; this introduces odorant molecules to the male's vomeronasal apparatus (Gillingham and Clark, 1981; Halpern and Kubie, 1980; Meredith and Burghardt, 1978). Males may also follow species-specific pheromonal trails in locating mates (Devine, 1977b; Ford, 1978, 1981, 1982). Females are recognized by a non-volatile, contact pheromone present on at least their dorsal skin (Burghardt, 1970; Crews, 1976; Garstka and Crews, 1981; Garstka et al., 1982; Gillingham and Dickinson, 1980; Kubie, 1977; Kubie et al., 1978b; Noble, 1937); however, ovariectomized females of any size are sexually unattractive (Crews, 1976; Garstka and Crews, 1982; Garstka et al., 1982; Kubie, 1977; Kubie et al., 1978b), suggesting that an ovarian product may be responsible for female attractivity. If the male's tongue is removed, access to the vomeronasal system blocked, or the vomeronasal nerve severed, the male will exhibit neither chin-rubbing nor any subsequent portion of the courtship sequence (Kubie, 1977; Kubie et al., 1978a). Chin-rubbing is an important and unambiguous component of sexual behavior in garter snakes. This highly stereotyped behavior is specific to a sexual context; the male presses his mental (chin) and labial areas tightly against the dorsal or dorsolateral skin of the female and may move forward and backward, traversing her length rapidly. Tactile-sensitive structures on the chin areas of snakes (Hiller, 1971; Landmann, 1975) may be involved in this behavior.

There are also tactile-sensitive areas dorsal to the male's cloaca (Noble, 1937); the male uses these in searching movements to locate the female's cloaca (Noble, 1937; Pisani, 1976). Once aligned with the female, with his head in the direction of her head and his cloaca apposed to hers, the male begins caudocephalic and cephalocaudal contractile waves along the length of his body; these contractions continue until intromission but then cease (Noble, 1937; Pisani, 1976). The male attempts intromission by rolling the tail region of the female upwards with his tail. Sexually receptive females must then at least remain stationary, and some authors describe female tail lifting and cloacal gaping as sexually receptive behaviors (Ball, 1979; Carpenter, 1977; Crews, 1976; Noble, 1937). Following intromission, the male ceases the contractile waves, relaxes, and the female may drag him around as she moves.

Copulation usually lasts from less than 15 minutes (*T. sirtalis*: Crews, 1976; Devine, 1977a; Fitch, 1965) to more than 90 minutes (*T. radix*: Ross and Crews, 1978). During that time, other males that may have been courting the female disperse from the mating pair and subsequent investigating males rarely begin chin-rubbing; mated females continue to be sexually unattractive and become intolerant of any further courtship (Devine, 1975, 1977a Fitch, 1965; Ross and Crews, 1977, 1978; Whittier et al., 1985). Recently mated *T. sirtalis*, *T. butleri*, and *T. radix* exhibit a translucent gelatinous plug in the cloaca (Devine, 1977a; Fitch, 1965). In *T. melanogaster* the plug appears in the oviducts just anterior to the cloaca (Garstka and Crews, unpubl. data). The plug material, which is a secretion of the male's renal sex segment (Devine,

1975), in at least one species contains a pheromone that inhibits the sexual activity of subsequent males (Ross and Crews, 1977, 1978). Ovarian follicular maturation with ovulation and fertilization in approximately 40-60 days may follow mating (Garstka et al., 1985). Alternatively, sperm may be stored by the female in a viable condition, with fertilization occurring at an appropriate time up to several years after mating (Fox, 1956; Halpert et al., 1982; Rahn, 1940; Stewart, 1972). Garter snakes are viviparous and form a placenta that provides nutritive support to the embryo in addition to the yolk (Hoffman, 1970; see also Guillette, 1982).

RESULTS: THE FEMALE SEXUAL ATTRACTIVENESS PHEROMONE

It has been understood for some time (Noble, 1937) that there is a contact sexual attractiveness pheromone on the skin of female garter snakes and that perception of that pheromone by a male garter snake is a necessary prerequisite for courtship. We use the term "attractiveness" here with the intent of implying the functional significance of the pheromone. Use of the terms "sex attractant" or simply "sex pheromone" would confuse the issue with their connotations of movement - of approach or luring. Sex attractant pheromones have been described in *Thamnophis* snakes (Devine, 1977b; Ford, 1981, 1982) that are functionally important in location of females by males and may be synonymous with the sexual attractiveness pheromone. However, at this time we do not feel that sufficient data exists to equate the contact pheromone with the trail-following pheromone. The chemical nature of these pheromones has been little studied and those few studies have produced conflicting results. Devine (1977b) has argued that the trail-following (sex-attractant) pheromone is in the cholesterol ester fraction of skin extracts, while we (Crews and Garstka, 1982; Garstka and Crews, 1981) have presented data that the contact (sexual attractiveness) pheromone is chemically related to the yolk precursor lipoprotein vitellogenin. In this section we will review the data relevant to the chemical nature and source, mechanism of release, and functional significance of the sexual attractiveness pheromone of *Thamnophis* snakes.

Nature and Source of the Pheromone

There are no glandular structures in the skin of garter snakes. Cloacal glands, the only exocrine glands in *Thamnophis*, seem not to be involved in sexual attractivity (Munro, 1948; Noble, 1937). However, vertebrate pheromones are not necessarily of glandular origin; urine and feces, for example, may contain pheromonal substances (Muller-Schwarze and Mozell, 1977). Thus, when we learned of a specialized network of blood vessels in the skin of *Thamnophis* snakes (Rauch, 1978), we hypothesized that the circulation could be the source of the pheromone. Although in retrospect it seems obvious that the skin would have a blood supply, and others had discussed it (White, 1957), Rauch's description served as a catalyst for our thinking. We reasoned that if the pheromone was present in the circulation it might reach the skin via Rauch's (1978) vessels and percolate out through the skin. The implications of this hypothesis are: 1) the pheromone, or a precursor of it, will be found in the blood, 2) some internal organ in the body will be the site of pheromone synthesis, 3) some process in the skin or at the level of the skin will be responsible for pheromone release onto the skin surface, and 4) either males produce pheromone with male skin limiting its release, or male skin is fundamentally like female skin, but males do not produce pheromone.

In order to test these predictions, a suitable bioassay was necessary. We had to determine whether the test preparation did or did not contain pheromone; that is, to know whether it was capable of inducing males to court. Previous investigations of chemical cues in snakes had largely relied on either counting tongue-flicks to material offered on cotton swabs (Burghardt, 1969, 1970) or trail-following of conspecifics (Devine, 1977b; Ford, 1978, 1981, 1982; Gehlbach et al., 1971). However, Crews (Crews, 1976; Ross and Crews 1977, 1978) and Gillingham and Dickinson (1980) had used courtship to a manipulated stimulus animal as a bioassay. Further, Noble (1937) had noted that male garter snakes rubbed with an attractive female were sometimes courted; that is, the pheromone was transferable. We were also aware that chin-rubbing was displayed in a sexual context but not at other times, and that the presence and degree of chin-rubbing could be used to form an ordinal scale of courtship intensity (Camazine et al., 1980). Therefore, to test our predictions, a test substance (0.5 ml volume) was applied to the back of a large male and that male placed into the isolation cages of two sexually active males (Garstka and Crews, 1981). If at least one male courted the coated stimulus male by chin-rubbing, a score of "positive" was given to the sample; if no chin-rubbing was directed at the stimulus male by either sexually active male, a score of "negative" was given to the sample. To be sure that the males used in these tests were indeed sexually active, each was daily tested with an estrogen-treated female and an untreated uncoated male before any tests with coated males. Only males courting the estrogen-treated female were used to test the putative pheromone substances. No male in our testing courted an untreated uncoated male. In order to be sure that a negative result to a putative pheromone was not the result of declining interest, each male was tested again with an estrogen-treated female and an untreated uncoated male following each day's testing. Only results from those tests in which the male courted the last estrogen-treated female were reported. No male was tested with more than two coated males daily.

We first tested blood serum for pheromonal activity with our behavioral bioassay (Garstka and Crews, 1981). We found that sera from males did not elicit courtship from males (n = 21; none courted), but sera from sexually attractive, courted females did (n = 26; 15 courted). Our sexually attractive females had been treated with estradiol (Crews, 1976; Kubie, 1977; Kubie et al., 1978b). Therefore, we treated males with the same paradigm of estradiol treatments we had used with the females (40 ug/75 g body weight each day for 7 days; Camazine et al., 1980; Kubie et al., 1978b). Although these males were not courted, their blood serum contained pheromone (n = 9 tested; 6 of the 9 sera elicited courtship when applied to the backs of other males).

Using the same bioassay we next tested skin extracts to determine which chemical fraction contained the pheromone (Garstka and Crews, 1981). Because the pheromone was non-volatile, we extracted lipids (chloroform:methanol 3:1 with washing), soluble proteins (Phosphate Buffered Saline, pH 7.2 = PBS), and membrane proteins (1 M NaCl in PBS), from the skins of 4 estradiol-treated, courted females. All of the lipid extracts, but none of the protein extracts, contained pheromone, indicating that the pheromone was a lipid.

Devine (1977b) had previously found that esterified cholesterol partitioned from his trichloroethylene or chloroform:methanol extracts of shed skins was responsible for trail-following in garter snakes, and we were curious whether our results were due to a specific lipid or if it was a quantitative phenomenon only. We found that the sera of estrogen-treated females contained much more lipid than the sera of males, and that estrogen treatment of males resulted in significantly elevated serum lipid (Garstka and Crews, 1981). Therefore we used our bioassay to test lipid non-specificity by testing lard, an animal lipid, applied to males as a stimulus. Lard did not result in courtship, nor did estradiol, our other lipid control, even though estradiol treatment was capable of inducing sexual attractivity (Camazine et al., 1980; Crews, 1976; Kubie, 1977; Kubie et al., 1978b). These data suggested that we were dealing with a specific lipid that was not estradiol.

A lipid present in the blood could come from two sources; it could either be released from the fat-bodies which in reptiles are gonad-associated, developing from the same embryonic genital-ridge tissue as the gonads, or it may be synthesized by and released from the liver. The fat-bodies and the liver are the vertebrate organs involved in lipid metabolism. Futher, both fat-bodies and liver in reptiles contain estrogen receptors (Crews, 1975; Heisermann, 1982; Heisermann et al., 1980), as do mammalian fat deposits (Wade and Gray, 1978). We tested the pheromonal content of liver and fat-bodies from untreated females (because estradiol treatment not only induces attractivity but also increases vascularity). Liver, but not fat-bodies, contained pheromone. Further, liver of estradiol-treated, but not intact, males also contained pheromone.

Thus, we had tested our implications 1, 2, and 4 (see above). Our results so far were that female sexual attractiveness pheromone 1) was present in the blood serum; 2) was a specific (but as yet not identified) lipid; 3) was in the liver but not the fat-bodies; and 4) could be induced in the serum and liver, but not onto the skin, of males by estradiol treatment. Our interpretation at this point (Garstka and Crews, 1981) was that the female sexual attractiveness pheromone was chemically related to vitellogenin, the precursor of yolk.

Vitellogenin is a large, sex-limited lipoprotein molecule which is synthesized in the liver under the control of estradiol and is transported in the circulation. Vitellogenin normally is not present in males, but its synthesis can be induced with estradiol treatment (Wallace, 1979; Wallace and Bergink, 1974). The hypothesis that vitellogenin is related to the female sexual attractiveness pheromone was first tested by using yolk from yolking ovarian follicles as the test substance; yolk from two other garter snake species, from Anolis lizards, and from chickens served as controls. Stimulus males with a coating of conspecific yolk were courted with chin-rubbing; other coatings did not elicit courtship. We next tested purified vitellogenin (method of Ansari et al., 1971; see also Gapp et al., 1979; Garstka et al., 1982), using our bioassay and found it to induce courtship.

Secretion of the Pheromone

Clearly vitellogenin is a good candidate for a female sexual attractiveness pheromone; it is sex-limited and it has a direct connection with reproduction. But how can a yolk precursor lipoprotein end up in the skin? And why? We took two approaches in order to understand this problem. First, we examined histological sections of the skin of untreated and estradiol-treated males and females using Sudan Black B as the lipid stain (Garstka and Crews, 1981; Garstka et al., 1982). Second, we used vitellogenin and rabbit antibody to vitellogenin in in vitro immunodiffusion tests to investigate the conditions for skin binding and release of vitellogenin (Garstka, 1982; Garstka et al., unpubl. data).

Histological examinations of lipid-stained frozen sections of the skin revealed two areas of intense lipid staining (Garstka and Crews, 1981). Within the epidermis of the outer scale surfaces of all snakes examined, regardless of sex or treatment, was a band of stained lipid in the mesos and adjacent alpha keratin layers. The outermost beta keratin layer of the outer scale surfaces of all snakes contained no lipid. In addition to the mesos and alpha keratin area (common to all snakes), females and estrogen-treated males exhibited a second, intense, lipid-stained area. In the dermis, especially the dermal connective tissue and striated muscle adjacent to the vascular bed of Rauch (1978), but generally spread through the dermis, were cells containing large numbers of lipid-stained droplet inclusions. Untreated males either lacked this staining pattern or showed an extremely reduced amount of staining relative to the females.

Lipid and lipoprotein are essential to the processes of skin generation and shedding in snakes. Epidermal lipid is sequestered during skin generation (Jackson and Sharawy, 1978; Landmann, 1979, 1980; Landmann et al., 1981; Maderson, 1984) and forms a barrier to water loss (Landmann, 1980; Landmann et al., 1981; Roberts and Lillywhite, 1980). Thyroid hormones are known to be involved in determining shedding frequency in snakes (Maderson, 1984; Maderson et al., 1970a, 1970b) and thyroidectomy results in extensive disruption of lipid metabolism in snakes (Gupta et al., 1975). Lipid in the form of cholesterol esters may even play a role in epidermal keratinization in snakes (Jackson and Sharawy, 1978).

In additon to the lipid staining areas seen in the epidermis of all snakes and in the dermis of females and estradiol-treated males, some individual snakes, particularly females, exhibited lipid staining material on the outer skin surface, in the hinge region between the scales. In this hinge region the heavy keratinization of the outer scale surfaces is nearly absent; the outer beta keratin layer is reduced to simply the outermost single layer of keratinized cells, the oberhautchen (Maderson, 1965). Further, although nearly lacking keratin, isolated hinge regions show complete water impermeability (Stokes and Dunson, 1982), suggesting extensive lipoidal deposition. Many of our sections which were cut obliquely through the hinge region show the oberhautchen cells outlined by blue stain indicative of lipid.

Our second approach involved use of rabbit antibody to vitellogenin isolated from the serum of estrogen-treated female garter snakes. Our intention was first to use this antibody to quantitate vitellogenin in the serum of female garter snakes, and second, to use the antibody to determine if vitellogenin was present in and on the skin of sexually attractive females. Ouchterlony radial immunodiffusion tests of sera against rabbit anti-vitellogenin absorbed with male garter snake serum were used to roughly quantitate vitellogenin in serum. Surprisingly, only 3 of 11 intact, untreated, sexually attractive females showed any circulating vitellogenin. However, sera of all estradiol-treated females (n = 4) and estradiol-treated males (n = 2) tested were positive for vitellogenin; all sera of intact untreated males (n = 8) were negative.

The skin, not the blood, is where males perceive the pheromone, however, and since our previous results indicated that the dermis contained abundant lipid, we tested minced dermal tissue in immunodiffusion tests. Initially, we tested dermal tissue of 5 intact untreated sexually attractive females, 2 long-term ovariectomized females, and 2 estradiol-treated females. Only tissue of estradiol-treated females reacted with anti-vitellogenin; tissue of untreated females known to be sexually attractive did not react with anti-vitellogenin. However, untreated tissue did react with adjacent wells containing vitellogenin-positive control solutions. Further, the precipitation line of the reaction between untreated tissue and vitellogenin control was a line-of-identity with the antigen-antibody line and was apparent even in the absence of

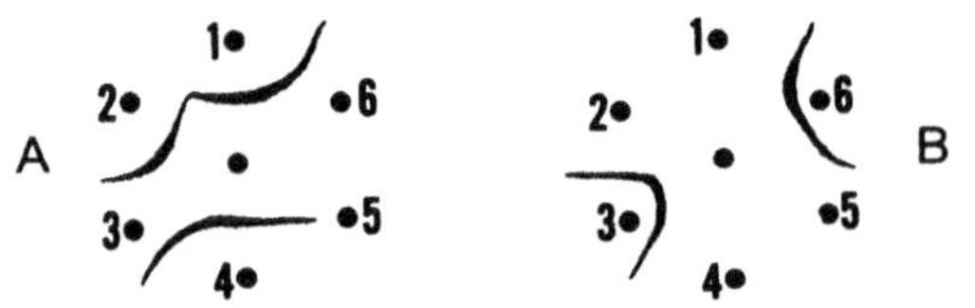

Fig. 2. Ouchterlony radial immunodiffusion assay of sera and tissues from intact and estrogen-treated female T. sirtalis. Center well assay A: rabbit 337 anti-vitellogenin. Center well assay B: vitellogenin, 100 mg/ml. Wells 1 and 4 both assays: estrogen-treated female serum. Well 2 both assays: untreated female serum. Well 3 both assays: dermal tissue from untreated female. Well 5 both assays: dermal tissue from untreated female, treated in vitro with estradiol in PBS. Well 6 both assays: dermal tissue from the same untreated female as the serum in well 2 and the tissue in well 5, treated as in well 5 but without estradion. Drawn from a plate in Garstka (1982).

anti-vitellogenin. We suggested that there is a soluble antibody-like receptor in the dermis and that pre-treatment in vivo with estradiol resulted in either a saturation of receptor by increasing vitellogenin synthesis or a decrease in binding affinity (Crews and Garstka, 1982; Garstka, 1982). These alternative interpretations were further tested by reacting previously frozen dermal tissue of intact, untreated females with estradiol in vitro. Tissues in buffer vehicle, and a mixture of vitellogenin and estradiol served as controls. After 24 hours incubation at 25°C, dialysis at 5°C and lyophilization, the samples were tested by immunodiffusion against both vitellogenin and anti-vitellogenin. The tissue controls without estradiol reacted with vitellogenin but not antibody, and the tissue plus estradiol samples as well as the vitellogenin controls reacted with antibody, but not with vitellogenin. These results are consistent with the second interpretation, that estradiol affects receptor affinity. Further support for this interpretation is the observation that estradiol inhibits skin shedding (Maderson et al., 1970b); that is, if lipid and lipoprotein are essential for skin proliferation and estradiol decreases skin lipoprotein receptor affinity, then estradiol would inhibit skin shedding.

From the above two lines of evidence we draw the following interpretation of how the vitellogenin-related pheromone may reach the skin. The dermal tissue surrounding the vascular network of Rauch (1978) contains abundant lipoprotein receptor. This receptor is not specific for vitellogenin, but appears to bind any lipoprotein and is found in both sexes. Lipoprotein is a critical resource; in snakes up to 11% of ingested energy is apportioned to the skin shedding process (Smith, 1976), and lipid is the most calorie-rich nutrient resource and therefore as much lipoprotein as possible is bound out of the circulation. Some lipid so acquired is diverted into the mesos-alpha keratin water barrier, some into the hinge region accumulation, and some into skin metabolic pathways. Females, who produce more lipoprotein than males (i.e., vitellogenin), and who characteristically have significantly higher serum lipid levels than males, accumulate more lipoprotein in the dermis than males. Estradiol treatment raises serum and dermal lipid and vitellogenin and causes the dermal lipid to be more loosely bound. Lipid from the dermal hinge regions may escape across the oberhautchen layer and accumulate under the dorsal scales.

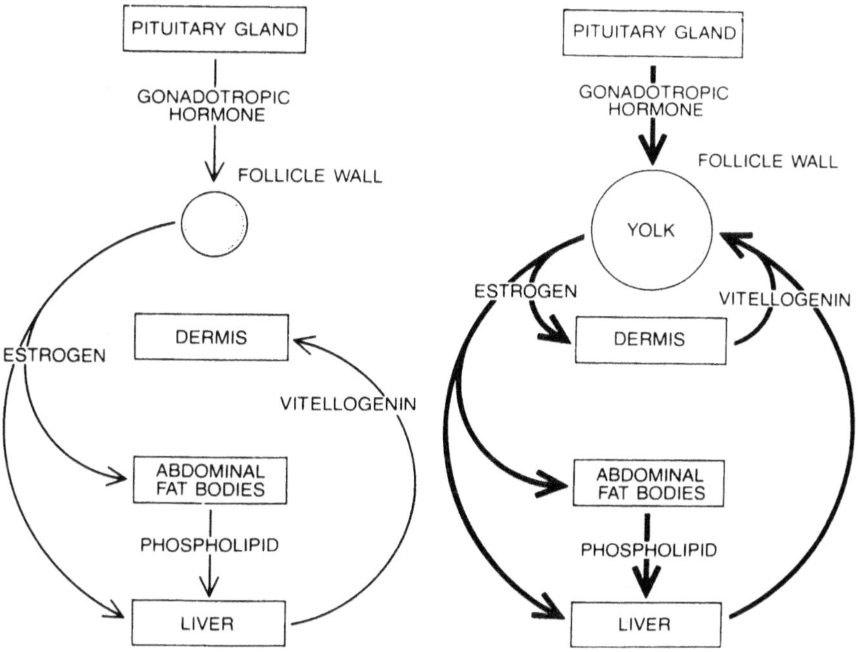

Fig. 3. From Crews, D. and W. R. Garstka. The ecological physiology of a garter snake. Copyright 1982 by Scientific American, Inc. All rights reserved. Hormonal changes induced in female T. sirtalis by mating in the spring may shunt vitellogenin from one role to another. Before mating (left), gonadotropic hormone, which is secreted from the adenohypophysis, causes the ovarian follicles to secrete a modest flow of estrogen. This in turn causes mobilization of precursors from the fat-bodies and synthesis of vitellogenin in the liver. The vitellogenin is rapidly bound into the dermis, some of it reaching the skin surface where it acts as a pheromone. Following mating (right), increased gonadotropin results in increased estrogen released by the ovary which then stimulates a higher level of vitellogenin synthesis and release of vitellogenin from the dermis into the blood. The vitellogenin is then rapidly incorporated into the developing eggs.

We feel this adequately describes what occurs prior to mating in female garter snakes. However, profound hormonal changes occur in female garter snakes after mating (Garstka et al., 1985). During the two days following mating, female T. sirtalis exhibit a rapid and substantial (more than 10 times) increase in circulating estradiol and decrease in serum progesterone. We suggest that these changes in circulating estradiol and progesterone following mating may facilitate the yolking response in female Thamnophis. Recall that garter snakes mate prior to follicular maturation; that is, in female Thamnophis, ovarian follicular yolk deposition follows mating. In the frog, Xenopus, ovarian follicular growth and maturation is dependent on a low ovarian estradiol to progesterone ratio; ovarian estradiol inhibits yolk uptake (Mulner and Ozon, 1981). Thus, we suggest that the dermis of female garter snakes may function as a vitellogenin store, and that the high levels of circulating estradiol following mating not only induce more rapid vitellogenin synthesis, but also simultaneously cause a release of dermal vitellogenin.

Communicative Value of the Pheromone

If there is a relationship between sexual attractivity and vitellogenin, males should be able to assess not only "femaleness," but also fecundity of a female. Vitellogenin content will reflect the size or number of ovarian follicles which can be yolked. Male garter snakes generally prefer larger females (Aleksiuk and Gregory, 1974; Joy and Crews, 1985), female garter snakes are larger than males (Crews et al., 1985; Fitch, 1965; Gregory, 1977), and offspring number is positively correlated with body length (Aldridge, 1979; Fitch, 1965; Ford and Ball, 1977; Gregory, 1977). Does the pheromone we have described serve to link these observations?

In the laboratory, we offered 27 intact, untreated females of varying body size individually to sexually active males; each male was tested with two females (Garstka et al., 1982). In this test larger females were preferred as expected. However, we also determined that males preferred females with significantly more ovarian follicles, even if body size was removed as a variable by analysis of covariance. We therefore suggest that larger females are preferentially courted not because of size, but as a consequence of the amount of pheromone - which we suggest directly predicts their higher fecundity.

The Pheromone and Mate Selection

Sensitivity of male garter snakes to female sexual attractiveness pheromone was initially the most useful variable in the elucidation of the identity of the pheromone: estradiol treatment enhanced sexual attractivity of females, and ovariectomy decreased sexual attractivity (Crews, 1976; Kubie, 1977; Kubie et al., 1978b). Further, every step in our tracing the source and nature of the sexual attractiveness pheromone required males sensitive and responsive to the pheromone. However, when we ovariectomized very large females to verify our hypothesized size-independent nature of the pheromone, we found that under certain conditions these large, long-term ovariectomized females were courted (Garstka et al., 1982; Garstka and Crews, 1984). This seems paradoxical in terms of our previous result - that attractivity was size independent.

In order to study this inconsistency, we compared the responses of males of two garter snake species to females of differing size and hormonal condition through natural and induced breeding periods. These species were *Thamnophis sirtalis* from Manitoba, Canada, and *Thamnophis melanogaster* from Michoacan, Mexico. They were chosen because they exhibit a marked contrast in natural histories: *T. sirtalis* inhabits a strongly seasonal environment (see Aleksiuk, 1975, 1976; Gregory, 1977), whereas *T. melanogaster* inhabits a relatively aseasonal environment (Ball, 1979; Garstka and Crews, 1982). Although both species exhibit synchronous breeding seasons, *T. sirtalis* breed on emergence from hibernation in early spring (Aleksiuk and Gregory, 1974; Camazine et al., 1980; Garstka et al., 1982; Gregory, 1974, 1977; Hawley and Aleksiuk, 1975), while *T. melanogaster* may breed at irregular intervals during the year (Alvarez and Huacuz E., pers. comm.; Ball, 1979; Garstka and Crews, 1982). In both cases, breeding is brief, lasting in the field or in the laboratory approximately 3-4 weeks in *T. sirtalis* and 7-10 days in *T. melanogaster*. Males cease to respond to females after the breeding season, even if fresh, estradiol-treated females which are highly sexually attractive during the breeding season are offered (Camazine et al., 1980; Crews et al., 1984; Garstka and Crews, 1982; Garstka et al., 1982).

A priming period of low-temperature conditions simulating hibernation, followed by exposure to warm temperatures, is necessary to stimulate male sexual activity in Canadian T. sirtalis (Aleksiuk and Gregory, 1974; Camazine et al., 1980; Crews et al., 1984; Hawley and Aleksiuk, 1975). T. melanogaster males do not appear to require a low-temperature dormancy before breeding, and will court estradiol-treated females at any time of the year (Garstka and Crews, 1982).

In sequential presentations of ovariectomized and estradiol-treated females in the laboratory, we found that 1) male T. sirtalis initially courted ovariectomized females, but after 5-9 days all males tested preferred estradiol-treated females, yet 2) male T. melanogaster initially courted only estradiol-treated females, but after 3-5 days most males courted ovariectomized females as well as estradiol-treated females (Garstka and Crews, 1985). These results were not an artifact of habituation or other form of learning, as indicated by tests in which males were presented with ovariectomized females after the critical time period of preference change. We found essentially similar results in the field using sequential and simultaneous presentations, with the exception that the time interval to cessation of courtship of ovariectomized females was variable in T. sirtalis.

Why do males change in their sensitivity to females? It is apparent that some physiological cause is responsible, perhaps the actual physiological cause of courtship (Camazine et al., 1980; Crews, 1983; Crews et al., 1984). Male T. sirtalis change preference in the laboratory at approximately the time when they are beginning to become sporadic in their courtship to estradiol-treated females. However, we believe that there is a clear link between the results observed with the two species: male T. sirtalis initially court and male T. melanogaster initially do not court ovariectomized females as selective responses to the inversely changing demographic situation in the two species. In nature, Canadian T. sirtalis breed in the immediate vicinity of the hibernation dens (Aleksiuk, 1975, 1976; Aleksiuk and Gregory, 1974; Garstka et al., 1982; Gregory, 1977) and female garter snakes usually mate only once (Blanchard and Blanchard, 1941a, 1941b; Devine, 1975, 1977a; Fitch, 1965; Ross and Crews, 1977, 1978; Whittier and Crews, 1985). Males emerge early in the breeding season and remain in the vicinity of the den for the duration of the breeding season. Initially, there may be hundreds or even thousands of males emerged and present at the den when the first females emerge. Thus, the effective sex-ratio is strongly male-biased. As the breeding season progresses, some males disperse, more females emerge, and females emerge earlier in the day, changing the effective sex-ratio to a much less biased or unbiased condition. In contrast, the semi-tropical T. melanogaster do not hibernate, so all females are present in the population at the onset of a breeding period (Garstka and Crews, 1982). Thus, as females mate and are removed from the pool of available mates, the effective sex-ratio becomes male-biased. We argue that at the selective level, male garter snakes are responding to mate availability. When they can choose among females, males choose fecund females; when no such choice exists (as when the effecitve sex-ratio is strongly male-biased), males court any female. The female sexual attractiveness pheromone which we have described serves in both cases to identify femaleness and fecundity.

CRITICAL EVALUATION AND DISCUSSION

Our interpretation accounts for the following observations about the garter snake female sexual attractiveness pheromone: 1) the pheromone is sex-limited; 2) the pheromone is non-volatile; 3) the pheromone is present on skin areas that lack glands; 4) estradiol increases female sexual attractivity; 5) ovariectomy decreases female sexual attractivity; 6) shedding enhances sexual attractivity; 7) the perhomone is species-specific; and 8) female sexual attractivity increases with increasing female body size. In addition, the interpretation is consistent with other known physiological information including 1) there is lipid in snake skin; 2) lipid and lipoprotein are preferentially taken up into snake skin; 3) the hinge region between dorsal scales is especially lipid-rich and leaky to lipid; 4) estradiol inhibits skin shedding; 5) yolk deposition in the ovarian follicles rapidly follows mating, and 6) estradiol can induce vitellogenin synthesis in males.

We have avoided equating the pheromone with vitellogenin because certain key evidence is either lacking or contradictory: 1) chloroform: methanol 3:1 extracts contain pheromone, but this treatment would certainly disrupt at least quaternary and tertiary structure of a lipoprotein; 2) while immunoreactive vitellogenin can be recovered from dermal tissue, our antibody preparations cross-react with vitellogenin from other garter snake species (anti-*T. sirtalis* cross-reacts with *T. melanogaster* and *T. eques*, and anti-*T. melanogaster* cross-reacts with *T. sirtalis*). The antibody may then be recognizing some feature of vitellogenin that the male is not. To the male garter snake there is a high degree of species-specificity in the sexual attractiveness pheromone; males only respond to conspecific females and conspecific yolk cues. Finally, 3) we have yet to demonstrate that vitellogenin crosses the epidermis; so far we have only demonstrated its presence in the deep dermal connective tissue.

At this point our conclusion, based on these data, is that the pheromone is a lipid-rich fragment of vitellogenin, most likely a small fragment. We also feel there may be more than the one pheromone which we have described. Devine's (1977b) cholesterol-ester skin aggregation pheromone, the trailing pheromone described in Ford's work (1978, 1981, 1982), and possibly a pheromone of low volatility (Aldridge, pers. comm.), may all be different and important. Clearly, the highest priority for subsequent studies is to resolve the discrepancies surrounding vitellogenin solubility in solvent media, cross-reactivity of vitellogenin antisera, and male response to only conspecific pheromone, and to use that new knowledge to directly assay epidermal surfaces of sexually attractive females for the pheromone.

A further difficulty with these studies may be the role of the female in pheromone release. Pheromones are usually thought of as being actively released, with the releasing organism communicating to receiving organisms. What evidence do we have that female garter snakes are actively communicating?

Several species of the related natricine genera *Rhabdophis* (=*Natrix*) and *Macropisthodon* (Nakamura, 1935; Smith, 1938) possess ductless dermal poison glands in the anterior dorsal region. When these snakes are aroused they force, by muscular contractions, glandular exudate through ruptures in the skin in the hinge regions between the scales overlying the glands. Although we have previously (Garstka and Crews, 1981) alluded to this example, we no longer feel that it is directly relevant to the *Thamnophis* problem. However, our observation of the hinge region in *Thamnophis* may point to sex differences in this critical area. We have observed (Garstka and Crews, 1981) that females increase their ventilatory volume while being courted, thus increasing the exposure of the hinge regions and anything that may be trapped under the scales, and perhaps stretching the very thin hinge regions as well. We have also observed that treatment of males with estradiol, while inducing pheromone in the blood serum and liver, never made those males sexually attractive (Garstka and Crews, 1981; Garstka et al., 1982). Thus, we argue that the skin of male and female garter snakes differs in some fundamental manner, and further, that the female plays an active role in communication by releasing the sexual attractiveness pheromone.

We have provided evidence that strongly suggests that the female sexual attractiveness pheromone of garter snakes is chemically related to vitellogenin, the circulating precursor of yolk, that it is present in the skin because of a lipoprotein receptor system common to both males and females, that in addition to "femaleness" it can communicate fecundity to males, and that it may be actively released by females to communicate with males. Kittredge and Takahashi (1972) and Karlson and Schneider (1973) have discussed the conceptual and physical difficulties in the notion that pheromone emission and reception systems could arise together de novo. They argue that these systems must have evolved separately and for different reasons. We have discussed the female, and have concluded that a system for lipoportein storage related to skin function has been modified by selection into a pheromone release system. What about the male? Has the male garter snake chemoreception system arisen expressly for pheromone perception or has it wider significance?

Garter snakes possess two separate chemical senses, the nasal olfactory system and the vomeronasal system. The garter snake vomeronasal system is involved in trail-following and prey selection, as well as sex recognition (Burghardt, 1970; Kubie, 1977; Kubie et al., 1978a), and is crucial to the display of courtship (Kubie, 1977; Kubie et al., 1978a; Noble, 1937). Therefore, we suggest that in the garter snake female sexual attractiveness system, selection has operated to modify both male and female functions. In the male, the vomeronasal system has gained a preeminence not seen in other vertebrates over hormonal controls in the regulation of male sexual behavior (Camazine et al., 1980; Crews, 1983; Crews et al., 1984). In the female, the secretion of a material already present in the skin has come under selective modification, whereas even when males are artificially induced to make pheromone, they lack the ability to release it onto the outside of the skin.

Selection may act generally to fix, canalize, or modify response to chemical stimuli when those stimuli communicate information directly pertinent to reproductive success.

ACKNOWLEDGMENTS

We wish first to recognize the contributions of our collaborators in the studies that we have reviewed here; without their labors and without the stimulating intellectual environment which they provided these studies would not have been possible. We are speaking of Brian Camazine, Maire Anne Diamond, Andrew Halpert, Milos Novotny, Vicki Sato, Francis Schwende, and Richard Tokarz. We also wish to thank the wildlife agencies and officials of both Canada and Mexico, especially Mr. R. Goulden, Mr. M. Shoesmith, and Lic. Ignacio Ibarrola B. for their assistance. Support was provided by grants from NICHHD and NINCDS (to DC) and the Richmond Fund of Harvard University (to WRG). Word processing equipment and software for manuscript preparation were made available by Project Access of U.A.H. (to WRG).

REFERENCES

Aldridge, R. D., 1979, Female reproductive cycles of the snakes Arizona elegans and Crotalus viridis, Herpetologica, 35:256.

Aleksiuk, M., 1975, Manitoba's fantastic snake pits, Natl. Geogr., 147:715.

Aleksiuk, M., 1976, Reptilian hibernation: Evidence of adaptive strategies in Thamnophis sirtalis parietalis, Copeia, 1976:170.

Aleksiuk, M., and Gregory, P. T., 1974, Regulation of seasonal mating behavior in Thamnophis sirtalis parietalis, Copeia, 1976:681.

Ansari, A. Q., Dolphin, P. J., Lazier, C. B., Munday, K. A., and Akhtar, M., 1971, Chemical composition of an oestrogen-induced, calcium-binding glycolipophosphoprotein in Xenopus laevis, Biochem. J., 122:107.

Arnold, S. J., 1983, Sexual selection: The interface of theory and empiricism, in: "Mate Choice," P. Bateson, ed., Cambridge Univ. Press., Cambridge.

Arnold, S. J., and Houck, L. D., 1982, Courtship pheromones: Evolution by natural and sexual selection, in: "Biochemical Aspects of Evolutionary Biology," M. H. Nitecki, ed., Univ. of Chicago Press, Chicago.

Ball, R. L., 1979, "The Courtship and Mating Behavior of a Mexican Garter Snake, Thamnophis melanogaster," MS Diss., Univ. Oklahoma.

Bermant, G., and Davidson, J. M., 1974, "Biological Bases of Sexual Behavior," Harper and Row, New York.

Bishop, J. E., 1959, A histological and histochemical study of the kidney tubule of the common garter snake, Thamnophis sirtalis, with special reference to the sexual segment in the male, J. Morphol., 104:307.

Blanchard, F. N., and Blanchard, F. C., 1941a, Factors determining the time of birth in the garter snake Thamnophis sirtalis sirtalis (Linnaeus), Mich. Academic., 26:161.

Blanchard, F. N., and Blanchard, F. C., 1941b, Mating of the garter snake Thamnophis sirtalis sirtalis (Linnaeus), Mich. Academic., 27:215.

Burghardt, G. M., 1970, Chemical perception in reptiles, Adv. Chemorecept., 1:241.

Camazine, B., Garstka, W. R., Tokarz, R. R., and Crews, D., 1980, Effects of castration and androgen replacement on male courtship behavior in the red-sided garter snake (Thamnophis sirtalis parietalis), Horm. Behav., 14:358.

Carpenter, C. C., 1977, Communication and display of snakes, Am. Zool., 17:217.

Cieslak, E. S., 1945, Relation between the reproductive cycle and the pituitary gland in the snake Thamnophis radix, Physiol. Zool., 18:299.

Crews, D., 1975, Psychobiology of reptilian reproduction, Science, 189:1059.

Crews, D., 1976, Hormonal control of male courtship behavior and female attractivity in the garter snake (Thamnophis sirtalis sirtalis), Horm. Behav., 7:451.

Crews, D., 1983, Control of male sexual behavior in the Canadian red-sided garter snake, in: "Hormones and Behavior in Higher Vertebrates," J. Balthazard, ed., Plenum Press, London.

Crews, D., 1984, Gamete production, sex steroid secretion, and mating behavior uncoupled, Horm. Behav., 18:21.

Crews, D. and Garstka, W. R., 1982, The ecological physiology of a garter snake, Sci. Am. 247:158.

Crews, D., Camazine, B., Diamond, M., Mason, R., Tokarz, R. R., and Garstka, W. R., 1984, Hormonal independence of courtship behavior in the male garter snake, Horm. Behav., 18:29.

Crews, D., Diamond, M., Whittier, J., and Mason, R., 1985, Small male body size in the garter snake depends on the testes, Am. J. Physiol., (in press).
Cuellar, H. S., Roth, J. J., Fawcett, J. D., and Jones, R. E., 1972, Evidence for sperm sustenance by secretions of the renal sex segment of male lizards, Anolis carolinensis, Herpetologica, 28:53.
Darwin, C., 1871, "The Descent of Man and Selection in Relation to Sex," John Murray, London.
Devine, M. C., 1975, Copulatory plugs in snakes: Enforced chastity, Science, 187:844.
Devine, M. C., 1977a, Copulatory plugs, restricted opportunities, and reproductive competition among male garter snakes, Nature, 267:345.
Devine, M. C., 1977b, "Chemistry and Source of Sex-attractant Pheromones and Their Role in Mate Discrimination by Garter Snakes," Ph.D. Diss.. Univ. Michigan.
Emlen, S. T., and Oring, L. W., 1977, Ecology, sexual selection, and the evolution of mating systems, Science, 197:215.
Farner, D. S., and Follett, B. K., 1979, Reproductive periodicity in birds, in: "Hormones and Evolution," E. J. W. Barrington, ed., Academic Press, New York.
Fitch, H. S., 1965, An ecological study of the garter snake, Thamnophis sirtalis, Univ. Kans. Publ. Mus. Nat. Hist., 15:493.
Ford, N. B., 1978, Evidence for species specificity of pheromone trails in two sympatric garter snakes, Thamnophis, Herpetol. Rev., 9:10.
Ford, N. B., 1981, Seasonality of pheromone trailing behavior in two species of garter snake, Thamnophis (Colubridae), Southwest, Nat., 26:385.
Ford, N. B., 1982, Species specificity of sex pheromone trails of sympatric and allopatric garter snakes (Thamnophis), Copeia, 1982:10.
Ford, N. B., and Ball, R., 1977, Clutch size and size of young in the Mexican garter snake, Thamnophis melanogaster (Reptilia, Serpentes, Colubridae), Herpetol. Rev., 8:118.
Fox, H., 1979, The urinogenital system of reptiles, in: "Biology of the Reptilia," Vol. 7, C. Gans, ed., Academic Press, New York.
Fox, W., 1952, Seasonal variation in the male reproductive system of Pacific coast garter snakes, J. Morphol., 90:481.
Fox, W., 1954, Genetic and environmental variation in the timing of the reproductive cycle of male garter snakes, J. Morphol., 95:415.
Fox, W., 1956, Seminal receptacles of snakes, Anat. Rec., 124:519.
Friedman, D. W., and Crews, D., 1985a, Role of the anterior hypothalamus-preoptic area in the regulation of courtship behavior in the male Canadian red-sided garter snake (Thamnophis sirtalis parietalis): Intracranial implantation experiments, Horm. Behav., (in press).
Friedman, D. W., and Crews, D., 1985b, Role of the anterior hypothalamus-preoptic area in the regulation of courtship behavior in the male Canadian red-sided garter snake (Thamnophis sirtalis parietalis): Lesion experiments, Behav. Neurosci., (in press).
Gapp, D. A., Ho., S-m., and Callard, I. P., 1979, Plasma levels of vitellogenin in Chrysemys picta during the annual gonadal cycle: Measurement by specific radioimmunoassay, Endocrinology, 104:784.
Gardner, J. B., 1955, A ball of snakes, Copeia, 1955:310.
Gardner, J. B., 1957, A garter snake ball, Copeia, 1957:48.
Garstka, W. R., 1982, "Chemical Communication and the Control of Garter Snake Reproductive Cycles," Ph.D. Diss., Harvard Univ.
Garstka, W. R., and Crews, D., 1981, Female sex pheromone in the skin and circulation of a garter snake, Science, 214:681.
Garstka, W. R., and Crews, D., 1982, Female control of male reproductive function in a Mexican snake, Science, 217:1159.
Garstka, W. R., and Crews, D., 1985, Mate preference in garter snakes. Herpetologica, 41:9.

Garstka, W. R., Camazine, B., and Crews, D., 1982, Interaction of behavior and physiology during the annual reproductive cycle of the red-sided garter snake (Thamnophis sirtalis parietalis), Herpetologica, 38:104.
Garstka, W. R., Tokarz, R. R., Diamond, M., Halpert, A. P., and Crews, D., 1985, Behavioral and physiological control of yolk synthesis and deposition in the female red-sided garter snake (Thamnophis sirtalis parietalis), Horm. Behav., (in press).
Gehlbach, F. R., Watkins, J. F., and Kroll, J. C., 1971, Pheromone trail-following studies of typhlopid, leptotyphlopid, and colubrid snakes, Behaviour, 40:282.
Gillingham, J. C., and Clark, D. L., 1981, An analysis of prey searching behavior in the western diamondback rattlesnake, Crotalus atrox, Behav. Neural Biol., 32:235.
Gillingham, J. C., and Dickinson, J. A., 1980, Postural orientation during courtship in the eastern garter snake, Thamnophis s. sirtalis, Behav. Neural Biol., 28:211.
Guillette, L. J., 1982, The evolution of viviparity and placentation in the high elevation Mexican lizard Sceloporus aeneus, Herpetologica, 38:94.
Gregory, P. T., 1974, Patterns of spring emergence of the red-sided garter snake (Thamnophis sirtalis parietalis) in the Interlake region of Manitoba, Can. J. Zool., 52:1063.
Gregory, P. T., 1977, Life history parameters of the red-sided garter snake (Thamnophis sirtalis parietalis) in an extreme environment, the Interlake region of Manitoba, Natl. Mus. Nat. Sci. Ottawa Publ. Zool. No. 13.
Gupta, S. C., Thapliyal, J. P., and Garg, R. K., 1975, The effects of thyroid hormones on the chemical constituents of different tissues of the chequered water snake Natrix piscator, Gen. Comp. Endocrinol., 27:223.
Halliday, T. R., 1978, Sexual selection and mate choice, in: "Behavioural Ecology; An Evolutionary Approach," J. R. Krebs, and N. B. Davies, eds., Blackwell Scientific Publs., London.
Halliday, T. R., 1983, The study of mate choice, in: "Mate Choice," P. Bateson, eds., Cambridge Univ. Press, Cambridge.
Halpert, A. P., Garstka, W. R., and Crews, D., 1982, Sperm storage and transport and its relation to the annual sexual cycle of the female red-sided garter snake, Thamnophis sirtalis parietalis, J. Morphol., 174:149.
Hawley, A. W. L., and Aleksiuk, M., 1975, Thermal regulation of spring mating behavior in the red-sided garter snake (Thamnophis sirtalis parietalis), Can. J. Zool., 53:768.
Hawley, A. W. L., and Aleksiuk, M., 1976, The influence of photoperiod and temperature on seasonal testicular recrudescence in the red-sided garter snake (Thamnophis sirtalis parietalis), Comp. Biochem. Physiol., A53:215.
Heisermann, G. J., 1982, "An Estrogen Receptor in the Liver of the Turtle, Chrysemys picta," M.A. Diss., Boston Univ.
Heisermann, G. J., Ho, S-m., and Callard, I. P., 1980, Identification of an estrogen receptor in the liver cytosol of a turtle, Chrysemys picta, Am. Zool., 20:792.
Hiller, U., 1971, Form und Funktion der Hautsinnesorgane bei Gekkoniden, 1. Licht und Raterelecktronenmikroskopie Untersuchungen, Form. Funct., 4:240.
Hoffman, L. H., 1970, Placentation in the garter snake, Thamnophis sirtalis, J. Morphol., 131:57.
Jackson, M. K., and Sharawy, M., 1978, Lipids and cholesterol clefts in the lacunar cells of snake skin, Anat. Rec., 190:41.
Joy, J., and Crews, D., 1985, Social facilitation of courtship behavior in the red-sided garter snake, Thamnophis sirtalis parietalis, J. Comp. Psychol., (in press).

Karlson, P., and Schneider, D., 1973, Sexual hormone der Schmetterlinge als Modelle chemischer Kommunikation, Naturwissenschaften, 60:113.
Kirkpatrick, M., 1982, Sexual selection and the evolution of female choice, Evolution, 36:1.
Kittredge, J. S., and Takahashi, F. T., 1972, The evolution of sex pheromone communication in the arthropoda, J. Theor. Biol. 35:467.
Kubie, J. L., 1977, "The Role of the Vomeronasal Organ in Garter Snake Prey Trailing and Courtship,": Ph.D. Diss., S.U.N.Y.
Kubie, J. L., Vagvolgyi, A., and Halpern, M., 1978a, Roles of the vomeronasal and olfactory systems in courtship behavior of male garter snakes, J. Comp. Physiol. Psychol., 92:627.
Kubie, J. L., Cohen, J., and Halpern, M., 1978b, Shedding enhances the attractiveness of oestradiol treated garter snakes and their untreated penmates, Anim. Behav., 26:562.
Lande, R., 1980, Sexual dimorphism, sexual selection, and adaptation in polygenic characters, Evolution, 34:292.
Lande, R., 1981, Models of speciation by sexual selection of polygenic traits, Proc. Natl. Acad. Sci., 78:3721.
Landmann, L., 1975, The sense organs in the skin of the head of squamata (Reptilia), Israel J. Zool., 24:99.
Landmann, L., 1979, Keratin formation and barrier mechanisms in the epidermis of Natrix natrix (Reptilia: Serpentes): An ultrastructural study, J. Morphol., 162:93.
Landmann, L., 1980, Zonulae occulodentes in the epidermis of the snake Natrix natrix L., Experientia, 36:110.
Landmann, L., Stolinski, C., and Martin, B., 1981, The permeability barrier in the epidermis of the grass snake during the resting stage of the sloughing cycle, Cell Tissue Res., 215:369.
List, J. C., 1950, Observations on the courtship behavior of Thamnophis sirtalis sirtalis, Herpetologica, 6:71.
Maderson, P. F. A., 1965, Histological changes in the epidermis of snakes during the sloughing cycle, J. Zool., 146:98.
Maderson, P. F. A., 1984, Some developmental problems of the reptilian integument, in: "Biology of the Reptilia," (in press).
Maderson, P. F. A., Chiu, K. W., and Phillips, J. G., 1970a, Changes in the epidermal histology during the sloughing cycle in the rat snake Ptyas korros Schlegel, with correlated observations on the thyroid gland, Biol. Bull., 139:304.
Maderson, P. F. A., Chiu, K. W., and Phillips, J. G., 1970b, Endocrine-epidermal relationships in squamate reptiles, Mem. Soc. Endocrinol., 18:259.
Meredith, M., and Burghardt, G. M., 1978, Electrophysiological studies of the tongue and accessory olfactory bulb in garter snakes, Physiol. Behav., 21:1001.
Muller-Schwarze, D., and Mozell, M. M., 1977, "Chemical Signals in Vertebrates," Plenum Press, New York.
Mulner, O., and Ozon, R., 1981, The roles of the follicular envelopes in the initiation of Xenopus oocyte maturation, Gen. Comp. Endocrinol., 44:335.
Munro, D. F., 1948, Mating behavior and seasonal cloacal discharge of a female Thamnophis sirtalis parietalis, Herpetologica, 4:185.
Nakamura, K., 1935, On a new integumental poison gland found in the nuchal region of a snake, Natrix tigrina, Mem. Coll. Sci. Univ., Kyoto Ser. B., 10:229.
Noble, G. K., 1937, The sense organs involved in the courtship of Storeria, Thamnophis, and other snakes, Bull. Am. Mus. Nat. Hist., 73:673.
Orians, G. H., 1969, On the evolution of mating systems in birds and mammals, Am. Nat., 103:589.
Pisani, G. R., 1976, Comments on the courtship and mating mechanics of Thamnophis (Reptilia, Serpentes, Colubridae), J. Herpetol., 10:139.

Prasad, M. R. N., and Reddy, P. R. K., 1972, Physiology of the sexual segment of the kidney in reptiles, Gen. Comp. Endocrinol., Suppl. 3:649.
Rahn, H., 1940, Sperm viability in the uterus of the garter snake, Thamnophis, Copeia, 1940:109.
Rauch, J. C., 1978, Integumentary bloodvascular system in garter snakes (Thamnophis sirtalis parietalis and T. radix), Can. J. Zool., 56:469.
Regaud, C., and Policard, A., 1903, Variations sexuelles de structure dans le segment preterminal du tube urinifere de quelques ophidiens, C. R. Soc. de Biol. (Paris), 55:216.
Roberts, J. B., and Lillywhite, H. B., 1980, Lipid barrier to water exchange in reptile epidermis, Science, 207:1077.
Ross, P., and Crews, D., 1977, Influence of the seminal plug on mating behavior in the garter snake, Nature, 267:344.
Ross, P., and Crews, D., 1978, Stimuli influencing mating behavior in the garter snake, Thamnophis radix, Behav. Ecol. Sociobiol., 4:133.
Rossman, D. A., and Eberle, W. G., 1977, Partition of the genus Natrix, with preliminary observations on evolutionary trends in natricine snakes, Herpetologica, 33:34.
Ruthven, A. G., 1908, Variation and genetic relationships of the garter snakes, Bull. U.S. Natl. Mus., 61:1.
Smith, G. C., 1976, Ecological energetics of three species of ectothermic vertebrates, Ecology, 57:252.
Smith, H. M., 1942, The synonymy of the garter snakes (Thamnophis), with notes on Mexican and Central American species, Zoologica, 27:97.
Smith, M. A., 1938, The nucho-dorsal glands of snakes, Proc. Zool. Soc., Lond., 108:575.
Stewart, G. R., 1972, An unusual record of sperm storage in a garter snake (genus Thamnophis), Herpetologica, 28:346.
Stokes, G. P., and Dunson, W. A., 1982, Permeability and channel structure of reptilian skin, Am. J. Physiol., 242:f681.
Wade, G. N., and Gray, J. M., 1978, Cytoplasmic 17β-(3H) estradiol binding in rat adipose tissues, Endocrinology, 103:1695.
Wallace, R. A., 1979, Oocyte growth in nonmammalian vertebrates, in: "The Vertebrate Ovary," R. E. Jones, ed., Plenum Press, New York.
Wallace, R. A., and Bergink, E. W., 1974, Amphibian vitellogenin: Properties, hormonal regulation of hepatic synthesis and ovarian uptake, and conversion to yolk platelets, Am. Zool., 14:1159.
White, F. N., 1957, A functional interpretation of the integumentary vessels of snakes, Herpetologica, 13:127.
Whittier, J., and Crews, D., 1985, Variation in reproductive patterns of female red-sided garter snakes, Thamnophis sirtalis parietalis, Anim. Behav., (in press).
Whittier, J. M., Mason, R. T., and Crews, D., 1985, Sexual selection and the effects of mating on sexual behavior in the red-sided garter snake, Thamnophis sirtalis parietalis, Behav. Ecol. Sociobiol., (in press).
Zahavi, A., 1975, Mate selection - a selection for a handicap, J. Theor. Biol., 53:205.

THE ROLE OF PHEROMONE TRAILS IN THE SOCIOBIOLOGY OF SNAKES

Neil B. Ford

Department of Biology
The University of Texas at Tyler
Tyler, Texas 75701

INTRODUCTION: WHY STUDY SNAKE PHEROMONE TRAILS?

In his review of reptilian social systems, Brattstrom (1974) stated that "nobody has, in my opinion, designed the appropriate experiments to study social behavior in snakes properly." In the subsequent decade, studies of snake behavior have proliferated, but Brattstrom's comments regarding studies of snake sociobiology are essentially still true. Most reports remain observational accounts and descriptions of ritualistic behaviors involved in reproductive and agonistic encounters (reviewed by Carpenter, 1977; Carpenter and Ferguson, 1977). Very few biologists have attempted to experimentally examine the sociobiological functions of such displays. However, for a good example, see Gillingham et al.'s (1983) test of courtship activity of male diamondback rattlesnakes (*Crotalus atrox*), after combat dances.

The problem in examining the functions of social behavior in snakes lies not in a lack of interest in this group, but rather in a lack of understanding of the two main sensory modalities involved in ophidian communication. Although courtship and agonistic rituals have visual components that have been examined, the olfactory and tactile senses likely play more important roles in snake intraspecific communication (Carpenter and Ferguson, 1977; Madison, 1977). As these two senses are of secondary importance in humans, researchers have often had difficulty both designing and explaining the results of experiments that deal with behaviors that are heavily influenced and controlled by odors and tactile stimuli. What we have learned about the mechanism by which odors are perceived by snakes has come about through a series of studies of feeding behavior (See Burghardt, 1980 for review). These experiments succeeded essentially because snakes produce an easily quantified bioassay of chemosensory input. Since the tongue of snakes picks up and delivers odors to the vomeronasal system (which is involved in detection, localization and identification of odors; Cowles and Phelan, 1958; Halpern and Kubie, 1983), the activity of the tongue can be a measure of olfactory response. Visual analysis of this activity (tongue flick rate) has therefore been useful in various experimental situations as a dependent measure of chemosensory activity.

Tongue flicks (TFs) have been quantified in some analyses of snake social activity (i.e., courtship and combat rituals; Gillingham et al. 1983); however, because tongue flick rates increase under a number of

Table 1. Observation and Studies of Trailing Behavior in Snakes.

Species	Observed In	Age	Suggested Function	Trail[a] Source	Reference
VIPERIDAE					
Crotalus horridus	Lab	Juvenile	Den location	Natural	Brown and MacLean 1983
Crotalus viridis	Lab	Adult	Den location	Natural[b]	King et al., 1983
Vipera berus	Lab	Adult	Reproduction	Natural	Andren, 1982
Vipera aspis	Lab	Adult	Reproduction	Cloacal musk	Baumann, 1929
COLUBRIDAE					
Lampropeltis getulus	Lab	Adult	Reproduction	Cloacal[b] musk	Price and LaPointe, 1981
Masticophis taeniatus	Lab	Juvenile	Aggregation	Skin	Parker and Brown, 1980
	Field	Adult	Reproduction	Natural	Parker and Brown, 1980
Nerodia sipedon	Lab	Adult	Reproduction	Natural	Ford, personal observation
Opheodrys aestivus	Lab	Adult	Reproduction	Natural	Ford, personal observation
Pituophis melanoleucus	Lab	Juvenile	Aggregation	Skin	Parker and Brown, 1980
Regina septemvittata	Lab	Adult	Reproduction	Natural	Ford, 1979
Storeria dekayi	Lab	Adult	Reproduction Aggregation	Skin	Noble and Claussen, 1936
Sonora semiannulata (episcopa)	Lab	Adult	Reproduction Aggregation	Natural	Gehlbach et al., 1971
Tantilla gracilis	Lab	Adult	Reproduction Aggregation	Natural	Gehlbach et al., 1971
Thamnophis butleri	Lab	Adult	Reproduction	Skin	Devine, 1977
	Lab	Adult	Reproduction	Natural	Ford, 1982
	Field	Adult	Reproduction	Natural	Finneran, 1949
Thamnophis couchi	Lab	Juvenile	Aggregation	Natural	Heller and Halpern, 1981
Thamnophis elegans	Lab	Juvenile	Aggregation	Natural	Heller and Halpern, 1981
Thamnophis marcianus	Lab	Adult	Reproduction	Natural	Ford and Schofield, 1984
Thamnophis proximus	Lab	Adult	Reproduction	Natural	Ford, 1978
Thamnophis radix	Lab	Adult	Reproduction	Natural	Ford and Low, 1984 Ford and Schofield, 1984
	Field	Adult	Reproduction	Natural	Davis, 1936
Thamnophis sirtalis	Lab	Juvenile	Aggregation	Natural	Heller and Halpern, 1981
	Lab	Adult	Aggregation	Skin	Noble, 1937
	Lab	Adult	Reproduction	Skin	Devine, 1977
				Natural	Ford, 1981
				Natural	Ford, 1982
				Skin	Noble, 1937
	Field	Adult	Reproduction	Natural	Bishop, 1927 Carpenter, 1952 Gardner, 1955 Munro, 1948 Noble and Claussen, 1936 Truitt, 1927 Wright and Wright, 1957
Virginia striatula	Lab	Adult	Reproduction Aggregation	Natural	Gehlbach et al., 1971
ELAPIDAE					
Dendroaspis polylepis	Field	Adult	Reproduction	Natural	FitzSimons, 1932
LEPTOTYPHLOPIDAE					
Leptotyphlops dulcis	Lab	Adult	Aggregation	Natural	Gehlbach et al., 1971
TYPHLOPIDAE					
Typhlops pusillus	Lab	Adult	Reproduction Aggregation	Natural	Gehlbach et al., 1971

[a]Natural - trail-laying snake allowed to move naturally.
Skin - body of snake rubbed on substrate.
Cloacal Musk - cloacal secretions rubbed on substrate.

[b]Trailing did not occur in these tests.

sociobiological situations, the results have been ambiguous. For example, elevated TF rates exhibited by a male snake that encountered a female could imply sexual interest, and agonistic response or just a reaction to a novel stimulus. Pheromone trailing is a visually quantifiable behavior, more restrictive than tongue flicking in its occurrence in social activity, and so studies of trailing behavior can give increased insight into the role of olfaction in the sociobiology of snakes.

PHEROMONE TRAILING BEHAVIOR

The following of odor trails is known to be important in several aspects of snake behavior such as locating prey (Kubie and Halpern, 1979) or following prey that has been envenomed and released (Chiszar et al., 1983). How ever, it is with respect to snake sociobiology that trailing may be most useful. Ophidian locomotion is particularly suited to laying continuous odor trails. As a snake undulates, the ventral and much of the lateral surface of the body contacts the substrate and pheromones from either the skin or cloacal glands can be deposited. (Most experimental evidence indicates, however, that cloacal musk is not the source for snake trails - Table 1: Noble, 1937; Price and LaPointe 1981. Trailing behavior has been documented in a number of species of snakes from five families, either from field observations or experiments in the laboratory (Table 1). Although many of the field accounts were anecdotal and some of the laboratory experiments were not well controlled, a summary of the pheromone trailing activity can be useful, particularly, in helping design future research on the phenomenon. Trails are used by males to locate females during the breeding season, and by young or adults of either sex to form aggregations (Table 1). The aggregations may be for control of environmental conditions (i.e., humidity or temperature in refugia or dens) or they may decrease predation (i.e., aggregations of gravid snakes just prior to parturition). In this paper, I review the role of trailing in these social activities, discuss how snakes orient to pheromone trails and describe some aspects of the evolution of trail pheromones.

THE ROLE OF TRAILING IN SEXUAL ACTIVITY

Snakes, as a group, are visually cryptic and auditorially impoverished. Therefore, it is probable that chemical cues are quite important in mate location. The utilization of pheromone trails in the reproductive activity of snakes has been examined to some extent in temperate zone colubrids. In these snakes, sexual behavior occurs primarily in the spring. Males leave the hibernacula first and remain in the vicinity to court the females when they emerge. At this time, female snakes leave trails for the males to follow. The pheromone involved is likely the same lipoprotein (vitellogenin) which the females produce in the liver and secrete through the skin to stimulate male courtship activity (Garstka and Crews, 1981, this volume), although the only evidence for this is that the sexual pheromone trails are also produced by skin secretions and, like a lipoprotein, are non-volatile and persistent (Ford and Low, 1984).

Male snakes are able to extract several pieces of information from the pheromone trail. There is evidence that they can tell both the species and the sexual receptivity of the female laying the trail.

Species-specific pheromone trails would be a useful prezygotic reproductive isolating mechanism as they would decrease the chance of hybrid matings. Such specificity would also save the male the energy of trailing a nonreceptive heterospecific female and reduce the male's exposure to predators. It seems likely that all species of snakes using trails in a

Table 2. Species specificity in sexual pheromone trails of sympatric snakes. Males were tested for trail preference in Y mazes and arm choice was analyzed with one-tailed binomial tests.

Species Tested	# Choosing Conspecific Female's Trail	# Choosing Heterospecific Female's Trail	P	Reference
Thamnophis sirtalis on				
T. butleri	13	4	0.025	Ford, 1982
	51	3	0.000	Devine, 1977
T. proximus	5	1	-	Ford, 1978
Thamnophis butleri on				
T. sirtalis	15	4	0.010	Ford, 1982
	28	5	0.000	Devine, 1977
Thamnophis proximus on				
T. sirtalis	5	1	-	Ford, 1978
Thamnophis radix on				
T. sirtalis	30	12	0.004	Ford and Schofield, 1984

particular area should produce different pheromones and indeed the four groups of sympatric species that have been examined can all distinguish conspecific trails (Table 2).

The evidence that pheromone trails contain information regarding female sexual receptivity is not as conclusive. A comparison of male trailing activity with the species' reproductive cycle is the obvious test to examine this hypothesis. But in the only seasonal examination of trailing, I found that garter snake trailing responses did not correlate well with the spring breeding period. Male Thamnophis sirtalis continued trailing females into June, and T. proximus males seemed to trail all summer (Fig. 1; Ford, 1981). While I suggested that mating might be occurring later than most other researchers believed, particularly in southern species like T. proximus, methodological problems (such as an inadequate removal of previous trails and inexact evaluation of the placement of the trail) made this conclusion premature.

Recent experiments in my laboratory have utilized a 250 cm long arena in which trailing activity can be more carefully evaluated. In this apparatus trails are laid on butcher paper on a removable floor (Fig. 2). Pegs are pushed through the paper in holes spaced every ten cm (peg spacing can be adjusted for smaller or larger species), and pieces of plastic soda straws are placed over each peg. The pegs give the animals traction in the arena, and the straws and paper can be changed to completely remove previous trails. The trail-laying animal is videotaped as it moves through the arena between two vertical guides. A map is made of all areas contacted by the snake's body. The guides are removed and the trailing snake allowed to enter the arena at its own discretion from a box placed at the beginning of the

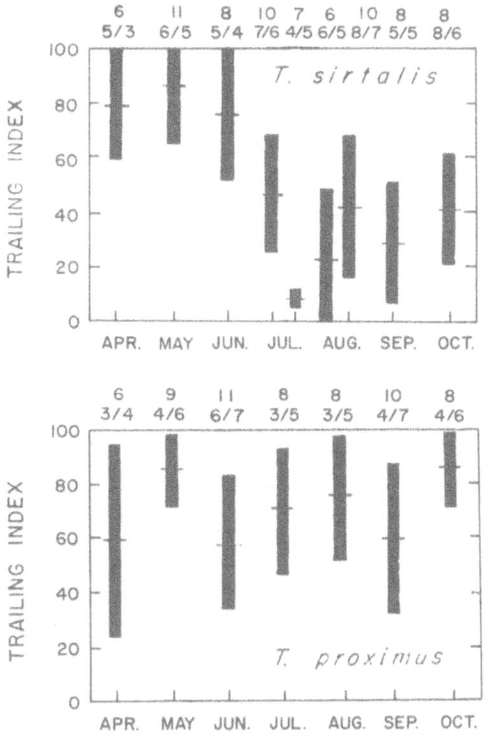

Fig. 1. Seasonal changes in trailing inclination of males of two species of Thamnophis. Mean trailing index (%) and 95% confidence limits. Top line: number of trials for each month. Second line: number of males and females used. From Ford (1981).

trail. The exterior room is darkened so that movements outside the arena are not visible to the snake inside. The trailing animal is also videotaped, and an index of trailing "ability" or "inclination" is produced by measuring all areas of the trail contacted by the head of the trail-following animal and dividing that by the length of the trail (see Fig. 3). This technique allows pheromone trailing to be reliably detected with very few trials (non-trailing animals tend to wall-seek and rarely cross the trail area). When this apparatus was used to test Thamnophis marcianus, 67% of males had trailing indices on female paths of nearly 100% in spring and none had indices over 10% in midsummer. Additionally, no trailing occurred when males were tested on male paths, females on female paths, and females on male paths. In this species it appears that trailing is primarily a spring activity and related to reproduction. Individuals of other species of temperate zone snakes have been tested (Table 1, personal observations), and all exhibited trailing during their courtship period but very little trailing at other times of the year. Since reproduction is the most important aspect of an organism's evolutionary survival, this widespread strong sexual trailing response is not surprising. The lack of evidence of nonsexual trailing in my laboratory does not negate the possibility of important roles for trails in nonsexual context; however it may indicate that it is a less intense response that is difficult to obtain in an artificial environment. Indeed, some researchers through careful experimentation have managed to observe pheromone trails used in several other social situations.

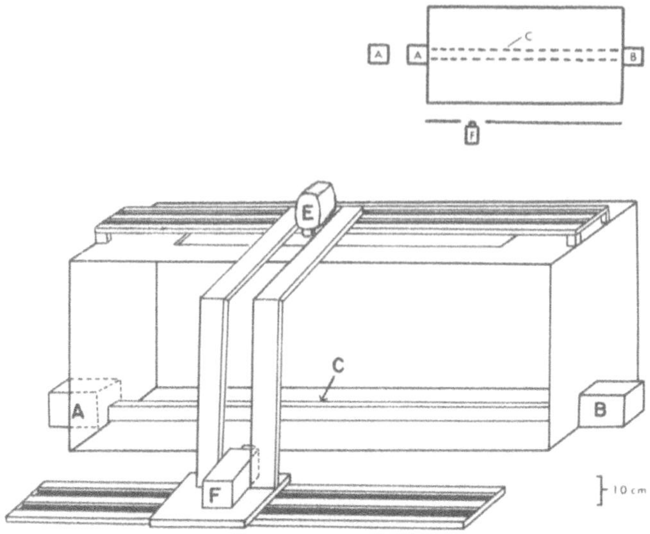

Fig. 2. Trailing arena. A - release box. B - holding box for trail-laying snake, removed prior to tests. C - removable guides. Insert - arena (dorsal view). The trail is laid within C. E - movie or video camera. F - video camera.

THE ROLE OF PHEROMONE TRAILS IN NONSEXUAL AGGREGATION

Trailing behavior has been suggested to play a role in the formation of various types of nonsexual aggregations of snakes. Researchers have indicated trails may lead snakes to dens, help form aggregations of gravid snakes, and be involved in aggregations that reduce water loss.

Denning

It was originally reported that pheromone trails might guide rattlesnakes to their dens in the fall (Klauber, 1972). However, it now appears that only juveniles and yearlings use trails and that adults may rely on other orientation mechanisms. Brown and MacLean (1983) documented juvenile timber rattlesnakes (Crotalus horridus) following adult pheromone trails in the fall (Table 3), but King et al. (1983) found that adult prairie rattlesnakes (C. viridis, would not trail conspecifics. Adult Crotalus apparently find dens by celestial cues (Landreth, 1973, Duvall et al., 1985) and

Fig. 3. Typical map of arena test (approximately 1/2 of total length and 1/3 of width). Dashed lines represent trail-laying snake's path (all areas touched). Solid line represents trailing snake's path (head movement).

Table 3. Laboratory tests of trail-following in newborn *Crotalus horridus* conducted in a Y-maze. Probability values for control condition were calculated using a two-tailed binomial test and with a one-tailed binomial test for experimental conditions. Modified from Brown and MacLean, 1983.

Test	Arm Choice		P
	L	R	
Control	11	6	0.332
	Trail	Blank	
Newborn trail	13	5	0.048
Female trail	17	3	0.001
Male trail	18	1	0.001

topographic landmarks (Brown and MacLean, 1983) but young snakes, with no knowledge of the location of dens, trail adults to find these aggregations. (For further discussion of this phenomenon see Graves et al., this volume).

Gravid Female Aggregation

Aggregations of female snakes that are nearing parturition have been recorded for several species (Duvall et al., 1985; Gregory, 1975; Henderson et al., 1980; Shine, 1979). Females may be aggregating at this time because of a scarcity of suitable refugia or to swamp predators with a mass crop of young. Whether pheromone trails are involved in the formation of these aggregations has not been examined.

Female *Thamnophis marcianus* maintained in a large circular tank (275 cm diameter) with 8 ceramic tiles displayed a tendency to aggregate in midsummer, and the aggregation sites were attractive to other females when the animals were placed singly in the arena. However, gravid females showed no inclination to trail other females in the trailing arena, and olfactometer tests indicated that the substrate attractant was not airborne (N. Ford and D. Holland, personal observation). It appears that chemical cues may indicate the suitability of a refuge, but in this species trailing is not utilized to locate the aggregation site.

Aggregations to Reduce Water Loss

Several species of small snakes aggregate in heterospecific groups to reduce cutaneous water loss (Burghardt, 1983; Dundee and Miller, 1968; Noble and Clausen, 1936). Several studies have indicated that trailing is involved in this behavior. Heller and Halpern (1981) found that juvenile *Thamnophis* of several species tended to follow the paths of any preceding animal regardless of species. Nonsexual trailing has been recorded in *Typhlops*, *Leptotyphlops*, *Virginia* and *Sonora* (Gehlbach et al., 1971), and *Thamnophis butleri* males will follow sympatric *T. sirtalis* trails (Ford, 1982). The common trait of these species is small size with a correspondingly large surface to volume ratio and, therefore, a water loss problem. Aggregation in numbers can decrease the rate at which dessication occurs (Duvall et al. 1985; Noble and Claussen, 1936; cf. Graves et al., this volume) and the aggregation is apparently facilitated by pheromone trails.

Fig. 4. Trailing posture of male Thamnophis sirtalis.

TRAIL ORIENTATION: WHY SNAKES HAVE FORKED TONGUES

Although we know the tongue and Jacobson's organ constitute the primary mechanism for the input of information from trails (Halpern and Kubie, 1983), we have little insight into how snakes utilize odors to orient spatially to locate either food or mates. To obtain some basic information on trail orientation, I filmed ten males of four species of snakes following female trails laid in the arena described previously (Fig. 2).

I examined three closely related species, two of which (Thamnophis sirtalis and T. butleri) live in primarily grassy microhabitats (Carpenter, 1952) and one (T. proximus) which occupies brushy areas in aquatic situations (Clark, 1974). In addition, I examined Regina septemvittata, another thamnophiine snake of somewhat confusing taxonomic status (Rossman, 1985) which occurs in habitats quite similar to those in which T. proximus is found (Branson and Baker, 1974). Differences in the evenness of the terrain of these species' preferred habitats should have an effect on both deposition of a trail and the mechanism of orientation to the trail.

From each film, two sections containing 20-30 tongue flicks (TFs) or about 200-400 frames each, were examined with a Lafayette motion film analyzer. The following data were collected: 1) duration of each TF, 2) time between TFs, 3) distance covered during each TF, 4) tongue tip spread (widest angle during oscillation phase), and 5) distance covered between TFs. From these data I calculated: 6) tongue flick rate, 7) speed during TF, and 8) speed between TFs. For each record, whether or not the male was on the trail was determined by examination of the maps of the trails. A measure of male trailing "ability" was also obtained from the maps by counting the number of times the male's head deviated from the female's trail. Speed between TFs was further examined by separating that period into three equal parts and calculating speed for each. Means for each of the factors were determined and species comparisons made by analysis of variance testing the effect of film segment, position on or off the trail, individuals, and species. Duncan's new multiple range tests of mean separation were conducted on all significant differences (Sokal and Rohlf, 1969).

Male Trailing Behavior

Trailing male snakes oriented closely to the narrow area contacted by the female (Fig. 3). When the edge of the trail was exceeded by one tongue tip during a TF, the snake reversed direction and swung his head back into the pheromone field during the period before the next TF. The tongue was extended approximately 1/2 of its length during most TFs and the tips were spread apart 55 to 180 degrees during its oscillation. The head was held level or angled slightly downward during trailing and at a height that allowed the tongue tips to just reach the substrate (Fig. 4), suggesting contact chemoreception much like that described during the prey trailing behavior of juvenile snakes (Kubie and Halpern, 1979).

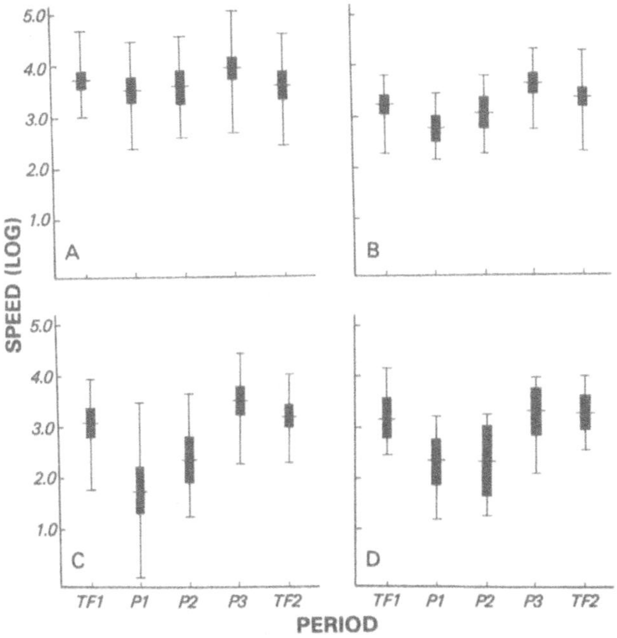

Fig. 5. Speed (mm/sec-converted to log) of trailing males during and between tongue flicks. TF1 - first tongue flick. P1 - first third after TF, P2 - second third, P3 - last third of the period between tongue flicks. TF2 - second tongue flick. A - T. sirtalis. B - T. butleri. C - R. septemvittata. D - T. proximus.

The males' precise orientation to the trail suggests the pheromone field is being simultaneously examined with two receptor organs (tropotaxis: Fraenkel and Gunn, 1940). Although not sense organs themselves, the tongue tips could bring in separate gradient information to the vomeronasal organ, allowing corrections toward the zone of highest pheromone concentration. The tongue tips were often widely spread during a TF, as much as twice the width of the snake's head (mean angle of approximately 125° for all four species). This would increase the concentration differential between tongue tips and could explain the ability of the males to keep close contact with the trail without extensive scanning movement. Could the forked tongue of snakes have evolved to increase the efficiency of trailing behavior?

Despite the generally accurate trailing of the males, snakes did periodically stray from the trail. This occurred when the male's head left the pheromone field between TFs. The snake would stop after the next TF for one or more subsequent TFs and then swing the head laterally between TFs until the trail was regained. This weaving behavior is more like klinotaxic orientation (Fraenkel and Gunn, 1940) and apparently dominates when the trail is lost and continues until the scent field is regained. Klinotaxis is often used by organisms in low-gradient pheromone fields as are found outside the active zone of a trail (Leuthold, 1968).

Male snakes briefly decelerated immediately after most TFs (Fig. 5 - period 1 and 2). In fact, the snakes often stopped completely, perhaps allowing analysis of information just collected by the tongue tips.

Table 4. Trailing behaviors of *Thamnophis* and *Regina*.

Tongue Flick Rate

Species	On Trail #/Min ± SE	Off Trail #/Min ± SE
T. butleri	83 ± 8.8	83 ± 10.8
T. sirtalis	77 ± 11.4	73 ± 10.6
T. proximus	70 ± 4.0	66 ± 7.6
R. septemvittata	62 ± 6.4	59 ± 6.2

Tongue Flick Duration

Species	On Trail Sec ± SE	Off Trail Sec ± SE
T. butleri	0.28 ± 0.023	0.30 ± 0.028
T. sirtalis	0.28 ± 0.023	0.29 ± 0.023
T. proximus	0.30 ± 0.025	0.35 ± 0.038
R. septemvittata	0.29 ± 0.021	0.35 ± 0.049

Time Between Tongue Flicks

Species	On Trail Sec ± SE	Off Trail Sec ± SE
T. butleri	0.51 ± 0.023	0.52 ± 0.074
T. sirtalis	0.69 ± 0.149	0.72 ± 0.149
T. proximus	0.58 ± 0.043	0.65 ± 0.127
R. septemvittata	0.78 ± 0.137	0.87 ± 0.197

Tongue Flick Duration/Total Time Trailing

Species	On Trail % ± SE	Off Trail % ± SE
T. butleri	38.2 ± 2.91	38.9 ± 2.66
T. sirtalis	35.3 ± 4.61	35.6 ± 4.52
T. proximus	35.3 ± 2.76	41.1 ± 3.11
R. septemvittata	30.6 ± 2.17	31.6 ± 2.13

Speed (during tongue flicking)

Species	On Trail mm/Sec ± SE	Off Trail mm/Sec ± SE
T. butleri	23.3 ± 3.73	34.6 ± 6.29
T. sirtalis	29.2 ± 6.07	33.8 ± 6.92
T. proximus	14.8 ± 2.21	25.9 ± 4.61
R. septemvittata	15.2 ± 2.57	17.9 ± 2.99

Speed (between tongue flicks)

Species	On Trail mm/Sec ± SE	Off Trail mm/Sec ± SE
T. butleri	29.2 ± 4.65	35.4 ± 6.97
T. sirtalis	40.6 ± 8.97	33.6 ± 9.23
T. proximus	24.2 ± 3.83	24.1 ± 3.83
R. septemvittata	19.4 ± 2.80	18.2 ± 2.35

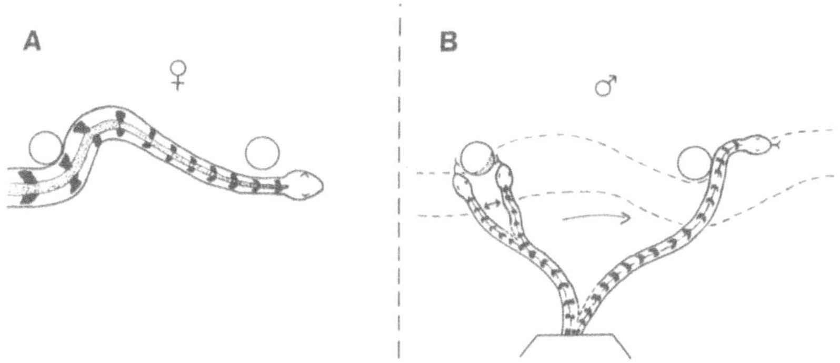

Fig. 6. Hypothesized mechanism for male snakes to determine direction of a female snake. (A) Female laying a trail. Note she pushes against the anterolateral surface of the pegs. (B) Behavior of a male snake when he encounters a female trail. Stippling on the first peg represents the probable deposition of the pheromone there, whereas the dashed lines represent the pheromone trail on the substrate. From Ford and Low (1984).

Generally, males did not change directions until after this assessment period. Then the animals' movements were rapid, causing the increase in speed noted in period 3.

Species' Differences in Trailing Behavior

The tongue flick rate varied with individual males, and although *T. proximus* and *R. septemvittata* had somewhat lower mean values, species' differences were not significant (Table 4). Tongue flick rate is directly correlated to stimulus concentration during prey trailing (Kubie and Halpern, 1979) and these pheromone trails probably had varying concentrations due to the use of different females and some differences in age of the trails.

Several other factors did correlate with the genera examined. *Thamnophis* differed from *Regina* in that they spent more time tongue flicking and covered a greater distance during the TF (Table 4). However, both *R. septemvittata* and *T. proximus* had significantly longer TFs (particularly when off the trail) and they were slower than *T. butleri* and *T. sirtalis*.

The similarities noted between *T. proximus* and *Regina* appear to be characteristics that would be influenced by having to search for trails in an uneven brushy habitat. A longer tongue flick and slower speed would allow increased chemosensory analysis and decrease the chances of losing the trail. Both species exhibited very accurate trailing and rarely left the trail boundaries. Both *T. proximus* and *R. septemvittata* are extremely cryptic, and a lower predation pressure may permit slower trailing in these species. The den sites of *T. butleri* and *T. sirtalis* contain sparse vegetation in the spring and are more conducive to the laying of unbroken trails; however trailing males may be more exposed to predation. Higher speeds during trailing would be of obvious advantage, and, since females are usually only a short distance away, the chance of losing the trail would be minimized.

Table 5. Ability of male plains garter snakes, Thamnophis radix to determine direction of a female pheromone trail under three different physical conditions. For experiments I and II a one-tailed binomial test of the hypothesis that the males would follow the trail in the direction of the female was conducted. For experiment III the hypothesis tested was that the males would go the direction the pegs "indicated". From Ford and Low (1984).

Test	N	# of males going correct direction	P
I Pheromone on substrate and vertical pegs	25	17	0.05
II Pheromone on substrate only	20	9	NS
III Pheromone on substrate and vertical pegs reversed	20	14	0.06

NS = not significant.

TRAIL ORIENTATION: WHICH WAY TO GO?

How males determine the direction in which the trail-laying female has gone has recently been elucidated (Ford and Low, 1984). We noted that males extensively tongue flicked pegs present in arena tests and thought the pegs might have some role in trail orientation. We tested male Thamnophis radix in situations in which the female had been allowed to contact pegs when laying the trail and also when she had not. The males could detect direction of the trail only when the female had pushed against pegs (Table 5). Noble (1937) originally suggested that the dorsal and lateral skin was the source of the pheromone. This would allow the female to mark the anterolateral surface of environmental objects she uses to obtain locomotor force (Fig. 6, A). When the male encounters a trail, he chemosensorily assays these objects (TFs), detects more pheromone on the side in the direction the female went, and then uses chemotactic orientation to follow the substrate trail (Fig. 6, B).

THE EVOLUTION OF PHEROMONE TRAILS

Pheromone trails can be used to form nonsexual aggregations; however, the predominate role of pheromone trails in snakes is to allow males to locate females for reproduction. As indicated earlier, this would necessitate that sympatric species produce trails with species-specific pheromones. Vitellogenin is a large polypeptide (Bast and Gibson, 1985) and even if the trail pheromone is only a subunit of the molecule, it would be large enough to allow chemical variation that would permit species identification. Although polymorphism of the trail pheromone within a species could occur, there would be evolutionary pressure for such variability to be conservative. Molecular diversity might decrease the ability of the males to distinguish trails and locate conspecific females (Ford and Schofield, 1984). Therefore,

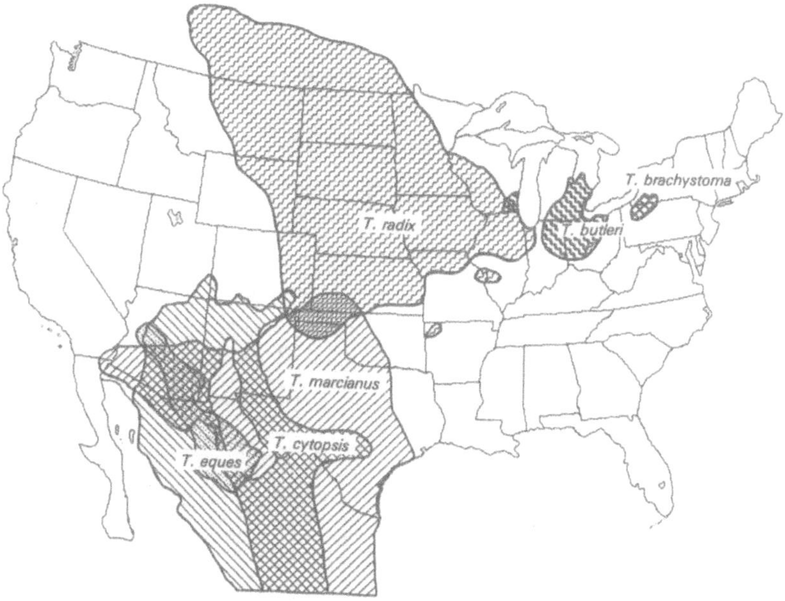

Fig. 7. Map of the distributions of members of the <u>Thamnophis radix</u> complex.

trail pheromones should not change (at least genetic drift would be slow) unless evolutionary pressure for specificity is present. This means that examination of trail specificity in recently evolved species of snakes can provide insight into the nature of their evolution. Particularly interesting is how evolutionary mechanisms such as adaptive divergence (allopatric speciation), clinical divergence, and habitat divergence (sympatric speciation) influence trail pheromone specificity (Templeton, 1981).

An optimal group in which to examine trail pheromone evolution is the garter snakes of the <u>Thamnophis radix</u> complex. In the United States, this complex consists of six recently evolved but distinct species: <u>T. radix</u>, <u>T. marcianus</u>, <u>T. cyrtopsis</u>, <u>T. butleri</u>, <u>T. brachystoma</u> and <u>T. eques</u> (Fig. 7). Over the last few years, my students and I have tested the specificity of the trails of four of the species in this complex and this information can be incorporated into the original explanation of <u>Thamnophis</u> evolution proposed by Ruthven in 1908.

Ruthven contended that the <u>T. radix</u> complex originally evolved in Mexico and followed the Pleistocene post-Wisconsinian glacial recession northward. Initially, the single wide ranging southern species evolved into three, <u>T. eques</u>, <u>T. marcianus</u> and <u>T. cyrtopsis</u>. One offshoot of the progenitor pushed across the deserts of northeastern Mexico and into the trans-Pecos and western Texas region where it became <u>T. marcianus</u>. Its eastward movement was limited by the transition of the desert grassland into forests, while in its subsequent westward movement it competed with the ancestral <u>T. eques</u> and so moved northward into Arizona and New Mexico where its range was limited by the high plateaus. <u>Thamnophis cyrtopsis</u> apparently evolved to cope with higher altitudes in the ranges of both of the other species. Pheromone trails of <u>T. cyrtopsis</u> and <u>T. marcianus</u> separated by these habitat differences are species-specific (Table 6). This indicates the chemical changes in the pheromones and discriminatory ability

Table 6. Y-maze tests of pheromone trail preference of members of the *Thamnophis radix* complex. P values were calculated using a one-tailed binomial test for preference of the conspecific trail. From Ford (1982), Ford and Schofield (1983), and Ford and O'Bleness (in press).

Male	Heterospecific Female	# Choosing Conspecific Trail/total	P
T. marcianus	*T. cyrtopsis*	15/20	0.02
T. radix radix	*T. marcianus*	16/23	0.05
T. radix haydeni	*T. marcianus*	16/21	0.02
T. butleri	*T. radix*	9/19	NS

NS = not significant.

of the males evolved either concurrently with the niche separation of these two species, or, alternatively, occasional heterospecific encounters over a long term produced enough selection pressure to cause the evolution of specificity.

The extension of the semiarid prairie-plains corridor from New Mexico and Texas to Canada furnished a highway to the north along which *T. marcianus* spread, becoming in Oklahoma and southern Kansas the form now known as *T. radix*. *Thamnophis radix* moved northward through the southern great plains corridor during interglacial xerothermic periods. Although widely separated in most of their range, *T. radix* and *T. marcianus* are sympatric in northern Texas, southwestern Kansas, and the panhandle of Oklahoma. Two subspecific populations of *T. radix* varying in distance form the area of sympatry with *T. marcianus* have been examined for their ability to distinguish conspecific trails (*T. r. radix* from Illinois, *T. r. haydeni* from Colorado; Table 6). Surprisingly, males of both subspecies could discriminate their females' trails. This implies that their area of sympatry was more extensive in the recent past, which could have occurred if, during one of the southern movements of a Wisconsinian glacier, *T. radix* was forced down into the range of *T. marcianus*. The evolution of reproductive isolating mechanisms, including species specific pheromone trails, would insure the retention of their specific ecological adaptations. When *T. radix* moved northward after the last glacial period, it retained the ability to discriminate between the trails, again suggesting the conservative nature of trail pheromone chemistry.

During one of the later interglacial xerothermic periods, *T. radix* invaded the eastern portions of the prairie peninsula. During this eastward dispersion, *T. butleri* evolved as a dwarf variety of *T. radix*. Again pheromone specificity gives some insight into this process. *Thamnophis butleri* and *T. radix* pheromone trails are not distinguishable by males (Table 6). It seems likely that during the dispersal of *T. "radix-butleri"* laterally, a southward movement of a glacier completely separated the group that became *T. butleri*. Although this is in disagreement with some previous authors (Schmidt, 1938; Conant et al., 1945), it appears that *T. butleri* evolved allopatrically in the post-glacial period and so still uses the same pheromone as *T. radix*.

In relatively recent times, *T. butleri* has moved westward into southeastern Wisconsin, where it is sympatric with *T. radix*. *Thamnophis radix* has also shown up in a very disjunct population in central Ohio within the range of *T. butleri* (Fig. 7). These populations are generally considered to be from very recent immigrations (Conant et al., 1945), and preliminary species specificity tests of the populations in Milwaukee, Wisconsin have indicated the pheromone trails are still indistinguishable by the males (Ford, unpublished data).

Relative to the other members of the complex, little information is available on the evolution of *T. brachystoma*. It is morphologically very similar to *T. butleri* and is geographically isolated from it. Discrimination tests for the species trails have not been conducted, but if the trails are indistinguishable by the males, then it is probable the species evolved as did *T. butleri*, during an extension of a Wisconsinian glacier.

It appears from these studies that pheromone trail specificity played an early role in the evolution of species reproductive isolation in *Thamnophis*. Behavioral assays of trail specificity of a variety of snakes would be valuable as they may be important clues to ancestral relationships of snakes and the ecological pressures during their evolution.

CONCLUSIONS

Although the database from experimental studies of pheromone trailing behavior in snakes is small, it is apparent from this review that such work is quite valuable in evaluating the role of olfaction in snake sociobiology. Trailing activity can be used as an easily quantified visual bioassay of chemosensory input relating to the various social responses of snakes. In addition, aspects of this activity may be useful in other disciplines of herpetology (i.e., evolution and taxonomy).

ACKNOWLEDGMENTS

I wish to thank William Lamar, Susan Novak, Joan Ford, David Duvall, Peg Dunklin, Melva Grisson, Winston Green and Martin Smith for their help with this manuscript. Several of the projects included were supported by University of Texas at Tyler faculty research grants. I also acknowledge the help of my students; Carolyn Schofield, James Low, Debera Holland and Michael O'Bleness.

REFERENCES

Andren, C., 1982, The role of the vomeronasal organ in the reproductive behavior of the adder, *Vipera berus*, *Copeia*, 1982:148.

Bast, R. E., and Gibson, A. R., 1985, Characterization of reptilian vitellogenin: Subunit composition and molecular weights of vitellogenin from the colubrid snake *Thamnophis sirtalis* (L.), *Comp. Biochem. Physiol.*, 80(B):409.

Baumann, F., 1929, Experimente über den Geruchssinn und den Beuteerwerb der Viper (*Vipera aspis* L.), *Z. Vergl. Physiol.*, 10:61.

Bishop, S. C., 1927, The amphibians and reptiles of Alleghany State Park, *Bull. N.Y. St. Mus.*, 3:1.

Branson, B. A., and Baker, E. C., 1974, An ecological study of the queen snake, *Regina septemvittata* (Say) in Kentucky, *Tul. Stud. Zool Bot.*, 18:153.

Brattstrom, B. H., 1974, The evolution of reptilian social behavior, *Am. Zool.*, 14:35.

Brown, W. S., and MacLean, F. M., 1983, Conspecific scent-trailing by newborn timber rattlesnakes, Crotalus horridus, Herpetologica, 39:430.
Burghardt, G. M., 1980, Behavioral and stimulus correlates of vomeronasal functioning in reptiles: Feeding, grouping, sex and tongue use, in: "Chemical Signals in Vertebrates and Aquatic Invertebrates," D. Muller-Schwarze, and R. M. Silverstein, eds., Plenum Press, New York.
Burghardt, G. M., 1983, Aggregation and species discrimination in newborn snakes, Z. Tierpsychol., 61:89.
Carpenter, C. C., 1952, Comarative ecology of the common garter snake (Thamnophis s. sirtalis), the ribbon snake (Thamnophis s. sauritus), and Butler's garter snake (Thamnophis butleri) in mixed populations, Ecol. Monogr., 22:235.
Carpenter, C. C., 1977, Communication and displays in snakes, Am. Zool., 17:217.
Carpenter, C. C., and Ferguson, G. W., 1977, Variation and evolution of stereotyped behavior in reptiles, in: "Biology of the Reptilia, 7," C. Gans, and D. Tinkle, eds., Acad. Press, London.
Clark, D., 1974, The western ribbon snake (Thamnophis proximus): Ecology of a Texas population, Herpetologica, 30:372.
Chiszar, D., Radcliffe, C. W., Scudder, K. M., and Duvall, D., 1983, Strike-induced chemosensory searching by rattlesnakes: The role of envenomation-related chemical cues in the post-strike environment, in: "Chemical Signals in Vertebrates 3," D. Muller-Schwarze, and R. Silverstein, eds., Plenum Press, New York.
Conant, R., Thomas, E. S., and Rausch, R. L., 1945, The plains garter snake, Thamnophis radix, in Ohio, Copeia, 1945:61.
Cowles, R. B., and Phelan, R. L., 1958, Olfaction in Rattlesnakes, Copeia, 1958:77.
Davis, D. D., 1936, Courtship and mating behavior in snakes, Field Mus. Nat. Hist. Zool. Ser., 20:256.
Devine, M. C., 1977, Chemistry and Source of Sex-attractant Pheromones and Their Role in Mate Discrimination by Garter Snakes, Ph.D. Dissertation, Univ. Michigan, Ann Arbor.
Dundee, H. A., and Miller, M.C., 1968, Aggregative behavior and habitat conditioning by the prairie ringneck snake, Diadophis punctatus arnyi, Tul. Stud. Zool Bot., 15:41.
Duvall, D., King, M. B., and Gutzwiler, K. J., 1985, Behavioral ecology and ethology of the prairie rattlesnake, Nat. Geog. Res., 1:80.
Finneran, L. C., 1949, A sexual aggregation of the garter snake, Thamnophis butleri (Cope), Copeia, 1949:141.
FitzSimons, F. W., 1932, "Snakes," Hutchison, London.
Ford, N. B., 1978, Evidence for species specificity of pheromone trails in two sympatric garter snakes, Thamnophis, Herpetol. Rev., 9:10.
Ford, N. B., 1979, Aspects of Pheromone Trailing in Garter Snakes (Thamnophis), Ph.D. Dissertation, Miami Univ. Ohio.
Ford, N. B., 1981, Seasonality of pheromone trailing behavior in two species of garter snake, Thamnophis (Colubridae), Southwest. Nat. 264:385.
Ford, N. B., 1982, Species specificity of sex pheromone trails of sympatric and allopatric garter snakes (Thamnophis), Copeia, 1982:10.
Ford, N. B., and Low, J. R., 1984, Sex pheromone source location by garter snakes: A mechanism for detection of direction in nonvolatile trails, J. Chem. Ecol., 10:1193.
Ford, N. B., and O'Bleness, M., in press, Species and sexual specificity of the pheromone trails of the checkered garter snake, Thamnophis marcianus, J. Herp.
Ford, N. B., and Schofield, C. W., 1984, Species specificity of sex pheromone trails in the plains garter snake, Thamnophis radix, Herpetologica 40:51.
Fraenkel, G. S., and Gunn, D. L., 1940, "The Orientation of Animals," Dover Publ., New York.

Gardner, J. B., 1955, A ball of gartersnakes, Copeia, 1955:310.
Garstka, W. R., and Crews, D., 1981, Female sex pheromone in the skin and circulation of a garter snake, Science, 214:681.
Gehlbach, F. R., Watkins, J. F., and Kroll, J. C., 1971, Pheromone trail-following studies of typhlopid, leptotyphlopid, and colubrid snakes, Behaviour, 40:282.
Gillingham, J. C., Carpenter, C. C., and Murphy, J. B., 1983, Courtship, male combat and dominance in the western diamondback rattlesnake, Crotalus atrox, J. Herpetol., 17:265.
Gregory, P. T., 1975, Aggregations of gravid snakes in Manitoba, Canada, Copeia 1975:185.
Halpern, M., and Kubie, J., 1983, Snake tongue flicking behavior: Clues to vomeronasal system functions, in: "Chemical Signals in Vertebrates 3," D. Muller-Schwarze, and R. Silverstein, eds., Plenum Press, New York.
Heller, S., and Halpern, M., 1981, Laboratory observations on conspecific and congeneric scent trailing in garter snakes (Thamnophis), Behav. Neur. Biol., 33:372.
Henderson, R. W., Binder, M. H., Sajdak, R. A., and Buday, J. A., 1980, Aggregating behavior and exploitation of subterranean habitat by gravid easter milksnakes (Lampropeltis t. triangulum), Milw. Publ. Mus. Cont. Biol. Geol., 32:1.
King, M., McCarron, D., Duvall, D., Baxter, G., and Gern, W., 1983, Group avoidance of conspecific but not interspecific chemical cues by prairie rattlesnakes, Crotalus viridis, J. Herp., 17:196.
Klauber, L. M., 1972, "Rattlesnakes", 2nd Ed. Univ. Calif. Press, Berkeley and Los Angeles.
Kubie, J. L., and Halpern, M., 1979, The chemical senses involved in garter snake prey trailing, J. Comp. Physiol. Psychol., 93:648.
Landreth, H. F., 1973, Orientation and behavior of the rattlesnake, Crotalus atrox, Copeia 1973:26.
Leuthold, K., 1968, A tibial gland scent-trail and trail-laying behavior in the ant (Crematogaster ashmeadi Mayr), Psyche, 75:223.
Madison, D. M., 1977, Chemical communication in amphibians and reptiles, in: "Chemical Signals in Vertebrates," D. Muller-Schwarze, and M. Mozell, eds., Plenum Press, New York.
Munro, D. R., 1948, Mating behavior and seasonal cloacal discharge of a female Thamnophis sirtalis parietalis, Herpetologica 4:185.
Noble, G. K., 1937, The sense organs involved in the courtship of Storeria, Thamnophis and other snakes, Bull. Am. Mus. Nat. Hist., 73:673.
Noble, G. K., and Clausen, H. J., 1936, The aggregation behavior of Storeria dekayi and other snakes with a special reference to the sense organs involved, Ecol. Monogr., 6:269.
Parker, W. S., and Brown, W. S., 1980, Comparative ecology of two colubrid snakes, Masticophis t. taeniatus and Pituophis melanoleucus deserticola, in northern Utah, Milw. Publ. Mus. Pub. Biol. Geol., 7.
Price, A. H., and LaPointe, J. L., 1981, Structure-functional aspects of the scent gland in Lampropeltis getulus splendida, Copeia, 1981:138.
Rossman, D. A., 1985, Liodytes resurrected, reexamined, and reinterred, J. Herp., 19:169.
Ruthven, A. G., 1908, Variations and genetic relationships of the garter snakes, Bull. U.S. Natl. Mus., 61.
Schmidt, K. P., 1938, Herpetological evidence for the post-glacial eastward extension of the steppe in North America, Ecology, 19:396.
Shine, R., 1979, Activity patterns in Australian elapid snakes (Squamata: Serpentes: Elapidae), Herpetologica, 35:1.
Sokal, R. R., and Rohlf, F. J., 1969, "Biometry," W. H. Freeman, San Francisco.
Templeton, A. R., 1981, Mechanisms of speciation - a population genetic approach, Ann. Rev. Ecol. Syst., 12:23.

Truitt, R. V., 1927, Notes on the mating of snakes, Copeia, 1927:21.
Wright, A. H., and Wright, A. A., 1957, "Handbook of Snakes of the United States and Canada," Comstock Publ. Assoc., New York.

PHEROMONE MIMICRY IN GARTER SNAKES

Robert T. Mason and David Crews

Institute of Reproductive Biology
The University of Texas At Austin
Austin, TX 78712

Female mimicry is found in many diverse taxa including insects and most vertebrate classes (Weldon and Burghardt, 1984) and occurs in two forms. In several species, males assume a female-like morphology or color pattern (Dominey, 1980; Gross, 1979), whereas in others, males mimic female behavior patterns (Arnold, 1976; Thornhill, 1979). In some species males mimic both female morphology and behavior (Gross, 1983).

We describe here an unusual form of female mimicry in a reptile (Mason and Crews, 1985). This form of mimicry is unique to the animal kingdom in that it is expressed not as a behavioral or morphological feminization but as a physiological feminization, the production of what was previously believed to be a sex-limited attractiveness pheromone.

The red-sided garter snake (Thamnophis sirtalis parietalis) is the most northerly living reptile in North America. During the winter months these animals hibernate in large communal dens throughout much of their range (Aleksiuk and Stewart, 1971). In the spring, males synchronously emerge from the dens and aggregate in areas immediately surrounding the den opening. Aggregations are comprised entirely of males and can number into the thousands. Females emerge singly over the next three to four weeks.

Upon emergence, a female will usually be courted by 10-20, but sometimes as many as 100 males forming a "mating ball" (Aleksiuk and Gregory, 1974). Males recognize and begin to court females by detecting a pheromone released by the female (Crews, 1976; Noble, 1937). Rapid tongue-flicking by the male delivers the pheromone cues from the female's dorsal surface to the male's vomeronasal organ (Garstka and Crews, 1981). During courtship, the male presses his chin along the female's dorsum and traverses the length of her body, eventually coming to rest along the female's back with the cloacas aligned. Only one male will be successful intromitting one of his hemipenes.

During the course of a five week field season in the spring of 1983, we captured entire mating balls and censused the participants. Each animal in the ball was sexed by cloacal probing and the hemipenes extruded (Laszlo, 1975). In 33 of the 200 mating balls censused no females were found. Rather, in each of these 33 mating balls, a single male was being courted by the other males. Because these males could be distinguished from other males by the fact that they elicited courtship, we referred to these males as she-males.

She-males are indistinguishable from normal males, each possessing hemipenes, testes, and accessory sex structures identical to those found in males. The mean snout-vent length and body weight of she-males does not differ significantly from that of other males. In addition, the weights of she-male livers, fat bodies, adrenals, and testes relative to body weight also do not differ significantly from those of other males. Histological examination of testes shows no difference in the stage of spermatogenesis between males and she-males. As is typical at this time of year, both she-males and males have regressed testes. Sperm is present in the ductuli deferentia in members of both groups, and the number of sperm and their motility and morphology do not appear to be different. Finally, blood samples have been taken from males ($n = 2$), females ($n = 2$), and she-males ($n = 2$) and stained to detect heterochromatin. In the garter snake, the condensed W chromosome, which is found only in females, is composed largely of heterochromatin (Becak, 1983). Only blood samples from females stain positively for heterochromatin; thus, she-males are genotypically as well as phenotypically male.

The mating behavior of she-males was investigated to determine whether they exhibit a normal sexual preference and court females or whether they court males. She-males were placed in a 40 cm x 28 cm x 10 cm clear plastic box with a recently emerged, unmated female for ten minutes. All 33 she-males courted vigorously for the duration of the test. Further, 3 of the she-males mated with the stimulus female in the time allotted. Although she-males would not court males, they did court each other.

The attractiveness of the she-males is not due to prior physical contact with recently emerged females. Males rubbed against a recently emerged, attractive female were not courted by other males. This agrees with field and laboratory observations that recently mated males that have been in contact with females for up to one hour have never been observed to be courted. The attractiveness of the she-males appears to be permanent as is the case in sexually mature females (Bona-Gallo and Licht, 1983). All of the she-males tested ($n = 33$) elicited courtship from sexually active males when tested at 3, 18, and 40 weeks following capture.

Both normal males and she-males prefer females to she-males. A simultaneous choice test was conducted in which a she-male and a female of similar size were placed in an arena with 10 males. After five minutes the number of males that courted either stimulus animal was recorded. A total of 12 simultaneous choice trials were run each having a duration of five minutes. Males showed a significant preference for females compared to she-males in all 12 simultaneous choice tests (binomial probability, $p < .005$). Similar results were obtained when males were presented with females and she-males in a sequential choice paradigm.

Production of the attractiveness pheromone may afford she-males an advantage in the mating ball. In observations of mating behavior in the field, she-males joining an established mating ball caused some males already courting the female to stop and begin to court the she-male, resulting in the she-male gaining a better position along the female's body. While in the field, a competitive mating test was run in which groups of 5 males and 5 she-males were selected at random and released into an arena (1 m x 1 m x 1 m) with an unmated, attractive female. The trials continued for one hour or until a mating occurred. Mating balls formed around the female in every trial. A total of 42 mating trials were conducted. Of these, 29 resulted in a she-male mating with the female and 13 with a male mating, making she-males more than twice as successful as males in the mating competition ($\chi 2 = 6.39$, $p < .025$). The advantage to the she-male's attractiveness thus appears to lie in its ability to gain a better position in the mating ball and at the same time effectively blocking males from mating, in essence causing sexual interference (Arnold, 1976).

To investigate the physiological basis of the she-males' attractivity, blood samples were taken from males, females, and she-males in the field and analyzed by radioimmunoassay (Camazine et al., 1980). Analysis of variance for each hormone revealed that females have significantly higher circulating levels of estradiol (E) than both males and she-males ($p < 0.05$, Student-Newman-Keuls Test). Conversely, she-males have significantly higher levels of testosterone (T) than both females and males. There were no significant differences in circulating levels of dihydrotestosterone (DHT) (Mason and Crews, 1985). Presumably, the higher levels of T in she-males are a result of either higher secretion or reduced clearance from the circulation.

The finding that a male animal with female-like aspects to its physiology has significantly higher levels of circulating T than do males lacking these qualities is intriguing. It is well known that T does not always exert its effect on target tissues as T directly but as one of two metabolites, DHT or E. We hypothesize that the high levels of T in the circulation of she-males may lead to physiologically significant levels of E in the skin. At this time, levels of E in the skin of she-males remain to be determined. However, in support of this hypothesis are those studies which have isolated aromatase, the enzyme responsible for converting T to E, in the skin and liver of both mammals and birds (George et al., 1981; Slaunwhite et al., 1965).

The female attractiveness pheromone may be related to vitellogenin and is transported in the blood and delivered through the skin (Crews, 1976). Further, exogenous estrogen injection will induce pheromone production in females and males (Kubie et al., 1978). Unlike females, estrogen-injected male garter snakes are never courted by males. However, serum from estrogen-injected males and females, when spread on the backs of control males, will elicit courtship from sexually active males (Crews, 1976). These observations suggest that the skin in the male and female differ in their transport of the attractiveness pheromone.

The pheromone produced by the she-males is the same as that produced by females. Samples obtained from the skin surface using the non-polar solvent toluene revealed that the skin of females and she-males, but not of males elicited courtship behavior from conspecific males at the hibernaculum (Camazine et al., 1980).

She-males may develop as a result of hormones experienced early in life. It is well-established that gonadal hormones organize morphological sex differences, including the brain, early in life, as well as activate sexual behavior in the adult (Feder, 1984; Goy and McEwen, 1980). It has been suggested further that the hormones associated with the heterogametic sex are responsible for determining the development of internal and external sex characters (vom Sall, 1983). In mammals in which the male is the heterogametic sex, T is the organizing hormone whereas in birds, in which the female is the heterogametic sex, it is the ovarian hormone estrogen that organizes sexual differentiation (Adkins-Regan, 1981). In those snakes that have sex chromosomes, the female is the heterogametic sex (Crews, 1985). Within this context of female heterogamety, it is significant that early estrogen treatment will induce production of the attractiveness pheromone in neonatal female and male *T. s. parietalis* (Crews, 1985).

ACKNOWLEDGMENTS

We wish to thank Michael Moore, Jim Blank, Joan Whittier, Eric Pianka, Mark Grassman, Mike Ryan, and Jim Bull for their suggestions which greatly improved this manuscript. We also wish to thank David Hinkley, Jim Bull, and Fred Marsteller for their assistance with the statistical analyses. Marie Anne Diamond and Yuki Morris provided excellent technical assistance. Special thanks to Bill Koonz and Merlin Shoesmith of the Manitoba Department of Natural Resources for their invaluable help. Supported by National Institute of Health grant NICHHD 16687 and by a National Institute of Mental Health Research Scientist Development Award to D. Crews.

REFERENCES

Adkins-Regan, E., 1981, Early organizational effects of hormones, *in*: "Neuroendocrinology of Reproduction," N. T. Adler, ed., Plenum Press, New York.

Aleksiuk, M., and Gregory, P., 1974, Regulation of seasonal mating behavior in *Thamnophis sirtalis parietalis*, *Copeia*, 1974:681.

Aleksiuk, M., and Stewart, K., 1971, Seasonal changes in the body composition of the garter snake, *Thamnophis sirtalis parietalis* at northern latitudes, *Ecology*, 52:485.

Arnold, S. J., 1976, Sexual behavior, sexual interference and sexual defense in the salamanders *Ambystoma maculatum*, *Ambystoma tigrinum*, and *Plethodon jordani*, *Z. Tierpsychol.*, 42:247.

Becak, W., 1983, Evolution and differentiation in vertebrates, *in*: "Mechanisms of Gonadal Differentiation in Vertebrates," U. Muller, and W. W. Franke, eds., Springer-Verlag, Berlin.

Bona-Gallo, A., and Licht, P., 1983, Effects of temperature on sexual receptivity and ovarian recrudescence in the garter snake, *Thamnophis sirtalis parietalis*, *Herpetologica*, 39:173.

Camazine, B., Garstka, W., Tokarz, R., and Crews, D., 1980, Effects of castration and androgen replacement on male courtship behavior in the red-sided garter snake (*Thamnophis sirtalis parietalis*), *Horm. Behav.*, 14:358.

Crews, D., 1976, Hormonal control of male courtship behavior and female attractivity in the garter snake (*Thamnophis sirtalis parietalis*), *Horm. Behav.*, 7:451.

Crews, D., 1985, Effects of sex steroid hormone treatment on courtship behavior and sexual attractivity in the red-sided garter snake, *Thamnophis sirtalis parietalis*, *Physiol. Behav.*, (In press).

Dominey, W., 1980, Female mimicry in bluegill sunfish--a genetic polymorphism? *Nature*, 284:546.

Feder, H., 1984, Hormones and sexual behavior, *Ann. Rev. Psychol.*, 35:165.

Garstka, W., and Crews, D., 1981, Female sex pheromone in the skin and circulation of a garter snake, *Science*, 214:681.

George, F. W., Noble, J. F., and Wilson, J. D., 1981, Female feathering in Sebright cocks is due to conversion of testosterone to estradiol in skin, *Science*, 213:557.

Goy, R. W., and McEwen, B. S., 1980, "Sexual Differentiation of the Brain," MIT Press, Cambridge.

Gross, M., 1979, Cuckoldry in sunfishes (*Lepomis*: *Centrarchidae*), *Can. J. Zool.*, 57:1507.

Gross, M., 1983, Sunfish, salmon, and the evolution of alternative reproductive strategies and tactics in fishes, in: "Fish Reproduction: Strategies and Tactics," R. J. Wootton, and G. Potts, eds., Academic Press, New York.
Kubie, J. L., Cohen, J., and Halpern, M., 1978, Shedding enhances the sexual attractiveness of oestradiol treated garter snakes and their untreated penmates, Anim. Behav., 26:562.
Laszlo, J., 1975, Probing as a practical method of sex recognition in snakes, Int. Zoo Year Book., 15:178.
Mason, R. T., and Crews, D., 1985, Female mimicry in garter snakes: Females, males and she-males, Nature, (In press).
Noble, G. K., 1937, The sense organs involved in the courtship of Storeria, Thamnophis, and other snakes, Bull. Am. Mus. Nat. Hist., 73:673.
Slaunwhite, W. R., Karsay, M. A., Hollmer, A., Sandberg, A. A., and Niswander, K., 1965, Fetal liver as an endocrine tissue, Steroids, 6:211.
Thornhill, R., 1979, Adaptive female-mimicking behavior in a scorpion-fly, Science, 205:412.
vom Saal, F., 1983, The interaction of circulating oestrogens and androgens in regulating mammalian sexual differentiation, in: "Hormones and Behaviour in Higher Vertebrates," J. Balthazart, E. Prove, and R. Gilles, eds., Springer-Verlag, Berlin.
Weldon, P. J., and Burghardt, G. M., 1984, Deception divergence and sexual selection, Z. Tierpsychol., 65:89.

INITIAL DEN LOCATION BY NEONATAL PRAIRIE RATTLESNAKES:
FUNCTIONS, CAUSES, AND NATURAL HISTORY IN CHEMICAL ECOLOGY

Brent M. Graves, David Duvall, Michael B. King, Stan L. Lindstedt, and William A. Gern

Department of Zoology and Physiology
University of Wyoming
Laramie, WY 82071

INTRODUCTION

The concept of adaptation has long been a central idea in the study of the survival and evolution of life on earth (e.g., Williams, 1966). The potential existence of biological adaptation is best thought of as a hypothesis to be tested or an idea to be approached, and not a phenomenon that is automatically inherent in each and every characteristic of a living system's biology. To be considered an adaptation, *sensu stricto*, potentially important mechanisms and processes should be rigorously studied to determine if they do in fact figure significantly into an organism's "fit" with its environment. Once such features of phenotype are determined to be genuinely critical to survival, subsequent analysis of the trait's mechanisms can lead to much more full explanations of how things actually work, and why they may have evolved into the structural and functional mosaics that we now see.

But which issues should be considered in attempting to test or examine the possibility that some biological mechanism that pleases the scientific and esthetic tests of an investigator is "more than just a mechanism," and, in fact, an adaptation? To expand on ideas of Bock (1980), Gould and Vrba (1982), Gould and Lewontin (1979), Williams (1966), Sober (1984) and many others, we see several questions that can be asked when attempting to sort out whether or not an interesting phenotypic trait is actually an adaptation. First, has the trait been shaped by natural selection for its current role? For example, and relevant to the larger subject of this volume, not all externally expressed exudates are chemical signals, pheromones, or semiochemicals. Indeed, most are simply end products of a wide variety of regulatory and metabolic processes and not "signals" of any sort (e.g., Duvall et al., 1983; Lucas, 1944; Weldon, 1983; Smith, 1977). Second, can environmental selection pressures be identified, either now or historically, that could have affected the derivation of traits as we now see them? We suspect, for example, that the extremely dry, warm, and windy conditions characteristic of the high deserts of Wyoming may have played a role in the evolution of the waxy, nonvolatile lipid pheromones used by prairie rattlesnakes when aggregating (see below). Third, does the operation of the trait in a popoulation affect fitness, perhaps even in a demonstrably positive way? Stated more simply, can it be demonstrated that the trait has "survival value" for organisms that exhibit it? Perhaps animals exhibiting the feature display faster growth or larger size, reproduce sooner or more frequently, produce more or larger offspring, have a higher survival rate, and so forth. And fourth, how variable is the trait's actual value in the population? Does it occur in most or all of the members of a population, and to the same extent or degree? Or does it occur so infrequently and with such variability that it is unlikely that the trait's operation has provided benefits sufficient to genetically fix the trait in the population? To the extent that these questions can be answered satisfactorally (and it is clear that their study would require a slow, cautious and interdisciplinary approach, see Bock, 1980), then the study of mechanism in behavioral biology is much more than mere description; indeed, it is the study of product and pattern in evolution. Further, when these criteria can be met, then a feature can fairly be called an adaptation.

In order to examine these criteria in the context of a living system's biology, access to a model system allowing gradual, interdisciplinary progress is necessary. In the context of chemical ecology and behavior, we believe that we are now studying such a potentially useful model system. Our goal in this paper is to discuss preliminary studies of aspects of the chemical ecology of the prairie rattlesnake, Crotalus viridis viridis, in hopes of describing evolutionary and ecological causes, functions and hypothesized adaptations characteristic of this taxon's biology. The problem specifically addressed here concerns the role(s) that conspecific chemical signals play in initial overwintering den (i.e., hibernaculum) location by neonatal snakes. The crux of this "problem in survival" lies in previous observations that (1) neonates are born only a few weeks before the arrival of the first snows of winter, (2) they are born far away from the den's entrance (and so do not know its location), and (3) their mothers do not actively lead them to it in any way (see Duvall et al., 1985). In the course of this discussion, we will consider those aspects of prairie rattlesnake life history, reproduction, ethology, habitat relations, and integumentary structure and function, that should be considered in order to construct a more complete explanation of both how and why these neonatal animals employ chemical signals as they do. Specifically, addressed herein are issues related to (1) integumentary production of chemical signals, (2) chemical signal chemistry, (3) lab and field studies of neonate and adult behavior, (4) environmental selection forces, and (5) relevant matters of natural history.

PRAIRIE RATTLESNAKES IN WYOMING: SOME RELEVANT NATURAL HISTORY

Because chemical ecology encompasses the interaction of many factors in the biotic and abiotic environment, we discuss several aspects of prairie rattlesnake biology that collectively produce heterogeneous selection pressures which together influence phenotypic expression of chemical signaling processes. Therefore, we begin by reviewing relevant life history features of prairie rattlesnakes in Wyoming that provide context within which to understand the chemical ecology of neonates of this species.

Because adaptation may be most evident under environmental conditions that are near the limit of a species' abilities to exist (e.g., at the edge of the species' range), we have chosen to study a population of prairie rattlesnakes in the Haystack Mountains of south-central Wyoming (about 25 km north of Sinclair). Winters there are long and cold, summers are short with hot days and cool nights, precipitation averages 30 cm annually, humidity is low, high winds are common, and at 2100 m elevation, solar radiation is relatively intense (Duvall et al., 1985). Indeed, this species does not occur just over the Continental Divide (about 5 km from our field site) in the eastern-most portion of the Great Divide Basin (Baxter and Stone, 1980). Therefore, we feel that this population has been exposed to challenges during its evolution whose "solutions" (i.e., adaptations) are especially evident in certain aspects of chemical signaling.

Much of the natural history and biology of prairie rattlesnakes is directly related to two critical events. Specifically, ectotherms in Wyoming must spend winters in thermally protected refugia (i.e., dens; Gregory, 1984) and, in the spring, must leave these overwintering sites and travel to summer foraging areas. Rattlesnakes accomplish these two tasks by overwintering in ancestral, communal hibernacula and undertaking extensive, vernal foraging migrations in search of dense populations of small rodents (Duvall et al., 1985).

The life history of prairie rattlesnakes revolves around this process of summer energy acquisition and winter hibernation. A striking example of the importance of these events is provided by their apparent effects on the reproductive cycle. While sexuallly mature male rattlesnakes are reproductively active annually (Aldridge, 1975; Bull and Shine, 1979; Duvall et al., 1982), female rattlesnakes have been reported to reproduce on annual, biennial and superennial cycles (Rahn, 1942; Fitch, 1949; Glissmeyer, 1951; Klauber, 1972; Tinkle, 1962; Keenlyne, 1972, 1978; Aldridge, 1979; Duvall et al., 1982; Reinert and Kodrich, 1982; Diller and Wallace, 1984; Gannon and Secoy, 1985). Such apparent "skipped opportunities for reproduction" may be a result of insufficient energy available to females. Hence, annual reproduction may be energetically impossible (Bull and Shine, 1979). In support of this hypothesis, Keenlyne (1978) found that female timber rattlesnakes (*Crotalus h. horridus*) reproduce biennially, whereas female eastern Massasaugas (*Sistrurus c. catenatus*) from an area with a longer growing season mature more quickly, have larger clutches, and reproduce annually. More convincingly, Diller and Wallace (1984) found that female *Crotalus viridis oreganus* from two dens reproduced biennially while females from a third (all were within a 40 km radius) reproduced annually. Interestingly, females from the annually reproducing population had a larger average clutch size relative to snout-vent length indicating that food availability was a contributing factor. Additionally, some females from all three populations reproduced annually and reproductive females had significantly greater fat stores than did nonpregnant snakes (cf. Tinkle, 1962; Keenlyne, 1972). These data concur with the hypothesis of Blem (1982) that larger females may reproduce annually while smaller females reproduce less frequently.

With regard to viperid snakes in temperate regions where prey is dispersed and patchy, such as in south-central Wyoming, long foraging migrations likely conflict with requirements for successful reproduction. Hence, during years in which females are pregnant, they do not make such extensive migrations, but instead exhibit an unusual strategy for efficient incubation and protection of developing embryos.

BIRTHING ROOKERIES: WHAT ARE THEY AND WHY DO THEY EXIST?

Associated with differences in reproductive cycles between male and female rattlesnakes are differences in behavior. In the spring, male and nonpregnant female prairie rattlesnakes in Wyoming make long, straight-line migrations away from hibernacula, apparently in search of dense patches of prey (Duvall et al., 1985). However, pregnant viperid snakes of several species have been observed to stay relatively close to the den and frequently are found in localized areas called "birthing rookeries" (Fitch and Shirer, 1971; Keenlyne, 1972; Klauber, 1972; Diller and Wallace, 1984; Reinert, 1984; Duvall et al., 1985; Gannon and Secoy, 1985; Martin, 1985), where pregnant snakes congregate prior to parturition and young are born. Pregnant rattlesnakes have been found in a rookery at our field site in 1983, 1984, and 1985, with 11, 14, and 6 individuals present, respectively.

Fitch and Shirer (1971) and Keenlyne (1972) reported that pregnant *C. v. oreganus* and *C. horridus* rarely if ever feed and postulated that, because of the mass of embryos in their body cavities, they are unable to consume prey. However, King (in preparation) has observed free-ranging pregnant prairie rattlesnakes in a rookery envenom and eat mice. Apparently, pregnant rattlesnakes can eat if prey are available. Why, then, would pregnant rattlesnakes aggregate in areas where food is not readily avaiable, rather than migrate to foraging areas as do nonpregnant snakes? This question seems especially pertinent since it appears that insufficient food is a major factor in preventing biennial reproducers from exhibitng annual parturition (see above). Many possible selection factors exist that may have influenced the derivation of this trait.

Birthing rookeries in Wyoming have been characterized as having large flat rocks beneath which pregnant female snakes take refuge from predators, and extreme night and day temperatures. These rocks may also act as solar collectors and provide thermal inertia for heat retention during night time or cloudy periods (Duvall et al., 1985; c.f., Graves and Duvall, 1987). Birthing rookeries described in other studies have occurred in open rocky areas (C. horridus and Agkistrodon contortrix mokeson in Pennsylvania; Reinert, 1984), a deep cleft in a limestone outcrop (A. contortrix in Kansas; Fitch and Shirer, 1971), a protected site near the summit of the hill above the den (C. v. viridis in Saskatchewan; Gannon and Secoy, 1985), and grassy southerly exposed slopes with large flat rocks (C. horridus in Wisconsin; Keenlyne, 1972). All of these sites would have provided basking sites and protection.

Hence, the first factor that may make rookery use an adaptive strategy is that refugia beneath large, flat objects may provide ideal thermal conditions for gestation (cooler than mid-day temperatures in the sun, yet warmer than night temperatures or those found in mammal holes and other underground resting places). Second, such refugia in rookeries may provide safety from predation. When humans approach rookery rocks while snakes are out basking, the startled animals rapidly retreat to holes beneath the rocks. Third, aggretations of several pregnant rattlesnakes could allow longer retention of body heat by effectively decreasing the surface-to-volume ratio and increasing the mass of the animals relative to when they are alone (Graves and Duvall, 1987). A fourth possible adaptive result of rookery use is that their location may function to insure that neonates get a meal before the first hibernation. Even though prey for adults is not abundant in rookeries, locations may be selected in areas of high population density of neonate prey (it is not known what or if neonate rattlesnakes eat prior to hibernation). Fifth, rookery use may involve the over-saturation of predators by prey (Janzen, 1967). For example, if two rattlesnakes have their young in the same place and a predator locates and attacks the neonates, assuming that the predator can only eat a certain number of neonates before the others escape, the probability that each female will lose offspring is decreased by one half. More litters born in the same place would further decrease the probability of individual females losing offspring to predation. Sixth, lengthy migrations may deplete lipid stores enough to delay embryo development and parturition until too late in the season for offspring survival. And seventh, and especially important to this discussion, it may be important that neonates are born relatively close to a hibernaculum so that they may find it before winter's arrival. If pregnant rattlesnakes embarked on foraging migrations, neonates might be born too far from hibernacula to locate one prior to winter. Parturition in the population of C. v. viridis that we study occurs in late August to early September and the onset of very cold weather usually occurs soon afterwards. Therefore, rapid den location by neonates seems critical.

THE PROTO-INTEGUMENT: ITS FUNCTIONS AND EFFECTS

With so short a time within which to locate a hibernaculum, neonatal prairie rattlesnakes ought to begin their search soon after birth. However, we have observed many litters of neonates in a rookery at our field site over three seasons and have found that they remain in the rookery for approximately two weeks after birth. In fact, neonates stay under the same rocks that their mothers used during the summer, seeking refuge from heat, cold and predators, and emerging in cool parts of the day to bask in the sun. We have also noticed that while sunning, neonates remain tightly aggregated in piles (Duvall et al., 1985). Termination of this behavior at one to two weeks post-partum coincides with the first shedding of the outer layer of epidermis, or "proto-integument," after which neonates disperse from the rookery to begin the migration to ancestral hibernacula. Consequently, we hypothesized that there are differences between the proto-integument and the second skin generation that influence aggregation in the rookery and dispersion for den location.

The proto-integument is a name that we have assigned to the outer layer of epidermis of snakes that is present at birth. Climatic conditions at our field site in August and September, when young are born, are such that evaporative potential is high and it is largely the integument of snakes that controls water loss (Bentley and Schmidt-Nielsen, 1966; Schmidt-Nielsen and Bentley, 1966). Therefore, the structure and permeability of the proto-integument and subsequent epidermal generations may have a critical impact on how, why and when neonatal rattlesnakes may move about in their microhabitat. As we shall discuss below, the skin of these animals appears to be the source of conspecific chemical signals that mediate important aggregative events as well.

Neonate Aggregation in the Laboratory

Our first step in investigating the functions and effects of the proto-integument involved a laboratory experiment to determine whether neonates actively aggregate with one another prior to shedding of this initial skin generation and disperse after it has been shed, or if our field observations were merely the result of attraction to localized environmental features by the young snakes that we observed. One group of four and two of five neonates were placed into clean plastic boxes measuring 39 x 26 x 8 cm and dispersed within. The boxes were then placed in a lighted incubator at 29.1 C and left there for 30 min. At the end of this period, the incubator was opened and the location of each snake was noted. If more than half of the animals (three in groups of four or five) were in bodily contact with one another, such a trial was scored as "aggregated." Otherwise, the trial was scored as "dispersed." The snakes were then immediately transferred to an identical clean plastic box and the procedure was repeated. Each group was used in 12 consecutive replications. All subjects were tested for aggregation tendencies in this manner on the day of birth and again approximately two weeks after shedding of the proto-integument which occurred at nine to ten days of age. Results from this experiment showed that a significantly greater proportion of groups of neonates exhibited aggregation before shedding relative to after shedding (29 of 36 before versus 7 of 36 after; $z = 3.99$, $P < 0.001$). This suggests that neonatal prairie rattlesnakes did indeed aggregate prior to shedding of the proto-integument, but a shift in behavior occurred after the first epidermal shed (at which time neonates are not attracted to one another). Although this experiment does not provide evidence of which sensory modalities and what kinds of cues mediate such behavior, it is likely that chemical signals play a role (Scudder et al., 1980; Brown and MacLean, 1983; Burghardt, 1977; see below).

In vivo Trans-cutaneous Water Loss

With this information indicating that neonates do actively aggregate just after birth, but less so after shedding of the proto-integument, we next attempted to obtain information that might explain why such behavior occurs. We hypothesized that the proto-integument may be adapted for operation within amniotic fluid and, hence, may not have developed the barrier to trans-cutaneous water loss (CWL) that is present in typical, adult squamate skin (Landmann, 1979, 1980). Some such means of controlling CWL may be critical for neonate survival due to their relatively high surface-to-volume ratio and the high evaporation potential of the environment into which they are born. Water loss may be minimized by remaining in highly humid microenvironments, such as under rookery rocks. Similarly, aggregation by neonates while basking in the sun would reduce CWL by minimizing the extent of exposed cutaneous surface area through which water could be lost (Noble and Clausen, 1936; Cohen, 1975). Concurrently, body temperature and metabolic rate may be elevated (via basking in groups and resting under rocks warmed by the sun). This could speed the shedding process, thereby allowing quicker dispersion for den location.

To test these hypotheses, in vivo water loss from neonatal prairie rattlesnakes was measured under four conditions. Specifically, water loss was measured before and after shedding, and in neonates alone or in aggregations. Water loss was quantified by passing completely dry air through a chamber holding three neonate rattlesnakes, then freezing moisture from the air excurrent from the chamber with an alcohol-dry ice bath. Water loss was quantified using this procedure after one hour trials. All trials were run at ambient temperatures which were within one degree of 23°C throughout the experiment.

Results from this experiment indicate that preshed water loss ($\bar{X}$ grams $\pm$ SE) was 5.20 $\pm$ 0.12 when neonates were grouped and 8.46 $\pm$ 0.31 when solitary. After shedding, water loss was 4.30 $\pm$ 0.10 when grouped and 5.61 $\pm$ 0.34 when solitary. Thus, neonates lost more water before shedding of the proto-integument than after, under grouped or isolated conditions ($F(1,6) = 176.35$, $P < 0.001$). Also, grouping of neonates reduced CWL both before and after shedding ($F(1,6) = 91.61$, $P < 0.001$). Hence, we propose that the proto-integument is more water permeable than subsequent skin generations and that neonates remain in the rookery until after the first shed, at least in part, to minimize CWL through minimizing exposed surface area via aggregation.

Total water loss was the dependent measure in this experiment and no allowances were made for possible differences in metabolic rate. Therefore, observed differences in water loss cannot be ascribed unquestionably to differences in CWL. It is possible that some of the differences are due to differences in respiratory water loss (RWL), though it is unlikely that aggregation would systematically cause a drop in metabolism. Thus, we believe that the results obtained from the present experiment reflect changes in CWL.

A replication of this experiment was performed using a different method for quantification of water loss and with measurements of metabolic rates to control for any changes in RWL. This experiment was performed the following year with different neonatal prairie rattlesnakes. Nine groups of three neonates each were subjects. The experiment was performed in the same manner as the previous experiment except that water loss was quantified using a dew point hygrometer (EG & B, Environmental Equipment Division, Waltham, MA, Model 911 Dew-all) and oxygen consumption was quantified using an oxygen analyzer (Applied Electrochemistry Inc., Sunnyvale, CA, Model S-3A) to correct for RWL. Additionally, temperature in the test chamber, flow rate of air through the test chamber, and weight of subjects was included in the analysis. The experiment was analyzed as a 9 X 2 X 2 (subjects X grouping X shedding condition) factorial with four covariates (weight, flow rate, temperature, and oxygen consumption) and water loss as the dependent measure.

Results from this experiment concurred with those of the previous study. Preshed water loss ($\bar{X} \pm$ SE in mg/g body weight/hr) was 0.94 $\pm$ 0.10 when separated and 0.79 $\pm$ 0.09 when grouped. Postshed water loss was 0.81 $\pm$ 0.05 when separated and 0.72 $\pm$ 0.04 when grouped. As before, there was a significant difference in water loss between aggregated versus separated snakes ($F(1,20) = 6.10$; $P < 0.02$). Also, water loss was higher for preshed snakes than for postshed snakes ($F(1,20) = 19.49$; $P < 0.0001$). Analysis of covariance removes the effects of the covariates (oxygen consumption, flow rate, temperature and subject weight) on the dependent measure (water loss) in order to assess the effects of the experimental variables (grouping and shedding condition) on the dependent measure (Hicks, 1973). Therefore, because oxygen consumption (a measure of respiration rate) was a covariate, this analysis indicates that effects of grouping and shedding condition on water loss are indeed due to changes in CWL and not to changes in RWL. Consequently, this experiment replicates our previous findings using different methods and provides additional evidence that differences in cutaneous water loss due to changes in permeability between the proto-integument and subsequent epidermal generations, as well as to reduction of exposed surface area via aggregation, are real phenomena.

How does ontogenetic change in cutaneous water permeability occur? Warburg (1966) and Jackson and Sharawy (1978) speculated that lipids in or on the skin may control CWL. Evidence for such a mechanism comes from observations that lipid extraction from shed skins increases water permeability greatly (Roberts and Lillywhite, 1980; c.f., Stokes and Dunson, 1982; Ljungman and Dunson, 1983; Burken et al., 1985a). Further, Landmann (1979, 1980) showed that the primary lipid barrier to CWL is located in the mesos layer of reptilian skin (see Maderson, 1965 for description of reptilian skin anatomy and histology). Thus, we hypothesized that one difference between the proto-integument and the second skin generation is the relative amount of lipid present in the skin and that this might account for differences in water permeability.

To determine whether the proto-integument contains less lipid than the second skin generation, shed skins from the first and second epidermal generations after birth (i.e., two shed skins from each individual) were collected from individuals of five litters of prairie rattlesnakes born in the lab by females recently captured at our field site. All neonatal rattlesnakes were fed one neonate mouse after their first shed. Shed skins from all neonates in a litter were extracted together. Thus, lipid content of shed neonate skins was compared for the first and second epidermal generations, with five replicates (litters) in each group. Lipids were extracted using a soxhlet apparatus with methylene chloride as a solvent in the manner of Schell and Weldon (1985; c.f., Graves and Duvall, 1983). Mean lipid content (mg/g shed skin after extraction) of the proto-integumental epidermal generation was 76.3 mg/g shed skin ($\pm$ 8.0 SE) and mean lipid content of the second epidermal generation was 130.5 mg/g skin ($\pm$ 18.2 SE) (c.f., Weldon and Schell, 1984; Burken et al., 1985b). A t-test for related samples indicates that this difference is significant ($t(4) = 3.76$; $P < 0.02$).

SKIN LIPIDS AS PHEROMONES: ADAPTATION AND PROTOADAPTATION

How do neonate prairie rattlesnakes find a hibernaculum of unknown location? They may follow adults that have hibernated there before and, hence, know its location (Klauber, 1972). However, rather than using visual and tactual cues to do so, we propose that neonatal rattlesnakes follow odor trails or markings of other snakes - principally adults - to the den. This idea is not novel(Klauber, 1972; Hirth, 1966), but no evidence has been available to support it until recently (Brown and MacLean, 1983; this paper). Furthermore, we hypothesize that, in addition to their role in controlling CWL discussed above, skin lipids also serve as pheromones mediating den location. It is probable that the structure and location of lipids within snake skin have been modified to enhance their efficacy in this dual role. Therefore, snake skin lipids that restrict CWL may have served as a proto-adaptation (Maderson, this volume), or trait that was modified through natural selection to function in a new role, namely as a source of chemical signals mediating adaptive aspects of social behavior. Another early role for these chemical may have been one of aiding the process of skin shedding (see Maderson, this volume). Appearance of semiochemical substances de novo seems unlikely; a coevolution of responses by the receivers and signal modification by the producers seems most parsimonious.

Available evidence indicates that aggregation among adults of several species of snakes, in a variety of situations, is mediated by chemical cues (Noble and Clausen, 1936; Dundee and Miller, 1968; Porter and Czaplicki, 1974; Burghardt, 1980; Heller and Halpern, 1981, 1982a,b; Ford, this volume). There is also evidence that neonatal snakes of a variety of species aggregate in response to chemical cues (Scudder et al., 1980; Brown and MacLean, 1983; Burghardt, 1983). Such aggregation capabilities suggest that similar orientation and approach behaviors could be employed for den location if conspecific odors were deposited near or around dens.

Certain squamate pheromones have been proposed to reside in the skin. Tongue flicking (associated with odor delivery to the vomeronasal organ, or VNO) directed at the skin of conspecifics has been noted in several social situations (Gillingham, 1974; Duvall, 1982; Cooper and Vitt, 1984). The sex attractiveness pheromone of garter snakes resides in the skin (Kubie et al., 1978; Gillingham and Dickinson, 1980; Garstka and Crews, 1981, this volume), and snakes have been observed to respond behaviorally to extracts from the skins of conspecifics and heterospecifics (Weldon and Burghardt, 1979; Weldon, 1982; Graves and Duvall, 1983; Weldon and Schell, 1984). Further, available evidence from the above citations indicates that squamate skin pheromones are lipoidal. Interestingly, Duvall (1981, 1982) and Smith et al. (1984) have hypothesized that squamate epidermatoglyphics (microscopic textural patterns on the outer surface of squamate skin) may aid in adherence of skin derived lipid pheromones (though their function, sensu stricto, is likely more complex).

Hence, if chemical signals mediate den location by neonatal prairie rattlesnakes, skin lipids would seem to be the best candidates. Odorous substances used in the experiments to be described next were obtained by extracting shed skins in a soxhlet apparatus with methylene chloride as a solvent. We assumed that our choice of solvent would provide us with extracts composed primarily of lipids. This assumption was later verified (see below).

But do neonatal prairie rattlesnakes respond to such skin extracts from conspecific adults in a manner suggestive of den location? We devised an experiment to determine if neonates detect and respond to skin lipids. Neonates were cooled to 6°C prior to testing as a behavioral anesthetic of sorts. This was intended to insure that all subjects were at the same "arousal state" at the start of the trial and minimize defensiveness due to handling and odor application (see Graves and Duvall, 1983). Subjects were immediately transported from the incubator to the test arena, a clean plastic container measuring 20 x 15 x 9 cm with a line drawn across the middle of its floor. Odor application involved touching a long glass rod that had been dipped in one of four extracts (test odors) to the snout of each subject. This procedure insured that, each time a neonate tongue flicked, it would encounter the test odor with its VNO. The four test odors were skin extracts

Table 1. Responses of neonatal rattlesnakes to skin lipid and control extracts ($\bar{X} \pm SE$) in small test arenas.

Dependent Measure	Odor Conditions: Conspecific Neonates	Conspecific Female adults	*B. constrictor*	CH_2CL_2
Tongue Flicks[a]	134 ± 24	132 ± 33	106 ± 18	72 ± 6
Activity Units[b]	12.0 ± 2.6	12.0 ± 3.6	4.2 ± 1.7	3.7 ± 0.9
Mouth Gapes	7.8 ± 2.7	9.8 ± 3.7	14.7 ± 6.2	5.2 ± 1.1

[a] $F\ (3,20) = 2.868;\ p < 0.06$.
[b] $F\ (3,20) = 3.636;\ p < 0.031$.

dissolved in methylene chloride (vehicle) from either (1) neonate conspecifics, (2) adult female conspecifics, (3) a male *Boa constrictor*, or (4) vehicle alone. The test arena was then placed on an electric heating block to warm subjects during the trial. Trials lasted 20 min during which subjects were observed from a blind while tongue flicks, activity units (the number of times that the subject's head crossed the center line), and mouth gapes (or yawns, a behavior which also is associated with VNO olfaction; Graves and Duvall, 1983, 1985) were quantified. Six different subjects were tested in each condition.

Table 1 presents the results of this experiment and statistical analyses. Although differences in numbers of tongue flicks exhibited by subjects in the respective groups were not significantly different at the classically accepted 0.05 probability level, they were very close. We feel that, given the small sample sizes and the observed effect, the findings are at least suggestive that conspecific chemical cues stimulated VNO investigation more than heterospecific skin lipids or the control did. More importantly, in this context, neonatal prairie rattlesnakes exposed to conspecific skin extracts locomoted significantly more than did subjects exposed to heterospecific skin extracts or the vehicle alone. Such a response is suggestive of a basic orientation and/or activity mechanism which, at a minimum, indicates that neonates can detect and respond to conspecific skin extracts.

Our next experiment expanded on the above findings and more directly investigated whether skin lipids might mediate den location by neonates. Neonates used in this experiment were housed individually within a few hours after birth. Subjects (N = 18) were introduced individually into the center of a large (122 x 79 cm) arena that had one rock pile at each end. The floor was covered with clean, plastic-backed bench paper prior to the start of each trial and a clean, circular piece of filter paper was placed under each rock pile. One min prior to the start of each trial, one ml of methylene chloride was poured onto the filter paper beneath one rock pile and the same amount of solvent with adult prairie rattlesnake skin extract was poured beneath the other rock pile. Several dependent measures (see Table 2) were quantified during continuous observations over the first 30 min of each trial and point observations of the subject's location were made at 0.5, 1, 12, 24, and 36 hr after the start of each test period. Additionally, each subject was pre-exposed to, or "primed" with, one of the

Table 2. Responses of odor-primed neonatal prairie rattlesnakes to skin lipid extracts or CH_2Cl_2 ($\bar{X} \pm SE$) in large test arenas.

Dependent Measures	Pre-test Prime	
	CH_2Cl_2	Skin lipids
Tongue flicks	560 ± 22	671 ± 23
Mouth gapes[a]	1.7 ± 0.6	9.8 ± 1.6
Accumulated locomotion duration (min and min/100)	15.51 ± 2.63	20.0 ± 2.28
Mouth gape-tongue flick sequences[b]	1.6 ± 0.6	7.9 ± 1.3
Tongue flick-mouth gape sequences[b]	0.4 ± 0.3	1.3 ± 1.2
Proportion occupying rattlesnake-marked arena zone at 36 hr[c]	9/9	8/9

[a] $F(1,16) = 21.19$, $p < 0.001$.

[b] $F(1,16) = 19.01$, $p < 0.001$.

[c] $Q(1) = 14.22$, $p < 0.001$.

two odors (solvent or solvent plus adult conspecific lipid) by touching a glass rod dipped into one of the two odors to its snout just prior to release into the arena. This preexposure was included because we have observed neonates tongue flicking, mouth gaping, head shaking, and rubbing their faces against their mother's bodies immediately after birth (Graves et al., unpublished data), suggesting the possibility that exposure to conspecific cues just subsequent to birth may facilitate an odor imprinting event of sorts. Further, these behaviors have been generally implicated in snake olfactory processes (Halpern and Kubie, 1980; Graves and Duvall, 1983, 1985; Scudder et al., 1980).

Several points are suggested by the results and statistical analyses in Table 2. First, as in previous experiments (Graves and Duvall, 1983), mouth gapes occurred more frequently after exposure to conspecific odors than after exposure to the control. Second, mouth gapes occur just prior to tongue flicks more frequently than vice versa, suggesting that mouth gaping may prepare the VNO for odor delivery via the tongue (additionally, it was noted that mouth gapes occur between bouts of locomotion; $F(1,17) = 13.06$, $P < 0.01$). Third, the priming procedure produced no differences in odor choice. And finally, and most important to our hypothesis, after 36 hr, 17 of 18 neonates were located on the half of the arena that contained conspecific skin lipids. To determine if this effect is mediated by conspecific skin lipids only, a control experiment was performed with different neonates ($n = 12$) identical to the above design except that extracts from the shed skin of a <u>B. constrictor</u> and methylene chloride alone were used as the two odor conditions. After 36 hr, neonates were found no more frequently on one side of the arena or the other. Thus, it appears that neonates not only detect and respond to conspecific skin lipids, but they are also attracted to such chemicals. Furthermore, while the above responses result from exposure to conspecific skin lipids, skin extracts from a heterospecific do not elicit similar behaviors.

Brown and MacLean (1983) showed that neonatal timber rattlesnakes (C. horridus) follow odor trails of both adult and neonate conspecifics. Our experiments support the hypothesis that chemical cues from adults may be used in den location by neonatal rattlesnakes, and strengthen this conclusion by extending it to another taxa. These experiments also have determined the locus of production of such chemical cues. Additionally, the findings of Brown and MacLean (1983) that neonate timber rattlesnakes follow trails of neonates as well as adults, and evidence in this paper that neonate prairie rattlesnakes exhibit a locomotory response to neonate skin extracts, suggests that neonates may follow skin-derived odor trails of other neonates, in addition to those of adults, when searching for hibernacula.

Other investigators have also presented evidence suggesting that squamate skin lipids that have derived pheromone function are likely located on the surface of the skin. Skin-shedding exposes pheromones on the outer skin surface of snakes (Kubie et al., 1978; Nilson, 1980) and these may be deposited as the lacunar and clear layer break down just prior to shedding (Lillywhite and Maderson, 1982). Indeed, Jackson and Sharawy (1978) found numerous lipid-filled globules within the lacunar layer of snake skin and these may be similar to those described here. Weldon and Schell (1984) have shown that lipoidal chemical signals can be obtained by wiping the dorsal surface of a variety of snakes with cotton soaked in methylene chloride. Roberts and Lillywhite (1980) found that wiping the surface of snake (Elaphe obsoleta) skin with an organic solvent does not increase CWL rate. Thus, substances, probably lipids, are present on the skin that may have pheromonal properties but do not influence CWL (the primary barrier to CWL is lipoidal, but located in the mesos layer; Landmann et al., 1981). We propose that lipids in squamate skin originally evolved to control CWL (and perhaps skin shedding; Maderson, this volume), and subsequently derived pheromonal properties through some interactive process of protoadaptation (Maderson, this volume) and exaptation (Gould and Vrba, 1982). Lipids may have originally been located more homogeneously throughout squamate epidermis, but chemical structure and spatial configurations of lipids may have been modified during the course of evolution to enhance the ability of lipids to perform different roles in different skin layers.

HISTOLOGY

Squamate epidermis is complex structurally (Maderson, 1965; Maderson et al., 1972) and functionally (Maderson, 1984). Therefore, histological examinations of prairie rattlesnake skin are needed to relate how lipid chemosignals are used to their distribution and arrangement in the integument. Further, microscopic observation may reveal relationships of lipids that act as pheromones and those that control CWL. That rattlesnake skin lipids have pheromonal properties is only half the story; how these substances and the skin that produces and contains them have been shaped for this role is important, and may also reveal patterns of adaptation.

Mature squamate skin is composed of at least six distinct cell layers. From the interior moving sequentially outward these are the clear, lacunar, alpha, mesos, beta, and Oberhäutchen layers (Maderson, 1965). Lipids are present on the surface of the Oberhäutchen (Crews and Garstka, 1982; Weldon and Schell, 1984) and in the beta, mesos, alpha (Roberts, 1980), and lacunar (Jackson and Sharawy, 1978) layers. Control of CWL is primarily accomplished by lipids in the mesos region (Landmann et al., 1981). Lipids are costly substances to produce. Hence, it seems likely that extra-mesos layer lipids

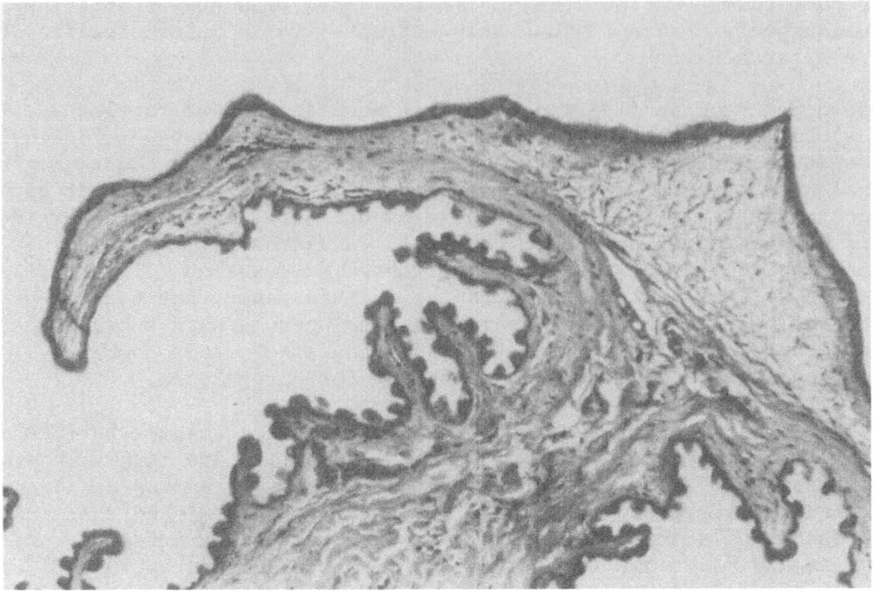

Fig. 1. Section through a scale of *Crotalus v. viridis* at 10 days post-shed (40x). Protuberances in the hinge region contain actively dividing, lipid-filled cells.

either serve accessory permeability barrier functions or have some other fole. We have proposed that squamate skin lipids play a dual role, both as modulators of CWL and as pheromones. However, lipids serving these two functions need not be compartmentalized in the same epidermal layers and, in fact, may be different chemically.

Although a considerable body of research concerning squamate skin anatomy and lipid distribution has been amassed (see review in Lillywhite and Maderson, 1982), crotalids have not been studied in this regard. Therefore, we prepared several integumental biopsies and pieces of shed epidermis for histological examination (c.f., Graves, in preparation). All specimens were fixed in neutral buffered formalin. These tissues were embedded in paraffin, sectioned at 10 µm, and stained with hematoxylin and eosin. Others were sectioned on a freezing microtome and stained with Sudan IV, which has an affinity for lipids (Humason, 1979).

Figures 1 and 2 show sections of epidermis from the nuchal region (dorsal, along the back of the neck) of one snake, a pregnant *C. v. viridis*. Figures 1 and 2 are sections of biopsied skin, 10 days post-shed.

In the hinge region of the scale shown in Figure 1 are a number of previously undescribed structures that appear as protuberances on the epidermal surface and are located only in this area. Figure 2 is a magnified view of these structures showing their location in the epidermis (note the outer, cornified layer). These epidermal protuberances contain actively dividing cells and stain positive for lipid. We hypothesize that these may be important sources of pheromone. It is not known how widespread among squamate taxa such lipid-containing protuberances in the hinge region are (but see Garstka and Crews, this volume), or if they are present in male, immature, or nonpregnant females. Furthermore, the existence and possible role of these structures as pheromone sources in no way contradicts other postulated pathways for expression of lipid pheromones on squamate skin (e.g., Duvall, 1981,

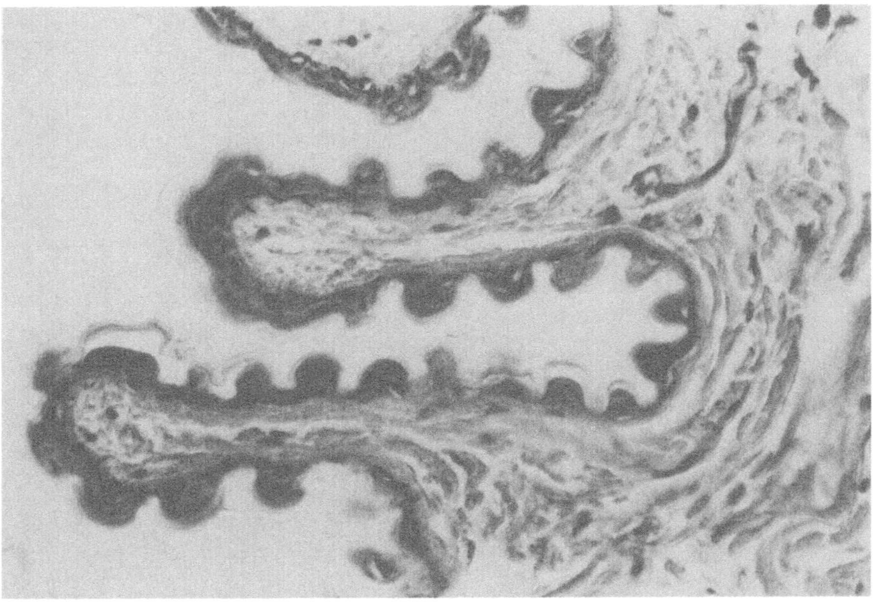

Fig. 2. Section through a scale of Crotalus v. viridis at 10 days post-shed (51x) providing a higher magnification of the epidermal protuberances in the hinge region.

1982; Garstka and Crews, 1981). However, it is possible that the hyperventilation and consequent body expansion observed in female garter snakes during courtship (Garstka and Crews, 1981) may, in addition to any other functions or effects, expose lipid pheromones in the hinge region. Finally, these lipid-containing protuberances may influence CWL patterns. It has been assumed that the hinge region is more permeable to water than the dorsal scale surface because of its thinness (Krakauer, 1970). However, Stokes and Dunson (1982) have shown that the hinge regions of Constrictor constrictor and Nerodia cyclopion floridana are completely impermeable to water. It is possible that the structures such as those described here play a role in this phenomenon.

Lipids also were stained in the shed skins of one pregnant female. Lipids in the skin could be exposed via the shedding process (see below) on both the outside of the new skin generation and the inside of the old skin generation. Hence, lipid pheromones could be deposited as continuous trails by snakes crawling along the ground (e.g., Brown and MacLean, 1983; Ford, this volume) or as point sources or "packets" of more concentrated and stable pheromone (i.e., where shed skins themselves are left). Interestingly, shed skins of pregnant female and neonates are found in the rookery area at our field site and often remain intact for several years. Hence, lipids in such skins may act as pheromonal marks, aiding females in locating the rookery and choosing suitable rocks under which to stay until parturition. Not only do many pregnant snakes use the same area as a rookery each year, but the same rocks are chosen by different females in consecutive years as well (Duvall et al., 1985; King, Duvall, and Graves, unpublished data). Rocks with odors from previous pregnant female and neonatal occupants must have provided adequate protection and microenvironmental conditions in previous seasons, and such rocks may be preferentially selected by pregnant females in subsequent years. Hence, lipids adhering to shed skins may facilitate rookery microhabitat selection. Shed skins located between

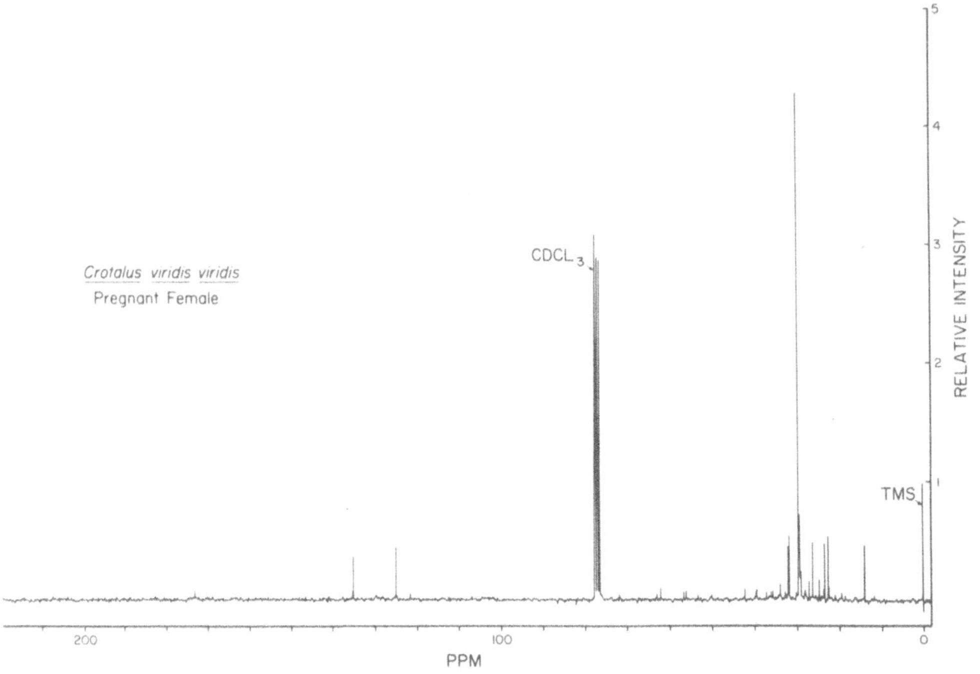

Fig. 3. ^{13}C Nuclear magnetic resonance spectrum of lipids extracted from the shed skin of a pregnant female Crotalus v. viridis.

rookeries and hibernacula similarly may aid neonates in locating overwintering sites, as described above. With regard to the ability of shed skins in the wild to retain lipid pheromones for long durations, sections of shed epidermis also have been examined, but after lipid extraction with methylene chloride in a soxhlet apparatus. Even after this procedure, lipid-specific dyes indicate that some lipid is still retained in the shed skin. Hence, some lipids are tightly associated with epidermal keratin (Lillywhite and Maderson, 1982), and these may remain in shed skins for extended periods.

SEMIOCHEMISTRY

Because we used methylene chloride as a solvent to extract pheromone-containing substances from shed snake skins, we hypothesized that resulting extracts were lipoidal in nature. And because our behavioral bioassays indicated that neonatal rattlesnakes responded differentially to extracts from conspecific and heterospecific sources, we hypothesized that there might be detectable differences between the chemical structures of such extracts. Additionally, knowledge of the extract's chemical structure might provide insight into their biological roles.

This idea was considered by using nuclear magnetic resonance (NMR) to characterize snake skin extracts. A JEOL FX 270 Superconducting NMR was used to obtain spectra of skin extracts. ^{13}Carbon (^{13}C) and hydrogen NMR were performed, both of which gave similar results. Concentration of lipid in each sample was standardized (1 mg lipid/1 ml solvent) and 500 scans of each sample were done with an interscan duration of 10 sec.

Figure 3 shows a ^{13}C spectrum of extracts from the shed skins of pregnant prairie rattlesnakes. Extracts from skins of nonpregnant and male prairie rattlesnakes, as well as B. constrictor, were also examined in this manner. Comparisons of spectra from these sources indicate that all of the skin extracts are long chain hydrocarbons that are relatively more aliphatic than olefinic, with much unsaturation, but generally lipoidal in nature. Though the spectra of extracts from all four sources are similar structurally, relative differences in peak height suggest potential differences in overall mixtures between these groups. However, an interscan duration of 25 sec, rather than 10 sec is needed for an accurate determination of quantitative differences.

Thus, although our analysis is preliminary, the differences that neonates detect and respond to behaviorally could very well be due to differences in relative quantities of constituents rather than their presence or absence. It is interesting that, even though B. constrictor and C. v. viridis are relatively distant taxonomically and live in vastly different habitats, there has been surprising conservativism in skin lipid structure during their evolution. It is possible that skin lipid chemical structure has remained relatively constant to maintain its apparent primary function, controlling CWL (Roberts and Lillywhite, 1980; Landmann, 1979, 1980), and that lesser changes have occurred to enhance pheromonal functions or effects in taxa where necessary selective pressures have existed.

Lipids are ideally suited for cooptation as pheromones in squamates. The squamate vomeronasal organ may be preferentially sensitive to nonvolatile substances (Cowles and Phelan, 1958; Kubie and Halpern, 1978; Burghardt, 1980; Duvall, 1981). Also, the lower relative volatility of lipids is ideal for producing pheromonal marks with persistence under hot, arid, windy, and intense solar radiation conditions, precisely the sorts of conditions that exist in the high plains desert of Wyoming. Such odorous substances must persist at least from spring until fall to be employed by snakes to locate the den from which they emerged. Similarly, chemical cues leading pregnant female rattlesnakes to rookeries must maintain their integrity from fall until the following spring to be effective.

CONCLUSIONS

At the outset of this paper, we outlined four criteria as sufficient or necessary to discuss the existence of biological adaptation at an organismal level. With regard to skin lipids of prairie rattlesnakes and their role(s) in chemical signaling, we believe that we have at least begun to address these ideas in our study of structure and function in adaptation among these animals. We first asked if the trait in question had been shaped by natural selection for its current role? Differences in skin lipid chemical composition between different taxa and the presence of unique lipid-containing structures in prairie rattlesnake epidermis suggest to us that it has. Second, can selection pressures be identified that could have affected the derivation of the trait? Because of the prominence of chemical senses in squamate behavior and the requirement for den location in this population of north-temperate region ectotherms, selection for aggregation-inducing properties of skin lipids seems likely. Third, can it be demonstrated in some way that the trait has survival value to the organisms that exhibit it? Neonates that do not respond to lipid trails likely will not find a den and, hence, may not survive the winter. Therefore, expression of skin lipids and deposition of trails leading to the den, as well as genetic transmission to offspring of sufficient orienting mechanisms to such cues, may be critical in ensuring survival of offspring. And fourth, we asked how variable is the trait in the population? Neonates in our experiments responded to conspecific skin lipids in a significant manner, suggesting the existence of mechanisms that could potentially mediate den location. Hence, this behavioral trait appears to be prevalent in this population. We are also beginning to assess variability in chemical structure of skin lipids and their distribution in the epidermis between various classes of snakes (i.e., male, female, juveniles and neonates) through NMR characterizations and histological study. Such investigations of the morphology underlying the adaptive behavior that we have seen will also help in this respect.

Of vital importance in attempting to identify true adaptation, and to provide complete explanations of causes and functions in behavioral biology, is the need for an interdisciplinary, integrative approach to empirical investigation. An ideal synthesis of this type would answer questions not only about what an animal does, but also how it responds, and why. Lillywhite and Maderson (1982) state that "Recent investigations...pose many questions for future research, particularly opportunities for integrating morphology, physiology, and ecology. Indeed, fuller understanding of the current questions may only be achieved through integrative approaches of research." And though we agree, we would add behavior to this list.

ACKNOWLEDGMENTS

This research was funded by grants from the National Geographic Society, the Wyoming Game and Fish Department, and the University of Wyoming to D. Duvall. We also thank Matt Goode and Al Redder for field assistance, Ed Clennen for aid in chemical analyses, Robert Cochran for help with statistical analyses, and Paul Maderson for help interpreting skin tissue samples.

REFERENCES

Aldridge, R. D., 1975, Environmental control of spermatogenesis in the rattlesnake Crotalus viridis, Copeia, 1975:493.

Aldridge, R. D., 1979, Female reproductive cycles of the snakes Arizona elegans and Crotalus viridis, Herpetologica, 35:256.

Baxter, G. T., and Stone, M. D., 1980, "Amphibians and Reptiles of Wyoming," Wyoming Game and Fish Dept., Cheyenne.

Bentley, P. J., and Schmidt-Nielsen, K., 1966, Cutaneous water loss in reptiles, Science, 151:1547.

Blem, C. R., 1982, Biennial reproduction in snakes: An alternative hypothesis, Copeia, 1982:961.

Bock, W. J., 1980, The definition and recognition of biological adaptation, Am. Zool., 20:217.

Brown, W. S., and MacLean, F. M., 1983, Conspecific scent trailing by newborn timber rattlesnakes, Crotalus horridus, Herpetologica, 39:430.

Bull, J. J., and Shine, R., 1979, Iteroparous animals that skip opportunities for reproduction, Am. Nat., 114:296.

Burghardt, G. M., 1977, Of iguanas and dinosaurs: Social behavior and communication in neonate reptiles, Am. Zool., 17:177.

Burghardt, G. M., 1980, Behavioral and stimulus correlates of vomeronasal functioning in reptiles: Feeding, grouping, sex, and tongue use, in: "Chemical Signals," D. Muller-Schwarze, and R. M. Silverstein, eds., Plenum, New York.

Burghardt, G. M., 1983, Aggregation and species discrimination in newborn snakes, Z. Tierpsychol., 61:89.

Burken, R. R., Wertz, P. W., and Downing, D. T., 1985a, The effect of lipids on transepidermal water permeation in snakes, Comp. Biochem. Physiol., 81A:213.

Burken, R. R., Wertz, P. W., and Downing, D. T., 1985b, A survey of polar and nonpolar lipids extracted from snake skin, Comp. Biochem. Physiol., 81B:315.
Cohen, A. C., 1975, Some factors affecting water economy in snakes, Comp. Biochem. Physiol., 51A:361.
Cooper, W. E., Jr., and Vitt, L. J., 1984, Conspecific odor detection by the male broad-headed skink, Eumeces laticeps: Effects of sex and site of odor source and of male reproductive conditions, J. Exp. Zool., 230:199.
Cowles, R. B., and Phelan, R. L., 1958, Olfaction in rattlesnakes, Copeia, 1958:77.
Crews, D., and Garstka, W. R., 1982, The ecological physiology of a garter-snake, Sci. Amer., 247:159.
Diller, L. V., and Wallace, R. L., 1984, Reproductive biology of the northern pacific rattlesnake (Crotalus viridis oreganus) in northern Idaho, Herpetologica, 40:182.
Dundee, H. A., and Miller, C. W., III, 1968, Aggregative behavior and habitat conditioning by the prairie ringneck snake, Diadophis punctatus arnyi, Tulane Stud. Zool. Bot., 15:41.
Duvall, D., 1981, Western fence lizard (Sceloporus occidentalis) chemical signals. II. A replication with naturally breeding adults and a test of the Cowles and Phelan hyyothesis of rattlesnake olfaction, J. Exp. Zool., 218:351.
Duvall, D., 1982, Western fence lizard (Sceloporus occidentalis) chemical signals. III. An experimental ethogram of conspecific body licking, J. Exp. Zool., 221:23.
Duvall, D., Guillette, L. J., and Jones, R. E., 1982, Environmental control of reptilian reproductive cycles, in: "Biology of the Reptilia, Vol. 13," C. Gans, and F. H. Pough, eds., Academic Press, New York.
Duvall, D., King, M. B., and Graves, B. M., 1983, Fossil and comparative evidence for possible chemical signaling in the mammal-like reptiles, in: "Chemical Signals in Vertebrates 3," D. Muller-Schwarze, and R. M. Silverstein, eds., Plenum, New York.
Duvall, D., King, M. B., and Gutzwiller, K. J., 1985, Behavioral ecology and ethology of the prairie rattlesnake, Nat. Geog. Res., 1:80.
Fitch, H. S., 1949, Study of snake populations in central California, Am. Midl. Nat., 41:513.
Fitch, H. S., and Shirer, H. W., 1971, A radiotelemetric study of spatial relationships in some common snakes, Copeia, 1971:118.
Gannon, V. P. J., and Secoy, D. M., 1985, Seasonal and daily activity patterns in a Canadian population of the prairie rattlesnake, Crotalus viridis viridis, Can. J. Zool., 63:86.
Garstka, W. R., and Crews, D., 1981, Female sex pheromone in the skin and circulation of a garter snake, Science 214:681.
Gillingham, J. C., 1974, Reproductive behavior of the western fox snake, Elaphe v. vulpina (Baird and Girard), Herpetologica, 30:309.
Gillingham, J. C., and Dickinson, J. A., 1980, Postural orientation during courtship in the eastern garter snake, Thamnophis s. sirtalis, Behav. Neural Biol., 28:211.
Glissmeyer, H. R., 1951, Egg production in the Great Basin rattlesnake, Herpetologica, 7:24.
Gould, S. J., and Lewontin, R. C., 1979, The spandrels of San Marco and the panglossian paradigm: A critique of the adaptationist programme, Proc. R. Soc. Lond., B205:581.
Gould, S. J., and Vrba, E. S., 1982, Exaptation - a missing term in the science of form, Paleobiology, 8:4.
Graves, B. M., In preparation, Behavioral and Chemical Ecology of Pregnant and Neonatal Prairie Rattlesnakes (Crotalus viridis viridis) in South-central Wyoming, Ph.D. Dissertation, University of Wyoming, Laramie.
Graves, B. M., and Duvall, D., 1983, Occurrence and function of prairie rattlesnake mouth gaping in a non-feeding context, J. Exp. Zool., 227:471.

Graves, B. M., and Duvall, D., 1985, Mouth gaping and head shaking by prairie rattlesnakes are associated with vomeronasal organ olfaction, Copeia, 1985:496.

Graves, B. M., and Duvall, D., 1987, An experimental study of aggregation and thermoregulation in prairie rattlesnakes (Crotalus viridis viridis), Herpetologica, in press.

Gregory, P. T., 1984, Communal denning in snakes, in: "Vertebrate Ecology and Systematics," R. A. Seigel, L. E. Hunt, J. L. Knight, L. Malaret, and N. L. Zuschlag, eds., Museum of Natural History, University of Kansas, Lawrence.

Halpern, M., and Kubie, J. L., 1980, Chemical access to the vomeronasal organs of garter snakes, Physiol. Behav., 24:367.

Heller, S. B., and Halpern, M., 1981, Laboratory observations on conspecific and congeneric scent trailing in garter snakes (Thamnophis), Behav. Neural Biol., 33:372.

Heller, S. B., and Halpern, M., 1982a, Laboratory observations of aggregative behavior of garter snakes, Thamnophis sirtalis, J. Comp. Physiol. Psychol., 96:967.

Heller, S. B., and Halpern, M., 1982b, Laboratory observations of garter snakes, Thamnophis sirtalis: Roles of the visual, olfactory, and vomeronasal senses, J. Comp. Physiol. Psychol., 96:984.

Hicks, C. R., 1973, "Fundamental Concepts in the Design of Experiments," Holt, Rinehart and Winston, New York.

Hirth, H. F., 1966, The ability of two species of snakes to return to a hibernaculum after displacement, Southwestern Nat., 11:49.

Humason, G. L., 1979, "Animal Tissue Techniques," Fourth Edition, W. H. Freeman, San Francisco.

Jackson, M. K., and Sharawy, M., 1978, Lipids and cholesterol clefts in the lacunar cells of snake skin, Anat. Rec., 190:41.

Janzen, D. H., 1967, Synchronization of sexual reproduction of trees within the dry season in Central America, Evolution, 21:620.

Keenlyne, K. D., 1972, Sexual differences in feeding habits of Crotalus horridus horridus, J. Herpetol., 6:234.

Keenlyne, K. D., 1978, Reproductive cycles in two species of rattlesnakes, Am. Midl. Nat., 100:368.

King, M. B., in preparation, Migratory and Foraging Ecology of Prairie Rattlesnakes in South-central Wyoming, M. S. Thesis, University of Wyoming, Laramie.

Klauber, L. M., 1972, "Rattlesnakes: Their Habits, Life Histories and Influence on Mankind," 2nd edition, Univ. Calif. Press, Berkeley.

Krakauer, I., 1970, The Ecological and Physiological Control of Water Loss in Snakes, Ph. D. Thesis, University of Florida, Gainesville.

Kubie, J. L., Cohen, J., and Halpern, M., 1978, Shedding enhances the sexual attractiveness of oestradiol treated garter snakes and their untreated penmates, Anim. Behav., 26:562.

Kubie, J. L., and Halpern, M., 1978, Garter snake trailing behavior: Effects of varying prey extract concentration and mode of prey extract presentation, J. Comp. Physiol. Psychol., 92:362.

Landmann, L., 1979, Keratin formation and barrier mechanisms in the epidermis of Natrix natrix (Reptilia: Serpentes): An ultrastructural study, J. Morph., 162:93.

Landmann, L., 1980, Zonulae occludentes in the epidermis of the snake Natrix natrix L., Experientia, 36:110.

Landmann, L., Stolinski, C., and Martin, B., 1981, The permeability barrier in the epidermis of the grass snake during the resting stage of the sloughing cycle, Cell Tis. Res., 215:369.

Lillywhite, H. B., and Maderson, P. F. A., 1982, Skin structure and permeability, in: "Biology of the Reptilia, Volume 12, Physiology C, Physiological Ecology," C. Gans, and F. H. Pough, eds., Academic Press, New York.

Ljungman, T. N., and Dunson, W. A., 1983, Integumentary water and sodium permeability of the yellow anaconda, Eunectes notaeus, Comp. Biochem. Physiol., 76A:51.

Lucas, C. E., 1944, Excretions, ecology and evolution, Nature, 153:378.
Maderson, P. F. A., 1965, The structure and development of the squamate epidermis, in: "Biology of the Skin and Hair Growth," A. G. Lyne, and B. F. Short, eds., Angus and Robertson, Sidney.
Maderson, P. F. A., 1984, The squamate epidermis: New light has been shed, in: "The Structure, Development, and Evolution of Reptiles," M. W. J. Ferguson, ed., Academic Press, London.
Maderson, P. F. A., Flaxman, B. A., Roth, S. I., and Szalo, G., 1972, Ultrastructural contributions to the identification of cell types in the lizard epidermal generation, J. Morph., 136:191.
Martin, W. H., 1985, Life history of the timber rattlesnake, Crotalus horridus, Investigator's Annual Report, U. S. Dept. Int., Natl. Park Serv.
Nilson, G., 1980, Male reproductive cycle of the European adder, Vipera berus, and its relation to annual activity patterns, Copeia, 1980:729.
Noble, G. K., and Clausen, H. J., 1936, The aggregation behavior of Storeria dekayi and other snakes, with especial reference to the sense organs involved, Ecol. Monogr., 6:269.
Porter, R. H., and Czaplicki, J. A., 1974, Responses of water snakes (Natrix r. rhombifera) and garter snakes (Thamnophis sirtalis) to chemical cues, Anim. Behav., 2:129.
Rahn, H., 1942, The reproductive cycle of the prairie rattler, Copeia, 1942:233.
Reinert, H. K., 1984, Habitat variation within sympatric snake populations, Ecology, 65:1673.
Reinert, H. K., and Kodrich, W. R., 1982, Movements and habitat utilization by the massassauga, Sistrurus catenatus catenatus, J. Herpetol., 16:162.
Roberts, J. B., 1980, The Role of Epidermal Lipids in Limiting Cutaneous Evaporative Water Loss in Squamate Reptiles, Ph.D. Thesis, University of Kansas, Lawrence.
Roberts, J. B., and Lillywhite, H. B., 1980, Lipid barrier to water exchange in reptile epidermis, Science, 207:1077.
Schell, F. M., and Weldon, P. J., 1985, ^{13}C NMR analysis of snake skin lipids, Agric. Biol. Chem., 49:3597.
Schmidt-Nielsen, K., and Bentley, P. J., 1966, Desert tortoise, Gopherus agassizii: Cutaneous water loss, Science, 154:911.
Scudder, K. M., Stewart, N. J., and Smith, H. M., 1980, Responses of neonate water snakes (Nerodia sipedon sipedon) to conspecific chemical cues, J. Herpetol., 14:196.
Smith, H. M., Duvall, D., Graves, B. M., Jones, R. E., and Chiszar, D., 1984, The function of squamate epidermatoglyphics, J. Maryl. Herp. Soc., 30:3.
Smith, R. J. F., 1977, Chemical communication as adaptation: Alarm substances of fish, in: "Chemical Signals in Vertebrates," D. Muller-Schwarze, and M. M. Mozell, eds., Plenum, New York.
Sober, E., 1984, "The Nature of Selection: Evolutionary Theory in Philosophical Focus," Massachusetts Institute of Technology Press., Cambridge, Mass.
Stokes, G. D., and Dunson, W. A., 1982, Permeability and channel structure of reptilian skin, Am. J. Physiol., 242:F681.
Tinkle, D. W., 1962, Reproductive potential and cycles in female Crotalus atrox from northern Texas, Copeia, 1962:306.
Warburg, M. R., 1966, On the water economy of several Australian geckos, agamids, and skinks, Copeia, 1966:230.
Weldon, P. J., 1982, Responses to ophiophagous snakes by snakes of the genus Thamnophis, Copeia, 1982:788.
Weldon, P. J., 1983, The evolution of alarm pheromones, in: "Chemical Signals in Vertebrates, 3," D. Muller-Schwarze, and R. M. Silverstein, eds., Plenum, New York.
Weldon, P. J., and Burghardt, G. M., 1979, The ophiophage defensive response in crotaline snakes: Extension to new taxa, J. Chem. Ecol., 5:141.

Weldon, P. J., and Schell, F. M., 1984, Responses by king snakes (*Lampropeltis getulus*) to chemicals from colubrid and crotaline snakes, *J. Chem. Ecol.*, 10:1509.
Williams, G. C., 1966, "Adaptation and Natural Selection," Princeton Univ. Press, Princeton.

EXPLORATORY TONGUE FLICKING BY GREEN IGUANAS IN LABORATORY AND FIELD

Gordon M. Burghardt, Barbara A. Allen, and Hannah Frank

Department of Psychology and Graduate Program in
Life Sciences, University of Tennessee
Knoxville, TN 37996

Tongue flicking (rapid tongue extrusion) is an integral component of the behavior of all snakes and many lizards (Gove, 1979; Simon, 1983). Such tongue flicking has been related to vomeronasal functioning in many snakes (e.g., Burghardt, 1970, 1980; Gillingham and Clark, 1981; Halpern and Kubie, 1981, 1983; Kubie and Halpern, 1979; Meredith and Burghardt, 1979; Chiszar, 1981; Graves and Duvall, 1985 and others). However, in lizards direct experimental evidence for vomeronasal involvement is sparse. Indeed a recent persuasive case has been made for taste being the function of much lizard tongue flicking, especially when contact with objects occurs (Schwenk, 1985). Lizards do often have well-developed taste buds on the tongue while no species of snake has been documented to have lingual taste buds. Surgical studies (c.f., Hammerman and Thomas, 1967) eliminating gustation in lizards may eventually be necessary.

In short, the sensory function of tongue flicking in lizards is still unclear. We also do not know much about the conditions in which tongue flicking occurs in lizards in natural and captive contexts. Certainly the use of the tongue in lizards is much more diverse than in snakes, having roles in prey capture, face cleaning, and food manipulation not found in snakes (Gove, 1979). The structure of the tongue also varies greatly in lizards, from long, thin, and deeply forked to thick and stubby; there are other morphological variants also, as in the elongated chameleon tongue. The frequency of tongue flicking has been shown to vary across lizard families in ways parallel to the degree of vomeronasal organ development (Bissinger and Simon, 1979). But regardless of the specific sense organs activated, tongue flicking is almost always considered chemosensory in function except in those cases where it may have a visual signal function secondarily imposed (Gove, 1979; Gove and Burghardt, 1983). The increased tongue flicking seen when snakes and lizards are introduced to novel environments or those containing social or food stimuli (Chiszar et al., 1976; Duvall, 1979; Simon et al., 1981) can, however, be related to information gathering (exploration), particularly if handling and home cage controls are imposed.

Members of the family Iguanidae have traditionally been considered to be primarily visual (e.g., Evans, 1959); they typically possess a fleshy nonserpent-like tongue with only moderate development of the vomeronasal organ and the associated accessory olfactory bulb. Contrary to expectations, however, some iguanid lizards have recently been demonstrated to show increases of tongue flicking in the presence of socially significant chemical stimuli (see review in Simon, 1983) as well as novel surroundings or disturbance (e.g., Greenberg, 1985). However, rates of tongue flicking are often very low in iguanids. Reported rates are virtually always less than 1.0/min, even in novel environments (c.f., De Fazio et al., 1977; Bissinger and Simon, 1979). Comparable rates are seen in the field (Simon et al., 1981); the rate for Anolis trinitatus was about 1 tongue flick (TF) every 100 min (Gravelle and Simon, 1980).

In contrast, preliminary observations on the green iguana (Iguana iguana), a large herbivorous lizard that has been the focus of extensive fieldwork on ecology and behavior (see references in Burghardt and Rand, 1982) revealed a high rate of tongue flicking at certain times in the field in two situations where observation conditions were ideal. These are at a communal nest site on the islet of Slothia adjacent to Barro Colorado Island in Panama where females migrating to nest (Rand, 1968) and hatchlings emerging months later (Burghardt et al., 1977) emit frequent tongue flicks to the substrate, emergent objects, air, and conspecifics.

It is now clear that all TFs are not the same; the most useful data come from at least some categorization of tongue flicks during the observational stage (Allen et al., 1984). Eventually, the single frame quantative approach (Gove and Burghardt, 1983) might be necessary. However, iguanid tongue flicks have never been observed to have multiple oscillations (Gove, 1979) and thus are more stereotyped. Further, air tongue flicks are less common, and data on the item touched by the tongue are easily obtained.

The specific questions addressed here are whether green iguanas show reliable patterns of tongue flicking related to novelty in both laboratory and field environments, whether the various kinds of tongue flicks are affected in the same way, whether individuals consistently differed in the number and kinds of tongue flicking engaged in, whether other response measures are correlated with them, and whether rates of tongue flicking (TFR) show habituation over time. Although partially addressed in previous studies on diverse iguanids (see review in Simon, 1983), this is the first to consider all of them with comparably gathered data on any lizard species.

STUDY 1

A group of iguanas being maintained in our laboratory in a wooden enclosure had clearly outgrown their facility, but no alternative was available until we moved to a new building. Here a room with a large one-way glass wall was converted into a walk-in iguanarium complete with large branches, sun lamp, etc. When the animals were transferred to the new facility, systematic observations were instituted focusing on tongue flicking and locomotion. Initial observations showed marked elevations and subsequent declines in several behaviors. Here we report observations on movements and the rate and kinds of tongue flicks emitted.

Methods

The subjects were four iguanas hatched in Panama in May, 1976 and captured in June near Frijoles, Canal Zone. They were brought to Tennessee shortly thereafter and maintained in wood cages approximately 1 m^3. For many months prior to the transfer to the large room in November, 1980 the iguanas were housed in a custom built reptile cage 87 x 57 x 57 cm and heated with a 60 w incandescent bulb on a 12/12 h on-off cycle. A ceiling mounted 15 w fluorescent Vitalite provided more balanced illumination.

Room ambient temperature was held to 23-27°C; temperature in the box would rise to about 30°C during the day but higher near the screen enclosed 60 w bulb where the lizards often basked (up to 38°C). At the time of the move, their mean snout-vent length was 22.2 cm (s.d. 3.03, range 19.5-25.5 cm). Based on femoral pores, two appeared to be males and two females.

The room the lizards were transferred to measured 240 x 172 x 245 cm. The walls and ceiling were white, and one narrow end contained the door and a one-way glass wall (63 x 200 cm). Lighting was provided on a 12/12 h cycle by 4 fluorescent 40 w bulbs: two Vitalites, one UV light, and one regular cool white. Heat was provided by a central system maintained at 27°C. A 250 w heat lamp was suspended over a basking perch and was on a 10/14 h on-off cycle beginning one h after the room lights came on. A KAZ Dyna-Steam model 175 vaporizer was used to raise the relative humidity to around 50%. The result was a temperature gradient from 24-27 to 30-33°C.

The iguanas were able to access a number of temperature locations because a small tree and several branches spread from floor to ceiling. The tree branches were festooned with plastic foliage providing traction, hiding areas, and condensation surfaces for water droplets. The room was misted once daily.

Food was provided daily in a bowl placed on a wood shelf affixed to the rear wall of the room. The same bowl and diet were used as when the animals were in the small cage. The food was primarily mixed vegetables (mostly greens) and occasionally mealworms (*Tenebrio molitor* larvae). The floor of the room was covered with about 5 cm of shredded paper.

Observations were recorded in 60 sec intervals on a checksheet using 15 min focal observation periods by the same person (HF) sitting in the darkened hallway outside the one-way wall. Each animal was observed in turn in a randomly determined manner once each day for 10 days. Head movements, locomotion, headbob displays, and other responses were recorded along with TFs directed at various objects (branch, leaf, wall, paper, food bowl), other iguanas, or air.

The animals were introduced into the room individually the first 2 days and placed back into the smaller home cage after the observations were completed on the first day. Observations began after a 5 min period. Animals were introduced and observed in turn; the earlier observed animal was not removed. One animal could not be observed the first day, and thus two 15 min blocks at several hour intervals were used on the following day. After the second day, the iguanas remained in the room and were observed without any disturbance to them.

Head movements were recorded when the iguana moved its head in any direction, as in head cocking. Barely noticeable jerks, quivers, or eyelid blinks were not recorded. *Locomotion* was recorded when the animal moved its body from one place to another as in climbing, walking, or chasing. Since in this study, our main interest was in tongue flicking, head and locomotory movements were checked once or twice in an interval depending on their frequency in a category ordinal scale. Tongue flicks were counted as *substrate* or *social* if the tongue tips actually contacted an inanimate object or other iguana, respectively, or if the head was so close to the target (< 2 cm) that contact could have been made. *Air* flicks were recorded when the animal's mouth opened and the tongue was visibly extruded and objects the snout faced were more than 2 cm distant. *Mouth open* was scored if the iguana opened its mouth without visibly extruding the tongue. All statistics were calculated using Statgraphics (v.1.1) on an IBM PC.

Results

Although sometimes longer samples were collected, only the first 15 min of observation each session were used for each animal. The inanimate object TFs were grouped together (most were to leaves and branches) for analysis and called substrate TFs here. Here and throughout the paper, a square root (+ .5) data transformation was used for all parametric tests if the variances differed significantly (Hartley's test).

Two-way analyses of variance (d.f. = 9, 27) revealed significant differences among days in substrate TFR ($F = 8.60$, $p < .0001$), air TFR ($F = 16.02$, $p < .0001$), and locomotion ($F = 4.82$, $p < .001$). Head movement ($F = 1.02$), mouth open ($F = 1.31$), and social TFR ($F = 1.66$) did not differ across days. There were no significant subject effects (d.f. = 3, 27) for any behavior.

There was a consistent decline in substrate TFR and air TFR the first three days [$p < .05$, Least Significant Difference (LSD) test], after which the level of these behaviors became rather uniform (Table 1). Locomotion declined also, but more slowly: a significant ($p < .05$) decline did not occur until Day 4, although locomotion continued to decline afterwards. Head movements were rather constant throughout all 10 days and showed no particular pattern except that on day 1 the variance was the least. Mouth open and social TFs were both emitted at low rates with most given the first three days.

The mean rates for substrate and air TFs over the ten days were highly correlated (product-moment $r = .995$, $p < .01$). Both of these were somewhat less correlated with locomotion (object, $r = .823$, $p < .01$; air, $r = .822$, $p < .01$). Head movements and locomotion were also correlated ($r = .684$, $p < .05$).

The above correlations imply that locomotion and both air and substrate TFRs are tied together in some way. Casual observation of lizards and snakes does often reveal that tongue flicking is related to locomotion. In the habituation design used here, however, such a correlation may be due to a common response of both TFRs and locomotion to a novel environment. To elucidate this issue, we took advantage of the fact that there were no subject effects and performed a correlation matrix on all six behaviors ($n = 40$). Although the correlations were lower, the significance levels were comparable. Nine of the 15 correlations were significant (6 at $p < .01$). Locomotion was correlated with substrate, air, and social TFRs, substrate TFR with air and social TFRs, social TFR with air TFR, mouth open, and head movement, and air TF with mouth open. But partial correlations that controlled for the remaining four variables showed a different result. Of all the partials, only two were significant: substrate and air TFRs ($r = .739$, $p < .01$) and mouth open and social TF ($r = .345$, $p < .05$).

After the first three days, most of the few tongue touches occurred when the lizards came to the ground or approached the food bowl. Displays were seen frequently involving head bobs, lateral body flattening, dewlap extension, and lunging at one another. These seem to involve dominance but are not further described here. In terms of the present behaviors, the 4 lizards were rather uniform.

Discussion

This set of observations demonstrates that iguanas transferred into a novel environment showed elevated rates of locomotion and object and air tongue flicking. Indeed, the mean total TFR the first day was close to 5.

Table 1. Mean responses by four green iguanas in a novel, much larger enclosure (responses/min[1]).

Day	Response					
	Substrate TF	Air TF	Social TF	Mouth Open	Head Move	Locomotion
Nov. 16	3.90^{a}	0.63^{a}	0.08^{a}	0.02^{a}	0.55^{a}	0.64^{a}
17	2.28^{a}	0.42^{b}	0.00^{a}	0.02^{a}	0.52^{a}	0.55^{ab}
18	0.47^{b}	0.07^{c}	0.03^{a}	0.02^{a}	0.72^{a}	0.45^{abc}
19	0.35^{b}	0.04^{c}	0.00^{a}	0.00^{a}	0.44^{a}	0.28^{bcd}
20	0.42^{b}	0.04^{c}	0.00^{a}	0.00^{a}	0.30^{a}	0.28^{cd}
21	0.29^{b}	0.04^{c}	0.02^{a}	0.00^{a}	0.54^{a}	0.29^{bcd}
23	0.08^{b}	0.00^{c}	0.00^{a}	0.00^{a}	0.20^{a}	0.04^{de}
24	0.09^{b}	0.02^{c}	0.00^{a}	0.00^{a}	0.32^{a}	0.00^{de}
25	0.30^{b}	0.00^{c}	0.00^{a}	0.00^{a}	0.43^{a}	0.18^{cde}
26	0.15^{b}	0.00^{c}	0.02^{a}	0.03^{a}	0.49^{a}	0.12^{cde}

[1]Over days, response means with no superscript in common are significantly different ($p < .05$, LSD test).

These responses then habituated so that after 4 days TFRs stabilized, although locomotion continued to decline.

A question arises as to the relationship among these behaviors. If substrate TFR is related to exploration and air TFR is related to stress or motivational conflict (Greenberg, 1985), then the close tie between the behaviors seen here is explicable even if the underlying mechanisms differ. In addition, locomotion and TFR are not as closely linked as descriptive studies (e.g., Duvall et al., 1985) or comparison of mean rates for subjects across days suggest. Furthermore, the rates of tongue flicking were much higher than those previously recorded in iguanid lizards.

STUDY 2

The previous experiment established that iguanas several years old when released into a large, rich environment showed a temporary increase in the rate of tongue flicking. The feasibility of recording different kinds of tongue extrusions was also shown. However, the study was based on only a few animals and other important factors intruded. The animals were not all together during the initial observations, the room differed significantly in many respects from the home cage, no home cage controls were run, and so on. The second study involved yearling iguanas reared from several weeks after hatching in a wooden cage similar to the animals above. Following baseline observations, the iguanas were either released into a similar, but novel, environment or were reintroduced into their home cage. This added a handling control. Further, refinements in the recording of head movements and locomotion were incorporated, as well as a much shorter (10 sec) interval.

Methods

The subjects were nine yearling iguanas approximately 14 months old in July, 1983. All the animals were from the same area in Panama near Barro Colorado Island; 8 were laboratory hatched and the ninth, the largest, was a hatchling marked on the Slothia nest site off BCI and recaptured several days later with the right front limb almost completely missing, presumably resulting from an encounter with a predator. This animal was agile and dominant by most measures.

The lizards arrived in Tennessee on June, 1982 and were raised in a box similar to the animals in Study 1 except that it was slightly smaller, measuring 72 x 57 x 57 cm. The novel environment was set up the same way except that it was the same size as the box in Study 1 and had the interior painted a light tan rather than green.

Both environments had a gravel substrate and a standard concrete block against the rear wall (40.6 x 20.3 x 20.3 cm) under a light fixture containing a 75 w incandescent bulb enclosed in a screen. A fluorescent fixture was mounted in front of this in both cages as in the box used in Study 1; however, UV bulbs rather than Vitalites were used as we had since established that the latter provided insufficient UV radiation. Both enclosures had branches approximately 2 cm in diameter crisscrossing the enclosure in a similar pattern, were on the same light/dark cycle, and were maintained at similar temperatures (approximately 29°C) as measured by a maximum - minimum thermometer in the left rear corner. Temperatures on the block were approximately 4°C higher and even more on the screen (where the animals often climbed during the morning and periodically throughout the day), up to a maximum of 38°C.

The observation procedure was to note the behavior of each iguana using 5 min focal animal sampling with 10 sec intervals. A total of 8 such sessions were made after preliminary sessions involving no disturbance to check out the behavior category system (Table 2).

The behaviors recorded were similar to those in Study 1 except that the place on an iguana where social TFs were received or delivered was noted. All head movements, posture changes, and locomotions were recorded as they took place in each 10 sec interval. All the responses were events except locomotion which often was continuous throughout the 10 sec interval. In these cases, a maximum of 5 checks were recorded in an interval (one per 2 sec). Thus a maximum rate of 30/min was possible for locomotion.

For the first three sessions, the order in which the 9 animals were observed was varied; after the third session (and during all the removal and reintroduction tests to both novel and home cage) the same randomised order was used. We decided to do this because the behavior of the animals seemed to shift slightly as time after reintroduction lengthened. The total time for a session was about an hour (5 min/subject plus a minute or two between subjects). Animals introduced to the novel cage were allowed to remain in it for a maximum of two hours before being moved back together to the home cage. There were two home cage and three new cage introductions.

In the new and home cage tests, the animals were quickly hand captured and placed in a 60 x 30 x 30 cm glass aquarium with a mesh screen cover. The sides of the aquarium were covered with brown paper to reduce visual input and thus more quickly calm the animals. Further, the room lights were off and the windowless lab was thus dark except for the light from the two enclosures themselves. Such illumination was adequate for data recording on checklists. The animals were held for 15 min in the aquarium at about 25°C before being again quickly transferred by hand into the home or novel cage. After introduction, 5 min elapsed before the start of focal observations.

As in Study 1, all observations were made primarily by one observer (GMB), and secondarily by another (BAA), who recorded the behavior as called out by the main observer and also observed the focal animal when possible. Agreement was very high on all behaviors except posture change, which was thus eliminated from subsequent analysis. The results were analyzed by converting the raw scores to rate means for each animal and each session.

Results

Two-way analysis of variance revealed that condition (d.f. = 7, 56) had a significant effect on all behaviors recorded. However, the magnitude of the effect did vary: locomotion ($F = 15.05$, $p < .0001$), head movement ($F = 8.31$, $p < .0001$), substrate TFR ($F = 4.07$, $p < .002$), air TFR ($F = 3.46$, $p < .01$), wall TFR ($F = 25.01$, $p < .0001$), and social TFR ($F = 3.28$, $p < .05$). There was no significant subject effect for any measure.

The TFRs are given for substrate, wall, and air in Table 2. The wall flicks are presented separately because they constituted a much larger percentage of total TFR than in Study 1. This appears related to the much smaller dimensions of the novel enclosure. Social TFR, head movements, and locomotion are also given in Table 2. Mouth open occurred too rarely to analyze.

During the two prefeeding undisturbed observations on July 13 and 14, virtually no behaviors were emitted except head movements. After food was

Table 2. Mean responses by nine yearling green iguanas observed in a social setting (responses/min). N = number of animals showing behavior at least once.[1]

RESPONSE	Day in July, 1983: 13 (Undisturbed)	14 (Undisturbed)	14 (After Feeding)	15 (Home Cage I)	18 (New Cage I)	20 (Home Cage II)	22 (New Cage II)	26 (New Cage III)
Substrate TF	0.00^{a}	0.02^{a}	0.36^{ab}	0.36^{ab}	2.40^{c}	0.49^{ab}	0.96^{bc}	0.40^{ab}
N =	0	1	4	7	7	5	4	4
Wall TF	0.00^{a}	0.16^{a}	0.18^{a}	0.16^{a}	2.27^{c}	0.04^{a}	0.49^{b}	0.11^{a}
N =	0	2	2	2	9	1	6	2
Air TF	0.02^{a}	0.04^{a}	0.13^{a}	0.33^{ab}	0.84^{c}	0.38^{ab}	0.51^{bc}	0.36^{ab}
N =	1	2	3	5	8	4	7	3
Social TF	0.00^{a}	0.00^{a}	0.04^{ab}	0.38^{bcd}	0.38^{cd}	0.13^{abc}	0.51^{d}	0.31^{abcd}
N =	0	0	1	3	8	3	7	3
Head movement	1.82^{a}	1.62^{a}	3.67^{ab}	6.00^{bc}	9.07^{c}	2.38^{a}	6.18^{bc}	2.58^{a}
N =	7	9	9	9	9	7	9	6
Locomotion	0.07^{a}	0.51^{a}	0.67^{a}	0.51^{a}	10.51^{c}	0.47^{a}	4.07^{b}	1.48^{ab}
N =	1	2	5	5	9	4	6	4

[1]Over days, response means with no superscript in common are significantly different ($p < .05$, LSD test).

introduced on July 14, there were slight but nonsignificant increases in a few responses. On the following day, when the lizards were removed from the home cage, held in an aquarium for 15 min, and then reintroduced, there was again little change. However, three days later when the lizards were introduced into the novel cage for the first time, there were increases in all behaviors except social TFR; rates of substrate, air, and wall tongue flicking, air and locomotion significantly so. Two days later, when the animals were reintroduced into the home cage, the rates of all responses were about the same as during the first home cage reintroduction except for head movements, which were significantly reduced. Two days later, during the second new cage reintroduction, all behavior rates were higher than both home cage introductions, although only locomotion and wall TFR on both days, and head movements on the second, were significantly different. On the other hand, all behaviors except social tongue flicks were lower in the second new cage introduction than the first one. The third new cage introduction four days later showed a further reduction in rates of all behaviors, and no behavior measure was significantly different from the second home cage introduction.

Comparing mean rates of behavior by condition showed that the rate of substrate tongue flicking was highly correlated with air TFR ($r = .929$, $p < .01$), wall TFR ($r = .965$, $p < .01$), head movements ($r = .872$, $p < .01$), and locomotion ($r = .977$, $p < .01$). Social TFR was not significantly correlated with either substrate TFR or locomotion. A correlation matrix of all six behaviors on the pooled data ($n = 72$) showed lower but significant positive correlations among all the measures: 14 out of 15 at $p < .01$). But a matrix of partial correlations among the six measures resulted in a greatly reduced number (7) of significant relationships. Locomotion was only correlated with wall TFR, by far the highest partical correlation ($r = .693$, $p < .01$). Substrate TFR was only correlated with air TFR ($r = .373$, $p < .01$) and head movement ($r = .327$, $p < .01$). Social TFR was correlated with air TFR ($r = .395$, $p < .01$), head movement ($r = .348$, $p < .01$) and negatively correlated with wall TFR ($r = -.267$, $p < .05$). Wall TFR was also correlated with head movement ($r = .317$, $p < .01$).

Discussion

This study replicated the habituation effects seen in Study 1 using more subjects and a home cage control. There was a tighter association between the various behaviors here, which may all have been influenced by a generalized arousal due to the handling as well as the more precise recording of head and locomotory movements. The nonassociation between locomotion and social tongue flicking was seen again. Furthermore, locomotion was only correlated with wall TF. The latter often seemed an escape response in this experiment. In the novel cage, the iguanas would often come to the mirrored front of the enclosure and walk back and forth tongue flicking the glass. However, they also initially climbed all over the cage, onto the screen sides and branches, in what appeared to be an exploratory fashion, especially the first two days. Head movement was associated with all TFs but air TF. If air TF is a sign of stress or conflict for *Anolis*, as postulated by Greenberg (1985), then it might be more prevalent when the animal is motionless.

The animals initiated movements earlier in the novel cage in this study than in Study 1, where they were transferred into a larger cage. This may be related to the more novel and inhibiting nature of the latter. Similarly, the negative relationship between social and wall TFRs may indicate that the social TFs are not likely to occur when a lizard is actively engaged in either locomotion or tongue flicking at the perimeter of its environment. Indeed, in the latter situation (wall TF), social TFs appeared inhibited.

Do the observations on tongue flicking in iguanas transferred to novel environments have any counterpart in the field? Is there even a meaningful parallel? Even being able to observe bouts of tongue flicking in the field is difficult in reptiles. Furthermore, such behavior in lizards and snakes is often related to foraging or courtship, the prior use of the area by the animals is unknown, and the observation of different animals in different habitats confounds interpretation.

Gravid female iguanas migrate to Slothia, a 0.3 hectare islet adjacent to Barro Colorado Island, Panama in order to nest, and have been observed doing so for many years. Tongue flicking has been noted (e.g., Rand, 1968), but no systematic observations of TFR have been made. During the nesting season in 1984, such observations were performed on known marked females to assess rates and kinds of TFs and changes in TFR as a function of phase of nesting: exploration, digging, or filling. It is important to realize that 1) iguanas do not live on Slothia; thus they have not been in the area for at least 11 months, 2) iguanas in Panama are primarily arboreal and thus rarely traverse open ground such as characterizes the clearing on Slothia, 3) iguanas do not eat during nesting and thus food-related TFs are not involved, 4) the prior nesting history of virtually all animals coming to Slothia for the prior four years was known (Bock, 1984), and 5) observation conditions were such that almost all details of behavior could be seen at close range.

Methods

Observations of females active in the approximately 6 x 8 m clearing on Slothia were made between February 14 and March 16, 1984. A permanently staked 5 x 5 m area in front of a wooden blind facilitated recording of iguana positions. Slothia has been used for communal nesting by iguanas (as well as by *Crocodylus acutus* and *Pseudemys scripta*) since at least the 1930's, and iguana behavior there has been studied since the 1960's. (Fuller descriptions are found in Rand, 1968; Rand and Rand, 1976; Drummond and Burghardt, 1982; and Bock, 1984).

Observations were made by the second author (BAA) from inside the blind through a one-way glass window and burlap curtain using close-focusing binoculars as needed. Most of the iguana activity took place within 8 m of the observer. As the observer was recording many facets of nesting behavior for a different project, the observations reported here were made when activity in the clearing was not intense. The time, shade temperature in blind, weather (cloudy, rain, partial or full sun), location of the iguana and whether she was in the direct sun, presence of other animals, possible disturbances (e.g., boat passing close by), and other data were recorded at the beginning of a session and periodically throughout. Being the dry season, rain rarely fell and the sun was normally out. Even during mid-day, however, the clearing and most iguanas were in dappled rather than in unblocked sunlight. Iguanas are active on the surface of the clearing only during daylight hours. The clearing was continually monitored every day during the entire nesting period and several days before and after.

Gravid iguanas tend to pause at the clearing edge for varying periods of time before entering the clearing, giving the observer time to identify the iguanas by means of numbers previously painted on their sides or from other distinctive characteristics. Only clearly identified animals were selected; consequently an animal was not observed during its first visit to the clearing, although sometimes on the first day. Iguanas may visit a clearing many times over a week or more before laying. Visits to the clearing, especially during the exploratory phase before active digging of burrows, may be numerous and brief. Animals may then not appear for several days or leave the islet entirely, presumably going elsewhere to lay (Bock et al., in press). For the animals observed in this study, it is known which ones eventually nested on Slothia.

Observations were started within a minute after an animal entered the clearing. Thirtyone 10 sec intervals, signaled by an automatically repeating countdown timer, were recorded using focal sampling on the selected animal. A checklist was used to record a series of behaviors chosen on the basis of reconnaissance observation and the preceding laboratory studies.

Head movements and locomotion were recorded and defined as in Study 2. Behavior included body movements such as standing, sitting, rearranging legs, digging, and filling that did not involve locomotion. Separate counts were kept of air, substrate (soil or leaf litter), and elevated object (usually scattered herbs, vines, small stumps, or stake markers) TFs. Mouth open, an agonistic act often accompanied by audible expulsion of air during encounters between nesting females, was rare as were social TFs. This is to some extent due to the fact that large numbers of iguanas were not in the clearing simultaneously; the logistics of the observer's need to observe general activity in the clearing limited her observations to those periods when only one or a few iguanas were present.

The three phases or stages of nesting were characterized in the following ways. During exploration, the female typically makes short visits to the clearing during which she walks about, basks, peers into the existing excavations, and engages in sporadic short digging bouts in one place then another. This phase may last up to a week. During digging, the female directs her attention almost entirely to lengthy digging bouts although digging sites may change. Typically this stage takes a day or two. Laying is underground in burrows; the female may stay underground overnight, emerging the following morning. The final phase is filling, which takes place immediately or within hours after emerging from laying. During this phase, the female intensively fills her own and even nearby holes by pushing dirt forward with her front legs and packing it down with her snout. During this phase, the female is remarkably oblivious to disturbances in general. When completed, the filled nest entrance is well hidden. Filling may last from a few hours to two days. It should be emphasized that during any stage the female may leave the clearing for hours. During exploration especially, the female may skip one or a few days with no appearances at all.

Results

A total of 35 samples were collected. Iguanas frequently failed to stay within observational range for the full specified period, but only two samples involved less than 6 intervals. Removing these, as well as two samples for an animal that did not move or engage in any behavior (e.g., digging) during the sampled period, left 31 samples on 22 different individuals. The median number of intervals in a sample was 30 (range of 7-31, mean = 26, s.d. = 7.5). Although several (8) animals were sampled more than once, on no response measure was there a significant difference based on sample number (Kruskal-Wallis one-way ANOVA), and thus the data were pooled in the following analyses. Since activity and also tongue flicking in lizards is sensitive to temperature (Cooper and Vitt, in press), shade air temperature differences across the three stages were also looked at but none were found (Kruskal-Wallis one-way ANOVA, $H = 2.32$, $n = 29$, $p = .31$). The median and mean temperature was 31°C (range of 28-34, s.d. = 1.6). Data on the samples as well as the behavioral results are given in Table 3.

Table 3. Nesting female iguana data on Slothia (mean rate/min). Number of samples = n.

Stage		n	Locomotion	Head Move	Behavior	Substrate TF	Object TF	Air TF	Shade Temp (°C)
Explore		19	3.99	7.02	1.86	2.10	0.54	0.57	31.0 (n = 18)
laid eggs in 1984	yes	11	5.12*	8.16	2.20	2.03	0.50	0.71	
	no	8	2.45	5.46	1.39	2.19	0.58	0.39	
previously laid eggs	yes	3	6.75	10.78*	2.12	2.25	0.44	1.35*	
	no	16	3.48	6.32	1.81	2.07	0.55	0.42	
days in clearing	1-2	12	4.59	7.40	1.75	2.78*	0.70	0.72	
	3-6	7	2.97	6.38	2.04	0.94	0.26	0.31	
Dig		7	3.11	8.31	3.35	1.62	0.47	0.23	30.7 (n = 6)
Fill		5	1.74	10.03	6.54	0.93	0.00	0.12	32.0

*$p < .05$, t-test.

A one-way Kruskal-Wallis analysis of variance was run on each of the 6 response measures as a function of stage (Table 3). Locomotion ($p = .12$) and head movement ($p = .14$) did not reach significance although locomotion decreased markedly while head movement increased over the three stages. Behavior, however, did show a significant increase ($p < .02$) in the digging and filling stages. Air ($p < .05$) and object ($p < .05$) TFRs declined significantly whereas stage differences in substrate TFR were not significant ($p = .25$). Substrate TFR was the highest TF response throughout, however.

Nesting iguanas in various stages of their nesting sequence thus show quantitative changes in two types of tongue flicking and behavior. Obviously the kinds of behavior engaged in differed across stages by definition. But the rate changes found here support the casual impression that filling females are most active. Whereas subdivided body movements and TFs showed significant stage effect, this was not true for pooled behaviors; total TFs ($p = .09$), combined object and substrage TFs ($p = .10$), nor total activity as given by the sum of locomotion and behavior ($p = .54$).

Many of the iguanas that explored the Slothia clearing, even remaining for several days, did not go on to lay their eggs on Slothia. There were sufficient samples in this stage to do a preliminary analysis of behavior during exploration as a function of either subsequent egg-laying or oviposition on Slothia in one of the four prior years. Using t-tests, controlling for equal variances, it was found that exploring iguanas that went on to lay in the clearing showed significantly (d.f. = 17) more locomotion ($t = 2.28$, $p < .05$), marginally more head movements ($t = 1.86$, $p = .08$) and no significant changes in behavior or air, substrate or object TFRs than those who left Slothia before laying (Table 3).

Samples from iguanas that had laid in the clearing in prior years (d.f. = 17) showed significantly more head movements ($t = 2.39$, $p < .05$), air TFs ($t = 3.44$, $p < .05$), marginally more locomotion ($t = 2.01$, $p = .06$), and no significant differences in behavior, substrate TFR, or object TFR than those on Slothia for the first time in at least four years (Table 3).

Since there were few repeated samples available for analysis, it is not possible to demonstrate habituation of responses during exploration in a manner similar to Studies 1 and 2. In one subject for which three samples were available, there was a marked decline in air and substrate TFRs from the first to the fourth day that she visited the clearing. Since we had records of which days the animals had visited the clearing, it was possible to evaluate the various response rates as a function of clearing day. Pooling clearing days 1 and 2 and days 3 through 6 (Table 3) showed that a significant decline was observed in substrate TFR and substantial but not significant declines in locomotion, object TFR, and air TFR.

A correlation matrix of the 6 response measures over all stages ($n = 31$) showed that only 3 of the 15 correlations were significant: locomotion with air TFR ($r = .701$, $p < .01$) and substrate TFR ($r = .475$, $p < .01$) and air TFR with substrate TFR ($r - .562$, $p < .01$). Temperature was not significantly correlated with any response measure; the strongest correlation was a negative one with locomotion ($r = -.343$, $p < .1$). A correlation matrix partialing out the effects of the other four response measures plus temperature resulted in only two significant partial correlations: air TFR with both locomotion ($r = .620$, $p < .01$) and substrate TFR ($r = .417$, $p < .05$). Two negative partial correlations were marginal: air TFR with behavior ($r =\ .343$, $p < .1$) and locomotion with temperature ($r = -.337$, $p < .1$).

Perhaps, however, more significant correlations would result if only a single stage was looked at. Explore was the only stage with sufficient observations. A correlation matrix of the 6 response measures ($n = 19$) led to only 2 out of 15 being significant: air TFR with locomotion ($r = .712$, $p < .01$) and substrate TFR ($r = .582$, $p < .01$). Behavior with locomotion ($r = .455$, $p < .1$) and head movement ($r = .433$, $p < .1$) were just short of significance. Head movement was negatively related to temperature ($r = -.468$, $p = .05$). A partial correlation matrix ($n = 18$) identical to that run with the entire data set led to strong correlations between locomotion and both air TFR ($r = .800$, $p < .01$) and behavior ($r = .705$, $p < .05$). Air and substrate TFRs had a larger partial correlation than in the complete data set, but it was no longer significant due to the smaller sample size ($r = .526$, $p < .1$). Head movement was marginally negatively correlated with temperature ($r = -.516$, $p < .1$).

Discussion

These results add to the conclusions drawn from the laboratory studies. Within the explore stage, substrate TFR declined over days the iguanas had been seen in the clearing; although the data were quite variable, suggestive declines were observed in locomotion and air and object TFRs as well. Locomotion in the clearing declined and behavior increased with successive nesting stages. Air TFR was strongly associated with locomotion; this might indicate that locomotion in the clearing is a potentially dangerous activity for nesting female iguanas and that air TFs help obtain some chemosensory information, result from stress or conflict, or act as a signal to other iguanas about their "security" in the clearing (c.f., Greenberg, 1985, on the role of foliage movement on air TFR in *Anolis*). These are not mutually exclusive possibilities, of course. Observations on social groups and more continuous observation of individuals would be valuable.

Iguanas who went on to lay in the clearing moved about somewhat more in the explore stage, but otherwise showed no significant differences from iguanas who left without laying. But iguanas that had previously laid on Slothia evidenced both more head movements and air TFs. It might be thought that iguanas nesting on Slothia for the first time would be the more cautious if any effects of prior experience were present. However, in the years prior to 1984, a nest guarding crocodile charged at and chased the iguanas repeatedly, even catching and killing them on occasion (Dugan et al., 1981). This may have led to the hesitation and alertness indicated by head movements, typically head cocks and head nods (not display bobs), occurring when the animal ceased locomotion. The increased air TFR in these animals supports the view that stress selectively potentiates or facilitates this form of TF (Greenberg, 1985), and suggests that the iguanas were in fact demonstrating a type of memory. During the period encompassed by the years 1979-1984, the numbers of iguanas coming to Slothia declined from those in earlier years when interference from the crocodile was either absent or infrequent (c.f., Rand, 1968). It should also be noted, however, that noosing and marking activities in prior years may also have led to the behavior changes in returning females (Bock et al., in press).

The partial correlation results in the explore stage show that locomotion was highly correlated with both behavior and air TFR while behavior and air TFR were themselves significantly negatively correlated. This seemingly paradoxical result may reflect the fact that the simple correlation between behavior and air TF was essentially nil ($r = .056$); removing the effect of other variables revealed a negative relationship. Perhaps both air TFs and behavior are related to differently elicited locomotion. That is, locomotion in the explore stage might, on the one hand, be associated with moving around the clearing, sampling possible nest sites and partially completed holes as well as holes being worked by other females, and on the other hand, be associated with cautious movement, frequent hesitation, and a high sensitivity to external sounds and other disturbances, leading to air TFs. This interpretation is supported by the clearing day analysis of the explore stage (Table 3): air TFR declined and behavior increased. This result points out the need for caution in interpreting partial correlations involving the same "response."

GENERAL DISCUSSION

The results for all three studies on different size and age groups of iguanas show that habituation to novel environments as measured by both activity and tongue flicking is rapid. However, the different kinds of tongue flicking diminish at different rates, and social tongue flicks show the least regular response. This is understandable since it is the environment that is new, not the conspecifics, in all but the last study.

Support for the role of the vomeronasal system in tongue flicking is supported by the findings of Distel (1978) that electrical stimulation of the nucleus sphericus in the brain of *Iguana iguana* leads to tongue touching behavior. The question is why does the increase occur. Can taste be the modality involved? The experiments here, along with others on the response of iguanid lizards to novel environments, document that TFR is related to exploratory behavior in lizards as well as snakes (Simon, 1983). In spite of Schwenk's (1985) evidence for the broader distribution of taste bud appearing structures in lizard tongues than was earlier appreciated, the fact remains that, 1) the evidence for taste in lizards is experimentally weak and involves ingestion only, and 2) taste appears unlikely in aerial tongue flicking.

A major conclusion we draw from these data is that there is no simple or necessary relationship among or between various degrees of body movement or kinds of tongue flicks in iguanas. Combined with our prior results on snakes (e.g., Allen et al., 1984), caution must be exercised in the collecting and analyzing of tongue flick data, especially outside of clearly defined contexts such as feeding or courtship. Our evidence showing a dissociation between locomotion and tongue touching is congruent with a study on rats showing a dissociation between locomotion and investigative behavior (snout and paw contact of stimuli or close orientation of the head towards a stimulus; Leyland et al., 1976).

In Study 1, locomotion did not correlate with substrate or air TFRs, but the two tongue flick types were related. In Study 2, where wall TFs were separated from substrate TFs, locomotion was correlated only with wall TFR but substrate and air TFRs were still correlated. In Study 3, where object TFs were separated from substrate TFs, locomotion was correlated with air TFR and substrate and air TFRs were still correlated. Such results indicate that locomotion may involve different functions and a variable relation with tongue flicking. The consistent positive correlations between air and substrate TFRs suggests a tighter relationship that deserves more careful scrutiny.

A final conclusion from these results is that tongue flicking in iguanas is a conspicuous and often frequent response systematically sensitive to different conditions in both laboratory and field, deserving of more study. If the responses to novel environments noted here are at all akin to "curiosity," then the reputed low levels of this phenomenon in reptiles (c.f., Glickman and Sroges, 1966) needs reassessment.

ACKNOWLEDGMENTS

We thank B. Dugan and B. Bock for obtaining the hatchling iguanas, A. S. Rand for essential field support and advice, and the students who assisted in the care of the captive iguana colony over many years. This work was supported by National Science Foundation grants BNS 78-14196 and BNS 82-17569 awarded to G. M. B. David Duvall, Neil Greenberg, A. Stanley Rand, and Gordon Rodda made useful suggestions on earlier drafts of this manuscript.

REFERENCES

Allen, B. A., Burghardt, G. M., and York, D. S., 1984, Species and sex differences in substrate preference and tongue flick rate in three sympatric species of water snakes (Nerodia), J. Comp. Psychol., 98:358.

Bissinger, B. E., and Simon, C. A., 1979, Comparison of tongue extrusions in representatives of six families of lizards, J. Herpetol., 13:133.

Bock, B. C., 1984, "Movement Patterns Relative to Nesting Site Locations in a Population of Green Iguanas (Iguana iguana) in Panama," Unpublished Ph.D. Dissertation, University of Tennessee, Knoxville.

Bock, B. C., Rand, A. S., and Burghardt, G. M., in press, Seasonal migration and nesting site fidelity in the green iguana, in: "Migration: Mechanisms and Adaptive Significance," M. A. Rankin, ed., University of Texas Press, Austin.

Burghardt, G. M., 1970, Chemical perception in reptiles, in: "Communiction by Chemical Signals," Vol. 1, J. W. Johnston, Jr., D. G. Moulton, and A. Turk, eds., Appleton-Century-Crofts, New York.

Burghardt, G. M., 1980, Behavioral and stimulus correlates of vomeronasal functioning in reptiles: feeding, grouping, sex, and tongue use, in: "Chemical Signals: Vertebrates and Aquatic Invertebrates," D. Muller-Schwarze and R. M. Silverstein, eds., Plenum, New York.

Burghardt, G. M., and Rand, A. S., eds., 1982, "Iguanas of the World: Their Behavior, Ecology, and Conservation," Noyes, Park Ridge, New Jersey.

Burghardt, G. M., Greene, H. W., and Rand, A. S., 1977, Social behavior in hatchling green iguanas: Life at a reptile rookery, Science, 195:689.

Chiszar, D., Carter, C., Knight, L., Simonsen, L., and Taylor, S., 1976, Investigatory behavior in the plains garter snake (Thamnophis radix) and several additional species, Anim. Behav. Learn., 4:273.

Cooper, W. E., Jr., and Vitt, L. J., 1985, Thermal dependence of tongue-flicking and comments on use of tongue-flicking as an index of squamate behavior, Z. Tierpsychol., in press.

De Fazio, A., Simon, C. A., Middendorf, G. A., and Romano, D., 1977, Iguanid substrate licking: A response to novel situations in Sceloporus jarrovi, Copeia, 1977:706.

Distel, H., 1978, Behavioral responses to the electrical stimulation of the brain in the green iguana, in: "Behavior and Neurobiology of Lizards: An Interdisciplinary Colloquium," N. Greenberg, and P. D. MacLean, eds., DHEW publ. (ADM) 77-491, Rockville, Maryland.

Distel, H., 1982, The behavioral inventory of the green iguana, Iguana iguana, in: "Iguanas of the World: Their Behavior, Ecology, and Conservation," G. M. Burghardt and A. S. Rand, eds., Noyes, Park Ridge, New Jersey.

Drummond, H. D., and Burghardt, G. M., 1982, Orientation in dispersing hatchling green iguanas, Iguana iguana, in: "Iguanas of the World: Their Behavior, Ecology, and Conservation," G. M. Burghardt and A. S. Rand, eds., Noyes, Park Ridge, New Jersey.

Dugan B. A., Rand, A. S., Burghardt, G. M., and Bock, B. C., 1981, Interactions between nesting crocodiles and iguanas, J. Herp., 15:409.

Duvall, D., 1979, Western fence lizard (Sceloporus occidentalis) chemical signals. I. Conspecific discriminations and release of a species-typical visual display, J. Exp. Zool., 210:321.

Duvall, D., King, M. B., and Gutzwiller, K. J., 1985, Behavioral ecology and ethology of the prairie rattlesnake, Nat. Geogr. Res., 1:80.

Evans, L. T., 1959, A motion picture study of maternal behavior in the lizard Eumeces obsoletus Baird and Baird, Copeia, 1959:103.
Gillingham, J. C., and Clark, D. L., 1981, Snake tongue flicking: Transfer mechanisms of Jacobson's organ, Can. J. Zool., 59:1651.
Glickman, S. E., and Sroges, R. W., 1966. Curiosity in zoo animals, Behaviour, 26:151.
Gove, D., 1979, A comparative study of snake and lizard tongue-flicking with an evolutionary hypothesis, Z. Tierpsychol., 51:58.
Gove, D., and Burghardt, G. M., 1983, Context-correlated parameters of snake and lizard tongue-flicking, Anim. Behav., 31:718.
Gravelle, K., and Simon, C. A., 1980, Field observations on the use of the tongue-Jacobson's organ system in two iguanid lizards, Sceloporus jarrovi and Anolis trinitatis, Copeia, 1980:356.
Graves, B., and Duvall, D., 1985, Avomic prairie rattlesnakes (Crotalus viridis) fail to attack rodent prey, Z. Tierpsychol., 67:161.
Greenberg, N., 1978, Ethological considerations in the experimental study of lizard behavior, in: "Behavior and Neurology of Lizards: An Interdisciplinary Colloquium," N. Greenberg and P. D. MacLean, eds., DHEW publ. (ADM) 77-491, Rockville, Maryland.
Greenberg, N., 1985, Exploratory behavior and stress in the lizard, Anolis carolinensis, Z. Tierpsychol., 70:89.
Halpern, M., and Kubie, J. L., 1980, Chemical access to the vomeronasal organs of garter snakes, Physiol. Behav., 24:367.
Halpern, M., and Kubie, J. L., 1983, Snake tongue flicking behavior: clues to vomeronasal functions, in: "Chemical Signals in Vertebrates," D. Muller-Schwarze and R. M. Silverstein, eds., Plenum, New York.
Hammerman, D. L., and Thomas, R. F., 1967, Lingual premetamorphic papillae as larval taste structures in frogs, Nature, 215:98.
Kubie, J. L., and Halpern, M., 1979, The chemical senses involved in garter snake prey trailing, J. Comp. Physiol Psychol., 93:648.
Leyland, M., Robbins, T., and Iverson, S. D., 1976, Locomotor activity and exploration: The use of traditional manipulators to dissociate these two behaviors in the rat, Anim. Learn. Behav., 4:261.
Meredith, M., and Burghardt, G. M., 1979, Electrophysiological studies of the tongue and accessory olfactory bulb in garter snakes, Physiol. Behav., 21:1001.
O'Connell, B., Greenlee, R., Bacon, J., and Chiszar, D., 1982, Strike-induced chemosensory searching in old world vipers and new world pit vipers at San Diego Zoo, Zoo Biol., 1:287.
Rand, A. S., 1968, A nesting aggregation of iguanas, Copeia, 1968:552.
Rand, W. M., and Rand, A. S., 1976, Agonistic behavior in nesting iguanas: A stochastic analysis of dispute settlement dominated by the minimization of energy costs. Z. Tierpsychol., 40:279.
Schwenk, K., 1985, Occurrence, distribution and functional significance of taste buds in lizards, Copeia, 1985:91.
Simon, C. A., 1983, A review of lizard chemoreception, in: "Lizard Ecology," R. B. Huey, E. R., Pianka, and T. W. Schoener, eds., Harvade, Cambridge, Mass.
Simon, C. A., Gravelle, K., Bissinger, B. E., Eiss, I., and Ruibal, R., 1981, The role of chemoreception in the iguanid lizard Sceloporus jarrovi, Anim. Behav., 29:46.

LIZARD PHEROMONES: BEHAVIORAL RESPONSES AND ADAPTIVE SIGNIFICANCE IN SKINKS OF THE GENUS *EUMECES*

William E. Cooper, Jr.,
Department of Biology
Auburn University at
Montgomery, Alabama

Laurie J. Vitt
Department of Biology
University of California
Los Angeles, California

INTRODUCTION

Chemical communication in lizards is a fascinating and theoretically important topic, but our knowledge in this area has until very recently been limited to merely suggestive anecdotal material. Experimental studies have begun to reveal hitherto unsuspected capacities for social discriminations in iguanid lizards (reviewed by Simon, 1983), yet little or no experimental evidence is available on other families, many of which appear to make much more extensive use of the chemical senses. In this article, we review a research program on pheromonal communication in three congeneric skinks. The major emphases have been on social discriminations based on chemical cues, including choice of mates and agonistic opponents, on selective forces underlying the evolution of pheromonal communication (especially sexual selection), and on the anatomical site of pheromone production.

Long before any experimental studies of social chemical discrimination were attempted, it was apparent to lizard natural historians that skinks of the genus *Eumeces* use the chemical senses in social contexts (e.g., Noble and Bradley, 1933). Because tongue-flicking in squamate reptiles provides an external sign that chemosensory processes, in particular vomeronasal olfaction, are in use (Burghardt, 1980; Halpern and Kubie, 1980; and Simon, 1983), chemosensory investigation is easily detected in these and other lizards. Several members of *Eumeces* have been reported to repeatedly tongue-flick conspecifics during encounters in the field and laboratory. Both sexes of *E. laticeps* tongue-flick conspecifics of either sex during initial stages of social encounters, but males tongue-flick much more than females. Males extrude the tongue more frequently in the presence of females than of males. The tongue touches various parts of the female, most frequently the cloacal region, tail, flanks, and neck (Cooper and Vitt, 1984a,b, unpublished observations).

Similar behavior has been reported in *E. egregius* (Mount, 1963), *E fasciatus* (Fitch, 1954), *E. inexpectatus* (Perrill, 1980), *E. multivirgatus* (Everett, 1971), *E. obsoletus* (Fitch, 1955), and *E. septentrionalis* (Breckenridge, 1943). Furthermore, several investigators have reported observations suggesting that skinks may be able to locate conspecifics by following scent trails (Fitch, 1954, for *E. fasciatus*; Goin, 1957 and Cooper and Vitt, in press e, for *E. laticeps*). The possibility that some skinks may accurately discriminate conspecifics from closely related heterospecific skinks is suggested by Perrill's (1980) report that, in a few laboratory encounters, members of *E. inexpectatus* and *E. fasciatus* ignored each other after a few mutual tongue-flicks. More prolonged social interactions are typical in meetings between conspecifics. Chemical identification of eggs and offspring by adult female *Eumeces* has been suggested on the basis of laboratory observations and improperly controlled experiments (Evans, 1959; Noble and Mason, 1933; Noble and Kumpf, 1936), but should not be uncritically accepted without further careful analysis. Female *Eumeces* are among the few reptiles which remain with the clutch of eggs until hatching. At least in one species, *E. obsoletus*, captive females remain with the young several days after hatching and frequently tongue-flick the cloacal areas of the hatchlings. Females of this species also exhibit chemical examination of young emerging from eggs (Evans, 1959). Groves (1983) has recently suggested that brooding females use chemical cues to detect addled eggs, which are consumed preferentially to viable ones in *E. fasciatus*. Female *E. fasciatus* and *E. laticeps* sometimes retrieve displaced eggs by rolling them back to the nest. In both species females retrieve eggs of other conspecific females and of congeners, but do not retrieve eggs of lizards belonging to other genera. To test the hypothesis that females identify their eggs by odor, Noble and his colleagues placed normal eggs or eggs coated with paraffin in the cages of brooding females. Females retrieved uncoated, but not coated, eggs (Noble and Mason, 1933; Noble and Kumpf, 1936). The conclusion that females failed to retrieve paraffin-coated eggs due to a lack of detectable conspecific chemical odors is not warranted because the paraffin coating altered textural as well as olfactory properties of the eggs. Identification of eggs and young by chemical cues remains an interesting but unproven hypothesis.

Among the first experiments on social responses to odors of adult skinks were olfactometric tests of the abilities of representatives of two different subfamilies to detect airborne conspecific odors, to respond differentially to such odors of conspecific males and females, and to discriminate between conspecific and heterospecific airborne odors (Duvall et al., 1980). In a y-maze olfactometer, lizards were observed until they passed a criterion point indicating approach to one of two airstreams, either or both of which might carry lizard odors. *Eumeces fasciatus*, a member of subfamily Scincinae (Greer, 1970), approached the source of airstreams which had passed through jars containing conspecifics significantly more often than they approached airstreams of equal velocity which had passed through empty jars. Although no differences in approach frequency based on sex of experimental lizards or odor source lizards were observed, males extruded their tongues at higher rates than did females, as in *E. laticeps*, and also lowered their snouts more often during locomotion. Such snout-dipping behavior is considered another possible sign that chemical stimuli are being sampled.

In similar experiments the ground skink *Scincella lateralis* (subfamily Lygosominae) also responded to conspecific odors, but the response pattern of this species differed dramatically from that of *E. fasciatus* (Duvall et al., 1980). Both sexes of *S. lateralis* significantly avoided odors of conspecific males. Females responded randomly to airborne odors of conspecific females, but males approached female odors with greater than chance frequency. A further interspecific difference was that female ground skinks exhibited higher snout-dip rates than did males. Tongue-flick rates of females were also considerably higher than those of males, but the difference was not significant. In further experiments, it was established that *E. laticeps* and *S. lateralis* did not significantly approach or avoid each other's airborne odors. Collectively, these studies show that two distantly related scincid lizards respond to conspecific chemical stimuli. The sharp contrast in reactions to conspecific odors and in visual indications of chemosensory activation presumably reflects social, ecological, and perhaps phylogenetic divergence between the two species.

Given this highly suggestive background, we began a research program in 1982 to assess the importance and determine functional characteristics of chemical communication in the social behavior of E. laticeps and its closest relatives. The primary findings of this continuing research are summarized here. The research began with attempts to determine whether adult E. laticeps could detect odors of conspecifics and discriminate between conspecific males and females solely by odor stimuli. Because vertebrate pheromonal systems typically convey messages of fundamental reproductive import, effects of sex steroids on responsiveness to and production of lizard odors were examined. These initial studies led to a series of experiments on the ability of skinks to follow scent trails on substrates and in moving air streams, on thermal dependence of responsiveness to conspecific odors, on the ability of E. laticeps and morphologically similar syntopic congeners to discriminate conspecific from heterospecific odors, and on the behavioral significance of chemical stimuli in sexual and aggressive behavior. The involvement of pheromones in maintenance of reproductive isolation, anatomical sources and chemical natures of pheromones, and function of the tongue-vomeronasal system are topics under current investigation.

TONGUE-FLICKING AS A MEASURE OF RESPONSE TO SOCIOCHEMICALS

Sensory Function

Study of responses involving vomeronasal olfaction is greatly facilitated in squamate reptiles by the presence of an easily observed action, tongue-flicking, which indicates that the vomeronasal system is in use (Burghardt, 1980; Halpern, 1983). The moist tongue in lizards and snakes is protruded through a volume of air and may be brought into direct contact with some surface before being retracted into the oral cavity (Gove, 1979). In garter snakes of the genus Thamnophis it has been shown that chemicals from the external environment adhere to the tongue's moist surface and, when the tongue is retracted, are transferred to the vomeronasal epithelium via the vomeronasal duct (Halpern and Kubie, 1980). Early experimental work suggests similar lingual function in lizards (Kahmann, 1932). Moreover, tongue-flick rates (in combination with attack latencies) have been used successfully by Burghardt and Greene (Burghardt, 1973, 1977) to measure relative responsiveness of lizards to cotton-tipped applicators saturated with odor extracts of food items and control substances.

Although taste is not a factor in ophidian studies, it cannot be excluded in cases of direct lingual contact in lizard studies. A recent report suggests, but does not demonstrate, that lizard tongues possess taste buds (Schwenk, 1985). Whether vomeronasal olfaction, gustation, or a combination of them contribute to tongue-flick rates in experiments on responses by lizards to chemical stimuli derived from conspecifics and heterospecific congeners is not certain. In any case, tongue-flicking directed toward an object indicates active chemosensory examination.

Most of the studies reviewed here rely on tongue-flicking rates to measure the relative intensities of chemosensory investigation elicited by various stimuli. A basic assumption in our interpretation of all such studies is that there is a direct correlation between tongue-flick rate and degree of chemosensory investigation of the stimulus. Differences in tongue-flick rates induced by various stimuli demonstrate a capacity for differential responsiveness to the stimulus sources. However, in defensive situations, many species of lizards, as well as snakes, perform tongue-flicks of longer duration and sampling a larger volume of air than tongue-flicks emitted by the same species in exploratory situations (Gove and Burghardt, 1983). This suggests that individual differences in responses of lizards to experimenters as potential predators might contribute to variance in tongue-flicking rates by reducing tongue-flick rates to a greater extent in more easily disturbed lizards than in bolder ones. Despite such complications, tongue-flick rates provide a rapid means of assessing reactions to potential social stimuli.

Experimental Procedures

In all measurements of tongue-flicking rate, stimuli were presented to lizards in a standardized manner. In experiments on *E. laticeps* or *E. fasciatus*, the cotton tip of a wooden applicator was first immersed in a control substance such as deionized water or peanut oil. Unless responses to the control substance were to be measured, experimental stimuli were then added to the cotton tip by rolling the cotton over skin surfaces and/or cloacal epithelia of lizards representing various sexual and taxonomic categories or by immersing the cotton in test substances.

To begin a trial, the experimenter approached the home cage of a skink, carefully removed the wire cage top, and slowly positioned the tip of the applicator 2 cm anterior to the lizard's snout. Numbers of tongue-flicks directed to the applicator were recorded for 20 s, 60 s, or both. In our earliest studies (Cooper and Vitt, 1984a,b), recording began immediately upon introduction of the stimulus. In subsequent studies rates were determined for intervals of the same durations but beginning at the time of the first tongue-flick directed to the applicator. If a skink failed to respond within 30 s, the applicator tip was brought into light contact with the lizard's snout and then withdrawn to its original position. This procedure induced tongue-flicking by many individuals which would otherwise have been immobile in the experimenter's presence. Trials in which lizards did not tongue-flick within 30 s after being touched were discarded. If a lizard retreated before the end of a trial, data for that trial were discarded. Lizards for which trials were aborted were retested on a later date. Any individual failing to complete a retrial was dropped from the experiment. To enhance comparability of responsiveness, data were collected only for fully active lizards in late morning and early afternoon.

Because individuals of *E. inexpectatus* often fled and failed to respond when confronted by the experimenter's hand, applicators were tied to the ends of 1 m wire rods which were held at the opposite end by the experimenter during trials. This procedure might also be helpful with *E. fasciatus* and other easily frightened species. We have successfully used wooden applicators to introduce stimuli to long-term captive *E. laticeps* thoroughly acclimated to laboratory conditions and accustomed to the daily introduction of human hands into their home cages. Other techniques will be required to study more timid species.

Other Experimental Considerations

Most of our early experiments measuring tongue-flicking directed to applicators employed repeated measures designs. However, it has become apparent that when sufficient lizards are available, independent groups designs are preferable. On the first presentation of an applicator, tongue-flick rate is presumably increased above the baseline rate in the given experimental setting by a novelty effect (see De Fazio et al.,

1977). Rapid habituation reduces response rates in subsequent trials. Therefore, randomization or counterbalancing of the sequence in which odor treatments are presented is mandatory for experiments with repeated measures designs. To minimize effects of habituation, an intertrial interval of at least one day was established in all but one experiment.

There is a strong functional relationship between temperature and tongue-flicking rate. Tongue-flicks emitted in 60 s by E. laticeps in response to conspecific cloacal odors increased roughly by a factor of 30 between 15 and 30°C. Above 30 degrees the tongue-flicking rate declined rapidly (Cooper and Vitt, in press d). The rapid acceleration of the tongue flick-temperature curve in the range of temperatures at which the lizards are normally active necessitates close attention to control of temperature in studies using tongue-flicking rate as a dependent variable. The wide variation in tongue-flicking rates elicited by identical stimuli at different temperatures also influences choice of experimental design because much tighter thermal control is required to achieve identical temperatures for repeated trials of the same individual on different days than to obtain similar temperature distributions across independent experimental groups (Cooper and Vitt, in press d).

Statistical analysis of tongue-flicking data was conducted by standard parametric procedures (Winer, 1971). Analyses of variance appropriate to the particular experimental design were conducted. Individual comparisons were made using Newman-Keuls or Duncan's multiple range tests. Hartley's test was run on all data sets to detect heterogeneity of variance. If significant heterogeneity was found, data were transformed (usually to square roots) and reanalyzed. Only when transformation failed to render variances uniform among groups were nonparametric procedures used (Siegel, 1956).

DETECTION OF CONSPECIFIC CHEMICAL STIMULI

The ability to approach or avoid airborne conspecific odors demonstrated by E. fasciatus and S. lateralis (Duvall et al., 1980), implies that these skinks can detect chemicals produced by or associated with conspecifics. Our initial studies of scincid chemical ecology were designed to determine whether E. laticeps can detect conspecific odors, and, if so, whether responsiveness differed with sex and reproductive condition of the experimental or odor-source lizards. Postreproductive adult males emitted significantly more tongue-flicks to odors of postreproductive conspecific males and females than to distilled water (Cooper and Vitt, 1984a). These results confirmed our expectation that males can detect conspecific odors (Fig. 1).

If broad-headed skinks use chemical signals in social contexts, particularly in sexual and/or aggressive situations during the spring breeding season (Vitt and Cooper, in press, unpublished data), differential response rates to male and female odors might occur and be accentuated in reproductively active individuals. Even when postreproductive, males protruded their tongues more rapidly in response to odors of postreproductive females than of males. Reactions of breeding skinks were simulated by testing lizards given steroid treatments known to induce orange head coloration, aggressive behavior, and courtship in males and sexual receptivity in females (Cooper and Vitt, 1984a). Males implanted subcutaneously with testosterone pellets showed the highest tongue-flick rates when responding to odors of females, as had postreproductive males. However, the mean rate of tongue protrusion by testosterone-treated males responding to cloacal odors of females treated with estrogen was 44.9/min, approximately fifty percent greater than the 30.6/min rate

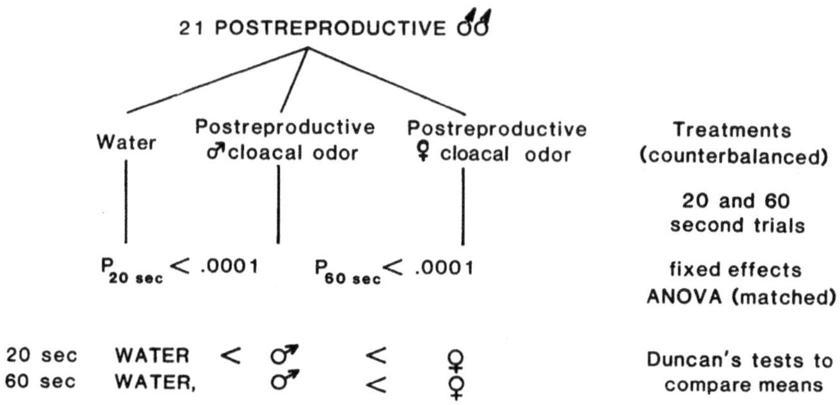

Fig. 1. Adult male *Eumeces laticeps* had higher tongue-flick rates in response to conspecific cloacal odors than to water and higher rates to female than male odors in the initial 20 seconds.

shown by postreproductive males responding to cloacal odors of postreproductive females. Because this difference might have been produced by a general increase in tongue-flick rates among reproductively active males rather than a specific reaction to female odors, a comparison of tongue-flicking rates by postreproductive and testosterone-treated males in response to cloacal odors of estrogen-treated females and testosterone-treated males was made. Results of this analysis clearly showed that testosterone treatment did not affect rates of tongue extrusion when the odor source was male. In striking contrast, when responding to cloacal odors of estogen-injected females, males given steroid treatments had tongue-flick rates 2.7 times greater than those of postreproductive males (44.9 and 16.9 per 60 s respectively).

Female *E. laticeps* also detect conspecific cloacal odors as revealed by the higher tongue-flick rates emitted when responding to either male or female cloacal odors than to distilled water. However, even females injected with estradiol-17β have similar tongue-flick rates regardless of the sex of the odor donor. Thus, whereas males exhibit clearcut differential responsiveness to odors based on sex, these experiments furnish no evidence that females discern any difference between male and female odors. Another marked sexual difference is that estrogen-injected females have lower 60 s tongue-flick rates to all stimuli tested in these experiments, even water, than do testosterone-treated males (Cooper and Vitt, 1984b). This finding is consistent with our observation that females protrude their tongues far less frequently during social encounters than do males (unpublished observations) and with the lower tongue-flick rates of female than male *E. fasciatus* in olfactometric experiments (Duvall et al., 1980). However, that the difference extends to water suggests either that females have lower baseline tongue-flicking rates or that female response rates show greater decrements in the presence of human experimenters than those of males. Females did appear to be more easily frightened, often moving away from the experimenter before completion of a trial.

These results from studies of tongue-flick rates demonstrate that *E. laticeps* can detect conspecific chemical stimuli, but do not prove that the odors are identified as arising from conspecifics. The major alternative hypotheses are that the skinks responded to potential prey odors or merely to unidentified odors. Predatory motivation is unlikely because very few individuals bit applicators, in contrast to the behavior of closely related skinks confronted with applicators bearing prey odors (Burghardt, 1973). Furthermore, tongue-flick rates even in response to male odors were much greater than the highest rate observed in earlier feeding studies. The striking difference in tongue-flick rates emitted by males to male and female chemical stimuli suggests a capacity for basing sexually appropriate social actions on conspecific chemical stimuli.

INTERSPECIFIC ODOR DISCRIMINATION IN THE FASCIATUS GROUP OF EUMECES

In the southeastern United States, three skinks in the fasciatus species group of Eumeces (Taylor, 1935) occur, E. fasciatus, E. inexpectatus, and E. laticeps. These skinks are so similar in appearance that they were considered a single species until the 1930s (Taylor, 1932a,b). Electrophoretic analysis has shown that these lizards are very closely related genetically (Murphy et al., 1983). Eumeces laticeps is considerably larger than the other two species, but even so, misidentification of all but large adults is common, even in museum collections. Distinguishing E. fasciatus from E. inexpectatus in the field is even more difficult. Within a broad range of sympatry, minor differences in habitat preference reduce contact to a variable degree between species pairs. Eumeces fasciatus occurs in mesic habitats; E. inexpectatus usually occupies drier areas, many of which have sandy soils. Eumeces laticeps has broader habitat preferences than the other species and is syntopic with them in many areas. Eumeces fasciatus and E. inexpectatus are found together less frequently, but are commonly syntopic along ecotones linking moist hardwood-dominated bottom land forests to well-drained upland forests. In such ecotones, all three species are sometimes syntopic (Cooper and Vitt, in press a,b).

The three species occur together, are reproductively active during a late spring breeding season, and are so similar in appearance that visual cues might not provide a reliable means of distinguishing conspecific mates and sexual competitors from heterospecific congeners. Given the strong response to conspecific odors by E. laticeps, it seemed possible that these skinks could discriminate odors of conspecifics from those of their closest heterospecific relatives. Previous work had shown that E. fasciatus and S. lateralis responded randomly to each other's odors although significantly approaching or avoiding conspecific odors (Duvall et al., 1980). However, these skinks are only distantly related, belonging to different subfamilies. Experiments were designed to determine whether members of the fasciatus group show higher tongue-flick rates to conspecific odors than to heterospecific odors or control stimuli when all are presented on applicators.

Tests for interspecific odor discrimination by E. laticeps were made using a repeated measures design in which each lizard responded in random sequence to cloacal odors of each of the three species, water, and a cologne used as a pungency control. In both sexes, the odors of conspecifics of opposite sex stimulated far higher tongue-flick rates than any of the other stimuli. Tongue-flick rates did not differ significantly among any of the other stimuli. Even the highly volatile pungency control, which the skinks showed obvious signs of detecting, did not induce elevated tongue-flick rates. Clearly, pungency alone does not account for responses to conspecific odors. That the odors of frequently syntopic, heterospecific congeners do not provoke detectable increases in tongue-flick rates demonstrates conclusively that the response to conspecific odors is highly specific. In males, the 60 s tongue-flick rate observed for conspecific female odor (26.7) was more than five times

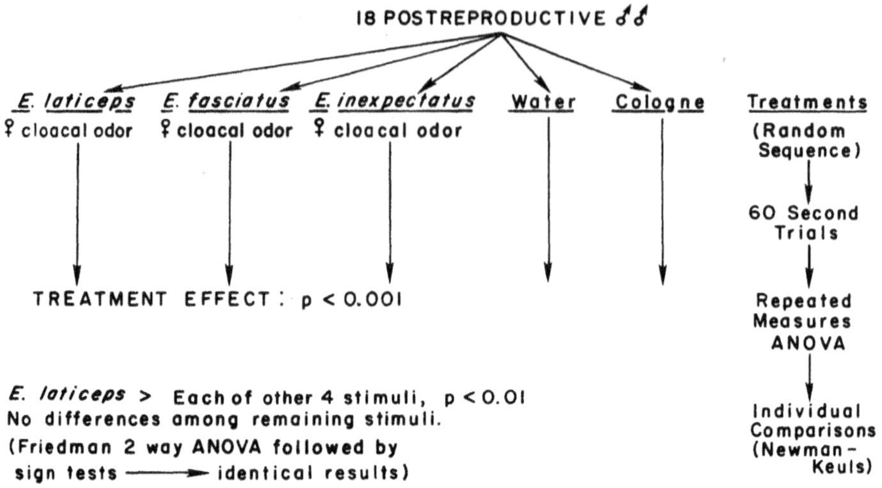

Fig. 2. Interspecific odor discrimination by male Eumeces laticeps. Lizards emitted significantly more tongue-flicks to female cloacal odors of conspecifics than to those of sympatric heterospecific congeners.

higher than the next highest rate, that for cologne (Fig. 2). As in earlier experiments, tongue-flick rates by females did not increase as much over baseline levels in response to conspecific odors as did those of males, the rate for conspecifics (10.6) being only 1.7 times that for cologne. Despite quantitative differences, both sexes unequivocally discriminated conspecific from heterospecific odors, showing prolonged chemosensory investigation only of the former (Cooper and Vitt, in press b).

Similar experiments were planned for males of the other southeastern members of the fasciatus group responding to female odors. Because the smaller species retreated or failed to respond much more often than E. laticeps, a repeated measures design was difficult to complete for E. fasciatus and impossible for E. inexpectatus. Limited availability of skinks and low responsiveness forced restriction of the experiment on E. inexpectatus to two odor treatments, conspecific odor and E. fasciatus odor. The pattern of results for E. fasciatus was identical to that for E. laticeps. That is, significantly more tongue-flicks were emitted in response to odors of female E. fasciatus than to any of the other stimuli, and no differences were observed in tongue-flick rates elicited by heterospecific or control stimuli. Likewise, male E. inexpectatus emitted more tongue-flicks in response to conspecific female cloacal odors than to those of E. fasciatus (Cooper and Vitt, in press a).

All three North American representatives of the fasciatus group are differentially responsive to chemical stimuli arising from conspecifics. Operationally, these lizards can discriminate members of their own species from their most closely related congeners solely on the basis of chemical cues. They appear to identify skinks as conspecific or heterospecific, but there is no evidence that they make any distinctions among heterospecific congeners pheromonally.

		fasciatus	inexpectatus	laticeps
Male-Male Agonistic Behavior	fasciatus	Most	None	None
	inexpectatus	None	Almost all	None
	laticeps	None	None	Almost all
Male Courtship of Females	fasciatus	Usually	None	None
	inexpectatus	Occasionally	Usually	Usually
	laticeps	None	None	All cases
Female Receptivity	fasciatus	All	—	—
	inexpectatus	None	Almost all	None
	laticeps	—	—	All

Fig 3. Both sexes of skinks in the fasciatus group of Eumeces from the southeastern United States recognize conspecifics as indicated by restriction of male agonistic behavior and sexual receptivity by females to conspecific lizards. Males of only one species, E. inexpectatus, courted heterospecific females.

PHEROMONES AND SOCIAL DISCRIMINATIONS IN THE FASCIATUS GROUP

The demonstration of ability to discriminate among syntopic congeners by chemosensory processes strengthened the possibility that skinks might use pheromones in making basic decisions regarding social behavior. During social encounters in the breeding season, skinks of the fasciatus group initially investigate each other by tongue-flicking, frequently near the cloacal margin (Fitch, 1954; Perrill, 1980; Cooper and Vitt, 1984a, in press a). Perrill (1980) observed that in a few staged encounters between E. inexpectatus and E. fasciatus, the lizards separated without further social interaction after a few mutual tongue-flicks. These preliminary observations intimated that the skinks could not only identify conspecifics by species-identification pheromones, but might also identify sexual partners and/or sexual competitors by some combination of sex and species-identification pheromones.

Reproductive Isolation

The potential involvement of chemical stimuli in mate choice across species within the fasciatus group was examined by observing responses of both sexes during staged encounters between reproductively active individuals of all species pairs and by transferring cloacal and skin odors between species (Cooper and Vitt, unpublished data). Postreproductive skinks were given steroid treatments, which reliably induce orange head coloration, aggressive behavior, and courtship by males and sexual receptivity by females (Cooper and Vitt, 1984a). Trials were conducted by placing a female in the home cage of a male.

At the outset of a typical encounter, a male would approach a female and touch her several times with his tongue prior to courting the female or subsequently ignoring her sexually. Females also tongue-flicked males, but much less frequently. In these experiments, male E. fasciatus and E. laticeps were highly selective, courting only conspecific females. Several male E. laticeps even attempted to prey upon heterospecific females. Male E. inexpectatus exhibited more "catholic" mate preferences, courting conspecific females and female E. laticeps with similar frequency. However, they courted female E. fasciatus in a significantly lower proportion of trials than they courted conspecific females (Fig. 3). Estrogen-treated females were universally receptive to courtship by conspecific males, but all females courted by heterospecific males rejected courtship. Of the few females not courted by conspecific males, over half (5 of 8) themselves initiated courtship. This contrasted markedly and significantly with the total absence of such behavior by uncourted females paired with heterospecific males.

The restriction of male courtship to conspecific females by two of the three species coupled with female rejection of courtship by heterospecific males in the third species may enforce strong reproductive isolation in the North American branch of the fasciatus group. An attempt was made to detect a possible role of species-identification pheromones in the premating isolating mechanism by observing interactions between heterospecific pairs consisting of a male or female E. fasciatus and an E. inexpectatus of opposite sex. In each encounter between steroid-treated lizards, the female bore odors of an estrogen-injected female belonging to the same species as the male. Odors were transferred by rubbing the skin and cloacal margin of one female on the skin of a heterospecific female. An encounter was staged by placing a female in the home cage of a male and observing ensuing activity from a blind.

Conspecific female odors stimulated male E. inexpectatus to court female E. fasciatus. The heterospecific females labeled by odors of sexually receptive conspecific females were courted at a significantly higher frequency than were similar females lacking conspecific odors. Conspecific and odor-treated heterospecific females were courted with equal frequency, but allospecific mating was rare because all female E. fasciatus courted by male E. inexpectatus vigorously rejected mating attempts. Male E. inexpectatus thus not only discriminate conspecific from heterospecific female odors, but also court sexually active females, even heterospecific ones, which bear conspecific chemical cues.

Ethological isolation between male E. inexpectatus and female E. fasciatus appears to be maintained in part by interspecific pheromonal differences. No female E. fasciatus were receptive when courted by male E. inexpectatus. However, the occurrence of two forced copulations involving such pairs (one female odor-treated, the other not) suggests that selective responsiveness by males to female species-identification pheromones could be adaptive. Individuals making errors in species-identification might suffer increased exposure to predation during misdirected courtship and copulation. For species in which males guard females from other males for some interval prior to and/or following copulation, heterospecific mating might substantially reduce mating opportunities, especially in skinks which have a short breeding season and produce a single clutch annually. Mate guarding has been reported in the fasciatus group only for E. laticeps (Vitt and Cooper, in press). Whether it occurs in the other species is unknown.

Quite different results were obtained for E. fasciatus. Conspecific female odors did not detectably alter responses of male E. fasciatus to female E. inexpectatus. No male E. fasciatus courted a female E. inexpectatus whether or not the female bore odors of sexually receptive female E. fasciatus. The absence of heterospecific courtship was not due to low sexual motivation because males readily copulated with conspecific females immediately after failing to court heterospecific females. Although conspecific female chemical cues did not elicit courtship of heterospecific females by male E. fasciatus, the presence of species-identifying or sex pheromones cannot be excluded. If chemical and visual signals operate conjointly, simultaneous exposure to species-typical stimuli in both modalities might be required to evoke courtship.

In aggregate, the experiments on courtship and transspecific odor effects in the North American members of the fasciatus group verify separation of species by an ethological premating mechanism including a pheromonal component. Participation of a pheromone in reproductive isolation has been shown only for the single mating combination of male _E. inexpectatus_ and female _E. fasciatus_. In the absence of other species-specific cues, conspecific female odors do not induce heterospecific courtship by male _E. fasciatus_ and may or may not in male _E. laticeps_.

Male Agonistic Behavior

Given the great similarity in appearance among males of all three species, their frequent syntopy, and their highly aggressive behavior toward conspecific males during the mating season, the existence of some mechanism for avoidance of potentially injurious heterospecific conflict was predicted. Preliminary evidence supporting this notion was reported by Perrill (1980), who observed no aggressive acts in a few staged encounters between male _E. fasciatus_ and _E. inexpectatus_. That the males exchanged mutual tongue-flicks before moving apart without any agonistic behavior suggested that males of these species might identify conspecific sexual competitors pheromonally.

To detect any heterospecific aggression in the fasciatus group and to test the hypothesis that conspecific males are recognized by species-identification pheromones, we staged encounters between androgen-treated males of all species pairs (Cooper and Vitt, unpublished data). In a large majority of heterospecific encounters, males tongue-flicked each other and then moved apart. No fighting occurred between heterospecific males despite the performance of aggressive head-tilting displays in a few cases prior to chemosensory investigation. However, large male _E. laticeps_, which are known predators of lizards including _Eumeces_ (Hamilton and Pollack, 1961; Cooper and Vitt, unpublished), attempted to prey on smaller heterospecific males in 3 of 11 encounters. In encounters between conspecific males, aggressive displays occurred subsequent to tongue-flicking in a large majority of trials for all species, as did biting and establishment of dominance.

Once it had been established that agonistic behavior following tongue-flicking was directed solely to conspecific males, an odor transfer experiment similar to that used to study the pheromonal isolating mechanism was conducted. Male _E. fasciatus_ were labeled with odors of male _E. inexpectatus_ by rubbing corresponding skin surfaces against each other, especially around the cloaca. A significantly higher proportion of male _E. inexpectatus_ displayed aggressive behavior to male _E. fasciatus_ bearing odors of male _E. inexpectatus_ than to males lacking those odors. There was no significant difference in frequency of encounters in which aggressive behavior was directed by male _E. inexpectatus_ to conspecific males or odor-treated male _E. fasciatus_. Because male _E. inexpectatus_ respond aggressively to male _E. fasciatus_ only when the latter have been labeled by odors of the former, it may be concluded that male _E. inexpectatus_ recognize conspecific males by chemical cues (Table 1). Species-specific pheromones thus allow males to discriminate sexual competitors from males of syntopic congeners not reliably distinguishable by visual cues.

Sexual selection has produced bright orange head coloration, large head size, and intense aggressive behavior toward conspecific males in the fasciatus group (Vitt and Cooper, in press). During the breeding season, almost all large adult males have fresh head wounds from bites by rival males. High levels of energetically expensive and potentially (and often

Table 1. Effects of interspecific odor transfers on responses by male *Eumeces inexpectatus* to *Eumeces fasciatus*.

sex of *E. fasciatus*	*E. fasciatus* without odor of *E. inexpectatus*	*E. fasciatus* with odor of *E. inexpectatus*
female	low probability of courtship	high probability of courtship
male	no aggression	frequent aggression

actually) injurious agonistic activity must be maintained by selection conferring greater mating success on winners of agonistic encounters. In the presence of syntopic congeners of similar appearance, there is frequent opportunity for misdirected aggression. That heterospecific agonistic behavior beyond preliminary display has not been reported suggests the operation of strong selection against it.

Conspecific males fight only during the breeding season, which suggests that competition for food is of little importance in maintaing agonistic behavior in these skinks. Instead, selection favoring agonistic behavior appears to be entirely sexual. There is no benefit to be gained from heterospecific agonistic behavior. The almost complete lack of interspecific copulation in the mating experiment shows that even in close confines heterospecific males are not serious sexual competitors. The major costs of heterospecific agonistic behavior are presumably injury inflicted during fights, increased exposure to predation due to lessened vigilance and greater detectability, and waste of time and energy which might have been used in other activities.

PHEROMONES AND SOCIAL RESPONSES TO CONSPECIFICS

Scent Trails

Field observations of *E. laticeps* during the breeding season strongly indicate that males sometimes follow chemical trails of conspecific females (Goin, 1957; Cooper and Vitt, in press e). While scent-trailing males repeatedly touch their tongues to the substrate over which the females has passed, testing both the female's path and the area to either side. Males have been observed to locate females initially invisible to them by following winding trails in this manner (Cooper, personal observations). To ascertain whether the skinks were following scent trails rather than some habitual meandering paths which coincidentally led to females, laboratory tests were made of the ability to follow a conspecific odor trail leading from the entry alley into one arm of a y-maze (Cooper and Vitt, in press e). Odor trails invisible to human observers were produced by dragging the ventral side of an adult male or female skink along a sheet of brown wrapping paper, maintaining contact between the paper and the lizard's cloacal area.

Adult male *E. laticeps* entered the trail-containing arm at well above chance frequency in trials with female trails, but responded randomly to male trails. Males directly contacted the female scent trails repeatedly with their tongues. In contrast, females did not follow chemical trails of either sex. Broad-headed skinks usually occupy structurally complex woodland habitats in which even nearby females may be out of view. Coupled with the necessity for males to locate females during a brief period of sexual receptivity and to guard the females against mating attempts by other males, restricted visibility in structurally complex habitats presumably provides strong selection favoring scent-trailing ability.

Airborne Odors

Although male E. laticeps followed female scent trails on solid substrates, neither sex approached or avoided sources of airborne conspecific odors in a 10 km/hr airstream moving through a y-maze olfactometer (Cooper and Vitt, in press c). Similar results were obtained for E. fasciatus. In these experiments, both temperature and air speed were well within the range encountered by active skinks during the reproductive season. Interestingly, both sexes of E. fasciatus approached sources of airborne conspecific odors of both sexes in similar experiments by Duvall et al. (1980). Factor(s) responsible for the disparity in results regarding responses to airborne conspecific odors are uncertain, but differences in air speed, ambient temperature, and reproductive state between experiments suggest avenues for further study. In social encounters, skinks appeared to respond only at very close range and usually to direct lingual contact with an odor-bearing surface, suggesting that social chemical signals are transmitted by substances of low volatility, i.e., contact pheromones.

Female Sex Pheromone

Tongue-flicking experiments demonstrated that male E. laticeps detected female odors from skin surfaces or the cloaca. Males also detected cloacal odors of conspecific males. Preliminary histological examination of the cloaca (with Stan Trauth and Steve Perrill) led to the discovery of an unreported sexual dimorphism in cloacal glands which will be presented elsewhere in detail. Males possess large dorsal glands overlying the dorsal and lateral aspects of the proctodaeum (Whiting, 1969), but lack prominent glandular structures in the urodaeal region. Females have large urodaeal glands which secrete directly into the urodaeal lumen.

It was hypothesized that the uredaeal gland might produce a female sex pheromone which elicits elevated tongue-flick rates and stimulates courtship by males. Because sex pheromones would presumably be produced maximally by and evoke strongest responses from reproductively active lizards, all skinks used in tests of these ideas were pretreated with appropriate steroids, except as contraindicated. Bill Garstka played an integral role in these as yet unpublished experiments. First, it was determined that macerated urodaeal products on cotton-tipped applicators elicited extremely high tongue-flick rates by males, even higher than those for cloacal swabs (Fig. 4). Presumably, this indicates that a major substance or mixture which stimulates chemosensory investigation is more concentrated in the urodaeal gland than in the cloaca. It has not been shown that the effective substance is a urodaeal glandular product which gains access to the skin and external environment via secretion into the urodaeal cavity, but this is our working hypothesis.

Another experiment was designed to examine the possible stimulation of courtship by urodaeal substances. In a repeated measures design, males were presented either a post-reproductive female or post-reproductive female painted with an aqueous suspension containing macerated urodaeal glands. Males courted females treated with urodaeal glandular

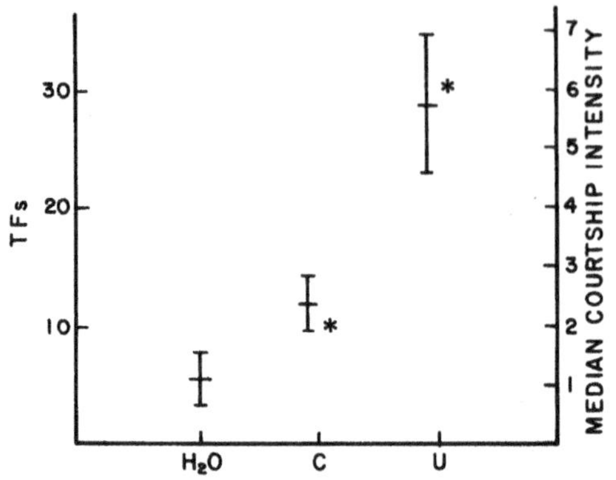

Fig. 4. Lingual and courtship responses of male Eumeces laticeps to conspecific female and control stimuli. Numbers of tongue-flicks emitted in 60s in response to applicators bearing water, female cloacal odors (C), or urodaeal glandular homogenate (U) are indicated (mean = horizontal bar, brackets indicate $\pm$ 1 s. e.). Asterisks indicate median courtship intensity directed by males to postreproductive conspecific females and to similar females painted with urodaeal pheromone.

suspension much more intensely than control females (Fig. 4). Whatever other functions it may serve, the urodaeal gland in E. laticeps is the source of a female sex pheromone which stimulates courtship by males. Virtually nothing is known about the chemical composition of the urodaeal pheromone. However, pilot studies have shown that males protrude their tongues at higher rates to the fraction of urodaeal glandular macerate extracted by a 2:1 chloroform-methanol solution than to peanut oil, a lipid control substance. The possibility that behaviorally active substances might be associated with other chemical fractions of urodaeal glandular homogenate has not been examined.

Male skinks in the fasciatus group could discriminate conspecific from heterospecific females by differences in urodaeal pheromones. However, only stimulation of courtship in male E. laticeps can be attributed to such substances on the basis of available data. It is entirely possible that the role of urodaeal pheromone is restricted to sexual arousal. Females might well produce additional substances identifying their species membership. Males, at least in E. inexpectatus, produce some other species-identifying chemical signals which cannot be of urodaeal glandular origin. The dorsal gland of the cloaca, which appears to be lacking in females, is a possible site of production, but so are numerous other structures. It is also possible that both sexes produce identical species-identification pheromones in an anatomical structure present in both sexes. The nature and sites of synthesis of species-identification pheromones are among the major subjects requiring future investigation before pheromonal communication in the fasciatus species group of Eumeces can be entirely understood.

Evolutionary Considerations

Pheromonal communication may be a familial trait in Scincidae and was very likely present in ancestral Eumeces. As it evolved in the fasciatus group, pheromonal communication was presumably modified and refined primarily by selection favoring production of and responsiveness to intraspecific signals. Each species is maximally responsive to conspecific odors and can direct appropriate social behaviors to conspecifics identified by chemical cues. Although mating and agonistic behavior are largely restricted to conspecifics, there is no direct evidence that such restriction evolved by selection against heterospecific mating or aggression (West Eberhard, 1983). Neither can the operation of such selection be excluded.

Production of a sex pheromone by female E. laticeps may have evolved to ensure the presence and/or sexual arousal of males during a brief interval of female receptivity. Although it is not known whether the female courtship-stimulating pheromone is identical to the substance sensed by scent-trailing males, this identity is a strong possibility. Females are fertilizable during the period of receptivity and unreceptive (and probably unfertilizable despite the possibility of sperm storage) at other times. In populations lacking pheromonal messages about sexual readiness, those males able to detect odors of substances produced exclusively or maximally during or immediately before sexual receptivity would be able to fertilize more females than would other males. The pheromone-detecting trait would presumably increase in frequency in subsequent generations until fixation. The presence of males responding to odors produced as by-products of physiological processes temporally associated with sexual receptivity would in turn favor females producing distinctive substances in sufficient quantity. Eventually, specialized glands for pheromone production might evolve. The urodaeal gland, which currently has no other known function, may have developed in this fashion, but even its phylogenetic origin is uncertain.

During the evolution of pheromonal communication, male sensitivity to urodaeal products and the amount and properties of the active urodaeal glandular molecules undoubtedly was refined by coevolutionary interaction between the sexes. If, as seems likely, males follow scent-trails by responding to urodaeal pheromone, the evolution of scent-trailing ability may have proceeded by such a process. Selection pressure on females to be located by males at the proper time may have acted to increase urodaeal glandular output to provide quantities of pheromone sufficient to allow reliable scent-trailing. Simultaneously, constraints on physical properties of molecules effective for scent-trailing, particularly those relating to volatility, may have selected for changes in molecular structure of urodaeal pheromone and thus induced evolution of changes in male sensitivity.

Except at extremely low population density, it is likely that virtually all females would be located and fertilized by males. Females producing more effective pheromones might be fertilized slightly earlier but because females produce a single clutch per year, the selective advantage might not be great. Another benefit might be the attraction of multiple males. If the female mated with the winner of agonistic contests among the males, she would be assured that her sexual partner had high relative fitness. While males were attracted to such females, other females which produced less effective pheromones might be discovered by fewer males, perhaps those less responsive to pheromones. These females would then copulate with males of lower mean fitness. Their female offspring would carry genes for low pheromone production, and their male offspring would not receive sex-linked genes for high male fitness.

From the above argument, it appears that pheromonal communication has evolved at least in part by sexual selection on pheromone production by females and on sensitivity to pheromones by males. Because chemical communication is of uncertain but wide-spread extent in Scincidae and other lizard families and because the phylogenetic extent and origin of the urodaeal gland and its products are unknown, the evolutionary history of pheromonal communication in E. laticeps and other species of the fasciatus group cannot now be reconstructed. Nevertheless, the dramatic sexual dimorphisms in head size, head color, and agonistic behavior argue for the operation of strong sexual selection on several traits in E. laticeps and its closest relatives (Vitt and Cooper; in press, unpublished data). The extreme anatomical and behavioral dimorphism in chemical communication involving urodaeal pheromones in Eumeces is consistent with these observations and with the evolutionary scenario.

ACKNOWLEDGMENTS

We wish to express appreciation to several colleagues who have generously contributed to this research. William R. Garstka performed skillful surgery and chemical separations in the study of female sex pheromone. Stanley E. Trauth and Stephen A. Perrill conducted a histological analysis of cloacal glandular structures. We thank Justin D. Congdon, J. Whitfield Gibbons, and Raymond D. Semlitsch for stimulating discussion. We are also grateful to Carolee Caffrey, Ruth Estes, Paul Heslip, Jeff Howland, Grady Knight, Jim Knight, Trip Lamb, Mary Mendonca, Steve Morreale, Sue Novak, Tim Owens, Joe Pechmann, and Rich Siegel for providing specimens and laboratory assistance. The research program on skink pheromonal communication has been conducted primarily at the Savannah River Ecology Laboratory (SREL) under contract DC-AC09-76SR00819 between the U.S. Department of Energy and the University of Georgia. The authors received support from the Faculty Research Participation Programs of SREL (WEC and LJV) and Oak Ridge Associated Universities (WEC). Partial support was provided by Auburn University at Montgomery and the University of California at Los Angeles.

REFERENCES

Breckenridge, W. J., 1943, The life history of the black-banded skink Eumeces septentrionalis septentrionalis (Baird), Amer. Midl. Nat., 29:591-606.

Burghardt, G. M., 1973, Chemical release of prey attack: Extension to naive newly hatched lizards, Eumeces fasciatus, Copeia, 1973:178-181.

Burghardt, G. M., 1977, The ontogeny, evolution, and stimulus control of feeding in humans and reptiles, in: "The Chemical Senses and Nutrition," M. Kare, and O. Maller, eds., Academic Press, New York.

Burghardt, G. M., 1980, Behavioral and stimulus correlates of vomeronasal functioning in reptiles: Feeding, grouping, sex, and tongue use, in: "Chemical Signals: Vertebrates and Aquatic Invertebrates," D. Müller-Schwarze, and R. M. Silverstein, eds., Plenum Press, New York.

Cooper, W. E., Jr., and Vitt, L. J., 1984a, Conspecific odor detection by the male broad-headed skink, Eumeces laticeps: Effects of sex and site of odor source and of male reproductive condition. J. Exp. Zool., 229:49-54.

Cooper, W. E., Jr., and Vitt, L. J., 1984b, Detection of conspecific odors by the female broad-headed skink, Eumeces laticeps, J. Exp. Zool., 299:49-54.

Cooper, W. E., Jr., and Vitt, L. J., 1986, Interspecific odor discriminations among syntopic congeners in scincid lizards (genus Eumeces), Behaviour, in press a.

Cooper, W. E., Jr., and Vitt, L. J., 1986, Interspecific odour discrimination by a lizard (Eumeces laticeps), Anim. Behav., in press b.
Cooper, W. E., Jr., and Vitt, L. J., 1986, Responses of five-lined and broad-headed skinks (Eumeces fasciatus and E. laticeps) to airborne conspecific odors: further appraisal, J. Herpetol., in press c.
Cooper, W. E., Jr., and Vitt, L. J., 1986, Thermal dependence of tongue-flicking and comments on use of tongue-flicking as an index of squamate behavior, Z. Tierpsychol., in press d.
Cooper, W. E., Jr., and Vitt, L. J., 1986, Tracking of conspecific odor trails by male broad-headed skinks (Eumeces laticeps), Z. Tierpsychol., in press e.
De Fazio, A., Simon, C. A., Middendorf, G. A., and Romano, D., 1977, Iguanid substrate licking: A response to novel situations in Sceloporus jarrovi, Copeia, 1977:706-709.
Duvall, D., Herskowitz, R., and Trupiano-Duvall, J., 1980, Responses of five-lined skinks (Eumeces fasciatus) and ground skinks (Scincella lateralis) to conspecific and interspecific chemical cues, J. Herpetol., 14:121-127.
Evans, L. T., 1959, A motion picture study of maternal behavior of the lizard, Eumeces obsoletus Baird and Girard, Copeia, 1959:103-110.
Everett, C. T., 1971, Courtship and mating of Eumeces multivirgatus (Scincidae), J. Herpetol., 5:189-190.
Fitch, H. S., 1954, Life History and ecology of the five-lined skink, Eumeces fasciatus, Univ. of Kansas Publ., Mus. Nat. Hist., 8:1-156.
Fitch, H. S., 1955, Habits and adaptations of the Great Plains skink (Eumeces obsoletus), Ecol. Monogr., 25:59-82.
Goin, O., 1957, An observation of mating in the broad-headed skink, Eumeces laticeps, Herpetologica, 13:155-156.
Gove, D., 1979, A comparative study of snake and lizard tongue-flicking with an evolutionary hypothesis, Z. Tierpsychol., 51:58-76.
Gove, D., and Burghardt, G. M., 1983, Context-correlated parameters of snake and lizard tongue-flicking, Anim. Behav., 31:718-723.
Greer, A. E., 1970, A subfamilial classification of scincid lizards, Bull. Mus. Comp. Zool., 139:151-184.
Groves, J. D., 1982, Egg-eating behavior of brooding five-lined skinks, Eumeces fasciatus, Copeia, 1982:969-971.
Halpern, M., 1983, Nasal chemical senses in snakes, in: "Advances in Vertebrate Neuroethology," J. P. Ewert, R. R. Caprianca, and D. J. Ingle, eds., Plenum Publ., New York.
Halpern, M., and Kubie, J. L., 1980, Chemical access to the vomeronasal organs of garter snakes, Physiol. Behav., 24:367-371.
Hamilton, W. J., Jr., and Pollack, J. A., 1961, The food of some lizards from Fort Benning, Georgia, Copeia, 1961:99-106.
Kahmann, H., 1932. Sinnesphysiologische Studien an Reptilien- I. Experimentalle Untersuchungen über das Jacobsonüsche Organ der Eidechsen und Schlangen, Zool. Jb. Abt. Allg. Zool. Phsyiol., 51:173-238.
Mount, R. H., 1963, The natural history of the red-tailed skink, Eumeces egregius Baird, Amer. Midl. Nat., 70:356-385.
Murphy, R. W., Cooper, W. E., Jr., and Richardson, W. S., 1983, Phylogenetic relationships and genetic variability of the North American five-lined skinks, genus Eumeces (Sauria, Scincidae), Herpetologica, 39:200-211.
Noble, G. K., and Bradley, H. T., 1933, The mating behavior of lizards: Its bearing on the theory of sexual selection, Ann. N. Y. Acad. Sci., 35:25-100.
Noble, G. K., and Kumpf, K. F., 1936, The function of Jacobson's organ in lizards, J. Genet. Psychol., 48:371-382.
Noble, G. K., and Mason, E. R., 1933, Experiments on the brooding habits of the lizards Eumeces and Ophisaurus, Amer. Mus. Novit., 619:1-29.

Perrill, S. A., 1980, Social Communication in Eumeces inexpectatus, J. Herpetol., 14:129-135.

Schwenk, K., 1985, On the occurrence, distribution, and functional significance of taste buds in lizards, Copeia, 1985:99-101.

Siegel, S., 1956, "Nonparametric Statistics for the Behavioral Sciences," McGraw-Hill, New York.

Simon, C. A., 1983, A review of lizard chemoreception, in: "Lizard Ecology: Studies of a Model Organism," R. B. Huey, E. R. Pianka, and T. W. Schoener, eds., Harvard University Press, Cambridge, Massachusetts.

Taylor, E. H., 1932a, Eumeces inexpectatus: A new American lizard of the family Scincidae, Univ. Kansas Sci. Bull., 20:251-258.

Taylor, E. H., 1932b, Eumeces laticeps: A neglected species of skink, Univ. Kansas Sci. Bull., 20:263-272.

Taylor, E. H., 1935, A taxonomic study of the cosmopolitan scincoid lizards of the genus Eumeces with an account of the distribution and relationships of its species, Univ. Kansas Sci. Bull., 23:1-643.

Vitt, L. J., and Cooper, W. E., Jr., 1985, The evolution of sexual dimorphism in the skink Eumeces laticeps: An example of sexual selection, Canadian J. Zool., in press.

West Eberhard, M. J., 1983, Sexual selection, social competition, and speciation, Quart. Rev. Biol., 58:155-183.

Whiting, A. M., 1969, Squamate Cloacal Glands: Morphology, Histology, and Histochemistry, Ph.D. thesis, Pennsylvania State University, University Park.

Winer, B. J., 1971, "Statistical Principles in Experimental Design, 2nd ed.," McGraw-Hill, New York.

CHEMORECEPTION IN THE HOMING AND ORIENTATION BEHAVIOR OF AMPHIBIANS AND REPTILES, WITH SPECIAL REFERENCE TO SEA TURTLES

David Owens[1], Diana Crowell Comuzzie[1], and Mark Grassman[2]

[1]Biology Department, Texas A & M University, College Station Texas, [2]Zoology Department, University of Texas Austin, Texas

INTRODUCTION

The importance of chemoreception in the life history of amphibians and reptiles has been reviewed on several occasions (Burghardt, 1970; Madison, 1977; Stoddart, 1980). Our task is considerably more specific as our interest is in the chemosensory components of orientation and homing.

One could broadly define chemical orientation to include food-finding and reproductive behaviors where chemoreception is involved; however, we have chosen not to discuss these better known areas of amphibian and reptile behavior. On the other hand, where semiochemicals or conspecific habitat marking appear to have a function in home range factors or homing to a specific site, we have chosen to include the available data in this discussion.

It is our general prejudice that the proportion of chemical sensing within the total sensory capabilities is greater in aquatic and semi-aquatic animals than in more terrestrial forms where vision in particular dominates. Thus it was somewhat of a surprise to us when we were unable to locate many recent publications on orientation and homing in the more aquatic amphibians. While our finding of an apparent lack of research on amphibians may be partially due to our own admitted historical bias towards the reptilian literature, we also believe that there indeed has been relatively less work on amphibians. It appears that since the important pioneering studies of Twitty (summarized in Twitty, 1966), very few scientists have had an interest in this area.

As in the far better documented avian and mammalian classes, it appears to be a gross over-simplification to think that an amphibian or reptile would rely on a single sensory system for homing or orientation. In fact, a major difficulty that researchers appear to have had is the experimental dissection of the separate sensory modes in preparing protocols using the various animal models. It is very tricky to test just for chemoreception for example. Even though other senses surely must be important, we have attempted to avoid digressions into visual, tactile or other sensory systems. In actuality, evolution is acting to integrate all these systems in animals that have developed unusually keen orientation and homing abilities.

Among the amphibia, homing to breeding ponds has been well documented in both anurans and urodeles. In these groups it is quite likely that chemosensory mechanisms alone are not responsible for homing abilities, but rather chemoreception in concert with vision and perhaps other senses allow animals to home successfully. However, there is apparently some dissension among scientists as to which senses are the most important and which are used secondarily. In addition, species differences may occur which compound the problem. Among the earliest research on chemosensory homing was Twitty's comprehensive study of homing in the newt, Taricha rivularis. Beginning in 1953 and continuing until 1967, Twitty and his co-workers captured and displaced newts from a stream divided into 58 sections, with each section approximately 50 yards long. Twitty's research was among the first to demonstrate the remarkable homing abilities of amphibians, including homing pathways across mountain ridges and streams. Twitty's newts homed with surprising accuracy to the same section of home stream for successive breedings over many years (Twitty et al., 1967a). Additionally, Twitty and co-workers consistently found that blinded newts were able to home successfully (Twitty, 1959) even after a long period of separation from the home stream (Twitty et al., 1964, 1966), when displaced to areas with which the newts had no previous familiarity (Twitty et al., 1967a), or when displaced over two miles (Twitty et al., 1967b). In these cases, chemoreception is believed to be the primary orienting mechanism. However, Grant et al. (1968) in the last year of Twitty's published study, found that anosmic newts did not distribute themselves randomly when displaced from their home streams although the newts' homing ability was clearly reduced. The authors speculated that other mechanisms may partially take over when chemoreception fails. In the green frog (Rana clamitans), ablation of the olfactory tracts resulted in a reduced ability of the frogs to locate the breeding site (Oldham, 1967). However, the frogs were not randomly distributed as would be expected if olfaction had been the only orienting mechanism. Oldham (1967) suggested that because of this distribution, other orienting mechanisms may be involved. Other authors have proposed that the toad Bufo boreas may use odor to note the direction of the breeding pond and then move to the goal using visual cues (Tracy and Dole, 1969). Conversely, Himstedt and Plasa (1979) suggested that for Salamandra, vision is the primary homing mechanism, and olfaction would only be effective near home.

There have been several suggestions as to what signal sources enable amphibians to locate their home breeding grounds. Grubb (1970), for example, felt that the pond itself may have its own unique odor qualities. Aquatic vegetation, terrestrial vegetation and soil biota may all combine to produce a unique odor. These same odors could be used as chemical landmarks to guide an amphibian along a homing route (Dole, 1968). Salamanders climb significantly more often on the night they are first released after capture than on all other nights. Madison and Shoop (1970), who observed this behavior, speculated that climbing could facilitate the detection of home-associated odors. Madison (1969) also suggested that salamanders may follow concentration gradients to reach localized odor sources. However, the presence of odors does not necessarily mean that the animal can discriminate them and will use them to home. Experiments by Martof (1962) illustrated that the frog Pseudacris triseriata, trained to discriminate odors in a T-maze, significantly chose odors from a breeding habitat (lowland muck and filamentous algae) over those from a nonbreeding, upland forest habitat (soil and decaying vegetation in distilled water).

In addition to habitat odors, odors produced from the animal itself may be used as homing signals. The salamanders Plethodon cinereus and Plethodon jordani can distinguish between their own markings and those of conspecifics (Tristram, 1977), between the substrate markings of familiar and unfamiliar conspecifics (McGavin, 1978) and between breeding and non-breeding neighbors and non-neighbors (Madison, 1975). Parzefall (1976) demonstrated that in the salamander Proteus anguineus, adults secrete a chemical substance on substrates which then become preferable to unmarked substrates among other salamanders. Jaeger and Gergits (1979) found that male salamanders (Plethodon cinereus) preferred their own marked substrates over those of other males. However, males showed no preference when exposed to their own marked substrates and substrates marked by females. Female salamanders preferred their own substrates over male-marked substrates, but showed no preference for their own substrate over female-marked ones. From these results, Jaeger and Gergits concluded that the salamander's movements were to avoid certain conspecifics rather than movements towards their own individual substrate odors. These semiochemicals would appear to function in dispersal rather than homing. Another species in this genus, Plethodon jordani, has been described as having "a well-developed homing response" (Madison, 1972). If P. cinereus likewise has a well-developed homing response, then Jaeger and Gergits' conclusions contradict what we would expect for an olfactory-based homing response. This dilemma merits further study in this species. Among anurans, Grubb (1973) has stated that the attraction to the smell of another animal is probably not the mechanism for homing since frogs and toads breed after heavy rains and these rains would eliminate odors left by the animals. However, this later assertion has not been substantiated by the limited research on anurans.

REPTILES

While vision is commonly rated as the dominant sensory mode in reptiles (i.e., Madison, 1977), we believe this is an impression based primarily on work with terrestrial lizards, snakes and tortoises which may not apply as well to the aquatic members of the class. Much more research needs to be done with aquatic and marine snakes and turtles whose habitats provide poor visual, but excellent chemical cueing characteristics.

The majority of the olfactory homing and orientation research on reptiles has been performed using snakes. Rattlesnakes (Crotalus spp.) overwinter in dens and then migrate out and disperse in spring. In the fall following breeding, rattlesnakes return to the hibernacula to await spring (Duvall et al., 1985). The mechanism by which rattlesnakes disperse and then return to the den has been the subject of research by several authors. Gregory (1982) stated that it would be unlikely for adult snakes migrating back to the den to follow the odor trails of other snakes since the snake would have to discriminate between trails leading to the den and those leading elsewhere. In addition, the first snake arriving at the den would not have the benefit of such a cue. However, research by Ford and Low (1984) has shown that snakes are capable of determining the direction of a trail by noting on which sides of rocks or other objects the scent is placed. This would point the snake in the direction of the den rather than away from it. It is a valid criticism that a snake would encounter many trails of conspecifics, and in order to return to the hibernaculum using chemoreception (vomeronasal sense?), the snake would have to pick out the correct one. We can only suggest that perhaps a migrating snake produces a different odor from that of a non-migrating snake. Our suggestion could be tested rather easily, but to our knowledge this has not been done. For the spring migration away from the den, Duvall et al. (1985) stated that olfactory cues would probably not have lasted over the winter and therefore would probably not be available for the vernal migration. Instead, these animals probably migrate by celestial cues (Landreth, 1973). The most likely candidates among rattlesnakes to follow olfactory migration cues are the neonates since they are born near the den and are the last to enter for the winter hibernation (Brown and MacLean, 1983; Duvall et al., 1985; Gregory,

1982; Hirth, 1966). Aggregations of the neonates in the few days following birth may serve to familiarize the neonates with the odors of other rattlesnakes and enhance their abilities to follow scent trails (perhaps deposits of adult skin lipids released while shedding) and locate the den they have not previously occupied (Duvall et al., 1985; c.f. Graves et al., this volume). Trailing has also been observed in garter snakes (Thamnophis spp.). Heller and Halpern (1981) have shown that adult garter snakes deposit chemical substances on filter paper substrates and that other snakes discriminated between marked substrates and unmarked substrates. In a Y-maze, snakes tended to significantly choose the arm of the maze containing the marked substrate, and tended to follow the trail of the previous individual in the maze. Chemoreception (probably via the vomeronasal organ, see below) then, is likely to be the mechanism by which Thamnophis spp. find their winter hibernacula.

Other research indicates that snakes respond in different ways to olfactants produced by other snakes and it is conceivable that these odors could play a role in snake orientation (see Burghardt, 1970, 1980, for discussions). One of the first studies of aggregation in snakes (Noble and Clausen, 1936) indicated that brown snakes, Storeria dekayi, use chemoreception as a secondary aggregative guide. Another study, by Dundee and Miller (1968) showed that the ringneck snake, Diadophis punctatus arnyi, appears to aggregate under discs which have been previously occupied by other ringneck snakes. In these experiments, ringneck snakes were released into a seven foot diameter tank filled with sand. Ten twelve-inch circular discs were radially arranged in the tank. Dundee and Miller found that snakes were attracted to substances under discs that had been previously occupied by other ringneck snakes. Sexual attraction was not believed to be implicated in the aggregation since some aggregations lacked females entirely. The snakes frequently pressed their snouts to the substrate after release during exploratory movements, suggesting a chemosensory searching mechanism. Burghardt (1980), using an apparatus similar to that of Dundee and Miller, found that tongueless garter snakes (Thamnophis sirtalis) did not aggregate as did normal T. sirtalis, assumedly due to a nonfunctioning vomeronasal system. Additional work by Heller and Halpern (1982a,b) has documented laboratory aggregations of male garter snakes (T. sirtalis, and reduced aggregation tendencies in male snakes with sutured vomeronasal organ ducts. Thus, the vomeronasal system appears to be an important component of chemoreception in this species and would be a likely candidate in other reptilian species as well. Porter and Czaplicki (1974) demonstrated that garter snakes (Thamnophis spp.) show a significant preference for areas soiled by conspecifics, but adult water snakes (Nerodia) avoided such areas (Porter and Czaplicki, 1974). Neonate water snakes do not avoid conspecific odors (Scudder et al., 1980). As suggested by Porter and Czaplicki (1974), perhaps chemical signals influence both aggregation and dispersal. In another example, Crotalus viridis exhibit conspecific avoidance when maintained on a mid-summer photothermal regimen. This would be expected for animals that in nature would be dispersed and feeding to expand fat reserves at this time (King et al., 1983). Chiszar et al. (1980) tested the ability of captive Crotalus spp. to discriminate between the fecal odors of conspecifics and the rattlesnakes' own fecal odor. The authors found that fecal odors did not provide sufficient information to identify conspecifics, but that the absence of odors did induce defecations in shorter periods of time and induced greater exploratory behavior. Chiszar et al. (1980) suggested that these snakes may use their own odors as marking cues and that the absence of odors induces individuals to re-mark an area. Fecal odors may not be the primary marking medium, but for captive animals with limited options, fecal odors or other self-generated odors may be the only ones available.

Besides squamates, some chemosensory research has been done on lacertilians. Duvall et al. (1980) have shown that five-lined skinks and ground skinks each are significantly attracted to the odors of conspecifics in an olfactometer. Five-lined skinks have been noted to form small aggregations in hibernacula after the breeding season, and as such, this chemical attraction may be the mechanism by which individuals locate the group. The chemical signals of western fence lizards (Sceloporus occidentalis) have also been investigated and it is clear that males and females distinguish conspecific chemicals in predictable ways, but the functions of these signals are not completely known (Duvall, 1979, 1982). Surfaces may be marked with urine, feces or gland exudates (Duvall, 1981). Markings seem to affect spacing and territorial maintenance (Duvall, 1980), but could also be involved in aiding a lizard to find its home area. This latter question has not as yet been addressed as far as we were able to ascertain.

Among turtles, some research has been done on orientation and homing in freshwater and terrestrial species, but only a few of the studies have suggested chemoreception as an important orienting mechanism. In a study of turtle orientation, Ortleb and Sexton (1964) found that painted turtles (Chrysemys picta) did not orient in a Y-maze to currents carrying odors of aquatic plants. However, Ortleb and Sexton in dismissing chemoreception as an orientation cue, fail to consider that other olfactants such as those of conspecifics may play a role in orientation. Studying homing, Carroll and Ehrenfeld (1978) observed that Clemmys insculpta turtles were able to home, that magnetic fields and their anomalies did not provide sufficient homing information and that vision would have been of limited value. The authors concluded that chemoreception was the most likely basis for homing. In another study, Chelazzi et al. (1981) tested the ability of Testudo hermanni Gmelin to home after a treatment with zinc sulfate which rendered the turtles anosmic. Both experimental and control animals were anesthetized. Animals were then displaced 800 m from the capture point and allowed to home for seven and fourteen days. Control animals homed normally but anosmic turtles were diffused around the release point. The authors concluded that chemical cues were probably the major homing mechanism in these turtles.

As with amphibians and other reptiles, turtles may be capable of producing their own odors for use in social orientation. Evans and his co-workers (1973, 1974) examined the responses of sibling Pseudemys scripta and Chelydra serpentina to chemical conditioning. Pairs of juvenile turtles were housed in separate compartments of the same tank, with water and presumedly chemical cues exchanged between the two compartments. The pairs of turtles were unable to see each other. Subsequent tests showed that pairs of turtles placed in a circular experimental chamber with no barriers spent more time together than did siblings that had been kept in separate tanks and not exposed to each other's odors. Other results from this experiment were less clear. For example, the two turtles of a pair spent about the same amount of time in opposite sections of the experimental tank as they did in the same section. This is not indicative of a strong attraction between members of a pair. Chemoreception appears to be involved in the orientation of these turtles in some way, but further research is required before it can be better understood. Finally, Patterson (1971) observing Gopherus agassizi, found that aggregations of tortoises in winter may be related to the absence of fecal pellets from dominant males in the group. Patterson placed fecal pellets into a group sleeping area. All of the tortoises in this area moved to other sleeping areas. The crowding in other sleeping areas continued for eight days until the feces, had dried out and then the tortoises returned. In another incident, a large male defecated in the area where two subordinate males had slept. These two males subsequently moved out of the area and slept in the open for several nights. Patterson suggested that these pellets served as deterrents to aggregations and that only in the winter, when the number of pellets decrease, are the animals likely to aggregate. Although Patterson's work seems mostly observational, without adequate controls, we believe there is limited evidence that these tortoises exhibit a type of chemical orientation which functions to disperse animals rather than to cause attraction or homing. Further research on this species could prove interesting.

Sea Turtle Studies

Among the reptiles, the most phenomenal migrators are the marine reptiles. Several populations of the green sea turtle, Chelonia mydas, are the best documented in terms of distance between nesting and feeding grounds, site fidelity and number of years in the multi-annual cycling of reproductive migrations. In C. mydas, both males and females undertake the migration, with mating probably occurring near the nesting beach during a discrete period of just a few days. Approximately one month later, the female begins to nest. By the peak of nesting season, the males have left to begin their return migration to the feeding grounds (Owens, 1980). The females deposit one to several clutches of eggs at approximate intervals of two weeks. This general pattern applies to all of the seven (or eight) species of marine turtles with a few important modifications. The hawksbill sea turtle, Eretmochelys imbricata, is thought to be more sedentary with few documented migratory populations (Witzel, 1983). A curious added behavioral trait is characteristic of the two ridley species (Lepidochelys), where the nesting females can retain mature eggs in their oviducts and apparently wait for a critical mass of nesting individuals to develop. In the most endangered sea turtle species, L. kempi, a mass nesting (arribada) occurs in the daylight hours, usually during a storm. In L. olivacea, on the other hand, the arribada begins at night and may last well into the next day, especially during a very large arribada (> 50,000 turtles). It is our contention that chemoreception is centrally involved in many of these behavioral patterns.

Koch et al. (1969) have presented a reasonable argument for the importance of the chemical senses to marine turtle navigation. Although definitive evidence is still lacking, Carr's (1967) suggestion of olfactory imprinting in sea turtles has gained wide popular acceptance. Certainly, a similar model for salmonid imprinting is very well substantiated (Scholz et al., 1976). The nasal morphology of reptiles has been reviewed by Parsons (1970, 1971). Rather elaborate olfactory and vomeronasal epithelia exist in many reptiles including marine turtles (Fig. 1). Beuerman (1975, 1977), using electrophysiological techniques, has shown that these two chemical systems may be functionally distinct. Thus, the aquatic species in particular apparently have the necessary anatomical prerequisites to have evolved a more elaborate chemical sense, even approaching that seen in fishes.

New evidence indicates that chemical imprinting in sea turtles will have to be even more phenomenal than previously imagined if it indeed occurs. Studies on predicted age of sexual maturity suggest 15-35 or more years for earliest maturation in green and loggerhead sea turtles (Balazs, 1980; Limpus and Walter, 1980). If chemical imprinting to the natal beach occurs, it surely must be the most extreme case of imprinting in the animal kingdom. The hatchling would have to learn chemical cues available perinatally during a few days or weeks at most, and as an adult (15-35 years later) discriminate these cues for the first time from many others.

Vision, while beyond the scope of this paper, is certainly important in the sea-finding ability of marine turtles (Mrosovsky, 1978) as well as in their courtship (Booth and Peters, 1972; Comuzzie, unpublished data), however celestial navigation for sea turtles appears unlikely due to the relatively poor focusing ability of the sea turtle eye out of water (Ehrenfeld and Koch, 1967). Thus, it is more or less a process of elimination that has led students of sea turtle biology to look at chemoreception. Clearly, the senses of hearing, sight and even magnetic field detection could be involved in sea turtle homing.

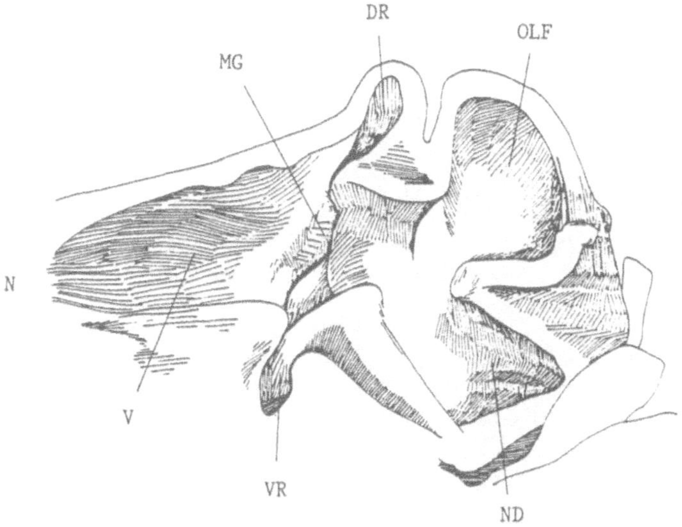

Fig. 1. Nasal morphology of the Kemp's Ridley Sea Turtle. Medial view showing the relative positions of the external nares (N), vestibulum (V), nasopharyngeal duct (ND), ventral recess (VR), dorsal recess (DR), medial groove (MG), and olfactory region (OLF).

Using small green sea turtles, Manton conducted a series of important operant conditioning experiments (Manton et al., 1972a, 1972b). These experiments have recently been reviewed by Manton (1979) and others (Owens et al., 1982) so they will not be discussed extensively here. Her experiments clearly indicate that:

1. C. mydas can learn to press a lever for a food reward upon recognition of the correct chemical cue at fairly low concentrations of 10-5 molar phenylethylalcohol, isopentyl acetate, triethylamine and cinnamaldehyde.

2. C. mydas is using chemoreception and not taste in this discrimination since zinc sulfate ablation of the chemosensory epithelia removed the capability demonstrated above.

3. C. mydas can remember these odors for at least one year since the same turtles exposed one year after initial training immediately initiated correct responses.

Ireland conducted experiments in Bermuda where immature C. mydas were displaced up to 6.8 km from the site of capture (Ireland, 1979a, 1979b). These animals demonstrated strong homing tendencies although there was no indication of the relative importance of chemoreception in these experiments. Behaviorally, sea turtles have been observed to plow their noses into the sand when they first approach the beach. This has been termed "sand smelling" and may aid the female in using chemoreception to locate the correct nesting beach (Fig. 2). An anecdotal observation by Mora and Robinson (1982) implicates chemoreception more strongly. At the major arribada nesting beach in Ostional, Costa Rica, a completely blind female successfully nested and returned to the sea. Since many of these females appear to migrate from feeding grounds several hundred kilometers away in Equador,

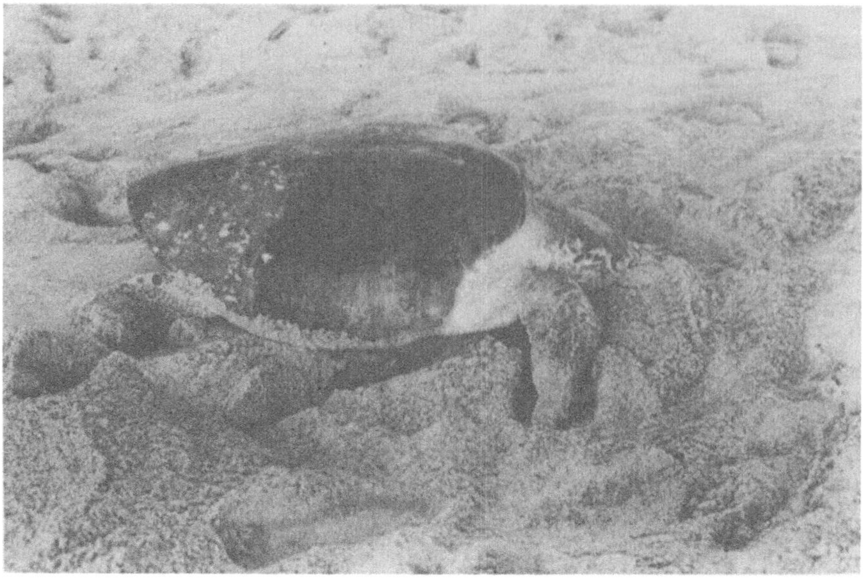

Fig. 2. A female olive ridley sea turtle, Lepidochelys olivacea, with head thrust down into the sand as she crawls up the beach to nest. Researchers have thought that this may be an olfactory behavior important in the individual's identification of the currect nesting beach.

Mora and Robinson felt this observation strongly supported olfactory imprinting as proposed by Carr. These authors also felt that their observation ruled out an alternative hypothesis (Hendrickson, 1958; Owens et al., 1982) which is based on an animal learning the location of the nesting beaches as an adult. While Mora and Robinson's observation does lend itself to a chemoreception-based interpretation, we feel its primary value is to reduce vision from the status of the critical sense for homing in these animals. The observation does not rule out the adult learning hypothesis, since chemoreception could be equally as important in this later model as it is to imprinting. We have reviewed the imprinting theory, discussed its implications and suggested the above alternative based on adults learning the migratory routes using chemical and possibly other cues (Owens et al., 1982).

In an attempt to extend the above mentioned work of Manton, where operant conditioning was used, our lab has conducted several experiments designed to look at olfactory behavior particularly as it might relate to imprinting (Table 1). It would be valuable to test imprinting in adults; however, the long age to sexual maturity in these species has made such a test quite problematical. As a consequence, we have chosen to look more closely at juvenile turtles. Two important points were considered in these experiments. First, a negative result with an immature animal can not disprove imprinting by chemoreception, since chemoreception may not be the only sensory system involved, and since we would not normally expect a juvenile to make a reproductive migration anyway. Rather, only a positive result might add credibility to the olfactory imprinting hypothesis. The second point is an experimental assumption which is both related to point one and apparently untested in olfactory imprinting research. We assume that a truly imprinted juvenile animal would retain an active receptor/CNS recognition system for the imprinted cues, but that the juvenile would not normally respond (migrate) to these cues even when the cues are present. The behavioral

Table 1. Aquatic Turtle Behavior as It Relates to Odor Imprinting.

	Experiment	Species	Animal Condition	Result	Reference
1.	Food Imprinting	<u>C. caretta</u>	Excellent	Only transitory preferences	Grassman and Owens, 1982
2.	Artificial Imprinting	<u>C. caretta</u>	Excellent	Suggestive preference but not clear-cut	Owens et al., 1982
3.	Padre Island Imprinting	<u>L. kempi</u>	Excellent	"Preferred" home	Grassman et al., 1984
4.	Padre Island Rancho Nuevo Imprinting	<u>L. kempi</u>	Poor	No preferences	Grassman, 1984
5.	Artificial Imprinting	<u>C. mydas</u>	Excellent	"Preferred" home water	Grassman, 1984

response threshold for imprinting cues would be much higher than for the more appropriate juvenile behaviors of feeding, seeking shelter, etc. However, we reasoned that if juveniles were placed in an arena where all normal age-appropriate behaviors were not possible, they might show some characteristic locomotory behavior indicative of responding or recognizing the imprinting cue. Of course, as adults, the threshold for the migratory imprint cue response would be reduced by the endocrine system to the point where all orientation behaviors would be dedicated to pursuing the correct migratory route. Supporting this, we have found that steroids will reduce feeding and initiate mating behavior in juvenile turtles (Owens, 1976). Our assumption is that the receptor system for imprinting is functional in the juvenile but not normally used due to its very low behavioral priority. In our studies, we had the juvenile turtles as well as the testing arena available (Kleerekoper, 1977) so why not look for a distinctive behavior.

Experiments 1 and 2, noted in Table 1, have been reviewed elsewhere (Owens et al., 1982). Briefly, we found that loggerheads would "imprint" in a transient way to early food exposures, that they could use chemoreception to locate the source of foods and that the "imprint" was not permanent or irreversible in these juveniles. In experiment 2, loggerheads were exposed to morpholine or phenylethanol in their nests (as eggs) and tanks (as hatchlings) or just in their tank water. The animals were tested in an aquatic arena with four compartmental choices. The results of this experiment were interesting, but not conclusive. One treatment of animals exposed to morpholine in their nest and water did show a significant elevation in cue compartment entries, but this observation was confounded by a control group which was equally "attracted" to the morpholine.

In experiment 3, Kemp's ridley juveniles from Mexico that had been artifically imprinted to Padre Island, Texas, were tested (Grassman et al., 1984). These turtles were borrowed from the international conservation program which is using "artificial imprinting" to attempt to establish a new nesting beach for this endangered species. Our four-compartment arena was used once again, and the animals were individually evaluated for number of entries, total time spent and time spent per entry into compartments treated with: 1) Padre Island seawater used to wash Padre Island sand; 2) Galveston Island seawater used to wash Galveston Island sand; or 3) unmodified experimental tank water (two compartments). Turtles entered Galveston-treated water more frequently than either Padre-treated or unmodified water. Although they did not enter the Padre-treated water as often as Galveston-treated water, the turtles spent a similar length of time in the two treated compartments. Thus, the turtles spent significantly more time per entry in the Padre Island seawater. That is, once they found the Padre Island seawater they stayed in it for longer periods. We feel that this observation is consistent with the likelihood that the turtles remembered the Padre Island exposure of several months earlier and were actually showing a preference to remain bathed in the solution from the proposed imprint environment. Unfortunately, this experiment lacked the reverse treatment in which turtles actually hatched at their Mexican home beach were evaluated in this same type of protocol. When we attempted such an experiment two years later, the results were discarded due to clearly unhealthy experimental animals (Grassman and Owens, in press).

In a more recent experiment (Grassman, 1984), green sea turtles were used. This experiment has proven to be the most interesting to date. Groups of hatchlings were "artificially imprinted" to morpholine and phenylethanol in a protocol similar to that of experiment 2 (Table 1). Treatment groups were exposed to one of the two chemicals at 5×10^{-5}molar concentrations either in the nest alone, in both the nest and tank water or in the tank water alone (total of six treatments). Using the four choice chamber once again, the turtles were evaluated two months after exposure to the chemicals. Only the turtles in the two treatments exposed to their "home" chemical in both the nest and later in their rearing tanks showed a significant and consistent response. In terms of average time spent in each compartment, the two groups each preferred the chemical to which they had been exposed (home) over the opposite chemical which served as a control.

Our interpretation of this experiment is that if turtles receive a relatively long exposure to an olfactory cue (e.g., both in the nest and tank water), they can learn (imprint to?) that cue and at a later time show a characteristic response to that cue. In this case, the response is a specific orientation pattern to the chemical used.

While this last experiment appears to satisfy the classical definition of imprinting (Lorenz, 1937), we have no way of knowing if these juvenile turtles could remember specific cues in two to three or more decades when, as adults, they would normally begin their reproductive migration in search of a breeding beach.

We suggest that the combined results of these laboratory experiments add to the results of Manton (see above) in the following ways:

1. Sea turtles can orient to specific chemical cues learned earlier in their lives without initial formal conditioning.

2. Sea turtles can orient to and distinguish between low concentrations of solutions prepared from natural beaches.

3. When no other stimuli are available, immature animals appear to show imprinting (e.g., an attraction) to chemicals to which they have had an early exposure.

4. In these experiments, exposure both in the nest (we did not separate pre-hatching from post-hatching in the nest) and in the early tank water was necessary for any consistent subsequent responses. This amounts to a fairly prolonged exposure of up to several weeks.

These experiments do not prove that chemosensory imprinting to the natal beach occurs in sea turtles. We do believe, however, that this research has added to the credibility of the olfactory (chemsensory) imprinting hypothesis as originally proposed by Carr (1967).

CONCLUSION

Researchers have not found it easy to design experiments that clearly indicate chemoreception to be the primary orientation system in amphibians and reptiles. In many cases (e.g., newts such as *Taricha* or chelonians such as *Clemmys* or *Chelonia*) chemoreception has been singled out as important for homing more by a process of elimination of the other senses than by sound laboratory and field research.

In some cases within the amphibia, there appear to be contradictory research results. For the most part, these contradictions are best explained by species-specific differences in what types of cues are actually available for the animals to use. Through evolution, a given species complex will certainly take advantage of whatever sensory mode has worked best in its habitat.

Chemoreception is probably important to sea turtle orientation; however, to date only the most preliminary laboratory studies have been conducted. Because of their awkwardly large size, their endangered species status, and the inaccessibility of these populations, good field studies of chemoreception have not been attempted.

On the surface one might think that the sense of smell in amphibian and reptile orientation and homing is a fairly obscure and useless corner of the zoological research world. We believe this is untrue for at least one very practical reason. Many amphibians and reptiles are threatened or endangered species. As man continues to dump the waste products of his high technology civilization into our aquatic systems, we run the increasing risk of sublethally altering the normal behavior of these animals. If chemoreception is of primary importance in the behavior of endangered aquatic species, increasing pollution levels may hasten the demise of these already depleted populations.

ACKNOWLEDGMENTS

We thank Molly Zachery and Cristina Germain who typed the manuscript. The original research cited in this paper was supported by the Texas A & M Sea Grant College Program under grant #NA81AA-D-0092. The authors were supported during the preparation of the manuscript by National Science Foundation grant #BNF-8418538 and by the Texas A & M Sea Grant College Program grant #NA83AA-D-00061.

REFERENCES

Balazs, G. H., 1980, "Synopsis of Biological Data On The Green Turtle In The Hawaiian Islands," NOAA Technical Memorandum NMFS.

Beuerman, R. W., 1975, Slow potentials of the turtle olfactory bulb in response to odor stimulation of the nose, Brain Res., 7:61.

Beuerman, R. W., 1977, Slow potentials in the turtle olfactory bulb in response to odor stimulation of the nose and electrical stimulation of the olfactory nerve, Brain Res., 128:429.

Booth, J., and Peters, J. A., 1972, Behavioural studies on the green turtle (Chelonia mydas) in the sea, Anim. Behav., 20:808.

Brown, W. S., and MacLean, F. M., 1983, Conspecific scent-trailing by newborn timber rattlesnakes (Crotalus horridus), Herpetologica, 39:430.

Burghardt, G. M., 1970, Chemical perception in reptiles, in: "Communication by Chemical Signals, Vol. 1," J. W. Johnston, Jr., D. G. Moulton, and A. Turk, eds., Appleton-Century-Crofts, New York.

Burghardt, G. M., 1980, Behavioral and stimulus correlates of vomeronasal functioning in reptiles: Feeding, grouping, sex and tongue use, in: "Chemical Signals, Vertebrates and Aquatic Invertebrates," D. Muller-Schwarze, and R. M. Silverstein, eds., Plenum Press, New York.

Carr, A., 1967, "So Excellente a Fishe," Natural History Press, New York.

Carroll, T. E., and Ehrenfeld, D. W., 1978, Intermediate-range homing in the Wood Turtle, Clemmys insculpta, Copeia, 1978:117.

Chelazzi, G., Calfurni, P., Graninetti, A., Carla, M., Delfino, G., and Callen, C., 1981, Modification of homing behavior in Testudo hermanii Gmelin (Reptilia:Testudinidae) after intranasal irrigation with zinc sulfate solution, Monit. Zool. Ital., 15:306.

Chiszar, D., Wellborn, S., Wand, M. A., Scudder, K. M., and Smith, H. M., 1980, Investigatory behavior in snakes. II. Cage cleaning and the induction of defecation in snakes, Anim. Learn. Behav., 8:505.

Dole, J. W., 1968, Homing in leopard frogs, Rana pipiens, Ecology, 49:386.

Dundee, H. A., and Miller, M. C., III, 1968, Aggregative behavior and habitat conditioning by the prairie ringneck snake, Diadophis punctatus arnyi, Tul. Stud. Zool. Bot., 15:41.

Duvall, D., 1979, Western fence lizard (Sceloporus occidentalis) chemical signals. I. Conspecific discriminations and release of a species-typical visual display, J. Exp. Zool., 210:321.
Duvall, D., 1980, Phermonal mechanisms in the social behavior and communication of the western fence lizard, Sceloporus occidentalis biseriatus, Dissertation Abstracts International, 41, 1981.
Duvall, D., 1981, Western fence lizard (Sceloporus occidentalis) chemical signals. II. A replication with naturally breeding adults and a test of the Cowles and Phelan hypothesis of rattlesnake olfaction, J. Exp. Zool., 218:351.
Duvall, D., 1982, Western fence lizard (Sceloporus occidentalis) chemical signals. III. An experimental ethogram of conspecific body licking, J. Exp. Zool., 221:23.
Duvall, D., Herskowitz, R., and Trupiano-Duvall, J., 1980, Responses of five-lined skinks (Eumeces fasciatus) and ground skinks (Scincella lateralis) to conspecific and interspecific chemical cues, J. Herpetol., 14:121.
Duvall, D., King, M. B., and Gutzwiller, K. J., 1985, Behavioral ecology and ethology of the prairie rattlesnake, Nat. Geogr. Res., 1:80.
Ehrenfeld, D. W., and Koch, A. L., 1967, Visual accommodation in the green turtle, Science, 155:827.
Evans, L. T., Preston, W., and Hardy, E., 1973, Imprinting in Chrysemys turtles, Am. Zool., 13:1269.
Evans, L. T., Preston, W. C., and McGeary, M. E., 1974, Imprinting in turtles, Am. Zool., 14:1278.
Ford, N. B., and Low, J. R., Jr., 1984, Sex pheromone source location by garter snakes: A mechanism for detection of direction in non-volatile trails, J. Chem. Ecol., 10:1193.
Grant, D., Anderson, O., and Twitty, V., 1968, Homing orientation by olfaction in newts (Taricha rivularis), Science, 160:1354.
Grassman, M. A., 1984, The Chemosensory Behavior of Juvenile Sea Turtles: Implication for Chemical Imprinting, Ph.D. dissertation, Texas A & M University.
Grassman, M. A., and Owens, D. W., 1982, Development and extinction of food preferences in loggerhead sea turtle, Caretta caretta, Copeia, 1982:965.
Grassman, M. A., and Owens, D. W., in press, An attempt at an experimental evaluation of imprinting in the Kemp's ridley sea turtle (Lepidochelys kempi), in: "Proceedings of the First International Symposium on Kemp's Ridley Sea Turtle Biology, Conservation and Management," N.M.F.S. Publications.
Grassman, M. A., Owens, D. W., McVey, J. P., and Marquez M., R., 1984, Olfactory-based orientation in artificially imprinted sea turtles, Science, 224:83.
Gregory, P. T., 1982, Reptilian hibernation, in: "Biology of the Reptilia, Vol. 13," C. Gans, and F. H. Pough, eds., Academic Press, London.
Grubb, J. C., 1970, Orientation in post-reproductive Mexican toads, Bufo valliceps, Copeia, 1970:674.
Grubb, J. C., 1973, Olfactory orientation in breeding Mexican toads, Bufo valliceps, Copeia, 1973:490.
Heller, S., and Halpern, M., 1981, Laboratory observations on conspecific and congeneric scent trailing in garter snakes (Thamnophis), Behav. Neural. Biol., 33:372.
Heller, S., and Halpern, M., 1982a, Laboratory observations of aggregative behavior of garter snakes, Thamnophis sirtalis, Comp. Physiol. Psychol., 96:967.
Heller, S., and Halpern, M., 1982b, Laboratory observations of aggregative behavior of garter snakes, Thamnophis sirtalis: Roles of the visual, olfactory, and vomeronasal senses, Comp. Physiol. Psychol., 96:984.
Hendrickson, J. R., 1958, The green sea turtle, Chelonia mydas (Linn.) in Malaya and Sarawak, Proc. Zool. Soc. Lond., 130:455.

Himstedt, W., and Plasa, L., 1979, Home-site orientation by visual cues in salamanders, Naturwissenschaften, 66:372.
Hirth, H. F., 1966, The ability of two species of snakes to return to a hibernaculum after displacement, Southwest. Nat., 11:49.
Ireland, L. C., 1979a, Homing behavior of immature green turtles, Chelonia mydas, Am. Zool., 19:952.
Ireland, L. C., 1979b, Homing behavior of juvenile green turtles, Chelonia mydas, in: "A Handbook on Biotelemetry and Radio Tracking," C. G. Amlander, and D. W. MacDonald, eds., Pergamon Press, Oxford.
Jaeger, R. G., and Gergits, W. F., 1979, Intra- and interspecific communication in salamanders through chemical signals on the substrate, Anim. Behav., 27:150.
King, M. B., McCarron, D., Duvall, D., Baxter G., and Gern, W., 1983, Group avoidance of conspecific but not interspecific chemical cues by prairie rattlesnakes (Crotalus viridis), J. Herpetol., 17:196.
Kleerekoper, H., 1977, Some monitoring and analytical techniques for the study of locomotor responses of fish to environmental variables, Am. Soc. Test. Materials Tech. Publ., 607.
Koch, A. L., Carr, A., and Ehrenfeld, D. W., 1969, The problem of open-sea migration: The migration of the green turtle to Ascension Island, J. Theor. Biol., 22:163.
Landreth, H. F., 1973, Orientation and behavior of the rattlesnake Crotalus atrox, Copeia, 1973:26.
Limpus, C. J., and Walter, D. G., 1980, The growth of immature green turtles (Chelonia mydas) under natural conditions, Herpetologica, 36:162.
Lorenz, K. Z., 1937, The companion in the bird's world, Auk, 54:245.
Madison, D. M., 1969, Homing behaviour of the red-cheeked salamander, Plethodon jordani, Anim. Behav., 17:25.
Madison, D. M., 1972, Homing orientation in salamanders: A mechanism involving chemical cues, in: "Animal Orientation and Navigation," S. Galler, K. Schmidt-Koenig, G. Jacobs, and R. Belleville, eds., N.A.S.A. Spec. Publ. 262:485.
Madison, D. M., 1975, Intraspecific odor preferences between salamanders of the same sex: Dependence on season and proximity of residence, Can. J. Zool., 53:1356.
Madison, D. M., 1977, Chemical communication in amphibians and reptiles, in: "Chemical Signals in Vertebrates, " D. Muller-Schwarze, and M. M. Mozell, eds., Plenum Press, New York.
Madison, D. M., and Shoop, C. R., 1970, Homing behavior, orientation, and home range of salamanders tagged with Tantalum-182, Science, 168:1484.
Manton, M. L., 1979, Olfaction and behavior, in: "Turtles: Perspectives and Research," M. Harless, and H. Morlock, eds., John Wiley and Sons, New York.
Manton, M., Karr, A., and Ehrenfeld, D. W., 1972a, Chemoreception in the migratory sea turtle, Chelonia mydas, Biol. Bull., 143:184.
Manton, M. L., Karr, A., and Ehrenfeld, D. W., 1972b, An operant method for the study of chemoreception in the green turtle, Chelonia mydas, Brain Behav. Evol., 5:188.
Martof, B. S., 1962, Some observations on the role of olfaction among salient amphibia, Physiol. Zool., 35:270.
McGavin, M., 1978, Recognition of conspecific odors by the salamander Plethodon cinereus, Copeia, 1978:356.
Mora, J. M., and Robinson, D. C., 1982, Discovery of a blind olive ridley turtle (Lepidochelys olivacea) nesting at Playa Ostional, Costa Rica, Rev. Biol. Trop., 30:178.
Mrosovsky, N., 1978, Orientation mechanisms of marine turtles, in: "Animal Migration, Navigation and Homing," K. Schmidt-Koenig, and W. T. Keeton, eds., Springer-Verlag, New York.
Noble, G. K., and Clausen, H. J., 1936, The aggregation behavior of Storeria dekayi and other snakes with especial reference to the sense organs involved, Ecol. Monogr., 6:269.

Oldham, R. S., 1967, Orienting mechanisms of the green frog, Rana clamitans, Ecology, 48:477.
Ortleb, E. P., and Sexton, O. J., 1964, Orientation in the painted turtle, Chrysemys picta, Am. Midl. Nat., 71:320.
Owens, D. W., 1976, The Endocrine Control of Reproduction and Growth in the Green Sea Turtle Chelonia mydas, Ph.D. Dissertation, U. of Arizona, Tuscon.
Owens, D. W., 1980, The comparative reproductive physiology of sea turtles, Amer. Zool., 20:549.
Owens, D. W., Grassman, M. A., and Hendrickson, J. R., 1982, The imprinting hypothesis and sea turtle reproduction, Herpetologica, 38:124.
Parsons, T. S., 1970, The nose and Jacobson's organ, in: "Biology of the Reptilia, Vol. II," C. Gans, and T. S. Parsons, eds., Academic Press, New York.
Parzefall, Von J., 1976, Die rolle der chemischen Information im Verhalten des Grottenolms Proteus anguineus Laru. (Proteidae, Urodela), Z. Tierpsychol., 42:29.
Patterson, R. G., 1971, Aggregation and dispersal behavior in captive Gopherus agassizi, J. Herpetol., 5:214.
Porter, R. H., and Czaplicki, J. A., 1974, Responses of water snakes (Natrix r. rhombifera) and garter snakes (Thamnophis sirtalis) to chemical cues, Anim. Learn. Behav., 2:129.
Scholz, A. T., Horrall, R. M., Cooper, J. C., and Hasler, A. D., 1976, Imprinting and chemical cues: The basis for home stream selection in salmon, Science, 192:1247.
Scudder, K. M., Stewart, N. J., and Smith, H. M., 1980, Response of neonate water snakes (Nerodia sipedon sipedon) to conspecific chemical cues (Reptilia, Serpentes, Colubridae), J. Herpetol., 14:196.
Stoddart, D. M., 1980, "The Ecology of Vertebrate Olfaction," Chapman and Hall, New York.
Tracy, C. R., and Dole, J. W., 1969, Orientation of displaced California toads, Bufo boreas, to their breeding sites, Copeia, 1969:693.
Tristram, D. A., 1977, Intraspecific olfactory communication in the terrestrial salamander Plethodon cinereus, Copeia, 1977:597.
Twitty, V. C., 1959, Migration and speciation in newts, Science, 130:1735.
Twitty, V. C., 1966, "Of Scientists and Salamanders," W. H. Freeman and Co., San Francisco.
Twitty, V. C., Grant, D. L., and Anderson, O., 1964, Long distance homing in the newt, Taricha rivularis, Proc. Nat. Acad. Sci. USA ., 51:51.
Twitty, V. C., Grant, D. L., and Anderson, O., 1966, Course and timing of the homing migration in the newt, Taricha rivularis, Proc. Nat. Acad. Sci. USA., 56:864.
Twitty, V. C., Grant, D. L., and Anderson. O., 1967a, Home range in relation to homing in the newt, Taricha rivularis, Copeia, 1967:649.
Twitty, V. C., Grant, D. L., and Anderson, O., 1967b, Initial homeward orientation after long distance displacement of the newt, Taricha rivularis, Proc. Nat. Acad. Sci. USA., 57:342.
Witzell, W. N., 1983, Synopsis of biological data on the Hawksbill turtle Eretmochelys imbricata (Linnaeus, 1766), FAO Fisheries Synopsis No. 137.

THE ECOLOGICAL AND EVOLUTIONARY CHALLENGES OF PROCELLARIIFORM OLFACTION

Bernice M. Wenzel

Department of Physiology
University of California
Los Angeles, CA 90024

INTRODUCTION

Once controversial, it is now well established that certain avian species use volatile chemical cues. This is the case even though no vomeronasal organ (Portmann, 1961) or accessory olfactory bulb has been found in adult birds. The existence of a septal organ and nervus terminalis in birds is unclear and needs direct study. It has been said, with no elaboration or references, that the terminal nerve occurs in all vertebrates (Ariëns Kappers et al., 1936) except cyclostomes (Larsell, 1950). Bang (1971) makes no mention of a septal organ. Peripheral olfactory structures - olfactory mucosa, nerve, and bulb - have been well documented in representatives of all 23 orders of living birds (Bang, 1971) and presumably exist in all species. The record shows, in addition, that these structures differ widely in size among the orders, and within the orders as well when enough specimens were available to provide good intraorder comparisons. Although the size of an organ is not irrefutable proof of its functional importance, it is customarily taken as strong evidence for its relative significance. A recent example is the use of the 4-fold increase in size of the male vocal control nuclei in the canary brain as a measure of its greater importance for male behavior (Nottebohm et al., 1981). In addition to the olfactory structures themselves, the avian nasal cavity is designed, sometimes very elaborately, to facilitate access of inspired air to the olfactory receptor cells and to maintain those cells in optimal condition (Bang and Wenzel, 1985). Were the nasal passages for respiration only, they would be far simpler, vascularized, mucociliated tubes. Based on anatomical features alone, therefore, some olfactory reliance would be expected in all birds, with greater reliance in those taxa having large amounts of olfactory tissue. Species with small amounts, in response to evolutionary pressures, may have retained or developed a single specialized olfactory function requiring a smaller number of olfactory cells than the more extensive uses favored by another species. Alternatively, fewer olfactory cells may indicate less critical use of them, perhaps like the human situation in which odors provide mainly affective, rather than critical cognitive, information. Whatever the facts may be, little research has been designed to discover them. This paper will deal with the challenging research questions raised by the one avian order, Procellariiformes, in which the amount of olfactory tissue is very large in almost all members, and the following paper with the order Passeriformes, in which almost all members have the smallest amounts of such tissue.

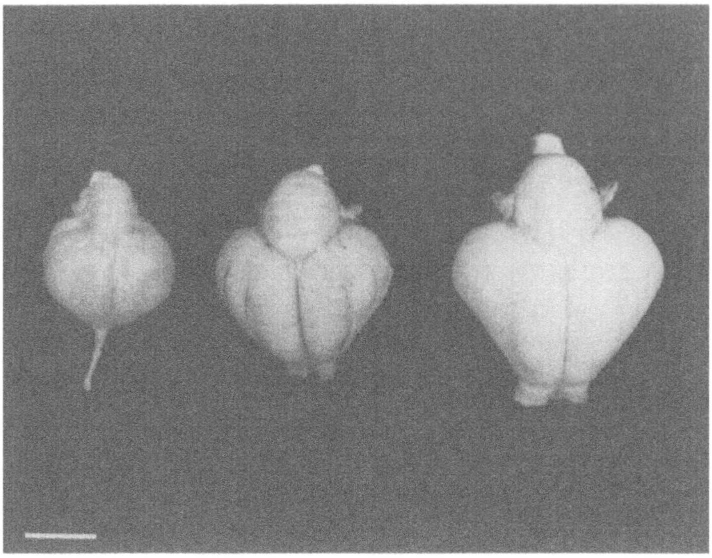

Fig. 1. From left to right, dorsal views of brains of Rock Dove, Cape Petrel, and Antarctic Fulmar. The olfactory nerve bundle remains attached to the pigeon's bulbs. Anterior is down. Calibration line is 1 cm.

Procellariiform birds include four families, viz., Diomedeidae - albatrosses; Procellariidae - fulmars, prions, petrels, and shearwaters; Oceanitidae - storm-petrels; and Pelecanoididae - diving-petrels. Because they all possess tubular nostrils, they are sometimes referred to collectively as the tubenoses. These tubes drain nasal salt glands, which aid excretion of the excess salt ingested with seawater. Peripheral olfactory structures have been examined in at least one representative of each family, specifically in 11 of the approximately 93 species in the order (Bang, 1971). In their classic work on olfactory bulb size in 108 avian species, Bang and Cobb (1968) calculated for each species the ratio of the largest diameter of one olfactory bulb to the largest diameter of the ipsilateral cerebral hemisphere. Of the 11 procellariiforms included, 10 ranked among the first 12 birds in order of bulb size, with bulb-hemisphere ratios of 27 to 37%. Further examination of nasal anatomy showed that these species had remarkably extensive olfactory mucosa spread over elaborately scrolled olfactory conchae (Bang, 1971). Other aspects of their nasal architecture favored efficient delivery of copious air streams to the mucosa. Figures 1 and 2 show the brains of several procellariids, all but one from high southern latitudes. Compared to the Rock Dove (Columba livia), whose ratio (18%) is average, the proportional size of their bulbs is clearly very large. In several cases the massive olfactory conchae are also shown.

The single known exception to the apparently uniform olfactory gigantism among the procellariiforms is the Georgian Diving-Petrel (Pelecanoides georgicus), which has only an average bulb ratio; presumably its other 3 family members would be similar. Their behavior also differs from their ordinal relatives in that they feed by pursuit diving and "fly" under water as do penguins and alcids. Murphy (1936, p. 781) commented on the many striking similarities in morphology between the Magellan Diving-Petrel (P. magellani) and the Little Auk (Alle alle) of Arctic seas, describing the resemblance as "one of the best examples of convergent evolution known among vertebrates." Indeed, Simpson (1976) has even characterized the diving-petrels as well on their way to becoming penguins. By two methods of

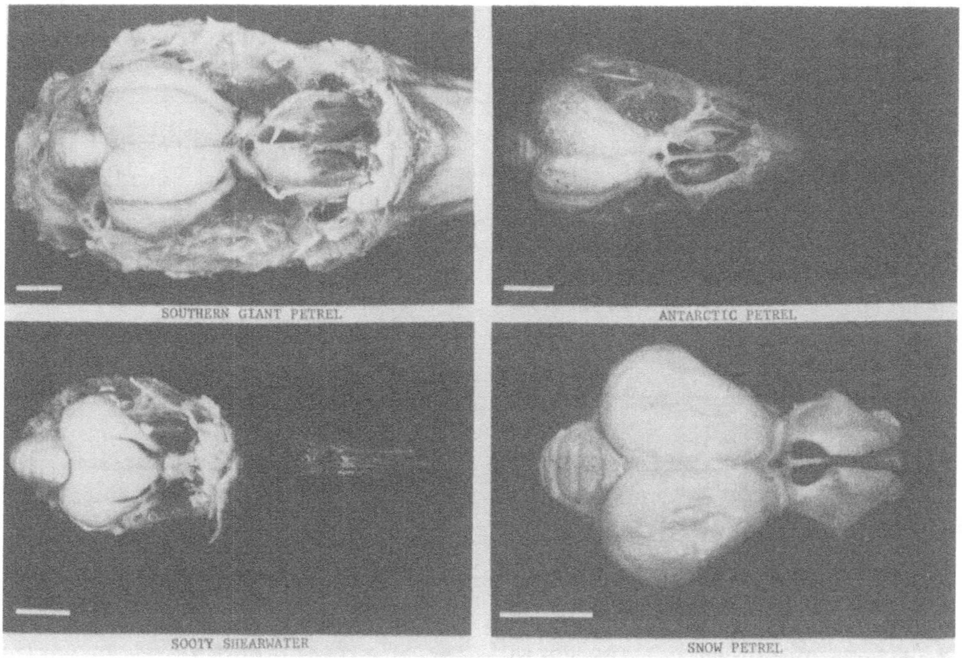

Fig. 2. Dorsal views of brains in situ of Southern Giant Petrel, Antarctic Petrel, Sooty Shearwater, and of the dissected brain of Snow Petrel. At least part of the olfactory conchae is shown in each case. Anterior is on the right. Calibration line is 1 cm.

chemical analysis, however, they are closely related to the other procellariiforms and are properly classified in that order (Harper, 1978; Jacob, 1976; Jacob and Ziswiler, 1982); no DNA-DNA hybridization data have been reported.

We have examined the internal structure of the olfactory bulb in two of the top-ranking species, the Northern Fulmar (Fulmarus glacialis) and the Sooty Shearwater (Puffinus griseus). As would be expected from the massive olfactory bulbs, the number of mitral cells in the bulb is high and their size is also large (Fox et al., 1983). Table 1 shows some values in comparison with the Rock Dove. Mitral cell counts for the rabbit and rat, unquestioned macrosmats, are 45,000 and 60,000, respectively (Allison and Warwick, 1949; Meisami, 1978), and our estimates for the Sooty Shearwater and Northern Fulmar are comparable. The bulb's internal structure (Fig. 3) shows the layers now well known for mammals and reptiles in the fulmar, shearwater, and pigeon but the greatly reduced, fused sparrow bulb shows scarcely a recognizable mitral cell let alone the regular layered organization. Modern bulbs in all of the higher vertebrate classes appear to have evolved along similar lines in their basic pattern. From olfactory morphology at both the gross and light microscopic levels, therefore, the expectation would be for a keen, heavily-used sense of smell in procellariiform birds.

Table 1. Estimated volumes (cc) of one cerebral hemisphere and one olfactory bulb in the Northern Fulmar, Sooty Shearwater, and Rock Dove, and the length (μm) of mitral cells in the olfactory bulb.

	Hemisphere	Bulb	Mitral Cell Length
Fulmar	1.87	.06	22
Shearwater	1.32	.02	23
Rock Dove	0.42	.0001	11

CHEMICAL SIGNALS

Prey

What signals might such a powerful system be responding to? An obvious possibility is odors associated with prey organisms. The food of pelagic predators usually contains odorous oils in the living state and becomes markedly odorous after death. The normal diet of procellariiform birds varies across species but includes zooplanktonic and nektonic organisms, such marine invertebrates as certain coelenterates, crustaceans and cephalopods, small

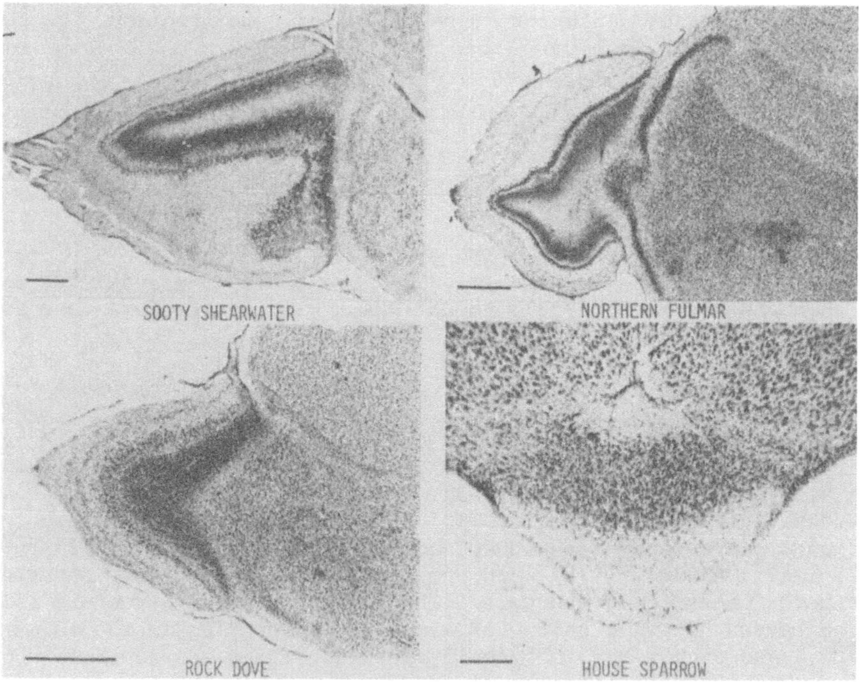

Fig. 3. Sagittal photomicrographs of olfactory bulb of Sooty Shearwater, Northern Fulmar, and Rock Dove; coronal view of House Sparrow olfactory bulb. Calibration line is 0.5 mm for the first three and 0.1 mm for the sparrow.

bony fishes, portions of cetacean carcasses, and other carrion and offal. The evidence is now strong that prey detection depends at least partly on chemical signals arising directly from the prey organisms themselves or from processes in association with them. Reports from many skilled observers of these birds' behavior patterns imply such dependence (see Wenzel, 1980 for summary). Perhaps the earliest is that of Collins (1882, p. 317), who described "tolling" up storm-petrels and shearwaters by using cod livers. "On many occasions, during the prevalence of a dense fog, when not a bird of any kind had been seen for hours, I have thrown out, as an experiment, pieces of liver to ascertain if any birds could be attracted to the side of the vessel. As the particles of liver floated away, going slowly astern of the schooner, only a short time would pass before either a Mother Carey chicken [storm-petrel] or a hag [shearwater]...could be seen coming up from the leeward out of the fog, flying backward and forward across the vessel's wake, seemingly working up the scent until the floating pieces of liver were reached...Should the first one to appear be a hag...with a greedy yell he pounces upon the pieces of liver and swallows them...[G]enerally...a flock of hagdons soon gather whatever may be the density of the fog, unless birds are very rare on the Bank or, perhaps, rendered indifferent to food by a recent feast." Murphy (1936) credits Rollo Beck with originating the practice of "luring" procellariiforms with oil and meat on the water but, according to Collins's report, it seems to have been a common part of fishermen's knowledge, at least in some parts of the world. Although Collins had been "assured" by Elliott Coues that smell was a rudimentary sense in birds, his observations were so compellingly dissonant with that advice that he deviated from the central theme of his report in order to describe them.

Sooty Shearwaters have been identified as a principal catalyst in the formation of mixed-species offshore feeding flocks in Alaskan waters (Hoffmann et al., 1981). From their descriptions, it is clear that the authors did not consider anything but vision in guiding the shearwaters to concentrations of prey. It would be helpful if future investigators working on feeding flock formation would treat olfactory guidance as a working hypothesis when making their observations.

Recent experiments have provided controlled observations and necessary support for earlier field reports suggesting olfactory guidance in procellariiform foraging. All have used cod liver oil as the standard attracting stimulus. Wilson's Storm-Petrel (Oceanites oceanicus) approached the lure in both its Antarctic breeding territory (Jouventin and Robin, 1984) and its North Atlantic non breeding ground (Grubb, 1972). In the Antarctic, Snow Petrels (Pagodroma nivea), Cape Petrels (Daption capense), and Southern Giant Petrels (Macronectes giganteus) were also attracted although Antarctic Fulmars (Fulmarus glacialoides) passed by the lure without showing any awareness of it (Jouventin and Robin, 1984). The brains of these four species are shown in Fig. 2.

In more extensive studies (Hutchison and Wenzel, 1980; Hutchison et al., 1984) under natural conditions at sea off central California, the release of odors of cod liver oil, squid homogenate, or krill homogenate was followed by the appearance of Northern Fulmars and Sooty Shearwaters from downwind, most of which flew very close to the buoy with its releasing wick and many of which made repeated approaches in an investigative manner. This behavior was intensified when the volatile fraction of cod liver oil was used as the stimulus. Other seabirds in the area showed no such systematic behavior. Virtually no fulmars or shearwaters appeared during tests with control substances. These data were collected at different times of year and under a wide range of weather conditions during the nonbreeding seasons of the species involved. There were no heavy concentrations uniformly nearby, therefore, and their routine foraging behavior was being expressed. Odor release was started only when no birds could be seen and the first fulmar or shearwater appeared minutes later. In its variable dispersal, the odor plume weakened like that of an insect pheromone. Detection in some cases very likely depended on a small number of molecules. Since no feeding occurred and there was no visual target material on the water, no flocks formed. The situation appeared to be a pure test of olfactory behavior.

Many of the smaller procellariiforms regularly feed at night, especially while caring for young, presumably to avoid predation. In our Pacific studies, Ashy Storm-Petrels (Oceanodroma homochroa), which were breeding about 90 km away, were attracted on nights with heavy overcast and no moonlight. No boat lights were used and observations were made with a night-viewing device (Wenzel, 1985). Storm-petrels, viz., Wilson's, Leach's (Oceanodroma leucorhoa), or both (species could not be identified), have also been attracted to cod liver oil at night, although light levels were not described (Grubb, 1972). These results suggest that night foraging does not depend solely on visual sighting of bioluminescent organisms. Although a study of squid intake for three albatross species (Imber and Berruti, 1981) found little material from nonluminescent varieties in the birds' castings, no direct work on the importance of luminescence as a signal has been done.

Secretory Products

Do procellariiforms, or any birds, produce chemicals that would be useful for intraspecific communication? Several external glands have been identified although almost no work has been done on most of them. All but a very few birds have a uropygial gland, a compact glandular organ at the base of the tail. It secretes a lipid substance, which the bird applies to its feathers. This coating is especially important for waterbirds because it aids water repellency. Proportional glandular size is largest in waterbirds and in terrestrial forms, like the Osprey (Pandion haliaetus), that enter water as normal behavior. Uropygial secretions have been analyzed extensively (Jacob and Ziswiler, 1982), especially for taxonomic purposes since their quantitative and qualitative characteristics vary across taxa. The waxes themselves are only weakly odorous; some constituents are very smelly, however, and probably contribute to the well-known powerful odor of many procellariiforms. The interspecific variation in constituent patterns implies a genetic component and, thus, a possible basis for individual recognition by chemical signals.

The preen substance contributes to feather hygiene and resilience as well, and may serve a pheromonal function in some species. Gland weight in the European Rook (Corvus frugilegus) doubles during the breeding season (Kennedy, 1971), for example. Chemical composition of the glandular secretion in the female Mallard Duck (Anas platyrhynchos) changes during the breeding period of February to June; the remainder of the year it does not differ from that of males and ducklings (Jacob et al., 1979). The sexual behavior of male Mallards given bilateral olfactory nerve section was significantly inhibited although their other behavior appeared normal (Balthazart and Schoffeniels, 1978). It seems likely, therefore, that the female's uropygial secretion normally stimulates the male.

Many birds also have internal or external anal glands, distinct from cloacal glands (Quay, 1967). They are mucus-secreting but their function is not understood and no chemical studies of the secretion have been reported. Both size and location vary across species, with the largest of Quay's samples in Columbidae, Cuculidae, and Scolopacidae. He included only one procellariiform, the Ashy Storm-Petrel, which had relatively small internal anal glands. These structures are obvious candidates for careful study in avian chemical signaling. In addition to anal glands, there are sebaceous glands in the orbit and external ear, as well as cells throughout the skin that secrete sebaceous material; little attention has been paid to these structures or their secretions. The Harderian gland, in the orbit, is said to be very large in "marine" birds (Prince, 1956, p. 305).

Stomach oil, produced in the proventriculus from dietary constituents, is another prominent odor source in all of these birds except the diving-petrels (Jacob, 1982; Warham, 1977). The oil is believed to play a nutritious role in slowing digestion and retaining nutritive material for later absorption. In addition, it is forcibly ejected for defense by several species, especially as nestlings. Although there is intra-specific variation in its chemical composition and its color because it originates from ingested food, which varies somewhat for any given species according to availability, interspecific variation is even greater so that it, too, could be under enough genetic control to permit individual recognition. Just as the smaller procellariiforms forage more safely in dim light, so must they return to their nest sites in burrows or cliff crevices in darkness. The smell of the crevice or burrow, which is a mix of odors from stomach oil, uropygial gland secretion, and feces, may facilitate homing at least when ambient illumination is very low. Snow Petrel nests in Antarctica are described as heavily encrusted with stomach oil residue, samples of which have been radiocarbon dated as 4-9,000 years old (Hiller and Wand, 1984). Leach's Storm-Petrels, homing to a heavily wooded island, proceeded upwind to their burrows if they were breeding; nonbreeders oriented less consistently into the wind (Grubb, 1974). Walking upwind to breeding burrows was not the mode for Wedge-tailed Shearwaters (Puffinus pacificus), however, on a near-shore island off Oahu with only low vegetation (Shallenberger, 1973), nor for Leach's Storm-Petrels on a similarly open island off Baja California (V. S. Fox, unpublished observations). In both of these cases, ambient illumination was much higher than in Grubb's study. After olfactory nerve section, none of Grubb's 23 birds and only 3 of Shallenberger's (1975) 12 returned to their burrows. Almost all control birds came back in both tests.

Reproductive Behavior

Increasing evidence supports a close relation between the mammalian vomeronasal system and the chemical signals of reproduction. Might the vomeronasal system be the obligatory route for such chemical signaling? If so, then its absence would imply the absence of sexual signaling by chemical stimuli. Duvall et al. (1983) have suggested that the reptilian ancestral stock had a vomeronasal system but that it disappeared in the development of modern crocodilians, birds, and certain mammalian groups including many primates. There is some evidence, however, that female primates of species without a vomeronasal system emit chemical signals about their reproductive status, which influence the behavior of conspecific males (Michael and Keverne, 1968). An alternative detection system for sex pheromones, the terminal nerve, has been proposed as significant across vertebrate classes (Demski and Northcutt, 1983) and presumably would be especially important in forms that have no vomeronasal system. Birds probably lack the terminal nerve also, although information is still very sparse. Functional sexual signaling by chemicals may be mediated entirely by the olfactory system in this class, an interesting possibility that deserves attention.

As suggested above, procellariiform behavior has components that could be interpreted as odor-dependent. The birds themselves are noticeably pungent to the human nose, unlike most other birds, and their nests are similarly odorous. Many burrow- and crevice-nesters return to the same sites each year, they rejoin the same mate each year, they home to their nesting island in darkness after foraging many miles away at sea, and they show little or no sexual dimorphism in plumage or size. The point has been made specifically about the Northern Fulmar that, despite nesting in large colonies with breeders spending much of the year at the site, there has been no "evolution of rich repertory of ritualized displays" (Cramp and Simmons, 1977, p. 125). No experimental studies have been done on how mates, chicks, or parents are recognized, there are virtually none on homing, and none at all on visual behavior or sensitivity. Shallenberger's (1975) report of increased heart rate in some Wedge-tailed Shearwaters when stimulated by mate odor as compared to nonmate odor is noteworthy as is Grubb's (1974) similar finding for Leach's Storm-Petrels to odors of their own burrows. A great deal more research of this kind is needed on procellariiform species and members of other orders for comparison. Pattern analysis of their glandular secretions, like that reported for two subspecies of tamarin monkey (Saguinus fuscicollis) (Smith et al., 1985) would provide critical information about individuality. Perhaps, like a bloodhound, the birds with very large olfactory systems are expert at recognizing unique identities by smell.

EVOLUTION

A complex of anatomical features in procellariiforms - structures in the nasal cavity, and the impressive size of the olfactory mucosa, nerve, and bulb - leads to the hypothesis that their olfactory system represents a specialized adaptation to selection pressures in adjusting to their pelagic habitat. As a group, their success has been outstanding. They have conquered all of the world's oceans, coming ashore solely for breeding, in some cases only biennially, usually on remote oceanic islands. Some of them migrate between deep southern and far northern latitudes, some circumnavigate the southern hemisphere, and some spend their entire lives at or below the Antarctic convergence. Their size variation is the largest of any avian order.

Procellariiforms are considered to be one of the oldest avian orders. Only penguins (Sphenisciformes), the flightless land birds (Apterygiformes), Rheiformes, and Casuariformes), loons (Gaviiformes), and grebes (Podicipeiformes) are listed ahead of them. Kiwis (Apteryx) are also macrosmats, both anatomically (Bang, 1971) and behaviorally (Wenzel, 1968). Loons may be somewhat above average in bulb size but grebes and penguins are not, based on the very limited evidence now available (Bang, 1971). Fossil data and several types of chemical analysis agree that procellariiforms and penguins are so closely related that they almost certainly arose from a common ancestral stock of aerial flyers. No record of such an ancestor has been found, however, and their divergence was complete by the early Eocene when most modern avian orders had appeared. We cannot say how much the modern penguin and procellariiform olfactory systems have changed from that of their progenitor. Indeed, the scarcity of reasonably intact fossil bird skulls from which brain casts could be formed makes it unlikely that a good estimate of bulb size can ever be made. In occupying their overlapping niches, the two groups adopted different sensory strategies. It may be that procellariiforms retained and perfected an existing system, exploiting it for aerial mastery of the sea as a feeding ground. Penguins are more like shorebirds, not at all as pelagic as the procellariiforms. The alcids, ecologically similar to the penguins in the northern hemisphere, are also more littoral in habit. They hunt from the surface of the ocean, or beneath the surface, and generally do not range above the water over the great distances regularly traveled by many procellariiforms. Vision would be useful for finding prey from the penguin's position in the water but less so from the procellariiform's above the water, and hardly at all at a distance or in poor visibility. Hearing would likewise contribute little information about underwater food sources. Of the major distance receptors, therefore, the nose is the only one left to rely on for information about remote points and regardless of weather. To rely on chemical stimuli in a frigid climate, however, would require extraordinary sensitivity because volatility is very low indeed at low temperatures. At the time of the great avian radiation, in the early Cenozoic era, the earth's climate was more tropical than in the Quaternary period so that the usefulness of a keen olfactory system in the cold was probably not an original selection factor. More likely, the system developed as an adaptation for aerial foraging, and possibly homing, at sea and then became an exaptation for life in colder climates as temperatures dropped. Only one of the approximately 17 modern tropical species of procellariiforms has been examined for olfactory anatomy, viz., the Wedge-tailed Shearwater. Its bulb-hemisphere ratio is in the same range as the 10 that have been measured from subtropical and polar regions. A quantitative comparison of tropical, subtropical, and polar species is needed.

The probably universal occurrence of large olfactory mucosae and bulbs in three of the four procellariiform families comprising over 90% of the species compels careful behavioral study. These birds unquestionably are reacting to odors in significant ways. How does such dependence interact with reliance on visual and other sensory cues? The prominence of their olfactory equipment implies some degree of functional dominance for smell. Our present understanding of olfactory mechanisms does not tell us what specific advantages reside in more, rather than less, olfactory tissue. The assumption is that finer, more numerous discriminations can be made, and that greater redundancy may be available for critical discriminations. When the general benefits of a large olfactory bulb are clarified, the implications for procellariiform behavior will suggest useful hypotheses for experiments. Until then, field biologists working on such species should constantly ask what the birds might be smelling and how the odors might be shaping their behavior. Answering these questions has hardly begun.

SUMMARY

The once controversial concept of functional olfaction in birds is now firmly established on the basis of anatomical, behavioral, and physiological evidence. Olfactory structures are present in all birds examined and their general characteristics are the same as those of reptilian and mammalian olfactory systems. A notable difference in chemoreceptive systems appears to be the absence in birds of both the vomeronasal system and the terminal nerve.

Procellariiform birds have an especially large number of olfactory receptor cells with correspondingly large olfactory bulbs. This group includes albatrosses, petrels, fulmars, prions, shearwaters, storm-petrels, and diving-petrels. Only the latter's olfactory system seems to be average in size. All species in the order are highly pelagic. Except when breeding, usually on remote oceanic islands, they spend all of their time at sea. Following up on many accounts of apparent olfactory guidance to feeding locations, field experiments at sea have confirmed that procellariiform species, unlike other seabirds, can be attracted from downwind by the smell of cod liver oil and other food odors. Furthermore, they exhibit investigative behavior around the odor source. In addition to utilizing volatile cues in foraging, such cues may also aid homing. It is characteristic of these species to return each year to the same breeding site, mate, and burrow in the case of burrow-nesters. The smaller forms forage mainly at night when the role of vision would be diminished.

The possibility of conspecific chemical signals has scarcely been considered for birds. Given the prominent olfactory anatomy of procellariiforms, however, it should be worth considering. Not only do they possess a uropygial gland, whose lipid secretion has a unique composition at least at the species level, but as in most birds there are also anal glands as well as sebaceous glands in the ear and skin. The composition of the mature female mallard's uropygial secretion changes during the breeding period as does the sexual behavior of olfactory nerve-sectioned mature males, indicating a cueing effect of the female's secretion. A special procellariiform feature is a strong-smelling stomach oil, which is ejected by some species in defense and is fed to chicks. Nest sites and the birds themselves are very odorous.

According to present fossil and chemical evidence, modern procellariiforms and penguins are closely related and are assumed to share a common flying ancestor. The olfactory system has been measured in only one penguin and proved to be average in size. It is possible that the very large procellariiform system is an adaptation for the strongly pelagic life these species lead, and has been of special benefit for adapting to life in extremely cold climates where volatility is low and where these birds are plentiful.

Many unanswered questions are raised by the facts presented here. The following are a sample of the research topics suggested.

Are the vomeronasal and terminal nerve systems completely absent in birds? If so, what is the significance of their disappearance or failure to develop in this class?

Do all procellariiform species, with the possible exception of the diving-petrels, have extremely large olfactory mucosae and bulbs? Are those of tropical species smaller than those of subantarctic and antarctic species? If diving-petrels are different, why?

What odors provide the cues followed in natural foraging?

How do olfaction and vision interact in foraging? How sensitive is the visual system?

Are procellariiforms a common catalyst for mixed-species feeding flocks at sea?

What complex of distance cues serves to guide homing to the breeding site and to the individual nest or burrow? Is olfaction a dominant one, especially for those species that return in darkness or dim light?

Do the anal and sebaceous glands secrete odorous material? Does any of it serve as signals for reproductive behavior or as bases for individual recognition?

Were the olfactory systems of fossil procellariiforms and penguins more similar in size than those of modern representatives? How might we speculate about that of their progenitor, who is still unidentified?

ACKNOWLEDGMENTS

Some of the research described was supported by National Science Foundation grant no. BNS-8013049 and National Geographic Society grant no. 2224-80 to B. M. Wenzel and L. V. Hutchison. Dr. David Duffy, Percy Fitzpatrick Institute of African Ornithology, University of Cape Town, and Brian Obst, Department of Biology, University of California, Los Angeles, contributed heads of Antarctic specimens. Boat transportation for collection of local specimens was provided by Eldon Jeffrey and the specimens were collected by Dr. Kenneth Stager.

REFERENCES

Allison, A. C., and Warwick, R. T. T., 1984, Quantitative observations on the olfactory system of the rabbit, Brain, 72:186.

Ariëns Kappers, C. U., Huber, G. C., and Crosby, E. C., 1936, "The Comparative Anatomy of the Nervous System of Vertebrates, Including Man," Macmillan, New York.

Balthazart, J., and Schoffeniels, E., 1979, Pheromones are involved in the control of sexual behaviour in birds, Naturwiss., 66:55.

Bang, B. G., 1966, The olfactory apparatus of tubenosed birds (Procellariiformes), Acta Anat., 65:391.

Bang, B. G., 1971, Functional anatomy of the olfactory system in 23 orders of birds, Acta Anat., Suppl. 58, 79:1.

Bang, B. G., and Cobb, S., 1968, The size of the olfactory bulb in 108 species of birds, Auk, 85:55.

Bang, B. G., and Wenzel, B. M., 1985, The nasal cavity and olfactory system, in: "Form and Function in Birds," Vol. 3, J. McLelland, and A. S. King, eds., Academic Press, London.

Collins, J. W., 1882, Notes on the habits and methods of capture of various species of sea birds that occur on the fishing banks off the Eastern coast of North America, and which are used as bait for catching codfish by New England fishermen, U. S. Bur. Fisheries Rept., 311.

Cramp, S., and Simmons, K. E. L., Eds., 1977, "Handbook of the Birds of Europe, the Middle East and North Africa. The Birds of the Western Palearctic," Vol. 1, Oxford University Press, Oxford.

Duvall, D., King, M. B., and Graves, B. M., 1983, Fossil and comparative evidence for possible chemical signaling in the mammal-like reptiles, in: "Chemical Signals in Vertebrates 3," D. Muller-Schwarze, and R. M. Silverstein, eds., Plenum Press, New York.

Demski, L. S., and Northcutt, R. G., 1983, The terminal nerve: A new chemosensory system in vertebrates?, Science, 220:435.

Fox, V. S., Davidian, H., Hutchison, L. V., and Wenzel, B. M., 1982, Olfactory neuroanatomy of the Sooty Shearwater and Northern Fulmar, Pac. Seabird Grp. Bull., 9:65.

Grubb, T. C., Jr., 1972, Smell and foraging in shearwaters and petrels, Nature, 237:404.

Grubb, T. C., Jr., 1974, Olfactory navigation to the nesting burrow in Leach's Petrel (Oceanodroma leucorrhoa), Behaviour, 22:192.

Harper, P. C., 1978, The plasma proteins of some albatrosses and petrels as an index of relationship in the Procellariiformes, N.Z.J. Zool., 5:509.

Hiller, A., and Wand, U., 1984, Radiocarbon dating of breeding places of petrels in the Antarctic, Z.F.I. Mitteilungen, 89:103.

Hoffman, W., Heinemann, D., and Wiens, J. A., 1981, The ecology of seabird feeding flocks in Alaska, Auk, 98:437.

Hutchison, L. V., and Wenzel, B. M., 1980, Olfactory guidance in foraging by procellariiforms, Condor, 82:314.

Hutchison, L. V., Wenzel, B. M., Stager, K. E., and Tedford, B. L., 1984, Further evidence for olfactory foraging by Sooty Shearwaters and Northern Fulmars, in: "Marine Birds: Their Feeding Ecology and Commercial Fisheries Relationships,", D. N. Nettleship, G. A. Sanger, and P. F. Springer, eds., Canadian Wildlife Service Spec. Pub., Ottawa.

Imber, M. J., and Berruti, A., 1981, Procellariiform seabirds as squid predators, in: "Proceedings of the Symposium on Birds of the Sea and Shore," J. Cooper, ed., African Seabird Group, Cape Town.

Jacob, J., 1976, Chemotaxonomical relationships between penguins and tubenoses, Biochem. Sys. Ecol., 4:215.

Jacob, J., 1982, Stomach oils, in: "Avian Biology," Vol. 6, D. S. Farner, and J. R. King, eds., Academic Press, New York.

Jacob, J., and Ziswiler, V., 1982, The uropygial gland, in: "Avian Biology," Vol. 6, D. S. Farner, and J. R. King, eds., Academic Press, New York.

Jacob, J., Balthazart, J., and Schoffeniels, E., 1979, Sex differences in the chemical composition of uropygial gland waxes in domestic ducks, Biochem. Sys. Ecol., 7:149.
Jouventin, P., and Robin, J. P., 1984, Olfactory experiments on some Antarctic birds, Emu, 84:46.
Kennedy, R. J., 1971, Preen gland weights, Ibis, 113:369.
Larsell, O., 1950, The nervus terminalis, Ann. Otol. Rhinol. Laryngol., 59:414.
Michael, R. P., and Keverne, E. B., 1968, Pheromones in the communication of sexual status in primates, Nature, 218:746.
Meisami, E., 1978, Influence of early anosmia on the developing olfactory bulb, in: "Maturation of the Nervous System," Prog. Brain Res., M. A. Corner, R. E. Baker, N. E. van de Poll, D. F. Swaab, and H. B. M. Uylings, eds., 48:211.
Murphy, R. C., 1936, "Oceanic Birds of South America," Macmillan, New York.
Nottebohm, F., Kasparian, S., and Pandazis, C., 1981, Brain space for a learned task, Brain Res., 213:99.
Portmann, A., 1961, Sensory organs: Skin, taste and olfaction, in: "Biology and Comparative Physiology of Birds," Vol. 2, A. J. Marshall, ed., Academic Press, New York.
Prince, J. H., 1956, "Comparative Anatomy of the Eye," Charles C. Thomas, Springfield, IL.
Quay, W. B., 1967, Comparative survey of the anal glands of birds, Auk, 84:379.
Shallenberger, R. J., 1973, "Breeding biology, homing behavior, and communication patterns of the Wedge-tailed Shearwater, Puffinus pacificus," Unpublished Ph.D. dissertation, University of California, Los Angeles.
Shallenberger, R. J., 1975, Olfactory use in the Wedge-tailed Shearwater (Puffinus pacificus) on Manana Is. Hawaii, in: "Olfaction and Taste 5," D. A. Denton, and J. P. Coghlan, eds., Academic Press, New York.
Simpson, G. G., 1976, "Penguins. Past and Present, Here and There," Yale University Press, New Haven.
Smith, A. B., III, Belcher, A. M., Epple, G., Jurs, P. C., and Lavine, B., 1985, Computerized pattern recognition: A new technique for the analysis of chemical communication, Science, 228:175.
Warham, J., 1977, The incidence, functions and ecological significance of petrel stomach oils, Proc. N. Z. Ecol. Soc., 24:84.
Wenzel, B. M., 1968, Olfactory prowess of the kiwi, Nature, 220:1133.
Wenzel, B. M., 1980, Chemoreception in seabirds, in: "Behavior of Marine Animals," Vol. 4, J. Burger, B. L. Olla, and H. E. Winn, eds., Plenum Press, New York.
Wenzel, B. M., 1985, Olfactory behavior in procellariiform birds, Nat. Geo. Res. Repts., 18:779.

CHEMORECEPTION AND THE SELECTION OF GREEN PLANTS AS NEST FUMIGANTS BY STARLINGS

J. Russell Mason[1,2] and Larry Clark[1]

[1]Monell Chemical Senses Center and [2]Department of Biology
University of Pennsylvania
Philadelphia, PA 19104

INTRODUCTION

Although once controversial, it is now well-established that many avian species use volatile chemical cues. This is the case even though birds lack a vomeronasal organ (Portmann, 1961). Chemoreception is not restricted to those species with anatomically well-developed olfactory and/or trigeminal systems. Songbirds and other birds (e.g., pigeons) with relatively undeveloped olfactory anatomy nonetheless use volatile cues for a variety of purposes, including feeding (Archer and Glen, 1969; Stager, 1967; Snyder and Peterson, 1979; Wenzel, 1968), homing (Benvenuti et al., 1973; Grubb, 1974; Papi et al., 1973; Wallraff, 1980), and, possibly, communication (Frings and Jumber, 1954). Because comprehensive reviews of these topics are available (e.g., Kare and Mason, 1985; Wenzel, 1973, 1980), we choose not to review them here. Instead, we will suggest that some birds may use chemical cues for a quite different purpose. We will present natural history, and behavioral, chemical, and electrophysiological data consistent with the hypothesis that birds that reuse nest sites use green plants as nest fumigants, and that they choose plants for incorporation into the nest on the basis of chemical cues. This use of green vegetation represents the first documented case of an air-breathing vertebrate using the chemical defenses of another organism for its own purposes.

NATURAL HISTORY

Many species of birds incorporate fresh green vegetation into their nests. Several hypotheses have been advanced to account for this behavior. Early naturalists ascribed to birds an esthetic sensibility for seeking brightly colored flowers (Welty, 1975). More recent interpretations have suggested selective benefits, including a form of sexual selection analogous to the courtship-related activities of bower birds (Collias and Collias, 1984). Fresh vegetation also might alter the insulative properties of nests (Kern, 1984), or maintain a high vapor pressure within the nest environment, thereby preventing desiccation of eggs (Mertens, 1977). In addition, nest material may closely match surrounding vegetation, concealing the nest from predators visually searching for prey (Lack, 1954; Skutch, 1967). Finally, chemicals contained in fresh vegetation may act as repellents, contact toxicants, or fumigants, effective at controlling parasite and/or pathogen

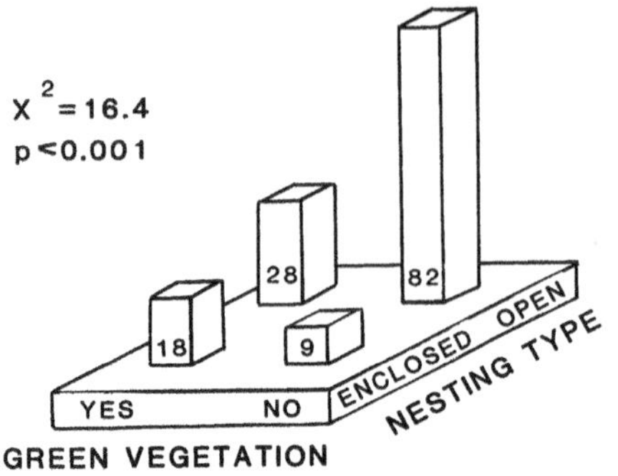

Fig. 1. The frequency distribution of nests of 137 eastern North American passerines as a function of nesting type (open versus closed), and whether or not the species incorporates fresh green vegetation into its nest matrix (Clark and Mason, 1985).

populations (Meyers, 1922, Johnston and Hardy, 1962; Sengupta, 1981; Wimberger, 1984; Clark and Mason, 1985).

The long term use of nest sites during a single breeding effort, or over several breeding efforts within a single season, increases the risk of incurring large parasite or pathogen loads (Stoner, 1936; Wasylik, 1971; Powlesland, 1977). Also, viruses, fungi and bacteria can lie dormant in nest debris and feces for several months, withstand freezing temperatures (Davies et al., 1971) and thereby increase the risk of infection to birds breeding at historically active sites. The inability to cope with parasite or pathogen infestation has led to nest desertion, egg spoilage, and juvenile mortality among several species (Neff, 1945; Szybalski, 1950; Florian and Trussell, 1957; Davies et al., 1971; Oliphant et al., 1976; Feare, 1976; Duffy, 1983) and is known to decrease vigor of nestlings which may diminish their future survivorship probabilities (Moss and Camin, 1971; Powlesland, 1977; Gold and Dahlsten, 1983; Whiteman and Bickford, 1983).

The case for plant compounds acting as a chemical defense against their own herbivores and pathogens is well established (e.g., Rosenthal and Janzen, 1979). We set out to test the hypothesis that some species of birds use attributes of fresh green vegetation to increase their probability of reproductive success. As an initial step, we reasoned that if fresh plants contain chemical compounds that act as contact toxicants or natural fumigants to the birds' ecto-parasites or pathogens, then species more likely to nest in previously used sites should be more prone to include fresh plant material into their nests. A review of the nesting behavior of 137 passerine species breeding in eastern North America was consistent with this hypothesis. Passerines nesting in enclosed spaces (e.g., secondary cavity nesters or species using crevices) were more likely to incorporate green plants into their nests, while passerines that infrequently reused nest sites and nested in open, cup-like nests were less likely to incorporate green vegetation into their nests (Fig. 1). Notable exceptions occurred. Some secondary cavity nesting species such as Nuthatches (e.g., *Sitta canadensis* and *S. carolinensis*) did not place fresh plant material into their nests. These species did, however, place pine pitch around their entrance holes, or squeeze and rub insects on the surface of the cavity (Bent, 1965c).

SPECIFIC QUESTIONS

We addressed specific questions regarding the use of nest material by a single species, the European Starling (Sturnus vulgaris). Four questions were asked. First, did Starlings select particular species of plants from the available vegetation for incorporation into their nests? If so, then were plants selected by Starlings more effective at controlling ecto-parasites or bacterial pathogens than a random subset of vegetation available in the habitat? Third, were the chemical profiles of volatiles found in plants used by Starlings different from profiles of a random subset of vegetation available in the habitat? Finally, we asked whether Starlings could discriminate among plants on the basis of chemosensory information alone.

Study Site and Breeding Behavior

The study site was a Starling nesting colony of about 100 nest boxes located at the Stroud Water Research Center of the Philadelphia Academy of Natural Sciences, Avondale, Pennsylvania. The habitat lies within the White Clay Creek watershed and is characterized by a patchwork of old fields (maintained via semi-annual mowing), pastureland, and secondary growth woodland.

Detailed accounts of Starling nesting behavior have been provided by several authors (e.g., Dunnet, 1955; Feare, 1984; Kessel, 1959). Observation of banded birds at the Stroud Water Research Center revealed the following patterns. Males intermittently visit the nesting area from August to March, show interest in cavity sites beginning in December or January, and throughout February, spend increasing amounts of time during the morning in or around one or two nest boxes. At this time, they may begin to clean out the previous season's nest material and debris. By March, males from the local population are well established at a cavity site, vigorously singing at the nest box during the early morning, and defending the nest site against newly arrived birds. Males also begin to place bits of fresh vegetation in the nest cavity. These bits are the tips of new growth and range between 1 and 3 cm in length. The behavior of placing only fresh vegetation in nest cavities lasts throughout March and into the first week of April. The number of species of plants found in the cavity increases with the availability of vegetation.

Males initiate nest construction in early April, and are aided subsequently by females. Nests consist of bulky masses of dried grasses of the previous year's growth. Woven into the matrix of dried grasses are bits of fresh green plant material. Bits of fresh vegetation are added continually to the nests until the eggs hatch, and then the behavior falls off dramatically.

(1) Do Starlings Select Particular Plants for Incorporation Into Their Nests?

Nest material. We checked nests for green vegetation every three days, from the initial stage of nest building to the hatching of young. We removed the nests from the boxes, picked apart the matrix by hand, and removed all green vegetation. The green vegetation was labeled and pressed for later identification. Thirty-eight herbaceous species were recorded, although a small fraction of green vegetation (< 5%) could not be identified to species. Dried nest matrix was reassembled, and placed back into the nest box. No birds abandoned their nests as a result of this manipulation.

Vegetation survey. During peak nest-building in the last week of April and the first week of May 1984, we sampled the green herbaceous vegetation

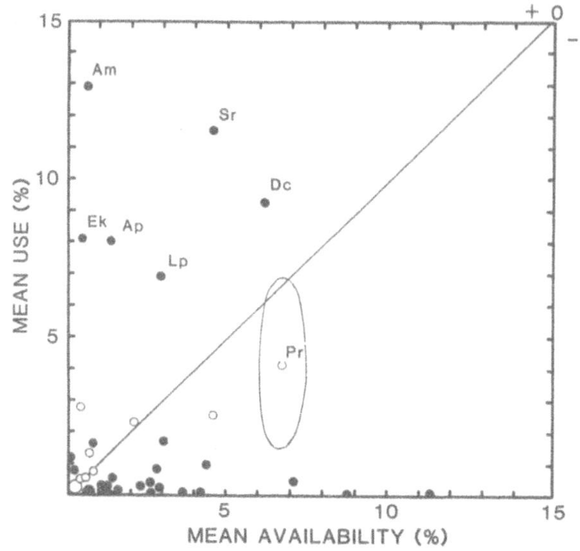

Fig. 2. Mean % use in nests versus mean % availability in habitat for 66 herbaceous species. The diagonal line represents a hypothetical selection scheme by Starlings, in which plants are selected in proportion to their abundance. Species selected in proportion to their abundance are represented by open circles and are classified as those species whose sampling errors along either axis intersected the diagonal line (e.g., Pr.) (Clark and Mason, 1985). Solid circles above the diagonal represent species selected in excess of their proportional abundance. Solid circles below the diagonal represent species selected in numbers less than proportional abundance. The large open circle is for 22 overlapping points, while the large solid circle is for 5 overlapping points. Species abbreviations are Am: *Achillea millefolium*; Sr: *Solidago rugosa*; Dc: *Daucus carota*; Ek: *Erigeron* sp.; Ap: *Agrimonia paraflora*; Lp: *Lamium purpureum*; Pr: *Polemonium reptans*.

in plots randomly placed along transects throughout the breeding colony. A total of 66.5 m^2 plots were sampled along 24 transects. This sampling effort was adequate, because starlings usually fly less than 50 m to choose material for their nests (Kessel, 1959; Bent, 1965).

Eighty-four herbaceous species were collected (samples of each plant were frozen for later use). Because the asymptotic number of plant species found in the habitat was approached after sampling 30 plots, we concluded that our sampling effort had in fact provided an adequate description of the available vegetation. Starlings selected nine species of plants in numbers greater than their proportional availability, 34 in proportion to their availability, and 23 species in numbers less than their availability from the habitat (Fig. 2). Such a pattern of vegetation use was not expected from a chance utilization of available vegetation (one-tailed Kolomogorov-Smirnoff test, $p < 0.005$). Too few plants were classified as being preferred by Starlings. For subsequent experiments, we used the six most frequently preferred species and six randomly selected species that Starlings either avoided or selected in proportion to their availability (Table 1).

Table 1. Plants chosen in excess of their availability ("preferred") versus plants chosen either in proportion to or below availability ("other").

Preferred Plants		Other Plants	
Scientific Name	Common Name	Scientific Name	Common Name
Agrimonia paraflora	Agrimony	*Potentilla recta*	Cinquefoil
Daucus carota	Wild carrot	*Brassica nigra*	Black Mustard
Erigeron sp.	Fleabane	*Polemonium reptans*	Greek Valerian
Achillea millefolium	Yarrow	*Claytonia virginica*	Spring Beauty
Lamium purpureum	Red-dead Nettle	*Conium maculata*	Poison Hemlock
Solidago rugosa	Rough Goldenrod	*Viola* sp.	Violet

(2) Are Plants Selected by Starlings More Effective at Controlling Ecto-parasites and Bacterial Pathogens than a Random Subset of Other Available Vegetation?

Ecto-parasites. Eggs of the louse, *Menacanthus* sp., and adult fowl mites, *Ornithonyssus sylviarum* were collected from adult birds and nests, respectively. This choice of ecto-parasite species was based on their potential for negative impact on nestling growth and survivorship (Moss and Camin, 1971; Powlesland, 1976). Logistical constraints determined which instar of each ecto-parasite was used.

Eggs of the louse were found attached to the base of feathers on the heads and necks of adult birds (Boyd, 1951). These eggs were removed and placed in covered, humidified petri dishes. Adult fowl mites were collected from infested nest boxes by placing a strip of paper amidst a swarm of mites, allowing the mites to board the paper, then depositing the mites and paper into a vial. Upon return to the laboratory, the mites were transferred to covered, humidified petri dishes.

Experimental treatments consisted of placing 0.2 g of dried plant material into randomly selected petri dishes. These dishes, and control dishes containing eggs or mites but no plant sample, were incubated at 30 ± 1°C, which is within the range of temperature of feathers on live birds (Clark, pers. obs.). The viability of louse eggs (whether or not they hatched) and mortality among mites was checked every other day for eight days. At no time was contact permitted between plant samples and *Menacanthus* eggs during experiments.

Bacterial pathogens. Four standard laboratory strains of bacteria (*Streptococcus aurealis*, *Staphylococcus epidermis*, *Pseudomonas aeroginosa*, and *Escherichia coli*) were chosen on the basis of literature pertaining to

Table 2. The effectiveness of plant material at inhibiting bacterial growth on a nutrient medium: (++) represents strongly bactericidal; (+) moderate effectiveness; blank for ineffective plants. The asterisk denotes plant species used preferentially by Starlings. Codes are Se: *Staphylococcus epidermis*; Pa: *Pseudomonas aeroginosa*; Ec: *Escherichia coli*; Sa: *Streptococcus aurelia*; and NB from nests and nest boxes. Unk. 21 is an unidentified plant species.

Plant Species	Se	Pa	Ec	Sa	NB
*Solidago rugosa**				+	+
Solidago ulmiflora					
Conium maculatum					
*Agrimonia paraflora**	++			++	++
*Achillea millefolium**					+
Polemonium reptans					
Glechoma hederacea					
Geum virginianum					
Geum canadense				+	
Taraxacum officinale					
Lonicera japonica				+	
*Daucus carota**		+		+	
Alliaria officinalis					
Senecio obvatus					+
Unk. 21					
Viola sp.					
*Lamium purpureum**					

the pathogenicity of these strains to birds or eggs (Davies et al., 1971). In addition, unidentified wild strains of bacteria were collected from nest boxes by swabbing the nest and nest cavity with sterilized cotton swabs dipped in sterilized distilled water. Each contaminated swab was carried back to the laboratory in a sterilized tube. Sterilized nutrient medium was innoculated with the standard strains or wild strains, and then each experimental plate was incubated in the presence of a 1 cm^2 piece of sterilized leaf material. There were three replications for each plant tested and a control (i.e., a plate innoculated with bacteria but without leaf material). Our presumption was that any bacteriocidal compounds present in the leaves would diffuse into the nutrient medium and inhibit bacterial growth. Plants were scored as strongly bacteriocidal if there was a bacteria free zone around leaf pieces greater than 1 cm. Plants were scored as moderately effective if a bacteria free zone was apparent, but less than 1 cm. Plants were scored as ineffective if no apparent inhibition of bacterial growth was observed.

Results. Plants preferred by Starlings retarded the hatching of *Menacanthus* sp. (one-tailed $t = 3.39$, $p < 0.005$) whereas other randomly chosen plants did not. Also, plants effective at controlling the growth of pure strains of *S. aurealis*, *S. epidermis*, *P. aeroginosa*, and wild bacteria obtained from nest cavities were more frequently used by Starlings (one-tailed $t = 2.05$, $p < 0.05$). No plants were effective against *E. coli* (Table 2). Also, no plants were effective against adult fowl mites (*O. sylviarum*). In fact, after three days of exposure to the plants, no mites had died in either group, and after five days of exposure, only a few mites from any group had died.

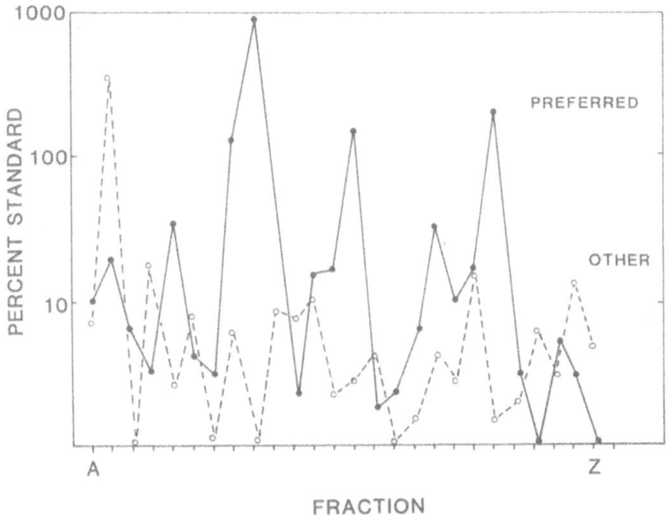

Fig. 3. A profile for the mean concentrations of volatile chemicals in preferred versus other plants. Each fraction represents either a single compound or a group of compounds based on relative retention times from a gas chromatograph.

(3) Are Chemical Profiles of Volatiles Found in Plants Used by Starlings Different From Profiles of a Random Subset of Vegetation Available in the Habitat?

The phytochemical analysis of plant volatiles was carried out in collaboration with Dr. John Labows, an analytical chemist at the Monell Center. Only large scale patterns of similarity among plants are reported here. Plant samples were thawed and blotted dry. Five grams of leaf tissue were placed in a 250 ml flask, and volatiles were collected from the headspace of the flask. Gas chromatographic and mass spectrographic techniques were used to identify potentially active fractions of plant volatiles and the identity of those fractions (Clark and Mason, in press). Plants preferred by Starlings contained significantly higher concentrations of volatile compounds (Wilcoxin matched-pairs test, $z = -1.73$, $p < 0.042$; Fig. 3)., although a cluster analysis did not reveal any other clear-cut dichotomy between preferred and other plants (Clark and Mason, 1985). GC/MS analyses indicated that the volatiles contained within preferred plants were mostly mono- and sesquiterpenes, such as myrecene, α- and β-pinene, cymene, ocimene, αphellandrene, αterpineol, Δ^3-carene, sabinene, αterpinolene and limonene. At least one of the preferred plants, *Solidago rugosa*, contained the sesqueterpenes 2-bornyl acetate and farnesol. These compounds have been implicated as potent insect juvenile hormone analogs (Rosenthal and Janzen, 1979). Such compounds generally suppress molting in arthropods. Any such action of compounds on ecto-parasites would delay the timing of terminal instars, and hence curtail the reproductive capacity of the parasites. As a consequence, the rate of population growth by parasites, and by implication, their adverse effect on nestling birds (Powlesland, 1978), would be reduced. Detailed analysis of the chemical composition of the plants is forthcoming.

(4) Can Starlings Discriminate Among Plant Species on the Basis of Volatile Chemical Characteristics?

Obvious sensory cues that the birds could use for discrimination among plants include visual characteristics. For example, passerines can make fine visual discriminations among leaf shapes (Real et al., 1984). Other possible cue sources mediating discrimination might be volatile or non-volatile chemical cues. Many of the used plant species are aromatic to human observers. Because we had already collected evidence suggesting that Starlings possess acute chemosensory abilities (e.g., Mason et al., 1985; Mason and Silver, 1983), we designed a series of electrophysiological and behavioral experiments to test the possibility that Starlings can discriminate among plants on the basis of some volatile cue.

Electrophysiology

Integrated multiunit responses were obtained from five adult Starlings. The birds were food deprived for approximately 12 hours before the experiment, anesthetized with urethane (4.6 ml/kg, ip), and placed into a head-holder. The feathers, skin, and bone between the eyes was removed. A small Teflon cap was fitted to the cleared area and fixed in place with dental cement. This cap provided a reservoir for mineral oil, which was pipetted into the cap to prevent desiccation. The olfactory nerves under the bony orbital walls were exposed, and electrical activity was recorded from the left olfactory nerve by dissecting a small twig free, and placing it on a pair of platinum-iridium wire electrodes. The preparation was grounded through the head-holder apparatus. The platinum-iridium hook electrodes were connected to the high impedance probe of an AC coupled preamplifier. The amplified neural activity was monitored with a storage oscilloscope and an audio analyzer, and stored on magnetic tape. Multiunit activity was summed using a short-time averaging circuit with a time constant of 0.5-1.0 sec (Kiyohara and Tucker, 1978), and displayed on a pen recorder.

The magnitudes of the neural responses were measured in arbitrary units from the baseline to the peak of the phasic response. Responses were reported as a percentage of the response to a standard stimulus, air saturated with *n*-butanol at 23°C.

Birds were tested for integrated neural activity in response to eight stimuli: the arbitrary standard (*n*-butanol), a blank that consisted of ambient air passed through the delivery apparatus, and volatiles from six plant species: *Agrimona paraflora*, *Daucus carota*, *Achillea millefolium*, *Aster prenanthoides*, *Aster praealtus*, and *Solidago rugosa*. Volatiles from 10 g of each plant were allowed to accumulate in 125 ml flasks for 15 min at 23°C. Fifty cc of headspace from each flask was then withdrawn by syringe, and injected into the left naris of each bird at a rate of 10 cc/sec. Five min interstimulus intervals were used, and as many series of the eight stimuli as possible were presented to each bird at 20 min intervals.

Results. Without exception, *n*-butanol, and volatiles from all the plant species elicited vigorous responses from the Starlings' olfactory nerves (Fig. 4). Room air also elicited responses, although to a considerably lesser extent. We took these results as evidence that Starlings have the olfactory capacity to detect volatiles from both preferred and unused plants. Experiments were then designed to determine whether Starlings could behaviorally detect and discriminiate among plant odors.

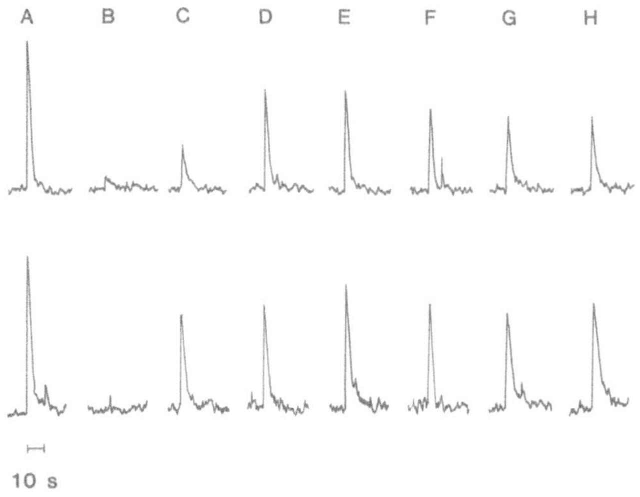

Fig. 4. Integrated multinuit responses from the olfactory nerves of two birds. Stimuli were: (a) _n_-butanol, (b) room air, (c) _Agrimonia paraflora_, (d) _Aster prenanthoides_, (e) _Solidago ulmiflora_, (f) _Solidago rugosa_, (g) _Aster praealtus_, (h) _Achillea millefolium_.

Behavior

Eight independent experiments (Table 3) were performed to determine whether Starlings could detect and discriminate among volatiles from plants. For each investigation, birds were individually housed and visually isolated eight days prior to training. After four days, the birds were adapted to a deprivation regime in which they were food deprived during the first

Table 3. Plant species used for adaptation and two-choice behavioral tests in eight independent experiments.

	Adaptation	Two-Choice Tests
1.	_Solidago canadense_	_Solidago rugosa_ vs. _Allaria officinalis_
2.	_Aster_ sp.	_Solidago rugosa_ vs. _Geum canadense_
3.	_Alliaria officinalis_	_Glechoma hederacea_ vs. _Ranunculus_ sp.
4.	_Ranunculus_ sp.	_Polemonium reptans_ vs. _Erigeron_ sp.
5.	_Allaria_ sp.	_Plantago major_ vs. _Lamium purpureum_
6.	_Potentilla recta_	_Achillea millefolium_ vs. _Brassica nigra_
7.	_Veronica_ sp.	_Agrimonia_ sp. vs. _Potentilla recta_
8.	_Solidago canadense_	_Barbarea vulgaris_ vs. _Daucus carota_

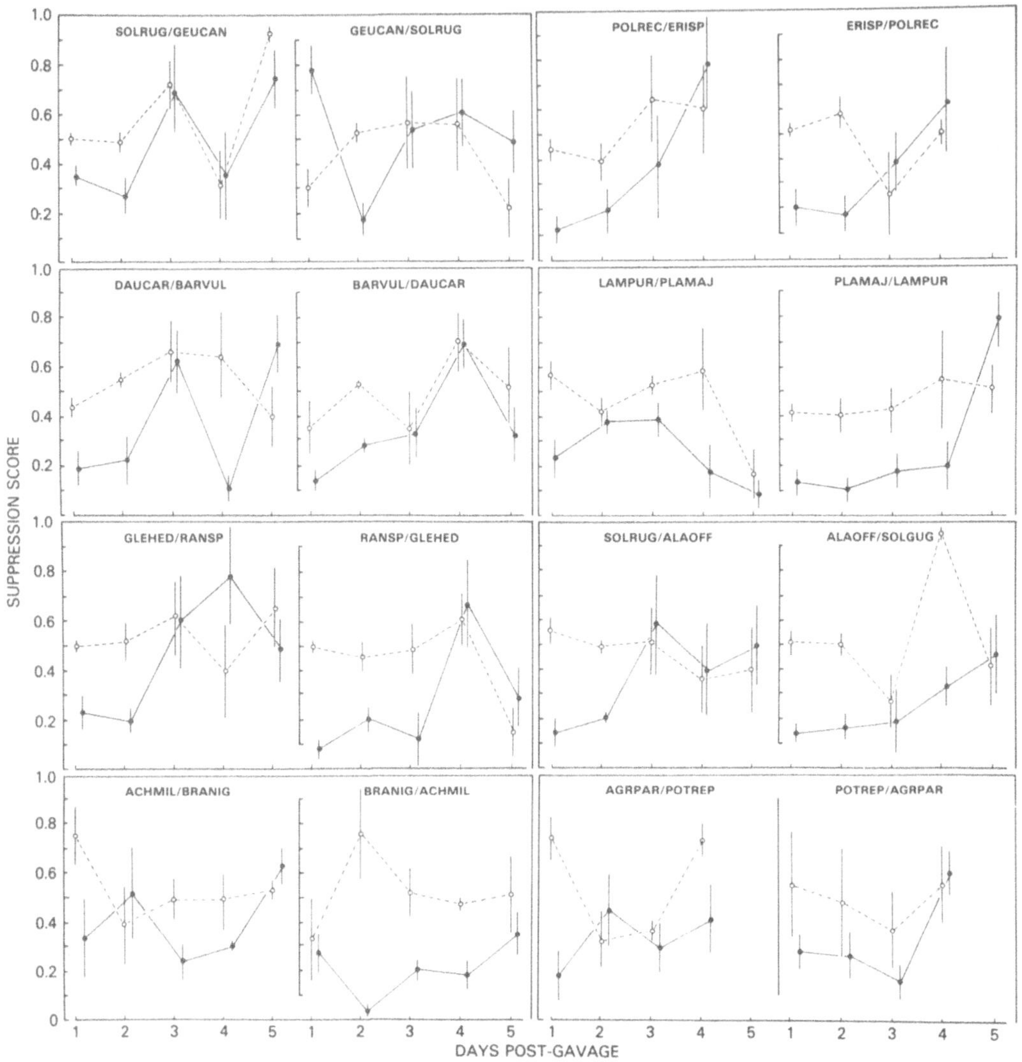

Fig. 5. Results of behavioral two-choice tests. Mean suppression scores were calculated by dividing CS+ by total consumption. Dotted lines represent control groups, while solid lines represent experimental groups. Vertical bars are standard errors of the means.

30 min, of light, and then presented with a cup containing 20 g of food. In the bottom of the food cup was a nylon mesh bag containing fresh leaves of one or another of the plant species. Volatiles from the plant material could diffuse freely into the headspace above the food. After one hour, food cups were removed, and consumption recorded. Plain food was then returned to the cages, and the birds were left undisturbed until the following day. Our aim in presenting fresh plant material during adaptation was to reduce the potential for neophobia during conditioning and test sessions.

Table 4. F values and probabilities for group main effects in eight independent odor discrimination experiments.

(1) Agrimonia paraflora vs. Potentilla recta
$F(1,16) = 23.23$, $p < 0.001$

(2) Achillea millefolium vs. Brassica nigra
$F(1,16) = 17.05$, $p < 0.001$

(3) Glechoma hederacea vs. Ranunculus
$F(1,16) = 7.19$, $p < 0.05$

(4) Geum canadense vs. Solidago rugosa
$F(1,16) = 0.23$, $p < 0.25$

(5) Lamium purpureum vs. Plantago major
$F(1,16) = 19.98$, $p < 0.001$

(6) Daucus carota vs. Barbarea vulgaris
$F(1,16) = 10.90$, $p < 0.01$

(7) Erigeron vs. Potentilla reptans
$F(1,16) = 6.78$, $p < 0.001$

(8) Solidago rugosa vs. Allaria officianales
$F(1,16) = 30.44$, $p < 0.001$

On the day of conditioning, birds were assigned to four groups (5 birds/group), food deprived for 30 min, and then each was presented with a cup containing 20 g of food. In the bottom of the food cup was one or another of two CS+ plant species different from the one presented during adaptation. After one hour, the birds in two experimental groups were gavaged with a toxicant methiocarb/propylene glycol solution (2 mg/kg; Mason and Reidinger, 1983) while the birds in the two control groups were gavaged with propylene glycol (2 mg/kg).

During the first hour of light on each of the five days immediately following the day of conditioning, all groups were food deprived for 30 min and then given 2-choice tests. These tests consisted of presenting two cups, each containing 20 g of food, and one or the other of the plant species presented on the day of conditioning. After one hour, the food cups were removed from the cages and consumption was measured.

Results. The data from each discrimination test were converted to suppression ratios (i.e., CS+ consumption/total consumption), and assessed using three-way analysis of variance (groups, plants, days). The results indicated that the Starlings could discriminate among volatiles from the plants (Fig. 5).

The probability of obtaining seven out of eight statistically significant results by chance was $p < 0.005$. Experimental groups consumed less food when it was associated with the species of plant presented on the day of conditioning. Control groups failed to exhibit such avoidance (Table 4). The one exception was for tests between Geum canadense and Solidago rugosa. This lack of differential consumption might mean that these plants have very similar odors, making discrimination difficult. Alternatively, it might reflect the failure of some experimental birds to learn. Half of

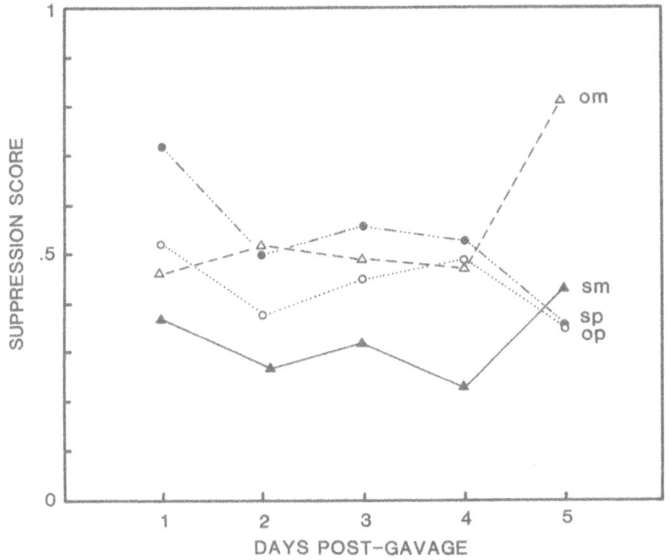

Fig. 6. Results of behavioral tests following bilateral olfactory nerve cuts or sham surgeries. Group om received olfactory nerve cuts and methiocarb intubations; Group sm, sham surgeries and methiocarb intubations; Group sp, sham surgeries and propylene glycol intubations; and Group op, olfactory nerve cuts and propylene glycol intubations. Suppression scores were calculated as previously described.

the experimental birds showed non-differential consumption, the other showed clear preferences for CS- over CS+ on all test sessions.

While birds could be trained to respond and discriminate among plants, we still had not demonstrated that such behavior was mediated by a nasal chemosensory cue. To answer this question, another series of behavioral experiments was undertaken with birds who had been given either bilateral olfactory nerve cuts or sham surgeries.

Starlings were assigned to two groups (n = 10/group). The birds in one group were lightly anesthetized with Equithesin, and placed in a headholder. The olfactory nerves under the bony orbital walls were exposed. The nerves were lifted slightly and a small section (approximately 1 mm) was removed. The cut ends of the nerves were folded back, the cavity packed with gelfoam, and the skin closed with cyanoacrylate glue. The birds in the other (sham) group were treated similarly, except that the nerves were not disturbed under the orbital walls. All of the birds recovered from surgery within 1 h of Equithesin injection. Beginning two days after surgery, the birds were given four days of adaptation to food deprivation. Five of the birds given olfactory nerve cuts, and five of those given sham surgeries were given a pairing of a plant species and methiocarb intubation. The other birds were given a pairing of the plant and propylene glycol intubation. On each of the four days following the day of conditioning all of the birds were given two-choice tests, as previously described.

Analysis of the results indicated that while sham operated methiocarb-gavaged birds exhibited avoidance of volatiles from the plant associated with sickness, methiocarb or propylene glycol gavaged olfactory nerve cut birds and propylene glycol gavaged sham birds failed to exhibit differential behavior (Fig. 6). We took these results as evidence, at least for the species of plants tested, that avoidance was mediated primarily by olfactory cues.

CONCLUSION

North American passerine species that reuse nest sites are more likely to incorporate fresh green vegetation into their nests than species who do not reuse nests. Several explanations for the use of green vegetation have been offered, including sexual selection (Collias and Collias, 1974), crypticity (Skutch, 1976), and the possibility that green plants possess structural or thermal characteristics which enhance nestlings' chances for survival (Kern, 1984). In the present paper, we have suggested an additional possibility; that the chemical defenses of green vegetation are being used by birds to control ecto-parasites and bacterial pathogens. Several lines of evidence support this possibility. For example, Sengupta (1981) reported that house sparrows (Passer domesticus) incorporate margosa (Azadirachta indica) leaves into their nests and that these leaves are used in preference to other available vegetation. Isolated fractions from leaves not only are repellent to arthropods, but also inhibit oviposition on treated surfaces. The suspected active chemicals are a steroid, β-sitosterol, and a phenolic, quercetion (Ambasta, 1980). Similarly, nuthatches smear nest cavities with pine resin (Bent, 1965), a material containing a variety of monoterpenes (e.g., alpha and beta pinene) known for their antiseptic and toxicologic properties (Rosenthal and Janzen, 1979). Finally, we have demonstrated that Starlings choose particular species of plants for inclusion in their nests, and that preferred plants exert fumigant effects in laboratory experiments. While the plant compounds exerting these effects remain unknown, the available evidence suggests that they may be mono- and sesqueterpenes, including known juvenile hormone analogs for ecto-parasites. We speculate that these analogs may act to control ecto-parasite populations by delaying maturation, and hence reproduction. The implied importance of plant derived juvenile hormone analogs can be appreciated when one considers that the per capita blood loss for a brood of four in a heavily infested nest can be 3.5 percent of the daily blood production of the chicks (Powlesland, 1978).

The underlying bases (e.g., unlearned preference, breeding experience, imprinting, observational learning) of plant selection and use remain unclear, although proximate mechanisms of discrimination likely include visual and chemical cues. We present electrophysiological and behavioral evidence that volatile cues could be used to detect and discriminate among plant species. In the field, we speculate that birds may visually orient towards patches of vegetation. However, once in a patch, vision alone probably would not serve for accurate discrimination, since the use of plant morphology as the sole criterion for plant selection would not guarantee a bird access to a particular chemical profile. Specifically, there is high variability of chemical profiles found within plants, even of the same species (Parks, 1974). An alternative is that, once in the patch, particular species of plants (and perhaps even particular individual plants) may be selected on the basis of a volatile cue. When foraging for green vegetation, Starlings pluck bits of plants, turn them in their bills, and then either accept or reject them (Clark, pers. obs.). These manipulations may facilitate retronasal access of plant volatiles to the olfactory epithelium (and possibly the detection of any salient taste cues that might be present). Welty (1975) has suggested that the large size of the internal nares in avians may serve to enhance access to the nasal cavity. From an ecological perspective, we suggest that chemicals represent a logical source of cues for plant selection. It would appear both logical and efficient to choose plant materials on the basis of cues (i.e., chemical volatiles) that may be directly responsible for insecticidal and anti-pathenogenic effects.

ACKNOWLEDGMENTS

This work was supported in part by a grant from the National Geographic Society. The authors thank Ms. Cheryl Boehm for her expert technical assistance.

REFERENCES

Ambasta, S. P., 1980, No title, Sci. Reporter, 17:210.
Archer, A. T., and Glen, R. M., 1969, Observations on the behavior of two species of honey-guides Indicator variegatus (Lesson) and Indicator exilia (Cassin). Los Angeles County Mus. Contrib. Sci., 160:1.
Bent, A. C., 1965, "Nuthatches, Wrens, Thrashers and Their Allies," Dover, New York.
Benvenuti, S., Fiaschi, V., Fiore, L., and Papi, F., 1973, Homing performances of inexperienced and directionally trained pigeons subjected to olfactory nerve section, J. Comp. Physiol. A, 83:81.
Boyd, E. M., 1951, A survey of the parasitism of the Starling Sturnus vulgaris L. in North America, J. Parasitol., 37:56.
Clark, L., and Mason, J. R., 1985, Use of nest material as insecticidal and anti-pathenogenic agents by the European Starling, Oecologia, in press.
Collias, N., and Collias, E., 1974, "Nest Building Behavior in Birds," Princeton Univ. Press, Princeton, N.J.
Davies, J. W., Anderson, A. C., Karstad, L., and Trainer, D. O., 1971, "Infectious and Parasitic Diseases of Wild Birds," Iowa State Univ. Press, Ames, Iowa.
Duffy, D. C., 1983, The ecology of tick parasitism on densely nesting Peruvian seabirds, Ecology, 64:110.
Dunnet, G. M., 1955, The breeding of the Starling (Sturnus vulgaris) in relation to its food supply, Ibis, 97:619.
Feare, C. J., 1976, Desertion and abnormal development in a colony of Sooty Terns (Sterna fuscata) infested by virus-infected ticks, Ibis, 118:112.
Feare, C. J., 1983, "The Starling," Cambridge University Press, Oxford.
Florian, M. L. E., and Trussell, P. C., 1957, Bacterial spoilage of shell eggs. IV. Identification of spoilage organisms, Food Technol., 11:56.
Frings, H., and Jumber, J., 1954, Preliminary studies on the use of specific sound to repel Starlings (Sturnus vulgaris) from objectionable roosts, Science, 11:318.
Gold, C. S., and Dahlsten, D. L., 1983, Effects of parasitic flies (Protocalliphora spp.) on nestlings of Mountain and Chestnut-backed Chickadees, Wilson Bull., 95:560.
Grubb, T. C., 1974, Olfactory navigation to the nesting burrow in Leach's petrel (Oceanodroma leucorrhoa), Anim. Behav., 22:192.
Johnston, R. F., and Hardy, J. W., 1962, Behavior of the Purple Martin, Wilson Bull., 74:243.
Kare, M. R., and Mason, J. R., 1985, The chemical senses in birds, in: "Avian Physiology," P. D. Sturkie, ed., Springer-Verlag, Berlin.
Kern, M. D., 1984, Racial differences in nests of white-crowned sparrows, Condor, 86:455.
Kessel, B., 1959, A study of the breeding biology of the European Starling (Sturnus vulgaris L.) in North America, Amer. Midl. Nat., 58:257.
Kiyohara, S. Y., and Tucker, D., 1978, Activity of new receptors after transection of primary olfactory nerves, Physiol. Behav., 21:987.

Lack, D., 1954, "The Natural Regulation of Animal Numbers," Oxford University Press, London.
Mason, J. R., and Reidinger, R. F., 1983, Importance of color for methiocarb-induced food aversions in red-winged blackbirds, J. Wildl. Manage., 47:383.
Mason, J. R., and Silver, W. L., 1983, Trigeminally mediated odor aversions in starlings, Brain Res., 269:196.
Mason, J. R., Glahn, J. F., Dolbeer, R. A., and Reidinger, R. F., 1985, Field evaluation of dimethyl anthranilate as a bird repellent livestock feed additive, J. Wildl. Manage., in press.
Mertens, J. A. L., 1977, Thermal conditions for successful breeding in Great Tits (Parus major major L.) II. Thermal properties of nests and nestboxes and their implication for the range of temperature tolerance of Great Tit broods, Oecologia, 28:31.
Meyers, L. E., 1922, The American swallow bug, Oeciacus vicarius Horvath (Hemptera, Cimicidae), Parasitol., 20:159.
Moss, W. W., and Camin, J. H., 1971, Nest parasitism, productivity and clutch size in purple martins, Science, 168:1000.
Neff, J. A., 1949, Maggot infestation of nestling mourning doves, Condor, 47:73.
Oliphant, L. W., Thompson, W. J. P., Donald, T., and Rafuse, R., 1976, Present status of the Prairie Falcon in Saskatchewan, Can. Field Nat., 90:365.
Papi, F., Fiore, L., Fiaschi, V., and Benvenuti, S., 1973, An experiment testing the hypothesis of olfactory navigation in homing piegons, J. Comp. Physiol A ., 83:93.
Parks, C. R., 1974, Chemical evidence, in: "Vascular Plant Systematics," A. E. Radford, W. C. Dickison, J. R. Massey, and C. R. Bell, eds., Harper and Row, New York.
Portmann, A., 1961, Sensory organs: Skin, taste, olfaction, in: "Biology and Comparative Physiology of Birds," A. J. Marshall, ed., Academic Press, New York.
Powlesland, R. G., 1977, Effects of the haematophagous mite Ornithonyssus bursa on nestling Starlings in New Zealand, New Zealand J. Zool., 4:85.
Real, P. G., Iannazzi, R., Kamil, A. C., and Heinrich, B., 1984, Discrimination and generalization of leaf damage by blue jays (Cyanocitta cristata), Anim. Learn. Behav., 12:202.
Rosenthal, G. A., and Janzen, D. H., eds., 1979, "Herbivores: Their Interaction With Secondary Plant Metabolites," Academic Press, New York.
Sengupta, S., 1981, Adaptive significance of the use of margosa leaves in nests of the house sparrow (Passer domesticus), Emu, 81:114.
Skutch, A. F., 1976, "Parent Birds and Their Young," Univ. Texas Press, Austin, Texas.
Stager, K. E., 1967, Avian olfaction, Amer. Zool., 7:15.
Stoner, D., 1936, Studies on the Bank Swallow (Riparia riparia riparia, L. in the Oneida Lake Region, Roosevelt Wildl. Annual., 4:126.
Snyder, G. K., and Peterson, T. T., 1979, Olfactory sensitivity in the black-billed magpie and in the pigeon, Comp. Biochem. Physiol. A., 62:921.
Szybalski, W., 1950, A comparative study of bacteria causing mustiness in eggs, Nature, 165:733.
Wallraff, H. G., 1980, Olfaction and homing in pigeons: Nerve section experiments, critique, hypotheses, J. Comp. Physiol., 139:209.
Wasylik, A., 1971, Nest types and abundance of mites, Ekol. Pol., 19:689.
Welty, J. C., 1975, "The Life of Birds," J. B. Saunders, Philadelphia, Pennsylvania.
Wenzel, B. M., 1968, The olfactory prowess of the Kiwi, Nature, 220:1133.
Wenzel, B. M., 1973, Chemoreception, in: "Avian Biology," D. S. Farner, and J. R. King, eds., Academic Press, New York.
Wenzel, B. M., 1980, Chemoreception in seabirds, in: "Behavior of Marine Animals," J. Burger, B. L. Olla, and H. E. Winn, eds., Plenum, New York.

Whiteman, C. E., and Bickford, A. A., 1983, "Avian Disease Manual," American Association of Avian Pathology, Univ. Pennsylvania, Kennett Square, Pennsylvania.

Wimberger, P. H., 1984, The use of green plant material in bird nests to avoid ectoparasites, Auk, 101:615.

ECONOMIC CONSEQUENCES OF SCENT MARKING IN MAMMALIAN TERRITORIALITY

L. M. Gosling

Coypu Research Laboratory
MAFF, Jupiter Road
Norwich NR6 6SP, UK.

INTRODUCTION

The mating strategies of male mammals consist of intrasexual competition for access to females and then attempts to maximise contacts with receptive females. The form that these attempts take is strongly influenced by female movements in response to food and their tendency to form social groups. Female reproductive success (RS) is primarily (although not entirely) determined by the number of offspring that can be produced and the factors that limit this are the availability of the nutrients needed for reproduction and the chance that offspring will be killed by predators. Many aspects of female behaviour are profoundly influenced by these considerations: for example, the movements of female antelopes in relation to spatial and temporal variation in grassland quality appear to be adaptations to optimise food quality while avoiding habitats that offer cover to predators. In addition, females can reduce the chance of predation by cryptic behaviour, alone or in small groups, or by the "selfish herd" advantages of joining larger social groups. Collectively, these female behaviours appear to dictate the mating strategy that is most profitable for males to adopt (Gosling, in press). Excluding lekking, these strategies fall into two main classes: (1) males follow one or more females, waiting until they become receptive, and (2) they defend part of the food resource that females need so that they can intercept females that are attracted to the resource (Gosling, in press). This second strategy is known as "resource defence territoriality" (Emlen and Oring, 1977).

Resource defence territories are probably always scent marked and in this paper I suggest a reason for this link. More specifically, I suggest that the costs of this form of territoriality (mainly the costs of area defence) would often outweigh the benefits (from increased reproductive success) in the absence of specialised scent glands and of scent marking. The argument is based on the interpretation of scent marking as a form of status advertisement and on the advantages of accurate competitor assessment in reducing the costs of territory defence and allowing larger areas to be defended. The examples used are drawn mainly from the African antelopes where resource defence territoriality is the most common male mating strategy (Jarman, 1974; Estes, 1974; Owen-Smith, 1977) and territory owners scent mark with faeces, urine, and the secretions of a number of specialised skin glands (Fig. 1; Gosling, 1985).

Fig. 1. Territorial male Thomson's gazelle, Gazella thomsoni, marking the tip of a grass stem with secretion from the antorbital gland (from Gosling, 1985; reprinted with permission from Oxford University Press).

MALE COMPETITION: INDIVIDUAL AND SPATIAL REFERENCES FOR DOMINANCE

In the case of both "following" and "resource defence" strategies, male antelopes first attempt to achieve high dominance status by repeated agonistic encounters with those males who are potential sexual competitors. Such contests may usually start when males are very young and they continue as males start to compete with non-breeding adult males, sometimes after they join groups of males (e.g., impala, Aepyceros melampus: Jarman, 1979; hartebeest, Alcelaphus buselaphus: Gosling, 1974). Stable dominance relationships are formed and these depend on the memory of the outcome of past encounters with individually known opponents: males thus acquire "an individual reference for dominance." When males achieve high dominance status, they may gain access to receptive females either by asserting their status directly in a mixed sex group (e.g., buffalo, Syncerus caffer: Mloszewski, 1983; greater kudu, Tragelaphus strepsiceros: Owen-Smith, 1984) or by using it to keep subordinate males away from a group of females which they thus monopolise (e.g., some oryx, Oryx beisa: Walther, 1978b). Alternatively, they may leave the male group and attempt to secure exclusive use of an area which contains resources that females need; in the mammals, this resource is usually food that females need for parental investment. When males succeed in getting such an area, or territory, the nature of their interactions with male competitors changes. Instead of contests being influenced by the memory of past encounters, the outcome now seems to depend mainly on the status of the male as the territory resident or owner. Owners nearly always win contests with intruders. Experiments with great tits (Parus major), which involved moving owners out of territories and then allowing them to compete with new occupants, show that ownership itself, rather than other possibilities (see below), is the main determinant of success (Krebs, 1982). The change from an individual to a spatial reference for dominance may not be instantaneous. A male's individual dominance status is probably important when it competes with other non-breeding males to acquire a territory and possibly also in the early days of occupancy when it will know many intruders. However, as time goes on, the status and identity of non-territorial males will become unfamiliar and the outcome of contests with intruders will increasingly depend on "owner advantage."

OWNER ADVANTAGE AND THE COSTS OF COMPETITION

But why do owners generally win? Theoretical analyses suggest three main possible reasons (Parker, 1974; Maynard Smith, 1976; Maynard Smith and Parker, 1976). These are (1) owners have higher "resource holding potential" (RHP; Parker, 1974), that is, better fighting ability, larger body size, etc., (2) owners have more to gain by retaining the territory than the intruder has in winning it and will thus be prepared to fight harder. Under these circumstances, most intruders should withdraw from the contest at an early stage, and (3) the contest is settled by an arbitrary rule such as "first arrival wins, later arrival retreats" (Maynard Smith and Parker, 1976); such rules are independent of the pay-off to each contestant and the contests thus depend on an "uncorrelated asymmetry." Of these various possibilities, most evidence supports the idea that owners generally win because they are prepared to fight harder (Reichert, 1978; Hyatt and Salmon, 1978; Krebs, 1982).

The reason that owners fight harder may be that the territory is more valuable to them because they have better knowledge of the resources in the area than an intruder. This knowledge might include the location of seasonally optimum food and water, the behaviour of predators which use the territory, likely ambush points and escape routes, the behaviour of potential mates and intruders, and so on (e.g., Davies and Houston, 1981; Metzgar, 1976). Another selection pressure that may be more important, is that the owner is more likely to be killed by a predator if it is defeated and expelled into a completely unfamilar area than an intruder which simply continues with its non-territorial existence. This danger is most acute in a short period after defeat when the owner risks its life. This is in contrast to the loss of knowledge about the territory's resources where the owner may temporarily lose foraging efficiency.

Whatever the basis of owner advantage, it certainly exists. Some hartebeest territories are continuously occupied by the same male for over three years (Gosling, 1974) and during this time they probably have thousands of agonistic encounters with intruders, many presumably with males of higher RHP, and all of which they win. Furthermore, nearly all of these encounters end quickly with the intruder withdrawing either as the owner approaches them or after a very brief encounter. There may thus be a large saving in the costs of territory defence so long as the owner is recognised and thus gains owner advantage in contests with intruders. The potential costs of competition if intruders did not recognise owners can be inferred from contests in which intruders are prepared to escalate to the same level as the owner. A classical case is where the intruder is a high status male that is trying to take over a territory and such contests are most likely to be severe when the intruder has occupied a territory which the owner has left temporarily. In this context, very long fights occur and, apart from their considerable energetic costs, they may end in serious injury or death (e.g., hartebeest: Gosling, 1974).

SCENT MARKING AND COMPETITOR ASSESSMENT

Owner advantage in contests with intruders depends on the identification of the owner, not as an individual (see below) but as the owner of the territory. Elsewhere I have argued that the function of scent marking in territories is to advertise the owner's status and thus to reduce the cost of territory defence (Gosling, 1982). If this is true then the proper context for scent marking is competitor assessment theory which is an offshoot of the ideas about animal contests mentioned in the previous section. Briefly, animals would be expected to adjust their behaviour to the probability and benefits of winning a contest, and the costs of fighting. Zahavi (1975, 1977) further suggested that animals may use information in the displays of opponents to assess the chance of winning a contest and the costs of fighting, and that they should avoid fighting if the costs were high (in relation to the expected benefits). It is implicit in this body of theory that animals can increment their fitness not only by increasing their reproductive effort, but also by reducing the costs of intrasexual competition since this will improve their survival chances and thus increase their residual reproductive value. Thus, there is a benefit to intruders in identifying a territory owner and, in a sense, owners subvert this benefit by advertising their identity to their own advantage.

A problem in considering the evolution of status advertisement signals is that low quality animals could cheat by using a signal to gain a competitive advantage. It may be that this is avoided because animals assessing fighting ability can discriminate between characters that closely reflect fighting ability and those that do not (although this could lead to improved deception), or by selection for signals that are very difficult to imitate, for example, as in the case of roaring by male red deer, *Cervus elaphus*, where the signal is energetically very expensive to produce (Clutton-Brock and Albon, 1979). Scent marking in territories may have evolved as a form of status advertisement because it takes some time to comprehensively mark a territory (one week for a newly arrived Thomson's gazelle male; Walther, 1978a) and, since a male that has done this must have successfully defended the territory, it must actually be the owner. The historical element in scent marking thus makes it uniquely cheat-proof.

THE MECHANISM: SCENT MATCHING VERSUS INDIVIDUAL RECOGNITION

Owners could advertise their identity using a "territory owner pheromone," but existing observations support an alternative idea which I have presented elsewhere (Gosling, 1982). This is that owners distribute scent marks in their territory where they are most likely to be detected by intruders and then provide an opportunity for intruders to compare the smell of the marks with their own smell. If the smells *match*, then the animal that the intruder has met must be the territory owner. This hypothesis provides an explanation for such behaviours as self annointing by owners and the high frequency of marking during agonistic encounters (which could demonstrate the link between the owner and the scent marks in the territory). If the hypothesis were true, one would expect owners to (1) mark their territories in a way that maximises the chance that marks will be found by intruders, (2) mark themselves with the substances used to mark the territory, (3) make themselves available for "scent matching" by intruders, and (4) remove or replace marks in the territory that do not match their own odour. There is some support for these predictions, and to a lesser extent, for corresponding predictions about the behaviour of intruders, from existing observations in the literature (Gosling, 1982) and subsequent testing (Kruuk et al., 1984); however, further tests are needed.

A remarkably similar functional explanation to that outlined in the previous section has subsequently been suggested by Gorman (1984); the main difference being the emphasis placed on the use of scent marks by intruders to individually recognise territory owners. This is a development of earlier work showing, in laboratory experiments, that mongooses and badgers can discriminate between the smells of individual donors (e.g., Gorman, 1976). However, the demonstration of a sensory capacity is not a demonstration of function, and the ability to resolve scents *at the level* of the individual is also an essential prerequisite of scent matching without individual recognition. If intruders could not discriminate at this level, they would not be able to identify the donor of a particular scent in a subsequent encounter, and since the scent matching hypothesis depends on the advantage to the owner in subverting this ability in intruders, scent marking would not evolve. But this does not mean that intruders must recognise the donor when they sample the smell. The information could simply be remembered and only used if an animal that matches the scent is subsequently encountered.

I labour this point because there are good functional reasons for intruders to use scent matching and not individual recognition. The interaction between intruder and owner is essentially competitive and the resource under dispute is the territory. As discussed before, the owner normally wins encounters because it has "owner advantage" or a "spatial reference for dominance." By scent matching, an intruder learns whether or not an animal that it meets is the territory owner and, if it is, it withdraws because such an animal would be prepared to escalate a contest to dangerous levels. But, if it decided that an animal was the owner by recognising its individually distinct smell, it could make critical mistakes. Firstly, the intruder could fail to recognise an animal that had just taken over a territory and escalate a contest with a potentially dangerous opponent (hartebeest males that have just taken over a territory are in good condition and very aggressive, Gosling, 1975, and in this species such a mistake could result in injury or death). Secondly, if an animal had recently lost or left its territory, the intruder would fail to recognise that the animal was no longer a serious competitor and that there might be a chance for it to occupy the territory. For these reasons a generalised form of status recognition, such as scent-matching, should be favoured by sexual selection. And the identification of territory owners by individual recognition, which gives no information about current status, should be selected against.

This is not to say that competing males would never learn each other's identity. In two contexts this seems particularly likely. The first is during the process of establishing an individual reference for dominance. Individuals must be recognised as dominance relationships are established and odours are a prime candidate. However, for the reasons outlined above, this information would be secondary when a male encountered another whose smell matched that of marks in the immediate environment. Secondly, neighbouring territorial residents probably learn each others identity through repeated social contact. Neighbouring territorial male wildebeest, *Connochaetes taurinus* (Estes, 1969) and hartebeest (Gosling, 1974) have prolonged interactions at common boundaries every day with scent marking as a conspicuous element and they could learn each others odour as well as other individual characteristics. In general, neighbours are a threat because of their tendency to expand their territory (Huxley, 1934; Petrie, 1984) rather than from taking over the territory, and scent marks at the boundary may define the spatial limit of the neighbour's "owner advantage." However, it is important for a male to defend its territory during ownership changes in the neighbouring territory when there is a danger of expansion by a new and vigorous male. Individual recognition is irrelevant here and there should be selection for a generalised form of status recognition, just as in contests with non-territorial intruders. However, most economic benefits from competitor assessment must come from the energy and risks saved in encounters with non-territorial animals since, numerically, these are the principal competitors for territory ownership. This competition pressure can sometimes be inferred from the ratio of non-territorial to territorial males: in hartebeest, this ratio is 1.63 (Gosling, 1974).

The relative importance of scent-matching and individual recognition as competing mechanisms for owner recognition becomes particularly clear in the various behaviours where owners annoint their bodies with foreign odours. Many ungulates mix mud and the sap of plants with their glandular secretions and excretory products (Gosling, 1985), but this sort of behaviour is best developed in carnivores where "scent-rubbing" in decaying corpses, rotting vegetation, the faeces of other species, strongly smelling fungi, etc., is common and conspicuous (Rieger, 1979). Since these odours are a feature of a particular locality, the owner of the territory in this place can gain exactly the same advantage by showing intruders that it matches these odours as it does when it demonstrates that it has the same odour as that of glandular secretions and excretory products that it places as scent marks in its territory. There is even a benefit for the owner in scent-rubbing over marking with glandular secretion since the substances used are energetically free; in this respect, scent-rubbing is reminiscent of marking with faeces and urine. There would be no advantage from scent-rubbing in terms of individual recognition because all such odours are ephemeral. In contrast, ephemeral odours are an advantage for owners which depend on scent-matching for status advertisement because they could usually find a scent source before intruders which visit the area less frequently. They would thus be more likely to match a smell originating in their territory and intruders would be less able to imitate them.

ECONOMIC CONSEQUENCES OF SCENT MARKING IN TERRITORIES

In resource defence territoriality, males defend a limited resource, usually food, and mate with females which use the resource. In general, the number of females attracted will be determined by the quantity of the resource and this quantity thus determines the number of matings that a male can achieve and, ultimately its fitness. Although the limited resource can be regarded as consisting of units, for example, discrete food items or unit areas of continuous grassland, the size of the area which contains the limiting resources is critical and the area that an owner should defend to maximise fitness can be considered using simple cost-benefit analysis (e.g., Davies, 1978). As a male gains increasing numbers of resource units by expanding its territory, it gains fitness units. However, independent of costs, the rate of fitness gain eventually slows down and then reaches an asymptote because the owner has an intrinsic finite ability to use resources which are distributed in a particular fashion. The pattern of resources that effect this asymptote will include the tendency for females to aggregate on food patches and to form social groups, as discussed earlier. In contrast, as the defended area increases, the costs of defence increase and, since these costs are determined by extrinsic factors, principally intrasexual competition, they could potentially continue to increase indefinitely. However, economic constraints intervene and the territory size just before the point at which the fitness costs outweigh the benefits defines the maxiumum area that can be defended (Tmax). But benefits are devalued by costs and to maximise its fitness a male should generally defend an optimum area, smaller than Tmax, where the excess of fitness benefits over costs is maximal (Davies, 1978).

The main costs of territory defence (i.e., negative contributions to fitness) are (1) the costs of finding and approaching intruders, (2) the costs of contests with intruders, and (3) the risk of losing a contest and thus the territory. In all cases there are energetic costs, with some risk of predation, and in (2) and (3) there is the risk of injury, and even death, when fighting. In (3) the cost is in the same currency as the benefits of territoriality since the male simply loses the chance to reproduce, either temporarily, or permanently, if for example it is near the end of its breeding life and competition pressure is high. All of these costs will tend to increase as territory size increases. In (1) this will be a consequence of the greater area to be covered and in (2) because the larger the amount of a limited resource that a male manages to monopolise, the greater the resistance from intra-sexual competition. This resistance could be either from higher numbers of excluded competitors or from neighbours whose territories were compressed; however, only the first of these is considered here. Where the number of competitors increase, there will generally be a proportionate increase in the costs of agonistic behaviour because of the higher number of encounters and because there would be a higher number of high status males which tend to escalate encounters. In addition to this increase, as the value of the defended resource increased, there would also be a higher pay-off for an intruder in taking over the territory and high status intruders might thus tend to escalate contests to a higher level.

These increased costs in larger territories would be compounded if owners which lacked a clear status signal were recognised less readily by intruders. Failure to recognise owners could be a direct result of the larger numbers of competitors, and correspondingly fewer encounters with each, and because of a higher number of membership changes within the pool of non-territorial males.

An owner might thus tend to be less well known by any individual competitor and there would be a higher chance of meeting a completely new intruder. Owner identification would thus be less certain and under these circumstances high status intruders would tend to delay their withdrawal from encounters. On average, as territory size increased, owners would thus need to escalate encounters to higher levels in order to expel, or assert dominance, over intruders. The chance of losing the territory (cost (3) above) would also increase under these circumstances, since in escalated encounters the relative RHP of the contestants becomes more important. Territorial males are known to lose condition faster than non-territorial males (e.g., hartebeest: Stanley-Price, 1974) and, additionally, the increased costs of agonistic encounters would make owners that defended large territories lose condition more quickly than average so that the chance of meeting an intruder with higher RHP would increase.

A number of these costs could be reduced if, as suggested in the scent-matching hypothesis, scent marking reduces ambiguity in status signaling so that territory residents clearly indicate their ownership status and readiness to escalate contests with an intruder, thereby giving the intruder the chance to withdraw early. However, it is not useful to compare the effect of advertising ownership status using scent marking with that of a complete absence of ownership signaling because this never occurs. All owners advertise their status in a variety of ways including particular postures, a direct and fast approach to intruders and vocalizations. Sometimes, the sight of an adult male hartebeest starting to walk towards an intruder is enough to elicit withdrawal even when it is some distance away (Gosling, 1975) and where there is presumably some ambiguity about its status. In this case, the intruders are generally of such low status that they have nothing to lose by very early withdrawal, even if the chance of the approaching male being a territory owner is only small. Hartebeest also scent mark and the function of this additional form of status advertisement may be to reduce any ambiguity remaining after other signals. High status intruders, including those which are serious competitors for ownership, are less likely to withdraw until after a close encounter where scent-matching may be important. It seems quite likely that the main costs of defence to an owner are from encounters with such males and, if so, the reduction of escalated encounters, and the consequent chance of losing the territory could be the main gain for an owner with unambiguous status signaling.

Status signaling must also have a cost and this will include both time and energy. The costs of postures and other displays must be mainly energetic and could include such factors as heat stress by remaining in a conspicuous exposed position, rather than seeking shade. Scent marking may involve considerable amounts of time, particularly in replenishing a pattern of marks every one or a few days and glandular secretion must be produced in substantial quantities. However, these costs can be reduced by combining scent marking excursions with feeding and by using energetically free marking substances such as urine, faeces, and, in scent-rubbing, completely foreign substances. In general, the costs of scent marking are probably low in relation to those of escalated encounters, including the chance of losing the territory.

If scent marking reduces the costs of important agonistic encouters with high status males then the overall defence costs of defending a territory of increasing size would increase more slowly for an owner which scent marks _in addition to_ other forms of status advertisement than one which relies on signals such as postures or on being individually recognised as an owner. Following the earlier economic argument, if the cost of defence is reduced, then an owner should be able to defend a larger territory. The argument is explicit in Fig. 2, where a graphical analysis predicts that as the rate of increasing defence costs is reduced by scent marking, the area that can be defended (abcissa) increases with a corresponding fitness increment (ordinate).

If scent marking favours the evolution of resource defence territoriality, then when it is absent, alternative strategies might often yield a higher fitness pay-off. In particular, the male mating strategy of following and defending a group of females might be more profitable (in terms of matings achieved) than defending a resource that females need. In the antelopes, following strategies are mainly restricted to the Tragelaphinae, and perhaps significantly, the species in this group are unusual in lacking both antorbital and pedal glands. However, this sort of correlation cannot resolve cause and effect. Future research should aim to explore the role of scent marking in species such as topi, _Damaliscus korrigum_, and oryx, _Oryx beisa_, where both strategies exist (Gosling, 1985).

With existing information it is not possible to determine the extent to which economic resource defence territoriality depends status on advertisement by scent marking. Without detailed studies of species with mixed strategies, the main empirical evidence that this may be the case is that despite good reasons for expecting following strategies to be common, where resources and potential mates are well dispersed, in practice they are rare. Resource defence territoriality is by far the most common male mating strategy amongst antelopes and occurs even where, from an ecological point of view, following strategies seem almost inevitable. For example, oryx feed on grasses in a very arid habitat and females thus forage on an unpredictable food supply over enormous areas. Following seems inevitable, and indeed it is a common strategy (e.g., Walther, 1978b). But where there is a slight increase in rainfall and some element of spatial predictability in the habitat, male oryx establish territories of about 4-5 km^2 (T. Wacher, quoted in Gosling, in press; see also Estes, 1974). They do not chase out intruding males, but assert their dominance over any that enter and comprehensively mark the territory with piles of faeces. The costs of maintaining dominance over such a large territory are potentially very large, but an area of this size is probably essential in such arid habitat if the male is to achieve more

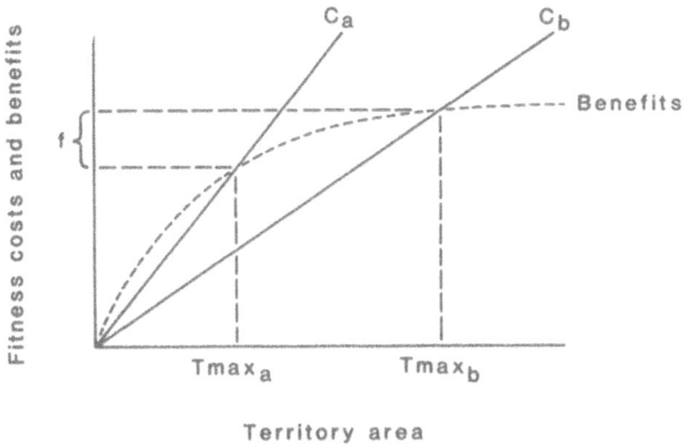

Fig. 2. Graphical analysis of the economic consequences of scent marking in area defence. The steep cost function (Ca) is the increasing cost of defending an area of progressively larger size when owners advertise their status by postures and calls, or rely on being recognised individually. The shallow function (Cb) is the cost of defending a progressively larger area with the additional use of scent marking which reduces the unambiguity of status signaling and thus reduces the costs of contests. These two cost functions define alternative native defence strategies. The broken line shows the increasing fitness benefits of monopolising an area of increasing size. The upper intercept between cost and benefit functions shows the upper limit of territory size for the Ca and Cb strategies (Tmaxa and Tmaxb respectively). The model predicts that males which adopt the Cb strategy can defend a larger territory. The fitness benefit (f) of adopting the Cb strategy instead of Ca is the difference between the intercepts of the cost and benefit functions on the ordinate. This benefit would accrue mainly because more females would be attracted to the larger area defended. The analysis is restricted to maximum territory size to keep the figure simple, but a similar result is obtained using optimum territory size (see text).

matings by resource defence than by following. In this species, and perhaps in many other cases, resource defence territoriality may be evolutionarily stable only because owners can reduce the costs of area defence by using scent-matching to advertise their status.

ACKNOWLEDGMENTS

I am grateful to Dr. Marion Petrie who helped to resolve a number of logical problems and who persuaded me to exclude a dubious passage on birds.

REFERENCES

Clutton-Brock, T. H., and Albon, S. D., 1979, The roaring of red deer and the evolution of honest advertisement, Behaviour, 69:145.

Davies, N. B., 1978, Ecological questions about territorial behaviour, in: "Behavioural Ecology: An Evolutionary Approach," J. R. Krebs and N. B. Davies, eds., Blackwell Scientific Publications, Oxford.

Davies, N. B., and Houston, A., 1981, Owners and satellites: The economics of territory defence in the pied wagtail, *Motacilla alba*, J. Anim. Ecol., 50:157.

Emlen, S. T., and Oring, L. W., 1977, Ecology, sexual selection, and the evolution of mating systems, Science, 197:215.

Estes, R. D., 1969, Territorial behaviour of the wildebeest, (*Connochaetes taurinus* Burchell, 1823), Z. Tierpsychol., 26:284.

Estes, R. D., 1974, Social organization of the African Bovidae, in: "The Behaviour of Ungulates and Its Relation to Management," F. R. Walther, and V. Geist, eds., IUCN, Morges.

Gorman, M. L., 1976, A mechanism for individual recognition by odour in *Herpestes auropunctatus* (Carnivora: Viverridae), Anim. Behav., 24:141.

Gorman, M. L., 1984, Scent marking and territoriality, Acta Zool., Fennica, 171:49.

Gosling, L. M., 1974, The social behaviour of Coke's hartebeest (*Alcelaphus buselaphus cokei*), in: "The Behaviour of Ungulates and Its Relation to Management," F. R. Walther, and V. Geist, eds., IUCN, Morges.

Gosling, L. M., 1975, The ecological significance of male bahaviour in Coke's hartebeest, *Alcelaphus buselaphus cokei*, Gunther, Ph.D. thesis, Univ. Nairobi.

Gosling, L. M. 1982, A reassessment of the function of scent marking in territories, Z. Tierpsychol., 60:89.

Gosling, L. M., 1985, The even-toed ungulates: Order Artiodactyla. Sources, behavioural context, and function of chemical signals, in: "Social Odours In Mammals," R. E. Brown, and D. W. Macdonald, Oxford Univ. Press, Oxford.

Gosling, L. M., (in press), The evolution of mating strategies in male antelopes, in: "Ecology of Social Evolution," D. I. Rubenstein, and R. W. Wrangham, eds., Princeton Univ. Press, Princeton.

Hamilton, W. D., 1971, Geometry for the selfish herd, J. Theoret. Biol., 31:295.

Huxley, J. S., 1934, A natural experiment on the territorial instinct, Brit. Birds, 27:270.

Hyatt, G. W., and Salmon, M., 1978, Combat in Fiddler crabs *Uca pugilator*: A quantitative analysis, Behaviour, 65:182.

Jarman, M. V., 1979, Impala social behaviour. Territory, hierarchy, mating, and the use of space, Adv. Ethol., 21:1.

Jarman, P. J., 1974, The social organisation of antelopes in relation to their ecology, Behaviour, 48:215.

Krebs, J. R., 1982, Territorial defence in the Great tit (*Parus major*): Do residents always win?, Behav. Ecol. Sociobiol., 11:185.

Kruuk, H., Gorman, M., and Leitch, A., 1984, Scent-marking with the subcaudal gland by the European badger, *Meles meles* L., Anim. Behav., 32:899.

Maynard Smith, J., 1976, Evolution and the theory of games, Am. Sci., 64:41.

Maynard Smith, J., and Parker, G. A., 1976, The logic of asymmetric contests, Anim. Behav., 24:159.

Metzgar, L. A., 1967, An experimental comparison of screech owl predation on resident and transient white-footed mice, J. Mammal., 48:387.

Mloszewski, M. J., 1983, "The Behaviour and Ecology of the African Buffalo," Cambridge Univ. Press, Cambridge.

Owen-Smith, N., 1977, On territoriality in ungulates, and an evolutionary model, Q. Rev. Biol., 52:1.
Owen-Smith, N., 1984, Spatial and temporal components of the mating systems of kudu bulls and red deer stags, Anim. Behav., 32:321.
Parker, G. A., 1974, Assessment strategy and the evolution of fighting behaviour, J. Theor. Biol., 47:223.
Petrie, M., 1984, Territory size in the moorhen (*Gallinula chloropus*): An outcome of RHP asymmetry between neighbours, Anim. Behav., 32:861.
Rieger, I., 1979, Scent rubbing in carnivores, Carnivore, 2:17.
Reichert, S. E., 1978, Games spiders play: Behavioural variability in territorial disputes, Behav. Ecol Sociobiol., 3:135.
Schaller, G. B., 1972, "The Serengeti Lion," Chicago Univ. Press, Chicago.
Stanley-Price, M. R., 1974, The feeding ecology of Coke's hartebeest, *Alcelaphus buselaphus cokei* Gunther, in Kenya, D. Phil. thesis, Univ. Oxford.
Walther, F. R., 1978a, Mapping the structure and the marking system of a territory of the Thomson's gazelle, E. Afr. Wildl. J., 59:243.
Walther, F. R., 1978b, Behavioural observations on oryx antelope (*Oryx beisa*) invading Serengeti National Park, J. Mammal., 59:243.
Zahavi, A., 1975, Mate selection - a selection for a handicap, J. Theor. Biol., 53:205.
Zahavi, A., 1977, Reliability in communcation systems and the evolution of altruism, *in*: "Evolutionary Ecology," B. Stonehouse, and C. Perrins, eds., Macmillan, London.

CHEMICAL SIGNALS AND KIN RECOGNITION

IN SPINY MICE (*Acomys cahirinus*)

Richard H. Porter

Department of Psychology and Human Development
George Peabody College
Vanderbilt University
Nashville, TN 37203

It follows from kin selection theory that the ability to influence the transmission of one's alleles into future generations would be enhanced in organisms who can readily distinguish between close relatives and unrelated conspecifics (Hamilton, 1964). Recognition of kin allows for preferential allocation of resources to such individuals (i.e., nepotism), which in turn increases the likelihood of survival and successful reproduction by the recipients--who share a large proportion of alleles with their related benefactor (e.g., Alexander, 1979). Likewise, optimal choice of mates, for example, reaching an appropriate balance between extreme outbreeding and incestuous mating, could be facilitated by discerning close kin (Bateson, 1983; Shields, 1983).

Kin recognition is believed to be a significant factor in the evolution of social behavior and has therefore become the subject of considerable current research (for recent reviews see Holmes and Sherman, 1983; Porter, in press). It appears that the sensory modalities involved in the recognition of kin are those that are otherwise the most salient for communication within a particular species. Accordingly, chemical cues have been implicated in the mediation of kin recognition in diverse taxa, including frog tadpoles (Blaustein and O'Hara, 1982; Waldman, 1984), coho salmon (Quinn and Busack, 1985), a variety of rodent species (e.g., Porter et al., 1978b; D'Udine and Partridge, 1981; Holmes, 1984), and even humans (Macfarlane, 1975; Russell, 1976; Porter and Moore, 1981; see also Colgan, 1983). The remainder of this chapter will focus upon kin recognition in spiny mice, *Acomys cahirinus* (a murid rodent indigenous to the eastern Mediterranean region); with particular emphasis on the role of chemical signals in parent-infant and sibling recognition.

PARENT-INFANT RECOGNITION

The most frequently studied version of kin recognition is that involving parents and offspring. Genetic benefits accruing to parents who respond preferentially to their own offspring are fairly obvious. Postnatal (or post-hatching) investment in progeny has direct positive consequences on the propogation of one's genes. In contrast, individuals who indiscriminately distribute their parental resources incur the genetic costs of wasted investment in unrelated young and, concomitantly, depriving their own offspring of their limited resources (Trivers, 1972). Thus, in contexts where biological offspring are likely to co-occur with unrelated young there would be considerable pressure favoring the evolution of mechanisms enabling parents to recognize their own offspring.

Within the first two days postpartum, laboratory-reared female spiny mice will indiscriminately nurse and retrieve their own offspring and unfamiliar unrelated pups of the same age (Porter and Doane, 1978). When two mothers, along with their eight-day-old pups are placed together, however, nursing interactions most often involve mothers and their own pups (Porter et al., 1980a). Related investigations indicate that the indiscriminate maternal responsiveness of one- to two-day postpartum females may be due (at least in part) to the similarity of chemical cues associated with newborn pups whose mothers are maintained on the same laboratory diet. Recently parturient females more rapidly retrieve unfamiliar pups born of females fed the same diet as the experimental mothers in comparison to simultaneously present pups born to females fed a distinctively different diet (Doane and Porter, 1978). Furthermore, the survival rate of pups fostered onto females fed the same diet as the mothers of the foster pups is greater than that of pups fostered onto different-diet females. Chemical cues that mediate maternal recognition of offspring thus appear to be dietary-dependent, so that the odors of neonates born to (and nursed by) mothers maintained on identical diets would be somewhat similar, whereas pups of mothers consuming different diets would have discernibly distinctive odors.

Spiny mice, like other muroid rodents, have eclectic feeding habits--taking advantage of whatever food sources are available. The diets of wild-living animals would presumably fluctuate and the types of food ingested, as well as the relative proportions of similar food items, differ somewhat across individuals. Therefore, reliance on chemical cues (labels) associated with the mother's unique diet would be a reliable strategy for early maternal recognition of offspring. Although the source of the dietary-dependent pup-recognition labels is not known at present, such cues could be deposited directly on the infants by the mother (e.g., through her saliva while grooming the pups, or through her urine), or result from the ingestion of mother's milk.

Unlike females, male spiny mice begin to respond discriminatively to their own familiar offspring vs. alien pups--all of whom were born of females maintained on the same diet--within 48 hours of birth (Makin and Porter, 1984). Such paternal recognition of offspring appears to involve chemosensory signals since males tested with two simultaneously present pups lick and sniff an alien pup more than their own. The increased responsiveness to the alien pups has been interpreted as heightened chemosensory investigation of an unfamiliar stimulus rather than a preference for such neonates over the fathers' own offspring.

Mammalian young are entirely dependent upon willing adult conspecifics (typically their mother) for the provision of nutrients and, possibly, other biological necessities such as warmth and protection from predators. As suckling young become capable of independent locomotion, they face the potential risk of wandering away from their mother or nest area and thereby losing access to parental resources. Alien lactating females are not adequate substitutes for the biological mother in many species since they are likely to reject or even respond aggressively toward unfamiliar young. Failure to distinguish between signals produced by their mother and alien females, and respond preferentially to the former, would jeopardize the continued survival of such young mammals.

Lactating females of several species of rodents emit chemical signals (maternal pheromone) to which preweaning young are attracted (Leon, 1980; Porter, 1983). Orientation to maternal pheromone is observed as soon as the pups' motor system is developed sufficiently to allow for independent locomotion. Corresponding with their advanced physical development at birth, the uniquely precocial spiny mouse pups begin to approach areas soiled with maternal pheromone within the first 26 to 36 hours postpartum (Porter and Doane, 1976); which is considerably younger than observed in other muroid rodents. Attraction to chemical cues produced by lactating females ceases as pups reach weaning age--at approximately 20 to 25 days (Porter et al., 1978b). Related neuroanatomical studies reveal a marked correlation between the age at which pups of different rodent species first approach maternal odors and the time when their olfactory bulbs attain "an adult level of structural organization" (Leon et al., 1984; see also Brunjes, 1983). One-day-old spiny mouse pups display mature structural organization of the olfactory bulbs as well as oriented movement towards maternal chemical signals. These anatomical and behavioral landmarks co-occur at approximately 14 days of age in Norway rats and 21 days in gerbils (Leon et al., 1984). Thus, across these three species, olfactory mediated approach responses begin when the olfactory bulbs reach a somewhat comparable stage of development.

Studies of recognition of specific maternal cues by _A. cahirinus_ pups have yielded results similar to those from investigations of mothers' recognition of offspring discussed above. In tests conducted one day after birth, pups are attracted to bedding soiled by their own mother _or_ by other lactating females fed the same diet, but not to bedding soiled by other classes of adult conspecifics (e.g., adult males, nulliparous females) (Porter and Ruttle, 1975; Porter and Doane, 1976). No preferences are evident when pups of this same age are tested with bedding from their own home cage vs. soiled bedding of a second suckling litter and parents (Porter and Ruttle, 1975).

Dietary influences on the attractiveness of maternal pheromone are evident from a further series of experiments. Pups prefer chemical cues from conspecific lactating females maintained on the maternal diet over such cues from females fed an unfamiliar diet (Porter and Doane, 1977). Lactating females of a different species, _Mus musculus_, fed the familiar maternal diet also produce chemical signals that attract _A. cahirinus_ pups (Porter, Deni, and Doane, 1977). Furthermore, chemical cues emanating from _Mus_ lactating females maintained on the maternal diet are preferred over cues from unfamiliar-diet conspecific females (Porter and Doane, 1977). It can be concluded that the maternal diet is more salient than are genetically determined species-specific characteristics in influencing the properties of chemical cues to which _A. cahirinus_ pups respond. Suckling spiny mouse pups rapidly become familiar with their mother's dietary-dependent odor and subsequently orient preferentially to such signals. To the extent that a mother's diet differs sufficiently from that of other lactating females, this process would result in her pups discriminating between maternal pheromone produced by their own mother and alien females.

SIBLING RECOGNITION

Whereas parent-infant recognition has long been a favorite topic of study by students of animal behavior, it is only recently that recognition among other classes of kin has become the subject of widespread empirical and theoretical interest. Hamilton (1964), in his now classic papers on the evolution of social behavior, pointed out that individual organisms need not produce their own offspring to have an impact on the representation of their alleles in future generations. It is also possible (at least in theory) to increase the reproductive success of other classes of kin by behaving to their advantage, and thereby exerting a positive influence on the propagation of alleles shared with those kin. This kinship, or indirect, component of inclusive fitness (e.g., West Eberhard, 1975), like the successful production of viable offspring, would benefit through an ability to recognize the appropriate individuals (or classes of individuals) to whom nepotistic behavior should be directed. Siblings are of special importance in this regard since the average coefficient of genetic relatedness among full siblings in diploid species is .50. Along with parents and offspring, who share 50% of their alleles through direct descent, full siblings are therefore the closest genetic kin (aside from monozygotic siblings who develop from a single fertilized egg and are genetically identical).

Evidence for Sibling Recognition in Spiny Mice

The initial investigation of sibling recognition in spiny mice involved observations of littermate full siblings who had been raised together with both parents prior to testing (Porter et al., 1978b). Each test group was composed of two pairs of full siblings (i.e., a total of four animals born to two females) who were placed together into a single observation cage when approximately one-month old. Spiny mice, like many rodent species, spend a great amount of time in physical contact with conspecifics. Consistent huddling preferences by the same pair of individuals would imply that these animals distinguish one another from other conspecifics--in other words, they recognize each other. Systematic observations were conducted on these groups of paired siblings over a period of several days (5 to 8 days, depending on the particular experiment). Social discrimination was assessed by recording the frequencies with which particular pairs of animals huddled together.

Regardless of the sex-combination of animals tested in the above manner, (familiar) sibling pairings were significantly more frequent than pairings by unfamiliar unrelated mice (Porter et al., 1978b). Replications of this experiment with pups of various ages reveal that preferential littermate huddling first occurs in pups approximatley 14- to 16-days old (Porter and Wyrick, 1979). When tested at a younger age, all four pups in a test group huddle together indiscriminately, with dyadic pairings (by either siblings or nonsiblings) only infrequently observed.

What benefits, if any, accrue to spiny mice as a result of an ability to recognize their littermate siblings? Detailed studies of free-living animals in their natural habitat, which would be necessary to provide ultimate answers to this question, have yet to be conducted. Nonetheless, the results of initial laboratory investigations afford insights into the possible functional significance of kin recognition in A. cahirinus.

In one such experiment, pairs of animals were food deprived and tested in novel environments containing a limited food supply (Porter et al., 1980b). Littermate pairs were more active in exploring their new environment than were pairs consisting of two unfamiliar agemates. Likewise, with food only available from a single small container, siblings frequently fed together--with such simultaneous feeding necessitating physical contact between the two animals. Paired unfamiliar animals, in contrast, displayed little simultaneous feeding. Rather, it was more common for the dominant animal to feed for a relatively long period of time and actively interfere with (by chasing or biting) the subordinate animal when the latter attempted to approach the food container. Agonistic behavior of this sort was never observed in any of the littermate pairs. A tendency to allow close kin to share one's food would enhance an animal's inclusive fitness providing that the available food was sufficient to sustain several individuals. Nepotistic food sharing might be especially important in the desert habitat of A. cahirinus where food is relatively scarce and sparsely distributed.

Olfactory Signatures

Recognition of kin, like other forms of social recognition, has two discrete components: (1) the phenotypic character (i.e., Signature) that establishes an individual's identity, or membership in a particular social category, and (2) decoding or recognition of that signature or its bearer (Alexander, 1979; Beecher, 1982). Within a given species, kin recognition might be mediated by overlapping signatures across two or more sensory modalities. Particular kin-recognition signatures could be genetically determined, such as the pattern of facial markings in Japanese quail (Bateson, 1980) and Bewick's swan (Bateson et al., 1980), or acquired from environmental sources, as for the dietary-dependent odor cues involved in parent-young recognition in spiny mice discussed above.

To a great extent, A. cahirinus weanlings rely on olfactory signatures for the recognition of littermate siblings. Weanlings rendered anosmic by intranasal injections of zinc sulfate display no behavioral evidence of recognizing their familiar siblings (Porter et al., 1978b). Rather, littermates, and unfamiliar agemates tend to huddle together indiscriminately in a single group comprising all four animals in a test cage (two pairs of siblings) following zinc sulfate treatment. Saline-treated control animals, on the other hand, huddle preferentially in sibling pairs. Indiscriminate responses among siblings and unfamiliar agemates by anosmic animals suggest that olfactory signatures are the basis for the selective interactions among littermates, as well as the avoidance of unfamiliar conspecifics, observed in intact animals. Related research currently in progress suggests that olfactory cues may also serve as signatures enabling animals to distinguish between individual members of their own litter. While intact animals respond differently to their littermates with whom they are familiar as opposed to unfamiliar, such discrimination between littermate full siblings disappears in anosmic animals. It would therefore appear that kin recognition is dependent upon discriminating individual odor signatures of particular kin, or, alternatively, that general signatures shared by members of the same kin class co-occur with unique individual signatures, with the latter being more salient for the mediation of huddling preferences.

The role of chemical signatures in A. cahirinus sibling recognition is further evident in experiments assessing responsiveness to chemical cues secreted by various classes of agemate conspecifics (Porter and Doane, 1979). Male, but not female, weanlings spend more time investigating bedding material soiled by a littermate sibling than that soiled by an unfamiliar animal of the same sex. Lack of discriminative responsiveness to littermate signals by female weanlings does not imply that females of this age are unable to recognize their littermates. As mentioned above, females, like males, show reliable preferences for huddling with littermates (of either sex) over strange agemates. These inconsistent responses by male vs. female weanlings to littermate-soiled bedding material cannot be explained further at the present time.

Processes Mediating Recognition of Siblings' Signatures

The mechanisms involved in recognition of kin (or kin signatures) has been the focus of a number of recent empirical and theoretical papers (e.g., Alexander, 1979; Holmes and Sherman, 1982; Porter, in press). Becoming familiarized with individual kin through direct association is probably the most obvious process for mediating kin recognition (e.g., Bekoff, 1981). According to this mechanism, individuals learn the signatures of kin with whom they have had direct contact and thereby recognize them during subsequent encounters. At the opposite extreme, it is theoretically possible for both the phenotypic signature and the ability to decode that signature to be genetically mediated--viz. encoded on recognition alleles (Blaustein, 1983; Dawkins, 1976). Individuals who share the same recognition allele complex would be capable of recognizing one another even though they have had no prior contact. A third recognition mechanism, phenotype matching, involves learning the signatures of close kin (or one's own signature) and afterwards comparing the signatures of others to this learned standard. Previously unfamiliar individuals could be recognized as kin, provided that their signature bears a sufficient resemblance to the kin standard (Holmes and Sherman, 1982; Lacy and Sherman, 1983).

Familiarity/Association. Research conducted in our laboratory over the last decade has determined that familiarization through direct association is the primary proximal mechanism involved in sibling recognition of spiny mice. In an attempt to assess the relative importance of genetic relatedness versus direct contact in the development of huddling preferences, single pups were exchanged between two litters born on the same day, with the families otherwise remaining intact (Porter et al., 1981). This cross-fostering manipulation resulted in the formation of litters, each containing one to three of the nursing mother's own offspring plus one unrelated foster pup. When the pups reached approximately 30 days of age, two pairs of foster littermates were placed together into a single cage. Each of the resultant groups of four animals observed in this experiment was composed of two pairs of biological siblings that had been reared separately following the fostering manipulation.

Regardless of whether pups were fostered on the day of birth, day 10, or day 20 postpartum, dyadic pairings by unrelated foster littermates (pups that had been housed together from the time of fostering) were significantly more frequent than were pairings by the separated (unfamiliar) siblings. Weanlings therefore recognize and evince social preferences for agemates with which they had been raised, whether or not such littermates are genetic relatives per se. In nature, of course, this strategy would result in the recognition of close kin since neonates raised in the same litter would presumably be biological siblings. The evolution of nepotism does not require an inborn ability to recognize kin, only that kin be discriminated from non-kin, regardless of the underlying mechanism. It should be pointed out that in this same series of experiments, pairings by unfamiliar siblings were no more frequent than pairings by unrelated unfamiliar animals--which is taken as further evidence that recognition alleles are not involved in the mediation of sibling recognition in A. cahirinus.

Once established, the recognition of siblings is not necessarily permanent; rather, its continued maintenance is dependent upon additional experience with those kin or their recognition signatures. Animals that are kept in isolation for eight full days following weaning subsequently display no signs of recognizing their littermates with whom they had been housed from birth until the beginning of the isolation period (Porter and Wyrick, 1979). If siblings are re-exposed to one another for four hours during the fifth day of an eight-day isolation period, however, they afterwards continue to engage in preferential huddling. Reinforcement of kin recognition through relatively brief periods of re-exposure would allow for long-term discriminative interactions among siblings--especially if individuals of a given species do not disperse widely from their place of birth so that littermates are likely to come intermittently into contact with one another's signatures.

Prior to weaning, suckling pups are exposed to a variety of cues across differing sensory modalities emanating from their littermates. Which of these modalities are most salient for developing an ability to recognize one's siblings? In particular, what evidence is there for the involvement of olfactory cues in the ontogeny of kin recognition through familiarization?

To elucidate further the types of experience necessary for the development of individual recognition, pairs of unfamiliar unrelated pups were exposed to one another in differing contexts for a period of nine days before being tested for social preferences (Porter et al., 1984). Pairs of animals that were housed in cages that allowed unrestrained interactions and full bodily contact developed striking preferences for huddling with each other, regardless of whether they had been maintained in complete darkness or under a recurring light/dark illumination cycle during the familiarization period. Likewise, the discriminative interactions among paired animals that had been separated by a wire-mesh partition, in either a dark or light/dark illumination condition, denote that they also recognize one another. Since huddling preferences were equally strong in the light vs.

light/dark conditions, visual cues appear to be of little importance for the development of social recognition. Nonetheless, huddling preferences among pairs in the wire-mesh partition condition were not as strong as those in the complete contact condition. While the wire-mesh partition did not eliminate visual, auditory, or distal odor cues, the possibility of proximal olfactory exploration of cagemates (e.g., of the anogenital area) was curtailed in this condition. Finally, animals that were separated by a wire-mesh partition covered with a thick layer of cotton gauze displayed no evidence of recognition in tests following the exposure period. Animals in this latter condition thus behaved similarly to weanlings that have been raised apart in entirely separate cages. Aside from restricting visual communication between the two halves of the exposure cage, the gauze barrier reduced airflow and thereby lessened the opportunity for olfactory familiarization of pups separated by that barrier. The gauze partition also reduced sampling of proximal olfactory cues (e.g., mutual sniffing by animals pressing their noses against the wire grid) more than the mesh partition. Accordingly, the development of social preferences indicative of recognition declines as the opportunity for olfactory familiarization is diminished. The combined results of this series of experiments attest to the role of olfactory cues--proximal and possibly more distal airborne odors--for the development of individual social recognition.

Phenotype matching. As mentioned above, relatedness of conspecifics might be assessed (at least in theory) by comparing their phenotypic signatures to that of a previously learned kin standard. Close kin could then be discerned by the degree of correspondance between their phenotype and the standard signature--providing that phenotypic similarity is indeed an accurate reflection of genetic similarity (Barash et al., 1978; Holmes and Sherman, 1982). The correlation between phenotype and genotype, not the underlying basis for this correlation, is critical for the evolution of kin recognition by phenotype matching. Therefore, the similarity in kin signatures could result from shared environmental influences or be genetically determined.

Matching chemical signatures of conspecifics, a putative kin recognition mechanism in a range of organisms from social insects through humans (for recent reviews see Holmes, in press; Porter, in press), apparently facilitates the development of sibling recognition in *A. cahirinus*. Earlier, it was pointed out that after nine days of isolation weanlings do not recognize their former littermates. If weanlings from litters originally containing four pups are housed in littermate pairs rather than in isolation, however, they subsequently huddle preferentially with their littermates with which they had no contact during the nine-day pair-housing treatment period (Porter et al., 1983). Exposure to a single littermate enables animals to recognize similar phenotypic signatures shared by other full siblings--even though the latter animals would otherwise have become unfamiliar. Self-matching, whereby an animal recognizes kin by their phenotypic resemblance to oneself, is not salient in this regard since isolation rearing, which entails only exposure to the isolate's own signature(s) obviates sibling recognition.

Topical application of artificial odorants has been employed as an experimental technique to investigate further phenotype-matching of olfactory cues. Groups of three littermates were caged together following weaning and all animals within a group treated with one of several artifical odors (e.g., musk oil, oil of cloves). Immediately after the final odor application (on the sixth day of odor exposure), four unfamiliar animals were placed together into an observation terrarium. In each instance, two of these animals (along with their littermates) had been treated with an identical odor, while the remaining two animals had been treated with a second odor. None of the four animals composing each test group had been exposed to each other prior to testing.

Over the three-day observation period, unfamiliar weanlings that had been treated with the same odor (applied directly to themselves and their littermates) paired significantly more often than did animals treated with two different odors. It was concluded that familiar social odors influence interactions among weanlings by serving as learned standards against which the odors of unfamiliar agemates are assessed. Conspecifics who match the phenotypic standard are recognized and preferred as huddling partners.

An additional example of phenotype matching suspected to occur in several mammalian species involves mothers (and possibly fathers) labeling their offspring with distinctive chemical stimuli that enable those young to be distinguished from others (e.g., Gubernick, 1980). In effect, mothers recognize their own familiar signature on their offspring through a process of self-matching. Comparison of shared maternal labels by spiny mouse pups has also been found to contribute to the development of sibling recognition (Porter et al., 1981).

Paired littermates were separated on the fourth day after birth and housed individually in isolation cages. The separated littermate siblings in one condition continued to feed from the <u>same</u> lactating female (their biological mother) over the next two weeks. Mothers alternated between the two cages (each containing one of her offspring) at 12-hour intervals. In a second condition, paired littermates were also reared apart in separate isolation cages, but each was nursed by a <u>different</u> female for 12 hours every day (i.e., one lactating female for each pup). Observations of social interactions within groups containing two pairs of separated siblings from the same feeding condition were conducted following the two-week treatment period.

Those littermates that had shared the same mother while separated from each other subsequently engaged in discriminative huddling indicative of recognition. There was no evidence of recognition among the separated siblings suckling from different females, however. Pups in the former condition apparently recognized one another through familiar maternal signatures that are transferred directly to the surface of the offspring (carried in the mother's urine, saliva, milk, or other glandular secretions), or indirectly through ingestion of the mother's milk.

The next experiment in this series investigated the robustness of maternal labels as recognition cues for spiny mouse siblings (Porter et al., 1984). That is, will pups anointed with the same maternal label still recognize one another if they are also labeled by two different females, or will the addition of dissimilar maternal labels reduce the effectiveness of the shared chemical signature?

On the fourth day after birth, littermate pairs were assigned to one of two treatment conditions. In the first condition, two siblings from the same litter were separately housed in individual isolation cages but suckled from the same female (their biological mother) for alternating 12-hour periods. During the intervals when the mother was not present, each pup was alone in its cage. Pups in the second condition were treated as above; however, during the 12-hour intervals when the biological mother was in the cage of one pup, its littermate was housed with another lactating female. For each littermate pair, therefore, the mother was shared by her two offspring, but those same pups also spent one-half of the time with another female who differed for each pup.

Pups in both treatment conditions displayed preferential littermate huddling during the subsequent recognition tests, beginning when they were 18 to 22 days of age. Furthermore, the frequency of discriminative pairing by siblings housed with two different females (as well as sharing their own mother) did not differ statistically from that of littermates who were not exposed to different lactating females. The addition of dissimilar maternal labels had no noticeable impact on the ability of pups to recognize one another by maternal labels that they received in common from the same lactating female.

Despite the apparent robustness of maternal labels as recognition cues, additional data reveal that phenotype-matching based upon such signatures has less of an effect than the process of familiarization on the development of social recognition (Porter et al., 1984). The relative salience of these two recognition mechanisms was assessed directly by testing groups, each containing four pups, in which pairings by the following categories of animals were possible:

1. Littermates who had been separated on the day after birth but continued to suckle from the same female (their biological mother) during an 18- to 19-day treatment period preceding testing.

2. Unrelated pups that were housed together for 12 hours per day throughout the treatment period but also suckled from _two different_ females during the remaining 12 hours each day.

3. Unfamiliar unrelated agemates that had _no_ prior contact with each other and suckled from two different females prior to testing.

Over the three-day social recognition tests, the familiar non-siblings (animals that had been housed together but nursed by two different females) paired significantly more frequently than did either the separated siblings who shared a common mother, or the unfamiliar non-sibling--reiterating the overriding importance of association and familiarity for the development of social preferences.

SUMMARY AND CONCLUSIONS

A variety of methodological approaches have been adopted in an attempt to elucidate the role of chemical cues in kin recognition in spiny mice (_Acomys cahirinus_): e.g., observations of unrestrained social interactions;

Table 1. Summary of Data Implicating Olfactory Cues in the Mediation of Kin Recognition in Spiny Mice (*Acomys cahirinus*)

I. Parent's Recognition of Offspring

Parturient females preferentially retrieve one-day-old pups born of same-diet females over pups from novel-diet females.

The survival rate among pups fostered onto females fed the same diet as the biological mother is greater than for pups fostered onto novel-diet females.

Males spend more time sniffing and licking (indicative of chemosensory investigation) novel pups as compared to their own offspring.

II. Recognition of Maternal Cues by Pups

Pups orient preferentially to olfactory signals produced by females maintained on the same diet as the biological mother of those pups.

III. Sibling Recognition

Intact animals huddle preferentially with familiar littermates; however, social huddling is indiscriminate among anosmic animals.

Male weanlings distinguish between bedding soiled by littermates vs. unfamiliar agemates.

Social recognition develops comparably among animals exposed to one another in complete darkness or under a light/dark cycle. As the opportunity for olfactory familiarization is reduced, evidence of recognition declines.

Animals huddle preferentially with unfamiliar conspecifics treated with the same odorant as themselves and their littermates.

Unfamiliar pups who suckled from the same lactating female display discriminative social interactions.

---

peripherally-induced anosmia; dietary manipulations of biological odors; treating animals with artificial odorants. Such studies conducted over the last decade provide both direct and indirect evidence that olfactory cues are crucial in mediating the development of discriminative interactions between parents and offspring, as well as amongst siblings. The results of these investigations are summarized in Table 1.

Sibling recognition, which has been investigated in considerably more detail than parent-offspring recognition for A. cahirinus, is apparently influenced by several proximal mechanisms. While interacting with one another, pre-weanling pups are exposed to their littermates' olfactory signatures and subsequently rely to a great extent upon such familiar odors to recognize those same siblings. Full siblings also share similar phenotypic characteristics (including common chemical labels provided by their mother) that facilitate recognition through the process of phenotype matching. Although matching of similar odor phenotypes on its own is a less effective recognition mechanism than familiarization, the former process would presumably reinforce the development of recognition among animals raised in the same litter.

The studies reviewed above are consistent in offering no evidence in support of hypothetical recognition alleles. Rather, the recognition mechanisms identified thus far are dependent upon exposure and learning, which might represent a more parsimonious solution to the problem of kin recognition than would recognition alleles. This latter recognition mechanism would necessitate a "supergene" (or gene complex) having pleiotropic influences on the phenotypic expression of the recognition signature; the ability to recognize that signature; and appropriate responsiveness to individuals bearing that phenotypic marker (Hamilton, 1964, but see also Blaustein, 1983; Yamazaki et al., 1976).

For spiny mice, conspecifics with whom (or with whose signatures) one has had experience in particular contexts (e.g., as members of one's own "litter"; as the source of milk and maternal care) are recognized and responded to accordingly--regardless of whether they are kin, per se. These mechanisms are likely to result in reliable recognition of kin in nature, and would guard against potential recognition errors, providing that non-kin do not usually co-occur with kin in such contexts. In contrast, for species wherein kin and nonkin overlap through all stages of the life cycle, familiarization (other than self-familiarization) would not be a viable mechanism for the development of kin recognition.

Finally, not all members of a given kin class (viz., full siblings in the present instance) are treated comparably. Separated biological littermates interact similarly to unfamiliar unrelated agemates, even though members of that same litter who had been raised together respond preferentially to one another. On the other hand, studies of phenotype matching suggest that individuals who are familiarized with one of their full siblings can recognize other members of that same kinship. It can be concluded that individual (olfactory) recognition cues are acquired through a process of association and familiarization, and that once a signature is learned, stimulus generalization would then account for the ability (albeit rather weak) to recognize unfamiliar close kin. However, since olfactory signatures produced by individuals within a particular kin class are likely to be similar but not identical (due to genetic differences between individuals), they are not responded to indiscriminately. Thus, even animals that share a common maternal label may also emit additional salient chemical cues that vary across individuals so that their complex recognition signatures are not identical. It appears, therefore, that the recognition of kin in *Acomys cahirinus* develops through a process of familiarization with individual olfactory recognition cues, and although there is some ability to generalize to other members of a kin class, there is no evidence for indiscriminate kin-group recognition (see Breed and Bekoff, 1981 for a further discussion of the distinction between individual recognition and generalized kin or group recognition).

ACKNOWLEDGEMENTS

The work summarized in this chapter was supported by grant HD 15051 from the National Institutes of Child Health and Human Development.

REFERENCES

Alexander, R. D., 1979, "Darwinism and Human Affairs," University of Washington Press, Seattle.

Barash, D. P., Holmes, W. G., and Greene, P. J., 1978, Exact versus probabilistic coefficients of relationship; some implications for sociobiology, *Amer. Natural.*, 112:355.

Bateson, P., 1980, Optimal outbreeding and the development of sexual preferences in Japanese quail, *Z. Tierpsychol.*, 53:231.

Bateson, P., 1980, Optimal outbreeding, *in*: "Mate Choice," P. Bateson, ed., Cambridge University Press, Cambridge.

Bateson, P., Lotwick, W., and Scott, D. K., 1980, Similarities between the faces of parents and offspring in Bewick's swan and the differences between mates, *J. Zool. London*, 191:61.

Beecher, M. D., 1982, Signature systems and kin recognition, *Amer. Zool.*, 22:477.

Bekoff, M., 1981, Mammalian sibling interactions: Genes, facilitative environments, and the coefficient of familiarity, *in* "Parental Care in Mammals," D. J. Gubernick, and P. H. Klopfer, eds., Plenum Press, New York.

Blaustein, A. R., 1983, Kin recognition mechanisms: Phenotype matching or recognition alleles?, *Amer. Natural.*, 121:749.

Blaustein, A. R., and O'Hara, R. K., 1982, Kin recognition cues in *Rana cascadae* tadpoles, *Behav. Neural Biol.*, 36:77.

Breed, M. D., & Bekoff, M., 1981, Individual recognition and social relationships, *J. Theoret. Biol.*, 88:589.

Brunjes, P. C., 1983, Olfactory bulb maturation in *Acomys cahirinus*. Is neural growth similar in precocial and altricial murids?, *Develop. Brain Research*, 8:335.

Colgan, P., 1983, "Comparative Social Recognition," John Wiley & Sons, New York.

Dawkins, R., 1976, "The Selfish Gene," Oxford University Press, Oxford.

Doane, H. M., and Porter, R. H., 1978, The role of diet in mother-infant reciprocity in the spiny mouse, *Develop. Psychobiol.*, 11:271.

D'Udine, B., and Partridge, L., 1981, Olfactory preferences of inbred mice (Mus musculus) for their own strain and for siblings: Effects of strain, sex and cross-fostering, Behaviour, 78:314.

Gubernick, D. J., 1980, Maternal "imprinting" or maternal "labelling" in goats?, Anim. Behav., 28:124.

Hamilton, W. D., 1964, The genetical evolution of social behavior: I,II, J. Theoret. Biol., 7:1:17.

Holmes, W. G., 1984, Sibling recognition in thirteen-lined ground squirrels: Effects of genetic relatedness, rearing association, and olfaction, Behav. Ecol. and Sociobiol., 14:225.

Holmes, W. G., in press, Kin recognition by phenotype matching in female Belding's ground squirrels, Anim. Behav.

Holmes, W. G., and Sherman, P. W., 1982, The ontogeny of kin recognition in two species of ground squirrels, Amer. Zool., 22:491.

Holmes, W. G., and Sherman, P. W., 1983, Kin recognition in animals, Amer. Scientist, 71:46.

Lacy, R. C., and Sherman, P. W., 1983, Kin recognition by phenotype matching, Amer. Natural., 121:489.

Leon, M., 1980, Development of olfactory attraction by young Norway rats, in: "Chemical Signals," D. Muller-Schwarze, and R. M. Silverstein, eds., Plenum, New York.

Leon, M., Coopersmith, R., Ulibarri, C., Porter, R. H., & Powers, J. B., 1984, Development of olfactory bulb organization in precocial and altricial rodents, Develop. Brain Research, 12:45.

Macfarlane, A., 1975, Olfaction in the development of social preferences in the human neonate, in: "Parent-Infant Interaction," Ciba Foundation Symposium 33, Elsevier, New York.

Makin, J. W., and Porter, R. H., 1984, Paternal behavior in the spiny mouse (Acomys cahirinus), Behavioral and Neural Biol., 41:135.

Porter, R. H., 1983, Communication in rodents: Adults to infants, in: "Parental Behaviour of Rodents," R. W. Elwood, ed., John Wiley & Sons, Chichester, Sussex.

Porter, R. H., in press, Kin recognition: Functions and mediating mechanisms, in: "Sociobiology and Psychology: Ideas, Issues and Findings," C. B. Crawford, M. F. Smith, and D. L. Krebs, eds., Lawrence Erlbaum Associates, Hillsdale.

Porter, R. H., Cavallaro, S. A., and Moore, J. D., 1980a, Developmental parameters of mother-offspring interactions in Acomys cahirinus, Z. Tierpsychol., 53:153.

Porter, R. H., Deni, R., and Doane, H. M., 1977, Responses of Acomys cahirinus pups to chemical cues produced by a foster species. Behav. Biol., 20:244.

Porter, R. H., and Doane, H. M., 1976, Maternal pheromone in the spiny mouse (Acomys cahirinus), Physiol. and Behav., 16:75.

Porter, R. H., and Doane, H. M., 1977, Dietary-dependent cross-species similarities in maternal chemical cues, Physiol. and Behav., 19:129.

Porter, R. H., and Doane, H. M., 1978, Studies of maternal behavior in spiny mice (Acomys cahirinus), Z. Tierpsychol., 47:225.

Porter, R. H., and Doane, H. M., 1979, Responses of spiny mouse weanlings to conspecific chemical cues, Physiol. and Behav., 23:75.

Porter, R. H., Doane, H. M., and Cavallaro, S. A., 1978a, Temporal parameters of responsiveness to maternal pheromone in Acomys cahirinus, Physiol. and Behav., 21:563.

Porter, R. H., Matochik, J. A., and Makin, J. W., 1984, The role of familiarity in the development of social preferences in spiny mice, Behav. Proc., 9:241.

Porter, R. H., and Moore, J. D., 1981, Human kin recognition by olfactory cues, Physiol. and Behav., 27:493.

Porter, R. H., Moore, J. D., and White, D. M., 1980b, Food sharing by sibling vs. nonsibling spiny mice (*Acomys cahirinus*), Behav. Ecol. and Sociobiol., 8:207.

Porter, R. H., and Ruttle, K., 1975, The responses of one-day old *Acomys cahirinus* pups to naturally occurring chemical stimuli, Z. Tierpsychol., 38:154.

Porter, R. H., Tepper, V. J., and White, D. M., 1981, Experiential influences on the development of huddling preferences and "sibling" recognition in spiny mice, Develop. Psychobiol., 14:375.

Porter, R. H., and Wyrick, M., 1979, Sibling recognition in spiny mice (*Acomys cahirinus*), Anim. Behav., 27:761.

Porter, R. H., Wyrick, M., and Pankey, J., 1978b, Sibling recognition in spiny mice (*Acomys cahirinus*), Behav. Ecol. and Sociobiol., 3:61.

Quinn, T. P., and Busack, C. A., 1985, Chemosensory recognition of siblings in juvenile coho salmon (*Oncorhynchus kisutch*), Anim. Behav., 33:51.

Russell, M. J., 1976, Human olfactory communication, Nature, 260:520.

Shields, W. M., 1983, Optimal inbreeding and the evolution of philopatry, in: "The Ecology of Animal Movement," I. R. Swingland, and P. J. Greenwood, eds., Oxford University Press, Oxford.

Trivers, R. L., 1972, Parental investment and sexual selection, in: "Sexual Selection and the Descent of Man," B. Campbell, ed., Aldine, Chicago.

Waldman, B., 1984, Kin recognition and sibling association among wood frog (*Rana sylvatica*) tadpoles, Behav. Ecol. and Sociobiol., 14:171.

West Eberhard, M. J., 1975, The evolution of social behaviour by kin selection, Q. Rev. of Biol., 50:1.

Yamazaki, K., Boyse, E. A., Mike, V., Thaler, H. T., Mathieson, B. J., Abbott, J., Boyse, J., Zayas, Z. A., and Thomas, L., 1976, Control of mating preferences in mice by genes in the major histocompatibility complex, J. Exper. Med., 144:1324.

GENETIC BASIS FOR INDIVIDUAL DISCRIMINATIONS:

THE MAJOR HISTOCOMPATIBILITY COMPLEX OF THE MOUSE

G. Beauchamp[1], A. Gilbert[1], K. Yamazaki[1] and E. A. Boyse[2]

[1]Monell Chemical Senses Center, Philadelphia, PA
[2]Memorial Sloan-Kettering Cancer Center
New York, NY

INTRODUCTION

The factors which permit altruistic behaviors to be differentially directed to kin and non-kin have aroused much interest. Hamilton's (1964) development of the concept of inclusive fitness allowed a rationale for the evolution of altruistic behavior. However, in order to evolve, such behaviors must be directed at kin, and thus kin and non-kin must be discriminated. Various means by which such discriminations take place have been elaborated by Alexander (1979), Blaustein (1983), Holmes and Sherman (1983) and others. One prominent issue is the extent to which phenotype matching (involving learning) and/or recognition alleles are involved in kin recognition. Here we consider this and other issues in the context of a specific set of studies that have explored a genetic basis for possible discriminations among individual mice.

Initial Observations

In 1976, Yamazaki, Boyse and colleagues (Yamazaki et al., 1976) reported that the same assembly of genes intimately involved in immune function (the Major Histocompatibility Complex or MHC, labeled H-2 in mice) also was involved in mate selection. Tests were conducted in which males of inbred strains were caged individually with two estrous congenic females which differed from each other only in the MHC region. Under these test conditions, matings were generally more frequent with females of one MHC type than with the other. These initial data established that genetic differences restricted to the MHC were somehow capable of providing the basis for discriminative behavior.

This observation had potential implications for evolutionary biology. Before discussing them, a brief overview of the MHC and MHC specification of olfactory signals will be presented. This will set the stage for a consideration of how this system may relate to issues in kin recognition.

The Major Histocompatibility Complex of Genes

The importance of the MHC genes can be judged from the fact that a similar set of genes is believed to operate in all vertebrates. The MHC of the mouse occupies a segment of chromosome 17 (Klein, 1975). It comprises many linked genes and can be divided into regions, notably the main regions called H-2K (K), H-2D (D) and Tla. The total set of variable alleles of all genes in the MHC region determines the animal's MHC type or H-2 type. The set of MHC alleles on a given chromosome 17 is called a haplotype, and a vast number of haplotypes is possible. The haplotype is conventionally denoted by a superscript letter or letters, such as $H\text{-}2^b$ (abbreviated b) and $H\text{-}2^k$ (abbreviated k) in homozygous animals or $H\text{-}2^{bk}$ (abbreviated bk) for heterozygotes.

The functional role of the MHC is best known from studies on tissue transplantation, because incompatibility of MHC types causes extreme rejection of grafts. The fate of kidney transplants in man depends mainly on MHC compatibility. Throughout the MHC region, there are also genes that determine the degree of response to particular antigens, and other genes that are selectively expressed in lymphocytes. Thus, the MHC is concerned with the many aspects of how immune cells -- lymphocytes equipped with specific receptors for antigen -- handle chemical information from the environment.

Genes in the MHC also influence phenotypes of no direct immunological significance. For example, the physiology of steroid hormones (testosterone and cortisone) is modulated by genetic variation in the MHC region (Gupta and Goldman, 1982; Ivanyi, 1978).

In order to identify specific genes or gene complexes with specific effects, inbred congenic mice have been essential. An H-2 congenic strain is produced by crossing two selected inbred strains of different H-2 types and by backcrossing many times to one of these strains, serologically selecting for the donor H-2 type in each generation (a simple outline is given in Beauchamp, Yamazaki and Boyse, 1985). Consequently, the final congenic strain is genetically identical to the base strain except for both copies of a segment of chromosome 17 bearing the H-2 haplotype introduced from the donor strain. Any difference between the base strain and its congenic partner strain, if proved to be genetic, must be due to genes in the H-2 region, because this is the only genetic difference between the inbred base strain and its congenic partner.

H-2 Control of Olfactory Signals

Following the demonstration of differential mating among mice according to H-2 type, the next step taken was to determine whether odor alone could be involved and, further, whether odor differences due to H-2 would be evident when a variety of paradigms were employed. Odor was thought a likely candidate for individual discrimination since odors are known to play significant roles in mouse social and sexual behaviors (Bronson, 1979). A series of studies demonstrated (Table 1) that both mice and rats could be trained to discriminate between urine odors of mice differing only at H-2 and that H-2 differences were sufficient to elicit pregnancy block (interruption of pregnancy by the odors of a male different from the stud).

Further studies revealed that independent regions within H-2 separately conferred individuality of odor (Yamazaki et al., 1982). At this point in the research it was still conceivable that within the H-2 complex of genes there were actual "recognition" alleles which had evolved for that function (c.f., Dawkins, 1976). More likely, however, genes known to have major roles in immunity were thought, perhaps secondarily, to direct production of olfactory signals. This latter idea was supported when it was shown that a mutation in a single gene,

Table 1. Evidence that individual identity is transmitted via signals arising from differences in the genes of the major histocompatibility complex of the mouse.

Paradigm	Results
Mating Preference	Males differentially mated with females differing only at the MHC[a]
Y-maze Training	Mice were trained to discriminate other mice or their urines when odor donors differed only within the MHC[b]
Olfactometer Training	Rats similarly were trained to discriminate between urines collected from mice differing only in the MHC[c]
Neuroendocrine Effects	In newly inseminated female mice, the odors of male mice differing from the stud only at the MHC were capable of inducing a pregnancy block[d]

[a]Yamazaki et al., 1976; Yamaguchi et al., 1978; Andrews and Boyse, 1978; Yamazaki et al., 1978.

[b]Yamazaki et al., 1979; Yamaguchi et al., 1981; Yamazaki et al., 1982; Yamazaki et al., 1983.

[c]Beauchamp et al., 1985.

[d]Yamazaki et al., 1983.

H-2K, altered urine odor. Since this gene is known to be involved in cell-cell recognition internally, it is likely that the odor-producing aspects are not its primary function, although it is possible that the production of individual odor has been a selective force in maintaining the gene in the population. In passing, it is important to stress the relatively minute genetic difference between the mutant and non-mutant mice whose urine odors can be discriminated. The protein coded by the mutant gene differs from that of the non-mutant by having amino acid differences at only 3 locations corresponding to a 7 nucleotide change in the DNA (Schulze et al., 1983). In all other respects, mutant and non-mutant mice are thought to be genetically identical.

While these data have shown that the MHC can underlie individual differences in odor, the question of the primacy of its role is as yet unanswered. In unpublished studies, it has been found that mice differing only at other autosomal loci while being identical at all others, are not easily discriminable. This suggests that H-2 may play a special role in individual recognition. However, other loci are clearly involved, since mice identical at H-2 but of different strains, thereby differing at many other loci throughout the genome, can be discriminated based on urine odors (unpublished results). Recently, it has been shown that different alleles at the T locus (also on chromosome 17 of the mouse) confer differences in body odor (see Lenington and Egid, 1985 and references cited therein). In the establishment of individual chemosensory identity, the relative importance of the H-2 region thus remains in question. For theoretical as well as practical reasons, we hypothesize that H-2 is paramount, but not exclusive.

To this point, genetic control of odor production has been discussed. There is no evidence that the MHC is involved in control of odor perception. Recent studies have demonstrated that rats can be trained to discriminate odors of mice differing at the H-2 implying species-specific receptors are not involved in perception (Beauchamp et al., 1985).

ODOR CONSTITUTION

How the MHC confers distinctive odors is not known. The odorants need not be structural derivatives of MHC products, although they could be. MHC genes may be involved in normal metabolic variation which in turn results in different patterns or profiles of odorous materials in the urines of animals of differing H-2 types (see Boyse et al., 1983). The observation that the urine odor of F_1 animals is discriminably different from urine odor of a 50:50 mixture of individuals of the respective inbred strains (Yamazaki et al., 1984) is consistent with this view. This metabolic variation hypothesis of H-2 odor determination suggests that several tissues, rather than one, contribute to the H-2 odor phenotype. Except for the intact mouse and urine, no highly potent source of H-2 odors has been identified, probably because the odorants, whatever their source, are concentrated in secretions and excretions.

The use of cell chimeras (cells of different genetic origin) permits an approach to the question of whether cells of particular lineages contribute to chemosensory identity. To test this, we have employed radiation chimeras where virtually the entire hematopoietic (blood-forming) system, but no other, is first destroyed by irradiation and is then replaced by donor cells. The red cells and lymphocytes of the chimeras are invariably and permanently of donor type, as indicated by reaction with antiserum to the alien H-2 haplotype of the donor.

In a typical experiment (Yamazaki, et al., in preparation), inbred mice of one H-2 type (e.g., $H\text{-}2^b$) were irradiated to destroy the entire hematopoietic system. Such mice would quickly succumb to infection if this system were not reconstituted with donor bone marrow and spleen cells. One group of irradiated animals was injected with these cells from a heterozygote (F_1) donor (see Fig. 1). Another (control) group was injected with donor cells from individuals of the same H-2 type as itself. The success of the reconstitution was determined subsequently by cytotoxicity assay of lymph nodes. The question was, would the reconstituted animal produce odor more like his own genetic type or more like the donor type (an F_1 heterozygote)? In short, did cell replacement change the chemosensory identity of the subject in the direction of the donor type?

In two experiments of this type, the answer has been that the recipients of the heterozygote donor cells came to smell like the donors (Yamazaki et al., in preparation). These results demonstrated that H-2 differences, manifest only in hematopoietic cells, are sufficient to confer sensory individuality, and the possibility that H-2 related odorants are made rather than concentrated by the kidneys is excluded.

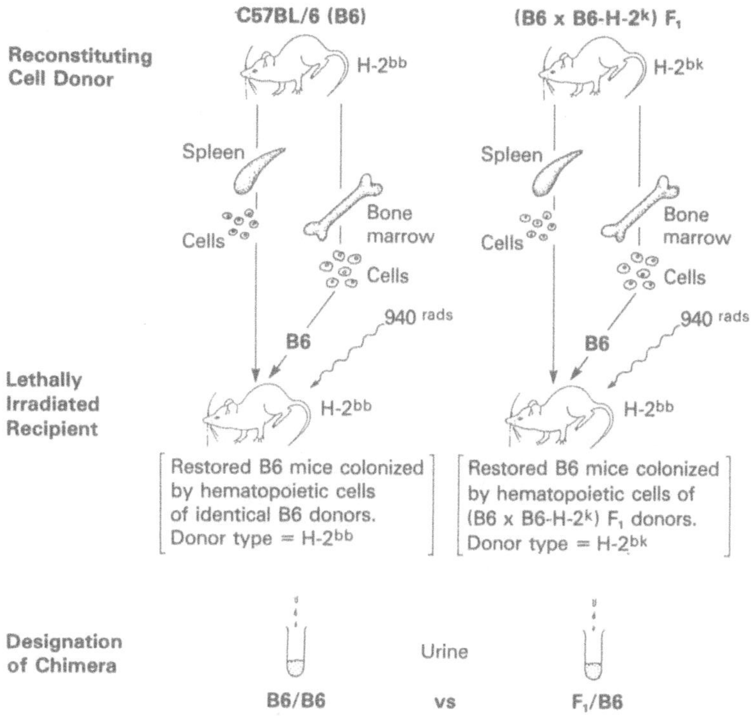

Fig. 1. Constitution of radiation chimeras. Inbred homozygous mice (H-2 type bb) were given 940 rads of gamma radiation, destroying the hematopoietic (blood-forming) system. Shortly thereafter, the mice were intravenously injected with 45-75 x 10^6 bone marrow and spleen cells, from either identical donors (left) or heterozygous donors (H-2 type bk). Injected cells re-established the hematopoietic system as shown by cytotoxicity tests conducted 5-11 weeks later.

MATING PREFERENCES

Several points concerning the initial observation of mating preferences according to H-2 type are pertinent.

1) Because highly inbred homozygous mice were studied, mating biases could be assigned to specific gene complexes. But the extent to which a similar set of biases would be expressed on a more varied and complex genetic background characteristic of natural mouse populations remains to be determined.

2) The most dramatic example of mating bias was the tendency for mice with different H-2 types to mate more frequently than expected by chance (Yamazaki et al., 1976; crosses 2 and 3). This was not always the case, however; among some strains, no choice was exhibited, and in other instances, males and females of the same H-2 type mated more frequently than expected by chance (Yamazaki et al., 1976, cross 1; Yamazaki et al., 1978; Andrews and Boyse, 1978). An explanation for these variations is not currently available (c.f., Bateson, 1979).

3) The expressed mating biases did not necessarily remain consistent under varying rearing conditions. For example, Yamazaki et al. (1978), reported that mice which were raised in a general colony room differed in mating bias from mice reared in a room with only a single strain. The responses of F_2 animals (whose parents were F_1 heterozygotes), did not always resemble those of the inbred or F_3 generation (whose parents were homozygotes) implicating rearing history as a factor in establishing mating biases (Yamazaki et al., 1978; see D'Udine and Alleva, 1983, for a listing of other research indicating a role for early experience in determining sexual preferences of rodents).

4) Recall that the test paradigm included two estrous females and one male. Yamazaki et al. (1976) argued that it was most likely that actual choice was being made by the male. Statistically, they found no consistency of choice (see Yamazaki et al., 1976 for the meaning of this term) among females which they believed implicated the male as the chooser. In subsequent studies, however, consistency of choice among females was sometimes found (Yamazaki et al., 1978), and it was suggested that the choice may be mutual, with both males and females involved in determining the ultimate pairing (Andrews and Boyse, 1978). Additionally, since mating preferences of inbred and F_2 males were not identical and environmental factors under which males grew up appear to influence mating preference, male choice appeared likely. It is important to determine the extent to which the female is involved in the choice. Theoretically, it is of more importance for females to be discriminating since they have the larger parental investment (Trivers, 1972). Further studies designed to explore the roles of males and females in mate choice are assuredly warranted.

Taken together, it can be seen that a system of biased mating exists where genetically dissimilar animals often mate more frequently than expected, where the mating biases may be partially influenced by rearing history, and where members of both sexes may play a role in establishing the bias.

RECOGNITION SYSTEMS

Halliday (1983) stated: "The question of whether animals can choose their mates on the basis of the quality of their genotype is the most controversial issue in the mate choice literature" (p. 9). One area where mate choice may be involved in determining genetic quality of offspring would be a system which would tend to limit the damage due to increased homozygosity which accompanies inbreeding. A selective advantage to MHC heterozygosity likely arises from the immune functions of the MHC molecules (Dausset, 1981). Where infections are a risk to fitness, hybrid mice would possess the advantage of a wider range of immune defenses (Yamazaki et al., 1976). In this regard, it is of interest that the hamster, a solitary-living species and hence, less at risk for spread of epizootic disease, may have little or no variation at the MHC (Darden and Streilein, 1984). More striking, perhaps, is the report that the cheetah exhibits very little MHC variation which may explain its particular vulnerability to disease and its difficulty in reproducing (O'Brien et al., 1985). Thus, a mating preference system which facilitates maintenance of heterozygosity at the MHC would appear adaptive at least among social species.

It has been argued on theoretical grounds that there may be an optimal level of outbreeding (Bateson, 1978, 1979, 1983). According to Bateson, one means of achieving optimal outbreeding would be for animals to learn the characteristics of close kin, and based on that standard, to choose to mate with an individual that is somewhat different, but not too different, from that standard. To support this mechanism, Bateson (1982) has demonstrated that Japanese quail spend more time near opposite-sexed first cousins than more closely related birds (siblings) or more distantly related birds (third cousins and unrelated individuals). A possible basis for avoiding matings with very closely related individuals could lie in the MHC genes, which through their products, provide information on genetic relatedness. It would be of interest to explore this in birds (see Simonsen, 1980).

In discussing the possibility of choice based on a general heterozygous advantage, Parker (1983) stated that choice (in his case, female choice) would not pay unless the chooser knew its own genotype; recognition of heterozygous vs. homozygous partners would not be sufficient. He doubted whether a fine-grained choice of this type was possible. In the case of the mouse, MHC individuals appear to "know" their MHC type, and this "knowledge" influences their choice of mates. Thus, at least under strict experimental conditions, choice based on mutual knowledge of genotype (or some correlate of it) is possible. What is the basis for this knowledge?

Non-random mating can arise from a variety of causes, not all of which necessarily imply recognition, discrimination or choice. Demographic factors could account for many cases of non-random mating, although it is difficult to see how demographic factors alone could facilitate avoidance of inbreeding without additional forces for dispersal of individuals. Where actual choice is involved, and thus where recognition and discrimination must be employed, two bases for discrimination have been hypothesized (see Alexander, 1979; Dawkins, 1982; Holmes and Sherman, 1982 for further discussion of these issues). First, recognition could occur through what has been termed phenotypic matching. Here, learning of the phenotype of relatives or self forms a template against which other individuals are compared. Second, recognition alleles could exist. These alleles would code for both the phenotype marker, and the ability to recognize it.

First, consider phenotype matching as it concerns mate choice based on H-2. Mice live in relatively stable social groups or demes (Bronson, 1979) and it is possible that, during development, the mice learn the characteristics of H-2 determined odors of surrounding mice (parents, siblings, others). These mice, who are kin, are then treated differently from unknown or more novel mice. While this scenario could underlie differential responsiveness to kin and non-kin, it is independent of the fundamental genetic definition of kinship. Thus, a simple crossfostering manipulation would disrupt it, since familiar non-kin, with which the animal was raised, would be treated as kin and vice versa.

A second route incorporates a comparator process whereby mice evaluate their own genetic make-up via H-2 related self-odor and then compare it with the odor of other mice. Assuming that genetically more similar mice smell more similar, it should be possible for a mouse to quantify the relatedness of other mice. Results of studies with ground squirrels (Holmes, 1984; Holmes and Sherman, 1982), macaques (Wu et al., 1980), frog tadpoles (Blaustein and O'Hara, 1981, 1982) and young coho salmon (Quinn and Busack, 1985) may be interpreted to support this process.

For H-2 and mating preference, we cannot now discriminate between these hypotheses. Ongoing cross-fostering studies will establish whether the bias favoring outbreeding can be disrupted by altered post-natal environments; preliminary evidence suggests that it can. Differential patterns of mating preference observed earlier (Yamazaki et al., 1976,

1978) certainly indicate a role for experience. The technique of ovary transplants between inbred strains permits tests of the role of the prenatal environment. These issues also may be clarified further in studies with radiation chimeras. In these mice we apparently have individuals who grew up with one chemosensory identity which was permanently altered at some point (via irradiation and reconstitution) to another such identity. This system may permit a rigorous test of the comparator alternative for discriminating relatedness.

Consider next the possibility of recognition alleles. In earlier studies (Yamaguchi et al., 1978; Andrews and Boyse, 1978), such recognition genes were postulated to underlie the H-2 determined mate choice. The possible existence of recognition alleles has been debated with the bulk of the theoretical discussion concluding that they are unlikely to exist (see Blaustein, 1983 for references). Blaustein has argued that while the available evidence is consistent with either phenotype matching or recognition alleles, the existence of recognition alleles could only be supported by studies which experimentally masked the ability of an individual to perceive his own phenotypic marker through ontogeny. In this way, phenotypic matching would be impossible and, if recognition still occurred, a recognition allele hypothesis would be supported. While Blaustein viewed such studies as difficult to perform, it may be possible to make use of radiation chimeras to investigate this issue. If it transpired that mate choice was not disrupted by cross-fostering or altering of chemosensory identity in adulthood, then a subsequent set of studies where phenotypic alteration occurred during early development could bear on this issue.

CONCLUSION

Inbred, congenic mice exhibit natural mating preferences, generally favoring matings between individuals dissimilar at the MHC. If one were to design a genetic system which would permit an organism to gauge genetic relatedness, the characteristics of MHC would most likely be selected. The MHC, the most variable locus known, is subject to high mutation or gene conversion rates (Pease et al., 1983; Weiss et al., 1983) and the genes themselves are involved in provisioning the animal with a chemical signal characteristic of the animal's genotype. Exploitation of this system in the laboratory may allow us to explore a variety of issues concerning mechanisms of mate selection and perhaps other aspects of kin recognition.

ACKNOWLEDGMENTS

This work was supported in part by NIH grants GMCA-30296 and 5F32 NS07353 and NSF grant BNS 8201759. We thank Dr. O. Matsuzaki, D. Kupniewski, H. Okada, Y. Okada and K. Suzuki for excellent technical assistance.

REFERENCES

Alexander, R. D., 1979, "Darwinism and Human Affairs," University of Washington Press, Seattle.

Andrews, P. W., and Boyse, E. A., 1978, Mapping of an H-2-linked gene that influences mating preference in mice, Immunogenetics, 6:265.

Bateson, P., 1978, Sexual imprinting and optimal outbreeding, Nature, 273.

Bateson, P., 1979, How do sensitive periods arise and what are they for? Anim. Behav., 27:470.

Bateson, P., 1982, Preferences for cousins in Japanese quail, Nature, 295:236.
Bateson, P., 1983, Optimal outbreeding, in: "Mate Choice," P. Bateson, ed., Cambridge University Press, Cambridge.
Beauchamp, G. K., Yamazaki, K., and Boyse, E. A., 1985, The chemosensory recognition of genetic individuality, Sci. Am., 253:86.
Beauchamp, G. K., Yamazaki, K., Wysocki, C. J., Slotnick, B. M., and Boyse, E. A., 1985, Chemosensory recognition of mouse major histocompatibility types by the rat, Proc. Nat. Acad. Sci. USA, in press.
Blaustein, A. R., 1983, Kin recognition mecahnisms: Phenotypic matching or recognition in alleles, Am. Nat., 121:749.
Blaustein, A. R., and O'Hara, R. K., 1981, Genetic control for sibling recognition? Nature, 290:246.
Blaustein, A. R., and O'Hara, R. K., 1982, Kin recognition cues in Rana cascadae tadpoles, Behav. Neural Biol., 36:77.
Boyse, E. A., Beauchamp, G. K., and Yamazaki, K., 1983, The sensory perception of genotypic polymorphism of the major histocompatibility complex and other genes. Some physiological and phylogenetic implications, Hum. Immunol., 6:177.
Bronson, F. H., 1979, The reproductive ecology of the house mouse, Q. Rev. Biol., 54:265.
Dardin, A. G., and Streilein, J. W., 1984, Syrian hamsters express two monomorphic classes. I. Major histocompatitility complex molecules, Immunogenetics, 20:603.
Dausset, J., 1981, The major histocompatitiliby complex in man, Science, 213:1469.
Dawkins, R., 1976, "The Selfish Gene," Oxford University Press, New York.
Dawkins, R., 1982, "The Extended Phenotype," W. H. Freeman, San Francisco.
D'Udine, B., and Alleva, E., 1983, Early experience and sexual preferences in rodents, in: "Mate Choice," P. Bateson, ed., Cambridge University Press, Cambridge.
Gupta, C., and Goldman, A., 1982, H-2 histocompatibility region: Influence on the murine glucocorticoid receptor and its response, Science, 216:994.
Halliday, T. R., 1983, The study of mate choice, in: "Mate Choice," P. Bateson, ed., Cambridge University Press, Cambridge.
Hamilton, W. D., 1964, The genetical evolution of social behavior. II, J. Theo. Biol., 7:17.
Holmes, W. G., 1984, Ontogeny of dam-young recognition in captive Belding's ground squirrels (Spermophilus beldingi), J. Comp. Psychol., 98:246.
Holmes, W. G., and Sherman, P. W., 1982, The ontogeny of kin recognition in two species of ground squirrels, Am. Zool., 22:491.
Ivanyi, P., 1978, Some aspects of the H-2 system, the major histocompatibility system in the mouse, Proc. R. Soc. London, Ser. B., 202:117.
Kirby, D. R. S., 1970, The egg and immunology, Proc. Roy. Soc. Med., 63:59.
Klein, J., 1975, "Biology of the mouse histocompatibility-2-complex," Springer-Verlag, Berlin and New York.
Lenington, S., and Egid, K., 1985, Female discrimination of male odors correlated with male genotype at the T locus: A response to T locus or H-2 variability? Behav. Gen., 15:53.
O'Brien, S. J., Roelke, M. E., Marker, L., Newman, A., Winkler, C. A., Meltzer, D., Colly, L., Evermann, J. F., Bush, M., and Wildt, D. E., 1985, Genetic basis for species vulnerability in the cheetah, Science, 227:1427.

Parker, G. A., 1983, Mate quality and mating decisions, in: "Mate Choice," P. Bateson, ed., Cambridge University Press, Cambridge.
Pease, R., Schulze, D. H., Pfaffenbach, G. M., and Nathenson, S. G., 1983, Spontaneous H-2 mutants provide evidence that a copy mechanism analogous to gene conversion generates polymorphism in the major histocompatibility complex, Proc. Nat. Acad. Sci. USA, 80:242.
Quinn, T. P., and Busack, C. A., 1985, Chemosensory recognition of siblings in juvenile coho salmon (Oncorhynchus kisutch), Anim. Behav., 33:51.
Schulze, D. H., Pease, L. R., Geier, S. S., Reyes, A. A., Sarmiento, L. A., Wallace, R. B., and Nathenson, S. G., 1983, Comparison of the cloned $H\text{-}2^{bm1}$ variant gene with the $H\text{-}2K^b$ gene shows a cluster of seven nucleotide differences, Proc. Nat. Acad. Sci. USA, 80:2007.
Simonsen, M., 1981, The major histocompatibility complex in a bird's eye view, in: "Immunobiology of the major histocompatibility complex," Karger, Basel.
Trivers, R. L., 1972, Parental investment and sexual selection, in: "Sexual Selection and the Descent of Man," B. Campbell, ed., Chicago University Press, Chicago.
Weiss, E. H., Mellor, A., Golden, L., Fahrner, K., Simpson, E., Hurst, J., and Flavell, R. A., 1983, The structure of a mutant H-2 gene suggests conversion-like events, Nature, 301:671.
Wu, H. M. H., Holmes, W. G., Medina, S. R., and Sackett, G. P., 1980, Kin preference in infant Macaca nemestrina, Nature, 285:225.
Yamaguchi, M., Yamazaki, K., Beauchamp, G. K., Bard, J., Thomas, L., and Boyse, E. A., 1981, Distinctive urinary odors governed by major histocompatibility locus of the mouse, Proc. Nat. Acad. Sci. USA, 78:5817.
Yamaguchi, M., Yamazaki, K., and Boyse, E. A., 1978, Mating preference tests with the recombinant congenic strain BALB.HTG, Immunogenetics, 6:261.
Yamazaki, K., Beauchamp, G. K., Bard, J., Thomas, L., and Boyse, E. A., 1982, Chemosensory recognition of phenotypes determined by the Tla and H-2K regions of chromosome 17 of the mouse, Proc. Nat. Acad. Sci. USA, 79:7828.
Yamazaki, K., Beauchamp, G. K., Egorov, I. K., Bard, J., Thomas, L., Boyse, E. A., 1983, Sensory distinction between $H\text{-}2^b$ and $H\text{-}2^{bm1}$ mutant mice, Proc. Nat. Acad. Sci. USA, 80:5685.
Yamazaki, K., Beauchamp, G. K., Thomas, L., and Boyse, E. A., 1984, Chemosensory identity of H-2 heterozygotes, J. Mol. Cell Immunol., 1:79.
Yamazaki, K., Beauchamp, G. K., Thomas, L., and Boyse, E. A., The hematopoietic system is a source of odorants that distinguish major histocompatibility types. In preparation.
Yamazaki, K., Beauchamp, G. K., Wysocki, C. J., Bard, J., Thomas, L., and Boyse, E. A., 1983, Recognition of H-2 types in relation to the blocking of pregnancy in mice, Science, 221:186.
Yamazaki, K., Boyse, E. A., Mike, V., Thaler, H. T., Mathieson, B. J., Abbott, J., Boyse, J., Zayas, Z. A., and Thomas, L., 1976, Control of mating preferences in mice by genes in the major histocompatibility complex, J. Exp. Med., 144:1324.
Yamazaki, K., Yamaguchi, M., Andrews, P. W., Peake, B., and Boyse, E. A., 1978, Mating preference of F_2 segregants of crosses between MHC-congenic mouse strains, Immunogenetics, 6:253.
Yamazaki, K., Yamaguchi, M., Baranoski, L., Bard, J., Boyse, E. A., and Thomas, L., 1979, Recognition among mice: Evidence from the use of a Y-maze differentially scented by congenic mice of different major histocompatibility types, J. Exp. Med., 150:755.

THE SUPPRESSION OF OVARIAN FUNCTION BY CHEMOSIGNALS

John G. Vandenbergh

Department of Zoology
North Carolina State University
Raleigh, NC 27695-7617

Endocrine regulation of ovarian function has been well worked out in recent years. While many mysteries remain for the reproductive endocrinologist to solve, the basic role of pituitary gonadotropins and the releasing factors regulating their levels have been well described (Kalra and Kalra, 1985). One system of internal regulating mechanisms is the pineal gland and its secretions. Interest has quickened recently in the role that this organ and its primary secretion, melatonin, play in suppressing reproductive functions (Tamarkin et al., 1985). We must postpone discussion of these internal regulating mechanisms because so little is known about how social signals influence suppressive physiological functions. The purpose of this paper is to review a selection of examples in which chemosignals have been shown to play a role in suppressing reproductive function.

A biological act or function often results from the removal of an inhibitor rather than the presence of a stimulator. The expression of a gene or the coordinated contraction of muscles depends upon selective disinhibition (Watson, 1976; Eccles, 1977). This concept is difficult to incorporate into our thinking but it is clear that inhibition plays an important role in both reproductive development and adult gonadal functioning. For example, the typical effect of hypothalamic releasing factors is to stimulate gonadotropin release from the pituitary, however, one releasing factor, that for prolactin, acts as an inhibitor and only its absence results in the release of prolactin (Krulich, 1979). In this paper I will explore reproductive phenomena as they are influenced by inhibitory chemosignals from conspecifics. Space prevents an exhaustive review so I have chosen some examples from among those that most clearly demonstrate chemosignal involvement.

SUPPRESSION OF ADULT OVARIAN FUNCTION

About forty years ago information was gathered indicating that social factors could play an important role in the regularity of ovarian cycles in rodents when Andervont (1944) and Merton (1938) showed that house mice, *Mus musculus*, living in a group displayed prolonged estrous cycles. The social suppression of estrus was further investigated and quantified by van der Lee and Boot (1955) in small groups of females and by Whitten (1959) in larger groups. Whitten demonstrated that airborne communication was involved when he showed that the suppressive effect was present in blinded females separated by wire mesh barriers. Champlin (1971) later confirmed

Table 1. Effect of prolactin inhibitor bromocriptine on the percent of grouped female mice showing estrus 3 or 4 days after treatment. The data were taken from Milligan (1980).

Treatment	Percent
No Treatment	20%
Control Injection	15%
Bromocriptine	60%
Presence of Male	63%

this conclusion when he showed that estrus could be suppressed in isolated females by exposure to soiled bedding material from groups of females. The chemical(s) involved in suppressing house mouse ovarian cycles remains unknown. Nor do we have any information on the physiological effects of this presumed pheromone.

Recent experiments have suggested a role for prolactin in the social suppression of ovarian cyclicity. Milligan (1980) suppressed estrus by grouping female mice 8 to a cage. Injection of suppressed mice with bromocriptine, a potent inhibitor of prolactin, restored estrus in 60% of the females in comparison to 15% of the females given a control injection (Table 1). Another example, this time in the talapoin monkey, Miopithecus talapoin, also links prolactin to reproductive suppression. Subordinate female monkeys living in a social group display suppressed ovarian cyclicity which is associated with elevated prolactin levels and the lack of a normal ovulatory surge of luteinizing hormone (Bowman et al., 1978). Although a beginning has been made in understanding the endocrine changes associated with ovarian suppression, as yet we have no information linking chemosignals to these endocrine changes. This gap in our knowledge is surprising in view of the high level of interest in the reproductive biology of the house mouse and the fact that the phenomenon of ovarian suppression appears to occur in several other species of rodents.

In the prairie deer mouse, Peromyscus maniculatus bairdii, for example, only about 10% of the females in a dense population reproduce. When nonreproductive females are isolated from the population they breed readily (Terman, 1973; Haigh, 1983). Such suppression of females who chronologically should be mature has not yet been proven to be based on a chemosignal. There is, however, good evidence that a delay in the onset of puberty in juvenile female prairie deermice is due to a pheromone. Haigh et al. (1985) showed that both soiled bedding and urine from females housed two per cage delayed the onset of puberty in isolated juvenile females. How much concordance there is between suppression of adult females and that of puberty delay is unknown. Other species in which ovarian cyclicity is known to be suppressed by social stimulation, presumably pheromonal, are the Indian field mouse, Mus booduga (Dominic and Pandey, 1978), and the wild hopping mouse, Notomys alexis (Breed, 1976).

SYNCHRONY OF OVARIAN CYCLES

The synchronization of ovarian cycles is a topic beyond the scope of this review but aspects that relate to the chemical suppression of ovarian function are relevant. For a more complete review of the topic of ovarian synchrony see the reviews of McClintock (1983) and Quadagno et al. (1981).

The clearest relationship between social chemosignals and ovarian inhibition comes from the work of McClintock (1978, 1981, 1983). She found that estrous synchrony occurred among female rats sharing the same air supply. From this result she hypothesized that female rats produce at least two airborne pheromones that have opposing effects on the ovarian cycle: one that phase-advances ovulation and one that phase-delays ovulation. To test this, odors from each part of the estrous cycle were presented continuously for 18 to 24 days to test females. Constant exposure to an air supply from rats in the follicular phase of the cycle shortened and regularized the cycle. Constant exposure to odors from rats in the ovulatory phase had the opposite effect; both the length and the variance in the estrous cycle were greater. Odors from the luteal portion of the cycle were without effect on the cycle of other females. The house mouse apparently differs from the rat in this regard because ovariectomy had no effect on the ability of grouped females to suppress cyclicity among other adult females (Marchlewski-Koj et al., 1983). In a recent study, McClintock (1983) demonstrated that odors from lactating rats or their pups lengthened the estrous cycle but odors from pregnant rats shortened the cycle. To date, the actual chemicals involved and their specific effect on the recipient female remain unknown.

Evidence for pheromonal suppression of primate ovarian cycles is far more suggestive than conclusive, but more and more reports support the possibility. Reports from a variety of field and captive social groups of nonhuman primates demonstrate that social factors are involved in reproductive synchrony of females. There are reports of reproductive synchrony in harems but not in the troops of hamadryas baboons, Papio hamadryas (Kummer,1968) and in female gelada baboons, Theropithecus gelada (Dunbar, 1980). Close contact also seems to influence the synchrony of the menstrual cycle in cynomolgus monkeys, Macaca facicularis (Wallis et al., 1984). A remarkably synchronized breeding season occurs in the talapoin monkey (Rowell and Dixon, 1975), and Rowell (1977) has suggested that the synchrony could be due to social signals, perhaps olfactory or visual. Recent evidence presented for our near relative the chimpanzee also suggests that ovarian synchrony results from social cues, perhaps pheromonal (Wallis, 1985). In her study of 10 captive female chimpanzees living in two social groups, there was a strong concordance of the menstrual cycles among the females living in a group. In Figure 1, I have redrawn her data to show the percent discrepancy between the onset of perineal swelling between females living together and those living in separate groups. These differences were statistically significant. A second measure of ovarian cyclicity, the data of detumescence differed but did not reach statistical significance. Another measure, however, the female's first postpartum estrus occurred in close temporal association to the estrus onset of cagemates. In each of these cases a role for chemosignals has not been confirmed and, if such a signal is involved, it is not known whether it is stimulatory or inhibitory in nature. It is possible that the progression of events leading to breeding is regulated by an inhibitory mechanism, and the removal of this mechanism by chemosignals results in reproduction.

An excellent opportunity to search for such an inhibitory system is provided by the recent work on the tamarin, Callithrix jacchus jacchus, by Abbott (1984). He established peer groups of pubertal and post-pubertal tamarins. One male and one female dominated each group. Ovulation was suppressed in all subordinate females as a result of low gonadotropin levels. Cortisol and prolactin levels were unaffected but control over gonadotropin release was markedly affected. There was no difference in basal levels of gonadotropins between dominant and subordinate females, but subordinate

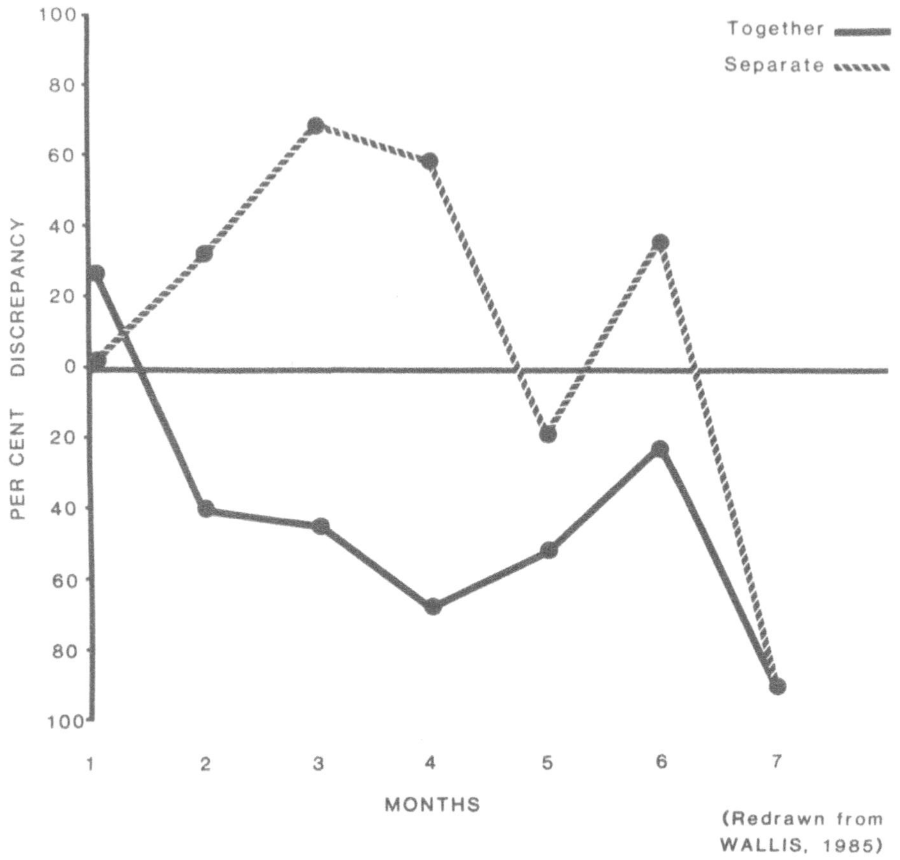

Fig. 1. The discrepancy in the onset of perineal swelling among female chimpanzees living together or separate. The data were taken from Wallis (1985).

females failed to show a significant LH response when injected with luteinizing hormone releasing hormone (LHRH) or a low dose of estradiol benzoate. This suggests a general impairment of LHRH secretion by the hypothalamus. Abbott suggests that these endocrine changes in subordinate females may be due to physical and visual contact with the dominant female or to pheromonal stimuli. Dominant females scent mark frequently (Epple et al., 1979). Tamarins are known to have at least a vestigial vomeronasal organ (Maier, 1980), a sensory organ known to be involved in detecting pheromonal signals (Meredith, 1983; Keverne and de la Riva, 1982). Thus, priming pheromones could have strong effects on the reproductive physiology of the female.

Epple et al. (this volume) will provide another interesting account of ovarian suppression in other small South American primates.

SUPPRESSION OF PUBERTY

Ascribing an age of puberty to a species is becoming more and more difficult. Not long ago one could comb the literature for the age at which an animal was first noted to breed and that date became enshrined as the age at puberty. With new information at hand we must now ask a number of questions related to the prior and current condition of the animal. Nutrition, genetic, and, most of all, social factors must be known before one can make an accurate prediction of when puberty is to be expected. Considerable attention has been paid to the acceleratory factors affecting puberty (Drickamer, 1981; Vandenbergh, 1983; Vandenbergh and Coppola, in press). Here I will focus on factors that inhibit puberty.

In the house mouse, considerable evidence exists linking pheromones to the social inhibition of female puberty. It was first noted that females matured later in groups than when housed singly (Vandenbergh et al., 1972). Bedding material soiled by a group of females, and later the urine of grouped females, was shown to carry the inhibitory message (Drickamer, 1977; McIntosh and Drickamer, 1977). Elsewhere in this volume Drickamer discusses in considerable detail some of his findings on puberty inhibition in mice. Here I will only touch on the highlights to set the stage for some other work of an ecological nature that I would like to discuss.

To help understand how a puberty inhibiting chemosignal may operate in nature, it is useful to have information on its source in the body, its physiological control, and its potency. Work by Drickamer and his colleagues have given us answers to each of these questions. The vehicle carrying the puberty inhibiting chemosignals from grouped females to juvenile females is excreted urine (Drickamer, 1977). Regulation over its presence in excreted urine seems to be the result of an unknown mechanism involving the urethra (McIntosh and Drickamer, 1977). Control over the production or presence of the puberty inhibiting pheromone in the bladder apparently depends upon adrenal function because removal of the adrenals from females eliminates the puberty inhibiting pheromone from their urine (Drickamer and McIntosh, 1980). The ovaries are apparently not involved because the presence of the puberty inhibiting pheromone was not affected by ovariectomy (Drickamer et al., 1978). The puberty inhibiting pheromone is remarkably potent. A daily dose of 0.001 ml urine produced the same delay as undiluted urine from grouped females (Drickamer, 1984). These results become even more remarkable when one considers that the active ingredient probably constitutes a very small fraction of the urine.

An important test of the significance of a laboratory finding is to determine whether it can be applied under field conditions. If the results can be replicated in the complex and unpredictable environment of a natural situation, we have much more confidence in its validity, and speculations on its ecological or evolutionary importance can be made with greater confidence. Two tests have been made of the possible role of puberty regulating pheromones in mice. These studies and their implications have been discussed at length elsewhere (Vandenbergh and Coppola, in press) but deserve mention here.

In each of the tests for the presence and action of puberty regulating pheromones, segments of highway interchanges, termed "highway islands," were used to contain freely growing populations of marked house mice. In the first study, populations were established on two highway islands (Massey and Vandenbergh, 1980). One became dense and peaked at an estimated level of 74 mice whereas the other remained sparse and peaked at about 16 mice. Voided urine was collected from trapped mice, both males and females, when populations were low and again when each population was at its peak, one dense and one sparse. Urine was then tested in the laboratory for its potency to influence the onset of puberty in juvenile females. Bioassay of the urine involved exposure of juvenile female mice to the test urine followed by daily vaginal smearing to determine the onset of puberty. Urine collected from male mice residing on the highway islands induced accelerated puberty in juvenile females at both high and low densities in both populations. Urine collected from females when populations were low had no effect on the puberty of juvenile females. When populations increased in density, only urine from females in the most dense population inhibited puberty.

A correlation of a priming pheromone with natural population changes suggests that a cause and effect relationship exists. An even more powerful test is to demonstrate alteration of pheromones after experimental manipulation of the population. Such a test has been conducted (Coppola and Vandenbergh, in prep.). Details of the study are reported in a poster presented at this conference (Coppola, this volume).

Four populations of house mice were established on different but similar highway islands. In a "flip-flop" design, pairs of highway island populations were abruptly made dense by the overnight release of 40 females. Urine was collected from resident females before and after the populations were made dense. The urine was tested in the laboratory for its effect on the onset of puberty in juvenile females. Urine collected before populations were made dense had no delaying effect on puberty, but that collected 3 weeks after increasing density caused a delay of approximately 5 days, and that collected 7 weeks later resulted in a delay of almost 3 days. These results provide experimental confirmation of the role of the puberty inhibiting pheromone as a density dependent factor regulating populations of house mice.

In view of the potentially important role played by this pheromone, discovering its identity becomes essential. Some work in progress may be yielding positive results (Jemiolo et al., this volume; Novotny and Jemiolo, personal communication). Further work in progress in my laboratory will explore the endocrine correlates of exposure to urine containing the puberty inhibiting pheromone.

In addition to the house mice, considerable work has implicated a role for pheromones in the suppression of puberty in voles. Carter et al. (1980) has shown that exposure to urine from male prairie voles stimulates the onset of puberty as measured by uterine weight increase. The stimulatory effect of the male can be blocked by the presence of another female or by exposure to female urine (Getz et al., 1983). Such blockage of an activating system is another form of ovarian suppression. Similar blockage of male stimulation of ovarian function has been reported in the house mouse (Drickamer, 1982) and should be sought for in other species. The physiological basis of such a blocking mechanism remains unknown and uninvestigated.

A similar suppression of ovarian development occurs in the California vole, _M. californicus_. Although the precise origin of the signals remains unknown, continued association with littermates suppresses reproduction (Batzli et al., 1977) and male puberty was delayed by chemosignals of family origin (Rissman et al., 1984). Meadow voles, on the other hand, fail to show suppression of maturation, at least physical growth, as a result of remaining with family members (Batzli et al., 1977). In recent work, we have obtained results indicating that the pine vole, _M. pinetorum_, responds like the California vole to family chemosignals (Lepri and Vandenbergh, in preparation; Lepri, this volume). Lepri showed first that the uteri of juvenile females more than doubled in weight after being paired with a stud male for two days. If the cage contained bedding material from the family group, almost half of the male-effect was suppressed.

The results of studies on pheromonal suppression of sexual development in house mice and several species of voles permit some speculation on the possible correlation between pheromonal mechanisms and habitat utilization (Table 2). In a preliminary attempt to relate habitat parameters to pheromonal systems, it appears that reproductive suppression by family chemosignals correlates with a stable and continuous habitat type. Batzli et al.

Table 2. Speculations on the relationship between type of reproductive suppression and habitat type in selected rodents.

Species	Reproductive Suppression	Habitat: Time	Habitat: Space
House mouse *Mus musculus*	Female density	ephemeral	patchy
Meadow vole *Microtus pennsylvanicus*	None	ephemeral	patchy
Prairie vole *M. ochrogaster*	Family	stable	continuous
Pine vole *M. pinetorum*	Family	stable	continuous
California vole *M. californicus*	Family	stable	continuous

(1977) have suggested that voles living in stable and continuous habitats may be more successful if they disperse prior to reaching puberty and becoming pregnant. Suppression of puberty would allow a longer period of time for dispersal. The possible evolutionary and ecological correlates of pheromonal suppression of puberty have been discussed at length by Bronson (1979) and Vandenbergh and Coppola (in press).

SUMMARY

The onset of puberty is such an important event in the life history of an animal that several regulating mechanisms have evolved to ensure its occurrence at the most propitious time for a specific individual in its specific environment. Photoperiod and nutrition are well known physical factors in the environment that regulate puberty. Social stimuli, particularly pheromones, have joined the list of factors regulating puberty and it seems that, among the pheromonal stimuli, inhibition plays the most important role in a natural setting. Perhaps the more conservative approach, to slow something down, is the most successful strategy.

REFERENCES

Abbott, D. H., 1984, Behavioral and physiological suppression of fertility in subordinate marmoset monkeys, Am. J. Primatol., 6:169.

Andervont, H. B., 1944, Influence of environment on mammary cancer in mice, J. Nat. Cancer Inst., 4:579.

Batzli, G. O., Getz, L. L., and Hurley, S. S., 1977, Suppression of growth and reproduction in microtine rodents by social factors, J. Mammal., 58:583.

Bowman, L. E., Dilley, S. R., and Keverne, E. B., 1978, Suppression of oestrogen induced LH surges by social subordination in talopoin monkeys, Nature, Lond., 275:56.
Breed, W. G., 1976, Effect of environment on ovarian activity of wild hopping mice (Notomys alexis), J. Reprod. Fert., 47:395.
Bronson, F. H., 1979, The reproductive ecology of the house mouse, Quart. Rev. Biol., 54:265.
Carter, C. S., Getz, L. L., Gavish, L., McDermott, J. L., and Arnold, P., 1980, Male-related pheromones and the activation of female reproduction in the prairie vole (Microtus ochrogaster), Biol. Reprod., 23:1038.
Champlin, A. K., 1971, Suppression of oestrus in grouped mice: The effects of various densities and the possible nature of the stimulus, J. Reprod. Fert., 27:233.
Dominic, C. J., and Pandey, S. D., 1978, Pheromonal influences on estrous cycle of the Indian field mouse Mus booduga Gray, Indian J. Exptl. Biol., 16:1134.
Drickamer, L. C., 1977, Delay of sexual maturation in female mice by exposure to grouped females or urine from grouped females, J. Reprod. Fert., 51:77.
Drickamer, L. C., 1981, Pheromones, social influences and population regulation in rodents, in: "Environmental Factors in Mammalian Reproduction," D. Gilmore, and B. Cook, eds., MacMillan Publ. London.
Drickamer, L. C., 1982, Acceleration and delay of first vaginal oestrus in female mice by urinary chemosignals: Dose levels and mixing of urine treatment sources, Anim. Behav., 30:456.
Drickamer, L. C., 1984, Effects of very small doses of urine on acceleration and delay of sexual maturation in female house mice, J. Reprod. Fert., 71:475.
Drickamer, L. C., and McIntosh, T. K., 1980, Effects of adrenalectomy on production and excretion of a pheromone delaying sexual maturation in grouped female mice, Horm. Behav., 14:146.
Drickamer, L. C., McIntosh, T. K., and Rose, E. A., 1978, Effects of ovariectomy on the presence of a maturation-delaying pheromone in the urine of female mice, Horm. Behav., 11:131.
Dunbar, R. I. M., 1980, Demographic and life-history variables of a population of gelada baboons (Theropithecus gelada), J. Anim. Ecol., 49:485.
Eccles, J. C., 1977, "The Understanding of the Brain," McGraw-Hill, New York.
Epple, G., Golob, N. F., and Smith, A. B., 1979, Odor communications in the tamarin Saguinus fuscicollis (Callithrichidae): Behavior and chemical studies, in: "Chemical Ecology: Odor Communication in Animals," F. J. Ritter, ed., Elsevier/North Holland, Biomedical Press.
Getz, L. L., Dluzen, D., and McDermott, J. L., 1983, Suppression of reproduction maturation in male-stimulated virgin female Microtus by a female urinary chemosignal, Behav. Proc., 8:59.
Haigh, G. R., 1983, Reproductive inhibition and recovery in young female Peromyscus leucopus, J. Mammal., 64:706.
Haigh, G. R., Lounsburg, D. M., and Gordon, T. A., 1985, Pheromone induced reproductive inhibition in young female Peromyscus leucopus, Biol. Reprod., in press.
Kalra, P. S., and Kalra, S. P., 1985, Control of gonadotropin secretion, in: "The Pituitary Gland," H. Imura, ed., Raven Press, New York.
Keverne, E. B., and de la Riva, C., 1982, Pheromones in mice: Reciprocal interaction between nose and brain, Nature, 296:148.
Krulich, L., 1979, Central neurotransmitters and the secretion of prolactin, GH, and LH and TSH, Ann. Rev. Physiol., 41:603.
Kummer, H., 1968, "Social Organization of Hamadryas Baboons," Univ. of Chicago Press, Chicago.

Lepri, J. J., and Vandenbergh, J. G., in preparation, The influence of chemosignals on reproductive activation in the female pine vole, Microtus pinetorum.
Maier, W. C., 1980, Nasal structures in Old and New World primates, in: "Evolutionary Biology of the New World Monkey and Continental Drift," R. L. Ciochon, and A. B. Chiarelli, eds., Plenum Publ. Co., London.
Marchlewska-Koj, A., Kruczek, M., and Toch, E., 1983, Suppression of estrous cycle of female mice by ovariectomized females, Horm. Behav., 17:233.
Massey, A., and Vandenbergh, J. G., 1980, Puberty delay by a urinary cue from female house mice in feral populations, Science, 209:821.
McClintock, M. K., 1978, Estrous synchrony and its mediation by airborne chemical communication, (Rattus norvegicus), Horm. Behav., 10:264.
McClintock, M. K., 1981, Social control of the ovarian cycle and function of estrous synchrony, Am. Zool., 21:243.
McClintock, M. K., 1983, Pheromonal regulation of the ovarian cycle: Enhancement, suppression and synchrony, in: "Pheromones and Reproduction in Mammals,", J. G. Vandenbergh, ed., Academic Press, New York.
McIntosh, T. K., and Drickamer, L. C., 1977, Excreted urine, bladder urine, and the delay of sexual maturation in female house mice, Anim. Behav., 25:999.
Meredith, M., 1983, Sensory physiology of pheromone communication, in: "Pheromones and Reproduction in Mammals," J. G. Vandenbergh, ed., Academic Press, New York.
Merton, H., 1938, Studies on reproduction in the albino mouse, Proc. Roy. Soc. Edinburgh., 58:80.
Milligan, S. R., 1980, Pheromones and rodent reproductive physiology, Symp. Zool. Soc., Lond., 45:251.
Quadagno, D. M., Shuberta, H. E., Deck, J., and Francoer, D., 1981, Influence of male social contacts, exercise and all-female living conditions on the menstrual cycle, Psychoneuroendocrinology, 6:239.
Rissman, E. F., Sheffield, S. D., Kretzmann, M. B., Fortune, J. E., and Johnston, R. E., 1984, Chemical cues from families delay puberty in male California voles, Biol. Reprod., 31:324.
Rowell, T. E., 1977, Reproductive cycles of the talapoin monkeys (Miopithecus talapoin), Folia. Primatol., 28:188.
Rowell, T. E., and Dixon, A. F., 1975, Changes in social organization during the breeding season of wild talapoin monkeys, J. Reprod. Fert., 43:419.
Tamarkin, L., Baird, C. J., and Almeida, O. F. X., 1985, Melatonin: A coordinating signal for mammalian reproduction? Science, 227:714.
Terman, C. R., 1973, Recovery of function by prairie deer mice (Peromyscus maniculatus bairdii) from asymptotic populations, Anim. Behav., 21:444.
Vandenbergh, J. G., 1983, Pheromonal regulation of puberty, in: "Pheromones and Reproduction in Mammals," J. G. Vandenbergh, ed., Academic Press, New York.
Vandenbergh, J. G., and Coppola, D., 1986, The physiology and ecology of puberty modulation by primer pheromones, in: "Advances In The Study of Behavior," J. S. Rosenblatt, C. Beer, and M. Busnel, eds., Academic Press, New York, in press.
Vandenbergh, J. G., Drickamer, L. C., and Colby, D. R., 1972, Social and dietary factors in the sexual maturation of female mice, J. Reprod. Fert., 28:397.
van der Lee, S., and Boot, L. M., 1956, Spontaneous pseudopregnancy in mice II, Acta Physiol. et Pharm. Neerl., 5:203.
Wallis, J., 1985, Synchrony of estrous swelling in captive group-living chimpanzees, (Pan troglodytes), Internat. J. Primatol., in press.
Wallis, J., King, B. J., and Roth-Meyer, C., 1985, Cyclical variation in social behavior and the influence of physical contact on the menstrual cycle of cynomolgus monkeys (Macaca facicularis), Am. J. Primatol., in press.

Watson, J. D., 1976, "Molecular Biology of the Gene," W. A. Benjamin, Inc., Menlo Park Calif., 3rd edition.
Whitten, W. K., 1959, Occurrence of anoestrus in mice caged in a group, J. Endocrinol., 18:102.

THE EVOLUTIONARY SIGNIFICANCE OF THE OLFACTORY BLOCK TO PREGNANCY

Eric B. Keverne and Anne E. Rosser

Department of Anatomy
University of Cambridge
Downing Site, Cambridge, England

INTRODUCTION

Mammalian pheromones influence a wide variety of behaviours and physiological states, few of which have attracted interest and stimulated the imagination more than the olfactory block to pregnancy, or the "Bruce effect" as it is more colloquially termed, after its discovery by Hilda Bruce in the early 60's. This pheromonal effect, classed as "primer" to distinguish it from the "releasing" action on behaviour of signalling pheromones, is but one of a number of physiological influences which urinary pheromones have on rodent reproduction. Indeed, murine rodents are unique among mammals for the extent to which this aspect of their chemical environment influences reproduction (Bronson, 1979). Over the past twenty years several different effects from primer pheromones have been demonstrated, particularly in laboratory female mice, following exposure to male urine. These include the acceleration of puberty (Vandenbergh, 1969), induction of oestrus in grouped females (Whitten, 1956) and the blocking of pregnancy in newly mated females (Bruce, 1959). Exposure of female mice to female urine has the converse effect of delaying puberty (Cowley and Wise, 1973), inducing anoestrus (Whitten, 1959) and protecting against the pregnancy block (Bruce, 1963). Some of these effects have also been demonstrated in wild mice, prairie deer mice, the laboratory rat, and several microtine rodents, although the latter appear to require the presence of the male in addition to urinary pheromones, these being ineffective in themselves (Milligan, 1976).

The adaptive advantages of the olfactory block to pregnancy have received so much discussion. Early hypotheses looked at the pheromones as providing exogamy, (Bruce and Parrot, 1960) or promoting male competition (Bronson, 1968). However, since the physiological mechanism appears to have evolved in the female, the precise way in which this benefits the female is difficult to assess, and has led to the view that, olfactory block to pregnancy may be a laboratory artifact. Natural selection should theoretically produce resistance to pregnancy block since the loss of offspring appears to be unlikely to enhance the individual fitness of females.

Dawkins (1976) suggested that pregnancy block may be advantageous for a female that has been deserted by her mate, for although she forfeits her initial investment, the female benefits from remating quickly with a male that will provide parental care. However, since parental care is not the prerogative of male mice, others have suggested that the Bruce effect has evolved as part of the female's strategy to minimize the consequences of encountering infanticidal males (Labov, 1981). Loss of fertilised eggs or recently implanted embryos represent a relatively small loss in overall reproductive fitness compared with the loss of a newborn litter. However, the female has evolved a perfectly adequate way of deterring male infanticide, namely by maternal aggression, a factor conveniently ignored in these sociobiological appraisals of the Bruce effect (Wolff, 1985).

Although these hypotheses may be intellectually appealing, none have attempted to account for the mechanisms on which the evolutionary forces have acted, and therefore none provide a completely satisfactory explanation. The purpose of this paper is to consider the neural and endocrine mechanisms involved in primer pheromone action and incorporate these into an evolutionary framework.

Neural Basis of Primer Pheromone Action

Most mammals, the exception being higher primates and man, possess a dual olfactory system. In addition to the olfactory epithelium, a second set of olfactory receptors located in the epithelial lining of the vomeronasal organ and also send axons, the vomeronasal nerves, through the cribriform plate. Vomeronasal nerves run medially between the main olfactory bulbs and synapse in the accessory olfactory bulbs. The accessory olfactory bulb is spatially and histologically distinct from the main olfactory bulb and usually lies on the dorsal surface of the main bulbs. In contrast to those of the main bulb, the projection of the accessory bulbs is directly into the limbic brain terminating in the ipsilateral corticomedial nuclei of the amygdala (Scalia and Winans, 1975). From the amygdaloid complex, fibres run in the stria terminalis and ventral amygdalafugal pathway to the ventromedial hypothalamus and the medial preoptic area. Thus the olfactory pathway has a fairly direct projection to those limbic structures involved in sexual and neuroendocrine regulation (Keverne, 1983).

Our studies have shown that it is this accessory olfactory system which is needed for the actions of primer pheromones in mice. Lesions to the vomeronasal organ or to the accessory olfactory bulb prevent female odours from inducing a state of pseudopregnancy (Reynolds and Keverne, 1979) and male odours from inducing oestrus, accelerating puberty (Lomas and Keverne, 1982) and blocking pregnancy (Bellringer et al., 1980). Although I use the term "odours," it is extremely unlikely that these pheromones are volatile since they elute chemically as small peptides and hence the females must make contact with urine. Contact with urine is also essential to stimulate the receptors in the vomeronasal organ and it is thought that the pumping action of the vomeronasal organ during licking and nuzzling facilitates access of chemical stimulants (Meredith and O'Connell, 1979).

Neuroendocrine Consequences of Primer Pheromone Action

In addition to the common neural mechanism, primer pheromones have a common endocrine effect expressed principally by changing prolactin secretion from the anterior pituitary. Our own studies have shown that lowering prolactin by injections of bromocriptine (a dopamine agonist) is able to reproduce the actions of all the described primer pheromones in female mice with identical timing (Keverne, 1982). Direct measurement of prolactin has also demonstrated that exposure of female mice to males results in rapid decreases of serum prolactin (Ryan and Schwartz, 1980). Such decreases in prolactin are probably brought about by increases in the release of hypothalamic dopamine from the tuberoinfundibular dopaminergic (TIDA) neurons into the hypophyseal portal circulation. Indeed, our recent studies show there to be an increase in the synthesis of dopamine in this area of the hypothalamus following exposure of recently mated females to strange males. However, the close anatomical relationship between dopamine and β-endorphin neurons in the mediobasal hypothalamus and the evidence for increased tonic activity of the β-endorphin neurons at times of reproductive quiescence (seasonal anoestrus, pregnancy, lactation, amenorrhea), do not exclude the possibility of an inverse relationship between these two neural systems (Grossman, 1983). Regardless of such interactions this common neuroendocrine mechanism can account for all the reproductive effects described as a consequence of pheromone stimulation, their difference being the time of life and reproductive status of the individual when exposure occurs. Hence, in the case of the anovulatory female, pheromones restore cyclicity by lowering prolactin which, in the mouse, is luteotropic. As a secondary consequence, progesterone secretion from the corpus luteum decreases, releasing the hypothalmopituitary axis from any inhibitory action of progesterone, and cyclicity is restored. In the case of pregnancy block, lowering prolactin in the first three days after mating prevents implantation, causing the developing blastocyst to enter a state of diapause and be lost at the ensuing oestrus. The acceleration of puberty by induction of early oestrus may not be governed by prolactin per se, but by the hypothalamic dopaminergic or β-endorphin which also regulates luteinizing-hormone (LHRH) release. Premature activation of these TIDA neurons is thought to lower prolactin and simultaneously permit (LHRH) release (Wuttke et al., 1980).

Thus, by stimulating the vomeronasal receptors, pheromones in male mouse urine activate the TIDA neurons, inhibit β-endorphin release, prolactin falls and luteinizing hormone (LH) may rise. In contrast, in the presence of female pheromones the converse is true; the hypothalamic dopamine system is inhibited, β-endorphin may be released, prolactin rises and hence cycles are suppressed. On the basis of such findings the need to hypothesize several types of pheromone is redundant, all that is required is a pheromone signal for "maleness" which probably competes for the same receptor as the signal for "femaleness." The olfactory block to pregnancy appears to be an exception to this simple formulation. Here maleness requires the further qualification of "strangeness," in that only males of a strain different to that of the stud will block pregnancy. If the pheromone signal were indeed simply maleness, then the block to pregnancy would indeed be induced by all males. Clearly an element of strangeness is an integral part of the olfactory block to pregnancy.

The Olfactory Block to Pregnancy by Strange Males: How is Familiarity Recognized?

Although strangeness is an integral part of the olfactory block to pregnancy, this has no absolute value. Thus if a female of strain 'A' is mated with a male of strain 'B', urine from strain 'A' will block pregnancy but not that of 'B', while the converse is true if female 'A' is mated with male 'A'. Clearly the act of mating "imprints" a recognition of the male that mates. Such recognition is essential if this male's pheromones are not to subsequently activate the neural and endocrine mechanisms which could result in pregnancy block. Presumably males of a strain different to that of the stud retain this capacity because the spectrum of their pheromones is sufficiently different from that of the stud male to activate nonimprinted pathways leading to pregnancy block.

Although pheromones from a male of a strain different to that of the stud are necessary to block pregnancy, it has not been possible to readily train female mice to discriminate these two strains of male bedding for food reward. However, female mice do show a preference for bedding from intact males compared with that from castrated males. It is the urine-soiled bedding from intact males and not castrates that contains the pheromones which produce a block to pregnancy (Bruce, 1965). Even when the ability of females to distinguish this bedding is prevented by selective lesions to the main olfactory system, the olfactory block to pregnancy is not impaired. Conversely, selective lesioning of the accessory olfactory system prevents the olfactory block to pregnancy without impairing either female discrimination or preference for the intact male bedding (Lloyd-Thomas and Keverne, 1982). These findings would imply that conscious olfactory recognition is not a prerequisite for the olfactory block to pregnancy by strange male odours. Hence, the kind of recognition or imprint that occurs is achieved entirely by the accessory olfactory system and its projection to the hypothalamus via the amygdala.

A Noradrenergic 'Print-now' for Recognition of Olfactory Pheromones

The brain's noradrenergic projections have been described as a subsystem which identifies situations of great survival value and hence instruct the storage of relevant incoming sensory information (Kety, 1970). In the olfactory systems there are clearly occasions for which the imprinting of olfactory information is of survival value; for example, a mother's recognition of her own offspring as opposed to those of strangers and, of particular relevance to primer pheromones, the recognition of the stud male's pheromones to prevent these from blocking pregnancy. In the mouse, the noradrenergic projection to the olfactory and accessory olfactory bulbs is via the medial olfactory stria. If this projection is selectively lesioned by small infusions of the neurotoxin, 6-hydroxydopamine, then the accessory olfactory bulbs are severely depleted (> 70%) of noradrenaline. In this condition females still respond to pheromones producing the olfactory block to pregnancy, but fail to recognise or become imprinted to pheromones of the male that mated, and consequently the stud male's pheromones block such a pregnancy (Keverne and de la Riva, 1982). Hence, females with noradrenaline depleted from their accessory olfactory bulbs would fail to become pregnant unless they were isolated from all male odours after mating.

For such a mechanism to be effective it is essential that these noradrenergic pathways are activated by the process of mating. Our recent studies have shown that this is indeed the case. Hence, coitus activates the brain's noradrenergic neurons, and those centrifugal projections which terminate in the female's olfactory bulb are essential to the recognition of the firing patterns of that population of receptors which code for male pheromones. Noradrenaline has been postulated to enhance signal-to-noise ratio in sensory systems, including the main olfactory system (Jahr and Nichol, 1982). Such a mechanism would permit sensory storage elsewhere in the brain to be called upon and gated should the same pheromones appear after mating during the critical period prior to implantation.

DISCUSSION

The olfactory block to pregnancy has provoked sociobologists to speculate at length on its evolutionary significance. The logic which prevails is based on the assumption that it must be of selective advantage to the female in the context of parental care (Dawkins, 1976), or as a female strategy to reduce infanticide (Labov, 1981), or to promote outbreeding (Parkes and Bruce, 1962). Based on such assumptions, experiments have been devised which further illustrate the failure of close kin to block pregnancy (Lendrem, 1984), the importance of dominance status for pregnancy block (Huck, 1982; Labov, 1981), and the linking of a subordinate's failure to block pregnancy with a lower incidence of infanticide (Huck et al., 1982).

None of these theories are entirely satisfactory and although appealing as hypotheses, represent armchair opinions lacking support from decisive experimental or ecological observations. Furthermore, they represent a rather blinkered view of evolution since they fail to take into account the mechanisms whereby such pheromones provoke their effects. Let us examine then, what are the potential benefits and costs to the female of possessing such a pheromonal mechanism.

The ecological lifespan of mice has been estimated at about 80 days. First oestrus normally occurs about day 44, and with a gestation period of about 18 days and weaning of a further 21 days, this would probably permit only one litter in the lifetime of most females. Accelerating the onset time of first oestrus to 36 days by pheromonal priming, and prematurely terminating lactational anoestrus by a similar mechanism as weaning wanes, could theoretically double the reproductive success of a species with such a relatively short lifespan. The evolution of such a neural and endocrine mechanism carries obvious selective advantage to the female. However, the possession of such a mechanism also carries certain risks, since such a neuroendocrine response would be extremely disadvantageous to the female following fertilisation; lowering the prolactin at this stage is known to prevent implantation. In order therefore to offset such risks to pregnancy the female has adapted the noradrenergic neural system to "imprint" recognition and hence offset the effect of male pheromones. Obviously, imprinting will be restricted to whatever pheromones are around at the time of mating. However, resident males are usually the only males which pose a threat to pregnancy, since they have such a strong advantage over intruders, and, therefore access by strange males is unlikely to be a frequent event. Hence, natural selection has favoured the recognition of familiar males. The olfactory block to pregnancy by strange males is simply circumventing this recognition, an event unlikely to occur at the critical period anywhere other than in the laboratory and by human intervention.

While it is essential to consider the olfactory block to pregnancy in the broader context of other primer actions, it is equally important to remember that primer pheromones are but one of many environmental stimuli which influence the reproduction of female mammals. The neuroendocrine mechanisms described, particularly the inverse relationship between dopamine and β-endorphin neurons, may be of fundamental evolutionary significance to female reproduction. These neural systems probably represent the final common pathway for suppression of oestrus during pregnancy and lactation -- nature's own means of contraception. This period of postpartum infertility has been extended as a result of photoperiodic stimuli acting on these neuroendocrine mechanisms in seasonal breeders, and circumvented in those species, which experience post-partum oestrus. In the latter, breeding success is optimised since delayed implantation is without undue risk to the present litter, while seasonal breeders optimise their lifetime reproductive success by producing offspring at the most favourable time of the year. In other words, sensory neural events can impinge upon this contraceptive neuroendocrine mechanism in an inhibitory or facilitatory manner. Tactile cues in the form of parturition or mating can provoke oestrus and ovulation in some species, while photoperiodic cues may delay the onset of oestrus and ovulation after parturition in others. On moving up the phylogenetic scale, then the social environment is all important in a species like the monkey, and evidence is accumulating which implicates this same neuroendocrine mechanism as the modulator of social influences on reproduction. Hence, olfactory pheromones may be viewed as yet another aspect of the environment which impinges on this common neuroendocrine mechanism, the evolution of which has been fundamental to female mammalian reproduction.

REFERENCES

Bellringer, J. F., Pratt, H. P. M., and Keverne, E. B., 1980, Involvement of the vomeronasal organ and prolactin in pheromonal induction of delayed implantation in mice, J. Reprod. Fert., 59:223.

Bronson, F. H., 1968, Pheromonal influences on mammalian reproduction, in: "Reproduction and Social Behaviour," M. Diamond, ed., Indiana University Press, Bloomington.

Bronson, F. H., 1979, The reproductive ecology of the house mouse, Q. Rev. Biol., 54:265.

Bruce, H. M., 1959, An exteroceptive block to pregnancy in the mouse, Nature, 184:105.

Bruce, H. M., 1965, Effect of castration on the reproductive pheromones of male mice, J. Reprod. Fert., 10:141.

Bruce, H. M., and Parrot, D. M. V., 1960, Role of olfactory sense in pregnancy block by strange males, Science, 131:1526.

Cowley, J. J., and Wise, D. R., 1972, Some effects of mouse urine on neonatal growth and reproduction, Anim. Behav., 20:499.

Dawkins, R., 1976, "The Selfish Gene," Oxford University Press, Oxford.

Grossman, A., 1983, Brain opiates and neuroendocrine function, Clinics in Endocrinol. Metab., 12:725.

Huck, U. W., 1982, Pregnancy block in laboratory mice as a function of male social status, J. Reprod. Fert., 66:181.

Huck, U. W., Soltis, R. L., and Coopersmith, C. B., 1982, Infanticide in male laboratory mice: Effects of social status, prior sexual experience, and basis for discrimination between related and unrelated young, Anim. Behav., 30:1158.

Jahr, C. E., and Nichol, R. A., 1982, Noradrenergic modulation of dendrodendritic inhibition in the olfactory bulb, Nature, 297:227.

Kety, S. S., 1970, "The Neurosciences, 2nd Study Program," F. O. Schmidt, ed., Rockefeller Univ. Press, New York.

Keverne, E. B., 1982, The accessory olfactory system and its role in pheromonally mediated changes in prolactin, in: "Olfaction and Endocrine Regulation," W. Breipohl, ed., IRL Press, London.

Keverne, E. B., 1983, Pheromonal influences in the endocrine regulation of reproduction, Trends Neurosci., 6:381.

Keverne, E. B., and de la Riva, C., 1982, Pheromones in mice: Reciprocal interactions between the nose and brain, Nature, 296:148.

Labov, J. B., 1981, Pregnancy blocking in rodents: Adaptive advantages for females, Am. Nat., 118:361.

Lendrem, D. W., 1985, Kinship influences the Bruce effect, Anim. Behav., in press.

Lloyd-Thomas, A., and Keverne, E. B., 1982, Role of the brain and accessory olfactory system in the block to pregnancy in mice, Neuroscience, 7:907.

Lomas, D. E., and Keverne, E. B., 1982, Role of the vomeronasal organ and prolactin in the acceleration of puberty in female mice, J. Reprod. Fert., 66:101.

Meredith, M., and O'Connell, R. J., 1979, Efferent control of the stimulus access to the hamster vomeronasal organ, J. Physiol., 286:301.

Milligan, S. R., 1976, Pregnancy blocking in the vole, Microtis agrestis. Effect of the social environment, J. Reprod. Fert., 46:91.

Parkes, A. S., and Bruce, H. M., 1962, Pregnancy block in female mice placed in boxes soiled by males, J. Reprod. Fert., 4:303.

Reynolds, J., and Keverne, E. B., 1979, Accessory olfactory system and its role in pheromonally mediated suppression of oestrus in grouped mice, J. Reprod. Fert., 57:31.

Ryan, K. D., and Schwartz, N. B., 1980, Changes in serum hormone levels associated with male mice - induced ovulation in group-housed adult female mice, Endocrinology, 106:959.

Scalia, F., and Winans, S. S., 1975, The differential projections of the olfactory bulb in mammals, J. Comp. Neurol., 161:31.
Vandenbergh, J. G., 1969, Male odour accelerates female sexual maturation in mice, Endocrinology, 84:658.
Whitten, W. K., 1956, Modification of the oestrus cycle of the mouse by external stimuli associated with the male, J. Endocrinol., 13:399.
Whitten, W. K., 1959, Occurrence of anoestrus in mice caged in groups, J. Endocrinol., 18:102.
Wolff, J. O., 1985, Maternal aggression as a deterrent to infanticide in *Peromyscus leucopus* and *P. maniculatus*, Anim. Behav., 33:117.
Wuttke, W., Homma, K., Lamberts, R., and Hohn, K. G., 1980, The role of monoamines in female puberty, Fed. Proc. Am. Soc. Exp. Biol., 39:2378.

PUBERTY-INFLUENCING CHEMOSIGNALS IN HOUSE MICE:

ECOLOGICAL AND EVOLUTIONARY CONSIDERATIONS

Lee C. Drickamer

Biology Department
Williams College
Williamstown, Massachusetts 01267 U.S.A.

INTRODUCTION

It is my purpose here to review the experimental evidence to date regarding urinary chemosignals excreted by house mice (Mus musculus) that affect puberty in young females. For many rodents, including house mice, the age of puberty may generally be equated with generation time. That is, attaining sexual maturation often means that the female rodent also begins to reproduce. The rate of growth or decline of populations of a variety of rodent species can be influenced significantly by changes in generation time.

A number of factors can influence the timing of female mouse puberty, including genetics (Drickamer, 1981), social conditions and non-social environmental factors (e.g., diet, temperature, photoperiod) (Laurie, 1946; Barnett and Coleman, 1959; Drickamer, 1975a,b; Bronson, 1979). Investigations of the social conditions that affect puberty in house mice have yielded evidence for a series of chemosignals contained in the urine that retard or accelerate sexual development in young females.

There are at least four urinary chemosignals that influence puberty in young female mice; (1) urine from males, (2) urine from pregnant or lactating females, and (3) urine from females in estrus, all of which accelerate puberty, and (4) urine from group-caged females which delays puberty. For each of these chemosignals, a variety of questions have been tested pertaining to both donors and recipients (Table 1). For several of the signals, the current state of our knowledge is more advanced than for others. For each chemosignal, I will summarize the available data and then speculate on the possible ecological significance and evolution of the signal. Lastly, I will discuss several aspects of the functional biology of the chemosignals together. For details of the methods used the reader should consult individual references.

MALE URINE

Urine from adult intact male mice accelerates puberty in young females (Vandenbergh, 1969; Kennedy and Brown, 1970; Colby and Vandenbergh, 1974). This acceleratory effect is even greater when the male is actually present with the young female mouse (Andervont, 1944; Vandenbergh, 1967) due to

Table 1. Outline of questions concerning the effects of urinary chemosignals on puberty in rodents.

I. Timing Effects
  A. Age at chemosignal treatment
  B. Duration of chemosignal treatment
    1. Number of days of treatment
    2. Hours/day of exposure or treatment
    3. Pulsatile treatment
  C. Daily patterns
    1. Donor patterns of chemosignal excretion
    2. Recipient patterns of sensitivity

II. Hormonal Involvement
  A. Gonadal hormones
  B. Adrenal hormones
  C. Reproductive history
  D. Time course patterns during pregnancy and lactation

III. Social Effects
  A. Density
    1. Density-dependent effects
    2. Number of days under a particular social condition necessary for chemosignal production/release
    3. Number of days after removal from a particular set of social conditions before cessation of release of a chemosignal or commencement of release of a chemosignal
  B. Dominance Status
  C. Kinship
    1. Mother, father
    2. Siblings - male or female
    3. Cross-fostering
  D. Actual social contact between donor and recipient

IV. Behavioral Issues
  A. Donors
    1. Pattern/location of urine deposition
    2. Timing of urine deposition
  B. Recipients
    1. Odor preferences - changes with age
    2. Activity Effects

V. Other Aspects
  A. Volatility
    1. Methods of chemosignal delivery
    2. Anosmia - zinc sulfate
    3. Anosmia - transection or ablation
  B. Excreted and bladder urine - presence of chemosignal
    1. Urethra involvement - singly-caged females
  C. Doses/microdoses
  D. Mixing urine from different donor sources
    1. Equal proportions
    2. Variable proportions

VI. Non-social Environmental Effects on Chemosignals
  A. Food deprivation
  B. Photoperiod
  C. Temperature

contact stimulation from the male (Drickamer, 1974a, 1975c; Bronson and Maruniak, 1975). The substance is effective with only three days of treatment when that treatment occurs prior to day 29 of age (Colby and Vandenbergh, 1974). Either two hours/day of urine stimulation or one hour/day of exposure to an intact adult male mouse is sufficient to produce acceleration (Drickamer, 1983a). The urinary substance is most effective in producing accelerated sexual development when it is collected at the onset of the light portion of the daily cycle, and it is also most effective when used to treat young females at the onset of the light portion of the cycle (Drickamer, 1982a). Because of seasonal variation in the sensitivity of young females to the signal, and not changes in the quantity or quality of chemosignal excreted by the males, puberty generally cannot be accelerated during the mid-winter months (Drickamer, unpub.).

Prepubertal and castrated males do not release the substance; hormone replacement therapy in castrated males restores the presence of the signal in the urine (Vandenbergh, 1969; Lombardi, Vandenbergh and Whitsett, 1976; Drickamer and Murphy, 1978). Adrenalectomy does not affect the male urinary chemosignal that accelerates female puberty (Drickamer, 1983a).

Dominant males release the acceleratory substance in their urine, but low-ranking or subordinate males do not (Lombardi and Vandenbergh, 1977; Drickamer, 1983a). Increasing the density of males does not alter these patterns of chemosignal release (Drickamer, 1983a). There are no differential acceleratory effects on female puberty based upon kinship between donors (father, brother) and recipient females (Drickamer, 1983a, 1984b).

The male chemosignal substance is effective in extremely small (0.0001 cc of urine/day) doses (Drickamer, 1982b, 1984c), appears to be quite nonvolatile (Vandenbergh et al., 1975; Vandenbergh et al., 1976; Drickamer, in press b), and is apparently received by the vomeronasal-accessory olfactory system (Keverne, 1969; Meredith, 1983; Drickamer and Assmann, 1981). The acceleration effect is not enhanced by simultaneous treatment of test females with urine from pregnant or lactating females or females in estrus, but the acceleration effect is blocked and delays in puberty occur when young females are treated simultaneously with urine from males and grouped females (Drickamer, 1982b).

The functional aspects of the male chemosignal must be considered from both the perspectives of the donor males and the recipient females. Adult male mice that are dominant within a deme (and presumably males that are living alone) can accelerate the sexual development of any young females they encounter; only a few days of stimulation for short periods each day suffices. The males are thus capable of acquiring potential additional or new mates. The relative nonvolatility of the substance may insure that it will remain present in the environment for some time after the male urinates in a particular location. However, this characteristic of the signal may simply be a function of the nature of the chemical substance. Young females apparently must make direct contact with the urine to receive acceleratory stimulation. It is interesting to note in this regard that male urine is most effective in accelerating puberty when it is collected at the onset of the light portion of the daily activity cycle, when presumably, the mice are about to become relatively quiescent. I have preliminary data on urine deposition patterns by male mice that bear on this point. Male mice urinate more frequently near or on soiled bedding from young females than on bedding that is clean (control) or soiled bedding from diestrous adult females. Could it be that male mice urinate on the bedding of or near nests containing young females, thus enhancing their chances of accelerating females' sexual development?

The male urinary chemosignal is influenced by dominance status and, presumably, by related testosterone levels. Under constant environmental conditions, there are no seasonal changes, no density effects, and no differential kinship effects associated with the male signal. However, I have recently reported that shortened photoperiods result in an absence of the acceleratory chemosignal from the urine of adult males (Drickamer, 1984d); this phenomenon may be related to a seasonal diminution of the chemosignal in mouse populations at some latitudes or locations. In addition, food deprivation of males results in some decrease in the amount of chemosignal in the urine (Drickamer, 1984d). Also, there is no special male-young female relationship necessary for acceleration; using urine from the same or different males each day to treat young females produces the same degree of puberty acceleration (Drickamer, 1983a). Male mice apparently release the maturation-accelerating chemosignal in their urine much of the time. Bronson (1976) has previously noted the larger broadcast patterns of urine deposition for dominant compared to subordinate males.

The chemical nature of the signal substance has received some attention (Vandenbergh et al., 1975; Vandenbergh et al., 1976). The substance may be a metabolic by-product of ongoing physiological processes in the male mouse; clearly these processes are, in part, testosterone dependent. The presence of the acceleratory substance in male urine may be epiphenomenal. It is possible that natural selection has influenced either the chemical substance itself or the manner in which it affects young females. What seems more likely, particularly in light of the foregoing comments on behavior(s) related to urine deposition by males, is that the behaviors of males associated with chemosignal release have been influenced by selection.

What about the young recipient females? Certainly, the physiological priming effect of the urinary substance plays a role in the onset of hormonal and related events culminating in ovulation and estrus. That the young females must make direct contact with the substance and apparently receive the signal via the accessory olfactory system may simply be a consequence of the nature of the nonvolatile substance.

The seasonal variation in sensitivity of the young females to the male urinary chemosignal may account for our inability to accelerate puberty in female mice during winter months (Drickamer, 1984a). The functional significance of the seasonal shift in sensitivity may be related to the breeding ecology of the mice. In climates and locations where there are reduced resources and/or increased risks related to weather conditions, house mice may breed seasonally (Bronson, 1979; Pelikan, 1981). In those seasonal environments we might expect to discover the evolution of one or more mechanisms to slow down or shut down breeding during periods when conditions are less than adequate for successful reproduction. Rather than risking energy expenditures on what may be unsuccessful attempts to reproduce, it may be advantageous for the mice to conserve their energies and utilize the available resources to survive. Females that delay puberty and/or reproduction until existing conditions are more favorable could be at an advantage. I think that the seasonal shift in sensitivity to the puberty-acceleating chemosignal from males is one such mechanism.

What of the long-range consequences of early sexual maturation in terms of lifetime reproduction and related parameters? I am currently testing this question; it appears from that portion of the data that has been collected and analyzed to date that females whose puberty is accelerated suffer higher mortality relative to control females during pregnancy and lactation, though not necessarily with their first litter. These puberty-accelerated females conceive their first, and sometimes their second, litters prior to the completion of normal morphological development. Even though they are accelerated, they produce normal sized litters with pup birth and weaning weights comparable to those for control females.

Lastly, consider the behavioral responses of young females to the male chemosignal. Unpublished data from an ongoing experiment in my laboratory indicate that young females, 20-35 days of age (with no acceleratory cues present), avoid the odor of a male mouse. However, by the time these females are postpubertal, aged 40-55 days, they exhibit some preference for male odor. Adult females exhibit a distinct and sustained preference for male odor under the same test conditions. Could it possibly be that the avoidance of the male odor by prepubertal females has evolved as a mechanism to avoid accelerated puberty and the likelihood of becoming pregnant at an early age before the completion of morphological development? Further research on the odor preferences of the mice is now in progress.

URINE FROM PREGNANT/LACTATING FEMALES

The urine of pregnant and lactating females accelerates the sexual development of young female mice (Drickamer and Hoover, 1979). To date, after a number of experiments exploring several aspects of the signal(s) from the two types of donor females, no clear differences have been discerned between the effects of the urine from pregnant and lactating females. Hence, I will consider them to be the same substance. Like the male chemosignal, three days of exposure to the pregnancy/lactation chemosignal beginning prior to day 20 of age is sufficient for producing accelerated puberty (Drickamer, 1984e). Two hours of stimulation per day is necessary for the effect to occur (Drickamer, unpub.). The patterns of daily excretion of the chemosignal by donor females and sensitivity by recipients reveal a complex set of interactions (Drickamer, 1984f). In general, the chemosignal is effective in producing accelerated sexual maturation regardless of when during the daily cycle it is collected, and regardless of when during the daily cycle it is used to treat young females. In a manner similar to the male chemosignal, the substance in the urine of pregnant and lactating females cannot be used to accelerate puberty in females during the winter months; this effect appears to be due to a seasonal shift in sensitivity to the chemosignal on the part of the young females (Drickamer, unpub.) and not to any changes in the chemosignal substance released by donor females.

The chemosignal from pregnant/lactating females is present in both excreted and bladder urine. The substance is present in the urine only during the last two-thirds of pregnancy and the last two-thirds of lactation (Drickamer, 1983b); it is not yet known what set of hormonal and related physiological conditions within the mice may account for this time course of chemosignal release.

There are no differential acceleratory effects from the pregnancy/lactation chemosignal with respect to related versus unrelated young females (Drickamer, 1984b). When pregnant or lactating females are caged in groups or housed with non-reproducing females, their urine neither accelerates nor delays puberty onset in recipient test females (Drickamer, 1983b). Possibly, the acceleratory chemosignal due to pregnancy/lactation and the delay signal that appears in the urine of grouped females are both present in the urine. Alternatively, the group-caged pregnant or lactating females may not be releasing either chemosignal.

The pregnancy/lactation signal is effective in moderately small (0.03 cc of urine/day) doses (Drickamer, 1982b, 1983c). The substance is rather volatile (Drickamer and Hoover, 1979; Drickamer, in press b), and is not effective with young females rendered anosmic with zinc sulfate (Drickamer, unpub.). Urine from pregnant or lactating females that is exposed to the air remains effective for only one to three days (Drickamer, unpub.). Thus, it appears that the chemosignal from pregnant and lactating females may be received at the olfactory epithelium. The acceleratory effect is not enhanced by simultaneous treatment of test females with either male urine or urine from females in estrus, but young females treated with urine from pregnant or lactating females and urine from group-caged non-reproducing females are delayed in attaining puberty (Drickamer 1982b, unpub.).

In the social deme structure of house mice, the young females most likely to be affected by the chemosignal(s) from pregnant or lactating females are the progeny of the female(s) depositing the urine. It is possible that other young females in the same deme, in nearby or even communal nests, could also be affected, but these reproducing females are quite possibly related to one another. Thus, the overall effect of the substance may be to accelerate the sexual development of a female's own progeny or those of a sister or daughter. I have not yet tested the urine deposition patterns of pregnant or lactating females; it will be interesting to determine whether such females urinate differentially near cues from young females, and, in particular, depending upon whether those cues are from their own or unrelated progeny. It may also be important to remember that house mice exhibit a postpartum estrus and quite frequently adult females are both lactating and pregnant simultaneously.

The chemical substance(s) involved for the pregnancy/lactation chemosignal is not yet known; like the male signal, I suspect that the substance is a metabolic by-product of physiological processes ongoing within the mice. The presence of the substance in the urine is likely to be a function of events resulting from reproduction (pregnancy, lactation) and has not evolved specifically as a communication signal. The substance has the additional feature of acting as a signal to all females that may encounter the urine regarding conditions for successful reproduction; the presence of the substance signifies that conditions are adequate or better. Not only are young females accelerated in their sexual development by the chemosignal, but also, adult females are induced by the presence of urine from pregnant or lactating females, into more frequent estrous cycles and more prolonged estrus (Hoover and Drickamer, 1978).

Under constant environmental conditions it appears that pregnant and lactating females release the signal throughout the year. We are currently testing the effects of various non-social environmental factors on the production and release of this signal. It is possible that some diminution of the substance in urine will be observed when pregnant or lactating female mice are housed under conditions of shortened photoperiods or reduced food supplies. These latter conditions may represent, directly or indirectly, less than adequate conditions for successful reproduction. Pregnant or lactating females housed at densities of three or more mice per cage apparently either cease to release the acceleratory signal, or they release both the acceleratory signal and the puberty-delaying chemosignal associated with grouped females.

The pregnancy/lactation chemosignal is relatively volatile and thus does not remain in the environment for more than a few days. This volatility characteristic has the effect of leaving the "conditions are adequate" message in the environment only as long as reproductively active female mice are present at low densities. The volatility may simply be a consequence of the chemical nature of the chemosignal substance.

Young females, prior to the time they leave the natal site to disperse (if they do), will be exposed to the substance from pregnant and lactating females. I am currently testing the possibility that preweaning stimulation as well as postweaning stimulation, by this chemosignal will result in accelerated sexual development. Prepubertal females exhibit some preference for the odor of pregnant and lactating females relative to some other mouse odor sources (Drickamer, unpub.). By the time the mice are postpubertal (ca. 50 days of age) and later as adults, they neither prefer nor avoid the odor from pregnant and lactating females. As is the case with the sensitivity of young female mice to the male signal, there is a distinct pattern of seasonal variation in the acceleration of puberty by the pregnancy/lactation substance. Thus, young females are not responsive to the signal for those months when the probability is higher that conditions for successful reproduction are marginal or less than adequate.

URINE FROM FEMALES IN ESTRUS

The urine of singly-caged females in estrus accelerates the onset of puberty in female mice (Drickamer, 1982c, 1984b), but urine from singly-caged females in diestrus does not. To date, considerably less is known about this chemosignal than about the other three substances. Like the two previously discussed acceleratory signals, the estrous urine signal is effective in inducing earlier puberty with only three days of treatment commencing prior to about day 29 of age (Drickamer, unpub.). Small doses (0.001 cc of urine/day) are sufficient for the acceleration effect (Drickamer, unpub.). Also, there are no differential acceleratory effects upon related, compared to unrelated, young females (Drickamer, 1984b). Urine from females that are housed in groups and are in estrus has no acceleratory or delaying effect upon puberty in young females; the subjects reach sexual maturity at the same age as water-treated control females (Drickamer, 1982c). This effect may result from the presence in the urine of both the acceleratory chemosignal and the delay signal associated with grouped females; alternatively, group-caged females in estrus may not release either of the puberty-influencing signals.

The simultaneous treatment of young females with urine from females in estrus and either male urine or urine from pregnant or lactating females does not enhance the acceleratory effect. However, the simultaneous treatment of subjects with the urine of females in estrus and urine from grouped females results in delayed puberty (Drickamer, unpub.).

The presence of this urinary chemosignal appears to be related to the reproductive cycles of the females and hence its presence is correlated with hormone changes that take place during the estrous cycle. The primary effect is as a signal to other mice that conditions are adequate for reproduction; young females that encounter the urine of females in estrus and that are affected by the signal may join the pool of reproducing females. The acceleratory effect is modulated by density conditions; females that are living in a population where density is above a certain level will not be accelerated and could be delayed in attaining puberty if exposed to the urine of non-reproducing females containing the delay chemosignal. No data are yet available on patterns of urine deposition by females in estrus compared to females in diestrus or other phases of the estrous cycle. Nor are there any data on the behavioral responses of young females to the urine from females at different stages of the cycle. Also, the effects of non-social environmental factors on the release of the estrous urine signal have yet to be investigated.

URINE FROM FEMALES HOUSED IN GROUPS

Females housed in groups reach sexual maturity later than females that are caged alone (Castro, 1967; Vandenbergh et al., 1972; Drickamer, 1974b). Young females treated with urine from females housed in groups are also delayed in attaining puberty and to the same extent as the group-caged mice (Drickamer, 1982c). Singly-caged females that are exposed to urine from grouped females will themselves excrete the puberty-delaying chemosignal in their urine (Drickamer, 1982c). At least four, and possibly up to seven, days of exposure to the substance beginning before about day 25 of age are necessary to produce delays in puberty (Drickamer, 1977a). A minimum of one hour/day of exposure is needed to produce delayed maturation (Drickamer, 1983c).

Females of all ages release the delay signal when caged at sufficient density for at least ten days (Drickamer, 1983c); females continue to release the delay signal in their urine for about ten days after being separated from grouped housing into individual cages (Coppola and Vandenbergh, 1985). Urine collected from grouped females at all hours of the day produces the delay effect and there is no difference in the sensitivity to the signal on the part of young females at different times of the day (Drickamer, 1982a). There is a pattern of significant seasonal variation in the delay of puberty - the chemosignal is generally not effective during the mid-summer months (Drickamer, This seasonal variation appears to be attributable to a change in sensitivity to the signal by young females and not to any shift in the potency of the delay substance as released by grouped females (Drickamer, unpub.).

The delay substance is present in the excreted urine of grouped females only, but it is contained in the bladder urine of all females regardless of age or caging density (McIntosh and Drickamer, 1977). In singly caged females, some process that occurs along the urethra apparently blocks or deactivates the delay substance (McIntosh and Drickamer, 1977). Ovariectomy does not alter the presence of the delay substance in either excreted or bladder urine (Drickamer, McIntosh and Rose, 1978). However, after adrenalectomy, the substance is no longer present in the excreted urine of grouped females or the bladder urine of all females (Drickamer and McIntosh, 1980). Replacement therapy given to grouped adrenalectomized females using hydrocortisone or corticosterone (gluccocorticoids) brings about a reappearance of the delay substance in the urine (Drickamer and Shiro, 1984). The reproductive history (virgin, primiparous, multiparous, bred unsuccessfully) of a female and any associated hormone changes are not factors in the production or release of the delay signal (Drickamer, 1982c).

Dominant female mice in groups do not release the delay substance; in diestrus the urine of dominant females has no effect on female puberty relative to water-treated control mice, while urine from dominant females in estrus accelerates puberty in test females (Drickamer, in press a). Urine from subordinate females in diestrus delays puberty, but urine collected from subordinate females within groups that are in estrus does not accelerate or delay puberty (Drickamer, in press a). There are no differential delays in puberty for mice that are related or unrelated to the donor females (Drickamer, 1984b). Nor is there any differential effect upon puberty when grouped siblings are compared to grouped non-siblings, or when female pups are cross-fostered from one dam to another. In all cases females exposed to other females, or to the delay signal, are retarded in their sexual development (Drickamer, 1982c).

The delay signal is effective in doses as small as 0.0001 cc of urine/day (Drickamer, 1982b, 1984c). The substance is apparently only slightly volatile; it loses its effectiveness after five days exposure to air (Drickamer, in press b). The mice need not make direct contact with the urine for puberty delays to occur (Drickamer and Assmann, 1981), but data on both peripheral and central anosmia suggest that the delay substance is received by the vomeronasal-accessory olfactory system (Reynolds and Keverne, 1979; Drickamer, unpub.). When young females are treated simultaneously with any pair or threesome of the chemosignals, one of which is the substance from grouped females, they are delayed in attaining sexual maturity; that is, the delay signal takes precedence over the other substances (Drickamer, 1982b). When singly-caged females are housed under shortened photoperiods, or placed on a food-deprivation regime, they release the delay chemosignal in their excreted urine (Drickamer, 1984d).

The functional significance of the puberty-delaying signal must be examined in some detail from several perspectives. The delay signal may function as a general communication to conspecifics that conditions for reproduction are less than adequate or poor. This mechanism may be effective for both social conditions; the signal is at least partly density-dependent and is certainly released by mice caged at moderate to high densities, and for non-social environmental factors. The delay signal is released in the excreted urine of all females caged under conditions involving shortened photoperiods or food deprivation. Females must experience the conditions that lead to the release of the delay chemosignal for a moderately prolonged period before the substance appears in their urine; this may be part of an evolved mechanism like a "built-in safeguard" to protect females from releasing the substance unless there is a substantial period of conditions that are less than adequate for successful reproduction. There is also a lag time of about ten days after the improvement of the conditions that bring about chemosignal release prior to the cessation of its release by the females. I wonder whether the same lag time would prevail if the females that are separated from a group are also, at the same time, exposed to a male, or male urine. Would their urine cease to contain the delay substance sooner?

Dominant females do not release the delay substance and are not apparently affected by the social conditions involving the moderate to high densities thus far tested. Dominant females continue to exhibit more regular estrous cycles than subordinates (Drickamer, in press a), and thus, may continue to reproduce even under less than adequate conditions. Presumably the dominant female has command of a greater than average share of the available resources, and thus, may be able to successfully carry a litter to term and wean the pups even when conditions are poor.

There are several interaction effects involving grouping females and either pregnancy/lactation or females in estrus, as has been noted in previous subsections. When the delay substance from grouped females is present in roughly equal proportions with any or all of the three acceleratory signals, females are delayed in attaining puberty. This finding makes sense if we consider that the delay signal indicates that the necessary conditions for successful reproduction are lacking. The three acceleratory chemosignals are apparently subject to some modulation by social and non-social environmental conditions, but the most critical mechanism that has evolved to safeguard against wasting reproductive energies involves the delay signal produced by grouped females and by singly-caged females exposed to social and non-social cues that indicate the existence of less than adequate conditions. Also, when females are delayed in attaining puberty, and when there are fewer adult females exhibiting regular estrous cycles (and hence fewer females either in estrus or pregnant or lactating) there will presumably be less of the acceleratory substances excreted by these mice present in the environment.

Grouped females deposit their urine in a general broadcast pattern throughout the environment when no other cues are present (Drickamer, unpub., but see Powell and Wolff, 1982). Thus, the delay signal would likely be present throughout a particular habitat. Though the substance must be contacted directly by young females to affect sexual development, very small daily doses are sufficient to produce the delay effect.

The puberty-delaying chemosignal has been only partially characterized in several unpublished investigations. The substance is clearly dependent, directly or indirectly, upon one or more substances from the adrenal glands. The release or non-release of the substance involves some mechanism at the level of the urethra. Thus, there are several possible locations or places in a pathway at which external and internal cues may influence the chemosignal system.

To be affected by the chemosignal, young females need more exposure to the delay chemosignal (four-seven days) than are needed for effects from the various puberty-accelerating substances (about three days). Furthermore, to be effective, the stimulation by the delay substance must begin by about day 25 of age rather than day 29 as for the acceleratory substances. Both the duration of stimulation and the relatively earlier critical age for onset of signal exposure could be mechanisms that have evolved to prevent young female mice from being incorrectly influenced by a false "delay puberty" message when conditions are actually adequate for reproduction.

Young female mice are not delayed in attaining puberty in the summer months even when exposed to the delay chemosignal, because of a lack of sensitivity to the substance. This finding may be related to the fact that, in most climates and locations, the summer months represent a period when resource conditions are generally adequate for successful reproduction. Thus, even though social density may be relatively high, the food, nesting, and other resources needed for reproduction are present in adequate amounts and the risks associated with cold temperature stress are not a factor.

An experiment in progress in my laboratory has yielded results that indicate that prepubertal female mice prefer the odor of grouped females relative to other mouse odors; the pattern shifts to avoidance of the odor from grouped females when the test subjects are 40-55 days of age, at and just after puberty. As adults, these mice do not exhibit a preference for, or avoidance of, the odor of grouped females.

Bronson (1979) hypothesized that the delay-of-puberty phenomenon is related to dispersal. Females may be delayed in attaining puberty until after they have departed from their natal deme. After dispersing, they may be influenced by various signals that hasten the onset of estrus. My foregoing summary of data on the delay chemosignal and related behavioral responses is in substantial agreement with Bronson's hypothesis. There are, however, many females that do not leave the general area where they were born, and yet certainly some of these females reproduce. Also, what of the effects of the puberty-accelerating chemosignal from pregnant and lactating females and the possible effects of that signal on their own progeny? Further studies of wild

Mus musculus populations could yield valuable data on dispersal patterns and provide further evidence on the functional significance of the delay chemosignal.

JOINT CONSIDERATIONS AND SUMMARY

Most of the data summaries in the foregoing sections have been based upon studies conducted utilizing laboratory stocks of house mice. It is therefore quite important to note that all of the chemosignals affecting puberty discussed here have also been demonstrated with wild house mice under laboratory conditions and that many aspects of the mechanisms discussed here have also been tested with wild Mus. No differences between the wild and laboratory stocks have been detected (Drickamer, 1979; 1984b, 1984g). In addition, the male urinary chemosignal that accelerates female puberty and the substance in the urine of grouped females that delays puberty have been tested under field conditions using populations of Mus musculus established in the cloverleaf islands of superhighways (Massey and Vandenbergh, 1980, 1981; Massey, 1982); the effects are the same as for laboratory stocks.

It is noteworthy that, in terms of the effects of altering the age of puberty on generation time, there is more of a potential spread from early to late puberty for wild Mus than for laboratory mice. Accelerated females with a male present attain puberty at 27-28 days of age and chemosignal-delayed females attain puberty at about 40 days of age for laboratory stocks (a spread of 12-13 days), whereas the accelerated wild Mus attain puberty at 42 days of age with a male present and are delayed up to 67-68 days of age for first estrus when caged in groups or exposed to the chemosignal from grouped wild females (a spread of 25-26 days). There may be some variation within stocks of laboratory mice and between populations of wild house mice, but it does appear that the potential effects on population growth or decline could be quite significant in wild Mus musculus populations.

The sexual maturation of young female mice is influenced by a hierarchy of cues as outlined by Bronson (1979). One of these sets of cues is the series of chemosignals summarized here. The chemosignals influence puberty directly, but are also themselves affected by non-social environmental cues such as food and photoperiod. The latter cues can also affect maturation directly. In a ranked hierarchy of cues affecting puberty, I think it unlikely that the chemosignals are the dominant or critical cue system, but rather they probably serve as a modulating influence. When the non-social environmental conditions are advantageous for successful reproduction, the urinary substances play key roles in hastening or retarding reproductive output. Also, almost as a corrective or checking influence, the usually socially mediated chemosignals are also sometimes influenced by factors like food and photoperiod. It is also possible that the chemosignals play a more prominent role in controlling the reproductive biology of house mice when conditions result in an outbreak or rapid growth of a population.

In general, I suggest that the signals have been influenced or shaped by natural selection only minimally. Rather, it is the behaviors of the donors and recipients in depositing and approaching or avoiding the signals, and the sensitivity of the young females to the signals that most likely have been altered by natural selection.

Three of the four urinary chemosignals thus far explored in house mice have acceleratory effects on puberty and/or enhancing effects on the estrous cycles of adult females. It may be that these urinary signals are more of a positive influence on reproduction than a negative or restraining influence. For many factors (e.g., food - nutrition and calories, temperature extremes) that affect reproduction, the tendency is to portray their effects as negative or restraining. For many mouse populations, the general conditions for successful reproduction may be present, and thus the acceleratory chemosignals may play important roles in sustaining reproduction. The negative, modulating effect of the delay chemosignal from females living in crowded conditions above a threshold density will become important only if conditions deteriorate.

Lastly, I should note that I have not attempted here to summarize the available literature on chemosignal effects on puberty in young male mice (Vandenbergh, 1971; McKinney and Desjardins, 1973) - though there are only scanty data for house mice, and considerable additional research effort is required in this regard. Also, I have not presented more than a few of the many references available on the effects of chemosignals and related social phenomena on the reproductive biology of adult house mice, nor any of the numerous references on these chemosignal phenomena in other rodent taxa. The chemosignal effects on puberty and reproductive biology in rodents appear to be relatively widespread across various taxa, though data are insufficient to conclude that the phenomena are universal or identical across all taxa (see chapters in Vandenbergh, 1983, for recent general reviews).

ACKNOWLEDGMENTS

The research conducted in my laboratory has been generously supported by the National Institutes of Health Grant No. HD-08585. Additional support has been provided by the Bronfman Science Center Discretionary Fund at Williams College. I would like to thank John King, Donald Dewsbury, C. Richard Terman, Edwin Banks, Douglas Meikle, Stephen H. Vessey and especially John Vandenbergh for their helpful comments and discussions of matters presented in this paper. I also would like to thank my several technicians and many Williams College undergraduate-thesis students who have participated in this research, many of whose names appear as co-authors of articles cited in this paper.

REFERENCES

Andervont, H. B., 1944, Influence of environment on mammary cancer in mice, J. Nat. Cancer Inst., 4:579.

Barnett, S. A. and Coleman, E. M., 1959, The effect of low environmental temperature on the reproductive cycle of female mice, J. Endocr. 19:232.

Bronson, F. H., 1976, Urine marking in mice: Causes and effects, in: "Mammalian Olfaction, Reproductive Processes and Behavior," R. L. Doty, ed., Academic Press, New York.

Bronson, F. H., 1979, The reproductive ecology of the house mouse, Quart. Rev. Biol., 54:265.

Bronson, F. H. and Maruniak, J. A., 1975, Male-induced puberty in female mice: Evidence for a synergistic action of social cues, Biol. Reprod., 13:94.

Castro, B. M., 1967, Age of puberty in female mice: Relationship to population density and the presence of adult males, Anais de Acad. Brasil de Ciencia, 39:289.

Colby, D. R. and Vandenbergh, J. G., 1974, Regulatory effects of urinary pheromones on puberty in the mouse, Biol. Reprod., 11:268.

Coppola, D. M. and Vandenbergh, J. G., 1985, Effects of density, duration of grouping and age of urine stimulus on the puberty delay pheromone, *J. Reprod. Fert.*, 51:77.
Cowley, J. J. and Wise, D. R., 1972, Some effects of mouse urine on neonatal growth and reproduction, *Anim. Behav.*, 20:499.
Drickamer, L. C., 1974a, Contact stimulation, androgenized females and accelerated scxual maturation in female mice, *Behav. Biol.*, 12:101.
Drickamer, L. C., 1974b, Sexual maturation of female house mice: Social inhibition, *Dev. Psychobiol.*, 7:257.
Drickamer, L. C., 1975a, Female mouse maturation: Relative importance of social factors and daylength, *J. Reprod. Fert.*, 44:147.
Drickamer, L. C., 1975b, Daylength and sexual maturation in female house mice, *Dev. Psychobiol.*, 8:561.
Drickamer, L. C., 1975c, Contact stimultion and accelerated sexual maturation of female mice, *Behav. Biol.*, 15:113.
Drickamer, L. C., 1977, Delay of sexual maturation in female house mice by exposure to grouped females or urine from grouped females, *J. Reprod. Fert.*, 51:77.
Drickamer, L. C., 1979, Acceleration and delay of first estrus in wild *Mus musculus*, *J. Mammal.*, 60:216.
Drickamer, L. C., 1981, Selection for age of sexual maturation in mice and the consequences for population regulation, *Behav. Neural Biol.*, 31:82.
Drickamer, L. C., 1982a, Acceleration and delay of sexual maturation in female mice via chemosignals: Circadian rhythm effects, *Biol. Reprod.*, 27:596.
Drickamer, L. C., 1982b, Acceleration and delay of first vaginal oestrus in female mice by urinary chemosignals: Dose levels and mixing urine treatment sources, *Anim. Behav.*, 30:456.
Drickamer, L. C., 1982c, Delay and acceleration of puberty in female mice by urinary chemosignals from other females, *Dev. Psychobiol.*, 15:433.
Drickamer, L. C., 1983a, Male acceleration of puberty in female mice (*Mus musculus*), *J. Comp. Psychol.*, 97:191.
Drickamer, L. C., 1983b, Chemosignal effects on puberty in young female mice: Urine from pregnant and lactating females, *Dev. Psychobiol.*, 16:207.
Drickamer, L. C., 1983c, Effect of period of grouping of donors and duration of stimulus exposure on delay of puberty in female mice by a urinary chemosignal from grouped females, *J. Reprod. Fert.*, 69:723.
Drickamer, L. C., 1984a, Seasonal viariation in acceleration and delay of sexual maturation in female mice by urinary chemosignals, *J. Reprod. Fert.*, 72:55.
Drickamer, L. C., 1984b, Urinary chemosignals from mice (*Mus musculus*): Acceleration and delay of puberty in related and unrelated young females, *J. Comp. Psychol.*, 89:414.
Drickamer, L. C., 1984c, Effects of very small doses of urine on acceleration and delay of sexual maturation in female house mice, *J. Reprod. Fert.*, 71:475.
Drickamer, L. C., 1984d, Urinary chemosignals and puberty in female house mice: Effects of photoperiod and food deprivation, *Physiol. Behav.*, 33:907.
Drickamer, L. C., 1984e, Acceleration of puberty in female mice by a urinary chemosignal from pregnant or lactating females: Timing and duration of stimulation, *Dev. Psychobiol.*, 17:451.
Drickamer, L. C., 1984f, Acceleration of puberty in female mice by a chemosignal from pregnant and lactating females: Circadian rhythm effects, *Biol. Reprod.*, 31:104.
Drickamer, L. C., 1984g, Acceleration of puberty by a urinary chemosignal from pregnant and lactating *Mus musculus*, *J. Mammal.*, 65:697.

Drickamer, L. C., in press a, Dominance effects among female mice (Mus musculus): Influences upon urinary chemosignals and reproduction, J. Comp. Psychol.

Drickamer, L. C., in press b, Acceleration and delay of puberty in female mice by urinary chemosignals: Age of the urine stimulus, Dev. Psychobiol.

Drickamer, L. C. and Murphy, R. X., 1978. Female mouse maturation: Effects of excreted and bladder urine from juvenile and adult males, Dev. Psychobiol., 11:63.

Drickamer, L. C., McIntosh, T. K. and Rose, E. A., 1978, Effects of ovariectomy on the presence of a maturation-delaying pheromone in the urine of female mice, Horm. Behav., 11:131.

Drickamer, L. C. and Hoover, J. E., 1979, Effects of urine from pregnant and lactating female house mice on sexual maturation of juvenile females, Dev. Psychobiol., 12:545.

Drickamer, L. C. and McIntosh, T. K., 1980, Effects of adrenalectomy on the presence of a maturation-delaying pheromone in the urine of female mice, Horm. Behav., 14:146.

Drickamer, L. C. and Assmann, S. M., 1981, Acceleration and delay of puberty in female house mice: Methods of delivery of the urinary stimulus, Dev. Psychobiol., 14:487.

Drickamer, L. C. and Shiro, B. C., 1984, Effects of adrenalectomy with hormone replacement therapy on the presence of a sexual maturation-delaying chemosignal in the urine of grouped female mice, Endocrinology, 115:255.

Hoover, J. E. and Drickamer, L. C., 1978, Effects of urine from pregnant and lactating female house mice on oestrous cycles of adult females, J. Reprod. Fert., 55:297.

Kennedy, J. M. and Brown, K., 1970, Effects of male odor during infancy on the maturation, behavior and reproduction of female mice, Dev. Psychobiol., 3:179.

Keverne, E. B., 1979, The dual olfactory projections and their significance for behavior, in: "Chemical Ecology: Odour Communication in Animals," F. J. Ritter, ed., Elsevier, Amsterdam.

Laurie, E. M. O., 1946, The reproduction of the house mouse (Mus musculus) living in different environments, Proc. Roy. Soc. London, 133:248.

Lombardi, J. L., Vandenbergh, J. G. and Whitsett, J. M., 1976, Androgen control of the sexual maturation pheromone in house mouse urine, Biol. Reprod., 15:179.

Lombardi, J. L. and Vandenbergh, J. G., 1977, Pheromonally induced sexual maturation in females: Regulation by the social environment of the male, Science, 196:545.

Massey, A. and Vandenbergh, J. G., 1980, Puberty delay by a urinary cue from female mice in feral populations, Science, 209:821.

Massey, A. and Vandenbergh, J. G., 1981, Puberty acceleration by a urinary cue from male mice in feral populations, Biol. Reprod., 24:532.

Massey, A., 1982, Ecology of house mouse populations confined by highways, J. Elisha Mitchell Sci. Soc., 98:135.

McIntosh, T. K. and Drickamer, L. C., 1977, Excreted urine, bladder urine, and the delay of sexual maturation in female house mice, Anim. Behav., 25:999.

McKinney, T. D. and Desjardins, C., 1973, Intermale stimuli and testicular function in adult and immature mice, Biol. Reprod., 9:370.

Meredith, M., 1983, Sensory physiology of pheromone communication, in: "Pheromones and Reproduction in Mammals," J. G. Vandenbergh, ed., Academic Press, New York.

Pelikan, J., 1981, Patterns of reproduction in the house mouse, in: "Biology of the House Mouse," R. J. Berry, ed., Academic Press, New York.

Powell, A. J. and Wolff, P. R., 1982, Sex differences in mouse urination patterns, Anim. Behav., 30:1207.
Reynolds, J. and Keverne, E. B., 1979, The accessory olfactory system and its role in the pheromonally mediated suppression of oestrus in grouped mice, J. Reprod. Fert., 57:31.
Vandenbergh, J. G., 1967, Effect of the presence of a male on the sexual maturation of female mice, Endocrinology, 81:345.
Vandenbergh, J. G., 1969, Male odor accelerates female sexual maturation in mice, Endocrinology, 84:658.
Vandenbergh, J. G., 1971, The influence of the social environment on sexual maturation in male mice, J. Reprod. Fert., 24:383.
Vandenbergh, J. G. , ed., 1983, "Pheromones and Reproduction in Mammals," Academic Press, New York.
Vandenbergh, J. G., Drickamer, L. C. and Colby, D. R., 1972, Social and dietary factors in the sexual maturation of female mice, J. Reprod. Fert., 28:397.
Vandenbergh, J. G., Whitsett, J. M. and Lombardi, J. L., 1975, Partial isolation of a pheromone accelerating puberty in female mice, J. Reprod. Fert., 43:515.
Vandenbergh, J. G., Finlayson, J. S., Dobrogosz, W. J., Dills, S. S. and Kost, T. A., 1976, Chromatographic separation of puberty accelerating pheromone from male mouse urine, Biol. Reprod., 15:260.

THE PUBERTY DELAYING PHEROMONE OF THE HOUSE MOUSE: FIELD DATA AND A NEW EVOLUTIONARY PERSPECTIVE

David M. Coppola

Worcester Foundation for Experimental Biology
Shrewsbury, Massachusetts
01545

Puberty delay occurs in natural, semi-natural, and laboratory populations of house mice (Mus musculus) in response to increasing densities (Christian et al., 1965; Crowcroft and Rowe, 1959; Lidicker, 1976; Southwick, 1958). Delayed puberty at high population densities is a factor in dampening further population growth (Christian, 1978). Urinary chemosignals may be part of the mechanism modulating puberty of females in dense house mouse populations. Studies in the laboratory reveal that housing females in groups or exposing them to urine from group-housed females delays puberty in wild and domestic female mice (Drickamer, 1977, 1979). These studies suggest that a primer pheromone inhibiting sexual maturation is produced by grouped females. A descriptive study of two feral mouse populations provided the first evidence that wild female house mice living under natural conditions produce the puberty delaying pheromone (Massey and Vandenbergh, 1980). The puberty delaying pheromone appeared in the urine of wild females coincident with density increases. Density-dependent release of a puberty delaying pheromone in female mice may play a significant role in regulating natural rodent populations (Drickamer, 1974). This report provides the first experimental evidence that female house mice living in wild populations produce the puberty delaying chemosignal in response to acute increases in population density and briefly outlines a new evolutionary perspective from which to study primer pheromones.

FIELD DATA

Methods

Four mouse populations were studied for eight months in 1983. Each resided on one of two highway cloverleaf interchanges along Interstate 40 near Raleigh, N.C. These old-field habitats, which have been termed "highway islands" (Massey and Vandenbergh, 1980), are excellent sites for the study of rodent populations owing to their small size (0.3-0.7 hectares) and virtual insularity.

This study utilizes these populations to provide a test of the causal relationship between population density and delay-pheromone release. The design was constrained by my ability to work with only four island populations at one time because of the large amount of labor involved in mark and recapture studies. The small number of islands available to me, along with

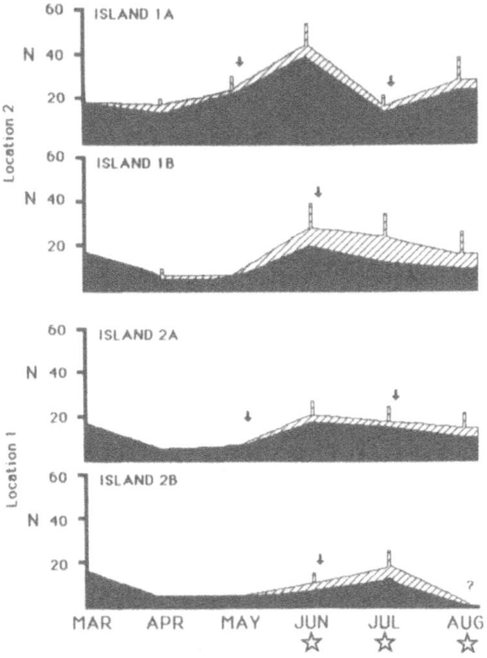

Fig. 1. Minimum number alive (solid area) and population estimates with their 95 percent confidence limits (cross-hatched area) for four populations of house mice confined to highway cloverleaf sections. Arrows represent the time of introduction of 40 female interlopers onto each cloverleaf. However, interlopers are not included in the population estimates. Stars represent the months in which urine was collected from females living on the cloverleafs to assess its puberty-delaying potency.

the great floral diversity between islands, required the use of repeated measures. Two adjacent islands at each of two locations were studied by monthly mark and recapture trapping over a six month period (Fig. 1). Acute population explosions were created on these islands by the introduction of 40 second or third generation female wild house mice that had been raised in the laboratory. Animals were introduced onto one island at each location three times during the study. A different island at each location received the interlopers for each introduction (Fig. 1). This allowed me to compare treated and untreated islands at the same point in time and also to examine an individual island across time.

Urine samples from resident females were collected for each island at monthly intervals during the six-to-eight-day trapping sessions by lining the floor of the live-traps with filter paper. Urine-soiled filter paper was removed from the traps and stored at -4°C under nitrogen. Three weeks separated the introduction of foreign females and the collection of urine samples from each island. This schedule was chosen because laboratory females do not release the delay pheromone until after the tenth day of grouping (Drickamer, 1973). Moreover, recent studies have shown that increasing the number of females per group will not shorten the length of time necessary for the initiation of pheromone release (Coppola and Vandenbergh, 1985). The ability of the urine samples to delay puberty was assessed in three bioassays, one for each month in which samples were collected. The assay of each sample entailed the delivery of 1 gm of urine-impregnated filter paper to the cages of 22-day-old female albino mice (n =

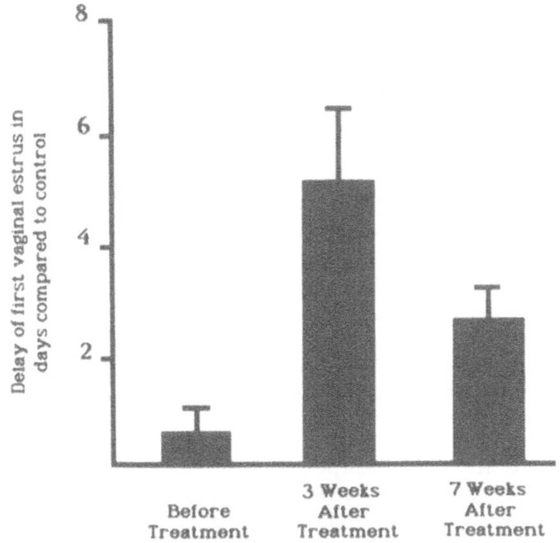

Fig. 2. Mean delay of first vaginal estrus (± 1 SEM) in days compared to the control for female mice treated with urine from wild females living on highway cloverleaf sections. The SEMs represent the variation between cloverleaf sections treated alike. Urine collected before the artificial population explosion did not cause a significant delay in puberty compared to controls. However, urine collected 3 weeks or 7 weeks after the explosion did cause a significant delay (3 weeks, $t[5] = 3.92$, $p < 0.02$; 7 weeks, $t[2] = 5.51$, $p < 0.05$). The delay produced by urine collected 3 or 7 weeks after the explosion were not significantly different.

18) for one week. The sample was taken from a pool of urine-impregnated papers for each collection from a highway island. Only paper impregnated with urine from adult (> 10 gms) females that were neither visibly pregnant nor lactating was included in the pool of samples. For each assay, a group of individually-caged females (n = 15) received clean filter paper to serve as controls. The assay females were examined daily for vaginal introitus, after which vagina were lavaged daily to determine the age at first estrus (Vandenbergh, 1969).

Results

Urine samples collected from the treated islands during the first trapping session after the introduction of foreign females caused an average delay of 5.3 days (Fig. 2). Foreign females that were captured during this trapping session were selectively removed from the islands. This was done to bring the population density back down to the pre-introduction levels. Urine samples collected from the islands at the next trapping session after the selected removal of foreign females also significantly delayed the puberty of test females albeit to a lesser extent than samples taken three weeks after the introductions. These samples were collected seven weeks after the introduction of foreign females onto an island and three weeks after the selected removal. The average delay in first estrus was 2.7 days for this group of females. In contrast to the urine samples taken from the islands after introductions, the urine from wild female mice residing on untreated islands did not cause a significant increase in the age at first estrus of test

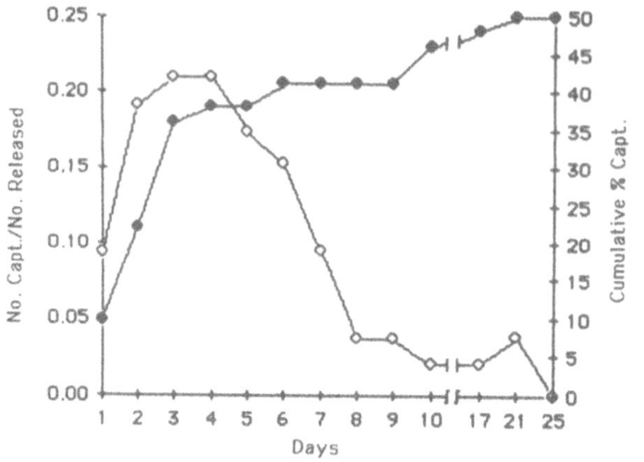

Fig. 3. The incidence of recapture for female house mice introduced to one highway cloverleaf population. The ratio of the number captured versus the number released and the cummulative percent captured are shown for each day traps were open.

females (Fig. 2). Density estimates for these untreated islands were low (Fig. 1). The association between low population density and lack of delay pheromone in the urine of wild female mice is consistent with the first study of highway island populations (Massey and Vandenbergh, 1980).

Of the 240 interlopers involved in the six introductions, only 21 were subsequently recaptured. In an effort to determine the fate of animals introduced to existing island populations, a separate study was undertaken. Fifty-two female mice from the same stock as the interlopers in the first experiment were released onto the center of a highway island. The movement of the interlopers on the island were monitored by live trapping over the following 21 days. Only 50 percent of the interlopers were ever captured, and the majority of these captures occurred within the first six days (Fig. 3). However, some interlopers remained on the island for at least three weeks after their introduction. These trapping results indicate the high mortality or emigration rate of the interlopers but also demonstrate that interlopers persisted on the island long enough to influence, transiently at least, the female interaction rate.

A NEW EVOLUTIONARY PERSPECTIVE

Future studies of the highway-island populations should further clarify the role of the puberty pheromones in the biology of the house mouse. This experimental system will allow us to choose between alternate hypotheses of pheromonal influences on population dynamics, given the appropriate manipulations of demographic parameters on the islands. It is critical that hypotheses concerning the puberty pheromones are based not only on empirical evidence but also on theoretical results pertaining to the age of first reproduction and how this trait interacts with other life-history characters.

I propose that life-history theory provides the proper evolutionary footing from which to study the function of primer pheromones in house mice and possibly primer pheromone function in general (Vandenbergh and Coppola, 1985). It is now well established that chemical cues from conspecifics alter the timetable for puberty in the young of many species. This cueing system, at least regarding the delay of puberty, seems to serve the signal recipients while the signal senders are little affected. Unwitting release of metabolic correlates of physiological state rather than altruism may explain the presence of the delay pheromone. The benefit gained by the recipients of the puberty pheromones are most easily understood in terms of their life-history. Life-history theory provides predictions of the environmental circumstances, social or otherwise, favoring early or delayed maturity. The puberty pheromones appear to act as cues to the social environment that determine the maturational rate most appropriate for the conditions under which they are released.

If these speculations are valid, then we should look for alterations in other life-history traits such as litter size and size of young in response to primer pheromones since the optima for these traits may also depend on the social milieu. Another area that is in need of work, not only with respect to the puberty pheromones but also with respect to primer pheromones in general, is behavior related to pheromonal communication. We should know, for instance, whether a young female will investigate or avoid urine marks that will influence her puberty and if her choice is influenced by the social milieu or other factors in the environment. In the field, new techniques must be developed to study secretive species such as the house mouse. More information about demography, social structure, and dispersal of feral and commensal mice will be required to evaluate adequately the hypotheses discussed here.

Life-history adaptations and behavioral adaptations undoubtedly interact to promote fitness despite the vagaries of the environment. Discovering the role of priming pheromones in this interactive process will require the melding of empirical and theoretical points of view.

ACKNOWLEDGEMENTS

I would like to gratefully acknowledge the contributions and guidance of Dr. John Vandenbergh. This project was supported by NSF grant BRS-8214558A02 to him. The preparation of this manuscript was supported in part by NIH grant NS-14453 to Dr. Robert J. O'Connell.

REFERENCES

Christian, J.J., 1978, Neurobehavioral endocrine regulation of small mammal populations, in: "Populations of Small Mammals Under Natural Conditions," D.P. Snyder, ed., University of Pittsburgh, Pittsburgh.

Christian, J.J., Lloyd, A., and Davis, D.E., 1965, The role of endocrines in the self-regulation of mammalian populations, Recent Prog. Horm. Res., 21:501.

Coppola, D.M., and Vandenbergh, J.G., 1985, Effect of density, duration of grouping and age of urine stimulus on the puberty delay pheromone in female mice, J. Reprod. Fert., 73:517.

Crowcroft, P., and Rowe, F.P., 1959, The growth of confined colonies of the wild house mouse (Mus musculus L.), Proc. Zool. Soc. London, 129:359.

Drickamer, L.C., 1974, Sexual maturation of female house mice: social inhibition, Devel. Psychobiol., 7:257.

Drickamer, L.C., 1977, Delay of sexual maturation in female house mice by exposure to grouped females or urine from grouped females. J. Reprod. Fertil., 51:77.

Drickamer, L.C., 1979, Acceleration and delay of first estrus in wild Mus musculus, J. Mammal., 60:215.

Drickamer, L.C., 1983, Effect of period of grouping of donors and duration of stimulus exposure on delay of puberty in female mice by a urinary

chemosignal from group females, J. Reprod. Fert., 69:723.
Lidicker, W.Z., 1976, Social behavior and density regulation in house mice living in large enclosures, J. Anim. Ecol., 45:677.
Massey, A., and Vandenbergh, J.G., 1980, Puberty delay by a urinary cue from female house mice in feral populations, Science, 209:821.
Vandenbergh, J.G., Male odor accelerates female sexual maturation in mice, Endocrinol., 84:658.
Vandenbergh, J.G., and Coppola, D.M., 1985, The physiology and ecology of puberty modulation by primer pheromones, in: "Advances in the Study of Behavior," J.S. Rosenblatt, ed., Academic Press, New York.

EFFECTS OF GENOTYPE, SOCIAL INTERACTION, AND TESTOSTERONE ON ESTERASES IN MALE MOUSE URINE

Jay Labov, Daniel Marra, Philip Allen, and Diane Zavotsky

Department of Biology
Colby College
Waterville, ME 04901

INTRODUCTION

Pregnancy blocking (Bruce effect) may result when a recently inseminated female mouse is introduced to an unfamiliar male or to pheromone(s) from his urine. Depending on the species, females may terminate pregnancies before (Bruce, 1959) or after implantation (Stehn and Richmond, 1975; Schadler, 1981). Pheromone production can be influenced by androgen levels, genotype, and social environment. C57BL/6J males do not release sufficient pheromone to induce estrus or the Bruce effect (e.g., Hoppe, 1975). Social subordination suppresses pheromone production in strains otherwise capable of pregnancy block (Lombardi and Vandenbergh, 1977; Huck, 1982). Although its chemical identity has not been established, the pheromone may be a polypeptide, a small protein, some moiety attached to a small protein, or a lipid (reviewed by Marchlewska-Koj, 1983).

Androgen-dependent urine esterases produced in both kidney and sex-accessory organs (Shaw and Koen, 1963; McPartland et al., 1981) of male rodents may be associated with these pheromones. Like pheromones, concentrations of urine esterases vary among mouse strains (McPartland et al., 1981). Esterases are less concentrated in urine from subordinated Swiss albino males than from their dominant counterparts (Labov, 1979).

The present experiments assessed the role of genotype, social factors and androgens on excretion of urine esterases by male house mice (*Mus domesticus*). Numbers and concentrations of electrophoretically separated urine esterases were compared for strains reported to be capable of blocking pregnancies (BALB/cByJ, CBA/J, DBA/2J, SWR/J, labeled BALB, CBA, DBA, and SWR, respectively) and for strains supposedly unable to do so (C57BL/6J, C57BL/6ByJ, C57BL/10J, C57BL/KsJ, labeled 6J, 6By, 10J, and KsJ, respectively) (Furudate and Nakano, 1980; Svare and Mann, 1981). Individual males within each strain were analyzed for differences in urine esterase levels when physically isolated and when grouped. Numbers and concentrations of urine esterases were also compared in SWR males following castration and androgen replacement.

METHODS AND MATERIALS

Subjects were 8 weeks old when purchased from the Jackson Laboratory, Bar Harbor, ME, and were maintained in 17 X 28 X 12 cm polypropylene cages with stainless steel tops. Wood shavings used as bedding were changed approximately weekly. Purina lab chow and tap water were available ad libitum. All strains were maintained in a colony room at 20-23°C and a 14:10 LD cycle. Males and females of other strains were also maintained in this room.

Each male was caged either singly or with 1 or 2 other males of the same strain upon arrival. Mice were isolated or grouped at least one week before the first urine samples were collected. Animals originally isolated were then grouped (2 or 3 per cage) and grouped males were isolated. This procedure was repeated for several strains. A total of 350 samples was collected from 105 subjects (90 of these males provided urine samples both when grouped and when isolated).

Urine was collected overnight (2100h-1300h) from individual males in 23 X 10 X 12.5 cm stainless steel metabolism cages (food and water were not provided during this time) and was subjected to electrophoresis 1-3 h later. Samples held more than 1 h were refrigerated in vials sealed with Parafilm. Unfiltered urine from an individual was mixed with an equal volume of 40% sucrose and bromphenol blue tracker dye in distilled water. One hundred microliters of this mixture were placed on top of a 2-cm stacker gel. Esterases were separated through 10-cm polyacrylamide disc gels (10% acrylaminde) in Tris-glycerine buffer (pH 8.5) at 100V and at 3.5 mA/sample until the tracker dye reached the lower end of the separation gels. Gels were sliced at the tracker lines, placed in idividual tubes containing esterase-specific substrate and stain (Selander et al., 1971), incubated for 0.5h at 37°C, and stored individually in 7% acetic acid. Gels were assayed with a scanning densitometer (λ = 575 nm) for total number of esterases, migration distances (distance from the beginning of the densitometer tracing to a line drawn perpendicularly through the apex of each peak), and relative concentrations as measured by the height of each peak. Individual peaks were labeled such that peak 1 represented the lowest molecular weight esterase (i.e., the esterase that migrated furthest).

After completing the experiments described above, numbers and concentrations of urine esterases were determined from ten grouped SWR males who were then castrated under Nembutal anesthesia and allowed to recover in isolation. Urine samples were assayed for numbers and concentrations of esterases two weeks after castration. Subjects were then injected daily with 500 μg of testosterone propionate in sesame oil for two weeks and their urine was collected and reanalyzed.

RESULTS

Males were caged for varying lengths of time (1-46 weeks) before their urine was sampled. However, correlation coefficients indicated that with the exception of Peak 2 migration distances, the length of time that males spent isolated or grouped did not significantly influence the other variables examined. Therefore, the amount of time spent isolated or grouped prior to sampling are not considered in subsequent analyses.

The number of detectable esterases per gel ranged from 0 (in several samples from 6J males) to 6 (means $\pm$ standard errors and sample sizes are in Table 1). Males in 7 strains released a greater number of esterases while grouped than while isolated. This effect was significant for 3 of 4 blocker strains [SWR: $F(1,34) = 6.5$, $p < .025$; CBA: $F(1,55) = 2.87$, $p < .001$; DBA: $F(1,20) = 7.7$, $p < 0.25$; BALB males showed a similar trend

Table 1. Numbers of Esterase Peaks Observed in the Urine of Males from Eight Inbred Mouse Strains After Being Grouped 2 or 3 per Cage or Physically Isolated

Nonblocker (N) or Blocker (B)	Strain	Number of Esterases When Grouped Mean ± SE (N)[a]	Number of Esterases When Isolated Mean ± SE (N)[a]	Number of Males[b] Showing Lowest Molecular Weight Esterase (%)
N	C57BL/KsJ	2.8 ± 0.1 (27)	2.8 ± 0.2 (32)	8/15 (53)
N	C57BL/6J	1.4 ± 0.1 (24)	1.1 ± 0.1 (8)	1/8 (12)
N	C57BL/10J	2.7 ± 0.1 (29)	2.4 ± 0.1 (25)	3/15 (20)
N	C57BL/6ByJ	3.0 ± 0.1 (34)	2.9 ± 0.1 (27)	11/11 (100)
B	SWR/J	2.7 ± 0.1 (24)	2.0 ± 0.0 (12)	5/12 (42)
B	BALB/cByJ	2.9 ± 0.1 (19)	2.4 ± 0.3 (11)	4/11 (45)
B	CBA/J	3.1 ± 0.1 (35)	2.1 ± 0.1 (22)	5/12 (42)
B	DBA/2J	4.0 ± 0.2 (13)	3.1 ± 0.1 (9)	7/7 (100)

[a]Total number of observations in each strain.

[b]Number of males that provided urine samples both while grouped and while isolated.

F(1,28) = 3.84, p = .06], but only for one non-blocker strain [10J: F(1,52) = 4.1, p < .05] (Table 1).

This difference in numbers of urine esterases was attributed primarily to the presence or absence of the lowest molecular weight esterase (lmwe) (Table 1). Of those 90 males whose urine was analyzed during both grouping and isolation, 10 males released lmwe in both environments, 23 males did so only while grouped, and lmwe was found in the urine of 12 others only during isolation. Forty-five males never released this esterase. Lmwe was detected in urine samples from 6J, SWR, and CBA males only during grouping (only one 6J male released lmwe, and only in one sample). Males in the other 5 strains excreted lmwe in both environments. In contrast, Peak 2 esterase was present in 100% of urine samples. Peak 3 was detected in urine from 84 of 90 grouped individuals and from 82 of 90 isolated males. Four 6J males did not release Peak 3 esterase in either environment.

Migration distance of the lmwe was unaffected by strain or caging. Caging significantly influenced migration distances of Peak 2 esterase in urine from KsJ: F(1,28) = 7.0, p < 0.25, 10J: F(1,28) = 15.2, p < .001, and SWR: F(1,22) = 15.5, p < .001 males. In two strains (KsJ, SWR), Peak 2 migrated further when animals were grouped than when they were isolated; the opposite was true for 10J males. Caging effects for Peak 3 were found for 10J: F(1,28) = 4.4, p < .05 and SWR: F(1,22) = 6.3, p < .025 urine. Peaks four through six were highly variable and difficult to categorize both within and among strains, and are not considered further.

Concentrations of lmwe (when expressed) were higly variable in different urine samples from individual males, among males within a strain, and among strains. No significant strain or caging differences were observed for lmwe concentrations. In contrast, concentrations of the other esterases were significantly influenced by strain and caging.

Castration and testosterone replacement significantly influenced the concentrations of lmwe and Peak 2 esterase in the urine from SWR males (Table 2). All ten subjects excreted both esterases prior to castration and after testosterone replacement, but only three males did so two weeks after castration. While concentrations of these two esterases did not differ from each other prior to or following castration, lmwe concentrations were significantly greater than Peak 2 concentrations following testosterone replacement.

DISCUSSION

Although males within each strain used in these studies are essentially isogenic, intrastrain variation in both numbers and concentrations of urine esterases was apparent. Males within a strain released lmwe in one caging environment but not in the other. This heterogeneity was especially conspicuous among non-blockers. 6J males excreted fewer numbers and lower concentrations of urine esterases than did mice from any other strain, but the percentage of KsJ and 6By males releasing lmwe was similar to that found among blocker strains. Possibly males from these two strains have higher titers of circulating testosterone or their tissues may be more sensitive to the effects of this hormone compared to other C57BL strains (Leiter et al., 1981; Leiter, pers. commun.). KsJ also differs from the other C57BL strains at the major histocompatability locus, H-2. 6J, 10J, and 6By males share the H-2b haplotype. KsJ haplotype (H-2d) is identical to that of BALB and DBA (Heiniger and Dorey, 1980). The H-2 locus in mice has been correlated with androgen-dependent physiological processes (Ivanyi, 1978) and with individuality of male odors which may be important for recognition of unfamiliar males by females (Yamazaki et al., 1983).

Table 2. Effects of Castration and Testosterone Replacement on Concentrations of Esterases in the Urine of SWR/J Male Mice

	Mean Concentration[a] ± SE (N)[b]			
	Prior to Castration	Two Weeks After Castration	Two Weeks After Testosterone Replacement	P[c]
Esterase Peak 1	22.7 ± 2.9	9.3 ± 4.9	61.8 ± 6.9	<.001
(Lowest Molecular Weight Esterase)	(10)	(3)	(10)	
Esterase Peak 2	25.2 ± 3.3	4.3 ± 3.4	42.2 ± 1.9	<.001
	(10)	(3)	(10)	
P[c]	0.58 (ns)	0.39 (ns)	.01	

[a]Distance (in mm) from the base of an esterase peak to its apex on densitometer scan.

[b]Number of males that excreted urine esterases.

[c]Probabilities as calculated from one-way analyses of variance.

Although other investigators have reported that 10J males are androgen-deficient and unable to block pregnancies (Furudate and Nakano, 1980; Stalvey and Payne, 1983), males in this strain showed patterns of esterase release intermediate to that of 6J and the other two C57 strains. However, like 6J males, a low percentage of 10J subjects excreted detectable concentrations of lmwe (Table 1). Because of these inconsistencies, it may be incorrect to assume that all C57BL-derived strains are unable to block pregnancies.

The significant strain and caging differences found for migration, distances of Peaks 2 and 3 may have resulted from our separation procedures. We did not remove sediment or standardize the concentration of individual urine samples before subjecting them to electrophoresis. Thus, more concentrated samples may have migrated through gels more slowly than dilute samples. When migration ratios of Peak 1 to Peak 2 are analyzed, significant strain differences are eliminated.

To our knowledge, no published reports specify which of the esterases found in other mouse tissues are excreted in the urine. Of the five somatic esterase loci for which all strains employed in this study have been typed, nonblockers are isogenic at all of them. Nonblockers and blockers show allelic differences at four loci and are completely isogenic at Es-2 (Heiniger and Dorey, 1980). Significant differences in migration of Peak 2 in blockers but not in nonblockers may be related to this allelic heterogeneity.

The pheromone(s) responsible for pregnancy blocking have been linked both to protein (MW 17,500) and polypeptide (MW 860) fractions of male urine (Marchlewska-Koj, 1981). Our data suggest that urine esterases, and particularly the lmwe, may be associated with this pheromone. Esterase concentration in male urine is strain-dependent. 6J and 10J males produce insufficient amounts of pheromone to induce the Bruce effect (Shire and Bartke, 1972; Hoppe, 1975) and also release less esterase in their urine (and virtually no lmwe) compared to males from other strains. The amount of esterase secreted by a male may also depend on social environment (e.g., dominance hierarchies). In six strains (KsJ, 6J, 10J, BALB, CBA, and DBA), one male per cage typically released lmwe when grouped. Because male social status and release of urine esterases are both androgen-dependent, future work should establish whether there is any relationship between male social status and the numbers and concentrations of esterases in the urine of male mice.

Although the role of urine esterases in pregnancy blocking has not yet been demonstrated, we suggest that these enzymes (particularly the lmwe) may activate pheromonal precursor(s). Testosterone surges resulting from dominance status or exposure to the urine of an estrous female (McKinney and Desjardins, 1973) may increase esterase excretion which, in turn, may enhance the biological activity of protein-bound lipids or sterols in the urine. Our collection, separation, and staining procedures demonstrate that urine esterases are enzymatically active for at least 18 hours (16h of urine collection and 1.5 to 2h of electrophoretic separation). Prolonged esterase activity may explain why extremely small volumes of male urine are sufficient for inducing estrus and blocking pregnancies (Novotny et al., 1980). If urine esterases are associated with pregnancy block or other types of estrus induction, we hypothesize that it should be possible to demonstrate this effect using males from any strain whose urine contains the lmwe, including strains supposedly unable to induce the Bruce effect. KsJ and 6By males are likely candidates for testing this hypothesis. Preliminary data from ongoing experiments suggest that KsJ males can block pregnancies in BALB/cByJ females (Labov, unpublished). Understanding the physiological and behavioral bases for the release of urine esterase may reveal new insights about the pheromone responsible for pregnancy blocking.

ACKNOWLEDGMENTS

This work was supported by the Natioal Science Foundation (PRM-8109719), by a Grant-in-Aid of Research from Sigma Xi, and by grants from the Andrew Mellon Foundation and Colby College to the senior author. The Jackson Laboratory graciously provided the animals used in these studies.

REFERENCES

Bartke, A., and Shire, J. G. M., 1972, Differences between mouse strains in testicular cholesterol levels and androgen target organs, J. Endocrinol., 55:173.

Bruce, H. M., 1959, An exteroceptive block to pregnancy in the mouse, Nature, 184:105.

Furudate, S., and Nakano, T., 1980, Studies on the pheromonal pregnancy block in the mice. I. Strain differences of mice in pheromonal pregnancy block, Exp. Anim., 29:341.

Heiniger, H., and Dorey, J. J., eds., 1980, "Handbook on Genetically Standardized Jax Mice, 3rd ed.," The Jackson Laboratory, Bar Harbor.

Hoppe, P. C., 1975, Genetic and endocrine studies of the pregnancy-blocking pheromone of mice, J. Reprod. Fert., 45:109.

Huck, U. W., 1982, Pregnancy block in laboratory mice as a function of male social status, J. Reprod. Fert., 66:181.

Ivanyi, P., 1978, Some aspects of the H-2 system, the major histocompatability system in the mouse, Proc. R. Soc. Lond., 202:117.

Labov, J. B., 1979, "Pregnancy Blocking in House Mice (Mus musculus) and Other Mammals: Sociobiological Implications and Adaptive Advantages for Females," Ph.D. Dissertation, University of Rhode Island, Kingston, R.I.

Leiter, E. H., Coleman, D. L., and Hummel, K. P., 1981, The influence of genetic background on the expression of mutations at the diabetes locus in the mouse. III, Effect of H-2 haplotype and sex. Diabetes 30:1029.

Lombardi, J. R., and Vandenbergh, J. G., 1977, Pheromonally induced sexual maturation in females: Regulation of the social environment of the male, Science, 196:545.

Marchlewska-Koj, A., 1981, Pregnancy block elicited by male urinary peptides in mice, J. Reprod. Fert., 61:221.

Marchlewska-Koj, A., 1983, Pregnancy blocking by pheromones, in: "Pheromones and Reproduction in Mammals," J. G. Vandenbergh, ed., Academic Press, New York.

McKinney, T. D., and Desjardins, C., 1973, Intermale stimuli and testicular function in adult and immature house mice, Biol. Reprod., 9:370.

McPartland, R. P., Sustarsic, D. L., and Rapp, J. P., 1981, Evidence for an androgen-dependent urinary arginine esterase in the rat: Separation from other urinary arginine esterases including kallikrein, Endocrinology, 108:1634.

Novotny, M., Jorgenson, J. W., Carmack, M., Wilson, S. R., Boyse, E. A., Yamazaki, K., Wilson, M., Beamer, W., and Whitten, W. K., 1980, Chemical studies of the primer mouse pheromones, in: "Chemical Signals: Vertebrates and Aquatic Invertebrates," D. Muller-Schwarze and R. M. Silverstein, eds., Plenum Press, New York.

Schadler, M. H., 1981, Postimplantation abortion in pine voles (Microtus pinetorum) induced by strange males and pheromones of strange males, Biol. Reprod., 25:295.

Selander, R. K., Smith, M. H., Yang, S. Y., Johnson, W. E., and Gentry, J. B., 1971, Biochemical polymorphism and systematics in the Genus Peromyscus. I. Variation in the old-field mouse (Peromyscus polionotus), Studies in Genetics VI, University of Texas Publication, Austin.

Shaw, C. R., and Koen, A. L., 1963, Hormone-induced esterase in mouse kidney, Science, 140:70.

Shire, J. G. M., and Bartke, A., 1972, Strain differences in testicular weight and spermatogenesis with special references to C57BL/10J and DBA/2J mice, J. Endocrinol., 55:163.

Stalvey, J. R. D., and Payne, A. H., 1983, Luteinizing hormone receptors and testosterone production in whole testes and purified Leydig cells from the mouse: Differences among inbred strains, Endocrinology, 112:1696.

Stehn, R. A., and Richmond, M. E., 1975, Male-induced pregnancy termination in the prairie vole, Microtus ochrogaster, Science, 187:1211.

Svare, B., and Mann, M., 1981, Infanticide: Genetic, developmental and hormonal influences in mice, Physiol. Behav., 27:921.

Yamazaki, K., Beauchamp, G. K., Wysocki, C. J., Bard, J., Thomas, L., and Boyse, E. A., 1983, Recognition of H-2 types in relation to the blocking of pregnancy in mice, Science, 221:186.

THE MAMMALIAN VOMERONASAL SYSTEM:

ITS ROLE IN LEARNING AND SOCIAL BEHAVIORS

Charles J. Wysocki[1], N. Jay Bean[1,2] and Gary K. Beauchamp[1]

[1]Monell Chemical Senses Center, 3500 Market Street
Philadalphia, PA 19104 and [2]Department of Psychology
Vassar College, Poughkeepsie, NY 12601

INTRODUCTION

Over the past several years, we have conducted behavioral work with mice and guinea pigs to define the functions of the mammalian vomeronasal system. This chapter will review the status of some of our ongoing research by summarizing the results of experiments designed to determine the role of the vomeronasal system in chemosensory investigation of female guinea pig urine by male guinea pigs, male mouse reproductive behavior, aggression among mice, and chemosensory-evoked ultrasonic vocalizations in mice. We will show that the vomeronasal system appears to be critical for maintaining certain behaviors; in its absence, chemosensory-mediated responses diminish or extinguish. Hence, we suggest that stimulation of the vomeronasal system is itself inherently rewarding. We will also demonstrate that deafferentation of the vomeronasal system neonatally or prior to experience is more detrimental to an animal than deafferentation subsequent to experience and will propose that the importance of stimuli detected by chemosensory systems other than the vomeronasal system, most likely olfaction, is learned during social encounters. These learned cues can later elicit behaviors in the absence of a functional vomeronasal system.

Vomeronasal Organ Structure and Neural Connections

The vomeronasal organ contains sensory afferents of a system that appears to be especially important in the detection of socially relevant chemical signals. The organ is physically separate from the olfactory epithelium and resides at the base of the nasal septum or in the hard palate. In most mammals it is a bilateral structure, containing two blindly ending fluid-filled lumens. Surrounding each lumen are two types of epithelium; pseudostratified non-sensory cells form the lateral wall and neuroepithelial microvillous sensory cells line the medial wall. Chemical stimuli have access to the sensory epithelium via a narrow duct into the nose, into the roof of the mouth or into the nasopalatine canal (which connects the nasal and oral cavities), depending upon the particular species. Stimuli, including nonvolatile substances (Ladewig and Hart, 1980; Wysocki et al., 1980), apparently can reach the receptors via mucus transport in the nose following direct contact between nostril and stimulus, suction into the organ lumen by a vascular pump (Meredith and O'Connell, 1979, Eccles, 1982), transport by accessory structures (the tongue and/or nasopalatine canal), and possibly as vapors in the air. The neural signals from the receptors pass via the

Table 1. Some Effects of Lesion of the Vomeronasal System.[a]

Response	Brief Description
Puberty	Accelerated onset of puberty in female mice and rats cannot be obtained upon exposure to male chemosignals.
	Group housed female mice that lack their vomeronasal organs do not produce the expected maturation inhibiting substance.
Female Cyclicity	Estrous synchrony in group housed female mice is disrupted.
Hormone Surges	Male mice and female rats will not show a surge in LH (males and females) or testosterone (males) upon exposure to chemosignals of the opposite sex.
Pregnancy Block	Recently inseminated female mice exposed to chemosignals from males will not show the expected resorption of embryos.
Courtship and Investigation	Vocalizations to female chemosignals from experienced and inexperienced male mice are reduced and eliminated, respectively.
	Headbobbing to female chemosignals by male guinea pigs extinguishes after repeated presentations.
Mating	Many male hamsters and mice fail to mate.
Maternal Behavior	Nulliparous rats more readily accept neonates, but primiparous mice are unaffected.
Aggression	Aggression is substantially reduced.

[a]Also see Johns (1980), Keverne (1983), Meredith (1983), and Wysocki (1979).

vomeronasal nerve and the accessory olfactory bulb to the amygdala, bed of nucleus of the stria terminalis, preoptic area and hypothalamus (for references see Wysocki, 1979), areas of the brain previously shown to be critical for reproductive success (see Wysocki, 1979).

Functions of the Vomeronasal System

Information about sex, reproductive condition and social status is all coded in chemical messages secreted or excreted by members of various species (see contributions in the three previous publications in this series). In addition to their information content, chemicals released by animals also affect physiological responses, and the vomeronasal organ seems to be particularly involved in reception of these chemical signals (pheromones). Many pheromonal effects depend on an intact vomeronasal system (see Table 1); they do not occur or are much reduced if the vomeronasal system of the recipient animal is rendered non-functional. Indeed, the organ has been implicated in the control of reproduction, especially in the onset of sexual maturity, the synchronization of mating behavior and the regulation of reproduction by population factors, e.g., density.

Experimental work has revealed a complex relationship between the vomeronasal and olfactory systems. Although the vomeronasal system mediates most of the classical primer effects, olfaction appears to regulate, in part, the expression of at least one. Normally, the reproductive systems of rats do not respond to shortened daylengths; however, removal of the olfactory and accessory olfactory bulbs or intranasal zinc sulfate treatment, but not removal of the vomeronasal organ, induces photoperiodism in rats (Nelson et al., 1985). Furthermore, many behavioral responses to chemical cues are affected by lesion of the vomeronasal system (see Table 1). Hence, the vomeronasal system and olfaction modulate both primer and signaller effects, and may interact depending upon past experience. This latter contention has been pursued in our laboratories in studies of animals lacking the sensory afferents of the vomeronasal system.

REMOVAL OF THE VOMERONASAL ORGAN

We deafferented the vomeronasal system by surgical extirpation of the vomeronasal organ (VNX). The methods have been published (Wysocki et al., 1982; Beauchamp et al., 1982); hence only a brief summary is provided.

Adults

Prior to surgery, each animal was injected with a general anesthetic chosen for the species under study. The subjects were then secured in a stereotaxic head-holder in a supine position, the jaws were retracted, and a midline incision was made in the palate from the incisors to the first palatal ridge. The soft tissues were then retracted and, as needed to free the organ, the incisive bones were drilled with a dental bit, or pressure was applied to the vomeronasal organ. With the use of forceps and the aid of a dissecting microscope, the entire vomeronasal organ complex was then removed, and the incision was closed with cyanoacrylate glue or sutures.

Neonates

Recently, we modified our surgical technique for use with one day mice. On the first day following parturition, pups were individually removed from the nest, weighed, marked for identification and placed upon crushed ice for approximately 2 minutes for anesthetization. The pup was then placed supine in a specially designed hollow holder and its head fitted into a modeling clay headrest to maintain a consistent orientation. To maintain adequate anesthesia throughout surgery, the holder was packed with crushed ice, the pup was then secured with dermicil tape, and the holder, with pup, was attached onto a tilted platform. Access to the vomeronasal organ was achieved through the palate and the organ was removed with microforceps with the aid of a 25X dissecting microscope. Following removal of the organ, the wound was sealed with cyanoacrylate glue and the pup was placed on a heating pad for at least one-half hour. Pups were then returned to the nest and growth rates were monitored (see Figure 1).

LEARNING: EXTINCTION, REINFORCEMENT AND CONDITIONING

Guinea Pig Chemoinvestigation

Male guinea pigs will approach a small amount of female guinea pig urine and investigate it while engaging in a distinctive behavior - headbobbing (Beauchamp, 1973). After contacting the female urine, the male apparently

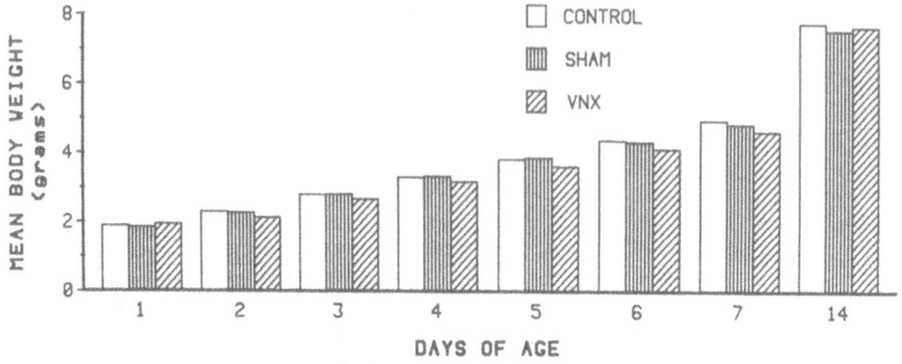

Fig. 1. Weight gains for neonatally-treated mice. One third of the pups had their vomeronasal organs removed (VNX), one third received a sham (SHAM) manipulation that included all but organ removal and the remaining pups served as anesthetic controls (CONTROL). Every attempt was made to assign 2 pups/sex/condition/litter. The surgical manipulation of the pups appeared to be quite benign; the lesioned pups maintained their presurgery body weights and exhibited a growth rate that was equivalent to the rates of the two control groups (Bean and Wysocki, 1985).

samples the urinary constituents with his vomeronasal organ; a nonvolatile dye placed in the urine was later found in the male's vomeronasal organ (Wysocki et al., 1980). Once contact is made with the urine, the male begins nodding his head at a rate of 3-5 cycles/sec. The behavior is present at the first presentation of female urine irrespective of the male's experience with females and is resistant to extinction provided that i) the behavior is elicited no more frequently than every 2-3 days and ii) urine from different female donors is used (Beauchamp and Wellington, 1984). These characteristics suggest that stimulation by female chemical cues may be inherently rewarding to male guinea pigs.

The vomeronasal organ is essential for maintenance of the guinea pigs's behavior. After removal of the vomeronasal organ, responses by male guinea pigs abate over repeated trials. Figure 2 summarizes the results obtained from SHAM and VNX guinea pigs on the first two tests after surgery. On the intial test, there is no difference between the groups; however a significant diminution in response was detected as rapidly as the second trial, conducted 3 days after the first.

Although unlikely, these observations could reflect surgical sequelae (e.g., changes in brain organization). However, the results of another study (Beauchamp et al., 1985), led us to conclude that the decline in response was extinction, resulting from repeated testing. Three groups of male guinea pigs were formed: one SHAM and two VNX. The SHAM and VNX_1 groups were tested repeatedly over a 15 week period. As in Figure 2, responses of the VNX_1 group declined during the period. Again, comparisons between these two groups alone cannot discriminate between sequelae or extinction. A second VNX group (VNX_2) was created. These males had their vomeronasal organs removed at the same time as the VNX_1 group, but they were not tested during the 15 weeks. The response of the VNX_2 group was critical; if it was low, then passage of time alone (sequelae) could account for the results. However, if the animals responded at control levels and the behavior

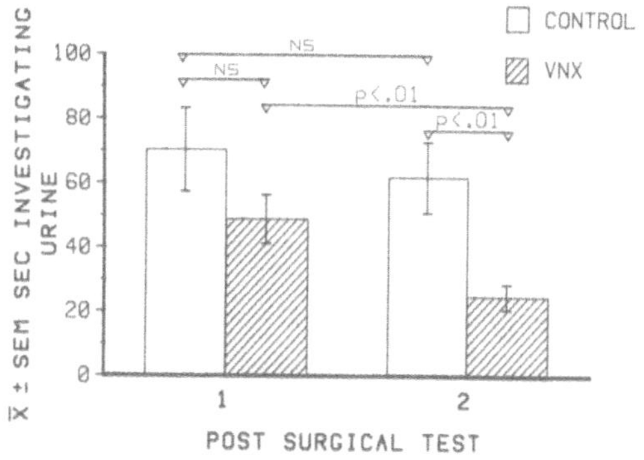

Fig. 2. Time spent investigating female urine by male guinea pigs on the first two trials subsequent to surgery. Note that on the first test, the groups did not differ. However, on the second test, the VNX group, but not the SHAMs, decreased their level of responding.

declined much like that of the VNX_1 group, then this would support an extinction hypothesis. After 15 weeks of trials, members of the VNX_2 group began post-surgical testing. Their initial responses were virtually equivalent to those of the SHAM group; however, like the VNX_1 group, the response declined over repeated trials, i.e., it extinguished. Although the response waned, we determined that males with nonfunctional vomeronasal systems could still smell the urine and could differentiate urines based on sex, species and individual identity of the donors (Beauchamp et al., 1983).

Extinction of the behavioral response to female guinea pig urine in males with a nonfunctional vomeronasal system lead us to conclude that stimulation of the vomeronasal system is inherently rewarding; removal of the vomeronasal organ eliminated reception of a chemical cue having reinforcing properties. Although the behavior, presumably mediated by conditioned olfactory cues, was initially at control levels, it rapidly dissipated upon repeated presentations of the conditioned cues in the absence of reception of the reinforcing cue.

A similar argument has been presented by Halpern and Kubie (1984) who interpreted the results of their earlier work to support a reward/extinction hypothesis. In their experiments on prey trailing by garter snakes, Kubie and Halpern (1979) used bits of worm as a reward for snakes following an appropriate trail. Rendering the vomeronasal system nonfunctional decreased trailing accuracy to chance levels. However, after vomeronasal nerve lesions, the snakes also extinguished attacking the bits of food presented in the goal box of the maze (Kubie and Halpern, 1979); and resistance to extinction was greater in snakes having had more experience with the bits (Halpern and Kubie, 1984). The authors hypothesized "that the vomeronasal stimulation provided by worm bit cues is inherently rewarding." Unfortunately, as a food, worm bits may also provide primary reinforcement, hence, dissociation of food reinforcers and other inherent reinforcers may be difficult with this bioassay.

Mouse Reproductive Behavior

We also observed that sex behavior appeared to wane in VNX mice subsequent to repeated exposures to females. The neonatally treated mice (Figure 1) underwent tests of reproductive behavior during adulthood (Bean and

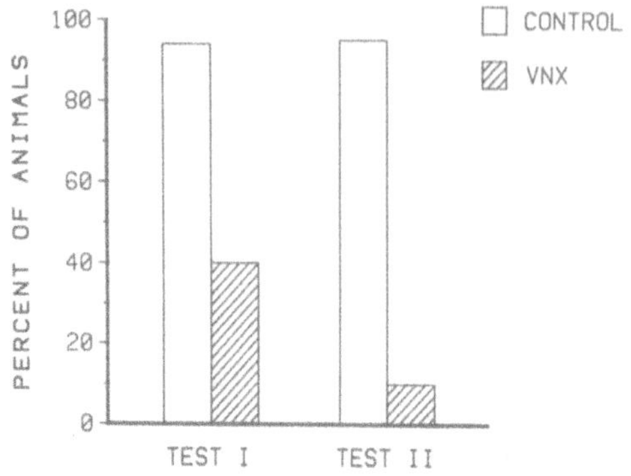

Fig. 3. Percent of males exhibiting reproductive behavior, or mounting a receptive female in a 20-min test session. Each animal was tested twice; the second test occurred after 2 weeks of cohabitation with a female. Note the further decrease in the VNX group during the second test, suggestive of extinction.

Wysocki, 1985). Not surprisingly, since these mice had not had adult experience with females, removal of the vomeronasal organ had a significant effect; only 40% of the VNX animals mounted a receptive female in the first 20-min test session, while over 90% of the control animals mounted (Test I in Figure 3). This deficit was even more pronounced during a second test session which followed 2 weeks of cohabitation with a female. During this test, only 10% of the VNX animals attempted to mount the receptive female, while 95% of the controls mounted (Test II in Figure 3). Furthermore, the VNX animals that responded did so with much longer latencies than did the animals in the other groups. This further significant reduction in behavior by the VNX mice also supports the extinction hypothesis elaborated above.

Aggression in Mice

Learning and chemosensory cues both play significant roles in aggression among mice. To date, only two reports on the effects of deafferentation of the vomeronasal system on aggression have been published. Bean (1982) found profound effects of deafferentation of the vomeronasal system on all measures of agonistic behaviors in mice. Fights, attacks, bites, tail-rattles, and total fight time all were reduced in lesioned males. Clancy et al. (1984) also reported that removal of the vomeronasal organ had a significant impact on intermale aggression. While tail-rattling occurred in half of the VNX males, the majority of them neither bit nor chased the intruder. In our laboratory we continue to see profound effects on aggressive behavior after removal of the vomeronasal organ; but the extent of the deficit depends upon the extent of experience prior to surgery. Learning, with an intact vomeronasal system, plays a significant role during subsequent potentially agonistic encounters.

In each test of aggressive behavior with inexperienced males, we have seen marked deficits following removal of the vomeronasal organ (unpublished observations). When tested in adulthood, neonatal organ removal resulted in a significant decrease in each measure of aggression. Aggression tests were conducted using a castrated intruder whose flanks were smeared with urine from a gonadally intact male. Only 20% of the lesioned animals either fought or attacked while over 85% of the SHAM animals were aggressive

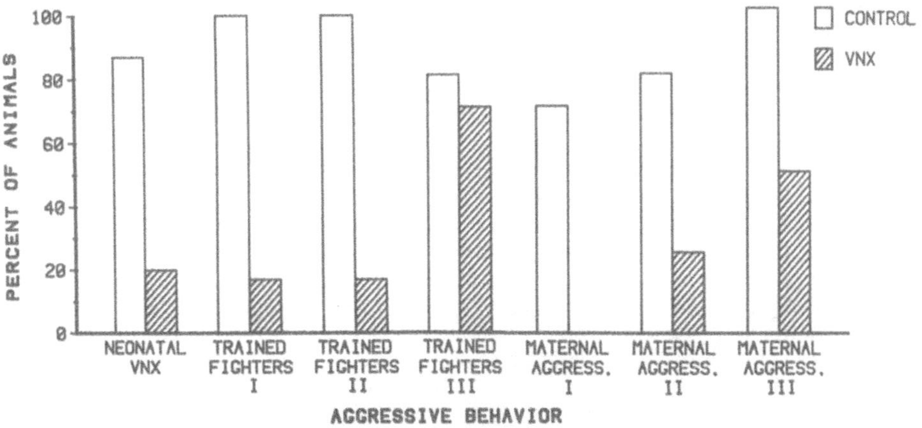

Fig. 4. Percent of animals responding with at least one attack or fight during a 5-min test session. Males in Trained Fighter I and Trained Fighter II are the same; those in Trained Fighter III are different. The females in Maternal Aggression III were the aggressive SHAM females from Maternal Aggression I. See text for details.

(Neonatal VNX in Figure 4; unpublished observations). Because of the magnitude of this deficit coupled with our observations that experience prior to vomeronasal organ removal appeared to spare behavioral responsivity in other situations (see Mouse Ultrasonic Vocalization) we decided to assess the effects of vomeronasal organ removal on aggression following extensive fighting experience. Sexually experienced, individually-housed ICR outbred male mice (Perfection Breeders) were trained to fight using the dangling method (Scott, 1966). All males received 3 consecutive days of training and a 5-min pretest with an intact male prior to surgery. Approximately one week after surgery, the animals were again tested for aggressiveness towards a castrated adult male mouse whose flanks were smeared with urine from a gonadally intact male. Each of the SHAM animals was highly aggressive, but only 17% of the VNX animals attacked or fought (Trained Fighters I in Figure 4; unpublished observations). These same subjects were again tested following 3 additional days of training. The same surprisingly low number of lesioned animals fought, while each of the controls further increased their aggressive behavior (Trained Fighters II in Figure 6, below; unpublished observations). The results of this second test were even more surprising than the first because all of the animals were highly aggressive toward the dangled animal during the training sessions. We were also puzzled by the lack of aggression from what we considered to be highly experienced males.

Although repeated training after surgery did not increase levels of intermale aggression, additional experience prior to removal of the vomeronasal organ appears to facilitate aggression once the vomeronasal organ has been removed. CF-1 outbred male mice (Charles Rivers) were trained by the dangling method for 8 pre-surgical days, rather than 3. After recovery, the males were presented with a castrated male painted with urine from a gonadally intact male. The VNX males were as aggressive as the SHAM males (Trained Fighters III in Figure 4; unpublished observations). Apparently the extended training period provided the sufficient conditions to maintain aggression subsequent to vomeronasal organ removal, however, the source of the mice cannot be ruled out as a confounding factor. Whether learning of olfactory cues or perhaps possible genetic differences account for our observations remains to be determined.

Removal of the vomeronasal organ does not affect basal levels of sex-related hormones (Wysocki et al., 1983; c.f., Sanchez-Criado et al., 1984), however, hormonal responses to chemosensory stimuli are not normal. Male mice do not exhibit a surge in LH (Coquelin et al., 1984) or testosterone (Wysocki et al., 1983) upon exposure to female urine or to anesthetized females respectively, and female rats do not show the expected surge in LH upon exposure to a male rat (Beltramino and Taleisnik, 1983). Since aggression in males is sensitive to circulating androgens, and we did not determine plasma levels of testosterone in all of our aggression studies, the hypothesis that changes in plasma testosterone were responsible for our observed results cannot be discounted. However, this appears unlikely; the SHAM, CONTROL and VNX groups formed during neonatal treatment had equivalent levels of plasma testosterone soon after their aggression test (unpublished observations).

In summary, removal of the vomeronasal organ has dramatic impact on intermale aggression, but its effects are conditional; experience apparently modulates the vomeronasal system's impact upon the expression of aggression. Reduction or elimination of aggression in males with little or no experience most likely results from sensory loss, while the maintenance of aggression in extensively trained males may reflect conditioning, perhaps with olfactory cues. Preliminary evidence suggests that repeated testing of agonistic VNX males produces less aggressive mice while SHAM mice continue to fight. This extinction effect (if replicated) lends additional support to a learning hypothesis.

The effects of vomeronasal organ removal on aggression by lactating females were also assessed and we observed a similar pattern of deficits. Lactating female mice normally exhibit a high incidence of aggression toward intruder males that are introduced into the female's cage. We tested primiparous females who had undergone vomeronasal organ removal prior to mating. After 10 days to recover from surgery, each female was paired with a male. After 2 weeks of cohabitation, 85% of the SHAM and 80% of the VNX females were pregnant (these data agree with other investigations on the role of the vomeronasal organ in female reproduction; Lepri et al., 1985).

Tests of aggression were conducted 3 days after parturition. None of the VNX females attacked or fought with a gonadally intact male, while most of the SHAM females attacked and fought at high levels with short latencies to the first attack (Maternal Aggression I in Figure 4; unpublished observations). To determine whether this effect resulted from the VNX females' inability, prior to parturition, to experience important cues mediating the appearance of aggression, we allowed another group of females to mate prior to surgery. The females were paired with males for 7 days, underwent surgery, and were returned with the male for an additional 7 days. Tests of aggression were conducted 3 days after parturition. The results of this experiment were very similar to the previous experiment. Females in the SHAM group were highly aggressive, while few of the VNX females were observed to fight or attack the intruders (Maternal Aggression II in Figure 4; unpublished observations). Apparently, experience with males prior to surgery was not sufficient to induce post-partum aggression directed at intruder males. We therefore assessed the effects of experience with aggression on subsequent occurrences of aggression.

The aggressive primiparous females in the SHAM group of the first maternal aggression study were split into two groups matched for aggressiveness and either had their vomeronasal organ removed or underwent the sham manipulation. Following a one week recovery period, the females were remated and maternal aggression was again assessed following parturition. All of the SHAM females were highly aggressive toward the intruders, but now, half of the VNX females fought or attacked the male (Maternal Aggression II in Figure 4; unpublished observations). When compared with the SHAM females, however, the VNX females that fought or attacked did so at a much reduced level on all measures. Aggression from females having had experience with aggression prior to removal of the vomeronasal organ supports the hypothesis that the vomeronasal system is involved in learning of cues which later elicit social behaviors.

Mouse Ultrasonic Vocalization

Chemosensory cues and learning also mediate the expression of male-generated ultrasonic vocalizations. In general, male mice emit a 70 kHz vocalization in the presence of a female or female-generated chemical cues, but not in the presence of male associated stimuli (see review by Nyby and Whitney, 1978). Females, on the other hand, rarely produce the vocalizations, but, in the presence of other females, they will if they are pretreated with testosterone (Nyby et al., 1977), or if they have not yet had experience with male mice (Maggio, personal communication). The male's vocalization is thought to play a role in courtship (Whitney et al., 1973), especially during post-partum estrus, but its exact function remains unclear. Females are more attracted to males that produce the vocalizations than to mute males and will select the side of a cage in which vocalizations are electronically produced (Pomerantz et al., 1983). However, attracting females may also be costly to males; vocalizing males elicit more aggression from lactating females than do mute males (Bean et al., 1985). Vocalizations occur upon the first presentation of a female (Nyby et al., 1977; Wysocki, 1982) and can be readily elicited by the female's urine, but only after the male has had 1-3, 3-min encounters with females (Dizinno et al., 1978), suggesting that the urinary cues are learned. [Nyby et al. (1983) suggested that lack of arousal accounted for non-responsiveness to urinary cues by heterosexually inexperienced male mice. These males vocalized, albeit at reduced levels, after they had had social interactions with castrated males that were anointed with female urine.] Additional support for a learning hypothesis comes from two investigations on the hormonal control of the female vocalization-eliciting cue. Males that have had experience with normal females emit high levels of vocalizations when given urine from normal females. These same males respond to the urine from hypophysectomized females at control levels (Nyby et al., 1979), implicating the pituitary in the production of the cue. However, males that are provided social interactions with hypophysectomized females emit high levels of vocalizations upon presentation of chemical cues from the hypophysectomized females (Maggio et al., 1984), suggesting the occurrence of cue-learning during the encounters. Using ultrasonic emissions as an assay, we also have explored the role of the vomeronasal organ in cue-learning during social encounters among mice.

The vomeronasal organ has been implicated in the detection of the female cue that elicits ultrasounds. We previously determined that removal of the vomeronasal organ had a detrimental effect upon the production of high frequency vocalizations by male mice in response to female-generated chemical cues (Bean, 1982; Wysocki et al., 1982); however, the extent of the deficit appeared to depend upon prior heterosexual experience. Males having had encounters with females prior to vomeronasal organ removal produced vocalizations in response to anesthetized females, albeit at reduced levels relative to their SHAM counterparts (Wysocki et al., 1982). The VNX males also vocalized more than SHAM males upon presentation of an anesthetized male. A much greater deficit, in many instances a lack of response, was observed in the behavior of males that were heterosexually naive prior to surgery (Bean, 1982; Wysocki, 1982; Wysocki et al., 1982).

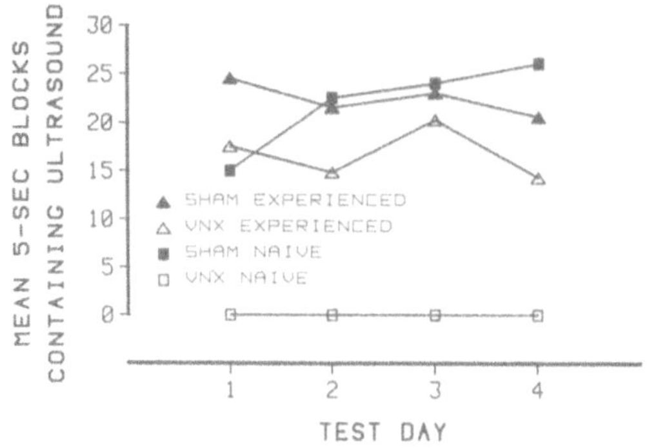

Fig. 5. Mean 5-sec blocks of ultrasonic vocalization produced by heterosexually experienced and naive male mice over four trials. Trial 1 represents the first contact since weaning with a female for the inexperienced males (modified from Wysocki, 1982).

Figure 5 summarizes the effects of vomeronasal organ removal in adulthood on responses to anesthetized females by heterosexually experienced or inexperienced males. Half of the males had their vomeronasal organs removed and the remainder were intact. As shown, inexperienced males lacking a vomeronasal organ suffered a signficant deficit in their vocalization response in the presence of normally adequate stimuli.

In many situations, neonatal insult to neural tissue is less traumatic than damage in adulthood (however, see Brunjes, 1985, and cited references). Neural plasticity is robust in neonates, and removal of the vomeronasal organ in newborn mice might have little effect upon behaviors that are normally mediated by the vomeronasal system in adulthood. Plasticity and reorganization might allow other systems to mediate the behavior. Alternatively, if the vomeronasal organ is critical for learning and expression of the behavior, especially during the initial encounters with females, then neonatal vomeronasal organ removal might be expected to mimic removal of the organ from inexperienced adult males.

The neonatally treated mice were tested in adulthood for ultrasonic vocalizations in the presence of an anesthetized female. Significantly fewer of the VNX males responded to the female than did the animals in the other two groups (SHAM and CONTROL; see Anesth. Female in Figure 6). Upon repeated testing, this deficit became even more apparent. During the first of two 20-min tests for reproductive behavior with a receptive female, the number of animals in the VNX group vocalizaing to the females during the first 3 min of the test was significantly less than the number responding in the other groups (see Awake Female I in Figure 6). This deficit further increased when the males were tested following 2 weeks of cohabitation with a female (see Awake Female II in Figure 6). During this third test only 10% of the males in the VNX group emitted high frequency vocalizations to females. In each of these tests the males showed interest in the females and typically groomed and sniffed the female for the majority of the test session, but the vocalizations apparently extinguished.

If female-generated olfactory cues are learned during the male's encounters with females, then, subsequent to removal of the vomeronasal organ, much like guinea pigs, the response by male mice to the conditioned cues should extinguish. This hypothesis was tested by forming two groups of mice,

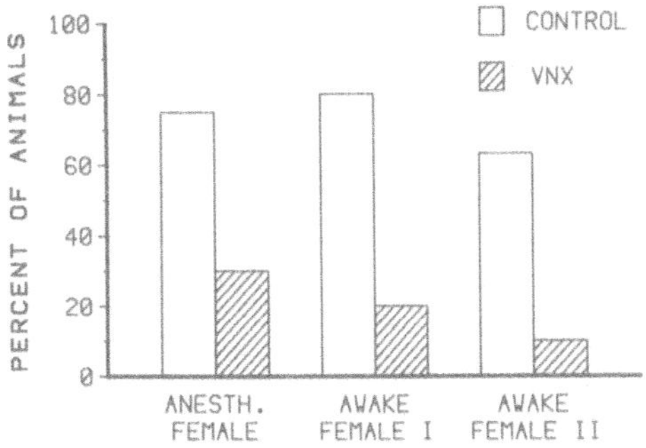

Fig. 6. Percent of males responding with ultrasonic vocalizations in more than 5 of the 5-sec blocks of the 3-min trial with the stimulus female indicated. The same males were tested in all three conditions.

VNX and SHAM, each having had experience with females. Subsequent to the surgical treatment, each group was repeatedly tested for their responses to anesthetized females (urine was not used because intact males extinguish responses to urinary cues; Dizinno et al., 1978). Unfortunately, both groups of mice responded in parallel fashion; their presurgical mean high scores (SHAM = 30, VNX = 31) declined over the seven postsurgical trials (SHAM = 16, VNX = 11). Since the response declined in normal mice, it was difficult to evaluate the possibility of extinction in VNX males. Learning, however, remains an important component in the expression of ultrasonic vocalizations (Nyby and Whitney, 1983).

The effects of vomeronasal organ removal in adulthood or in neonates suggest that the vomeronasal organ is important in the detection of female-generated, unlearned chemical cues that elicit ultrasonic vocalizations. However, the data from experienced males show that the vomeronasal organ is not always necessary for the expression of the behavior. During encounters with females, males apparently learn the importance of other cues that come to elicit responses from them. We suggest that the vomeronasal organ detects an unconditioned stimulus, and during encounters with females (and stimulation of the vomeronasal system) odors, detected by the olfactory system, become conditioned stimuli. If the vomeronasal organ is removed after conditioning has occurred, then the conditioned stimuli will maintain the behavior, at least initially. However, if the vomeronasal organ is removed prior to conditioning, then conditioning does not take place and odors do not acquire the status normally attained in males with a functional vomeronasal system. [The unconditioned and conditioned cues need not be different chemicals. When detected via the vomeronasal system the substance may act as an unconditioned stimulus, and when detected by experienced males via the olfactory system the substance may act as a conditioned stimulus.] This hypothesis was put to a more direct test.

In two separate studies (one testing experienced males, the other inexperienced; unpublished observations) 6 groups of male mice were tested with 3 different stimuli: female mouse urine, male mouse urine, and a perfume (Coty Wild Musk Spray) previously shown to acquire vocalization-eliciting properties, presumably via conditioning (Nyby et al., 1978). Three groups had heterosexual experience prior to surgery, three after surgery. Each set of three contained one SHAM group that experienced ethanol-annointed females (a stimulus control group), one SHAM group that experienced perfume-annointed females (a surgery control group), and one VNX

Table 2. Mean 5-sec Blocks Containing Ultrasounds From Males Having Undergone Treatment and Experience Indicated.

	Stimulus		
Group[a]	Male Urine	Female Urine	Perfume
ETOH/SHAM	0	13	0
Perfume/SHAM	0	14	11
Perfume/VNX	2	0	11
SHAM/ETOH	0	22	0
SHAM/Perfume	3	16	13
VNX/Perfume	1	4	0

[a]The sequence of the experience and surgical treatment is indicated by the order of presentation. For example, mice in the ETOH/SHAM group experienced ethanol-annointed females and then underwent a sham surgical treatment.

group, the members of which experienced perfume-annointed females. It is important to note the sequence of events. For one of the VNX groups (perfume/VNX), experience with the odorized females occurred prior to removal of the vomeronasal organ; for the other (VNX/perfume), experience with the odorized females occurred after removal.

As expected, the SHAM groups exposed to ethanol-annointed females responded to female urine but not to male urine or perfume (Table 2). Hence, the perfume does not act as an unconditioned stimulus. Also as expected (Nyby et al., 1978), the perfume on a cotton swab did elicit responses from the members of the two SHAM groups exposed to the perfume-annointed females. The two critical VNX groups responded differently. The perfume/VNX group, which experienced perfume-annointed females prior to removal of the vomeronasal organ, responded to the perfume in the absence of a vomeronasal organ. Surprisingly, they did not respond to female urine. We suggest that the volatile compounds in the urine that are normally detected by the olfactory system were masked by the perfume during heterosexual encounters and conditioning did not occur; when tested alone, the volatile components of the female urine were ineffective. The responses of the VNX/perfume group are most interesting, especially when compared with those of the perfume/VNX group. The perfume did not elicit significant amounts of vocalizations from the males that lacked their vomeronasal organ during experience with the odorized female. We suggest that, in the absence of the vomeronasal organ, the unconditioned stimulus was never detected as such, and no conditioning occurred; the perfume did not acquire the status of a conditioned stimulus.

CONCLUSIONS

Fewer than 15 years ago, after the demonstration of independent vomeronasal and olfactory system projection pathways (Winans and Scalia, 1970), the mammalian vomeronasal organ was a structure in search of a function. Clearly, our knowledge of the vomeronasal system has increased dramatically. This chemosensory system now appears to be critical for learning chemosensory cues important in normal social functioning. Furthermore, stimulation of the vomeronasal system appears to provide inherent rewards. We speculate that this latter provision mediates the conditioning of other odor cues during social encounters.

ACKNOWLEDGMENTS

This research was supported by the National Science Foundation (BNS 83-16437 to CJW, BNS 82-01759 to GKB, and supplemental sabbatical funding to NJB) and institutional funds from the Monell Chemical Senses Center and Vassar College. We thank Linda M. Wysocki, Ronald Bernhard, Christine Wojciechowski-Metzler and Susan Yocum for technical assistance.

REFERENCES

Bean, N. J., 1982, Olfactory and vomeronasal mediation of ultrasonic vocalizations in male mice, Physiol. Behav., 28:31.

Bean, N. J., 1982, Modulation of agonistic behavior by the dual olfactory system in male mice, Physiol. Behav., 29:433.

Bean, N. J., Nunez, A. A., and Wysocki, C. J., 1985, 70kHz vocalizations by male mice do not inhibit aggression in lactating mice, Behav. Neural Biol., under review.

Bean, N. J., and Wysocki, C. J., 1985, Behavioral effects of removal of the vomeronasal organ in neonatal mice, Chem. Senses, in press.

Beauchamp, G. K., 1973, Attraction of male guinea pigs to conspecific urine, Physiol. Behav., 10:589.

Beauchamp, G. K., Martin, I. G., Wysocki, C. J., and Wellington, J. L., 1982, Chemoinvestigatory and sexual behavior of male guinea pigs following vomeronasal organ removal, Physiol. Behav., 29:329.

Beauchamp, G. K., and Wellington, J. L., 1984, Habituation to individual odors occurs following brief, widely spaced presentations, Physiol. Behav., 32:269.

Beauchamp, G. K., Wellington, J. L., Wysocki, C. J., Brand, J. G., Kubie, J. L., and Smith, A. B., III, 1983, Chemical communication in the guinea pig: Urinary components of low volatility and their access to the vomeronasal organ, in: "Chemical Signals: Vertebrates and Aquatic Invertebrates," D. Muller-Schwarze, and R. M. Silverstein, eds., Plenum Press, New York.

Beauchamp, G. K., Wysocki, C. J., and Wellington, J. L., 1985, Extinction of response to urine odor as a consequence of vomeronasal organ removal in male guinea pigs, Behav. Neurosci., in press.

Beltramino, C., and Taleisnik, S., 1983, Release of LH in the female rat by olfactory stimuli, Neuroendocrinology, 36:53.

Brunjes, P. C., 1985, Unilateral odor deprivation: Time course of changes in laminar volume, Brain Res. Bull., 14:233.

Clancy, A. N., Coquelin, A., Macrides, F., Gorski, R. A., and Noble, E. P., 1984, Sexual behavior and aggression in male mice: Involvement of the vomeronasal system, J. Neurosci., 4:2222.

Coquelin, A., Clancy, A. N., Macrides, F., Noble, E. P., and Gorski, R. A., 1984, Pheromonally induced release of luteinizing hormone in male mice: Involvement of the vomeronasal system, J. Neurosci., 4:2230.

Dizinno, G., Whitney, G., and Nyby, J., 1978, Ultrasonic vocalizations by male mice (Mus musculus) in response to a female-produced pheromone: Effects of experience, Behav. Biol., 22:104.

Eccles, R., 1982, Autonomic innervation of the vomeronasal organ of the cat, Physiol. Behav., 28:1011.

Halpern, M., and Kubie, J. L., 1984, The role of the ophidian vomeronasal system in species-typical behavior, Trends NeuroSci., 7:472.

Johns, M. A., 1980, The role of the vomeronasal system in mammalian reproductive physiology, in: "Chemical Signals in Vertebrates and Aquatic Invertebrates," D. Muller-Schwarze, and R. M. Silverstein, eds., Plenum, New York.

Keverne, E. B., 1979, Dual olfactory projections and their significance for behavior, in: "Chemical Ecology: Odour Communication in Animals," F. J. Ritter, ed., Elsevier, Amsterdam.

Keverne, E. B., 1983, Pheromonal influences on the endocrine regulation of reproduction, Trends NeuroSci., 6:381.

Kubie, J. L., and Halpern, M., 1979, Chemical senses involved in garter snake prey trailing, J. Comp. Physiol. Psychol., 93:648.

Ladewig, J., and Hart, B. L., 1980, Flehmen and vomeronasal organ function in male goats, Physiol. Behav., 24:1067.

Lepri, J., Wysocki, C. J., and Vandenbergh, J. G., 1985, Mouse vomeronasal organ: Effects on chemosignal production and maternal behavior, Physiol. Behav., in press.

Maggio, J. C., Maggio, J. H., and Whitney, G., 1983, Experience-based vocalization of male mice to female chemosignals, Physiol. Behav., 31:269.

Meredith, M., 1983, Sensory physiology of pheromone communication, in: "Pheromones and Reproduction in Mammals," J. G. Vandenbergh, ed., Academic Press, New York.

Meredith, M., and O'Connell, R. J., 1979, Efferent control of stimulus access to the hamster vomeronasal organ, J. Physiol., 268:301.

Nelson, R. S., Fleming, A. A., Wysocki, C. J., Shinder, T. W., and Zucker, I., 1985, Chemosensory and neural influences on photoperiodic responsiveness of laboratory rats, Neuroendocrinology, 40:285.

Nyby, J., Bigelow, J., Kerchner, M., and Barbehenn, F., 1983, Male mouse (Mus musculus) ultrasonic vocalizations to female urine: Why is heterosexual experience necessary? Behav. Neural Biol., 38:32.

Nyby, J., Dizinno, G., and Whitney, G., 1977, Sexual dimorphism in ultrasonic vocalizations of mice (Mus musculus): Gonadal hormone regulation, J. Comp. Physiol. Psychol., 91:1424.

Nyby, J., and Whitney, G., 1978, Ultrasonic communication of adult myomorph rodents, Neurosci. Biobehav. Rev., 2:1.

Nyby, J., and Whitney, G., 1983, Experience affects behavioral responses to sex odors, in: "Chemical Signals: Vertebrates and Aquatic Invertebrates," D. Muller-Schwarze, and R. M. Silverstein, eds., Plenum Press, New York.

Nyby, J., Whitney, G., Schmitz, S., and Dizinno, G., 1978, Postpubertal experience establishes signal value of mammalian sex odor, Behav. Biol., 22:545.

Nyby, J., Wysocki, C. J., Whitney, G., and Dizinno, G., 1977, Pheromonal regulation of male mouse ultrasonic courtship (Mus musculus), Anim. Behav., 25:333.

Nyby, J., Wysocki, C. J., Whitney, G., Dizinno, G., and Schneider, J., 1979, Elicitation of male mouse (Mus musculus) ultrasonic vocalizations: I. Urinary cues, J. Comp. Physiol. Psychol., 93:957.

Pomerantz, S. M., Nunez, A. A., and Bean, N. J., 1983, Female behavior is affected by male ultrasonic vocalizations in house mice, Physiol. Behav., 1983, 31:91.

Sanchez-Criado, J. E., Fernandez-Galaz, C., Vaticon, M. D., and Mora-Novard, O. A., 1984, Niveles plasmaticos de LH y testosterona en ratos macho sometidas a deaferentacion periferica del sistema vomeronasal, Rev. Esp. Fisiol., 40:359.

Scott, J. P., 1966, Agonistic behavior of mice and rats: A review, Am. Zool., 6:683.

Whitney, G., Coble, J. R., Stockton, M. D., and Tilson, E. F., 1973, Ultrasonic vocalizations: Do they facilitate courtship of mice? J. Comp. Physiol. Psychol., 84:445.

Winans, S. S., and Scalia, F., 1970, Amygdaloid nucleus: New afferent input from the vomeronasal organ, Science, 170:330.

Wysocki, C. J., 1979, Neurobehavioral evidence for the involvement of the vomeronasal system in mammalian reproduction, Neurosci. Biobehav. Rev., 3:301.

Wysocki, C. J., 1982, Non-olfactory and vomeronasal influences on reproduction, in: "Olfaction and Endocrine Regulation," W. Breipohl, eds., IRL Press, London.

Wysocki, C. J., Katz, Y., and Bernhard, R., 1983, Male vomeronasal organ mediates female-induced testosterone surges in mice, Biol. Reprod., 28:917.

Wysocki, C. J., Nyby, J., Whitney, G., Beauchamp, G. K., and Katz, Y., 1982, The vomeronasal organ: Primary role in mouse chemosensory gender recognition, Physiol. Behav., 29:315.

Wysocki, C. J., Wellington, J. L., and Beauchamp, G. K., 1980, Access of urinary nonvolatiles to the mammalian vomeronasal organ, Science, 207:781.

OLFACTORY COMMUNICATION AMONG RATS: INFORMATION CONCERNING DISTANT DIETS

Bennett G. Galef, Jr.

Department of Psychology
McMaster University
Hamilton, Ontario, Canada

INTRODUCTION

Animal Communication in the Absence of Display

The study of animal communication has traditionally focused on analyses of conspicuous displays, ritualized (Blest, 1961), or formalized (Smith, 1977) over evolutionary time for the function of intraspecific communication (see McKay, 1972, for discussion of problems of definition). However, particularly among vertebrates, communication need not involve such specialized signals.

As any organism engages in its routine daily activities, it coincidentally provides usable information to others. Stimuli emitted as incidental by-products of behaviors necessary for individual survival can serve communicative functions similar to those of ritualized displays. For example, the rasping sounds an agouti makes while gnawing on a nut, carry dozens of meters through the tropical forest and attract other agouti, as well as the occasional potential predator, to a promising feeding site (Smythe, 1970); a vulture, descending to scavenge from a carcass lying on the Serengeti plain, draws both others of its kind and competing species to a recent kill (Houston, 1979; Schaller, 1972). It does not seem reasonable to suppose that in such cases the discoverer of a resource is emitting signals specialized for communication to announce its find. Yet the feeding behavior of an agouti or vulture provides unambiguous messages to those sufficiently alert and behaviorally flexible to detect and exploit them. Stimuli generated by individuals engaging in life-sustaining activities can be utilized by others to facilitate their own acquisition of resources. Although such unspecialized signals have been discussed at length as contexts modifying the meaning of formalized displays (Smith, 1977), they have not been treated as primary forms of communication. Hence, the potential importance of unspecialized signals in mediating interaction among organisms has been little studied.

In the present chapter, I describe recent work from my laboratory, the results of which are consistent with the view that incidental by-products of ingestion are important media of communication among rats. Olfactory signals, passively emitted by rats following feeding, provide conspecifics with information that both facilitates and orients subsequent foraging by its recipients.

A number of behavioral ecologists have suggested that in social species that forage from a central location, such as a burrow, roost, or nesting area, unsuccessful foragers could enhance their own subsequent foraging efficiency by acquiring information from successful foragers encountered at the central site (DeGroot, 1980; Ward and Zahavi, 1973; Erwin, 1977; Bertram, 1978, Waltz, 1982). Wild Norway rats are social, central-place foragers. In natural circumstances, each rat lives as a member of a colony that inhabits a fixed system of burrows. When foraging, colony members disperse from their burrow, feed, and then return to it (Calhoun, 1967; Telle, 1966).

As part of a long-term study of the behavioral processes underlying social learning in Norway rats (see Galef, 1977; 1983; 1984 for reviews), my students and I have been examining the possibility that rat burrows serve as 'information-centres' (Ward and Zahavi, 1973), i.e., as aggregation sites at which information concerning current availability of foods in the larger environment is exchanged among foragers. As will be seen below, our data indicate that one Norway rat can use information acquired during interaction with conspecifics in orienting its own subsequent foraging excursions. One rat can exploit diet-identifying cues passively emitted by recently-fed others both to determine what to eat and where to seek food.

THE LABORATORY PARADIGM

The procedures used in the studies described below were designed to mimic situations in which a foraging rat ingests a food at some distance from its burrow, returns to its burrow, and then interacts with a burrow-mate. Our purpose was to discover whether, as a result of such interaction: (1) the burrow-mate could acquire information concerning the food the forager had eaten and (2) whether the burrow-mate would subsequently use that information either in selecting foods for ingestion or in orienting its own foraging behavior.

Of course, any laboratory analogue of a natural foraging situation will fail to reflect many of the environmental complexities faced by free-living animals. Experimental control frequently requires both reduction in the number of alternative courses of action available to subjects and replacement of spontaneous behavior by experimenter-initiated manipulations. Consequently, experiments such as those described below, though adequate to reveal behavioral capacitites that might be employed in natural circumstances, do not provide evidence that these capacities are actually used by free-living animals. Demonstration that rat burrows actually function in nature as information-centres requires field studies that have not yet been undertaken (Galef, 1984; Galef and Wigmore, 1983).

The Basic Experiment

During all experiments described below, our subjects were housed in same-sex pairs in cages divided in half by screen partitions. For purposes of exposition I will refer to one member of each pair as a "demonstrator" and the other as an "observer".

The basic experiment, schematized in Figure 1, was carried out in five steps:

Step 1. Demonstrator and observer were maintained together with *ad lib* access to Purina Laboratory Rodent chow and water for a 2-day period of familiarization with both apparatus and cage-mate.

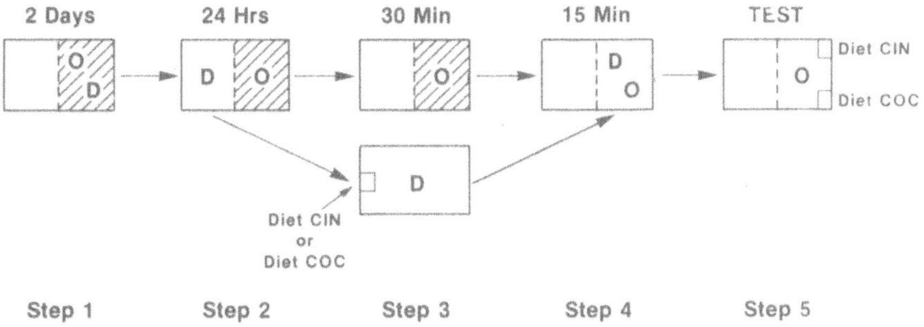

Fig. 1. Schematic diagram of the procedure of the basic experiment. 0 = observer; D = demonstrator; hatching indicates maintenance diet present in cage. (Galef and Wigmore, 1983. Copyright 1983 by Baillière Tindall. Reprinted by permission of the publisher and authors.)

Step 2. The demonstrator was moved to the opposite side of the screen partition from its observer and food-deprived for 24 hr to ensure that the demonstrator ate when given the opportunity to do so.

Step 3. Chow was removed from the observer's side of the cage (in preparation for testing) and the demonstrator was moved to a cage in a separate room and allowed to feed for 30 min on either cinnamon-flavored diet (Diet Cin) or cocoa-flavored diet (Diet Coc).

Step 4. The demonstrator was returned to the observer's side of the cage and demonstrator and observer were allowed to interact for 15 min.

Step 5. The demonstrator was removed from the experiment and the observer was offered a choice between two weighed food-cups, one containing Diet Cin and one containing Diet Coc.

Figure 2 shows the mean amount of Diet Coc, as a percentage of total amount eaten, ingested during testing (Step 5) by observers whose demonstrators had eaten either Diet Coc or Diet Cin during Step 4 of the experiment. As can be seen in Figure 2: (1) Those observers whose demonstrators ate Diet Coc ate a far greater percentage of Diet Coc than did those observers whose demonstrators ate Diet Cin and (2) effects of demonstrators' diet on observers' diet preference were still observable 48-60 hr after interaction of demonstrator and observer. The results of this first experiment clearly show that an observer rat can extract from a demonstrator information identifying the diet that demonstrator had eaten at a time and place distant from the locus of demonstrator-observer interaction. The data also show that this information is sufficient to bias its recipient's subsequent selection of diet.

Variations on a Theme

We have repeated the basic experiment described above many times: with a variety of different diets (Galef and Wigmore, 1983), with hungry and replete observers, with male demonstrator-observer pairs and female ones, with wild and domesticated rats, with demonstrator-observer pairs familiar with one-another and with pairs that had never met prior to their interaction during Step 4 of the experiment, with both old demonstrators and observers and young ones, and with observers selecting distinctively flavored fluids rather than solids for ingestion (Galef, Kennett and Wigmore, in press). In every case, we have seen robust preferences by

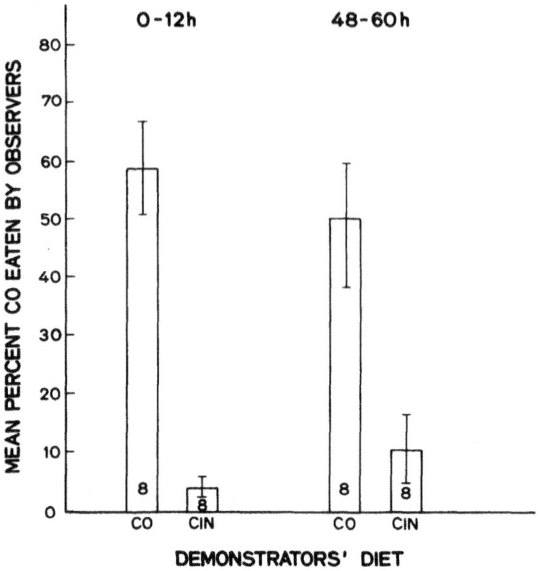

Fig. 2. Mean amount of cocoa-flavored diet ingested, as a percentage of total amount eaten, by observers whose demonstrators ate either cocoa- or cinnamon-flavored diet. CO = Cocoa-flavored diet; Cin = Cinnamon-flavored diet. (Galef and Wigmore, 1983. Copyright 1983 by Baillière Tindall. Reprinted by permission of the publisher and authors.)

observers for their respective demonstrators' diets. Similarly, Posadas-Andrews and Roper (1983) and Strupp and Levitsky (1984), using rather different paradigms, have repeatedly observed demonstrator influence on subsequent observer diet preference. The phenomenon of demonstrator influence on observer diet preference seems a general one in Norway rats, not dependent on some restricted set of experimental parameters for its expression.

Effects of the Passage of Time

In the basic experiment diagrammed in Figure 1, observers and demonstrators interacted immediately after demonstrators had eaten a diet. Observers had the opportunity to choose between diets immediately following extraction of information from their respective demonstrators. Free-living rats must expend time in returning from a feeding site to their burrows. In the field, foragers departing from their burrows must expend further time in reaching a feeding site. If the capacity of rats to transmit information concerning a food eaten at a distance from their burrow is to function in information exchange in natural settings, communication must occur even if there are delays both between a successful forager's ingestion of a food and its return to its burrow and between the interaction of a successful forager with other rats and the latters' arrival at a potential feeding site. In terms of the laboratory analogue illustrated in Figure 1, rats must be able to tolerate delays between Steps 3 and 4 and between Steps 4 and 5 and still successfully exchange information.

The results of experiments in which independent groups of subjects experienced varying delays (1) between a demonstrator feeding and its interaction with an observer and (2) between an observer interacting with a demonstrator and its choosing between diets are presented, respectively,

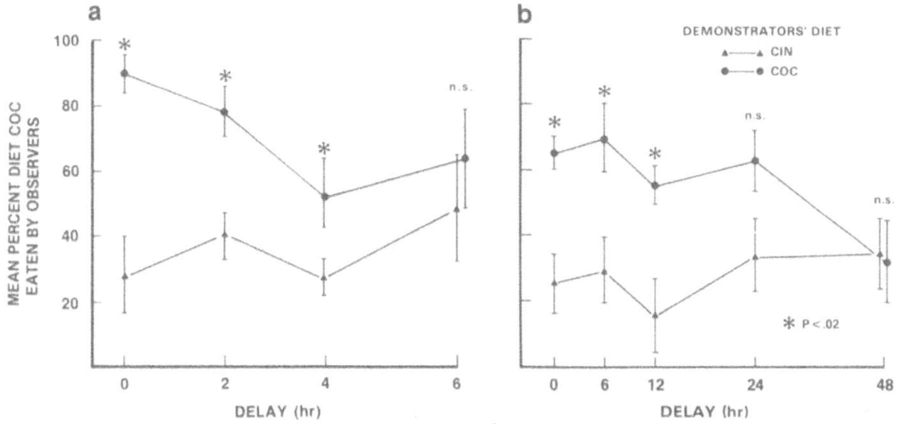

Fig. 3. Mean amount of cocoa-flavored diet ingested, as a percentage of total amount eaten, by observers whose demonstrators ate Diet Cin or Diet Coc: (3a) as a function of time between ingestion by demonstrator and interaction with observer (3b) as a function of time between interaction of demonstrator and observer and initiation of testing. (Galef and Wigmore, 1983. Copyright 1983 by Baillière Tindall. Reprinted by permission of the publisher and authors.)

in Figures 3a and 3b. As is clear from examination of the figures, and as statistical tests confirmed (Mann-Whitney U tests, see Figure 3 for p values), rats could tolerate considerable delays between Steps 3 and 4 or between Steps 4 and 5 of the basic experiment and still successfully exchange information.

As can be seen in Figure 3a, demonstrators continued to emit, for at least 4 hr after feeding on a diet, cues sufficient to permit an observer to identify its demonstrator's diet (Galef and Kennett, in press). Data presented in Figure 3b indicate that observers can use diet-identifying information obtained from demonstrators for 12 to 24 hr after receiving it (Galef, 1983). Both the time course of emission of diet-identifying information obtained from demonstrators seem appropriate to permit use of the information-transmission system under investigation in natural environments.

Handling of Multiple Messages

In natural circumstances, each rat lives within a social group or colony (Telle, 1966). While there is relatively little information available concerning social life in wild rat colonies, it seems reasonable to suppose that each colony member, prior to departing from its colony's burrow system on a foraging expedition, might have an opportunity to acquire information from several conspecifics about foods they had recently ingested. It is possible that an individual rat, remaining in its burrow and interacting with a succession of colony-mates returning from successful foraging trips, could collect information concerning the entire range of foods the returning foragers had exploited.

If rats in their burrows are to make use of information received from a succession of returning colony-mates, they must be able to distinctively encode and store information extracted from each informant. The results of several studies indicate that rats have such a capacity (Galef, 1983).

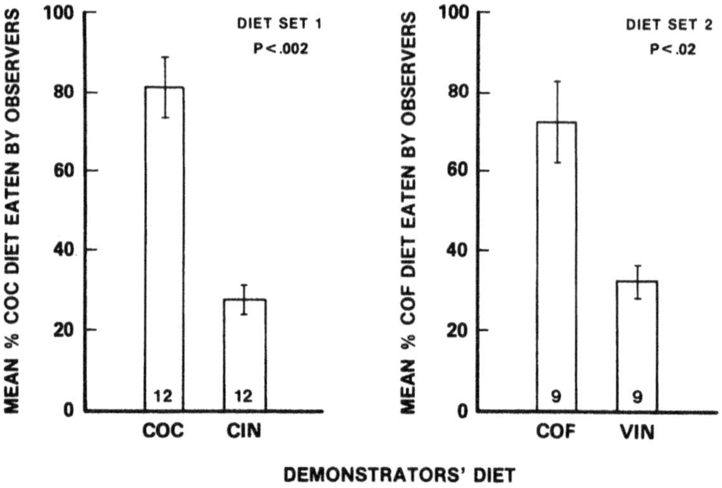

Fig. 4. Mean percent Diet Coc (left-hand panel) or Diet Cof (right-hand panel) eaten by observers one of whose demonstrators ingested, respectively, either Diet Cin or Coc, or either Diet Cof or Vin. (Galef, 1983. Copyright 1983 by the American Psychological Association. Reprinted by permission of the author and publisher.)

Our method was similar to that outlined in Figure 1. However, in the present experiment, each observer, instead of interacting during Step 4 with a single observer that had eaten either Diet Cin or Coc, interacted for 15 min with each of a series of four demonstrators, one of which had eaten vinegar flavored-diet (Diet Vin), one of which had eaten coffee-flavored diet (Diet Cof), one of which had eaten a casein and cornstarch based diet (Diet NPT), and one of which had eaten _either_ Diet Coc _or_ Diet Cin. During testing of observers (Step 5 of Figure 1), each observer was offered a choice between Diets Cin and Coc. As can be seen in the left-hand panel of Figure 4, those observers one of whose four demonstrators had eaten Diet Coc prefereed Diet Coc, while those observers one of whose four demonstrators had eaten Diet Cin, preferred that diet. Of course, it might have been the case that the two diets selected for testing (Diets Cin and Coc) were simply the most salient of those offered to demonstrators. To control for this possibility, the entire experiment was repeated using Diets Cin, Coc, and NPT, as irrelevant diets and Diets Cof and Vin as critical test items. As can be seen in the right-hand panel of Figure 4, those observers one of whose four demonstrators ingested Diet Cof preferred Diet Cof, while those observers one of whose four demonstrators ingested Diet Vin preferred Diet Vin.

Such preferences for the diet eaten by one demonstrator, embedded in a series of three masking demonstrators, indicate that observers are able to distinctively encode a number of diet-identifying signals sequentially extracted from demonstrators. These findings are consistent with the notion that a rat remaining in its burrow and interacting with a succession of returning successful foragers is able to construct an inventory of foods currently available in the larger environment and exploited by its fellows.

Use of Extracted Information in Orientation of Foraging

Although the experiments described above indicate that a successful

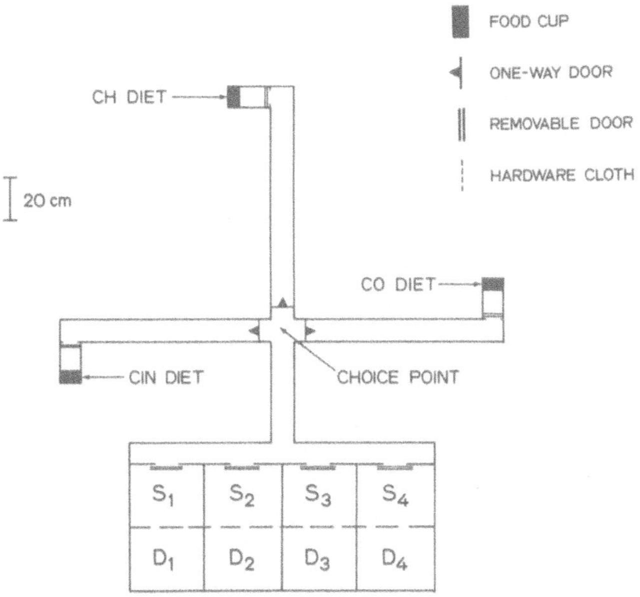

Fig. 5. Plan view of apparatus. Ch = cheese-flavored diet; Cin = cinnamon-flavored diet; Co = cocoa-flavored diet; S = subject; D = demonstrator. (Galef and Wigmore, 1983. Copyright 1983 by Baillière Tindall. Reprinted by permission of the publisher and authors.)

forager can provide information about diets it has ingested far from the locus of information transfer, these studies do not provide evidence that such information can be used by its recipients to facilitate their foraging. To investigate the usefulness of socially-transmitted information in increasing foraging efficiency, we introduced our subjects into the environment depicted in overhead schematic in Figure 5. The rule here was that each of three discriminable foods was available at a different, fixed location: cheese-flavored diet (CH Diet) in the central arm of the maze, cocoa-flavored diet (CO Diet) in the right arm, and cinnamon-flavored diet (CIN Diet) in the left. Only one of the three diets was accessible to each subject on any given day, and the particular diet available to a subject on any day was randomly selected.

Each subject (S_n in Figure 5) was given four trials/day using a correction procedure. On the first trial of each day, each subject had no information as to which food was available, and therefore had only one chance in three of selecting the correct arm of the maze. If it chose the correct arm, it ate for a few minutes. If it did not, it was locked in the arm it had chosen for a few minutes, and the first trial was repeated until the subject found the food. Trials 2, 3, and 4 of each day were run in the same fashion.

Each subject could, in effect, tell the experimenter when it understood this little world by exhibiting more or less perfect performance in its first choices on trials 2, 3, and 4 of each day. Once a given subject had reached the necessary criterion of performance on trials 2, 3, and 4, testing was instituted.

On each test day, for 15 min prior to Trial 1 of that day, each subject was allowed to interact with a demonstrator rat (D_n in Figure 5) that

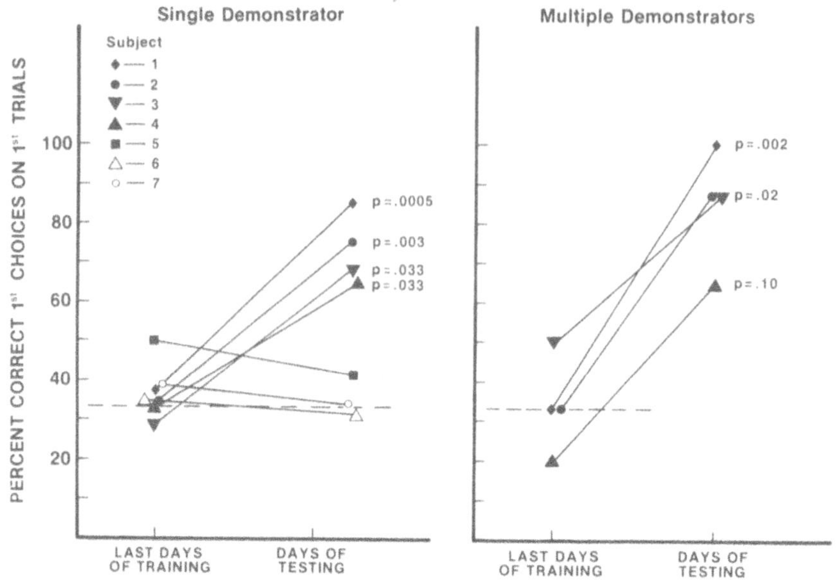

Fig. 6. Percentage of correct first choices on first trials by subjects at the end of training and during testing in the apparatus illustrated in Figure 5. (Galef, 1983; Galef and Wigmore, 1983. Copyrights 1983 by Baillière Tindall and the American Psychological Association. Reprinted by permission of the authors and publishers.)

had eaten the diet that was going to be available to that subject on that day. That is, if cheese-flavored diet was going to be available to S_3 on a given day, S_3's demonstrator, D_3, was fed cheese-flavored diet for 30 min and then allowed to interact with S_3 for 15 min prior to initiating Trial 1 of testing of S_3.

In order to determine whether subjects were capable of using information acquired from demonstrators to enhance foraging efficiency, we compared the probability of a correct response on the first choice of the first trial of each day of testing (when information from a demonstrator was available to subjects) with the probability of a correct response on the first choice of the first trial of each of the last days of training (when noninformation from a demonstrator was available to subjects). As can be seen in the left-hand panel of Figure 6, four of our seven subjects were able to use the information provided by their respective demonstrators to facilitate location of unpredictable foods (Galef and Wigmore, 1983).

As indicated in the immediately preceding section, rats can distinctively encode diet-identifying information extracted from a series of conspecifics. We have also found that they can use such diet-identifying information, embedded in a series of like messages, to orient foraging trips. During testing, instead of allowing each subject to interact with a single demonstrator predicting the food to be available on a given day, we had each subject interact with four demonstrators, three of which had eaten irrelevant foods (Diets NPT, Vin, and Cof) and one of which had eaten the food to be available to the subject on that day. As can be seen in the right-hand panel of Figure 6, subjects were still able to extract and use the relevant information in selecting an arm of the maze for initial exploration on test days (Galef, 1983).

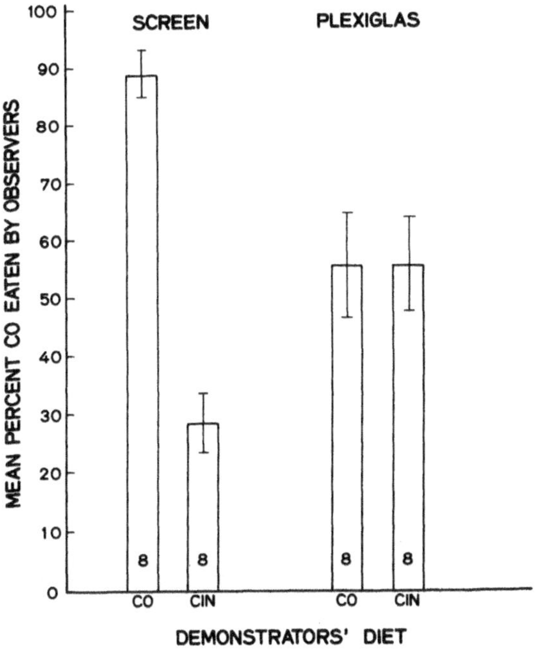

Fig. 7. Mean amount of cocoa-flavored diet ingested, as a percentage of total amount eaten, by observers whose demonstrators ate cinnamon- or cocoa-flavored diet. Left-hand bars, observer and demonstrator separated by a screen partition during interaction. Right-hand bars, observer and demonstrator separated by a Plexiglas partition during interaction. (Galef and Wigmore, 1983. Copyright by Baillière Tindall. Reprinted by permission of the publisher and authors.)

Taken together, the results of the experiments described above indicate that rats returning from a successful foraging trip can provide information to conspecifics concerning foods they have exploited. Further, the data indicate that information acquired by one rat from another can be used to facilitate foraging in situations in which a food is available intermittently at fixed locations.

THE NATURE OF MESSAGES PASSING FROM DEMONSTRATORS TO OBSERVERS

Implication of Olfactory Signals

An obvious question arising from the observations described above concerns the means by which an observer rat acquires information from a demonstrator as to the diet that demonstrator has been eating. We have developed several converging lines of evidence, each consistent with the hypothesis that olfactory cues passing from demonstrator to observer are sufficient to allow observer identification of demonstrators' diets.

In order to examine the mode of communication of diet-identifying information from demonstrator to observer, it was necessary to gain some control over their interaction. We employed a procedure similar to that depicted in Figure 1, but with one important modification. During the period of demonstrator-observer interaction (Step 4 of Figure 1), the members of each demonstrator-observer pair were on opposite sides of the

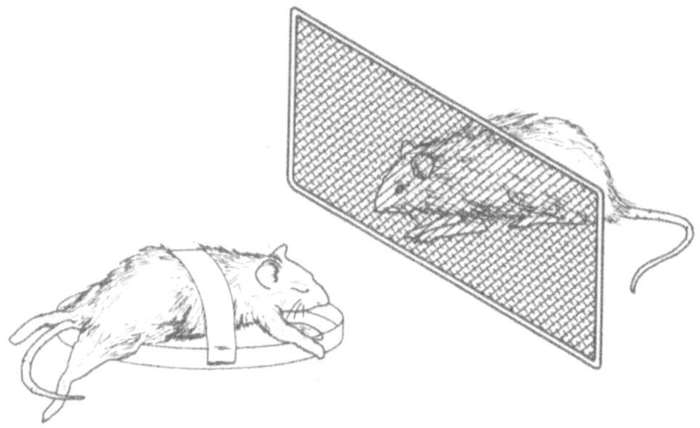

Fig. 8. Illustration of procedure during interaction of anesthetized demonstrator and observer.

screen partition dividing their cage.

As can be seen in Figure 7, observers had no trouble in developing a preference for their respective demonstrators' diets when separated from their demonstrators by a screen during interaction. However as can also be seen in Figure 7, if the screen partition was replaced by a clear Plexiglas partition, demonstrator influence on observer diet preference was completely abolished (Galef and Wigmore, 1983).

Further, we have conducted an experiment in which each demonstrator, after eating either Diet Cin or Diet Coc, was anesthetized and placed 5 cm from and facing a screen partition, with its observer on the other side of the screen (see Figure 8). During subsequent preference testing, observers still exhibited (Figure 9) a robust preference for their respective demonstrators' diets (Galef and Wigmore, 1983). Also, as one would expect if olfactory cues play an important role in information transfer between demonstrator and observer, observers rendered anosmic by passing Zinc-sulfate solution through their nares (Alberts and Galef, 1973) failed to exhibit a preference for their respective demonstrators' diets. Control rats whose nasal passages had been rinsed with saline solution continued to exhibit a preference for their respective demonstrators' diets (See Figure 10).

Last, but not least, humans, as well as rats, can use olfactory cues emitted by a previously-fed rat to tell what diet that rat has been eating. A human observer presented with a dozen rats in random sequence, half of which had eaten Diet Coc and half of which had eaten Diet Cin, could, by sniffing their breaths, tell with better than 85 percent accuracy which rat had eaten which diet (Galef and Wigmore, 1983).

CAUSES OF DEMONSTRATOR INFLUENCE ON OBSERVER DIET PREFERENCE

The simplest behavioral explanation of the observed influence of demonstrators on observers' subsequent diet selection would be something like the following. Rats are always somewhat hesitant to ingest unfamiliar foods (Barnett, 1958; Galef, 1970). An observer rat that, as the result of interacting with a demonstrator that has eaten a Diet X, has been exposed to cues associated with Diet X, should be familiar with Diet X. Therefore, an observer rat that has interacted with a Diet-X-fed demonstrator should eat Diet X in preference to other roughly equipalatable,

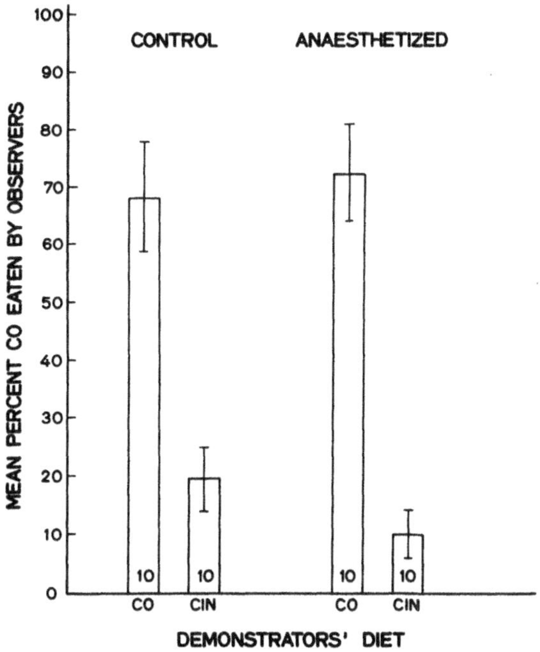

Fig. 9. Mean amount of Diet Coc eaten by observers, as a percentage of total amount ingested. Left-hand bars, observers interacting with intact demonstrators. Right-hand bars, observers interacting with anesthetized demonstrators, as depicted in Figure 8. (Galef and Wigmore, 1983. Copyright 1983 by Baillière Tindall. Reprinted by permission of the publisher and authors.)

but totally unfamiliar, diets. Demonstrator influence on observer diet preference could, thus, be explained as the result of a simple increase in observer familiarity with a demonstrator's diet.

We have conducted a number of experiments designed to test the adequacy of explanations of demonstrator influence on observer diet preference in terms of demonstrator-induced diet familiarity of the sort described above (Galef, Kennett, and Stein, in press). In every case, the results of our studies have been contrary to the most straight-forward predictions from the familiarity hypothesis.

For example, if reduced diet novelty, resulting from observer exposure to diet-identifying cues during interaction with a demonstrator, were responsible for subsequent demonstrator influence on observer diet preference, one would expect observers choosing between two familiar diets for be relatively immune to demonstrator influence in their diet selection. Any additional familiarity with one test diet, resulting from a brief period of interaction with a demonstrator fed that diet, should be overwhelmed by observers' extensive previous experience with both test diets.

We conducted an experiment much like that outlined in Figure 1 except that during the 2-day period of familiarization (Step 1 of Figure 1), observers in a Cin/Coc Pre-exposure Group were left alone in their cages and allowed ad lib access to two food bowls, one containing Diet Cin and one containing Diet Coc. Following two days of feeding on Diets Cin and Coc, each observer was exposed for 15-min to an unfamiliar demonstrator that had eaten either Diet Coc or Diet Cin (Step 4 of Figure 1). Each

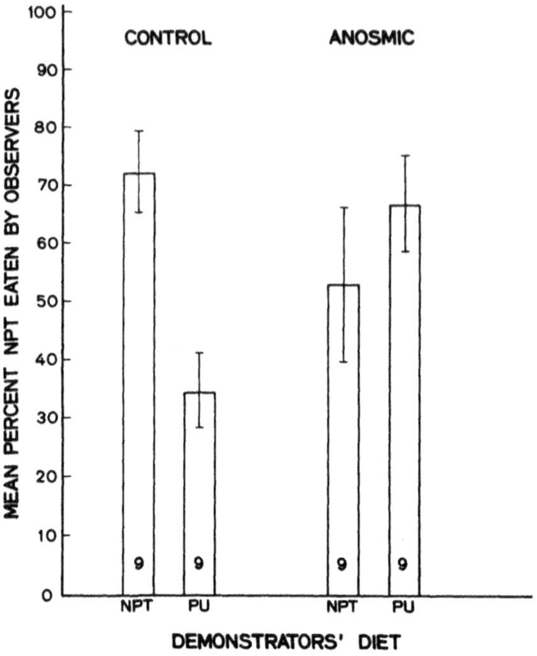

Fig. 10. Mean amount of Diet NPT ingested, as a percentage of total amount eaten, by observers whose demonstrators ate either Diet NPT or Pu. Left-hand bars, observers' nasal cavities rinsed with saline. Right-hand bars, observers' nasal cavities rinsed with $ZnSO_4$, i.e., anosmic. (Galef and Wigmore, 1983. Copyright 1983 by Baillière Tindall. Reprinted by permission of the publisher and authors.)

observer was then tested for its preference between Diets Coc and Cin (Step 5 of Figure 1). Observers in the Control group were treated identically to those in the Cin/Coc Pre-Exposure Group except that during familiarization (Step 1 of Figure 1) observers in the Control Group had access to two food bowls containing a powdered form of their standard maintenance diet.

The main results of the experiment are presented in Figure 11 which shows the mean amount of Diet Coc, as a percentage of total amount eaten, ingested by observers in Cin/Coc Pre-exposure and Control Groups. As can be seen in the figure, the diet eaten by demonstrators profoundly affected the food choice of observers in both groups. This finding renders unlikely interpretation of the effects of demonstrator influence on observer diet preference as resulting solely from familiarity with the taste or small of a diet experienced during 15 min of interaction with a demonstrator. Further, simply feeding a rat either Diet Cin or Coc for 15 min (or, for that matter, 24-hr) prior to offering it a choice between Diets Cin and Coc had no effect on rats' subsequent choice of diet (Galef, Kennett, and Stein, in press).

On the basis of such data, it is difficult to maintain the hypothesis that demonstrator influence on observer diet preference is the result of a simple increase in observers' familiarity with their respective demonstrators' diets. An obvious alternative is that observer experience of diet-identifying cues within a context provided by the presence of a demonstrator is necessary, if experience of diet-identifying cues is to

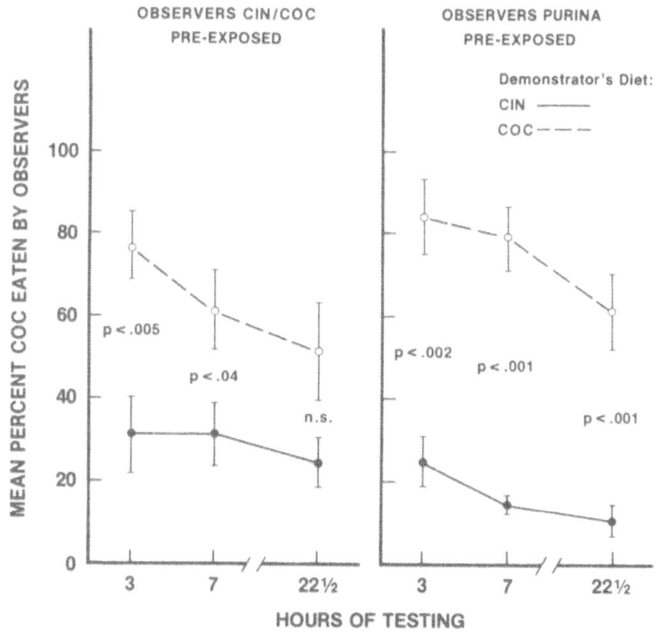

Fig. 11. Mean amount of Diet Coc ingested, by observers either pre-exposed or not pre-exposed to Diets Cin and Coc and whose demonstrators ate either Diets Cin or Coc. (Galef, Kennett and Stein, 1985. Copyright 1985 by the Psychonomic Society, Inc. Reprinted by permission of the authors and publishers.)

alter observers' subsequent diet preference. It is this hypothesis that has been the guiding principle in our recent research.

Analysis of Olfactory Cues

Assume, for the sake of argument, that I am correct in asserting that the effects of demonstrator rats on the subsequent dietary preferences of their observers are the result of observers experiencing the smell of a diet within the sensory context provided by a demonstrator, rather than the result of simple exposure to diet-related cues during interaction with a demonstrator. If so, then analysis of the message passing from demonstrator to observer presents two separable problems: (1) determination of the source of the diet-identifying cues emitted by demonstrators and (2) determination of the source and nature of the contextual cues, also emitted by demonstrators, that act in concert with the diet-identifying cues to alter observers' subsequent diet preference.

Diet identifying cues. In order to look more closely at the cues involved in demonstrator influence on observer diet preference, we again changed our experimental procedures slightly. The new procedure was similar to that described in Figure 1, but differed both in the way in which demonstrators were made to emit diet-identifying cues (Step 3) and the treatment of demonstrators and observers during the period of their interaction (Step 4). Rather than feed all demonstrators during Step 3 of the procedure, we employed a variety of techniques for attaching diet-identifying cues to demonstrators. Some demonstrators were allowed to eat Diet Cin or Diet Coc for 30 min, as was done in the basic experiment. Other demonstrators were anesthetized and their faces dusted with either

Fig. 12. Illustration of apparatus used to analyze diet-identifying and contextual cues. (Galef and Stein, 1985. Copyright 1985 by the Psychonomic Society, Inc. Reprinted by permission of the authors and publisher.)

Diet Cin or Diet Coc. Yet other demonstrators were anesthetized and tube-fed one of two distinctively flavored fluids. Some observers, instead of interacting with a demonstrator during Step 4 of the procedure, spent 30 min interacting with a surrogate rat (constructed of cotton-batting and surgical gauze) one end of which had been dusted with either Diet Cin or Coc. Further, instead of permitting demonstrator and observer to interact freely during Step 4, each demonstrator was placed, following anesthetization, in the screen tube of the apparatus illustrated in Figure 12. Observers were introduced into the bucket-shaped area of the enclosure, left there for 30 min, then moved back to their respective home-cages for testing (Step 5 of Figure 1).

As can be seen in Figure 13, 30 min observer interaction with a surrogate rat, dusted with either Diet Coc or Cin, failed to affect observer diet preference during testing (Step 5), providing further evidence of the inadequacy of simple exposure to a diet to produce alterations in observer diet preference. In contrast, exposure to a fed demonstrator, an anesthetized demonstrator powdered with diet, or an anesthetized demonstrator tube-fed a flavored solution each had the capacity to alter observers' diet selection during testing.

The finding that diet applied to the face of a demonstrator enhanced diet preference in their observers indicates that ingestion of a diet by a demonstrator is not critical in demonstrator production of diet-identifying cues. The finding that demonstrators stomach-loaded with a flavored solution also induced observers to increase their preference for the solution placed in the stomach of demonstrators indicates that particles of food clinging to the fur and vibrissae of demonstrators are not necessary for transmission of diet-identifying information to observers. Taken together the results of the present study (Galef, Kennett, and Stein, in press; Galef and Stein, in press) indicate that both particles of food on the exterior of rats and portions of diet in the stomach of rats provide cues sufficient to permit observers to identify their respective demonstrators' diets.

<u>Contextual cues</u>. The results of the studies presented in Figure 13 also suggest that a demonstrator rat provides a context within which exposure to diet-identifying cues alters observers' subsequent diet preference. Observer preference was not affected by exposure to a diet

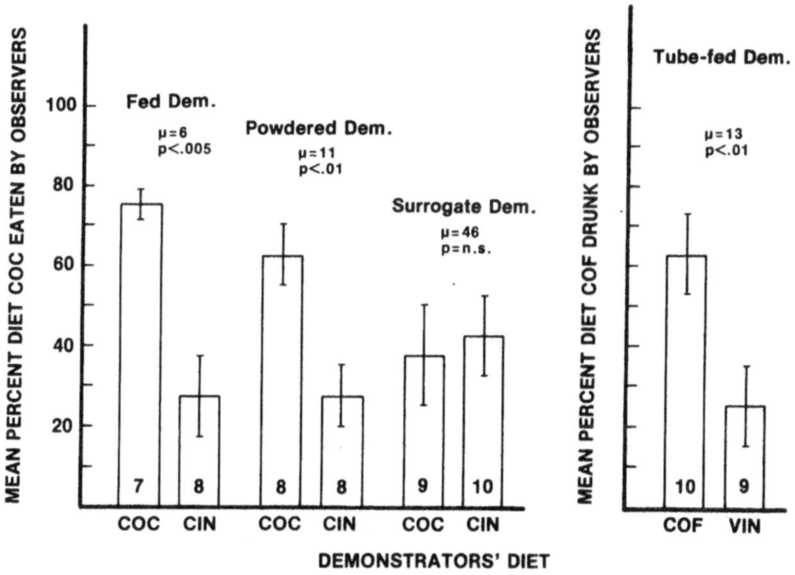

Fig. 13. Left-hand panel: Mean amount of cocoa-flavored diet eaten, as a percentage of total amount ingested, by observers interacting with demonstrators or surrogates. Right-hand panel: Mean amount of coffee-flavored solution, as a percentage of total amount drunk, ingested by observers whose demonstrators were tube-fed with either coffee- or vinegar-flavored solution.

presented on a surrogate, but was affected by exposure to the same diet presented on the face of a rat. Further, the procedure employed in these studies provides an opportunity to define more precisely the nature of the contextual cues which, in combination with diet identifying cues, produce demonstrator influence on observer diet preference.

We allowed observers to interact for 30 min in the apparatus illustrated in Figure 12 with demonstrators treated in one of four ways: (1) Observers in the Powdered-face Group interacted with anesthetized demonstrators whose faces had been rolled in either Diet Cin or Diet Coc. (2) Observers in the Dead-powdered-face Group interacted with demonstrators that had been sacrificed by anesthetic overdose and had their faces rolled in either Diet Cin or Diet Coc prior to the demonstrators introduction into the apparatus. (3) Observers in the Powdered-rear Group, interacted with anesthetized demonstrators whose rear ends were rolled in Diet Cin or Diet Coc. These demonstrators were introduced into the screen tube of the apparatus illustrated in Figure 12 with their rear ends inside the bucket and their heads outside of it. Last, (4) observers in the Surrogate Group were allowed to interact in the apparatus with a rat-size cotton-batting stuffed length of tubular gauze one end of which had been rolled in either Diet Cin or Diet Coc.

Figure 14 presents a measure of the degree of influence of the various sorts of demonstrators on their respective observers' subsequent diet preferences during testing (Step 5 of Figure 1). The greater the Cocoa-demonstrator/Cinnamon-demonstrator ratio the greater the influence of demonstrators' diets on observers' subsequent diet preference.

To summarize the results of a series of statistical analyses discussed in detail elsewhere (Galef and Stein, in press): (1) Observers in

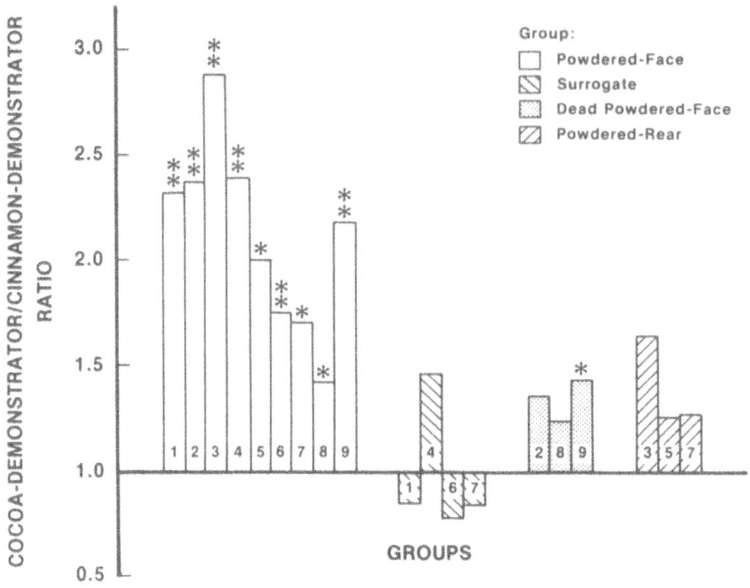

Fig. 14. Cocoa-demonstrator/Cinnamon demonstrator ratios of groups of observers interacting with demonstrators and surrogates in the apparatus illustrated in Figure 12. Observers were randomly assigned across groups labeled with the same integer. * = $p <$ .05, ** = $p < .01$. (Galef and Stein, 1985. Copyright 1985 by the Psychonomic Society, Inc. Reprinted by permission of the authors and publisher.)

the Powdered-face Group consistently exhibited a significant tendency to choose for ingestion the diet applied to their respective demonstrators. (2) Observers in the Surrogate Group exhibited no tendency to select the same diet for ingestion that their respective demonstrators had been fed. (3) Observers in the Dead- powdered-face and Powdered-rear Groups were both significantly less affected in their diet selection by demonstrators than were observers in the Powdered-face Group, and significantly more affected by demonstrators than were observers in the Surrogate Group. Taken together these findings suggest: (1) that simple exposure of an observer rat to the smell of a diet is not sufficient to enhance observer preference for that diet, and (2) that the contextual cues emitted by demonstrator rats, producing preference for a diet in their observers, are both widely distributed and most concentrated at the anterior end of live rats. These findings do not, of course, satisfactorily resolve the issue of the nature or origins of the effective contextual cues emitted by demonstrator rats. Determination of the active chemicals involved in potentiating observer preference for demonstrators' diets must await biochemical analyses. We will attempt such analyses in the future, but I suspect that identification of the critical agent or agents will prove as difficult in the present case as it has in other attempts to chemically define mammalian pheromones.

SUMMARY AND CONCLUSIONS

The series of studies described above provide compelling evidence that naive rats have the capacity to extract information from recently-fed conspecifics, permitting identification of the food those fed individuals have eaten. Recipients of such information are biased in their subsequent food selection by the diet-related cues they encounter during interaction with a fed individual, and, in a restricted set of circumstances, can use the information acquired during social interaction to orient subsequent foraging activities.

Our data further indicate that information concerning distant diets is contained in olfactory signals passing from recently-fed rats to naive ones. These olfactory cues both permit recipient identification of the diet eaten by a conspecific and provide an as yet undefined social context that results in subsequent enhanced preference for diets eaten by informants.

It does not seem appropriate to conceive of the signals passing from a recently-fed rat to its fellows as ritualized or formalized over evolutionary time for a communicative function; anesthetized rats lacking the capacity to initiate or execute skeletal responses are as effective in signaling as are intact individuals. It seems to be the case that in rats the life-sustaining processes of ingestion, digestion, and respiration produce as by-products olfactory diet-identifying signals and contextual cues that are exploited by recipients to guide their future behavior. Thus, the results of the present experiments, in addition to demonstrating the possibility of communication of information concerning distant diets by rats, call attention to the potential importance in vertebrate communication of information extraction as a complement to formalized signaling. Life sustaining activities, such as feeding, not only promote the fitness of their initiators, but also provide a source of potential survival-enhancing information to others capable of extracting relevant information from their social environments.

ACKNOWLEDGEMENTS

The research described in the present paper was supported by grants from the Natural Sciences and Engineering Research Council of Canada and the McMaster University Research Board.

I thank my collaborators, Stephen Wigmore, Moni Stein, Deborah Kennett, Sandra Vegeris, Cathy Maskell, and Cecilia Malinski for their enthusiasm for and dedication to this research program. I also thank Mertice Clark and Harvey Weingarten, whose thoughtful critiques of earlier drafts have made this manuscript far more "user friendly" than it would otherwise have been.

REFERENCES

Alberts, J. R., and Galef, B. G., Jr., 1971, Acute anosmia in the rat: A behavioral test of a peripherally-induced olfactory deficit, Physiol. Behav., 6:619-621.

Barnett, S. A., 1958, Experiments on "neophobia" in wild and laboratory rats. Brit. J. Psychol., 49:195-201.

Bertram, B. C. R., 1978, Living in groups: Predators and prey, in: "Behavioural Ecology", J. R. Krebs and N. B. Davies, eds., Sinauer, Sunderland, Mass.

Blest, D. A., 1961, The concept of ritualization, in: "Current Problems in Animal Behavior", W. H. Thorpe and O. L. Zangwill, eds., Cambridge University Press, Cambridge.

Calhoun, J. B., 1962, "The Ecology and Sociology of the Norway Rat," U.S. Department of Health, Education, and Welfare, Bethesda, Md.

DeGroot, P., 1980, Information transfer in a socially roosting weaver bird (Quelea quelea: Ploceinae): An experimental study, Anim. Behav., 28:1249-1254.

Erwin, M. R., 1977, Foraging and breeding adaptations to different food regimes in three seabirds: the common tern, Sterna hirundo, Royal tern, Sterna maxima, and black skimmer, Rynchops niger, Ecology, 58:389-397.
Galef, B. G., Jr., 1970, Aggression and timidity: Responses to novelty in feral Norway rats, J. Comp. Physiol. Psychol., 70:370-381.
Galef, B. G., Jr., 1977, Mechanisms for the social transmission of food preferences from adult to weanling rats, in: "Learning Mechanisms in Food Selection," L. M. Barker, M. Best, and M. Domjan, eds., Baylor University Press, Waco Tex.
Galef, B. G., Jr., 1982, Studies of social learning in Norway rats: A brief review, Dev. Psychobiol., 15:279-295.
Galef, B. G., Jr., 1983, Utilization by Norway rats (R. norvegicus) of multiple messages concerning distant foods, J. Comp. Psychol., 97:364-371.
Galef, B. G., Jr., 1984, Social learning in wild Norway rats, in: "Issues in the Ecological Study of Learning," T. D. Johnston and A. T. Pietrewicz, eds., Lawrence Erlbaum, Hillsdale, N.J.
Galef, B. G., Jr., in press, Direct and indirect behavioral pathways to the social transmission of food avoidance, in: "Experimental Assessments and Clinical Applications of Conditioned Food Aversions." P. Bronstein and N. S. Braveman, eds., Ann. N.Y. Acad. Sci.
Galef, B. G., Jr., in press, Reciprocal heuristics: A discussion of the relationships of the study of learned behavior in laboratory and field, Learn. Motiv.
Galef, B. G., Jr. and Kennett, D. J., in press, Effects on communication about distant diets of events occurring between ingestion and information transmission, Anim. Learn. Behav.
Galef, B. G., Jr., Kennett, D. J., and Wigmore, S. W., in press, Transfer of information concerning distant food in rats: A robust phenomenon, Anim. Learn. Behav.
Galef, B. G., Jr., Kennett, D. J., and Stein, M., in press, Demonstrator influence on observer diet preference: Effects of familiarity and exposure context in R. norvegicus, Anim. Learn. Behav.
Galef, B. G., Jr., and Stein, M., in press, Demonstrator influence on observer diet preference: Analyses of critical social interactions and olfactory signals, Anim. Learn. Behav.
Galef, B. G., Jr., and Wigmore, S. W., 1983, Transfer of information concerning distant foods: A laboratory investigation of the 'information-centre' hypothesis, Anim. Behav., 31:748-758.
Galef, B. G., Jr., Wigmore, S. W., and Kennett, D. J., 1983, A failure to find socially mediated taste aversion learning in Norway rats (R. norvegicus), J. Comp. Psychol., 97:458-463.
McKay, D. M., 1972, Formal analysis of communicative processes, in: "Non-verbal Communication," R. A. Hinde, ed., Cambridge University Press, Cambridge.
Houston, D. C., 1979, The adaptations of scavengers, in: "Serengeti, Dynamics of an Ecosystem," A. R. E. Sinclair and M. Norton-Griffiths, eds., University of Chicago Press, Chicago.
Posadas-Andrews, A., and Roper, T. J., 1983, Social transmission of food-preferences in adult rats, Anim. Behav., 31:265-271.
Schaller, G. B., 1972, "The Serengeti Lion," University of Chicago Press, Chicago.
Smith, W. J., 1977, "The Behavior of Communicating," Harvard University Press, Cambridge, Mass.
Smythe, N., 1970, Ecology and Behavior of the Agouti (Dasyprocta punctata) and Related Species on Barro Colorado Island, Panama, Ph.D. Thesis, University of Maryland, College Park, Maryland.
Strupp, B. J. and Levitsky, D. A., 1984, Social transmission of food preferences in adult hooded rats (Rattus norvegicus), J. Comp. Psychol., 98:257-266.
Telle, H. J., 1966, Beitrag zur Kenntnis der Verhaltensweise von Ratten,

vergleichend dargestellt bei Rattus norvegicus und Rattus rattus, Z. Angew. Zool., 53:129-196.
Waltz, E. C., 1982, Resource characteristics and the evolution of information centers, Amer. Nat., 119:73-90.
Ward, P., and Zahavi, A., 1973, The importance of certain assemblages of birds as 'information-centres' for food finding, Ibis, 115:517-534.

ALARM CHEMOSIGNALS IN A MERIONES UNGUICULATUS: PREY-PREDATOR INTERACTIONS

Del D. Thiessen and Robert Cocke

Department of Psychology
University of Texas
Austin, Texas

INTRODUCTION

A number of chemical exudates originating from invertebrates and vertebrates are released under stressful environmental conditions. They are variously defined, but are known generically as "alarm signals," chemosignals released by one or more individuals to affect the behavior and physiology of conspecifics (von Frisch, 1938; Muller-Schwarze et al., 1984; Pfeiffer, 1963). These chemicals can cause exploration and investigation (Jaffe and Puche, 1984; Sonenshine, 1985), hesitation and physiological signs of frustration (Courtney et al., 1968; Herreid and Mooney, 1984), immobilization and hiding (Rottman and Snowdon, 1972; Carr et al., 1970), flight and dispersion (Stevens and Gerzog-Thomas, 1977; Hornbuckle and Beall, 1974; Ressler et al., 1968), and social mobilization and attack (Cherix, 1983). The environmental features which provoke the release of such chemicals are diverse, but usually involve stimuli of potential danger, such as novelty, social intrusion or physical insult. Generally, among higher vertebrates the body source of these chemicals is unknown and no compounds have been isolated and identified.

Our own studies were prompted by an interest in the importance of olfactory communication within and between species and the possibility that airborne "alarm signals" may provoke different responses in prey and predator species. Thus, a chemical exudate may cause individuals of a prey species to disperse and seek cover; whereas that same chemical may act as a target signal for a predator. Thus, what may be an adaptive "pheromone" reaction in prey, leading to dispersion and hiding, may result in an equally adaptive "kairomone" response in predators, leading to the selection and attack of vulnerable prey. Specifically, the experiments reported here demonstrate that individuals of Meriones unguiculatus show physiological evidence of arousal and stress when exposed to odors from stressed conspecifics, especially odors from the blood of stressed individuals. Gerbils avoid areas, such as simulated burrows and nesting chambers, that are contaminated with the blood of stressed gerbils. Natural predators of rodents, on the other hand, such as domestic cats tend to prefer dry cat food containing blood from stressed gerbils. We also discuss more limited evidence that ferrets (Mustela putorius furo) and foxes (Vulpes fulva) show an olfactory preference for the blood from stressed gerbils. Our contention is that gerbils react with physiological arousal when exposed to odors of stressed conspecifics, and avoid sources of these odors as a means of reducing the probability of predatory attack. Predators, on the other hand, use olfactory substances from stressed prey as signs of vulnerability and as signals for tracking their targets.

A second intent of this paper is to offer a tentative categorization of "alarm signals" in prey species based on ecological criteria. It is our impression that several apparently diverse reactions can be understood on a continuum running from brief negative reactions to conspecific chemicals to intense avoidance and flight reactions. Cutting across this continuum are species-typical behaviors that add to defensive maneuvers. The diversity of chemical and behavioral reactions are expected to have their counterparts in the perceptual and tracking abilities of predators. Thus, relative to highly stressful situations, less stressful physical or social situations should evoke less intense conspecific reactions, result in fewer or qualitatively different chemosignals, and reduce the intensity of predation. This particular classification may help bring together observations that have been treated as unlike in the literature.

METHOD

Subjects, Apparatus, and General Method

Male *Meriones unguiculatus* were obtained from Tumblebrook Farms (West Brookfield, MA). They were housed individually prior to the experiment under a 12L-12D light-dark cycle at 23°C and approximately 40% relative humidity. Animals were at least 120 days of age at the start of the experiment. Purina laboratory chow and tap water were continuously available. All testing was conducted during the light half of the cycle in a room different from that of the colony.

The apparatuses used in the experiments are schematically illustrated in Figure 1. The details can be obtained by writing the authors, or consulting the paper by Thiessen and Kittrell (1980). Between each discrete trial of each experiment the apparatus was washed with 50% ethanol and toweled dry prior to the introduction of the next animal and stimuli.

Experiment I (Activity Suppression)

In order to generate chemosignals from stressed gerbils, animals were individually placed in a foot-shock chamber and exposed to 60 sec of scrambled 600 V 2.5 ma shock. Following this procedure, each animal was placed in a 500 ml chamber (donor chamber) from which the odor could be directed to a recipient animal via an air stream (7.5 l/m) through 1/4 in tubing. These animals in this and subsequent experiments will be referred to as Stress Odor (SO) animals. Six SO animals were used in Experiment I.

Control animals were handled similarly, except they were not shocked in the shock chamber. Additionally, they were individually adapted to the donor chamber for 2 hr a day for 2 days prior to observing responses of recipient animals. These animals in this and subsequent experiments will be referred to as Control Odor (CO) animals. Six CO animals were used in Experiment I.

The odors from the donor chamber containing SO or CO animals were directed into a Plexiglas activity apparatus (34 x 34 x 50 cm) from the center of the ceiling roof (Figure 1a). Six recipients received odors from SO animals and six received odors from CO animals. The position of the donor animals was alternated between chambers. The activity of the recipient animal was recorded for 5 min, consisting of the number of times an animal crossed two lines on the floor which separated the floor into thirds.

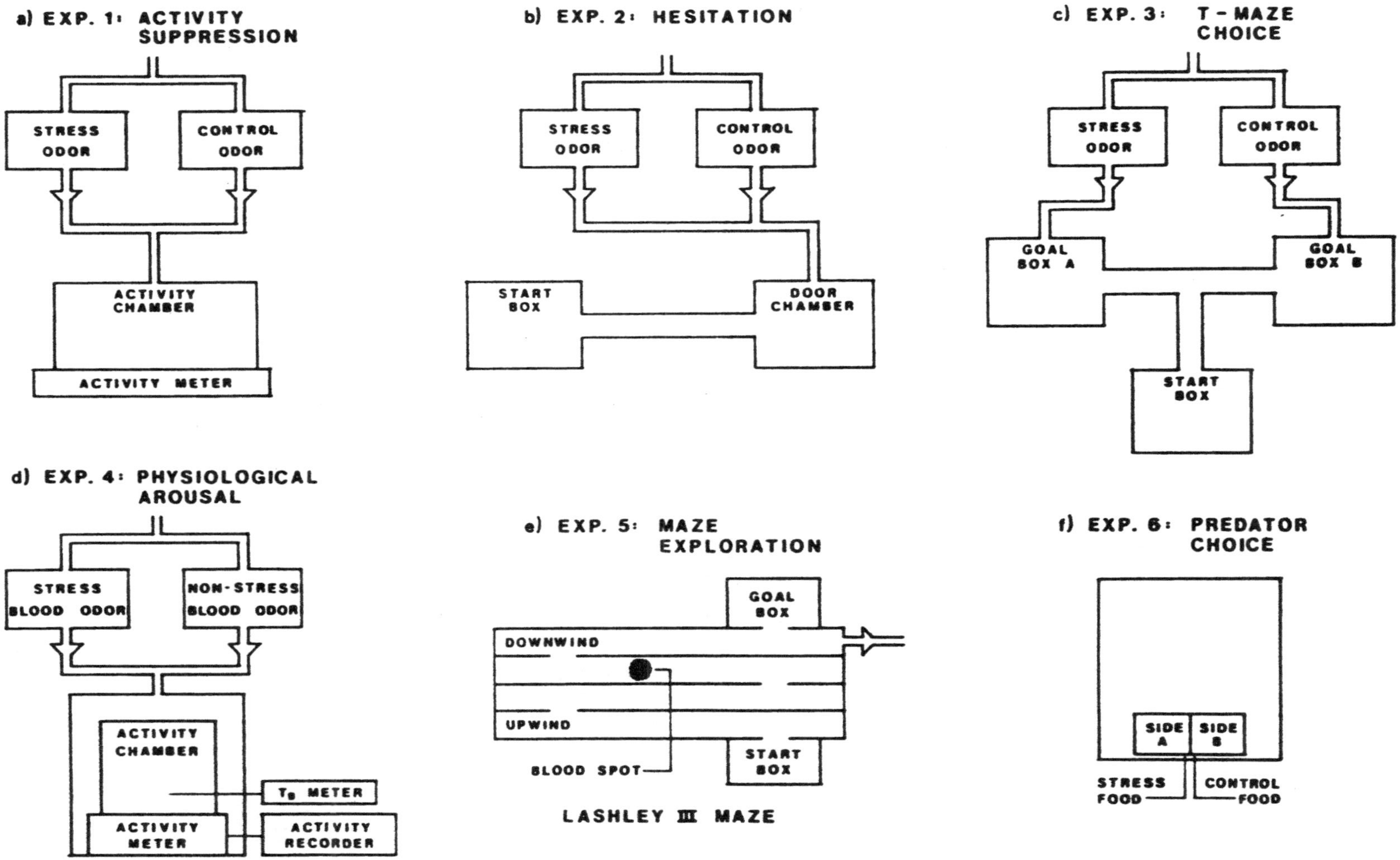

Fig. 1. Schematic diagrams of testing devices used in Experiments 1 - 6.

Figure 2a illustrates the results. Recpients receiving odors from SO animals were significantly suppressed in their activity, relative to those receiving odors from CO animals ($t = 6.75$; $df = 10$; $p < .002$). These animals were observed to be more thigmotaxic and tended to remain immobilized in a corner for longer periods of time.

Experiment II (Hesitation)

The odor of 12 SO and 12 CO animals were derived by the same procedure as in Experiment I and presented in the same manner using the same donor chamber. Experimental odor recipient animals were individually placed in a start box (11.5 x 7.5 x 5.0 cm), which was connected to an odor chamber of the same dimensions by a 2.5 in plastic (PVC) tube (Figure 1b). For one group of 12 animals the odor of SO was introduced into the odor chamber. The CO odor was presented to 12 animals in the second group. The order of odor presentation was randomized for individuals of the two groups.

The latency from initial placement in the start box until all four of the animal's paws were on the floor of the odor chamber was recorded for each animal. If an animal did not enter the second chamber after 2 min the trial was terminated.

Figure 2b illustrates an increased latency to enter the chamber odorized by SO animals ($t = 2.02$; $df = 22$; $p < .05$). Animals receiving stress odors were observed to initially enter the connecting tube and then withdraw more often than the animals which approached the control odor.

Experiment III (T-maze Choice)

In pilot studies we found that the odors from animals who had been involved in fights were avoided by conspecifics in various experimental paradigms. We therefore decided to test whether odors produced by animals involved in fight bouts resulted in avoidance in an olfactory T-maze.

Two animals which fought when paired in a fight chamber (11.5 x 7.5 x 5 cm) were used to generate odors used in a T-maze avoidance procedure. Approximately 2 min after the initiation of fighting the odors were introduced into one arm of a 2.5 in plastic pipe T-maze (Figure 1c).

In a control odor condition, two male animals that did not fight when paired were placed in an identical chamber. The air from this chamber was then exhausted at the rate of .75 l/m into one arm of the T-maze.

The odorized arm of the T-maze was randomly chosen on each trial (coin toss). Compressed air was introduced into the non-odorized arm of the maze at the same flow rate.

Forty-eight animals with no prior experience in the T-maze were individually tested as odor recipients. Twenty four of these encountered stress odors versus compressed air in the maze, and twenty four encountered odors from non-aggressive animals versus compressed air.

As can be seen in Figure 2c, the frequency with which animals approached odors from fighters or non-fighters was different ($X^2 = 10.02$; $df = 1$; $p < .001$). Not only did test animals avoid the alleys containing odors from stressed gerbils, but they appeared to be attracted to odors from animals not engaging in fights.

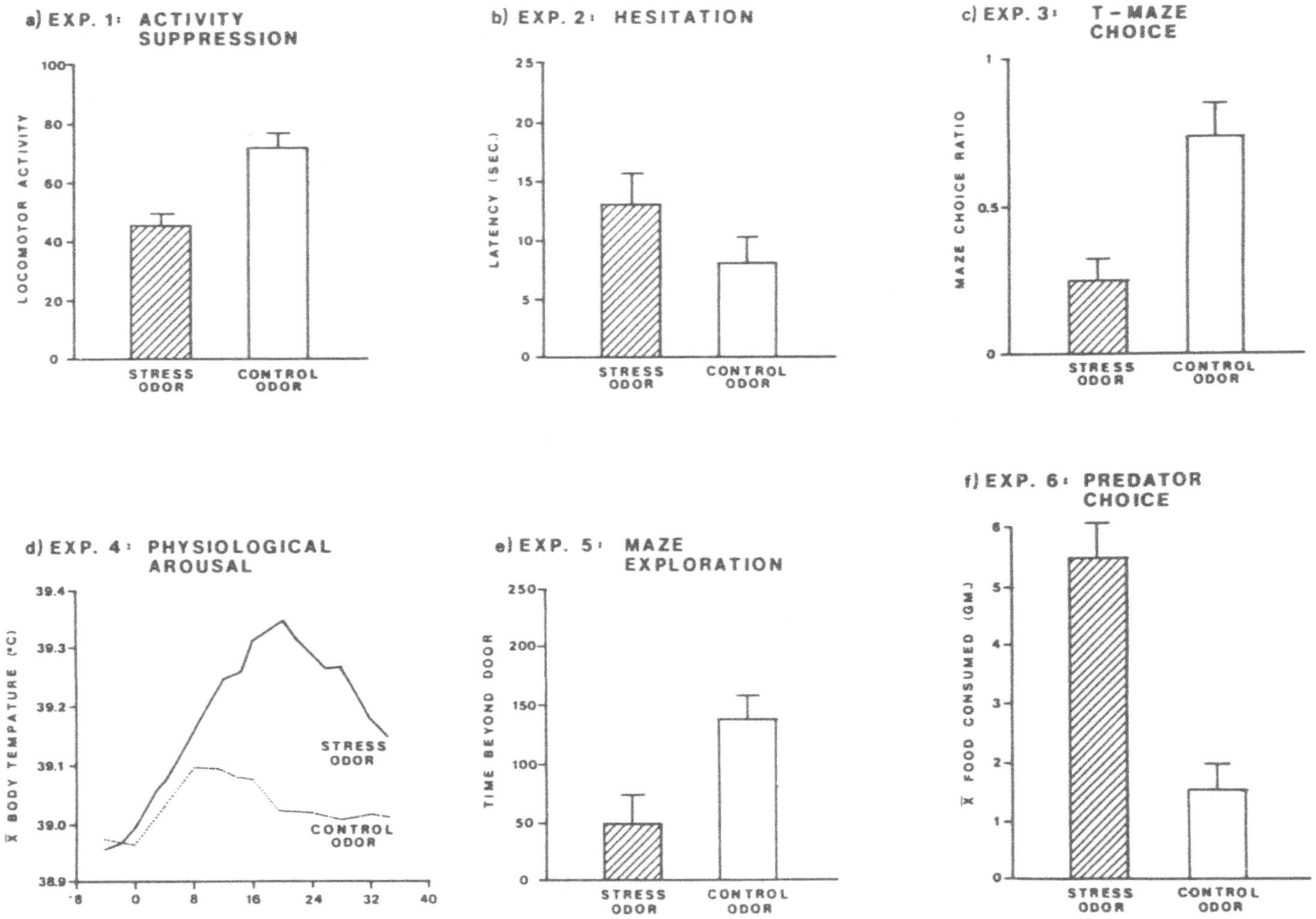

Fig. 2. Results of Experiments 1 - 6.

Experiment IV (Physiological Arousal)

In Experiment III, and other unpublished experiments with fight odors, more robust effects were apparent when trauma and bleeding resulted from the aggressive encounters. We speculated that blood from stressed animals might be more effective in producing avoidance and arousal in conspecifics. Experiment IV was intended to demonstrate that gerbils respond to presentation of stress odors with physiological arousal. A number of behavioral responses may be indicative of physiological arousal; animals respond to stress odors with freezing, piloerection, increased defecation, increased body temperature, thigmotaxis, and occasional tonic-clonic seizures (Stevens and Gerzog-Thomas, 1977). Because of its robustness and its ease of measurement, body temperature increases were used as an index of physiological arousal.

We adapted a procedure used by Mackay-Sim and Laing (1981a). Fifteen min after the foot-shock administration, blood from stressed animals was collected in a heparinized watch glass. Control odor blood was collected from donors immediately following removal from the home cage.

Body temperature was measured in 9 test animals with surgically implanted biotelemetry radio transmitters (Mini-mitter, Model M) according to the procedure described by Thiessen and Kittrell (1980). Two weeks following surgery, individuals were placed in a Plexiglas activity apparatus measuring 40 x 34 x 25 cm (Animex, Model Type O, Farad Electronics) enclosed in an unlighted and soundproof cabinet (Figure 1d). Either stress odors or control odors from the blood samples could be directed into the apparatus. The T_A ($23^{\circ}C$) of the chamber was monitored continually and the animal's body temperature and activity in the Animex apparatus were recorded on a Coulbourn printer (Model R22-10). Ten successive inter-signal intervals from the transmitters were summed and averaged by the Coulbourn instrument. This procedure generated a fourth significant figure which allowed for accurate measurement of relatively small temperature changes ($.05^{\circ}C$). Locomotion was scored as the number of times the animal passed over six electromagnetic sensors built into the floor of the Animex chamber. After the odor recipient animal had a 45 min habituation in the apparatus, a fresh blood sample from either the stress odors or control odors condition was introduced into the air stream of the chamber for 30 min. Each animal was exposed to both odors on alternate days in a counter-balanced sequence.

The average increase in body temperature shown in Figure 2d was significantly greater among animals exposed to stress odors than control odors ($\underline{t} = 4.18$; $\underline{df} = 17$; $\underline{p} < .01$) (Figure 2d). There was no significant difference in locomotor activity, although there was a trend toward lower activity during stress odors presentation, and there were no covariations between body temperature, locomotion and ambient temperature. Thus, a primary effect of exposure to the blood odors of stressed animals was an average increase of $.38^{\circ}C$ in body temperature. Stress odors therefore appear to physiologically arouse conspecifics.

Experiment V (Maze Exploration)

Experiment V was an attempt to demonstrate that presentation of blood-borne stress odors not only results in physiological arousal but also causes avoidance in conspecifics. We employed a multi-alley Lashley III maze in which stress odors or control odors were placed on the floor midway through the maze where the animals would encounter the stimuli as they traversed the alleys (Figure 1e). The apparatus measured 42 x 17 x 96 cm overall and consisted of four 96 x 17 x 18.5 cm runways involving 3 choice points. The maze was covered with Plexiglas and the floor was covered with butcher paper. Air was pulled from the start box and through the apparatus at approximately

.75 l/m. Thus the animals would progress through one-half of the maze before encountering the stimulus or its air-borne odors. Twenty male gerbils were placed individually in the start box and admitted to the maze via a guillotine door for 5 min of free exploration. Blood from either a stress odor or control odor donor was swabbed on the paper flooring midway through the maze immediately before each trial. The amount of blood placed on the floor was approximately 0.25 ml. Ten individuals encountered the stress odors and ten encountered the control odors. The running of the two conditions was alternated and the test animals were randomly assigned to the stress odor and control odor encounters. Video tapes of each trial were analyzed for the amount of time that animals spent up- or down-wind from the odor sources and for the frequency of crosses over the blood spots in the alleyway.

The average time spent down-wind of the odor source for the animals encountering the stress odor and control odor conditions was 49 sec and 139 sec, respectively ($\underline{t}$ = 6.74; $\underline{df}$ = 18; $\underline{p}$ .001), and the average number of times that animals crossed the blood spot for the stress odors and control odors conditions was .3 and 1.9, respectively (X^2 = 4.62; $\underline{df}$ = 1; $\underline{p} < .05$). Thus, both in terms of time spent down-wind of the odor source and the probability of crossing over the odor source, animals encountering the blood from stressed animals showed more avoidance (Figure 2e). In addition, two subjects encountering the stress odors immediately developed tonic-clonic seizures. Apparently odors from the blood of stressed gerbils physiologically arouse recipients (as indicated by elevations in body temperature) and also evoke evasive behaviors during exploration of a novel environment.

Experiment VI (Predatory Preference)

Our next question was whether the odors which stimulated arousal and avoidance were also attractive to rodent predators. In pilot investigations we found that foxes, ferrets and domestic cats seek out blood odors from stressed gerbils. In Experiment VI, therefore, we attempted to verify this orientation by testing the food preferences of domestic cats exposed to stress odors or control odors in their food. Twenty-three male and female cats were tested at a local cattery for food preferences. The ages of the cats were often uncertain but ranged from adolescence to maturity. Both sexes were represented in approximately equal numbers. The cats were individually housed in stainless steel cages (60 x 38 x 40 cm) exposed to natural lighting at approximately 23°C. Thirty minutes prior to their normal feeding time (1600) a two-compartment bowl with 20 g of dry Purina cat chow in each side was placed in the subject's home cage. One compartment of the bowl contained three additional food pellets saturated with the blood from stress odors and the other compartment contained three additional food pellets saturated with the blood from control odors. The food bowl remained in the cage for 5 min, with the position of the stress odors and control odors compartments rotated every 90 sec. The initial food preference and the amount of food eaten for each compartment was recorded.

The majority of cats initially oriented toward and selected the food from the compartment containing the stress odors (X^2 = 4.34; $\underline{df}$ = 1; $\underline{p}$ < .05), and the greater amount of food was eaten from this same compartment ($\underline{t}$ = 3.30; $\underline{df}$ = 22; $\underline{p} < .01$). It appeared that the cats were immediately attracted to the stress odors and then tended to persist in eating on that side even after the pellets saturated with stress odors were eaten. We view these findings as evidence that rodent predators appear to be attracted to the odors from the blood of stressed prey and may accept these odros as part of their food source.

DISCUSSION

Our experimental results confirm the many findings that rodent species avoid the odors of stressed conspecifics (Carr et al., 1970; Courtney et al., 1968; Mackay-Sim and Laing, 1981a; Rottman and Snowdon, 1972). When odors from foot-shocked gerbils are directed into an open field, recipients become more hesitant, are less active and generally become immobilized in a corner of the apparatus (Experiment I). When similar odors are funneled into one of two interconnecting chambers, recipients of stress odors are more hesitant to move from the first chamber into the chamber containing the odor source (Experiment II). They frequently investigate the connecting gateway but are reluctant to enter the odorized area. Test animals exposed to odors from fight-stressed individuals in a T-maze avoid the side of the maze containing the odor (Experiment III). In all three situations odors from handled but otherwise nonstressed individuals had relatively little effect on test animals. If anything, our impression is that test animals are attracted to the odors of animals who have experienced non-aggressive social interactions. In any case, our results indicate that odors from stressed animals cause hesitation and active avoidance during exploratory locomotion by nonstressed conspecifics. The type of behavior that occurs apparently depends upon the options available to the organism and not the differences among odors; if an active avoidance is possible animals will choose areas not contaminated by stress odors. If avoidance is not possible, the predominant response appears to be hesitation and immobilization. Animals appear to first investigate the odor source and then respond according to the availability of appropriate responses.

Further exploration suggested that the "alarm chemosignal" produced by an animal receiving foot-shock is blood-borne (Experiment IV). When the blood from stressed donor individuals is used as the odor stimulus, recipient animals respond to the odors with significant elevations in body temperature. Blood from nonstressed individuals is relatively ineffective. The body temperature increases upon exposure to odors from stressed blood are rapid and last for the greater part of the 30 min test, suggesting that the arousal caused by blood odorant is intense. The suggestion is amplified by the subsequent maze test in which it was found that gerbils are quite reluctant to move downwind and across a spot of blood from stressed animals (Experiment V). The negative reaction to blood from stressed gerbils is similar to those reported by other investigators (Hornbuckle and Beall, 1974; Mackay-Sim and Laing, 1980, 1981a, 1981b; Stevens and Saplikoski, 1973). There is little information in the literature to suggest how these volatile blood odorants are generated, or how they are released outside of the body in intact animals. Recently, however, McGlone (1985), reported that domestic pigs injected with ACTH produce a urinary signal which induces submissive behaviors in conspecifics, similar to the immobilization and hesitation we see in gerbils exposed to airborne signals from the stressed individuals. The parallel in the experiments suggests that a pituitary and/or adrenal reaction may be responsible for the production of the chemosignals found in the blood which are released as olfactory stimuli. Recently Faneslow (1985) has reported that odors from stressed conspecifics evoke another physiological alteration in rodents. Exposure to shock-induced stress odors produces profound analgesia in rats. The effect appears to be mediated by endorphins as it can be reversed with nalaxone.

In contrast to the negative reactions found in gerbils to odors of stress, potential rodent predators appear to be attracted to these odors. It has been suggested that rodent predators may be capable of tracking prey by attending to odors such as urine odors (Cushing, 1984). Our own preliminary investigations with ferrets (Mustela putorius furo) and foxes (Vulpes fulva) suggested that the odors from foot-shocked gerbils are attractive to these species. Thus, in a two-compartment chamber, ferrets spent more time investigating the side containing an odor from a stressed gerbil versus the control odor on the other side. When foxes were offered a choice between two cloth bags containing odors from stressed or nonstressed gerbils, they spent more time investigating and manipulating the bag containing the odor from stressed animals. A similar result occurred in our more extensive study with domestic cats (Experiment VI). Three food pellets soaked in the blood of stressed gerbils, but not the blood of nonstressed gerbils, were sufficient to influence their gustatory preference for one of two food dishes containing dry cat food. Once eating began in a dish containing the stress odor, it tended to continue, even after the stimulus pellets were consumed. We view these findings as evidence that predators are attracted to the same odors that repel conspecific prey.

The olfactory interaction between prey and predator species is even more complex than indicated in these studies. Vernet-Maury (1980) and Cattarelli (1981) have found that when rodents are exposed to odors from predators they respond with fright reactions, including freezing and active avoidance. Furthermore, Sullivan and Crump (1984) have reported that odors from weasels are disturbing to snowshoe hares and disrupt their feeding behaviors. These observations suggest prey might avoid the odors of predator species as well as those of injured conspecifics in order to reduce the probability of exposure to predators.

Much of the diversity of reactions to odors associated with stress appears to be related to the differential vulnerability of prey species compared to that of predators. Individuals who are constantly under the threat of predation may have evolved behavioral reactions to odors associated with stress as a means to avoid conspecifics who are vulnerable to attack and therefore attractive to predators. They also have seemingly developed the capacity to react to odors from predators. At the same time, predators of these species may have evolved abilities to perceive odors of stress or injury and use these odors during foraging. It is possible that predators learn that certain odors from prey denote vulnerability, or that predators acquire a preference for these odors because of an increased frequency of encounter. Gosling and Petrie (1981) suggest that predators may cull unhealthy individuals from the prey population; hence orientation toward stress odors might play a role in this process of prey selection. Whatever the proximate mechanism for the regulation of these behaviors, it does appear that the different reactions to odors of stress by prey and predators have adaptive significance.

In the course of our studies we were led to consider the possibility that "alarm chemosignals" were highly variable in origin and effect, yet rested on a common continuum of physiological and behavioral arousal. What is known as "frustration odor responses," (whereby an animal hesitates to run a maze alley toward the goal box on a trial following another animal which received unexpected nonreinforcement) (Davis et al., 1981), appears to be a weaker version of the more intense alarm reactions because odors from stressed animals mimic this effect (Courtney et al., 1968). It appears that, depending on the intensity of environmental stimuli, and the amount of exudate released, animals will show a wide range of behaviors. Weak stimuli apparently provoke the release of few exudates. Typical reactions of conspecifics include alert reactions, orientation toward the source of the stimulus, caution and rapid habituation (Stevens and Saplikoski, 1973). Intense stimuli, including injury, on the other hand, apparently exaggerate the release of exudates or cause the release of qualitatively different exudates. Reactions to these signals include avoidance, hiding, immobilization and slow habituation (Stevens and Saplikoski, 1973; Hornbuckle and Beall, 1974). Some of the influences of odors upon maturation rates in rodents may also be due to low levels of "alarm chemosignals" produced by population density stress (Jones and Nowell, 1973; Drickamer and Shiro,

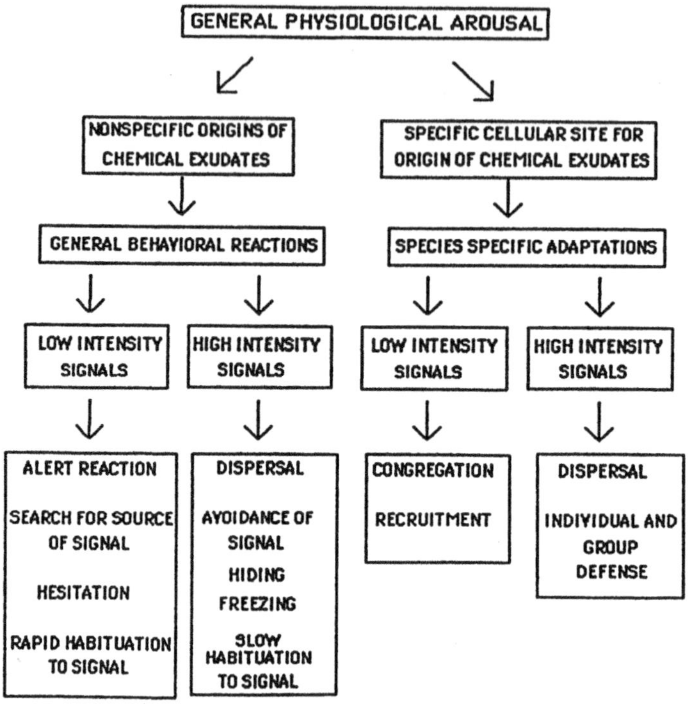

Fig. 3. Proposed variations in reaction to alarm chemosignals.

1984). Physiological reactions to these chemosignals presumably vary in intensity along this same continuum.

Features of alarm systems range from nonspecificity of both odor source and reactions, to high specificity of odor source and responses. Thus, many species appear to release general body odors when stressed and react to these signals with general arousal and variable behavior (Stevens and Gerzog-Thomas, 1977). Other species, such as fish and insects, have evolved specialized cells which release unique signals and provoke stereotyped reactions (von Frisch, 1938; Sonenshine, 1985). Therefore, both in terms of the tissue source of chemosignals and the specificity of physiological and behavioral reactions, alarm substances appear to vary in their degree of specificity. A rough scheme illustrating the range of variation in "alarm chemosignals" is given in Figure 3. We view this scheme as only heuristic. It could, nevertheless, promote a classification of chemosignals whereby similarities and differences among various olfactory responses can be seen. The classification could obviously be sharpened by the isolation and identification of chemical exudates released during environmental stress.

It seems possible that the long history of prey-predator interactions could have resulted in the evolution of specific cellular sources of alarm signals and adaptive reactions to these signals by prey as well as by predators. This may have been the basis for the coevolution of prey-predator interactions and the specialization (ritualization) of the intraspecific and interspecific chemical communication system. Alarm signals in social insects, generated in abdominal glands, and similar signals produced in fish by specialized club cells, appear to be illustrations of this tight interrelationship between prey and predators (Cherix, 1983; Pfeiffer, 1963).

In more complex species, facing variegated environments, alarm signals may have originated as responses to general metabolic exudates associated with stressful stimuli (Weldon, 1982).

REFERENCES

Carr, W. J., Martorano, R. D. and Krames, L., 1970, Responses of mice to odors associated with stress, J. Comp. Physiol. Psychol., 71:223.

Çattarelli, M. and Vigouroux, M., 1981, Evaluation des reactions émotionnelles déclenchees par des odeurs a signification biologique chez le rat au cours d'un conditionnement opérant, Physiol Behav., 27:445.

Cherix, D., 1983, Intraspecific variation of alarm pheromones between two populations of the red wood ant Formica lugubris (Zett.) (Hymenoptera, Formicidae), Schweiz. Entomol. Gesamte, 56:57.

Courtney, R. J., Jr., Reid, L. D., and Wasden, R. E., 1968, Suppression of running times by olfactory stimuli, Psychonom. Sci., 12:315.

Cushing, B. S., 1984, A selective preference by least weasels for oestrus vs. dioestrus urine in prairie deer mice, Anim. Behav., 32:1263.

Davis, S. F., Thomas, R. L., and Prytula, R. E., 1981, The development and maintenance of odor-based double-alternation responding under conditions of Thorazine and Elevil injection, Anim. Learn. Behav., 9:551.

Drickamer, L. C., and Shiro, B. C., 1984, Effects of adrenalectomy with hormone replacement therapy on the presence of a sexual maturation-delaying chemosignal in the urine of grouped female mice, Endocrinology, 115:255.

Fanselow, M. S., 1985, Odors released by stressed rats produce opioid analgesia in unstressed rats, Behav. Neurosci., 99:589.

Frisch, K. von., 1938, Zur Psychologie des Fischschwarmes. Naturwissenchaften, 26:601.

Gosling, L. M., and Petrie, M., 1981, The economics of social organization, in: "Physiological Ecology: An Evolutionary Approach To Resource Use," C. R. Townsend, and P. Calow, eds., Blackwell Scientific Publications, Oxford.

Herreid, C. F., II and Mooney, S. M., 1984, Color change in exercising crabs: Evidence for a hormone, J. Comp. Physiol. B, 165:207.

Hornbuckle, P. A., and Beall, T., 1974, Escape reactions to the blood of selected mammals by rat, Behav. Biol., 12:573.

Jaffe, K., and Puche, H., 1984, Colony-specific territorial marking with the metapleural gland secretion in the ant Solenopsis geminata (Fabr), J. Insect Physiol., 30:265.

Jones, R. B., and Nowell, N. W., 1973, Aversive and aggression-promoting properties of urine from dominant and subordinate male mice, Anim. Learn. Behav., 1:207.

Mackay-Sim, A., and Laing, D. G., 1980, Discrimination of odors from stressed rats by non-stressed rats, Physiol. Behav., 24:669.

Mackay-Sim, A., and Laing, D. G., 1981a, Rats' responses to blood and body odors of stressed and non-stressed conspecifics, Physiol. Behav., 27:503.

Mackay-Sim, A., and Laing, D. G., 1981b, The sources of odors from stressed rats, Physiol. Behav., 27:511.

McGlone, J. J., 1985, Olfactory cues and pig agonistic behavior: Evidence for a submissive pheromone, Physiol. Behav., 34:195.

Muller-Schwarze, D., Altieri, R., and Porter, N., 1984, Alert odor from skin gland in deer, J. Chem. Ecol., 10:1707.

Pfeiffer, W., 1963, Alarm substances, Experientia, 19:113.

Ressler, R. H., Cialdini, R. B., Ghoca, M. L., and Kleist, S. M., 1968, Alarm pheromone in the earthworm Lumbricus terrestris, Science, 161:597.

Rottman, S. J., and Snowdon, C. T., 1972, Demonstration and analysis of an alarm pheromone in mice, J. Comp. Physiol. Psychol., 81:483.

Sonenshine, D. E., 1985, Pheromones and other semiochemicals of the acari, Amer. Rev. Entomol., 30:1.

Stevens, D. A., and Gerzog-Thomas, D. A., 1977, Fright reactions in rats to conspecific tissue, Physiol. Behav., 18:47.

Stevens, D. A., and Saplikoski, N. J., 1973, Rats' reactions to conspecific muscle and blood: Evidence for an alarm substance, Behav. Biol., 8:75.

Sullivan, T. P., and Crump, D. R., 1984, Influences of mustelid scent gland compounds on suppression of feeding by snowshoe hares (Lepus americanus), J. Chem. Ecol., 10:1809.

Thiessen, D. D., and Kittrell, E. M. W., 1980, The Harderian gland and thermoregulation in the gerbil (Meriones unguiculatus), Physiol. Behav., 24:417.

Vernet-Maury, E., 1980, Trimethyl-thiazoline in fox feces: A natural alarming substance for the rat, in: "Olfaction and Taste VII," H. Van der Starre, ed., IRL Press, Ltd., Washington, D.C.

Weldon, P. J., 1982, The evolution of alarm pheromones, in: "Chemical Signals In Vertebrates 3," D. Muller-Schwarze, and R. M. Silverstein, eds., Plenum Press, New York.

AVOIDANCE RESPONSE OF POCKET GOPHERS (THOMOMYS TALPOIDES) TO MUSTELID ANAL GLAND COMPOUNDS

Thomas P. Sullivan[1] and Douglas R. Crump[2]

[1]Applied Mammal Research Institute, 23523 47th Avenue R. R. #7, Langley, B.C., Canada V3A 4R1
[2]Chemistry Division, Department of Scientific and Industrial Research, Private Bag, Petone, New Zealand

INTRODUCTION

Several species of gophers in the family Geomyidae inflict feeding damage to agricultural and forest crops in North America. These fossorial rodents inhabit burrow systems and are seldom active above ground. The northern pocket gopher (*Thomomys talpoides*) is a persistent problem in the fruit-growing regions of British Columbia, Canada (Anderson, 1980; Sullivan et al., 1985) and Washington and Oregon of the U.S.A. Similar feeding attacks by the fossorial vole (*Microtus pinetorum*) also occur in orchards of the eastern United States and Canada (Byers et al., 1976; Davis, 1976). These herbivorous rodents feed on the bark, vascular tissues, and roots of orchard trees, which may result in reduced growth and yield or even direct mortality from girdling. Similar feeding damage by pocket gophers to coniferous seedlings in areas of reforestation is considered a major problem to regeneration of western forests (Barnes, 1973; Barnes et al., 1982; Crouch, 1982).

Attempts to alleviate this damage problem usually involve the use of poison bait to reduce or eliminate the rodent population. However, the resiliency of gophers (and voles) to depopulation, hazard of poison to non-target species, and increased resistance of rodents to poison formulations make the development of new control techniques highly desirable. The use of semiochemicals (pheromones and kairomones) is one such avenue which might provide effective protection for fruit tree orchards.

Avoidance and reduced activity responses to intraspecific odors have been reported in fossorial yellow voles (*Lagurus luteus*) (Fan, 1983), bank voles (*Clethrionomys glareolus*) (Brinck and Hoffmeyer, 1984), and other rodents such as muskrats (*Ondatra zibethica*) (Van Den Berk and Muller-Schwarze, 1984) and beaver (*Castor canadensis*) (Muller-Schwarze and Heckman, 1980; Muller-Schwarze et al., 1983). The potential use of pheromones in biological control of rodents has been discussed by Christiansen and Døving (1975) and Christiansen (1976). Similar avoidance responses of voles (*M. agrestis*, *M. arvalis*) to predator odors, particularly those of the weasel (*Mustela nivalis*) and stoat (*M. erminea*), were reported in several studies (Stoddart, 1976, 1980; Gorman, 1984). These responses are innate and presumably adaptive since the chances of an individual vole suffering predation would be reduced (Gorman, 1984). Stoats (also called ermine or short-tailed weasels) and long-tailed weasels (*M. frenata*) prey on pocket gophers throughout their range in western North America (Criddle and Criddle, 1925; Simms, 1979). The use of stoat or weasel odor (anal gland secretion) as an area repellent within burrow systems could protect forest and agricultural crops from gopher depredations.

The chemical composition of some mustelid anal gland secretions has been reported for the stoat (Crump, 1978, 1980a; Schildknecht and Birkner, 1983), ferret (*M. putorius*) (Crump, 1980b; Schildknecht and Birkner, 1983), and mink (*M. vison*) (Schildknecht et al., 1976, 1981; Brinck et al., 1978; Anderson and Bernstein, 1980; Sokolov et al., 1980). A comparison of anal gland secretion in eight mustelid species was conducted by Brinck et al. (1983). The major sulfur-containing compounds identified in these studies were 2-propylthietane, 3-propyl-1,2-dithiolane, 2,2-dimethylthietane, and 3,3-dimethyl-1,2-dithiolane.

The impact of certain mustelid anal gland compounds on suppression of feeding by snowshoe hares (*Lepus americanus*) was reported by Sullivan and Crump (1984). This paper reports on the influence of the major anal gland compounds from the stoat and ferret in generating an avoidance response by northern pocket gophers. This study was designed to test the hypothesis that gophers would avoid areas "marked" by one or more mustelid compounds, thereby providing a preliminary assessment of the use of such semiochemicals in crop protection.

MATERIALS AND METHODS

GCMS Analysis of Anal Gland Secretion from Canadian Stoats

The composition of the anal gland secretion, with respect to sulfur-containing compounds, of Canadian *Mustela erminea* was analyzed for comparison with the introduced species in New Zealand. Male stoats were collected by a fur trapper at Prince George, B.C., Canada and frozen immediately after trapping. The anal glands were dissected later and were kept at -5°C prior to being vacuum sealed in glass ampules and shipped to New Zealand for analysis. The GCMS analysis followed that of Crump (1980a).

Anal Gland Compounds

Anal gland compounds from the stoat: 2-propylthietane and 3-propyl-1,2-dithiolane; and ferret: 3,3-dimethyl-1,2-dithiolane, were prepared according to Crump (1978, 1980a, b, 1982). The non-sulfur compounds, indole and o-aminoacetophenone, were available commercially. In an attempt to simulate the anal gland secretion of the Canadian stoat, a group of compounds: 2-propylthietane, 3-propyl-1,2-dithiolane, indole, and o-aminoacetophenone; was mixed in the following ratio: 7:1:0.1:0.6. In addition, 2-propylthietane and 3-propyl-1,2-dithiolane were mixed in a 1:1 ratio. Polymerized 3-propyl-1,2-dithiolane and partly polymerized 3,3-dimethyl-1,2-dithiolane (in field bioassays only) were also available for testing. Single sulfur-containing compounds and the mixtures of compounds were tested in laboratory and field bioassays. The sulfur-containing novel odor, 2-methylthiophene, was also tested in the laboratory bioassays. All compounds were dispensed in 140-ul capillary tubes (75 mm in length). The capillary tubes provided a point source of odor and prevented contamination of the arena or pocket gophers, themselves, in the laboratory bioassays. This encapsulation also protected the compounds from adverse environmental conditions in the field bioassays. Approximately 30 mg of each compound or mixture was placed in a given capillary tube using a 1-ml microsyringe with a 20-gauge needle. An empty capillary tube represented the control to complete the design of a given bioassay.

Laboratory Bioassays

The potential avoidance behavior of gophers to the various anal gland compounds, mixtures, and novel odor was assessed by tallying the number of "captures" in the control and treatment rooms of a bioassay arena (see Figure 1). The walls of the arena were 17 cm high and composed of 2-cm thick fir plywood which was covered with a smooth washable surface. The arena (no floor) was set on a sheet of wax paper (on a lab bench) which was changed each time a gopher was tested. This changing of wax paper and cleaning of the arena walls attempted to eliminate other sources of chemical communication (e.g., feet, hair, feces, urine) left by a given gopher, which might have affected the responses of subsequent individuals. The control and treatment rooms (A or B) and the neutral zone (except for opening for gophers to enter) were covered with light-proof aluminum foil. This helped keep the odor within the treatment room for optimum evaluation purposes. In addition, the trials were conducted under dim red light, thereby simulating darkness and, essentially, an underground environment for the gophers. Laboratory temperature was maintained at 22°C to provide for consistent volatility of compounds. Upon entering the neutral zone, access to the control or treatment rooms was through 6.5 x 6.5 cm openings. A Longworth live-trap was located directly opposite these openings with the tunnel of the trap extending 0.5 cm into the control or treatment room. These traps simulated a burrow system for a gopher to enter. One capillary tube with a given compound or mixture was attached with adhesive tape to the floor of the tunnel of the live-trap in the treatment room. An empty tube was attached in a similar manner to the trap in the control room to test whether or not the tube itself would prevent a gopher from entering the trap. All traps in which gophers had been "captured" were thoroughly cleaned with detergent and hot water, and dried prior to further use.

Within a given trial, an individual gopher was placed in the neutral zone and allowed to enter either room. Upon being "captured" in a live-trap, the animal was removed to its holding pen, a fresh trap and wax paper replaced, and another gopher was tested. The number of gophers "captured" in each room was tallied to assess the impact of each compound on avoidance behavior. The control and treatment rooms were alternated for each trial with individual gophers being tested only once in a given trial. Only one trial was conducted per given day with a new compound tested on each consecutive day. Each gopher was given a maximum of 10 min to be "captured." Four groups ($A = 25$, $B_1 = 45$, $B_2 = 25$, $C = 32$) of gophers were tested. Each group was tested initially with both arena rooms acting as controls to determine if there was a preference by individual gophers to enter and be "captured" in one room or the other. Sample size of each group varied depending on number surviving throughout the series of trials and exclusion from results of those individuals (usually $\leq$ 2 per trial) responding to neither control nor treatment conditions. Depending on availability, all compounds were tested with three different groups of gophers, except for 2-propylthietane (2 groups) and 2-methylthiophene (1 group). All gophers were collected from intensive study areas at Summerland and Vernon, B.C., Canada. Male and female gophers were pooled for all trials.

Field Bioassays

Gopher populations. Pocket gopher populations were monitored on control and treatment areas at the Agriculture Canada Research Station, Summerland, B.C. from Septemter 1983 to July 1984; and on control, pulse-removal control, and pulse-removal treatment areas in Kalamalka Lake Park, Vernon, B.C. from April 1984 to May 1985. These study areas were located in natural habitats near extensive blocks of orchards.

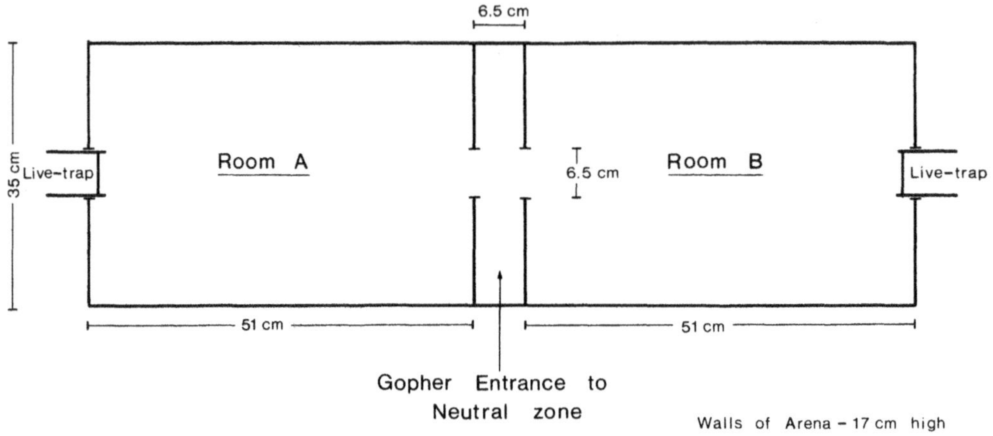

Fig. 1. Laboratory bioassay arena used for assessing the influence of mustelid anal gland compounds on avoidance behavior of pocket gophers.

Gophers were live-trapped on 1-ha checkerboard grids with Longworth live-traps. Each grid (except for Summerland treatment at 4 x 12) had 40 (7 x 7) stations located at 15.2-m intervals. Live-traps were set within recently used tunnels or burrows indicated by fresh excavations of soil. Thus, traps did not usually have permanent stations but were moved in response to gopher activity. Traps were baited with whole oats, peanut butter, and carrot; coarse brown cotton was supplied as bedding. Traps were set on day 1, checked on the morning and afternoon of day 2, and morning of day 3, and then locked open between trapping periods. The Summerland control and treatment grids had 30-35 traps and 20-25 traps, respectively, set in a given trapping period. An average of 40-45 traps was set on each of the Vernon grids during a trapping period.

All gophers captured were ear-tagged with serially numbered tags, breeding condition noted, weighed on Pesola spring balances and point of capture recorded. These populations were monitored at 3-week intervals during spring, summer, and fall, and at 4-week intervals in the winter (Summerland populations were not monitored from December 1983 to February 1984 nor Vernon populations from December 1984 to April 1985 because of snowfall and frozen ground conditions).

Experiment A. To test the initial field efficacy of those compounds which gophers avoided to a significant degree in the laboratory, capillary tubes containing a mixture of 2-propylthietane and 3-propyl-1,2-dithiolane were attached to the inside of 500-ml cans (open at one end) by a plastic twist-tie and put in gopher burrows in place of locked-open Longworth traps. This experiment tested the effectiveness of weasel odor in driving out the resident gopher population from the Summerland treatment grid. Totals of 25 and 24 (with empty tubes) cans were placed on the treatment and control grids, respectively, in September 1983. This experiment was repeated in October 1983 with 24 and 28 cans placed on the treatment and control grids, respectively. Cans were placed in burrows immediately after a trapping period and checked 3 weeks later, prior to the next trapping session.

Experiment B. To provide a somewhat more direct test of gopher avoidance of mustelid anal gland compounds, a capillary tube with one or more compounds was attached (by a plastic twist-tie) to the treadle of the tunnel of each Longworth trap during regular trapping periods. The open end of the tube faced the door of the trap such that a gopher could not enter and be caught in the trap without being exposed to the odor of the given compound. Empty tubes were attached to trap tunnels in an identical manner on the control grid. The following compounds or mixtures were tested in consecutive trapping periods from March to July 1984 on the Summerland treatment grid: 2-propylthietane, 3-propyl-1,2-dithiolane, mixture of 2-propylthietane and polymerized 3-propyl-1,2-dithiolane, 3,3-dimethyl-1,2-dithiolane, control, 3-propyl-12-dithiolane, and 3-propyl-1,2-dithiolane. In general, traps were not cleaned prior to these field bioassays except for the final trial which provided comparative information.

Experiment C. Prior to this field experiment at Vernon in August 1984, all resident gophers were permanently removed for two consecutive trapping periods from the control and treatment pulse-removal grids. This removal allowed the odor of the anal gland compounds to fumigate the burrow system of the treatment grid without disturbance from the digging activites of resident gophers. The mixture of four compounds was dispensed via one capillary tube attached by a plastic twist-tie to a 500-ml can (open at both ends). A total of 72 cans with the weasel odor was systematically placed in burrows on the treatment grid with 2 cans per each of the 36 squares on the 7 x 6 matrix. Cans with empty tubes were placed on the control pulse-removal grid in an identical manner. This experiment commenced immediately after completion of the second removal trapping period. Locked-open traps were left in place on both grids. This experiment tested whether or not new gophers would colonize or avoid the removal area which smelled of a weasel predator. Gophers colonizing a new area usually inhabit the already existing burrow systems. The third grid acted as a control for monitoring the population dynamics of the gophers at this study area.

Experiment D. This experiment was similar to C except that resident gophers were removed from the Vernon control and treatment pulse-removal grids for three consecutive trapping periods starting in late September 1984. Two capillary tubes were used in each of 48 cans: 24 cans containing polymerized 3-propyl-1,2-dithiolane and 24 cans with polymerized 3,3-dimethyl-1,2-dithiolane. Each group of 24 cans with the respective odor was systematically placed in the burrows on the treatment grid with 2 cans per each of the 12 squares on opposite sides of the 36-square matrix. Thus, 12 squares in the middle third of the grid did not have cans. Cans with empty tubes were placed on the control pulse-removal grid in an identical manner. Locked-open traps were left in place on both grids. This experiment commenced immediately after completion of the third removal trapping period and covered the overwinter period of 1984-85, with the results recorded in May 1985.

Statistical analysis. Results of the laboratory and field bioassays were statistically evaluated by chi-square with an acceptable significance level of at least $p < 0.05$.

RESULTS

GCMS Analysis of Anal Gland Secretion

Diethyl ether extracts of the anal gland secretion were analyzed on an AEI MS-30 mass spectrometer combined with a Varian 2800 gas chromatograph and a data system to obtain high-resolution mass spectra of the components. The compounds were identified by comparison with samples obtained by synthesis (Crump, 1980a) or from commercial sources. As shown in Figure 2, the analysis revealed an essentially identical chromatogram to that obtained in the results from New Zealand animals (Crump, 1980a).

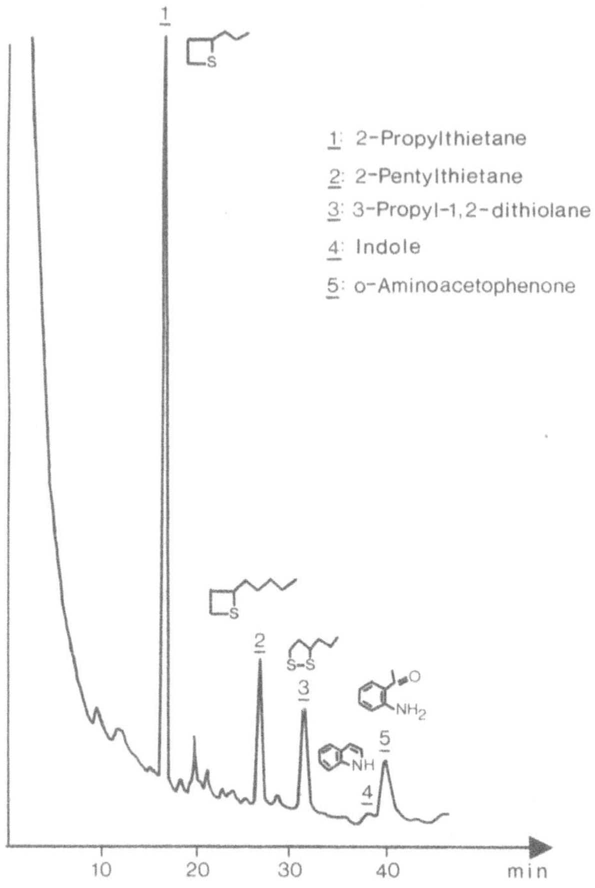

Fig. 2. Gas chromatogram of male stoat anal gland extract on 5% OV - 101 column, temperature 50°C, 10 min then 5°C/min to 120°C, flow 20 ml/min.

Laboratory Bioassays

The responses of gophers to the weasel mixture and single anal gland compounds are illustrated in Figure 3. There was no difference in responses between males and females in these trials. Gophers clearly avoided the 2-propylthietane, polymerized 3-propyl-1,2-dithiolane, and the weasel mixture. They also tended to avoid the 3,3-dimethyl-1,2-dithiolane from the ferret with the following results from 3 trials (control vs. treatment): 14-9; 14-9; 8-1. The control test, fresh 3-propyl-1,2-dithiolane, and 2-methylthiophene (novel odor) had little effect on gopher behavior.

Field Bioassays

Experiment A. The technique of placing a mixture of 2-propyl-thietane and 3-propyl-1,2-dithiolane in cans at the trapping points did not repel gophers from the Summerland treatment grid. The results in proportion of cans plugged with soil were: trial 1- control 14/24, treatment 14/25; trial 2- control 16/28, treatment 15/24. Between trapping periods, the usual response of gophers to locked-open traps (or cans) in their burrow system was to "block" the burrow by filling the trap or can with soil. Traps not

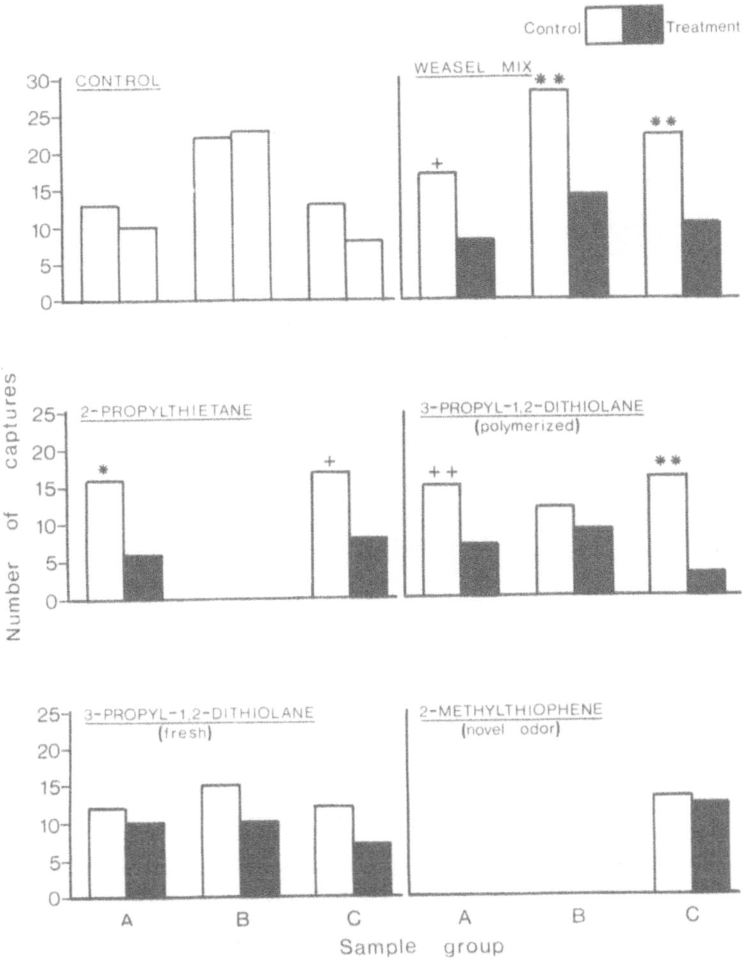

Fig. 3. Avoidance behavior responses ("capture" in live-traps) of pocket gophers to mustelid anal gland compounds and a novel odor in the laboratory bioassay arena. * $p < 0.05$; ** $p < 0.01$; significant difference by chi-square. + $p = 0.75$; ++ $p = 0.09$.

filled with soil were a reliable indication that the burrow system in the immediate vicinity of the trap was abandoned. Thus, if this technique had allowed the weasel odor to repel gophers from the treatment grid, presumably the majority of cans should have been empty indicating that the burrow systems were abandoned. That this was not the case was verified by trapping results, which showed little change in gopher numbers on either grid.

Experiment B. The response of gophers to anal gland compounds in traps are given in Table 1. Small sample sizes precluded statistical analysis, but fewer gophers tended to be captured in the initial trial with 3-propyl-1,2-dithiolane, the mixture of 2-propylthietane and polymerized 3-propyl-1,2-dithiolane, and 3,3-dimethyl-1,2-dithiolane. The control period (No. 5) showed an increase (from 5 to 9) in number of gophers captured on the treatment grid which provides some additional evidence that gophers were indeed avoiding certain odors. There was no difference in number of

Table 1. Number of pocket gophers captured on control and treatment grids at Summerland prior to and during trials of Experiment B.

Trapping Period or Trial	Compound	Control	Treatment
Prior to Experimental Trials		11	8
1	2-Propylthietane	11	7
2	3-Propyl-1,2-dithiolane	11	4
3	Mixture of 2-propylthietane and polymerized 3-propyl-1,2-dithiolane	9	3
4	3,3-Dimethyl-1,2-dithiolane	10	5
5	Control	13	9
6	3-Propyl-1,2-dithiolane	15	7
7	3-Propyl-1,2-dithiolane (clean traps)	14	7

gophers captured between trapping periods 6 and 7 (3-propyl-1,2-dithiolane and clean vs. field traps). The general increase in density of gophers on the control grid throughout this experiment was due to recruitment of juveniles into the population. Traps not catching gophers during these trials on the treatment grid were invariably plugged with soil. This response may be a typical anti-predator strategy by gophers which encounter potentially dangerous odors (and perhaps actual predators) in their burrow system.

Experiment C. The results of placing the weasel mix in gopher burrow systems in the field at Vernon are illustrated in Figure 4. Gophers readily colonized both the control and treatment pulse-removal grids in this trial. Again, the gopher response to odor in the cans was to plug them with soil. However, only 88.8% (16 of 18) of resident gophers was removed in the last trapping period prior to placement of the weasel mixture. Thus, some plugging of cans by the remaining two gophers probably occurred, thereby restricting the initial odor penetration of the burrow systems. Gophers colonizing the treatment grid also presumably plugged cans since 61.4% of cans were filled with soil at the end of this experiment on September 30.

Experiment D. This experiment began with complete removal of all gophers from the treatment grid which was essential to allow the 3-propyl-1,2-dithiolane and 3,3-dimethyl-1,2-dithiolane to permeate the burrow systems without disturbance. The results of this overwinter experiment are illustrated in Figure 4. Significantly ($p < 0.05$) fewer gophers colonized the treatment (3) than control (13) pulse-removal grid. A total of 13 gophers was present on the control grid during this initial post-winter trapping period in May. An influx of juvenile (young of the year) gophers occurred on all grids in the final trapping period (see Figure 4 and Table 2). Thus, the anal gland compounds deterred adult gophers from colonizing the treatment grid throughout the 5-month period December 1984 to April 1985. However, the annual reproduction produced a surplus of juvenile gophers which

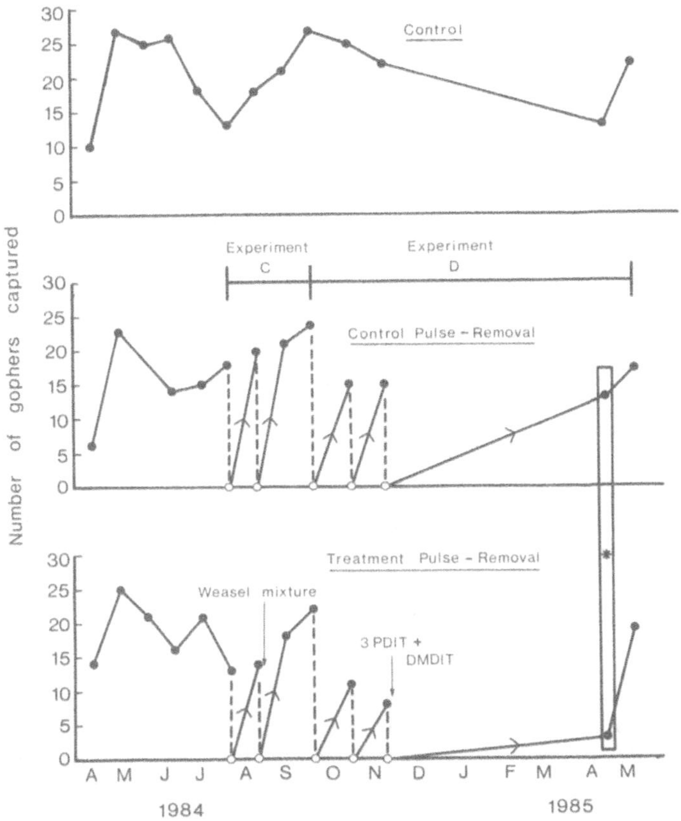

Fig. 4. Responses of pocket gophers to mustelid anal gland compounds in the field at Vernon study area. 3PDIT = 3-propyl-1,2-dithiolane. DMDI = 3,3-dimethyl-1,2-dithiolane. * $p < 0.05$; significant difference by chi-square.

filled in the open grid area. Evidence for this surplus was provided by the colonization of a permanent removal grid at the Vernon study area where juveniles constituted 13 of 17 gophers captured in the final trapping session (Sullivan and Sullivan, unpubl.). Further verification of the influence of the dithiolane odors on gophers during the overwinter period was the presence of voles (M. montanus) on both pulse-removal grids prior to the winter (Table 2). Voles persisted on the treatment grid until the final trapping period, but disappeared from the control pulse-removal during the winter. Voles will occupy only abandoned gopher burrow systems (Sullivan and Sullivan, unpubl.). This pattern was reflected by the control grid data (Table 2) where only 1 or 2 voles were captured in association with the resident population of gophers.

DISCUSSION

This study has demonstrated that certain mustelid anal gland compounds will produce an avoidance response by pocket gophers in both laboratory and field environments. As discussed for snowshoe hares (Sullivan and Crump, 1984), this biological activity is presumably a fear response by gophers to one or more components of mustelid (predator) anal gland odor.

Table 2. Ratio of adult to juvenile gophers and number of voles on grids during Experiment D at Vernon study area.

Trapping Period	Adult : Juvenile Gophers			Voles		
	Con	Con-P-R	Treat-P-R	Con	Con-P-R	Treat-P-R
1	27:0	24:0	22:0	0	0	0
2	25:0	14:0	11:0	2	4	4
3	22:0	15:0	8:0	0	8	9
Overwinter						
4	13:0	9:4	3:0	0	1	12
5	15:5	9:8	4:15	1	0	1

Con = control; Con-P-R = control pulse-removal; Treat-P-R = treatment pulse-removal.

This chemical scent communication may represent an interspecific signal between predator and prey. Similar avoidance responses by voles to mustelid odors (Stoddart, 1976, 1980; Gorman, 1984) and by rats to fox fecal odors (Vernet-Maury et al., 1984) suggest that this predator-prey communication may be well established among herbivorous small mammals and their principal predators. These semiochemicals are clearly of adaptive advantage to the prey if an individual animal avoids a given area, thereby lowering the probability of being preyed upon.

Pocket gophers are preyed upon by a wide variety of mammalian predators including weasels (Mustela spp.), coyotes (Canis latrans), foxes (Vulpes spp.), bobcat (Lynx rufus), and badger (Taxidea taxus) (Teipner et al., 1983). Gophers are also prey for avian and reptilian (snake) species. Weasels and snakes (e.g., Pituophis spp. and Crotalus spp.) are the only predators, in terms of body size, capable of entering subterranean burrow systems to prey upon pocket gophers. Weasels (both stoat and long-tailed), in particular, may invade gopher territory and prey heavily on these burrowing rodents (Criddle and Criddle, 1925; Hansen and Ward, 1966; Simms, 1979; Andersen and MacMahon, 1981). This invasion of burrow systems by weasels is particularly prevalent in winter (Criddle and Criddle, 1925; Simms, 1979) when above ground hunting may be poor for these mustelids. This provides a possible explanation for the significant avoidance of the Vernon treatment grid during the winter of 1984-85. Thus, the anal gland odor of the stoat and perhaps odors of the rattlesnake or gopher snake (see Henessy and Owings, 1978) are semiochemicals which have potential value as area repellents for pocket gophers. The results of our study have provided evidence that gophers will indeed avoid the mustelid compounds, 3-propyl-1,3-dithiolane and 3,3-dimethyl-1,2-dithiolane, in the field. The other mammalian predators must excavate the burrow, and the avian predators must catch an unwary gopher making a foray above ground.

The technique of placing a predator odor within the burrow systems appears to generate an avoidance response only if all resident gophers have been previously removed. The plugging of cans and traps (presumably an anti-predator strategy) with soil in Experiments A and B, respectively, clearly supports this conclusion. In addition, the failure of the weasel mix to repel gophers in Experiment C, even though this mixture was very effective in laboratory bioassays, may be due to plugging of cans before the odor could permeate the burrow systems. An odor delivery apparatus utilizing compressed air (Muller-Schwarze, 1983) may be a technique for rapid fumigation of burrows without disturbance by gophers. The plugging of traps in Experiment B suggests that Longworth live-traps may not be as suitable for field bioassays of the responses of gophers to various odors as they are for voles (see Stoddart, 1983).

Another aspect of gopher population dynamics which may help explain the poor field results with the weasel mix is the surplus of gophers still evident in September 1984 (Experiment C), and those colonizing the treatment grid in late May 1985 in Experiment D. As recorded by a permanent removal grid, juvenile and subadult (young of the year) gophers were available to fill in any open habitat during these periods (Sullivan and Sullivan, unpubl.). This colonization of removal areas by surplus gophers is presumably a response to the aggressive behavior of dominant adult animals who cause juveniles to disperse from their natal areas. In addition, as evidenced by the control population (see Fig. 4), few juvenile gophers recruit into and persist within an established adult population. Young subordinate gophers may fear and avoid intraspecific behavioral strife more than the prospects of living in an area with the odor of a potential predator. Intraspecific odors (e.g., compounds within the urine of dominant adult gophers) may elicit similar avoidance responses to those recorded in this study with mustelid anal gland compounds. If this is the case, both juvenile and adult gophers might avoid areas which have the odor of one or more intraspecific compounds. The repellent effect of anal gland odor of yellow voles (*Lagurus luteus*) on conspecifics has been reported by Fan (1983). It should be stressed that this intraspecific repellency was only temporary, whereas our study with interspecific compounds indicates that gophers avoided the treatment grid in Experiment D for a 5-month period.

Additional field trials with mustelid anal gland compounds and pocket gophers should be conducted to provide replication and investigation of the potential for practical use of these semiochemicals in crop protection. The polymerization of the dithiolanes deserves considerable attention since these compounds may provide some degree of persistence for naturally deposited anal scent marks. As discussed by Sullivan and Crump (1984), polymerization may be the key to a truly durable and effective repellent. After 5 months in the field, both dithiolanes on the Vernon treatment grid still had strong odors which could be detected above ground where tunneling by voles had perforated the abandoned gopher burrows and allowed the odors to escape. Thus, these odors had presumably permeated throughout most, if not all, of the burrow systems on the treatment grid during the 1984-85 winter, thereby rendering these burrows as unfavorable habitat for gophers.

ACKNOWLEDGMENTS

We thank the Research Branch of Agriculture Canada for financial support and the study areas and facilities provided by the Summerland Research Station; the British Columbia Ministry of Lands, Parks, and Housing for study areas in Kalamalka Lake Park; and the field assistance of D. S. Sullivan, J. Gareau, B. Jones, W. Kaiser, J. Krebs, and C. Robinson.

REFERENCES

Andersen, D. C., and MacMahon, J. A., 1981, Population dynamics and bioenergetics of a fossorial herbivore, *Thomomys talpoides* (Rodentia:

Geomyidae), in a spruce-fir sere, Ecol. Monogr., 51:179.
Anderson, G. G., 1980, Rodent control on agricultural land in British Columbia, B.C. Ministry of Agriculture, Victoria, B.C.
Anderson, K. K., and Bernstein, D. T., 1980, Sulfur compounds in mustelids, in: "Natural Sulfur Compounds," D. Cavallini, G. E. Gaull, and V. Zappia, eds., Plenum Press, New York.
Barnes, V. G., 1973, Pocket gophers and reforestation in the Pacific Northwest: A problem analysis, USDI Fish and Wildlife Service, Special Scientific Report, Wildlife No. 155.
Barnes, V. G., Anthony, R. M., Evans, J., and Lindsey, G. D., 1982, Evaluation of zinc phosphide bait for pocket gopher control on forest land, in: "Proceedings, Tenth Vertebrate Pest Conference," R. E. Marsh, ed., University of California, Davis.
Brinck, C., Gerell, R., and Odham, G., 1978, Anal pouch secretion in mink Mustela vison, Oikos, 38:68.
Brinck, C., Erlinge, S., and Sandell, M., 1983, Anal sac secretion in mustelids. A comparison, J. Chem. Ecol., 9:727.
Brinck, C., and Hoffmeyer, I., 1984, Marking urine and preputial gland secretion of male bank voles (Clethrionomys glareolus L.): Chemical analyses and behavioral tests, J. Chem. Ecol., 10:1295.
Byers, R. E., Young, R. S., and Neely, R. D., 1976, Review of cultural and other control methods for reducing pine vole populations in apple orchards, in: "Proceedings, Seventh Vertebrate Pest Conference," C. C. Siebe, ed., University of California, Davis.
Christiansen, E., and Døving, K., 1975, Pheromones in small rodents, occurrence and supposed usefulness in biolotical control, in: "Biocontrol of Rodents," L. Hansson, and B. Nilsson, eds., Swedish Natural Science Research Council, Stockholm.
Christiansen, E., 1976, Pheromones in small rodents and their potential use in pest control, in: "Proceedings, Seventh Vertebrate Pest Conference," C. C. Siebe, ed., University of California, Davis.
Criddle, N., and Criddle, S., 1925, The weasels of southern Manitoba, Can. Field-Nat., 39:142.
Crouch, G. L., 1982, Pocket gophers and reforestation on western forests, J. Forestry, 80:662.
Crump, D. R., 1978, 2-Propylthietane, the major malodorous substance from the anal gland of the stoat, Tetrahedron Lett., 1978:5233.
Crump, D. R., 1980a, Thietanes and dithiolanes from the anal gland of the stoat (Mustela erminea), J. Chem. Ecol., 6:341.
Crump, D. R., 1980b, Anal gland secretion of the ferret (Mustela putorius forma furo), J. Chem. Ecol., 6:837.
Davis, D. E., 1976, Management of pine voles, in: "Proceedings, Seventh Vertebrate Pest Conference," C. C. Siebe, ed., University of California, Davis.
Fan, Z., 1983, The ecological importance of the anal gland secretion of yellow voles (Lagurus luteus), in: "Chemical Signals in Vertebrates III," D. Muller-Schwarze, and R. M. Silverstein, eds., Plenum, New York.
Gorman, M. L., 1984, The response of prey to stoat (Mustela erminea) scent, J. Zool. Lond., 202:419.
Hansen, R. M., and Ward, A. L., 1966, Some relations of pocket gophers to rangelands on Grande Mesa, Colorado, Agriculture Experiment Station, Colorado State University, Tech. Bull. No. 88.
Henessy, D. F., and Owings, D. H., 1978, Snake species discrimination and the role of olfactory cues in the snake-directed behavior of the California ground squirrel, Behaviour, 65:116.
Muller-Schwarze, D., and Heckman, S., 1980, The social role of scent marking in beaver (Castor canadensis), J. Chem. Ecol., 6:81.
Muller-Schwarze, D., 1983, Chemical signals in alarm behavior of deer, in: "Chemical Signals in Vertebrates III," D. Muller-Schwarze, and R. M. Silverstein, eds., Plenum, New York.

Muller-Schwarze, D., Heckman, S., and Stagge, B., 1983, Behavior of free-ranging beaver (Castor canadensis) at scent marks, Acta Zool. Fenn., 174:111.

Schildknecht, H., Wilz, I., Enzmann, R., Grund, N., and Ziegler, M., 1976, Mustelan, the malodourous substance from the anal gland of the mink (Mustela vison) and polecat (Mustela putorius), Agnew. Chem. Int. Ed. Eng., 15:242.

Schildknecht, H., Birkner, C., and Kraub, D., 1981, Struktur und Wirkung der Musteliden-Odomone II. Erweiterte Analyse des Analbeutelsekretes des Nerzes Mustela vison L., Chemiker-Zeitung, 105:273.

Schildknecht, V. H., and Birkner, C., 1983, Analyse der Analbeutelsekrete Mitteleuropäischer Musteliden, Chemiker-Zeitung, 107:267.

Simms, D. A., 1979, North American weasels: Resource utilization and distribution, Can. J. Zool., 57:504.

Sokolov, V. E., Albone, E. S., Flood, P. F., Heap, P. F., Kagan, M. Z., Vasilieva, V. S., Roznov, V. V., and Zinkevich, E. P., 1980, Secretion and secretory tissues of the anal sac of the mink, Mustela vison. Chemical and histological studies, J. Chem. Ecol., 6:805.

Stoddart, D. M., 1976, Effect of the odor of weasels (Mustela nivalis L.) on trapped samples of their prey, Oecologia, 22:439.

Stoddart, D. M., 1980, Some responses of a free living community of rodents to the odors of predators, in: "Chemical Signals: Vertebrates and Aquatic Invertebrates," D. Muller-Schwarze, and R. M. Silverstein, eds., Plenum, New York.

Stoddart, D. M., 1983, Odor as a component of trap entry behavior in small rodents, in: "Chemical Signals in Vertebrates III," D. Muller-Schwarze, and R. M. Silverstein, eds., Plenum, New York.

Sullivan, T. P., and Crump, D. R., 1984, Influence of mustelid scent gland compounds on suppression of feeding by snowshoe hares (Lepus americanus), J. Chem. Ecol., 10:1809.

Sullivan, T. P., Krebs, J. A., and Kluge, H. A., 1985, Survey of mammal damage to tree fruit orchards in the Okanagan Valley of British Columbia, Can. Field-Nat., submitted.

Teipner, C. L., Garton, E. O., and Nelson, L., 1983, Pocket gophers in forest ecosystems, U.S.D.A. Forest Service, Intermountain Forest and Range Experiment Station, Gen. Tech. Rep, INT-154.

Van Den Berk, J., and Muller-Schwarze, D., 1984, Responses of wild muskrats (Ondatra zibethica L.) to scented traps, J. Chem. Ecol., 10:1411.

THE COMPARATIVE RESPONSES OF MICE AND VOLES TO CONSPECIFIC AND HETEROSPECIFIC ODORS IN THE FIELD - A LESSON IN SOCIAL BEHAVIOR

*D. Michael Stoddart
Department of Zoology
King's College London
Strand, London WC2R2LS

INTRODUCTION

Notwithstanding the very many population studies that have been conducted on small mammals in the field, remarkably little is known about their social behavior and organization under natural conditions. This is because of their small size and cryptic life style which render direct observation impossible. Small rodents are generally seen only when they enter live traps, which they do very readily but not all with equal frequency. Some aspects of social behavior have been deduced from the pattern of trap entry in a few species (e.g., Shillito (1963) for Microtus agrestis; Wuensch (1982) for Mus musculus and Peromyscus maniculatus; Boonstra and Krebs (1976) for Microtus townsendii; Daly et al. (1980) and Mazdzer et al. (1976) for Peromyscus leucopus; Rowe (1970) for Mus musculus), although the different likelihood of entry into scented traps shown by different members of a population (Stoddart, 1982a,b) cloud the general applicability of these deductions. Over a number of years I have conducted a series of studies on the responses of free living wood mice (Apodemus sylvaticus) and short-tailed voles (Microtus agrestis) to specific odors introduced into live-traps. Although these have been conducted under a variety of weather conditions and population densities, and not all at the same time of year, examination of the combined data yields important information on intra- and interpopulation interrelationships. The objective of this paper is to construct models of the involvement of conspecific and heterospecific odors in the social behavior of these two species of common European rodent, based on their responses to odor cues emanating from live-traps.

MATERIALS AND METHODS

Field work has been conducted in grassland and hedgerow habitats at King's College Field Station at Rogate, West Sussex, England (National Grid reference SU 807238), from 1975 until 1984. Longworth live-traps in pairs have been used on all occasions, provided with hay and baited with mixed poultry grains. Traps were invariably washed in hot water and left to dry in the air prior to each trapping session. Depending upon the specific objective of each trapping experiment, traps were either treated with a

*Present address: Department of Zoology, University of Tasmania, Box 252C, G.P.O. HOBART, Tasmania, Australia 7001.

Table 1. Table showing choices between scented and odorless traps by a) Apodemus sylvaticus and b) Microtus agrestis.

	Observed	Expected
a. Observed and expected choices by A. sylvaticus		
Scented (conspecific and heterospecific)	394	314.5
Odorless	235	314.5
	629	629.0
$\chi^2 = 40.2 \quad p = < 0.001$		
b. Observed and expected choices by M. agrestis		
Scented (conspecific and heterospecific)	28	40.5
Odorless	53	40.5
	81	81.0
$\chi^2 = 7.72 \quad p = < 0.01 \quad > 0.01$		

specific odorant or replaced with clean traps. Details of all trapping procedures are given in Stoddart (1976, 1980, 1982a, 1982b, 1983) and Stoddart and Smith (1984, in press). It is important to stress here that all trapping procedures ensured no spatial bias in the presentation of traps, randomization being assured by the toss of a coin, or by reference to listed random numbers. In all studies, great attention has been paid to the subject of whether a true choice existed for the rodents, and Stoddart and Smith (1984) described a data handling protocol to ensure that the only data selected for analysis were those which related to the existence of a real choice.

RESULTS

A number of questions have been posed, and these are listed below together with a presentation of relevant data. The composite nature of the data has resulted in not all questions applying to both species.

Can the Small Rodents Apodemus sylvaticus and Microtus agrestis Be Demonstrated to Be Able to Perceive Trap-borne Odor in the Field?

Apodemus sylvaticus. Table 1a shows that when confronted with a choice between odorless and scented traps, *A. sylvaticus* chooses to enter the scented traps significantly more often than the odorless traps. This choice is expressed throughout the year.

Table 2. Table showing choices between conspecific and heterospecific scented traps by a) *Apodemus sylvaticus* and b) *Microtus agrestis*.

		Observed	Expected
a.	*A. sylvaticus*		
	Conspecific	85	59.5
	Heterospecific*	34	59.5
		119	119.0
	$\chi^2 = 21.8$ p = < 0.001		
b.	*M. agrestis*		
	Conspecific	23	17
	Heterospecific*	11	17
		34	34
	χ^2 4.24 p = < 0.05 > 0.01		

*Heterospecific odor consists of trap scrapings from *A. flavicollis*, *Clethrionomys glareolus*, *Mus musculus*, *Sorex araneus* and *S. minutus*, as well as the two species considered here.

Microtus agrestis. Table 1b shows that a significantly smaller number of captures of *M. agrestis* are effected in traps scented with conspecific odor than in odorless traps. These data further confirm Stoddart's (1982a) data suggesting vole odor acts as an inhibitor to trap entry behavior.

The data in Table 1 indicates that *A. sylvaticus* and *M. agrestis* can both perceive residual trap odors and that this perception can be demonstrated under field conditions by observing the pattern of trap choice.

Can *A. sylvaticus* and *M. agrestis* Discriminate Between Different Odors In the Field?

To answer this question, it has been necessary to draw upon different experimental data for the two species in which different odor choices have been presented.

A. sylvaticus. *A. sylvaticus* discriminates clearly between members of a pair of traps, one scented with conspecific and the other with heterospecific odor, and chooses to enter the one scented with conspecific odor statistically significantly more frequently than the heterospecifically scented one (n = 119, $\chi^2 = 21.8$, p = < 0.001).

M. agrestis. Voles were presented with paired traps--one scented with their own body odor and the other with body odor of another vole. The number choosing the trap in which they themselves had been held previously was statistically significantly higher than those choosing the other trap (n = 204, $\chi^2 = 9.49$, o = < 0.005 > 0.001). I conclude from these two sets of data that odor discrimination occurs and can be demonstrated to occur under field conditions by observing the pattern of trap choice.

Table 3. Table showing choice of trap by sexually mature and immature A. sylvaticus, for each sex separately. This experiment was conducted between April and October 1983; i.e., wholly within the breeding season.

	Sexually Mature Male	Sexually Immature Male	Sexually Mature Female	Sexually Immature Female
Sexually mature male	22	15	11	28
Sexually mature female	25	11	10	36
Totals	47	26	21	64
	$\chi^2 = 0.8$ p = n.s.		$\chi^2 = 0.5$ p = n.s.	

Can A. sylvaticus and M. agrestis Distinguish Between Conspecific and Heterospecific Odors?

A. sylvaticus. The data for 119 straight choices are shown in Table 2a. A. sylvaticus clearly discriminates between these two odors, choosing to enter conspecific scented traps almost three times more frequently than heterospecific scented traps.

M. agrestis. When presented with a choice between equal numbers of heterospecifically scented and conspecifically scented traps, significantly more voles enter the conspecifically scented traps (Table 2b). Although the data are few, which influences the level of statistical significance, more than twice as many voles entered conspecific as heterospecific traps.

Can A. sylvaticus and M. agrestis Discriminate Between Individual Conspecifics By Odor?

A. sylvaticus. Our investigations indicated no ability to discriminate between "self" and "non-self." In 30 direct choices, $\chi^2 = 0.013$, p = n.s.

M. agrestis. Data have already been presented above which indicate a clearly marked ability to discriminate between "self" and "non-self."

Can Sex of Subject Be Discriminated From Odor Residue?

In an experiment conducted during spring and summer in 1983 and involving 158 choices, it was apparent that the responses of sexually mature and immature mice of both sexes to odor residue from sexually mature female mice showed no differential attraction or repulsion (Table 3).

M. agrestis. The results of 383 entries by adult voles into traps scented either with heterosexual of ipisexual odor were analyzed and a significant departure from the expected frequency of trap choice was detected (Table 4). More males than expected entered traps scented with female

Table 4. Table showing number of entries made by adult male and female *M. agrestis* into traps scented with either heterosexual or ipisexual odor.

	Trap Odor Type					
	Heterosexual Scent			Ipisexual Scent		
	Obs.	Exp.	χ^2	Obs.	Exp.	χ^2
Males	36	27.2	2.8	116	124.7	0.6
Females	33	41.7	1.8	200	191.2	0.4
Total	$\chi^2 = 5.68$ $p = < 0.02; > 0.01$					

odor, and fewer females than expected entered traps scented with male odor. These data were all obtained during the breeding season of the voles, but the precise sexual condition of female scent donors and responders was not assessed.

Do Both Sexes of A. sylvaticus and M. agrestis Respond to Odor in the Same Way?

A. sylvaticus. There exists a strong tendency for female mice to exercise a more conservative response to odor. Thus, when exposed to a direct choice between conspecific and heterospecific odors, females express a significant selection for conspecific odor ($\chi^2 = 24.9$, n = 61, p = < 0.001). Males do not make a statistically significant selection for conspecific odor in the same choice paradigm (n = 58, $\chi^2 = 3.4$, p = n.s.). However, this response must not be interpreted as an inability on the part of males to perceive heterospecific odor. When male mice occupying their "normal" habitat (woodland scrub) are presented with the choice between heterospecifically scented and odorless traps, they show a significant selection for the heterospecific odor (n = 81, $\chi^2 = 16.7$, p = < 0.01; > 0.001). These data pose a particularly interesting problem, for when male mice living in a grassland habitat are presented with a choice between heterospecifically scented and odorless traps, no selection is apparent (n = 128, $\chi^2 = 0.76$, p = n.s.). The most plausible interpretation is that heterospecific odor is perceived in both habitats, but only in the "normal" habitat is heterospecific odor actively investigated. It is interesting to note that female mice, in neither "normal" nor "abnormal" habitat, show any selection for or against heterospecifically scented traps.

M. agrestis. There appears to be tendency for the nature of the response of female voles to be similarly conservative, though the data are too few for rigorous statistical analysis. Stoddart (1980, 1983) indicated that heterospecific odor brought about a greater reduction in the numbers of female voles caught than males, but catches of both sexes were markedly depressed.

Do All Members of the Population Respond in an Identical Manner to Perceived Odor?

A. sylvaticus. During a trapping study designed to examine the effect of conspecific odor on the estimation of population size (Stoddart and Smith, in press), data were collected on the frequency of retrappability of mice. Mice were caught, on an average, on 71% of the maximum number of occasions possible, when traps were treated with conspecific odor. During this phase of the experiment, many more mice were caught than in a control period (of equal duration but employing no conspecific odor). Immediately after, inspection of the data revealed the presence in the first phase of a group of mice which were absent from the control period and which had a mean trappability of just 55%. By contrast, mice which were trapped in both phases had a mean retrappability of 90%. It thus appears as if there are (at least) two sections to the trap-revealed population; the larger part enters traps readily, irrespective of whether they carry conspecific odor, while the smaller part is far more reticent to enter traps and appears to do so only when they smell of conspecific odor and then with a much reduced frequency.

M. agrestis. In a study with a similar objective, Stoddart (1982a,b) treated live-traps for voles with tiger urine. A control period, when no tiger urine was used, preceded and followed an experimental period when this pungent odorant was used. A reduction in the number of voles caught occurred when the urine was used, as expected (Stoddart, 1976; 1982a), and this was caused by the absence of a group of voles from the traps during the experimental period which had been present in the controls both before and after. The mean retrappability of this group prior to the experimental period was 37.4%, while that of the voles which were undeterred by the tiger urine odor was 67%. This difference is significant ('t' = 3.46, 34 d.f., p = < 0.001).

DISCUSSION

The above data have been collected over a number of years and for a number of separate purposes. When pooled, what do they tell us about the manner in which odor signals are involved in the social behavior of each species?

There is little doubt that both species are able to perceive odors in the field, and base decisions on whether or not to enter a particular trap upon the perceived information. Both can discriminate between different odors and, when one of these is a familiar conspecific odor, choose to enter conspecifically scented traps more readily than either odorless or heterospecifically scented traps. When presented with a choice between heterospecifically scented and odorless traps, male A. sylvaticus living in their normal habitat exhibited a response pattern consistent with the hypothesis that, in this habitat, they are behaviorally dominant over M. agrestis. No such response was shown by female mice or by males living in a habitat from which they are normally excluded by voles. Consideration of these interactions suggests that woodmice are dominant to short-tailed voles in woodland, but not dominant in grassland. As woodmice normally do not inhabit grassland, it is possible that this pattern of spatial avoidance is maintained through differential response to heterospecific odor. The above data further suggest that it is only male woodmice which exert interspecific dominance; examination of the present data indicate that females are apparently inert.

It would not seem unreasonable to assume that the specific habitat exerts an influence on an individual mouse's response to the odor of another species. Calhoun (1963) and Blaustein (1980) have drawn attention to the fact that a relative dominance hierarchy exists between different species members of a community of small mammals; the data presented above might suggest that expressed dominance is a species characteristic which owes much to the habitat.

It is perhaps surprising that no identification of individuals could be demonstrated for A. sylvaticus, since individual variation in caudal organ secretion has been shown (Gorman, 1982), but the field data are few. In the vole, there is a clearly marked discrimination between the odors of "self" and "non-self," with voles choosing "self" significantly more often than "non-self." These data suggest that dispersion in burrow and tunnel systems might be maintained through the recognition of individual odor.

A major difference between the two species' odor responses appears to be that sex is encoded in the odor signals of M. agrestis, but is not encoded in the signals of A. sylvaticus. Voles live at much higher densities than mice (French et al., 1975), and one male is reportedly dominant in a group (Evans, 1977). The data herein presented are not inconsistent with the notion that males identify which females are sexually mature by olfactory cues. Mice, in contrast, live at lower density than voles (Flowerdew, 1977) and the relationship between male and female is closer. From the above data, it would seem reasonable to suggest that recognition of sex in A. sylvaticus is not dependent upon whole-body odor, though it would be wrong to imply that body odor plays no part in this process.

It is clear that in both species some members of the population are more influenced by odor than others, and this differential response can even be reflected in estimates of population size by Catch-Mark-Release-Recapture techniques (Stoddart, 1982b; Stoddart and Smith, in press). Heterospecific odor scented traps are avoided by a portion of the M. agrestis population which may be regarded as "trap shy" (although it is not known whether the group identified via odor screening is precisely the same as that originally defined as "trap shy" by Chitty and Kempson (1949)), and a group of mice of low trappability may be caught when conspecifically scented traps are employed. These data indicate the importance of standardization of trapping procedures in rodent population estimation.

In spite of the fact that the above-described models have been constructed from many studies on populations of unknown and various densitites, it appears that odors exert strong influences on trap entry behavior. It is logical to assume that it plays a similar role of importance in burrow and nest-site entry behavior, and upon all occasions in which an individual chances upon, or is drawn to, a place where another individual of the same, or a different species, has been. The models have the advantage of being constructed under field conditions, even if they suffer from the lack of comprehensiveness obtainable only in the laboratory.

REFERENCES

Blaustein, A. R., 1980, Behavioral aspects of competition in a three species rodent guild of coastal southern California, Behav. Ecol. Sociobiol., 6:247.

Boonstra, R., and Krebs, C. J., 1976, The effect of odours on trap response in Microtus townsendii, J. Zool. Lond., 180:467.

Calhoun, J. B., 1963, The social use of space, in: "Physiological Mammalogy, Vol. I," V. V. Mayer, and R. G. van Gelder, eds., Academic Press, New York.

Chitty, D., and Kempson, D., 1949, Prebaiting small mammals and a new design of live trap, Ecology, 30:536.

Daly, M., Wilson, M. I., and Behrends, P., 1980, Factors affecting rodents' response to odours of strangers encountered in the field. Experiments with odour-baited traps, Behav. Ecol. Sociobiol., 6:323.

Evans, D. M., 1977, Genus Microtus, in: "Handbook of British Mammals," G. B. Corbet, and H. N. Southern, eds., Blackwell Scientific Publications, Oxford.

Flowerdew, I. R., 1977, Apodemus sylvaticus, Blackwell Scientific Publications, Oxford.
French, N. R., Stoddart, D. M., and Bobek, B., 1975, Patterns of demography in small mammal populations, in: "Small Mammals, Their Productivity and Population Dynamics," F. B. Golley, K. Petrusewicz, and L. Ryszkowski, eds., C. U. P.
Gorman, M. L., 1982, Social differences in the subcaudal scent gland secretion of Apodemus sylvaticus (Rodentia:Muridae), J. Zool. Lond., 198:353.
Mazdzer, E., Capone, M. R., and Drickamer, L. C., 1976, Conspecific odors and trappability of deer mice (Peromyscus leucopus noveboracensis), J. Mammal., 57:607.
Rowe, F. P., 1970, The response of wild mice (Mus musculus) to live traps marked by their own and foreign mouse odour, J. Zool. Lond., 162:517.
Shillito, E. E., 1963, Exploratory behaviour in short-tailed vole, Microtus agrestis, Behaviour, 21:145.
Stoddart, D. M., 1976, Effect of the odour of weasels (Mustela nivalis L.) on trapped samples of their prey, Oecologia, 22:439.
Stoddart, D. M., 1980, Some responses of a free-living community of rodents to the odors of predators, in: "Chemical Signals in Vertebrates," D. Muller-Schwarze, and R. M. Silverstein, eds., Plenum, New York.
Stoddart, D. M., 1982a, Does trap odour influence estimation of population size of the short-tailed vole, Microtus agrestis? J. Anim. Ecol., 51:375.
Stoddart, D. M., 1982b, Demonstration of olfactory discriminiation by the short-tailed vole, Microtus agrestis L., Anim. Behav., 30:393.
Stoddart, D. M., 1983, Odor as a component of trap entry behavior in small rodents, in: "Chemical Signals in Vertebrates 3," D. Muller-Schwarze, and R. M. Silverstein, eds., Plenum, New York.
Stoddart, D. M., and Smith, P. A., 1984, Woodmice (Apodemus sylvaticus) can distinguish conspecific from heterospecific odors in the field, J. Chem. Ecol., 10:923.
Stoddart, D. M., and Smith, P. A., Recognition of odour-induced bias in the live trapping of Apodemus sylvaticus L., Oikos, in press.
Wuensch, K. L., 1982, Effect of scented traps on captures of Mus musculus and Peromyscus maniculatus, J. Mammal., 63:312.

MORPHOMETRIC PATTERNS AMONG MICROTINE RODENTS. I. SEXUAL SELECTION SUGGESTED BY RELATIVE SCENT GLAND DEVELOPMENT IN REPRESENTATIVE VOLES (MICROTUS)

Frederick J. Jannett, Jr.

Department of Biology
The Science Museum of Minnesota
30 East 10th Street
St. Paul, Minnesota 55101

INTRODUCTION

Chemical communication research is lacking in theoretical approaches (see, for example, Perrigo and Bronson, 1983; Sokolov et al., 1984). It should proceed within the context and theory of organic evolution, in which an active area of research has been sexual selection. Sexual selection depends on the advantage which certain individuals have over other individuals of the same sex and species in exclusive relation to reproduction, and it can affect an animal's glands for emitting odours (Darwin, 1871). Recent comments on sexual selection in mammals (Blaustein, 1981; Arnold and Houck, 1982) have stressed the possible role of scent communication without citing any comparative analyses of closely related species.

Voles and lemmings (Muridae : Microtinae) are ecologically diverse and there are about 18 recent genera currently recognized. The genus *Microtus* is the largest with about 45 species (e.g., Corbet and Hill, 1980). Upon morphological analyses, species in the genus can be considered highly evolved (Hinton, 1926), and there is increasing evidence for diversity in microtine social systems, which, as discussed below, include monogamy, polygyny, and promiscuity.

Voles also have a diversity of sebaceous scent glands, of which the largest complexes are the preputial glands, anal glands, and posterolateral glands. Aksenova (1973) described the paired preputial glands, which open on the prepuce, in nine species of *Microtus*. Vrtis (1929a) described the anal glands of *M. arvalis* as a series of glands encircling the rectum, and L. Skurat (pers. comm.) has found those of other species to be of similar structure. Quay (1968) reviewed the structure and positions of paired posterolateral skin glands on species of *Microtus* and concluded that a species may have them on the hips, or on the flanks, or not at all. Subsequent investigation indicates that there is a continuous zone of posterolateral glandular development among species from flanks to posterior hips and that some species have glands in an intermediate position (Jannett, unpubl.).

I hypothesized, first, that the development of the three major glandular regions would be correlated, that is, a species with relatively large preputial glands would also have relatively large anal and posterolateral glands, and, secondly, insofar as social systems have been identified, those species with well developed glandular areas would be polygynous.

MATERIALS AND METHODS

There were three sources of the voles analysed in this study, listed by species in Table 1: field-trapped specimens, laboratory colony animals, and field-trapped animals held in captivity for various lengths of time. Only voles with scrotal testes were used. All species determinations of Palaearctic forms were made by Soviet museum and laboratory colleagues.

Older specimens in the collections of the Museum of Zoology (Moscow State University) and the Institute of Zoology (USSR Academy of Sciences) were probably fixed in alcohol; later specimens were fixed in formalin and then preserved in alcohol. All North American specimens were fixed in 10% formalin. Of the 13 M. chrotorrhinus, 11 field-trapped specimens obtained in 1983 were shortly thereafter transferred to 35%→50%→70% isopropyl alcohol, from which organs were removed about 15 months later. Organs from M. richardsoni were never transferred to alcohol. To minimize possible loss of lipids from tissues in alcohol, all other material was left in formalin for periods ranging from three weeks to more than 11 years, until all were at hand. Carcasses were then transferred to 35% isopropyl alcohol for several days; organs were dissected and stored in 35% ethanol and weighed within about one week. Although tissues may have lost some weight in alcohol, this would seem of minor concern here inasmuch as this is a preliminary survey, and some of the heaviest glands were in old museum specimens.

For many of the Palaearctic museum specimens, body weight was unavailable. Organs were not weighed if they were damaged, except in the case of minor damage to a few large preputial glands of Palaearctic specimens. The weights of the posterolateral glands and testes were obtained by doubling one respective organ chosen at random so as to leave the remaining organ for future study. The weight of the paired preputial glands was obtained by doubling the weight of one gland in four instances because one gland was missing or damaged. Vesicular glands were weighed inasmuch as they may reflect androgen levels. Vesicular gland weight was never obtained by doubling the weight of one gland because many pairs appeared to be nonsymmetrical.

Hair was removed from the posterolateral gland by clipping it with scissors as close to the skin as possible. Inasmuch as the gland weight included skin weight, those glands that were lightest contained relatively less secretory tissue than those that were heaviest. After dissection of the proctodeal anal glands and the associated muscles (see Vrtis, 1929a), the rhabdosphincter encircling the glands was carefully teased away. A small strip of thin leiosphincter and rectum between the glands, to which the encircling glands were attached, was left intact so that the lobes of the glands remained intact.

Fixative on the surfaces of all glands was removed with blotting paper before weighing to the nearest 0.0001 g.

Because a test of the first hypothesis required relative weights of the scent organs, rank tests were used to analyse the data. Kendall rank correlation coefficients were obtained using the SAS (1982) program. In at least one species (M. montanus; Jannett, 1978b), all three scent glands are responsive to testosterone, so partial rank correlation coefficients were obtained wherein weight of the testes, presumably the primary source

Table 1. Species and sources of specimens. Numbers of specimens (n).

Species	Geographic locality of capture or of laboratory stock	Sources of specimens	Dates of capture or ages of laboratory voles
evoronensis	USSR, RSFSR, Primor'e, bank of Lake Evoron	L[a](1)	212 days
afganus	S113842[b], S113843, S113847	F[c](3)	
chrotorrhinus	USA, Minnesota, Cook Co. (several sites)	F(13)	Aug.8-Sept.13,1983; Aug. 18-31,1984
pinetorum	USA, New York, Ulster Co., New Paltz	L(11)	4-38 months
kirgisorum	USSR, Uzbek SSR, near delta of Amu Dar'ya	L(1)	1 year
socialis	USSR, Kazakh SSR, Jambul District, Betpakdala (2); S113886	F(3)	Oct.15,1970 (2)
longicaudus	USA, Wyoming, Teton Co., Moran (several sites)	F(6);C[d](3)	June1-5,1973;July24-27,1984
pennsylvanicus	USA, New York, Tompkins Co., Beam Hill	F(4)	Sept.1,1976
ochrogaster	USA, Illinois, Champaign Co.	L(9)	68 days
transcaspicus	USSR, Turkmen SSR, Kopetdag Mtns., Al-Dere (1); S129325	C(1);F(1)	
juldaschi	S132263, S132264	F(2)	
mongolicus	USSR, RSFSR, Chita Obl., Borzya (4); S123465	L(4);F(1)	271-413 days (4)
arvalis	USSR, Lithuanian SSR, near Trakai; USSR, RSFSR, Krasnodar	L(2)	
rossiaemeridionalis	USSR, RSFSR, Voronezh Obl., Chrenovoe	L(4)	
californicus	USA, California, Sonoma Co., Bodega Marine Laboratory	C(9);L(2)	55-60 days (2)
montanus	USA, Wyoming, Teton Co., Moran (several sites)	F(22)	May23-28,1984
brandti	S113850, S113851, S113852	F(3)	
maximowiczii	USSR, RSFSR, Primor'e, nw of city of Khabarovsk	L(1)	207 days
gud	USSR, RSFSR, Dagestan ASSR, Khvarshe; USSR, Georgian SSR, Bakureany	L(2)	242+334 days
gregalis	S4098, S4100	F(2)	
fortis	S127321	F(1)	
richardsoni	USA, Wyoming, Teton Co. (several sites)	F(18)	July29-Sept.17,1973; July17-Sept.7,1974

[a] L=born in laboratory.

[b] Catalogue numbers refer to specimens in the Museum of Zoology, Moscow State University.

[c] F=field-trapped.

[d] C=field-trapped, held in captivity as adults.

of testosterone, was statistically controlled (Siegel, 1956). Inasmuch as the vesicular glands of M. montanus exhibited a stronger correlation with testosterone dosage than the scent glands (Jannett, 1978b) and may therefore reflect androgen levels, partial rank correlation coefficients were obtained wherein weight of the vesicular glands was statistically controlled. Lastly, because species varied in body weight, partial rank correlation coefficients were obtained, statistically controlling for body weight.

RESULTS

Body, vesicular gland, and scent gland weights are listed in Table 2, where species are entered in order from that with the smallest mean sum of the three scent gland weights to the species with the largest mean sum, the exception being M. evoronensis, for which no anal gland weight was available. The sums were obtained from individual voles for which all three weights were had, not from the average weights of the three glands. Vesicular and scent gland weights are depicted in Figure 1.

The ranks of preputial gland mean weights were correlated with those of the posterolateral gland mean weights (Kendall rank correlation coefficient, $\tau = 0.62$, $p < .0001$, $N = 22$), and those of anal gland mean weights ($\tau = 0.48$, $p = .001$, $N = 21$). Ranks of posterolateral gland mean weights were correlated with ranks of anal gland mean weights ($\tau = 0.43$, $p < .005$, $N = 21$). The ranks of the mean weights of the three scent glands were correlated as measured by the Kendall coefficient of concordance ($W = 0.79$, $p < .001$).

For the 20 species for which all three scent gland, vesicular gland, and testes weights were obtained, Kendall τ was recomputed, and compared with corresponding partial correlation coefficients for which testes weight and vesicular gland weight were statistically controlled. Kendall τ was recomputed and partial correlations obtained for the 15 species for which body weights were available (Table 3). When testes weight and body weight were controlled, correlations remained high. Correlations dropped when vesicular gland weight was controlled. Nevertheless, some species with large vesicular glands had relatively small scent glands. This was most striking in M. pennsylvanicus.

Discordant patterns of scent gland development were apparent among some species. For example, M. arvalis had larger preputial glands but smaller anal glands than M. rossiaemeridionalis, and the former species had posterolateral glands whereas the latter did not.

The sum of the three scent gland weights expressed as a percentage of body weight, calculated on the basis of individual voles, is given in Table 2. It ranged from 0.09% in M. chrotorrhinus (n=9) to 0.89% in M. richardsoni (n=17) and 1.11% in M. gud (n=2).

DISCUSSION

The pattern of scent gland development in these species fulfills three criteria proposed by Darwin (1871) for characters suggesting evolution by sexual selection, namely, the glands are larger in males than in females, they are most highly developed at sexual maturity, and they are most highly developed during the breeding season. All three trends are evidenced most clearly by the patterns of posterolateral gland development (Quay, 1968). Vrtis (1929b) described the anal glands of M. arvalis as being larger on males than on females, and largest on adult males during

Table 2. Body weight, organ weights, sum of three scent gland weights, and sum of three scent gland weights as a percentage of body weight. Mean ± SE (n). Body weight, g; organ weights, mg.

SPECIES	BODY WEIGHT	TESTES WEIGHT	VESICULAR GLANDS WEIGHT	PREPUTIAL GLANDS WEIGHT	ANAL GLANDS WEIGHT	POSTEROLATERAL GLANDS WEIGHT	SUM OF SCENT GLANDS WEIGHTS	SUM OF GLANDS WEIGHTS AS PERCENTAGE OF BODY WEIGHT
EVORONENSIS	79.6 (1)	506 (1)	238 (1)	26 (1)	--	0 (1)	--	--
AFGANUS	--	66± 11 (3)	15 (1)	3± 1 (3)	21± 3 (3)	0 (3)	24± 3 (3)	--
CHROTORRHINUS	26.9±1.9 (10)	385± 41 (13)	102± 18 (13)	7± 2 (12)	14± 2 (13)	3± 2 (13)	25± 4 (12)	.09 (9)
PINETORUM	25.6±1.0 (11)	83± 5 (11)	35± 6 (11)	3± 1 (11)	27± 2 (11)	0 (11)	31± 3 (11)	.12 (11)
KIRGISORUM	35.5 (1)	276 (1)	70 (1)	5 (1)	37 (1)	0 (1)	42 (1)	.12 (1)
SOCIALIS	--	192± 40 (3)	96± 10 (2)	16± 5 (3)	42± 9 (3)	0 (3)	57± 14 (3)	--
LONGICAUDUS	32.0±3.5 (9)	439± 49 (9)	141± 21 (9)	21± 3 (9)	31± 2 (9)	9± 6 (9)	60± 9 (9)	.20 (9)
PENNSYLVANICUS	43.5±1.9 (4)	1604± 59 (4)	514± 89 (4)	18± 2 (4)	49± 5 (4)	0 (4)	67± 3 (4)	.15 (4)
OCHROGASTER	37.5±2.3 (9)	288± 15 (9)	144± 25 (9)	35± 3 (9)	33± 3 (8)	0 (9)	67± 5 (8)	.18 (8)
TRANSCASPICUS	60.5 (1)	523± 10 (2)	137±105 (2)	6 (1)	104±25 (2)	0 (2)	85 (1)	.14 (1)
JULDASCHI	--	359+ 48 (2)	202± 78 (2)	42± 5 (2)	80± 7 (2)	0 (2)	122± 12 (2)	--
MONGOLICUS	30.8±3.1 (4)	277± 42 (5)	127± 17 (5)	37±15 (4)	50±18 (5)	86± 20 (5)	147± 40 (4)	.46 (4)
ARVALIS	31.5±3.5 (2)	147±118 (2)	232± 45 (2)	35± 3 (2)	91± 3 (2)	23± 23 (2)	148± 23 (2)	.48 (2)
ROSSIAEMERIDIONALIS	47.2±1.9 (4)	327± 7 (4)	146± 28 (4)	8± 1 (4)	141±20 (4)	0 (4)	149± 21 (4)	.31 (4)
CALIFORNICUS	54.2±2.4 (11)	180± 10 (11)	220± 22 (11)	30± 5 (11)	133±12 (11)	109± 37 (11)	271± 37 (11)	.50 (11)
MONTANUS	36.5±1.1 (22)	394± 17 (22)	158± 14 (22)	97±10 (22)	96± 7 (22)	78± 10 (22)	271± 21 (22)	.74 (22)
BRANDTI	--	600± 61 (3)	392± 95 (3)	118±79 (2)	71±25 (3)	113± 15 (3)	331± 94 (2)	--
MAXIMOWICZII	66.5 (1)	407 (1)	271 (1)	61 (1)	106 (1)	251 (1)	419 (1)	.63 (1)
GUD	48.9±8.1 (2)	795± 76 (2)	303± 56 (2)	184±54 (2)	229±38 (2)	135± 18 (2)	548±110 (2)	1.11 (2)
GREGALIS	--	224± 44 (2)	288±132 (2)	132±56 (2)	117±64 (2)	392±130 (2)	641±250 (2)	--
FORTIS	--	1313 (1)	--	82 (1)	501 (1)	82 (1)	665 (1)	--
RICHARDSONI	80.0±5.6 (18)	979± 96 (18)	499± 85 (18)	67± 8 (18)	402±33 (17)	254± 47 (18)	713± 85 (17)	.89 (17)

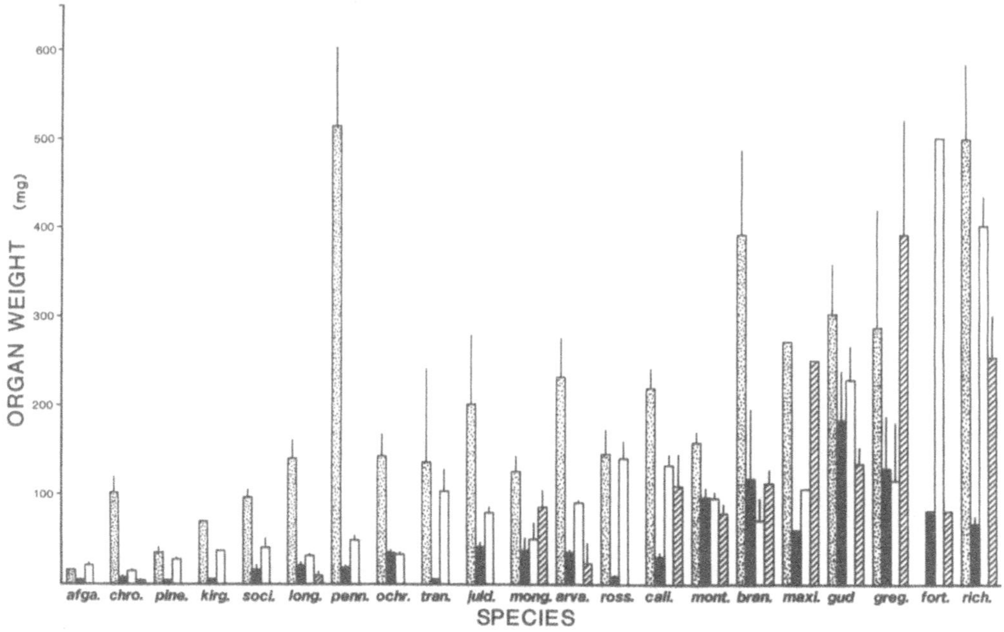

Fig. 1. Organ weights. Average for each species + SE. Sample sizes are given in Table 1. No vesicular gland weight was available for *M. fortis*. Where no weight is depicted for posterolateral gland weights, the specimen(s) lacked this gland. Stippled = vesicular glands; solid = preputial glands; open = anal glands; hatched = posterolateral glands.

the breeding season. Preputial glands are larger on males than on females in numerous species of *Microtus* and are also largest on mature breeding males (Jannett, unpubl.). That hypertrophy is correlated among the three scent organs is predicted in the context of sexual selection theory (e.g., Kodric-Brown and Brown, 1984).

The second hypothesis presented here, that species with well developed scent glands would be polygynous, is supported by field work, some of it including telemetry, on several species. Among the species with most highly developed scent organs were *M. richardsoni*, *M. montanus*, and *M. californicus*. *Microtus richardsoni* is polygynous (Ludwig, 1984a, 1984b), *M. montanus* is polygynous (Jannett, 1978a, 1981b, 1982), and one might conclude from field work on *M. californicus* (e.g., Lidicker, 1980) that it is also polygynous. Among those species with relatively small anal and preputial glands and no posterolateral glands, were *M. pennsylvanicus* which is promiscuous (Getz, 1972; Madison, 1980), *M. ochrogaster* which is monogamous (Getz and Carter, 1980), and *M. pinetorum*, which, in the highly structured orchard habitat, is found in groups of one adult female associating with up to several adult males (FitzGerald and Madison, 1983). Lastly, preliminary radiotelemetric study of *M. chrotorrhinus* which has small anal, preputial, and posterolateral glands indicates that it is neither monogamous, nor promiscuous, nor polygynous; most adults apparently nest solitarily at low densities around small isolated habitat patches (Jannett, unpubl.).

Table 3. Kendall correlation and partial correlation coefficients.

Variables			N	τ_{12}[a]	$\tau_{12.3}$[b]
1	2	3			
Preputial glands	Posterolateral glands	Testes	20	.64	.62
"	"	Vesicular glands	20	.64	.49
"	"	Body	15	.66	.63
"	Anal glands	Testes	20	.47	.43
"	"	Vesicular glands	20	.47	.24
"	"	Body	15	.43	.36
Posterolateral glands	"	Testes	20	.44	.41
"	"	Vesicular glands	20	.44	.26
"	"	Body	15	.45	.36

[a]Correlation coefficients.

[b]Partial correlation coefficients.

Scent glands probably aid in territorial maintenance by males in polygynous species. In laboratory terraria where they were established as residents and stimulated by constrained conspecific males, M. richardsoni males used two active marking patterns to deposit posterolateral glandular secretion; M. montanus males in the same test situation used one marking behavior (Jannett and Jannett, 1974, 1981). Both species had a graded behavioral pattern for actively marking with anal secretion (Jannett, 1977; Jannett, 1976). Both species had an additional posture that apparently served to deposit preputial secretion; vital dye injected into the preputial glands of M. montanus appeared in the scent mark (Jannett, 1977; Jannett, 1976). There is evidence for M. montanus in laboratory tests of field-trapped males of different social status (Jannett, unpubl.) that the posterolateral glandular secretion inhibits the movements of some males.

Glandular secretions of males may also facilitate mate choice by females in the polygynous species. Territorial males scent mark more than other males (Jannett, 1981b). Marked territories would allow females a basis for mate choice. Furthermore, lactating M. montanus in low and moderate density populations generally abandon their litters prior to subsequent parturition (Jannett, 1978a). Occasionally this shift in territory results in the female moving from the territory of one male to that of another (Jannett, 1982). At such times the female may be exhibiting mate choice, and the pattern of movement has been implicated for several species of Microtus (Jannett, 1980).

Two alternate explanations for the observed correlation of glandular development are that the pattern is largely the product of natural selection, and that development is dependent on an overriding mechanism of androgen responsiveness, which is not necessarily adaptive. Natural selection is not likely of primary importance. Hypertrophied glandular areas are costly: they direct the attacks of male voles of other species (Jannett, 1981a); they attract predators. The adaptive significance of the glands in the context of natural selection would perhaps be similar to that of smaller glands of other species in the context of natural selection, e.g., with respect to species recognition.

A single mechanism controlling androgen responsiveness of all tissues is not likely given the variability of control mechanisms (e.g., Ebling, 1977). Also, in *M. pennsylvanicus* the vesicular glands, presumably concerned primarily with ejaculate, are very large, whereas the preputial and anal glands are small and the posterolateral glands are lacking. Lastly, even though all of the scent glands are responsive to testosterone in *M. montanus* males, they are responsive to different degrees (Jannett, 1978b).

The pattern does not simply reflect taxonomic groupings. Species in the subgenus *Microtus* (Anderson, 1960; Hooper and Hart, 1962) include those with both large scent glands (*M. montanus* and *M. californicus*) and small scent glands (*M. pennsylvanicus* and *M. chrotorrhinus*). On the other hand, the taxonomic positions are less certain for the two species with the extremes of sum of scent gland weights. *Microtus richardsoni* had the largest average sum of scent gland weights. This species has been assigned to *Arvicola* and to *Microtus*, subgenus *Aulacomys* (see Hooper and Hart, 1962). *Microtus afganus* had relatively small glands. It has been assigned to the monotypic *Blandfordimys* (Ellerman and Morrison-Scott, 1951), to *Pitymys* (Corbet and Hill, 1980), and to *Microtus*, subgenus *Blandfordimys* (Gromov and Polyakov, 1977). The sizes of the scent gland assemblages of these species may be correlated with body size, but may in part reflect phylogeny. The posterolateral glands of *M. richardsoni* are also more specialized than those of other species of *Microtus* (Quay, 1968).

One aspect of scent gland development subject to natural selection may be species recognition. The disparate development of scent glands on *M. arvalis* and *M. rossiaemeridionalis*, until recently considered to be one species (Meyer et al., 1972), may be related to species recognition in areas of sympatry. Glandular development may subsequently be found to differ geographically. This species complex would seem to be an ideal model for future chemical investigation.

The consequences of scent organs evolving by sexual selection with respect to chemical communication theory, are that such structures may evolve rapidly (c.f., Fisher, 1930). I suggest there may be considerable redundancy in the messages; redundancy is a general phenonenon in animal communication (Markl, 1985), and its evolution and significance are poorly understood. I also hypothesize that the functional chemical constituents of the large glands of the polygynous species will be more diverse than the functional chemical constituents of the respective small glands of the non-polygynous species.

The consequences of evolution by sexual selection among voles with respect to demography, genic structure, and life history strategies may be immense. For example, some species with hypertrophied scent glands on males exhibit other patterns of sexual dimorphism, a greater disparity in the operational sex ratio than is shown by other species, and generally higher population densities than are achieved by other species (Jannett, in prep.).

Inasmuch as species of *Microtus* may exhibit different degrees of sexual selection, the genus will be an ideal group in which to investigate the evolution of sexual selection and the role of chemical communication in sexual selection.

ACKNOWLEDGMENTS

Collection of material was supported by the Ford Foundation Ecology of Pest Management Program at Cornell University, the New York Zoological Society, the Minnesota Non-Game Research Program, and the Science Museum of Minnesota. I thank the National Park Service for permission to collect in

Grand Teton National Park. Work at the Museum of Zoology (Moscow) and Institute of Zoology (Leningrad) was supported by the Academies of Sciences of the USA and USSR.

I thank V. E. Sokolov for hosting my visit to the USSR, and O. Rossolimo, P. P. Strelkov, and I. Ya. Pavlinov for access and assistance in collections. I thank J. Jannett for assistance at each stage of the project, S. Nowland for typing the manuscript, and J. Jannett, J. Moodie, and S. Pfeifer for critically reading the first draft. I especially thank L. L. Getz, F. N. Golenischev, V. Malikov, V. M. Malygin, M. N. Meyer, R. Ostfeld, and M. Schadler for animals from their respective colonies.

REFERENCES

Aksenova, T. G., 1973, The structure of accessory male glands in some species of the genus Microtus (Rodentia, Cricetidae), Zool. Zh., 52:1843.

Anderson, S., 1960, The baculum in microtine rodents, Publ. Mus. Nat. Hist. Univ. Kans., 12:181.

Arnold, S. J., and Houck, L. D., 1982, Courtship pheromones: Evolution by natural and sexual selection, in: "Biochemical Aspects of Evolutionary Biology," M. H. Nitecki, ed., University of Chicago Press, Chicago.

Blaustein, A. R., 1981, Sexual selection and mammalian olfaction, Amer. Nat., 117:1006.

Corbet, G. B., and Hill, J. E., 1980, "A World List of Mammalian Species," British Museum (Natural History), London.

Darwin, C., 1871, "The Descent of Man, and Selection in Relation to Sex," John Murray, London.

Ebling, F. J., 1977, Hormonal control of mammalian skin glands, in: "Chemical Signals in Vertebrates," D. Muller-Schwarze, and M. M. Mozell, eds., Plenum Press, New York.

Ellerman, J. R., and Morrison-Scott, T. C. S., 1951, "Checklist of Palaearctic and Indian Mammals 1758 to 1946," British Museum (Natural History), London.

Fisher, R. A., 1930, "The Genetical Theory of Natural Selection," Clarendon Press, Oxford.

FitzGerald, R. W., and Madison, D. M., 1983, Social organization of a free-ranging population of pine voles, Microtus pinetorum, Behav. Ecol. Sociobiol., 13:183.

Getz, L. L., 1972, Social structure and aggressive behavior in a population of Microtus pennsylvanicus, J. Mamm., 53:310.

Getz, L. L., and Carter, C. S., 1980, Social organization in Microtus ochrogaster populations, Biologist, 62:56.

Gromov, I. M., and Polyakov, I. Ya., 1977, "Microtinae," Fauna SSSR, Mammals (Nauka, Moscow - Leningrad), 3:1.

Hinton, M. A. C., 1926, "Monograph of the Voles and Lemmings (Microtinae) Living and Extinct," British Museum (Natural History), London.

Hooper, E. T., and Hart, B. S., 1962, A synopsis of recent North American microtine rodents, Misc. Publ. Mus. Zool. Univ. Mich., 120:1.

Jannett, F. J., Jr., 1977, "On the Sociobiology of the Montane Vole, Microtus montanus nanus (Rodentia: Muridae)," Ph.D. dissertation, Cornell University, Ithaca, New York.

Jannett, F. J., Jr., 1978a, The density-dependent formation of extended maternal families of the montane vole, Microtus montanusnanus, Behav. Ecol. Sociobiol., 3:245.

Jannett, F. J., Jr., 1978b, Dosage response of the vesicular, preputial, anal, and hip glands of the male vole, Microtus montanus (Rodentia: Muridae), to testosterone propionate, J. Mamm., 59:772.

Jannett, F. J., Jr., 1980, Social dynamics of the montane vole, Microtus montanus, as a paradigm, Biologist, 62:3.

Jannett, F. J., Jr., 1981a, Scent mediation of intraspecific, interspecific, and intergeneric agonistic behavior among sympatric species of voles (Microtinae), Behav. Ecol. Sociobiol., 8:293.
Jannett, F. J., Jr., 1981b, Sex ratios in high-density populations of the montane vole, Microtus montanus, and the behavior of territorial males, Behav. Ecol. Sociobiol., 8:297.
Jannett, F. J., Jr., 1982, Nesting patterns of adult voles, Microtus montanus, in field populations, J. Mammal., 63:495.
Jannett, F. J., Jr., and Jannett, J. Z., 1974, Drum-marking by Arvicola richardsoni and its taxonomic significance, Amer. Midl. Nat., 92:230.
Jannett, F. J., Jr., and Jannett, J. Z., 1981, Convergent evolution in the behavior of a shrew and a rodent, Mammalia, 45:473.
Jannett, J. A., 1976, "The Behavior of the Water Vole, Arvicola richardsoni DeKay, with Notes on Growth, Reproduction, and Scent Glands," M. S. thesis, Syracuse University, New York.
Kodric-Brown, A., and Brown, J. H., 1984, Truth in advertising: The kinds of traits favored by sexual selection, Amer. Nat., 124:309.
Lidicker, W. Z., Jr., 1980, The social biology of the California vole, Biologist, 62:46.
Ludwig, D. R., 1984a, Microtus richardsoni, Mammalian Species, 223:1.
Ludwig, D. R., 1984b, Microtus richardsoni microhabitat and life history, Spec. Publ. Carnegie Mus. Nat. Hist., 10:319.
Madison, D. M., 1980, Space use and social structure in meadow voles, Microtus pennsylvanicus, Behav. Ecol. Sociobiol., 7:65.
Markl, H., 1985, Manipulation, modulation, information, cognition: Some of the riddles of communication, in: "Experimental Behavioral Ecology and Sociobiology," B. Holldobler, and M. Lindauer, eds., Fischer Verlag, Stuttgart.
Meyer, M. N., Orlov, V. N., and Skholl, E. D., 1972, On the nomenclature of 46- and 54-chromosome voles of the type Microtus arvalis, Zool. Zh., 51:157.
Perrigo, G., and Bronson, F. H., 1983, Communication disparities between genetically-diverging populations of deermice, in: "Chemical Signals in Vertebrates 3," D. Muller-Schwarze, and R. M. Silverstein, eds., Plenum Press, New York.
Quay, W. B., 1968, The specialized posterolateral sebaceous glandular regions in microtine rodents, J. Mammal., 49:427.
SAS Institute, Inc., 1982, "SAS User's Guide: Basics," SAS Institute, Inc., Cary, North Carolina.
Siegel, S., 1956, "Nonparametric Statistics for the Behavioral Sciences," McGraw-Hill Book Co., New York.
Sokolov, V. E., Aleinikov, P. A., and Zinkevich, E. P., 1984, Chemical communication of mammals: A challenge to a biologist, a semioticist and a chemist, Acta Zool. Fennica, 171:35.
Vrtis, V., 1929a, The anal glands and the anus of the field-vole (Microtus arvalis [Pall.]), Biol. Spisy, Brno., 8(11):1.
Vrtis, V., 1929b, The development of the sebaceous anal glands of the field-vole (Microtus arvalis [Pall.]) and their changes during breeding season and aging, Biol. Spisy, Brno., 8(12):1.

FEMALE-INDUCED DELAY OF PUBERTY IN BANK VOLE AND EUROPEAN PINE VOLE FEMALES

Anna Marchlewska-Koj and M. Kruczek

Institute of Zoology
Jagiellonian University
30-060 Krakow, Poland

INTRODUCTION

In several species of rodents, the rate of sexual maturation and density of natural populations are negatively correlated (Krebs and Myers, 1974). Chemical signals discharged by females could be one of the important factors limiting reproductive activity.

Here we report evidence that ovarian activity of females from Microtidae family may be suppressed by the presence of conspecific females or their urine.

MATERIALS AND METHODS

All the experimental bank voles (*Clethrionomys glareolus*) and European pine voles (*Pitymys subterraneus*) came from outbred stock colonies reared in our laboratory. Animals were maintained in metal cages (40 x 20 x 25 cm) at 18-20°C, in a 10:14 photoperiod schedule and fed with natural food (oats, wheat, apples and red beets). Wood shavings were provided as bedding and nesting material.

Bank Voles (*Clethrionomys glareolus*)

Experiments were carried out from the beginning of April until the end of September in 1982 and repeated in similar conditions in 1983.

At the beginning of April, 30 adult, marked females and 10 males were introduced into a 36 m^2 enclosure. The natural grass covered the enclosure floor, but the animals were fed with the same diet as those maintained in laboratory. At the end of each month, bank voles were caught with live-traps. The number of animals and their marks were checked. Nonmarked, young animals were assumed to have been born in the month of trapping.

Males born in the enclosure were removed, but females were marked and randomly divided into 2 groups. They were either re-introduced free into the enclosure or were kept in cages (1 or 2 animals per cage) in the same enclosure (outdoor cages) until the end of September. If an old female (mother) or old male (father) died, the missing animal was replaced by the next one of proper age from laboratory stock.

Table 1. Body weight, uterine weight, ovarian weight, and number of large ovarian follicles in bank vole females born from June to August and reared in the enclosure, in outdoor cages or in the laboratory until killed on September 30

	Enclosure	Cages	
		Outdoor	Laboratory
No. of females	27	26	26
Body weight (g)	19.20 ± 0.25^{Aa}	17.79 ± 0.59^{A}	14.52 ± 0.42
Uterine weight (mg)	5.64 ± 0.25	8.92 ± 1.66^{b}	9.87 ± 0.89^{b}
Ovarian weight (mg)	1.48 ± 0.11	3.10 ± 0.28^{C}	3.45 ± 0.28^{C}
Ovarian follicles	2.92 ± 0.59	10.00 ± 0.93^{D}	12.69 ± 1.42^{D}

Mean ± s.e.m. not followed by the same letter within a line are significantly different at $P < 0.05$ for a, b and $P < 0.01$ for A, C, D (Duncan's New Multiple Range test).

For the experiment in the laboratory, 5 adult females and 5 adult males were used each year. One female and one male per cage were mated at the beginning of April. Their pups of both sexes were kept with parents until weaning at 21 days of age. Then they were separated and 1 or 2 females per cage were housed until the end of the experiment. All young females were killed on September 30.

European Pine Voles (*Pitymys subterraneus*)

Females at 21 days of age with a body weight of 9.9 ± 0.30 g (mean ± s.e.m.) were randomly allocated to 5 experimental groups and housed 1 to 2 animals per cage. They were reared in the presence of adult female urine during the next 3 weeks. A cage with juvenile females was connected with a metabolic cage in which an intact or ovariectomized female was maintained for 24 hrs. Urine donors were housed singly or as 6 animals per cage (grouped females). They were transferred to the metabolic cage once a week. Females were ovariectomized at least 4 weeks before use. All young females were killed when 42 days old.

Bioassay

Uterus and ovarian weights and number of large ovarian follicles were used as an indicator of hormonal activity of tested females.

Histology

Both ovaries were fixed in Bouin's fluid, serially sectioned (7 μm thick) and stained with haematoxylin and eosine. The large ovarian follicles (types: 6, 7, 8, atretic) were classified according to Pedersen and Peters (1968).

Data were analysed by using ANOVA, Duncan's New Multiple Range test and Dunnett's test (Zar, 1984).

Table 2. Uterine weight and number of ovarian follicles in European pine vole females exposed to the urine of adult intact and ovariectomized (OVX) females

	No. of	Uterine	Ovarian Follicle Number				
	Females	Wt. (g)	Type 6	Type 7	Type 8	Atretic	Total
Control	16	$6.8 \pm 0.38^{a,b}$	8.7 ± 0.93	4.4 ± 0.45	$3.3 \pm 0.42^{a-d}$	3.7 ± 0.58	$20.1 \pm 1.54^{a,b}$
Isolated Intact Females	9	5.6 ± 0.58	5.9 ± 1.34	5.0 ± 0.93	1.4 ± 0.29^{a}	3.6 ± 0.74	15.9 ± 2.64
Grouped Intact Females	9	7.2 ± 0.55	7.0 ± 0.92	4.1 ± 0.64	0.9 ± 0.43^{b}	3.7 ± 0.61	15.6 ± 1.94
Isolated OVX Females	9	4.9 ± 0.33^{a}	7.1 ± 1.14	3.2 ± 0.43	1.6 ± 0.24^{c}	3.0 ± 0.44	14.0 ± 1.23^{a}
Grouped OVX Females	9	4.6 ± 0.57^{b}	6.3 ± 1.05	3.6 ± 0.44	1.1 ± 0.20^{d}	3.1 ± 0.56	14.1 ± 1.80^{b}
ANOVA		$P < 0.01$	NS	NS	$P < 0.01$	NS	$P < 0.01$
Dunnett's test		a,b $P < 0.01$			a–d $P < 0.01$		a,b $P < 0.05$

RESULTS AND DISCUSSION

The data presented in Table 1 clearly show that the weights of uteri and ovaries and number of ovarian follicles (sum of ovarian follicle types: 6, 7, 8 and atretic) in bank vole females were lower when the females maturated in the presence of other females. Bank voles reared in the enclosure and outdoor cages were under the influence of the same climatic variations and fed with the same diet, so social cues are suspected as the main factors affecting gonadal activity of these females.

When working with the other species of Microtidae family, P. subterraneus, we found that hormonal activity of females could be suppressed by the urine of intact and ovariectomized females (Table 2). There is a strong similarity in the responses of house mice (Drickamer, et al., 1978) and European pine vole females. However, in house mice the maturation-delaying chemosignal was released only by grouped females. Increase in population density of house mice was correlated with the production by females of a substance that delayed the onset of puberty of other females in the laboratory colony (Drickamer, 1980) or in feral populations (Massey and Vandenbergh, 1980). Wild European pine voles do not live in high density populations (Wasilewski, 1960) and this may explain the fact that urine from singly reared females delayed the onset of puberty in juvenile females.

The results presented above indicate that sexual maturation of juvenile females of two species of the Microtidae family is inhibited by adult females. There is evidence that inhibition of puberty of P. subterraneus females is evoked by urinary chemosignals. We suggest that lower hormonal activity of C. glareolus females matured in high density populations is also controlled by the chemosignals released by conspecific females.

REFERENCES

Drickamer, L. C., 1980, Social cues and reproduction of rodents and primates, in: "Biosocial Mechanisms of Population Regulation," M. N. Cohen, R. S. Malpass, and H. G. Klein, eds., Yale Univ. Press, New Haven and London.

Drickamer, L. C., MacIntosh, T. K. and Rose, E., 1978, Effect of ovariectomy on production and excretion of a pheromone delaying maturation in female house mice, Horm. Behav., 11:131.

Krebs, C. J., and Myers, J. H., 1974, Population cycles in small mammals, Adv. Ecol. Rev., 8:267.

Massey, A., and Vandenbergh, J. G., 1980, Puberty delay: A urinary cue from female house mice in feral populations, Science, 209:555.

Pedersen, T., and Peters, H., 1968, Proposal for a classification of oocytes and follicles in the mouse ovary, J. Reprod. Fert., 17: 555.

Wasilewski, W., 1960, Angaben zur Biologie und Morphologie der Kurzohrmause, Pitymys subterraneus, Acta Theriol., 12:185.

Zar, J. H., 1984, "Biostatistical Analysis," Prentice-Hall, New York.

FAMILIAL CHEMOSIGNALS INTERFERE WITH REPRODUCTIVE ACTIVATION IN FEMALE PINE VOLES, MICROTUS PINETORUM

John J. Lepri

Department of Zoology
North Carolina State University
Raleigh, NC 27695-7617

Why do many populations of small mammals periodically build up to high numbers and then suddenly crash? Though the question seems straight-forward, humans have been unable to provide a straight-forward answer for centuries (Elton, 1942). The correct answer is probably neither straight-forward nor simple, because many factors influence reproduction in rodents, including food, weather, disease, predation, and social factors. Food availablity is probably the most important regulator of female reproduction due to the energetic constraints of gestation and lactation (Bronson, 1985), but social factors can fine-tune the timing of reproductive effort to insure that reproduction is attempted only in social environments that provide a high probability of success. This report focuses on a single social factor that appears to influence reproduction of female pine voles, *Microtus pinetorum*.

Social factors, notably chemosignals, affect reproduction in a variety of rodent species (see Vandenbergh, 1983). Stimulatory chemosignals from males activate female reproduction in many microtine species, including the pine vole (reviewed by Lepri, in preparation), but with the exception of pregnancy blockage by strange males, there are few reports of chemosignals that suppress reproduction in microtines. In the prairie vole, *M. ochrogaster*, urinary chemosignals from females interfere with male-induced activation of reproduction in other females (Carter and Getz, in press; Getz et al., 1983). Familial chemosignals suppress reproductive maturation in male California voles, *M. californicus* (Rissman et al., 1984). In this report, I describe how males activate reproduction in female pine voles, and demonstrate that continued contact with familial chemosignals can interfere with this process in female offspring.

METHODS AND RESULTS

The pine voles used in these experiments were 1st through 4th generation laboratory descendants of wild voles trapped in 1982 and 1983 in apple orchards in Henderson County, North Carolina. Breeding pairs and their offspring were maintained in 36 x 30 x 18-cm plastic cages containing peat moss substrate and fitted with wire lids. Stud males were at least 120 days of age and were individually housed. Water and Wayne Lab Blox were available ad libitum; a slice of apple, 3-5 g sunflower seeds, and 5-7 g Wayne Guinea Pig Diet were provided three times a week. Colony photoperiod was 14L:10D and room temperature maintained at 20 ± 3° C.

In the following two experiments, female reproductive state was assayed by measuring physical variables. Following exposure to one of the treatments, females were killed by cervical dislocation and body, ovarian, and uterine weights were determined. Presence of sperm in the vaginal smear was taken as evidence that mating had occurred. Statistical comparisons were made using one-way analysis of variance (ANOVA) and Duncan's new multiple range test by the General Linear Models procedures of SAS computer programs (SAS, 1982). Organ weights were divided by body weight and multiplied by 30 to produce standardized values of mg organ weight per 30 g body weight. An adult non-pregnant female pine vole typically weighs 26-30 g.

EXPERIMENT 1

A preliminary experiment determined the time course of reproductive activation in adult females taken from isolation and paired with stud males for the following number of hours: 0, 12, 24, 36, 48, 60, or 72. The average (± SEM) age of the females was 72.3 (± 2.5) days, and ages were equivalently distributed across all treatment groups, as were body weights, which had a mean (± SEM) value of 27.1 (± 0.4) g.

Uterine weight doubled within 12 hours of pairing whereas ovarian weight increased gradually across all treatment groups (Figure 1). The uterine weight differences were statistically significant, $F(6,77)=4.73$, $P<0.0004$, with uteri from unpaired females (zero hours) being lighter than all the others (Duncan's new multiple range test, $P<0.05$). The ovarian weight differences were also statistically significant, $F(6,77)=3.55$, $P<0.0038$, with an overlapping distribution in which only the zero and 72 hour groups were statistically different. Sperm was present in vaginal smears of some females that had been paired with males for 48 or more hours.

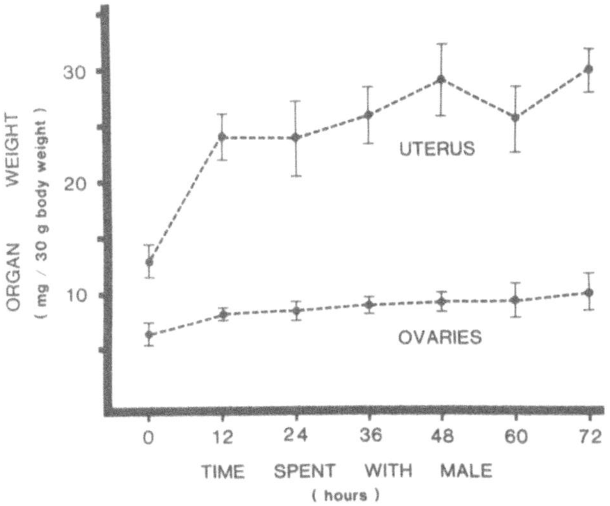

Figure 1. Changes in mean (± SEM) reproductive organ weights of female pine voles paired with stud males for varying periods of time (n = 12 females per treatment).

EXPERIMENT 2

In the second experiment, young females were randomly assigned to one of three treatment groups: i) taken directly from family cages; ii) taken from family cages and paired with stud males for 48 hours in a cage containing soiled peat moss from the female's family; and iii) taken from family cages and paired with stud males for 48 hours in a cage containing clean peat moss substrate. In the latter 2 treatments, an additional 100 ml aliquot of the appropriate substrate was placed in the pairing cage 24 hours after the initiation of the treatment. The reproductive state of the females was assayed as described above. Females in this experiment were of average (± SEM) age 56.4 (± 2.1) days, and of average (± SEM) body weight 25.5 (± 0.6) g. Ages and body weights were equivalently represented across all three treatment groups.

Females paired with stud males in the presence of familial chemosignals had lighter uteri, $F(2,38)=11.19$, $P<0.0002$, and ovaries, $F(2,38)=12.52$, $P<0.0001$, than females paired with stud males in the absence of familial chemosignals (Table 1). Females taken directly from their families and not paired with a male had the lightest uteri. None of the females had sperm in the vagina.

Table 1. Mean (± SEM) uterine and ovarian weights standardized to 30 g body weight of females paired with stud males for 48 hours in the presence or absence of familial chemosignals, and of females taken directly from family cages but not paired with stud males.

Treatment	n	Uterine weight (mg/30 g body weight)	Ovarian weight (mg/30 g body weight)
Paired with a stud male:			
clean pairing cage	14	27.71 (± 2.69)[a]	10.11 (± 0.49)[a]
pairing cage soiled with familial bedding	15	20.74 (± 2.10)[b]	8.31 (± 0.33)[b]
Not paired with a stud male:			
taken directly from family cage	12	13.18 (± 0.77)[c]	7.38 (± 0.29)[b]

[a]Means (± SEM) in the same column with different superscripts differ ($P<0.05$).

DISCUSSION

Reproduction in female pine voles was dramatically activated as a result of pairing with a stud male, but continued contact with familial chemosignals interfered with activation. However, familial chemosignals only partially suppressed reproductive activation, as uteri of females paired with stud males in the presence of familial chemosignals were larger than those of females taken directly from their family cages (Table 1). Prolonged observations beyond 48 hours would determine if the observed suppression of reproductive organ weight by familial chemosignals results in the delay of mating.

There are at least three possible mechanisms by which familial chemosignals could suppress female reproduction. First, continued contact with familial chemosignals might cause females to behave in a non-receptive manner to activational stimulation from males. For example, prairie vole siblings rarely mate with each other even if they are housed together for long periods of time. However, if urine is collected from the male sibling and applied to his sister's upper lip, sufficient stimulatory chemosignals will be transferred to activate her reproductive system (Carter et al., 1980). Hence, Carter et al. (1980) proposed that the inbreeding avoidance shown by brother-sister pairs of prairie voles is due to a behavioral block that prevents transfer of stimulatory chemosignals. Such a system might also function in families of pine voles whereby continued association with familial chemosignals causes females to react aggressively, rather than receptively, to strange males, thereby preventing the transfer of stimulatory chemosignals. Second, chemosensory processing of familial chemosignals by females might act neurologically to inhibit the release of reproductive hormones. Urinary chemosignals from males accelerate, whereas urinary chemosignals from group-housed females delay, sexual maturation in female house mice (reviewed in Vandenbergh, 1983). However, when the two types of chemosignals are mixed together, the suppressive effects of the grouped females override the stimulatory effects of the male (Drickamer, 1982). This indicates that neural processing of these conflicting cues is hierarchically arranged so that the inhibitory effect predominates. If chemoreception of familial chemosignals renders female pine voles physiologically insensitive to the stimulatory effect of the male, then removal of the appropriate chemosensory system from females should permit normal reproductive activation in spite of continued association with familial chemosignals. The vomeronasal organs (see Meredith, 1983) are likely candidates as the chemosensory organs involved. Third, familial chemosignals might inhibit the reproductive behavior of stud males. This could be tested by observing the reproductive behavior of stud males with hormonally-primed females on various odor substrates.

How does familial suppression of female reproduction relate to the ecology of pine voles? A primary consideration is the distribution of resources and how this affects opportunites for dispersal. Pine voles occur in stable woodland habitats (and more recently, in fruit orchards) throughout the eastern United States (see Smolen, 1981). Opportunities for dispersal may be limited to occasional openings as a result of deaths of individuals in neighboring colonies. Theoretically, then, young female pine voles that defer reproduction while remaining in the safety of the familial colony could gain additional body mass and social stature, which would increase the probability that future dispersal would be successful. Delayed maturation while in the familial colony also decreases opportunities for inbreeding, and non-dispersers may increase their inclusive fitness by "helping" to rear younger offspring (Fried and Powell, 1985).

The life history strategy of delayed maturation in pine voles is accompanied by an unusually low litter size, 2 or 3 pups per litter (Kirkpatrick and Valentine, 1970; Schadler and Butterstein, 1979), compared to other species of voles (Hasler, 1975). Small litters and delayed maturation both might result in the production of a few high-quality individuals that are competent to successfully disperse and recruit into the breeding population (see Stearns, 1976). However, delayed maturation is a facultative response to continued association with the family, as females can conceive as early as 32 days of age when isolated from the family (Lepri, in preparation). Thus, should an opportunity for dispersal occur, there is sufficient flexibility in the reproductive development of females to rapidly exploit it.

The social organization of pine voles provides the appropriate environment for familial suppression of female reproduction. Extended families consisting of one or more reproductively mature males, one reproductively active female, and many offspring of various ages, are the basis of social organization in field populations of this species (FitzGerald and Madison, 1983). Reproductive activation of females born into such groups has not been observed, but sibling pine voles rarely mate with each other, and the mere presence of a male sibling on the opposite side of a screen partition has an inhibitory effect on reproduction by females that are in direct contact with strange males (Schadler, 1983). Furthermore, reproductive development of young females reared in crowded cages is inhibited relative to what occurs in less-crowded conditions (Schadler, 1980). Further experiments are needed to determine whether reproductive suppression is dependent on density, implying population regulation, or familiarity, implying inbreeding avoidance.

Delayed maturation and dispersal occur seasonally in other species of voles. In spring and summer populations of field voles, M. arvalis, subadults usually disperse from their natal area shortly after weaning, but in autumn the older offspring remain at the natal area with their mother and siblings throughout the winter, effectively reducing the thermoenergetic requirements of individuals in these "great families" (Frank, 1957). Reproduction by offspring is not observed in such conditions. The extended families of field voles bear resemblance to the colonial organization of pine vole populations (FitzGerald and Madison, 1983). In many field populations of voles, more dispersing than resident females are in breeding condition (see Gaines and McClenaghan, 1980), indicating that dispersal may be a result of sexual maturation and that maturation is inhibited by continued association with the family group. The current experiment and that of Rissman et al. (1984) provide a chemosensory-based mechanism for suppression of reproduction in offspring.

In conclusion, familial chemosignals might play a biologically significant role in the timing of reproduction by suppressing it until after dispersal from the natal unit. Many other factors influence reproduction, however, and the ecological reality of population fluctuations under natural conditions most likely involves many factors, temporally and spatially interacting with the population and each other (Lidicker, 1978).

REFERENCES

Bronson, F. H., 1985, Mammalian reproduction: an ecological perspective, Biol. Reprod., 32:1.

Carter, C. S., and Getz, L. L., (in press). Social and hormonal determinants of reproductive patterns in the prairie vole. Proc. 1st

Internat. Comp. Physiol. Biochem. Conf., Hormones and Behavior, 1984, Liege, Belgium, Springer-Verlag.
Carter, C. S., Getz, L. L., Gavish, L., McDermott, J. L., and Arnold, P., 1980, Male-related pheromones and the activation of female reproduction in the prairie vole (Microtus ochrogaster), Biol. Reprod., 23:1038.
Drickamer, L. C., 1982, Acceleration and delay of first vaginal oestrus in female mice by urinary chemosignals: dose levels and mixing urine treatment sources, Anim. Behav., 30:456.
Elton, C. S., 1942, "Voles, Mice and Lemmings: Problems in Population Dynamics," Clarendon Press, Oxford.
FitzGerald, R. W., and Madison, D. M., 1983, Social organization of a free-ranging population of pine voles, Microtus pinetorum, Behav. Ecol. Sociobiol., 13:183.
Frank, F., 1957, The causality of microtine cycles in Germany, J. Wildl. Manage., 21:113.
Fried, J. J. and Powell, R. A., 1985, Helping behavior in the pine vole Microtus pinetorum, Abstract, Natl. Conv. Anim. Behav. Soc., 23-28 June 1985, North Carolina State University, Raleigh.
Gaines, M. S. and McClenaghan, Jr., L. R., 1980, Dispersal in small mammals, Ann. Rev. Ecol. Syst., 11:163.
Getz. L. L., Dluzen, D. and McDermott, J. L., 1983, Suppression of reproductive maturation in male stimulated virgin female Microtus by a female urinary chemosignal, Behav. Processes, 8:59.
Hasler, J. F., 1975, A review of reproduction and sexual maturation in the microtine rodents, Biologist, 57:52.
Kirkpatrick, R. L., and Valentine, G. L., 1970, Reproduction in captive pine voles, Microtus pinetorum, J. Mamm., 51:779.
Lepri, J. J., (in preparation), "Investigations on the Reproductive Ecology of the Pine Vole Microtus pinetorum," Ph.D. thesis, Department of Zoology, North Carolina State University, Raleigh.
Lidicker, Jr., W. Z., 1978, Regulation of numbers in small mammal populations- historical reflections and a synthesis, in: "Populations of Small Mammals under Natural Conditions," D. P. Snyder, ed., Spec. Publ. Series, vol. 5, Univ. Pittsburgh.
Meredith, M., 1983, Sensory physiology of pheromone communication, in: "Pheromones and Reproduction in Mammals," J. G. Vandenbergh, ed., Academic Press, New York.
Rissman, E. F., Sheffield, S. D., Kretzmann, M. B., Fortune, J. E., and Johnston, R. E., 1984, Chemical cues from families delay puberty in male California voles, Biol. Reprod., 31:324.
Schadler, M. H., 1983, Male siblings inhibit reproductive activity in female pine voles, Microtus pinetorum, Biol. Reprod., 28:1137.
Schadler, M. H., 1980, The effect of crowding on the maturation of gonads in pine voles, Microtus pinetorum, J. Mamm., 61:769.
Schadler, M. H., and Butterstein, G. M., 1979, Reproduction in the pine vole Microtus pinetorum, J. Mamm., 60:841.
Smolen, M. J., 1981, Microtus pinetorum, Mammalian Species, 147:1.
SAS Institute, Inc., 1982, "SAS User's Guide: Statistics, 1982 Edition," SAS Institute Inc., Cary, NC.
Stearns, S. C., 1976, Life-history tactics: a review of the ideas, Quart. Rev. Biol., 51:3.
Vandenbergh, J. G., 1983, Pheromonal regulation of puberty. in: "Pheromones and Reproduction in Mammals," J. G. Vandenbergh, ed., Academic Press, New York.

ODOR IMAGES: RESPONSES OF BEAVER TO CASTOREUM FRACTIONS

D. Muller-Schwarze, L. Morehouse, R. Corradi,
Cheng-hua Zhao, and R. M. Silverstein

College of Environmental Science and Forestry
State University of New York
Syracuse, NY 13210

INTRODUCTION

The ideal way to bioassay a mammalian pheromone is to experiment with freely moving animals in a natural setting, and intact social units. The year-round family territories of the North American beaver, Castor canadensis, provide such an opportunity. Being largely nocturnal, beaver have depended on their chemical senses for social communication, food selection, and possibly other behaviors such as orientation in space, and predator detection and avoidance.

There are examples of chemically simple mammalian pheromones, as in the golden hamster, Mesocricetus auratus, (Singer et al., 1976) or pig, Sus scrofa, (Melrose et al., 1970), but many others have proven to be quite chemically complex. Examples are the individually differentiated secretion from the circumgenital glands of the tamarin, Saguinus fuscicollis, (Epple, 1978; Yarger et al., 1977; Smith et al., 1985) and Flehmen-releasing urine constituents in black-tailed deer, Odocoileus hemionus columbianus (Crump et al., 1984).

We now report results from a field experiment with fractions of a territorial odor of the North American beaver, Castor canadensis.

Beaver not only scent mark their territories, but also build a supporting structure for the scent. With their front paws, they scoop up mud from the pond bottom and deposit it on the bank, usually within 50 cm of the water. This mud pile is then scented by secretion from the castor sacs and anal glands during cloacal contact with the substrate. By elevating the scent mark, the active space (volume of air containing above-threshold amounts of the putative pheromone) is greatly enlarged. Even humans can smell these scent marks from some distance, depending on air currents.

Lederer (1946, 1950) identified 45 of over 100 compounds isolated from castoreum, but was not clear on the species of beaver used. Nitrogen-containing compounds were identified by Maurer and Ohloff (1976). The only chemical study of anal gland secretion is that of Grønneberg (1978/79). He identified wax esters with chain lengths of 5-22 for the carboxylic acid parts and 14-19 for the alcohol parts. None of the identified compounds has been linked to any biological function.

Much has been speculated about the functions of the scent mounds. They may be important for the marking individual itself, for colony members, or for neighbors and strangers. An individual may use its scent mark for orientation in space, especially at night, and particularly for returning to a food item such as a felled tree into which considerable energy already has been invested. Colony members may be informed by scent mounds about the location of food items. Potential mates may be attracted by the scent of single individuals. Finally, intercolony functions may decrease the confidence of approaching strangers ("warning sign") or actually keep them out ("fence function").

Beaver are known to respond readily to artificial castoreum scent marks that are placed near the lodge (Hodgdon, 1978; Muller-Schwarze and Heckman, 1980; Svendsen, 1980; Bollinger, 1980). They sniff, paw, and mark such scent marks (Muller-Schwarze et al., 1983). This stereotyped and elaborate behavior provided the bioassay in the present study.

METHODS AND MATERIALS

Study site. The experiment was conducted at Allegheny State Park, Cattaraugus County, New York, an area of exceptionally high beaver population density with 1.03 colonies per stream kilometer (col/km). This compares with 0.9 col/km in an unexploited population in Grand Teton National Park (Collins, 1976) or 0.63 col/km in Alaska (also unexploited; Boyce, 1981). The saturation density in northeastern hardwood forests is estimated to be approximately 1 col/km (Nordstrom, 1972). Most colonies in our study area are located along streams. The vegetation consists of deciduous forest with large species diversity. At many colony sites, all woody vegetation has been cleared by the beaver for as much as 100 m from the pond. Even red pine (*Pinus resinosa*) and scotch pine (*P. sylvestris*) plantations are utilized in winter although they usually are not considered beaver habitat.

Odor stimuli. Odor samples were placed at the banks of seven different colonies, usually upwind from the lodge and main area of beaver activity in the pond, and within 50 cm from the water. The scent was applied to an artificial mud pile at about 1700 hr. Mud was scooped from the pond edge with rubber gloves or a small trowel. The mudpile was 12 cm in diameter and 6 cm high, and the samples were applied to the pile with a disposable pipette. Each application contained 0.033 gland equivalent (GE) of material in 0.25 ml methanol. The behavior at and near these scent mounds was observed directly from time of emergence of the beaver, usually from between 1800 and 1900 hr., until nightfall, which occurred at 2100 hr. at the latest. One observer was stationed at a particular colony, and 2 to 4 colonies were worked each night. The experiment lasted from May 21 to August 28, 1984.

The following odors were used, in sequence from the most to the least complex. A mixture of male and female castoreum and anal gland secretions ("family odor"), a mixture of male and female castoreum, a mixture of male and female anal gland secretion, single-sex castoreum and anal gland secretions separately, a single-sex distillate and residue of castoreum separately, three fractions of distillate, four and three subfractions respectively of fractions 2 and 3, and recombinations of the fractions at the two levels of fractionation. Methanol was used as control.

Chemical fractionation. Male and female castor sacs were macerated and extracted in methanol (200 ml per gland). The methanol extract was filtered through glass wool, and then concentrated to 50 ml by fractional distillation (water-bath 85°C, atmospheric pressure, glass column 25 cm x 1 cm I.D., filled with glass beads). Twenty-five g NaCl was added to the concentrated

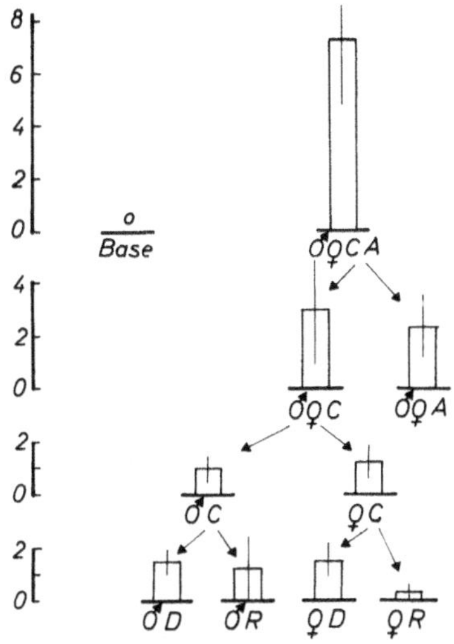

Fig. 1. Mean number of land visits ($\pm$ SEM) by all members of beaver colonies in response to odor samples. Ordinate: Number of visits. Base: Number of visits prior to any odor application. C: Castoreum; A: Anal gland secretion; D: Distillate; R: Residue.

extract, which was extracted with 50 ml ether four times. The combined ether extract was dried over anhydrous Na_2SO_4 and concentrated by fractional distillation (water-bath 55°C, column 25 x 1 cm, atmospheric pressure) to a volume of 20 ml.

Vacuum distillation. The volatile components (after ether removal) were distilled in a short-path distillation apparatus. The ether fraction was distilled (49 mm, oil bath temperature increased from room temperature to 80°C, then at 0.9 mm, oil bath 140-150°C for female sample and 0.12 mm and 110-120°C for male sample). The receiver was cooled by dry-ice/acetone. The distillate was washed from the condenser and the receiver with ether and concentrated to 1.5 ml of a slightly yellowish solution.

GC fractionation. The column was 10% Carbowax 20M on Chromosorb W-AW-DMCS 60-80 mesh, glass column 2 m x 9 mm I.D.; column temperature was increased from 100°C (after 6.6 min) to 192°C at the rate of 4°/min. F.I.D. detector temperature 200°C, injector temperature 180°C, N_2 flow rate 80 ml/min. Male and female castoreum distillates were separated into 3 fractions. GC fractions were collected in 3 mm O.D. glass capillaries with a Brownlee-Silverstein thermal gradient collector (Brownlee and Silverstein, 1968). For each injection, 0.1 GE of distillate was used. The glass capillaries were rinsed with 0.3 ml methanol. Each fraction and the total recombination were used in the bioassay. Male beaver GC fraction 3 was fractionated into 3 fractions (c.f., Fig. 3) under the following GC conditions: 10% Carbowax 20M on Chromosorb W-AW-DMCS 60-80 mesh, glass column 2 m x 9mm I.D., isothermal 190°C, detector 220°C, injector 180°C, N_2 flow rate 80 ml/min. The largest peak in the male extract occurred in fraction 2. Fraction 2 was refractionated with a 10% Carbowax 20M column (column temperature 140°C, other conditions as above).

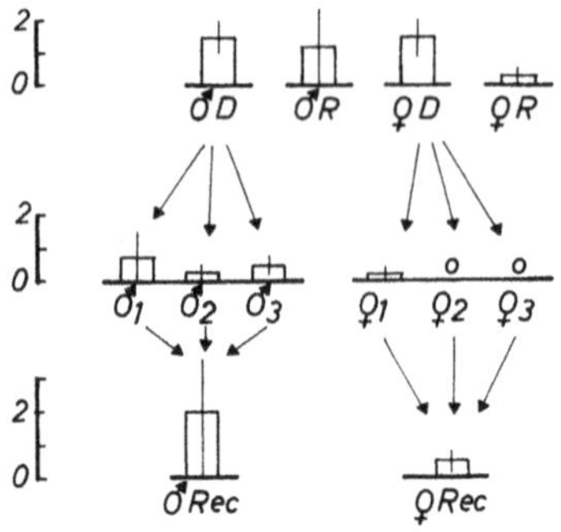

Fig. 2. As in Figure 1. Numbers denote fractions of distillate. Rec: Recombination of fractions.

RESULTS

Typically, a beaver, after emerging from its lodge in the evening, cruised through a pond, usually to the dam, floated, presumably to defecate, then turned its attention to the experimental scent mound. It swam back and forth in front of the scent mound and approached closer to the odor source while its lateral excursions grew shorter. Having reached shallow water, an animal typically raised itself on its hind-quarters and sniffed the air coming from the scent source.

In the second phase of the odor response, the beaver left the water, walked slowly to the scent pile, and sniffed it, with its nose 5 cm or less from the sample. After sniffing, the beaver pawed the mudpile with its front feet, usually flattening it, then stepped forward, straddled the mudpile, and rubbed its cloacal area over the mud. During this movement, secretions of the castor sacs and anal glands are applied to the substrate. Having thus scent marked the experimental mudpile, the beaver walked from the pile toward the water, slid into the pond, and swam away. With many samples, the beaver did not return after one bout of activity at the experimental scent mound, but other samples attracted several beaver or the same beaver several times. The most intense response consisted of 19 land visits by various members of the colony.

Of the several variables measured, number of land visits, and total time on land are most useful for characterization of the response intensity.

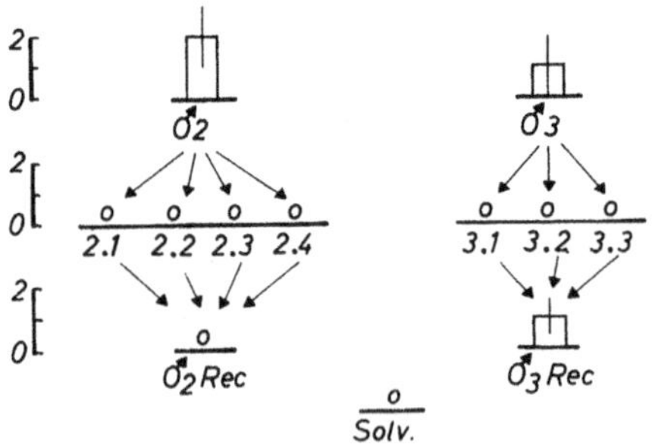

Fig. 3. As in Figure 1. Continued fractionation. Two-digit numbers denote subfractions. Solv: Solvent only (methanol) as control.

Table 1. Sniffing, Pawing and Scent Marking Frequencies (Mean ± SEM) in Response to Mixed Sex Secretions and Controls (M = Male, F = Female, C = Castoreum, A = Anal Secretion).

Stimulus	N	Sniffing	Pawing	Scent Marking
Baseline (No Odor)	6	0	0	0
M and F: C and A	6	9.5 ± 3.09	2.83 ± 0.83	3.67 ± 1.45
M and F: C	3	2.33 ± 1.45	0.67 ± 0.67	1.00 ± 1.00
M and F: A	3	1.00 ± 0.58	0.67 ± 0.67	0.33 ± 0.33
Methanol (Solvent only)	4	0	0	0

*: $p < 0.05$, all other differences NS.

As a general rule, the more complex the odor, the stronger the response. The most intense response was released by a mixture of male and female castoreum and anal gland secretion, the most complex odor used. A castoreum mixture from both sexes was half as active or less as the mixture of castoreum and anal gland secretion. Anal gland secretion from both sexes was also about half as active as the combination of castoreum and anal secretion (Fig. 1 for number of land visits, Fig. 4 for time on land, and Table 1 for sniffing, pawing and scent marking of odor mixtures). Single sex castoreum released still fewer responses, with secretions from both sexes about equally active (Figs. 1 and 4; Tables 2 and 3).

The three fractions of distillate of male castoreum were all mildly active, while two or three fractions of female castoreum distillate did not trigger any response (Figs. 2 and 5; Tables 2 and 3). None of the subfractions showed activity, although recombination restored the activity, at least

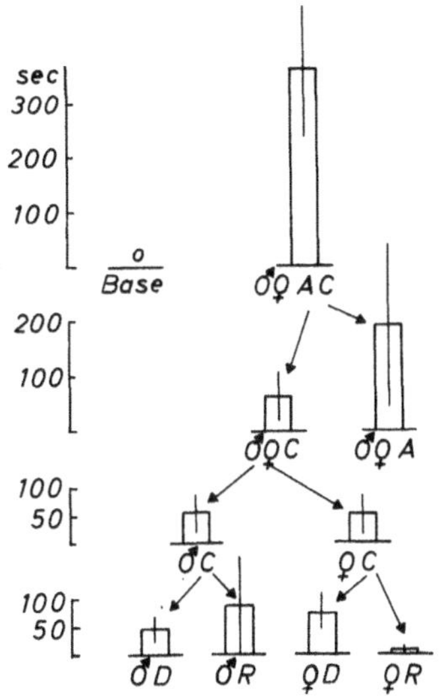

Fig. 4. Time on land (in sec; ordinate) in response to odor samples. As in Figures 1-3.

Table 2. Sniffing, Pawing and Scent Marking Frequencies in Response to Male Secretions, Fractions, and Controls. (As in Table 1.)

Stimulus	N	Sniffing	Pawing	Scent Marking
Baseline (No Odor)	6	0	0	0
Castoreum (C)	4	1.00 ± 0.58	0.67 ± 0.33	0.33 ± 0.33
C: Distillate (D)	4	1.75 ± 0.75	0.50 ± 0.29	0.25 ± 0.25
C: Residue	4	1.50 ± 1.19	0.25 ± 0.25	0.25 ± 0.25
D: Fraction 1	4	0.75 ± 0.75	0	0
D: Fraction 2	7	1.00 ± 0.53	0.57 ± 0.43	0.57 ± 0.43
D: Fraction 3	7	0.71 ± 0.29	0	0
Fractions 1-3 Recombined	4	2.00 ± 1.68	0.50 ± 0.29	0.50 ± 0.29
Subfraction 2.1	3	0	0	0
Subfraction 2.2	3	0	0	0
Subfraction 2.3	3	0	0	0
Subfraction 2.4	3	0	0	0
Fraction 2 Recombined	3	0	0	0
Subfraction 3.1	3	0	0	0
Subfraction 3.2	3	0	0	0
Subfraction 3.3	3	0	0	0
Fraction 3 Recombined	3	1.00 ± 0.58	0.67 ± 0.67	0.33 ± 0.33
Methanol (Solvent only)	4	0	0	0

in fraction 3 (Figs. 3 and 6; Tables 2 and 3). No response to methanol was ever noted (Figs. 3 and 6; Tables 1-3). Note the redundancy of the effects of the stimulus pairs, castoreum and anal gland secretion, male and female castoreum, distillate and residue, and also the three fractions of male castoreum (Figs. 1, 2, 4, and 5; Tables 1-3).

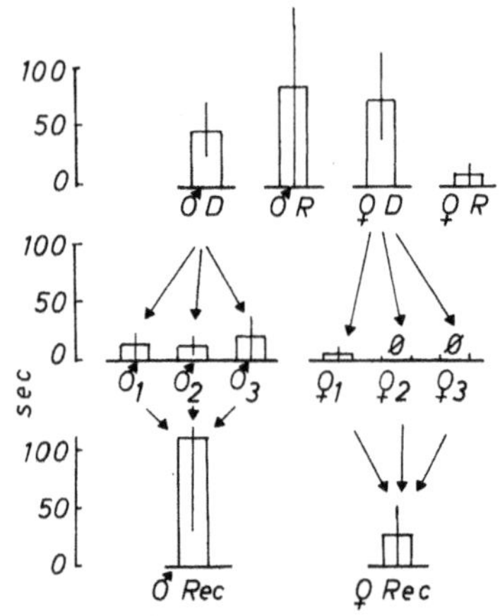

Fig. 5. Time on land in response to fractionated male and female castoreum. As in Figures 1-4.

Table 3. Sniffing, Pawing and Scent Marking Frequencies in Response to Female Secretions, Fractions, and Controls. (As in Tables 1 and 2.)

Stimulus	N	Sniffing	Pawing	Scent Marking
Baseline (No Odor)	6	0	0	0
Female Castoreum (C)	4	1.25 ± 0.63	0.50 ± 0.29	0.25 ± 0.20
Castoreum Distillate (D)		1.50 ± 0.65	0.50 ± 0.29	0.25 ± 0.20
Castorum Residue	3	0.33 ± 0.33	0.33 ± 0.33	0.33 ± 0.20
D: Fraction 1	4	0	0	0
D: Fraction 2	4	0	0	0
D: Fraction 3	4	0	0	0
Fractions 1-3 Recombined	4	0	0	0
Methanol (Solvent only)	4	0	0	0

DISCUSSION

Beaver in their natural setting are almost ideal experimental mammals for bioassays in the field with all its ecological complexity. Consistent results were obtained despite numerous uncontrollable intervening variables and the fact that their largely nocturnal behavior precludes prolonged detailed observation.

The responses of the beaver show that a number of components are necessary for a behavioral effect, and that the response grows more intense with increasing complexity of the chemical stimulus. Thus, what started out as an experiment using "response-guided strategy" to pin down some active components, came to demonstrate an "odor image" (Albone, 1984).

In mammalian pheromone studies, varying degrees of chemical complexity of the biologically active stiumlus have been encountered. Even insect pheromones, with a few exceptions, are multicomponent, and generally it has been possible to isolate several components that account for most of the activity. However there have been only a few examples in mammals in which

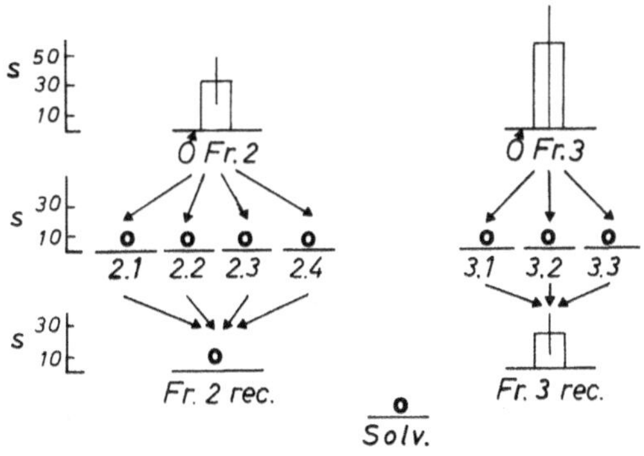

Fig. 6. Continued fractionation. As in Figures 1-5.

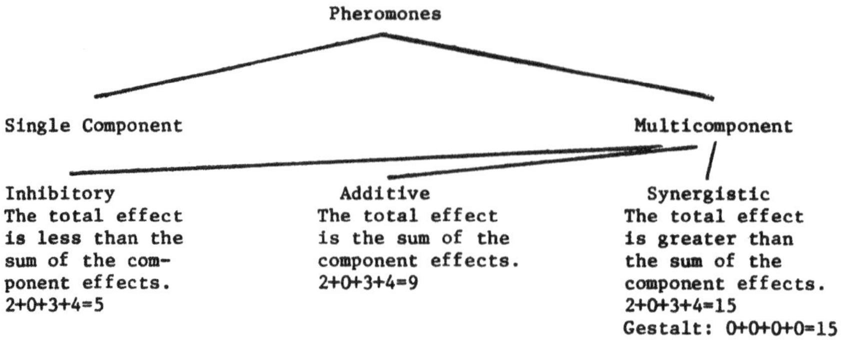

Fig. 7. Possible classification of pheromones according to interaction types of mixture components.

one or two compounds have accounted for a substantial part of the total activity [dimethyl disulfide in the golden hamster, *Mesocricetus auratus* (Singer et al., 1976); 5α-androst-16-en-3β-ol and 5α-androst-16-en-3-one in the pig, *Sus scrofa*, (Melrose et al., 1971); isovaleric acid in the pronghorn, *Antilocapra americana* (Muller-Schwarze et al., 1974); *cis*-4-hydroxydodec-6-enoic acid lactone in the black-tailed deer, *Odocoileus hemionus columbianus* (Brownlee et al., 1968); and phenylacetic acid in the Mongolian gerbil, *Meriones unguiculatus* (Thiessen et al., 1974)]. Most investigators have found, as we have in the present study and that of female urine of black-tailed deer (Crump et al., 1984), that the response disappeared completely before the fractionation had reached the level of single compounds. Claesson and Silverstein (1977) had noted that the preferred *modus operandi*--to monitor each fraction and combinations thereof with a bioassay--may not be feasible when most or all of the components are needed for substantial activity and when the field bioassay itself is complex, difficult to reproduce, and may yield only unspecific responses such as sniffing. Many responses occur or are valid only in the proper physical, social, and temporal contexts. Under these conditions, it may be possible to select significant compounds "because they are present, or predominate, in one sex, may appear at a specific time such as estrus, or may appear, increase, or decrease in response to manipulation such as castration or hormone treatment. The consistent appearance of a compound through wide dietary variation may be significant." Sometimes we have to identify all compounds present, synthesize them, and bioassay the total mixture with all components in the proper proportions. In the extreme case, removal of any single compound could result in total loss of the bioassay response. We may use the psychologist's term "Gestalt," and consider it the extreme version of Albone's (1984) "odor image." Smith et al. (1985) have developed a sophisticated computer approach to analyze complex chemical patterns to predict the cues used by the test animal. The risk with the "total analysis" approach is that expediency may dictate that minor components be disregarded, but these minor compounds may be biologically significant far out of proportion to the amount present.

The components of insect pheromones have been found to function additively or non-additively; the term "synergistic" has been used to describe a mixture that evokes a response that is greater than the sum of the responses to single components; the term "inhibitory" for a mixture that evokes a response that is less than the sum. These effects extend to the level of enantiomers (Silverstein, 1982) and are summarized in Figure 7. Carr and Derby (1986) applied the useful descriptor "interactive" to synergistic or inhibitory effects in fish and crustacean responses.

In our present study, the bioassay response is diminshed in the course of fractionation and the "response-guided strategy" (Albone, 1984) is not appropriate. Redundant components can further complicate the bioassay, but may function biologically to ensure the effectiveness of an important scent mark. In future studies, we propose to identify and synthesize as many of the odor components as possible in order to reproduce the odor image.

ACKNOWLEDGMENTS

We thank the staff of Allegany State Park, especially Hugh Dunne, Regional Director, and Dan Colvin, for their cooperation, Dr. John Berglund, for providing a camp in the park, and Christine Muller-Schwarze and Olive Silverstein, for help as volunteers.

REFERENCES

Albone, E. S., 1984, "Mammalian Semiochemistry: The Investigation of Chemical Signals Between Mammals," Wiley, Chichester.

Bollinger, K. S., 1980, Scent marking behavior of beaver (*Castor canadensis*), M. S. Thesis, Univ. Massachusetts.

Boyce, M. S., 1981, Habitat ecology of an unexploited population of beavers in interior Alaska, *in* "Proceedings Worldwide Furbearer Conf.," Joseph A. Chapman, and Wayne Pursley, eds., Vol. 1, 1981, Frostburg, MD.

Brownlee, R. G., and Silverstein, R. M., 1968, A micro-preparative gas chromatograph and a modified carbon skeleton determinator, *Anal. Chem.*, 40:2077.

Brownlee, R. G., Silverstein, R. M., Muller-Schwarze, D., and Singer, A. G., 1969, Isolation, identification and function of the chief component of the male tarsal scent in black-tailed deer, *Nature*, 221:284.

Carr, W. E. S., and Derby, C. D., 1986, Chemically stimulated feeding behavior in marine animals: The importance of chemical mixtures and the involvement of mixture interactions, *J. Chem. Ecol.*, 12:989.

Claesson, A., and Silverstein, R. M., 1977, Chemical methodology in the study of mammalian communication, *in*: "Chemical Signals in Vertebrates," D. Muller-Schwarze, and M. M. Mozell, eds., Plenum, New York.

Collins, T. C., 1976, Population Characteristics and Habitat Relationships of Beavers, *Castor canadensis*, in Northwest Wyoming, Ph.D. Thesis, Univ. Wyoming.

Crump, D., Swigar, A. A., West, J. R., Silverstein, R. M., Muller-Schwarze, D., and Altieri, R., 1984, Urine fractions that release Flehmen in black-tailed deer, *Odocoileus hemionus columbianus*, *J. Chem. Ecol.*, 10:203.

Epple, G., 1978, Studies on the nature of chemical signals in scent marks and urine of *Saguinus fuscicollis* (Callitrichidae, Primates), *J. Chem. Ecol.*, 4:383.

Grønneberg, T. Ø., 1978-79, Analysis of a wax ester fraction from the anal gland secretion of beaver (*Castor fiber*) by chemical ionization mass spectrometry, *Chemica Scripta*, 13:56.

Lederer, E., 1946, Chemistry and biochemistry of the scent glands of the beaver (*Castor fiber*), *Nature*, 157:231.

Lederer, E., 1950, Odeurs et parfums des animaux, *Fortschr. Chem. org. Naturstoffe*, 6:87.

Maurer, B., and Ohloff, G., 1976, Zur Kenntnis der stickstoffhaltigen Inhaltsstoffe von Castoreum, *Helv. Chim. Acta*, 59:1169.

Melrose, D. R., Reed, H. C. B., and Patterson, R. L. S., 1971, Androgen steroids associated with boar odour as an aid to the detection of oestrus in pig artificial insemination, *Brit. Vet. J.*, 127:497.

Muller-Schwarze, D., and Heckman, S., 1980, The social role of scent marking in beaver (*Castor canadensis*), *J. Chem. Ecol.*, 6:81.

Muller-Schwarze, D., Heckman, S., and Stagge, B., 1983, Behavior of free-ranging beaver (*Castor canadensis*) at scent marks, *Acta Zool. Fenn.*, 174:111.
Muller-Schwarze, D., Muller-Schwarze, C., Singer, A. G., and Silverstein, R. M., 1974, Mammalian pheromone: Identification of active component in the subauricular scent of the male pronghorn, *Science*, 183:860.
Nordstrom, W. R., 1972, Comparison of Trapped and Untrapped Beaver Populations in New Brunswick, M. S. Thesis, Univ. New Brunswick.
Silverstein, R. M., 1982, Chemical communication in insects: Background and application, *Pure and Appl. Chem.*, 54:2479.
Singer, A. G., Agosta, W. C., O'Connell, R. J., Pfaffmann, C., Bowen, D. V., and Field, F. H., 1976, Dimethyl disulfide: An attractant pheromone in hamster vaginal secretions, *Science*, 191:950.
Smith, A. B., III, Belcher, A. M., and Epple, G., 1985, Computerized pattern recognition: A new technique for the analysis of chemical communication, *Science*, 228:175.
Svendsen, G. E., 1980, Patterns of scent mounding in a population of beaver (*Castor canadensis*), *J. Chem. Ecol.*, 6:133.
Thiessen, D. D., Regnier, F. E., Rice, M., Goodwin, M., Isaacks, N., and Lawson, N., 1974, Identification of a ventral scent marking pheromone in the male Mongolian gerbil (*Meriones unguiculatus*), *Science*, 184:83.
Yarger, R. G., Smith, A. B. III, Preti, G., and Epple, G., 1977, The major volatile constituents of the scent mark of a South American primate (*Saguinus fuscicollis*, *Callitricidae*), *J. Chem. Ecol.*, 3:45.

SCENT-MARKING IN PIKAS (*Ochotona princeps*):

TEST OF A BREEDING-FACILITATION HYPOTHESIS

Carron A. Meaney

Zoology Department
Denver Museum of Natural History
City Park
Denver, CO 80205

INTRODUCTION

Pikas (*Ochotona princeps*) regularly scent mark rocks in their environment with cheek glands located at the angle of the jaw. Glandular activity is most pronounced during the breeding season (Harvey and Rosenberg, 1960). Previous work (Meaney, 1983a) has shown that the rate of scent marking is high in June, males mark more than females, and individuals mark more in potential mates' territories as a consequence of a greater degree of overlap. Pikas also can discriminate individual odors (Meaney, 1983b). Furthermore, pikas are aggressive mammals, although the level of aggression is lowest during the breeding season (Smith, 1981).

These components of pika behavior led me to hypothesize that, from the point of view of the female, familiarity with a potential mate's odors serves to facilitate mating with him. The purpose of this pilot study was to test the prediction that a female would be more likely to mate with a male whose odors are familiar. As a corollary, I predicted that there would be less aggression in pairs in which the female was familiar with the male's odors compared to pairs in which she was not.

Priming effects on the reproductive physiology of female mice exposed to male odors have been well documented. These include regulation of the estrous cycle (Whitten, 1956), puberty acceleration (Vandenbergh, 1967), and pregnancy block (Bruce, 1959). Various behavioral effects, on the other hand, have only been suggested. Synchrony of reproductive activity (Epple, 1976; Harrington, 1975) and attraction, stimulation, or arousal of the female (Bronson, 1976; Grau, 1976; Vandenbergh and Drickamer, 1974) are four such examples. In addition, Daly (1977) has suggested that the function of male scent marks is to reduce aggression by a female toward a male whose odors are familiar. He found that males marked the female territories they visited in his field work on two species of gerbil

(Meriones libycus and Psammomys obesus). Laboratory research on a third species (Meriones unguiculatus) revealed that females were, in fact, less aggressive towards males whose odors were familiar.

METHODS

The Laboratory Animals

The general laboratory and housing conditions are described in Meaney (1983b). Twenty-nine pikas were trapped from Monte Cristo, Summit County, Colorado, in May and June, 1982. One laboratory-born female, the sole survivor from Whitworth's (1984) previous established colony, was also included. Due to a high (57%) mortality, only 14 healthy animals remained for the study. The fatalities were due to a variety of causes. Possible contributing factors were bloating (possibly from eating too much dry alfalfa hay), not enough fresh vegetation, and possibly a Coccidia infection. As these problems were recognized they were dealt with by removing alfalfa from the bins, gathering fresh dandelions (Taraxacum officinale) daily, and treating for coccidiosis.

The animals were allowed two weeks' acclimation to the laboratory before initiation of a test. Only healthy animals were tested. Nine females and five males were used in the experiment.

Vaginal smears were collected from all females immediately after the second three-hour test. For this procedure, the animals were anesthetized with Metaphane gas (Methoxyfluorane manufactured by Pitman-Moore). Smear samples were obtained by inserting an eye dropper containing two ml of distilled water into the vagina. The water was expressed and then aspirated several times to collect cells from the vaginal wall. A drop of fluid was retained in the eyedropper and deposited onto a slide. When dry, each slide was dipped in absolute methyl alcohol for 5 seconds, dried, dipped into Giemsa solution (1:20) for 30 minutes, and rinsed in water.

Testing Paradigm

A table of random numbers was used to assign the animals into pairs. Each female was randomly assigned two males. One male, called familiar, had his cheek-gland odors presented to the female while the other male, called unfamiliar, did not. Due to the high mortality of males, some were paired with more than one female in order to test all nine females. Pairs were not housed together. On Day 1 of the test, cheek-gland secretions were collected from the familiar male by rubbing a small rock eight times across his cheek gland. The rock was then placed in the female's bin. This procedure was repeated on Day 4 and Day 7. On Day 8, the female's bin was connected with one of her partners' bins by a plastic tunnel. The bins remained connected for three hours after one partner had entered the other's bin. The tunnel was then disconnected for one hour, after which a clean tunnel was used to connect the female with her other partner for another three-hour period. The assignment of the familiar or unfamiliar male for the first connection of the day was consistently alternated between females. Frequencies of behavioral acts were recorded during the three-hour connection periods. The Wilcoxon matched-pairs signed-rank test was used in the comparisons of behavioral frequencies for familiar and unfamiliar pairs. Reported probabilities are one-tailed. A minimum of six samples is required for the Wilcoxon test. If the behavior in question did not occur in either pair, or if the difference was zero for more than two of the pair combinations, sign tests were used.

Table 1. Mean frequencies of aggressive acts, $\pm$ SE, during pairings of females with males whose odors were not familiar and males whose odors were familiar, n = 8.

	NOT FAMILIAR	FAMILIAR	
CHASE	14.1 ± 10.79	0.2 ± 0.16	*
BOX	4.6 ± 1.68	3.5 ± 2.15	NS
FIGHT	3.5 ± 2.37	0.2 ± 0.16	--

* = $P < 0.05$, Wilcoxon matched-pairs signed-rank test.
NS = $P > 0.05$.
-- = Cannot be analyzed, 6 pairs had 0 difference.

Descriptions of Behavioral Acts

Behavioral descriptions are adapted from Whitworth (1984).

Box..........Aggressive behavior between two animals, in which both maintain a full upright stance, with front legs pawing each other.

Chase..........Pursuit of another conspecific during an aggressive encounter.

Fight..........Highly aggressive behavior characterized by biting and rolling on the substrate, bodies often somersaulting.

Male Follows, Female Runs..........The female moves away before the male has made physical contact.

Male Follows, Female Waits..........The male makes physical contact before the female moves away.

Mount..........The male mounts the female but she moves away before pelvic thrusting occurs.

Copulate..........A mount accompanied by pelvic thrusts.

RESULTS

One pair combination was eliminated from the analyses because the animals did not interact. Data from the remaining eight pair combinations were analyzed. There were significantly more Chases in the unfamiliar than in the familiar pairs ($T = 2$, $n = 7$, $P < 0.05$) (see Table 1). In all cases, the male chased the female and not the reverse. The frequencies of Box and Fight were higher in the unfamiliar than in the familiar pairs. These results are not significant for Box ($T = 12$, $n = 8$, $P > 0.05$) and they could not be tested for Fight due to an extremely low frequency of occurrence.

There was no significant difference between Male Follows, Female Runs in the unfamiliar and familiar pairs ($T = 13$, $n = 7$, $P > 0.05$) (Table 2). The frequency of Male Follows, Female Waits was significantly lower ($T = 0$, $n = 7$, $P < 0.05$) in the unfamiliar versus familiar pairs. The frequencies of Mount and Copulate were both lower in the unfamiliar versus

Table 2. Mean frequences of following sequences and sexual acts, ± SE, during pairings of females with males whose odors were not familiar and males whose odors were familiar, n = 8.

	NOT FAMILIAR	FAMILIAR	
MALE FOLLOWS,[1] FEMALE RUNS	23.2 ± 10.09	19.6 ± 5.22	NS
MALE FOLLOWS,[1] FEMALE WAITS	21.1 ± 14.84	65.5 ± 28.46	*
MOUNT[2]	0.9 ± 0.64	15.0 ± 5.99	*
COPULATE[2]	6.5 ± 6.50	22.2 ± 8.44	NS

NS = P > 0.05
* = P < 0.05 sign test

[1]Wilcoxon matched-pairs signed-rank test.

[2]Sign test.

familiar pairs. This difference was significant for Mount ($T = 5$, $n = 5$, $P < 0.05$) but not for Copulate ($T = 4$, $n = 5$, $P > 0.05$).

Of the eight vaginal smears, half were characterized by nucleated epithelial cells and half contained at least some non-nucleated cornified squames. In rats, the former are typical of proestrus, and the latter are seen in true estrus (Gorbman and Bern, 1962). Two of the females in proestrus had sperm visible or had been observed in copula. Of the five females that copulated, two had sperm visible on the vaginal smear.

DISCUSSION

All three aggressive bahaviors -- Chase, Box, and Fight -- occurred more frequently between pairs in which the female was unfamiliar with the male's odors. However, only Chase was statistically significant. Of these aggressive behaviors, Chase was the most common, Box was infrequent, and Fight was rare. On one occasion, a fight was intense enough to warrant an early termination of the test, which was a pairing of a female and an unfamiliar male.

The sequences of following are notable because they show the behavioral changes associated with completion of sexual behavior. During breeding behavior, the male follows the female repeatedly. Initially, she runs off before he reaches her, regardless of whether or not she does eventually mate with him. But, concomitant with an increase in the probability of mating is an increase in her latency to run off; thus a prelude to mating is that the male makes physical contact with the female before she withdraws. This aspect of pika breeding behavior explains the lack of difference in the first following sequence (Male Follows, Female Runs). The signficantly greater propensity of a female to wait (in the second following sequence) when paired with a male whose odors are familiar suggests that she is more likely to mate with this male. Time may be an important factor here;

mating may simply be delayed in pairs where the male's odors are not familiar to the female. The three-hour time limit is a laboratory constraint not experienced by pikas in the field. However, it behooves a male to make use of any behavior that will speed up or facilitate the process of inseminating a female.

The importance of olfactory familiarity reveals itself in that there were more Mounts and Copulations in the familiar odor pairings. The fact that the difference was not significant for Copulations is probably a consequence of the sample size. The experiment indicates that this line of research is well worth pursuing.

Results from the vaginal smears suggest that pikas are receptive during both proestrus and estrus. Puget (1973), in his work on the Afghan pika (*Ochotona rufescens*), suggested that these animals have a pseudo-cycle, as in the rabbit. Such a pseudo-cycle is not characterized by spontaneous ovulation, rather, ovulation is induced by copulation.

The data presented in this pilot study do provide suggestive evidence for a mating-facilitation function of marking. The mechanism responsible for mating facilitation is unknown, but three possibilities are suggested. The first is that familiarity with a male's odor reduces a female's aggression towards him as Daly (1977) has proposed for gerbils. The problem here is that there is no evidence that females are more aggressive than males (but see Stewart et al., 1982), and it was the males who chased the females in the present study and never the reverse. A second possibility is that male odors prime a female for reproduction with that male. A priming effect of male odors on various aspects of female reproduction has been shown in other animals (Bronson, 1971; Bruce, 1959), and this mechanism may be a good candidate for further exploration. A third possibility is that familiarity with a male's odor reduces a female's fear in interactions with him. Such an effect may be similar to a preference for familiar nest odors, as found in rats (Gregory and Pfaff, 1971; reviewed in Geyer, 1981).

From the female's point of view, an evaluation of male scent marks addresses several important criteria. The first is that the male is in the vicinity and available. This is important because few immigrants are successful in establishing territories (Smith, 1983). The second is that the vigorousness of the scent-mark display may be an indication of sperm supply (Halliday, 1978). Thirdly, it may be an indication of testosterone levels (Thiessen et al., 1971; Turner, 1975; Yahr, 1981), which are often correlated with dominance. In pikas, dominant animals benefit in terms of food resources and territory (Kawamichi, 1976; Whitworth, 1982). Thus by choosing a male whose odors are familiar, a female may be investing wisely in terms of the reproductive success of her offspring, although the criterion for her choice is an indirect measure of selectively advantageous traits.

An interesting aspect of this study is that cheek-gland secretions had an effect on the interaction between two individuals. I refer to the aggressive behaviors (in Table 1), in which the male was always the aggressor. Familiarity with a male's odor, or lack of it, somehow affected a female in such a manner that the male's subsequent rate of aggression towards her was altered. What signals are involved, and how these cues are communicated, is not known. Perhaps postural signals from the female play a role.

REFERENCES

Bronson, F. H., 1971, Rodent pheromones, *Biol. Reprod.*, 4:344.
Bronson, F. H., 1976, Urine marking in mice: Causes and effects, *in*:

"Mammalian Olfaction, Reproductive Processes, and Behavior," R. L. Doty, ed., Academic Press, New York.
Bruce, H. M., 1959, An exteroreceptive block to pregnancy in the mouse, Nature, 184:105.
Daly, M., 1977, Some experimental tests of the functional significance of scent marking by gerbils (Meriones unguiculatus), J. Comp. Physiol. Psychol., 91:1082.
Epple, G., 1976, Chemical communication and reproductive processes in non-human primates, in: "Mammalian Olfaction, Reproductive Processes, and Behavior," R. L. Doty, ed., Academic Press, New York.
Geyer, L., 1981, Ontogeny of ultrasonic and locomotor responses to nest odors in rodents, Am. Zool., 21:117.
Gorbman, A., and Bern, H. A., 1962, "A Textbook of Comparative Endocrinology," John Wiley and Sons, Inc., New York.
Grau, G. A., 1976, Olfaction and reproduction in ungulates, in: "Mammalian Olfaction, Reproductive Processes, and Behavior," R. Doty, ed., Academic Press, New York.
Gregory, E. H., and Pfaff, D. W., 1971, Development of olfactory-guided behavior in infant rats, Physiol. Behav., 6:573.
Halliday, T. R., 1978, Sexual selection and mate choice, in: "Behavioural Ecology," J. R. Krebs, and N. B. Davies, eds., Blackwell Scientific Publications, Oxford.
Harrington, J. E., 1975, Field observations in social behavior of Lemur fulvus fulvus E. Geoffroy 1812, in: "Lemur Biology," I. Tattersal, and R. W. Sussman, eds., Plenum Press, New York.
Harvey, E. B., and Rosenberg, L. E., 1960, An apocrine gland complex of the pika, J. Mammal., 41:213.
Kawamichi, T., 1976, Hay territory and dominance rank of pikas, J. Mammal., 57:133.
Meaney, C. A., 1983a, Rate and location of scent marking by pikas during the breeding season, in "Chemical Signals in Vertebrates 3, D. Muller-Schwarze, and R. M. Silverstein, eds., Plenum Press, New York.
Meaney, C. A., 1983b, Olfactory Communication in Pikas (Ochotona princeps), Ph.D. Dissertation, Univ. of Colorado, Boulder.
Puget, A., 1973, Etude anatomique, physiologique et biochemique de l'Ochotone afghan (Ochotona rufescens rufescens) envue de son utilization comme animal de laboratoire, These Sci. Toulouse.
Smith, A. T., 1981, Territoriality and social behavior of Ochotona princeps., in: "Proceedings of the World Lagomorph Conference," K. Myers, and C. D. McInnes, eds., Guelph University Press, Guelph, Ontario.
Smith, A. T., 1983, Colonization in a pika population: Dispersal vs. philopatry, Behav. Ecol. Sociobiol., 13:37.
Stewart, N. J., Scudder, K. M., and Southwick, C. H., 1982, Investigative response of pikas (Ochotona princeps) to odors produced by conspecifics, J. Mammal., 63:690.
Thiessen, D. D., Owen, K., and G. Lindzey, 1971, Mechanisms of territorial marking in the male and female Mongolian gerbil (Meriones unguiculatus, J. Comp. Physiol. Psych., 77:38.
Turner, J. W., 1975, Influence of neonatal androgen on the display of territorial marking behavior in the gerbil, Physiol. Behav., 15:265.
Vandenbergh, J. G., 1967, Effect of the presence of a male on the sexual maturation of female mice, J. Endocrinol., 81:345.
Vandenbergh, J. G., and Drickamer, L. C., 1974, Reproductive coordination among free-ranging rhesus monkeys, Physiol. Behav., 13:373.
Whitten, W. K., 1956, Modifications of the estrus cycle of the mouse by external stimuli associated with the male, J. Endocrinol., 13:399.

Whitworth, M. R., 1982, Maternal Care, Infant Development and the Ontogeny of Social Behavior in Pikas, Ph.D. dissertation, Johns Hopkins Univ., Baltimore, Maryland.

Whitworth, M. R., 1984, Maternal care and behavioural development in pikas (Ochotona princeps), Anim. Behav., 32:743.

Yahr, P., 1981, Scent marking, sexual behavior and aggression in male gerbils: Comparative analysis of endocrine control, Am. Zool., 21:143.

HISTOLOGY AND GROSS MORPHOLOGY OF THE SEXUALLY DIMORPHIC STERNAL GLAND IN THE NORTH AMERICAN OPOSSUM, DIDELPHIS VIRGINIANA KERR

Donna Holmes Meisner

Department of Biological Sciences
Bowling Green State University
Bowling Green, Ohio 43403

INTRODUCTION

Cutaneous scent glands have been described for a wide taxonomic range of mammalian species, including the Marsupialia (for reviews, see Strauss and Ebling, 1970; Eisenberg and Kleiman, 1972; Thiessen and Rice, 1976; Stoddart, 1980; Muller-Schwarze, 1983). Pronounced sexual dimorphism of skin glands is a strong indication that their secretions are involved in social communication and, hence, are of evolutionary importance (Darwin, 1874; Thiessen and Rice, 1976; Blaustein, 1981).

Dimorphic sternal scent glands have been documented in several genera of the New World metatherian family Didelphidae, including Didelphis, Marmosa, and Monodelphis (Schaffer, 1940; Barnes and Barthold, 1969; Boggs, 1969; McManus, 1974; Pippitt, 1976; Barnes, 1977; Hunsaker and Shupe, 1977; Pine and Abravaya, 1978; Gardner, 1982; Trupin and Fadem, 1982; Meisner, 1983; Streilein, 1983). Few histological descriptions exist of cutaneous glands in this group, however. In general, the role of chemo-signalling in social organization and sexual selection in didelphid opossums has been sorely neglected.

Adult male Virginia opossums (Didelphis virginiana Kerr) have an odorous yellow-orange stain on the fur of the chest (Schaffer, 1940; McManus, 1974; Pippitt, 1976; Gardner, 1982; Meisner, 1983; Streilein, 1983). It is particularly striking just prior to, and during, the breeding season (personal observation). Opossum sternal gland secretions have been suggested to function in sexual recognition or advertisement (Pippitt, 1976; Streilein, 1983). The objective of this project was to characterize the apparent dimorphism of these glands in more detail, at both the gross morphological and histological levels, as part of an ongoing study of social behavior and chemocommunication in captive opossums.

MATERIALS AND METHODS

Gross Examination

The sternal integument of 40 opossums (15 adult females, 16 adult males, four subadult males, and five subadult females) was examined between July 1982 and March 1985. All subjects had either been live-trapped in northwest Ohio or born to members of a captive opossum colony that has been maintained at Bowling Green State University October 1983. Of 21 adults weighed to the nearest 0.1 kg, the mean weight for males was 2.46 kg, with a range from 1.7 to 3.8 kg. Females averaged 2.18 kg, ranging from 1.3 to 3.8 kg. Opossums of both sexes were examined at various phases of the reproductive cycle, including the nonbreeding period (November to early January), but most were examined during the breeding season or in December, just prior to the onset of estrus in early January. Of these 40 animals, 11 adult males and 10 adult females were anesthetized with ether to facilitate examination. The remaining opossums (five adult males, five adult females, and 10 subadults) were killed with ether or carbon dioxide and their sternal skin shaved and examined.

Histological Preparation

Skin for histological examination was collected from four adult males and two adult females in late December 1984 and February and March 1985, following gross examination. Several five-mm square skin specimens were excised from each of three areas on each animal: the sternal region (the area of visible staining in males), an area of homologous skin on the shoulder several cm adjacent to the sternal region, and on the dorsum between the scapulae.

After trimming off fat and connective tissue, skin specimens were fixed in 10 percent neutral buffered formalin overnight and transferred to 70 percent ethanol, or fixed in Carnoy's solution overnight. Specimens fixed in formalin were then passed through a standard series of dehydration alcohols, and all were cleaned for several days in cedar oil, cleared briefly in benzene, and embedded in paraffin under vacuum. Skin was sectioned at 8 to 10 μm on a rotary microtome and stained using a standard hematoxylin and eosin series. Upon microscopic examination, the general morphology and development of sudoriferous (apocrine) and sebaceous (holocrine) skin gland units were noted, as well as the relative thickness and appearance of the epidermis and dermis. Hair was also collected from the sternal and dorsal regions of several animals, whole mounted and examined.

RESULTS

Gross Morphology of the Sternal Integument

All adult males examined (n = 11), regardless of season, had a yellow-orange stain over the manubrium of the sternum, with a strong musky odor. It was triangular, with the apex pointed caudad. It averaged 7.18 cm in length and 4.36 cm at the widest part of the anterior end, thus averaged approximately three and one-half times longer than wide. In some breeding males, the stain extended in a thin line down the midline of the ventrum to the scrotum. Male sternal secretions were heaviest in specimens examined in December and January, when scent marking by males reaches its peak (Meisner, unpublished data). The fur in this area was matted and slightly damp to the touch, and oily orange flakes of exudate which collected could readily be rubbed off. The amount and color of secretion present varied greatly among individual males, even those examined at the same time of year, and began to wane in early summer.

Sternal secretions were lacking in females. During the breeding season, however, a thin, pale-yellow stripe appeared along the length of the ventral fur in some adult females. In no case did a female produce the musky odor characteristic of male secretions, or produce an amount of stain approaching five percent of the amount found on the least heavily stained males.

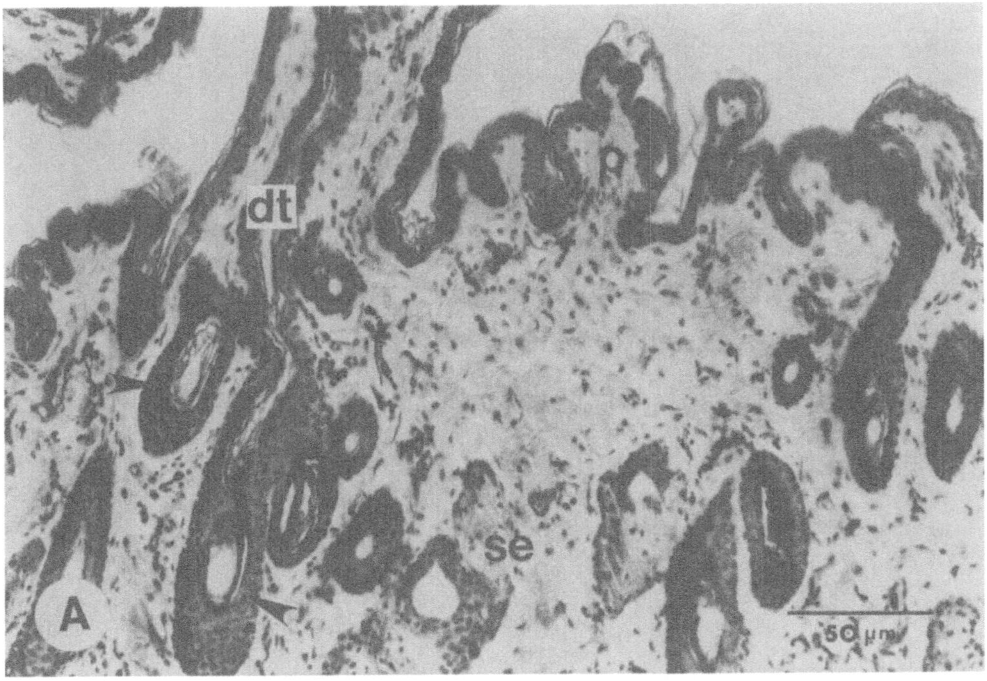

Fig. 1A. Photomicrograph of sternal skin of male Didelphis virginiana (p, dermal papilla, dt, duct, se, sebaceous gland). Arrows indicate apocrine gland units.

No complex glandular ducts were found opening to the surface of the sternal skin in males. In adults of both sexes, the sternal skin was darkly pigmented. The pigmented region underlay but did not directly correspond to the stained area, and was roughly butterfly-shaped, often with an unpigmented stripe down the midline. No consistent sexual dimorphism in pigmentation was noted, but its shape and extent varied much among individuals. The skin of the sternal region in all males was noticeably thicker, with more subcutaneous fat, than that of the surrounding region or sternal skin of females. In large males, this skin protruded as far as four cm from the chest to form a "dewlap."

There was no noticeable thickening or thinning of the chest fur in either sex, but the short, soft, white hairs characterizing the underfur are more prevalent on the ventrum than are the longer, coarser, black-tipped guard hairs of the dorsal fur in males and females.

Gross examination of known-aged subadults showed that the development of the sternal stain in males coincides closely with the time of independence of the young from the mother (Reynolds, 1952). By 100 to 110 days of age the musky odor is present, and at approximately 120 days the yellow stain is evident. Male opossums are not sexually mature until eight or nine months of age (Gardner, 1982), and cannot breed until the onset of estrus the following January in Ohio, but I have seen captive males as young as 150 days vigorously scent marking unfamiliar objects.

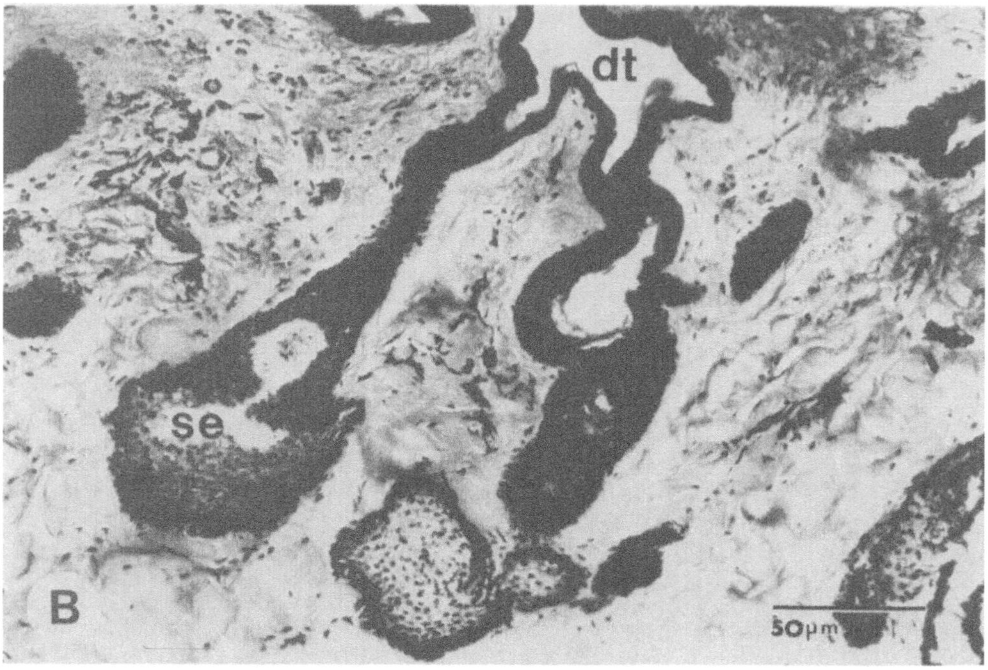

Fig. 1B. Photomicrograph of sternal skin of male Didelphis virginiana, showing hypertrophied sebaceous gland units (se, sebaceous gland, dt, duct).

Histology of the Sternal Integument

Microscopic comparison of skin from male and female opossums confirmed the suggestion, based on gross morphology, of greater cutaneous gland activity in the sternal region of males. The dermis was markedly thicker in male sternal skin than in female sternal skin or specimens from the adjacent shoulder region in either sex, and characterized by many infoldings or convolutions (dermal papillae) (Fig. 1A). Papillae were also present, to a lesser extent, in adjacent skin from males and females. They were best developed, however, in male sternal specimens, where they often formed elongate ducts to the epidermal surface.

No specializations of the hair of the sternal region of opossums of either sex were noted.

Holocrine sebaceous (oil) glands were more numerous, hypertrophied, and more active (based on the presence of exudate) in both pigmented and unpigmented specimens from sternal skin of males (Fig. 1B) than from the sternal area in females or the adjacent or dorsal skin in males or females. The degree of glandular development seen in female sternal skin was most similar to that in skin of the chest area adjacent to the stained area in males. Sebaceous glands from male sternal skin, which were usually associated with hair follicles, were teardrop-shaped, and extended into the deep layers of the dermis. The glands were characterized by cells with many darkly staining basophilic nuclei, and many cells were actively degenerating and emptying their contents into the lumina. Often several sebaceous acini were seen associated with a single large hair follicle. In a few cases, the ducts from two sebaceous glands converged to form a common secondary duct which emptied to the skin surface.

Apocrine sudoriferous (sweat) glands were also more prevalent and better developed in male sternal skin than in other skin types. The variability in size and configuration of apocrine units in male sternal skin exceeded the variability seen in male sternal sebaceous glands. Some of these were extremely complex, coiled structures located deep in the dermis, while others occurred much closer to the skin surface (Fig. 1A). Heightened activity of apocrine glands of all types was evident from the presence of eosinophilic contents of their lumina.

In general, the relative number and degree of hypertrophy of holocrine and apocrine gland units showed substantial individual variation, even among same-sex opossums of comparable age, weight, and time of collection. However, the variation in skin glands among females did not approach that seen among males.

DISCUSSION

The sexual dimorphism in cutaneous gland activity indicated by the sternal stain in male opossums is readily apparent at the histological level during the breeding season. The sternal skin of males is a site of hypertrophied sebaceous and sudoriferous gland units and well adapted for odor collection and dispersion. The folding of the skin surface provides a likely area for bacteria to collect and aid in the degradation of exudate and odor production, as well as an increased surface area for scent dispersal.

The fact that cutaneous glands in female sternal skin do not approach the degree of development or scent production seen in males suggests that opossums may be able to discriminate the sex of conspecifics on the basis of sternal odors alone. Moreover, the striking differences in relative numbers and types of sternal gland units among the few males examined thus far suggests that they are a source of variation in individual male scent. Olfactory discrimination tests are now in progress to test these hypotheses.

Virginia opossums do not have a well developed sternal scent marking behavior, although males occasionally rub their chests on objects or over the backs of females during courtship (Meisner, unpublished data). Adults of both sexes frequently cheek rub objects, however, often alternating this behavior with licking (Reynolds, 1952; McManus, 1970; Hunsaker and Shupe, 1977). The sternal region is pressed to the substrate in conjunction with this form of scent marking, and males sometimes deposit enough sternal gland secretion to be detectable by humans.

Social and communication systems in the Didelphidae have been described as "primitive," "poorly developed," and even "more reptilian than eutherian" (Reynolds, 1953; McManus, 1970; Tyndale-Biscoe, 1975; Hunsaker and Shupe, 1977; Gardner, 1982, Charles-Dominique, 1983). The sexual dimorphism in the sternal glands of this species is as pronounced as that in more recently evolved, ostensibly "advanced," mammalian groups, including placentals. While chemosignalling is admittedly an ancient mode of communication, the role of secretions like those of this gland in coordinating social interactions must be thoroughly explored before reaching firm conclusions about social complexity in New World marsupials.

ACKNOWLEDGMENTS

I am grateful for the technical assistance of P. McCann and D. Morse of Medical College of Ohio, F. Rabalais, M. E. Newport, the Electron Microscopy Facility of Bowling Green State University, and E. Csanady.

K. Foltz tended the opossums. S. Vessey and M. E. Newport reviewed the manuscript, and L. DeVenney typed it. Financial support was provided by Dissertation Improvement Grant BNS-8412441 from the National Science Foundation, the Association for Women in Science, Sigma Xi, the Scientific Research Society, and the Graduate College of Bowling Green State University.

REFERENCES

Barnes, R. D., 1977, Special anatomy of *Marmosa robinsoni*, *in*: "Biology of Marsupials," D. Hunsaker, II, ed., Academic Press, New York.

Barnes, R. D., and Barthold, S. W., 1969, Reproduction and breeding in an experimental colony of *Marmosa mitis* Bangs (Didelphidae), *J. Reprod. Fert.*, Suppl., 6:477.

Blaustein, A. R., 1981, Sexual selection and mammalian olfaction, *Am. Nat.*, 117:1006.

Boggs, J., 1969, The general and agonistic behavior of the mouse opossum, *Marmosa robinsoni*, M. S. Thesis, San Diego State Univ., San Diego.

Charles-Dominique, P., 1983, Ecology and social adaptations in didelphid marsupials: Comparison with eutherians of similar ecology, *in*: "Advances in the Study of Mammalian Behavior," J. F. Eisenberg, and D. G. Kleiman, eds., *Spec. Publ. Am. Soc. Mammal.*, 7:395.

Darwin, C., 1874, "The Descent of Man and Selection in Relation to Sex," Merrill and Baker, New York and London.

Eisenberg, J. F., and Kleiman, D. G., 1972, Olfactory communication in mammals, *Ann. Rev. Ecol. Syst.*, 3:1.

Gardner, A. L., 1982, Virginia opossum, *in*: "Wild Mammals of North America," J. A. Chapman, and G. A. Feldhammer, eds., Johns Hopkins Univ. Press, Baltimore and London.

Hunsaker, D., II, and Shupe, D., 1977, Behavior of New World marsupials, *in*: "Biology of Marsupials," D. Hunsaker, II, ed., Academic Press. New York.

McManus, J. J., 1970, Behavior of captive opossums, *Didelphis marsupialis virginiana*, *Am. Midl. Nat.*, 84:144.

McManus, J. J., 1974, *Didelphis virginiana*, *Mamm. Species*, 40:1.

Meisner, D. H., 1983, Psychedelic opossums: Fluorescence of the skin and fur of *Didelphis virginiana* Kerr, *Ohio J. Sci.*, 83:4.

Muller-Schwarze, D., 1983, Scent glands in mammals and their functions, in: "Advances in the Study of Mammalian Behavior," J. F. Eisenberg and D. G. Kleiman, eds., *Spec. Publ. Am. Soc. Mammal.*, 7:150.

Pine, R. H., and Abravaya, J. P., 1978, Notes on the Brazilian opossum, *Monodelphis scalops* (Thomas) (Mammalia: Marsupiala: Didelphidae), *Mammalia*, 42:379.

Pippitt, D. D., 1976, A radiotelemetric study of the winter energetics of the opossum *Didelphis virginiana* Kerr., Ph.D. Thesis, Univ. Kansas, Lawrence.

Reynolds, H. C., 1952, Studies reproduction in the opossum (*Didelphis virginiana virginiana*), *Univ. Calif Publ. Zool.*, 52:223.

Reynolds, H. C., 1953, The opossum, *Sci. Amer.*, 188:88.

Schaffer, J., 1940, "Die Hautdrüsenorgane der Säugetiere," Urban and Schwarzenberg, Berlin.

Stoddart, D. M., 1980, "The Ecology of Vertebrate Olfaction," Chapman and Hall, London and New York.

Strauss, J. S., and Ebling, F. J., 1970, Control and function of skin glands in mammals, *Mem. Soc. Endocr.*, 18:341.

Streilein, K. E., 1983, Behavior, ecology, and distribution of South American marsupials, *in*: "Mammalian Biology in South America," M. A. Mares, and H. H. Genoways, eds., *Spec. Publ. Pymatuning Lab. Ecol.*, Univ. Pittsburgh, Pittsburgh.

Thiessen, D., and Rice, M., 1976, Mammalian scent marking and social behavior, Psych. Bull., 83:505.
Trupin, G. L., and Fadem, B. H., 1982, Sexual behavior of the gray short-tailed opossum (*Monodelphis domestica*), J. Mammal., 63:409.
Tyndale-Biscoe, H., 1975, "Life of Marsupials," American Elsevier Publ. Co., New York.

CHEMICAL COMMUNICATION IN GRAY SHORT-TAILED OPOSSUMS (MONODELPHIS DOMESTICA) WITH COMPARISONS TO OTHER MARSUPIALS AND WITH REFERENCE TO MONOTREMES

Barbara H. Fadem

University of Medicine and Dentistry of New Jersey
New Jersey Medical School
Newark, NJ 07103

INTRODUCTION

The Importance of Chemical Communication in Marsupials and Monotremes

The Infraclass Metatheria contains only the single Order Marsupialia which includes some 250 species of living marsupials and at least 150 known fossil species (Kirsch and Calaby, 1977). Marsupials are found in two widely separated regions of the world, Australasia and the Americas (Stonehouse and Gilmore, 1977). The Australian marsupials are found in almost every habitat and include a wide range of insectivores, herbivores and carnivores, while the extant American marsupials are more limited in ecological range and are mainly small to medium sized carnivores (Kirsch and Calaby, 1977).

The monotremes are the only known order of the Subclass Prototheria and are found only in Australasia (Griffiths, 1978). There are two families, Ornithorhynchidae and Tachyglossidae. The former contains the single species *Ornithorhynchus anatinus*, the platypus; the latter contains two genera, *Tachyglossus aculeatus*, the echidna, and *Zaglossus bruijni*, the New Guinea long-beaked echidna.

Behavioral and physiological aspects of chemical communication in marsupials and monotremes are important for a number of reasons. First, these animals are different in many ways from the familiar eutherian mammals and are thus intrinsically interesting. The major difference between marsupials/monotremes and eutherians is the method of reproduction. In marsupials the young are born at a very early stage of development and are nurtured during a further period of development outside of the body of the mother, attached to teats of abdominal mammary glands. The pouch or marsupium, from which this order takes its name, is not, however, a universal feature of the group (Sharman, 1970; Stonehouse and Gilmore, 1977). Monotremes, although they carry two diagnostic characters of mammals, hair and mammary glands, lay eggs instead of bearing live young (Matthews, 1971; Collins, 1973). The degree of morphological development of the monotreme hatchling is much like that of the newborn marsupial (Griffiths, 1978). Thus, both groups have an extended period of early development and of maternal dependence. Both groups also must possess the means to gain maternal protection outside of the womb, i.e., to reach the pouch, while in a vulnerable, undeveloped condition. These characteristics have implications with regard to communication between mother and young (Eisenberg and Golani, 1977).

Chemical communication in marsupials is also of interest because communication systems in this group are often olfactory and may be extremely complex (Ganslosser, 1982). As in eutherian mammals, specific scent glands are present in a number of species and scent-marking behavior serves to deposit the secretions of these glands within the environment. Marsupials have cloacal (anal) glands, and frontal and sternal glands are present in some families (Muller-Schwarze, 1983). Scent-marking behavior in marsupials includes chin, sternal, ventral, cheek, flank and forehead rubbing and cloacal dragging (Eisenberg and Golani, 1977). Marsupials also scent-mark with saliva and urine (Muller-Schwarze, 1983). Monotremes scent-mark with the cloaca (Dobroruka, 1960). Although glands at the base of the neck and on the foot pads are present in this group, direct observation of their use has not been made (Griffiths, 1978).

Although suggested by location on the body and by the ecology of a species, the function of scent glands and of scent-marking in social communication has been studied systematically for only a few marsupial species (Schultze-Westrum, 1965, 1969; Ewer, 1968; Fadem and Cole, 1985). There are also very few studies of other forms of behavior in marsupials, perhaps because the behavioral repertoire of marsupials was not of major interest in the past since it was thought to be simple when compared with that of eutherians. This idea probably came about because many generalizations about marsupial behavior were based on studies of a very few, small, solitary species (Ewer, 1968). It has recently been noted that when a larger number of marsupial species are taken into account, the social behavior of marsupials is no less complex than that of ecologically or morphologically similar eutherians (Ganslosser, 1982).

A Model for Studying Chemical Communication in Marsupials

The Virginia opossum (*Didelphis virginiana*), a member of the Didelphidae or American opossums, is the only native North American marsupial. This species has probably been utilized more than any other marsupial for behavioral studies (Kirkby, 1977). There are also a number of didelphids in South America, although very little has been known heretofore about the ecology or behavior of these South American forms (Streilein, 1982).

Various aspects of chemical communication have recently been studied and reported for one South American didelphid, the gray short-tailed opossum (*Monodelphis domestica*), hereafter referred to as the gray opossum. Scent-marking behavior was described and conspecific responses to olfactory cues quantified in a recent study (Fadem and Cole, 1985). In two other studies, characteristics of the estrous cycle and influence of social factors on estrous cyclicity were assessed. It was found that estrous synchrony occurred in females (Fadem and Rayve, 1985), and that female reproduction was activated by olfactory contact with males (Fadem, 1985). It is of interest that these studies provide the first evidence for an effect of olfactory communication on female reproductive physiology in a marsupial.

The major objective of this chapter is to summarize existing information on chemical communication in marsupials. To do this, the gray opposum will be used as a model. Aspects of olfactory communication in a variety of other marsupials, and in monotremes, also will be addressed.

Scent Marking in Gray Opossums

The gray opossum. There are at least 11 species of *Monodelphis* distributed from eastern Panama to central Argentina. *Monodelphis* is nocturnal and semiarboreal, reportedly living near water and building nests among rocks and in hollow logs (Walker, 1968). Its short tail, typical of the genus, is slightly prehensile and is used for carrying nesting material and for support during climbing. The gray opossum (*M. domestica*) is found in the arid Chaco, Cerrado and Caatinga regions and in the remnant Atlantic rainforest on the eastern border of the Caatinga region in Brazil (Streilein, 1982). The animals used in the studies described here were descendants of 9 animals trapped in the Caatinga region by representatives of the National Zoological Park, Washington, D.C. Because laboratory breeding has been obtained through at least 7 generations, and because it is hardy, tractable and disease resistant, the gray opossum is likely to become the first practical marsupial model for laboratory research (Fadem et al., 1982; VandeBerg, 1983).

Adult female gray oppossums weigh 60-100 grams and adult males 90-150 grams. Head and body length of adults ranges from 170-200 mm with tail length an additional 60-80 mm. The coat color is gray-brown with lighter underparts. Sexual maturity is reached at 4-5 months; reproduction has occurred up to 28 months in females and 39 months in males, and breeding occurs throughout the year. The estrous cycle is approximately 32 days in length. Short, possibly anovulatory cycles of approximately 14 days are also seen (Fadem and Rayve, 1985). Individual females have had up to 4 litters per year (VandeBerg, 1983). The gestation period is 14-15 days, most litters contain between 5 and 12 spring and postpartum dependence is about 50 days (Maliniak and Taft, 1981; Fadem et al., 1982; VandeBerg, 1983).

Behavioral description of scent-marking in gray opossums. Scent-marking in gray opossums involves several body parts, objects and areas in the environment, and a number of different movements. In a recent laboratory study (Fadem and Cole, 1985), four body parts used in scent-marking were identified: the head, flank, chest and hip. In the most commonly observed type of scent-marking, head marking, the animal rubbed the side of its head and/or snout against an object, in a series of sharp, repeated motions. On occasion, either before or after head marking, the animal sniffed, bit or licked the object being marked, while holding onto it with the front paws. In the second type of marking, flank marking, the animal pressed the side of its body against an object and slid forward so that the length of its flank rubbed across the object. This motion often followed head marking. In the third type of marking, chest marking, the animal slid the slightly outspread front paws forward and upward and pressed the chest area over the object being marked. Again, this motion often followed head marking. After flank or chest marking, the animal sniffed the area just marked and occasionally groomed. The fourth type of marking, hip marking, although mainly seen in naturally and artificially estrous females (Fadem, unpublished observation), was also seen in males. When it occurred in males, it appeared almost frantic or paroxysmal in nature and was seen only after a considerable bout of head, flank or chest marking had taken place.

Sex and age relationships to scent-marking in gray opossums. Male gray opossums marked significantly more frequently than females ($P < 0.005$). Males marked with head, flank and chest, while diestrous females marked only with head and flank. Gray opossums tended to place

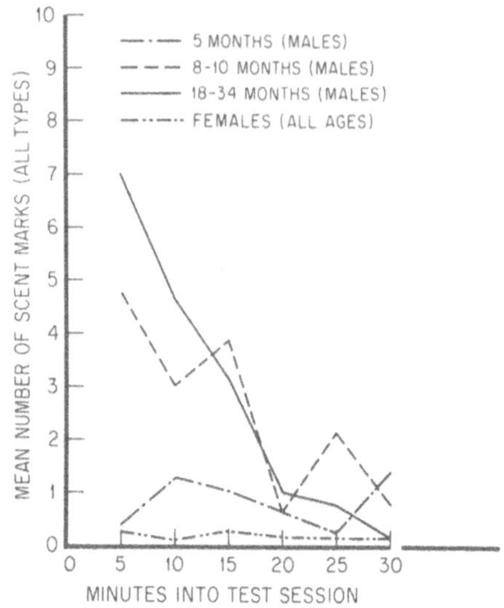

Fig. 1. Mean number of scent-marks of all types made over 30 min test period by gray opossums of different ages (Fadem and Cole, 1985).

their marks high in the environment and to mark the entrance to their nests. Scent-marking frequency was positively correlated with age; two month old males did not scent-mark, males 5 months old marked infrequently and older males 8-10 months and 18-34 months marked frequently. Also, there was a clear tendency of males 8 to 34 months of age to scent-mark within the first 10 minutes of their introduction into a testing chamber, while younger males scent-marked equally throughout a 30 min test ($P < 0.01$) (Fig. 1). Although 5 month old males scent-marked less frequently than older males, they self-groomed significantly more frequently ($P < 0.004$).

Similarities and differences of gray opossum scent-marking to scent-marking in well studied eutherian species. The primary models for experimental studies of scent-marking have been two laboratory rodent species, the golden hamster (Johnston, 1975), and the Mongolian gerbil (Thiessen et al., 1969). Like both hamsters and gerbils, objects within the environment are marked and males show a preference for marking where another male has marked (Johnston, 1975; Thiessen et al., 1971). Unlike hamsters and gerbils, which use one primary body part in scent-marking (the flank in hamsters (Johnston, 1975) and the ventrum in gerbils (Thiessen et al., 1969)), mature gray opossum males utilize three body parts in scent-marking, the head, flank and chest. The tendency to use a number of body parts in scent-marking is seen in other marsupial species (Eisenberg and Golani, 1977).

Functional significance of scent-marking in gray opossums. Although scent-marking in mammals often serves to define a home range (Muller-Schwarze, 1983), Hunsaker (1977) expressed a widely held view in maintaining that the didelphids are generally nomadic and have variable home ranges. Streilein (1982) reported that individual gray opossums were rarely trapped more than once in a particular area, and suggested that this species probably leads a nomadic existence because of the uneven rainfall and food resources in its native habitat. It is therefore not likely that scent-marking serves to delineate a home range for individual gray opossums.

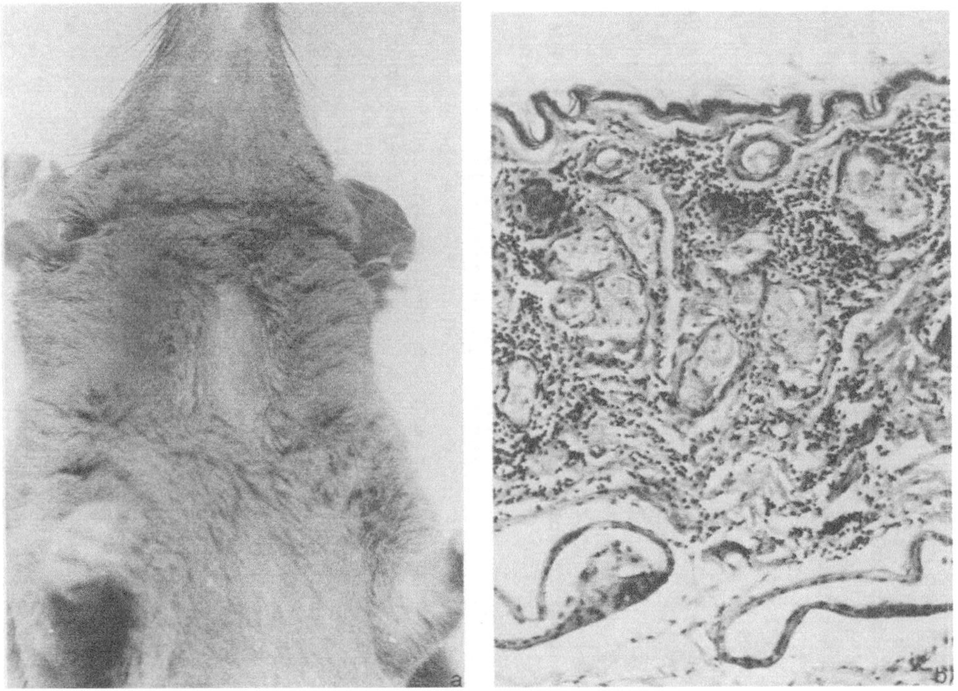

Fig. 2. The suprasternal scent gland in a male gray opossum (a). A section through this gland (b). (Fadem and Schwartz, 1985.)

Two observations of scent-marking in gray opossums support the possibility that, as in other species (Kleiman, 1966; Thiessen, 1968), a gray opossum uses scent-marking to personalize the immediate area with its own scent and to repel conspecifics from the immediate area. First, in mature males, scent-marking occurred at highest levels upon first introduction into the test chamber, and decreased as time went on as the area was fully marked and personalized. Second, the observed preference for marking high in the environment and the nest entrance indicate that a gray opossum may personalize the nest with its own scent and repel intruders; the effectiveness of these marks to an intruder being increased by placement at nose level (Hopp and Timberlake, 1983).

The suprasternal scent gland in gray opossums. Mammalian scent glands are often sexually dimorphic, occurring with greater frequency in males (Ebling, 1977), and are often seen in small, nocturnal species indigenous to arid environments (Regnier and Goodwin, 1977). Thus it is not unexpected that gray opossum males possess a specific scent-marking structure.

Physical examination of 40 adult male and 40 adult female gray opossums aged 30-122 weeks, and 25 juvenile males and 15 juvenile females aged 10-15 weeks revealed that all of the adult males and three juvenile males ages 13, 14 and 15 weeks of age possessed a suprasternal gland. None of the adult or juvenile females possessed a suprasternal gland. The gland appeared as a waxy area, devoid of hair along the midline of the body between the neck and rostral end of the sternum (Fig. 2a). The average length of the suprasternal gland was 13.8 mm $\pm$ 2.8 and ranged from 9 to 18 mm, while average width of the gland was 6.4 mm $\pm$ 1.8 and ranged from 3 to 10 mm. Histological investigation of the suprasternal scent gland revealed an intact epidermis, prominent sebaceous units in the upper to mid-dermis, and apocrine elements in the dermis (Fig. 2b), much like scent glands seen in eutherian species (Quay and Tomich, 1963; Kumari and Prakash, 1983; Sokolov and Skurat, 1966; Thiessen, Blum and Lindzey, 1969).

Hormonal aspects of scent-marking behavior in gray opossums. Testosterone stimulates and castration decreases the size and secretion of sebaceous glands in all mammalian species studied. Furthermore, the degree of stimulation by androgens of specialized sebaceous glands such as those used for scent-marking may be greater than the stimulation of sebaceous glands found on the rest of the body (Strauss and Ebling, 1970). The effects of estrogens on size and secretion of sebaceous glands are more equivocal. While in some eutherian species estrogens produce atrophy of sebaceous glands (Ebling, 1948), in others, estrogens have a stimulatory effect (Glenn and Gray, 1965). It is of interest that the first study of the hormonal control of a mammalian skin gland was conducted on a marsupial species, *Trichosurus vulpecula*. In this species, expectedly, castration led to a reduction and testosterone to an increase in the size of the sternal gland (Bollinger, 1944).

Scent-marking behavior first appears in gray opossums at the age of about 14 weeks, although most animals do not begin to mark until about 15 weeks of age. This correlates with the age of first appearance of the suprasternal gland in males and just prior to puberty, suggesting that, as in other species (Ebling, 1977), scent-marking behavior and the size of the suprasternal scent gland in gray opossums are related to androgen secretion. The androgen-dependence of the suprasternal gland was examined in an experiment in which seven adult male gray opossums were castrated; scent-marking behavior was observed and scent gland measurements taken over an 8 week period. It was found that the size of the suprasternal gland decreased significantly (Fig. 3, $P < 0.01$) in the 8 week period. Scent-marking behavior continued to be displayed, but at reduced frequency (Fadem, unpublished).

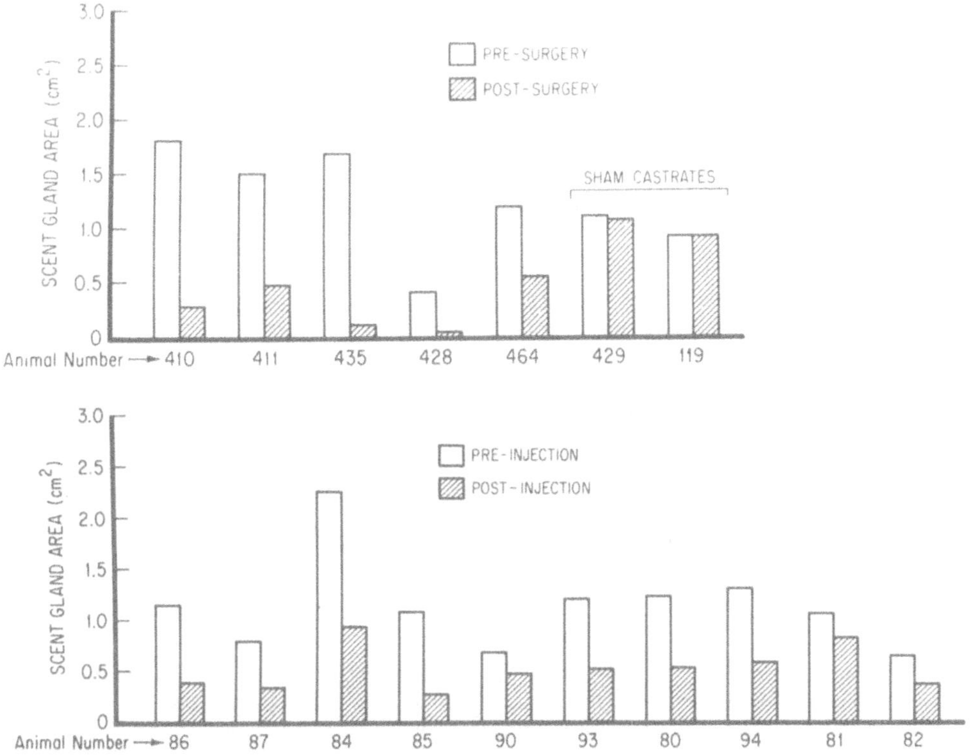

Fig. 3. The effect of castration on the size of the suprasternal scent gland of male gray opossums (Fadem, unpublished) (Upper panel). The effect of estradiol benzoate on the size of the suprasternal scent gland of male gray opossums (Fadem unpublished) (Lower panel).

In another experiment, intact male gray opossums were injected with 25 ug estradiol benzoate (EB) every other day for a period of 8 days (a total of 4 injections per animal). They were tested for scent-marking behavior prior to and at the end of this injection period. It was found that the size of the suprasternal gland decreased (Fig. 3) following estrogen injection, although scent-marking behavior continued. Chest-marking was the only type of scent-marking to show a decrease following EB injection. Thus it appears that the size of the suprasternal gland is a more sensitive monitor of the effects of gonadal hormones than is scent-marking behavior in this species. Currently, studies are being conducted in which crystalline estradiol and testosterone are being implanted subcutaneously in ovariectomized female gray opossums. Preliminary results indicate that treatment with either hormone results in increased scent-marking, particularly hip-marking.

Scent Glands and Scent-Marking Behavior in Monotremes and Marsupials

Scent glands and scent-marking in monotremes. Although physiological and ecological studies have been done, few behavioral studies have been conducted on the monotremes (Griffiths, 1978).

In a study of scent-marking in echidnas, Dobroruka (1960) noted that olfactory markings were applied with the extruded cloaca. Griffiths (1978) suggested that olfaction probably plays a role (1) in bringing the normally solitary echidnas together at breeding season and (2) in locating food. The echidna, which inhabits hot and arid areas, also uses olfaction to explore the surroundings since its sight is poor. Conspecifics are recognized olfactorily, and, if a member of another species smells of echidna, it is treated as a conspecific.

Platypuses are often found in fresh water creeks, rivers, and lakes in Australia, since they depend on fresh-water invertebrates for food. When not in water, platypuses spend time in their burrows, which protect them against changes in outside temperature. Since the olfactory epithelium is cut off from outside stimuli when the animal is feeding under water, olfaction is probably of lesser importance to the platypus than to the land-based echidna (Griffiths, 1978).

The male platypus has glands at the base of the neck that emit a "musky" odor. "Sweat glands" in the foot pads of the fore and hindlegs have also been described in this species (Klastsch, 1888, cited in Muller-Schwarze, 1983).

The long-beaked echidna occupies humid and often cold habitats and is restricted to the upland areas of Papua New Guinea and Irian Jaya. Unlike the other monotremes, members of this genus possess sweat glands (Griffiths, 1978).

Although little is known about their breeding biology, the monotremes normally produce one or two offspring per year. Female platypuses apparently incubate their eggs for approximately 10 days after laying and the young are suckled in burrows for over three and one-half months (Griffiths, 1978). Thus, as in marsupials, there is a prolonged period of maternal-young interaction.

One form of non-olfactory chemical communication, the crural system, is present in all three species of monotremes. The crural system consists of a kidney-shaped alveolar gland located on the dorsal aspect of the upper thigh muscles, connected by a duct to a spur on the heel. The gland secretes a venom which is passed into the spur and which can be injected into other animals by erection of the spur (Temple-Smith, 1973, cited in Griffiths, 1978). In platypuses, although females possess the structural components, only males have an active crural system. In contrast to the situation in the platypus, echidna and long-beaked echidna females, as well as males, possess crural systems (Griffiths, 1968). In the platypus, the activity of the crural system waxes and wanes seasonally in close parallel to the waxing waning of the reproductive system. In this species, the spur is apparently used in inter-male fighting during the breeding season (Temple-Smith, 1973, cited in Griffiths, 1978). Thus, the activity of the crural system, is probably related to androgen secretion in the platypus.

Scent-marking in marsupials. In addition to using specific anal-cloacal, and sternal glands, marsupials use saliva, urine and feces in chemical communication (Eisenberg and Golani, 1977). Other specific glands, such as frontal glands and parts of the body (such as the flank), are also used (Schultze-Westrum, 1965, 1969; Fadem and Cole, 1985). As in eutherian species, type of scent gland and scent-marking, as well as the degree to which the behavior occurs, are related to the eoclogy of the marsupial species (Muller-Schwarze, 1983). For example, while they possess specific scent-marking structures, it is likely that the large, grazing marsupials, such as the kangaroos, utilize olfactory communication to a lesser extent than do the small, nocturnal, arboreal forms such as the sugar glider in which visual communication may be limited. Also, as previously mentioned, scent-marking, particularly where an oily or waxy secretion remains, may be emphasized in species such as gray opossums that inhabit arid areas, since a deposited scent may be long-lasting; i.e., "released" from a mark only by moisture from the breath of another animal when the mark is sniffed (Regnier and Goodwin, 1977).

Anal glands and cloacal marking. Anal (cloacal) glands are present in most marsupial families. These include the Didelphidae, Dasyuridae, Phalangeridae, Phascolomyidae, and Macropodidae (reviewed in Muller-Schwarze, 1983). The secretions of these glands may be copious; anal gland secretions of female Virginia opossums and gray opossums often contaminate vaginal smears (Jurgelski and Porter, 1974; Fadem and Rayve, 1985). It has been suggested that since the anal glands of marsupials are used in stressful circumstance, they may function in defense (Bollinger and Whitten, 1948). Anal glands may prove particularly useful in interspecific defense, since many carnivores find the strong odor of marsupial anal scent glands distasteful (Gill, 1955).

The brush possum, T. vulpecula, has two sets of anal glands, one of which produces an apocrine secretion which is ejected as a reaction to extreme fear (Tyndale-Biscoe, 1973). When exposed to the anal gland secretions of other males, T. vulpecula and T. caninus males produce a threat display including agonistic vocalizations. This does not occur in response to the anal gland secretions of females (Thompson and Pears, 1962). In the Tasmanian devil, Sarcophilus harrisii, threat displays are often preceded, interrupted or followed by cloacal dragging in which the rump is lowered and the perianal area is pressed to the substrate. Dominant individuals often start to fight immediately after dragging, which is likely to accompany aggression over food (Buchmann and Guiler, 1977). Extensive cloacal dragging is also seen in Marmosa, Dasycercus, Sarcophilus (Eisenberg and Golani, 1977) and Phalanger gymnotis (Wemmer and Collins, 1978). In P. gymnotis, evacuation of the cloacal glands as the animal moves along occurs in both males and females and produces a line of scent on the substrate (Wemmer and Collins, 1978). In the "rat" kangaroo, Hypsiprymnodon moschatus, anal glands in both sexes emit a musky scent (Schaffer, 1940; Walker, 1968; Muller-Schwarze, 1983).

Virginia opossums introduced into a strange cage often taste or eat the feces of the cage occupant (Meisner, personal communication). Gray opossums do the same, particularly in the context of sexual encounters (Trupin and Fadem, 1982). It has been suggested that although fecal material may be a source of olfactory information, the fecal material itself may be of less importance than the deposits of the anal glands left on the feces (Schultze-Westrum, 1965).

Sternal glands. As in eutherian species, sternal glands in marsupials are characteristic of arboreal and semiarboreal forms such as the sugar glider (Schultze-Westrum, 1965, 1969), gray opposum (Fadem and Cole, 1985), mouse opossum, Marmosa robinsoni (Boggs, 1969), and brush possum, T. vulpecula (Wemmer and Collins, 1978). Sternal glands are also present in D. virginiana, Antechinus flavipes, Myrmecobius fasciatus (Muller-Schwarze, 1983) and T. vulpecula. Only male T. vulpecula have modified sebaceous glands on the chin; both males and females have these glands on the chest (Tyndale-Biscoe, 1973). In a number of species such as D. virginiana, M. robinsoni, M. domestica, Megaleia rufa, and Phascolarctos cinereus, the sternal gland is found only in males (McManus, 1974; Smith, 1980; Muller-Schwarze, 1983; Fadem and Cole, 1985). Apocrine glands have been found in the breast and axillary regions of male red kangaroos, M. rufa and in the cloaca-pouch area of females of this species (Mykytowycz and Nay, 1964). Similar glands are seen in the gray kangaroo, Macropus

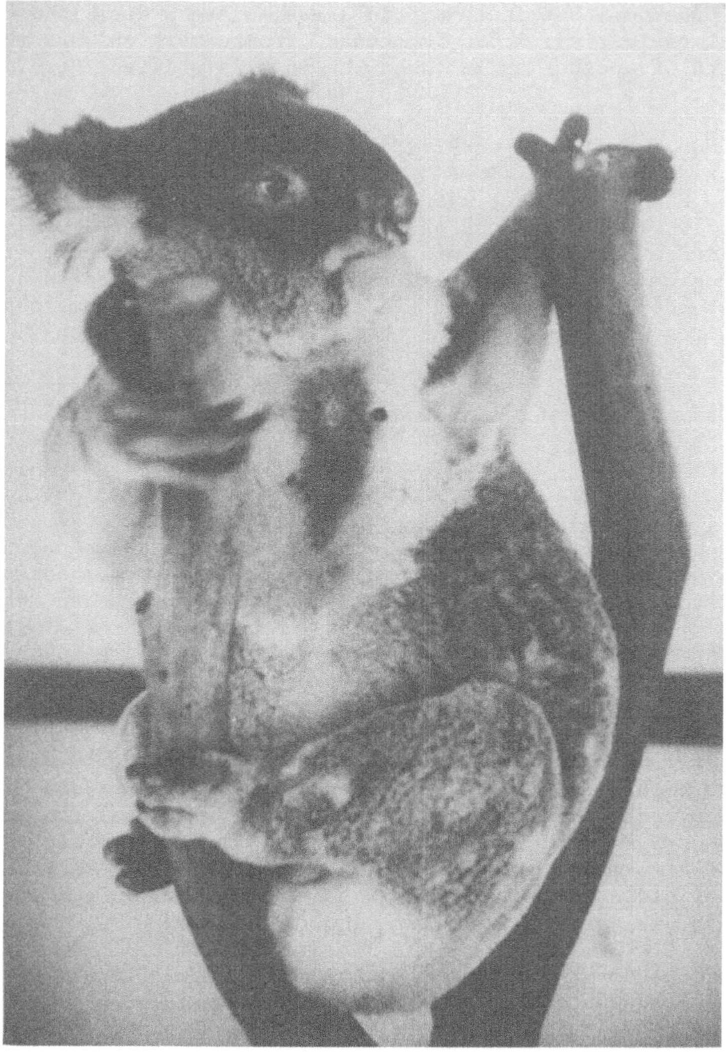

Fig. 4. The sternal gland in the koala (Phascolarctos cinereus) appears as a small, hairless area in the center of the darkly-stained chest fur (Thompson and Fadem, unpublished).

cangaru, Macropus robustus, and the swamp wallaby, Wallabia bicolor (Muller-Schwarze, 1983).

Like anal glands, sternal glands may also function in agonistic encounters. In the mouse opossum, the sternal gland is used by dominant males to mark their habitats (Boggs, 1969). The glands on the chest and abdomen of male red kangaroos are licked or bitten in aggressive encounters with other males (Sharman and Calaby, 1964).

In a recent study of scent-marking behavior in the koala, P. cinereus (Thompson and Fadem, unpublished observations) a conspicuous sternal gland, seen only in adult males, was used in chest-marking (Fig. 4). In chest-marking, the male koala stood upright, lowered its forelimbs to the side, placed the gland over the area to be marked, and flexed and extended

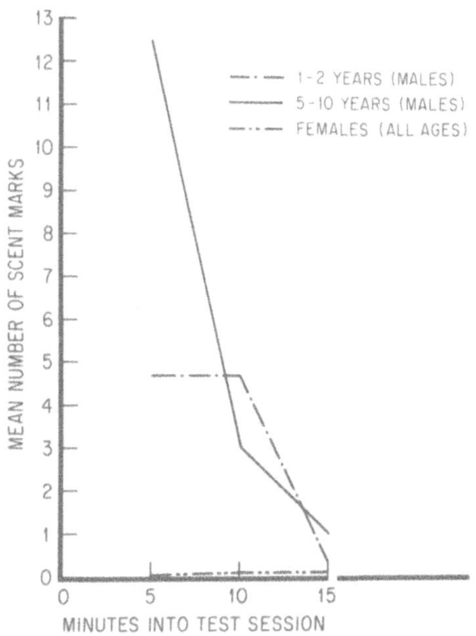

Fig. 5. Mean number of chest marks made in a 15 min test session by koalas (Thompson and Fadem, unpublished).

the hindlimbs so that the mark extended vertically. Sniffing of the area to be marked occurred prior to and immediately following the behavior. Occasionally, the area was marked again by a repeat of this sequence. A distinctive vocalization, "bellowing," often accompanied scent-marking.

Chest-marking in the koala occurred primarily upon introduction to an unfamiliar tree and declined thereafter. Females rarely chest-marked (only one chest mark was seen in over 20 15-min trials), and highest levels of chest-marking behavior were seen in males 5-10 years old; males 1-2 years old marked less frequently (Fig. 5). This is probably related to the fact that puberty occurs at between 1 and 3 years of age and the sternal gland first appears at about 1 year of age in this species (Thompson, personal communication). Thus, age, sex, and time-related aspects of scent-marking in the koala were remarkably similar to those seen in the gray opossum (c.f., Fig. 1).

Saliva and oral marking. The use of saliva and licking or biting in scent-marking occurs in the yellow footed marsupial mouse, *A. flavipes*, the crest-tailed marsupial mouse, *Dasycercus cristicauda* (Ewer, 1968), *M. robinsoni*, *T. vulpecula* and *P. gymnotis* (Eisenberg and Golani, 1977). Saliva is apparently used for identification of sex and for recognition of individuals in the fat-tailed marsupial mouse, *Sminthopsis crassicaudata* (Ewer, 1968).

Oral marking, like anal and sternal marking, is used to display aggression in a number of marsupial species. Male gray kangaroos, *Macropus giganteus*, and red kangaroos spread saliva on the chest and then rub the chest on the substrate during aggressive encounters (Frith and Calaby, 1969). In the Virginia opossum, adult males mark objects by licking and rubbing their head and neck region against them. These marks elicit an aggressive response in other males; upon encountering the mark of one male, another male will display an aggressive behavior pattern described as the "fighting dance" (Reynolds, 1952). Hunsaker and Shupe (1977) suggested that marking by the male Virginia opossum may identify his presence in an area, allow for recognition of individual males by females and repel other males.

Urine marking. Urine dribbling behavior is seen in *Marmosa*, *Sminthopsis*, *Dasyurus*, *Sarcophilus*, *Trichosurus*, *Phalanger* and *Macropus* (Tyndale-Biscoe, 1973; Eisenberg and Golani, 1977). Both male and female koalas urinate in a steady stream, or a few drops at a time (spot marking), upon introduction to an unfamiliar tree (Thompson and Fadem, unpublished observation). The wooly opossum, *Caluromys phalander*, also urinates in a trail, or only on a spot along a tree branch (Charles-Dominique, 1983). Charles-Dominique (1983) suggested that in *C. phalander*, urine marking may play a role in individual familiarization, although probably not in territoriality. Male and female Virginia opossums apparently use urine marking in social interactions (Meisner, personal communication).

Social determinants of scent-marking in marsupials. The social determinants of scent-marking can give clues to its functional significance for a species. For example, scent-marking might occur in response to the odor of a conspecific if it serves to personalize an environment with the individual's own odor (Kleiman, 1966). Alternatively, if the presence of an intruder has aversive qualities (Lorenz, 1966) scent-marking might be depressed in response to the odor of a conspecific. Also, preferentially marking where another animal has marked may be a response to an absence of one's own odor cues and an attempt to personalize the area (Thiessen et al., 1971). Although observations of scent-marking behavior have been made in some marsupial species, experimental studies of the social aspects of scent-marking behavior (i.e., behavioral responses to odors) have been conducted for only a few marsupials, among them the sugar glider, *Petaurus breviceps* (Schultze-Westrum. 1965 1969), and the fat-tailed marsupial mouse, *Sminthopsis crassicaudata* (Ewer, 1969). Recently, an experimental study of the social determinants of scent-marking in the gray opossum was conducted (Fadem and Cole, 1985). The results of studies on these three species are summarized below.

Social aspects of scent-marking in sugar gliders. The most extensive studies of chemical communication in marsupials were conducted by Shultze-Westrum (1965, 1969) on the sugar glider. This species, which is nocturnal and arboreal, lives in social nesting communities which contain up to six adult males and females together with their young. Sugar gliders use many forms of chemical communication; odors are produced by the sternal, frontal and anal glands and may be present in the urine. It is likely that the odor of urine may be used in individual recognition. The frontal and sternal glands are apparently used in intersexual communication, since both males and females rub their frontal glands against the sternal glands of group members of the opposite sex. Sugar gliders of both sexes produce strong odors from the secretion of anal glands when danger threatens.

Intraspecific agonistic behavior is only displayed between communities of sugar gliders and only one or two dominant males in each community mate, maintain the territory and distribute odors (Schultze-Westrum, 1965). The dominant male apparently is the source of a community odor which functions, in part, to keep the social activities of subordinate males at a low level. If the dominant male is removed, scent-marking, scent production, mating activity, patrol activity and body weight of other males in the community increase.

Social aspects of scent-marking in the fat-tailed marsupial mouse. The marsupial mouse uses two areas of the body, the mouth and the cloaca, in communication (Ewer, 1968). The smell of the mouth is apparently involved in individual recognition; when strangers are introduced, mutual mouth sniffing results. If a female is separated from her litter, mouth-sniffing occurs between mother and young upon reunion. Ewer (1968) suggested that although saliva may be used as a scent-marking substance in other species, this use is not apparent in the marsupial mouse, since one animal does not show a response to the mouth-mark of another.

Anal marking is easily elicited in the marsupial mouse by contact with a strange object, or by a strange smell on a familiar object. In contrast to effects of saliva, anal gland secretion is clearly perceptible to another animal. Both sexes produce anal gland secretions and both sexes mark, but when a male and female are housed together, marking is done only by the male. Odor emanating from the cloacal area may also serve to identify an estrous female to a male.

Social aspects of scent-marking in gray opossums. As previously noted, little is known about the natural social behavior of gray opossums. In a laboratory study of sexual behavior in this species, males placed into cages recently occupied by females sniffed and then scent-marked areas of the cage using the chest, flank, and head (Trupin and Fadem, 1982). In a recent laboratory study (Fadem and Cole, 1985), the scent-marking behavior of male and female gray opossums in a clean test chamber was compared with their behavior in a test chamber that had been previously marked by a male. It was found that although there was no significant difference in the overall frequency of marking of test males between control tests and tests in a previously marked testing chamber, there was a change in the pattern of marking. A male's marking pattern in a previously marked testing chamber was more similar to that of the male that had marked the chamber than to his own prior marking pattern in a clean testing chamber. Thus, male gray opossums preferentially marked objects and areas that had been scent-marked by other males. Although they vigorously sniffed areas that had been previously scent-marked by males, females that had not previously scent-marked in a clean testing chamber did not scent mark in a male-marked testing chamber.

Scent-marking in gray opossums may serve a reproductive function. Females of this species display intense aggression toward unfamiliar males, although these aggressive responses are reduced if the female has become familiarized with the male's scent prior to actual physical contact (Fadem et al., 1982). Thus, the tendency of a second gray opossum male to scent-mark in an area marked by a first male may confer a reproductive advantage to the second male. By replacing the odor of the first male (by marking over his marks), the second male can ensure that a female will become familiarized with his own scent. This would serve to reduce her aggression toward the second male and increase his chances over those of the first male of mating with the female. Recent studies also indicate that the odor of males alters the estrous cycle of female gray opossums (Fadem and Rayve, 1985; Fadem, 1985).

The use of chemical communication in maternal-young interactions in marsupials and monotremes. As previously noted, marsupials and monotremes are born (hatched) at a very early stage of development and have an extended period of development outside of the body of the mother. Despite its very immature condition at birth, the neonatal marsupial must climb to the pouch and attach to a teat if it is to survive. The means by which such an undeveloped neonate accomplishes this feat has been a subject of great interest. In 1806, Barton (cited in Griffiths, 1978) described the enormous nature of this task: "The young opossums, unformed and perfectly sightless as they are at this period, find their way to the teats by the power of an invariable, a determinate instinct, which may, surely, be considered as one of the most wonderful that is furnished to us by the science of natural history."

The migration of the neonatal marsupial to the pouch appears to be guided by olfactory cues rather than, as had been previously thought (Hartman, 1928), by negative geotropism or tactile responses. In marsupial species in which birth has been directly observed, just prior to and during birth, the mother licks the cloacal and pouch area extensively. Although she gives no assistance to the young as they climb toward the teats, the odor trail left by the maternal licking, apparently orients the young toward the pouch (Sharman and Calaby, 1964).

Although many sensory systems of the newborn marsupial are undeveloped at birth, the presence of a functional olfactory system at this time is likely. Anatomical studies indicate that the incisive ducts of Stensen, open at birth in the marsupial (McCrady, 1938), are connected to the vomeronasal organs whose nerves have formed by this time (Langworthy, 1928). It has been suggested (Hunsaker and Shupe, 1977) that the chemoreceptive ability of the vomeronasal organs could be sufficient to enable the neonatal marsupial to orient toward areas of greater concentration of chemical cues of the mother's saliva or of pouch secretions. Thus, the neonate would follow the path made by the mother's saliva, move in the correct direction, find a teat and begin suckling.

Like the marsupial neonate, the monotreme neonate possesses fully differentiated olfactory sense receptors when it emerges from the egg. Fibers from these olfactory sense cells which are present in the olfactory epithelium, enter the olfactory bulbs. In contrast to the absence of differentiated organs in other sensory systems, the differentiation of the olfactory system at hatching suggests that in monotremes as in marsupials, sense of smell plays a part in location of the maternal teat (Griffiths, 1978).

THE EFFECTS OF OLFACTORY COMMUNICATION ON FEMALE REPRODUCTIVE PHYSIOLOGY IN GRAY OPOSSUMS

In the first part of this chapter, the effects of chemical signals provided by the scent-marking of one animal in "releasing" immediate behavioral responses in another animal were discussed. Chemical signals can not only release such immediate behavioral motor responses, but can also cause relatively slow physiological changes, so called "priming" effects. The priming effects induced by chemical signals can lead ultimately to changes in behavior, size of scent-marking structures and occurrence and timing of ovulation and pregnancy (Whitten, 1966).

In a number of eutherian species, social-olfactory stimulation provided by males or by females influences reproduction by synchronizing reproductive cycles of females (Parkes and Bruce, 1961; Whitten, 1956; McClintock, 1971, 1978, 1984), or by inducing estrus and ovulation (Whitten, 1966; Hughes, 1964; Carter et al., 1980).

The hormonal control of the estrous cycle and ovulation is similar in eutherian and marsupial mammals (Sutherland et al., 1980). However, in marsupials, gestation is always shorter than the estrous cycle and pregnancy does not interrupt the cyclic recurrence of estrus (Sharman, 1970). Spontaneous ovulation appears to be the rule in marsupials and the estrous cycle of polyestrous species is approximately one month in length (Tyndale-Biscoe, 1973). The effects of social factors on reproduction in female marsupials have not heretofore been examined systematically, in part because it has been difficult to maintain and breed

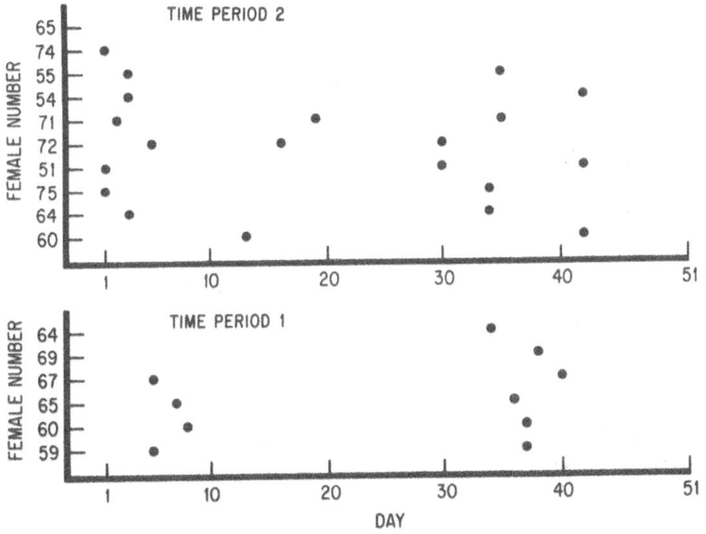

Fig. 6. Estrous synchrony in the gray opossum. Each symbol represents the first day of a period of estrus (Fadem and Rayve, 1985).

marsupials in the laboratory (Farris, 1950: Barnes, 1968; Barnes and Wolfe, 1971). The only reports of the influence of social factors on reproduction in marsupials were the observation that Virginia opossum females from the same litters came to estrus at the same time (Reynolds, 1952); also, as previously noted, reproduction in subordinate male sugar gliders was inhibited by odors produced by the dominant male (Schultze-Westrum, 1965).

Estrous Synchrony in Gray Opossums

In a recent study (Fadem and Rayve, 1985), it was found that social-olfactory factors influenced the estrous cycle of gray opossums. The estrous cycles of groups of females housed in one room tended to occur in synchrony ($P < 0.01$, Fig. 6). Also, females living in close proximity to males in a caging situation in which sensory contact between male and female was permitted, although mating was prevented, were more likely to show vaginal smears typical of estrus over the course of the estrous cycle than those housed in a caging situation in which both physical and sensory contact with males were prevented ($P < 0.01$).

Activation of Reproduction by Males

In another study (Fadem, 1985), the temporal pattern of births following pairing was examined over a one year period for 79 male-female gray opossum pairings. Males and females were paired in a random fashion, without regard to the estrous cycle of the female. To avoid the aggression common between females and unfamilar males (Fadem et al., 1982), a female was familiarized with a male by placing her in a small cage placed inside of a larger cage in which a male had free movement. At the end of this familiarization period, the female was released from the small cage and allowed to interact freely with the male in a large cage. Fourteen days following her release, the female was checked daily for a period of six weeks for the presence of young.

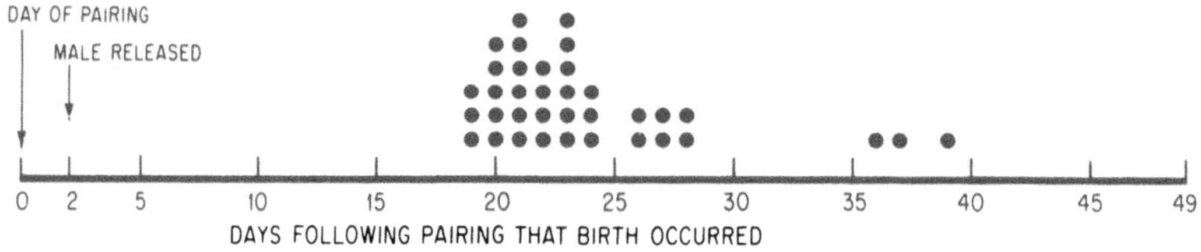

Fig. 7. The temporal pattern of births by female gray opossums following pairing with males on day zero. Each symbol represents one litter (Fadem, 1985).

Estrus, ovulation and mating were expected to occur in a random pattern over the 32 days ("ideal" estrous cycle length) following the familiarization period. Since the gestation period is approximately 15 days in gray opossums, and there was a two day familiarization period, births were expected to occur uniformly over the 17-49 days following pairing. A chi-square test for uniform occurrence of births between days 17 and 49 was therefore conducted.

Thirty-six of the 79 pairings produced litters. All litters were born within 39 days following pairing. The remaining 43 pairings did not result in litters although the animals were paired for a total of 56 days.

The occurrence of births was not uniform. Seventy-five percent of the litters were born 19-24 days following pairing and 16.6 % of the litters were born 26-28 days following pairing. Thus, a total of 91.6 % of litters were born 19-28 days following pairing. The remaining 8.3 % of litters were born 36-39 days following pairing (Fig. 7). No litters were born 29-35 days or 40-71 days following pairing. Thus, the occurrence of births over days 17 through 49 was not uniform ($\chi^2 = 35.9$, d.f. = 3, $P < 0.0001$). Also, significantly more births occurred on days 19-24 than at any other time following pairing ($\chi^2 = 6.26$, d.f. = 2, $P < 0.05$).

The results of this experiment indicated that a factor associated with males altered the expected occurrence of estrus in female gray opossums. To obtain information on the nature of this factor, the temporal pattern of the appearance of vaginal estrus following sensory exposure to males without actual mating stimulation was then examined.

The estrous cycles of 11 females living together in a room without males were monitored for one month. At the end of this period, 7 of these females (Group 2) were placed into small cages put into larger cages in which males had free access as in the previously described familiarization period. Thus, sensory contact was permitted, although mating was prevented. The remaining 4 females acted as controls (Group 1) and were not exposed to males.

Only 2 of the 11 females exhibited periods of estrus prior to the introduction of males to Group 2. Thus, the probability of a period of estrus occurring in the absence of males was estimated at 2/11. Three to 7 days following the introduction of males to Group 2, 4 of the 7 animals in this group exhibited a period of estrus; 1 of these 7 females was already in estrus when the males were introduced and 2 females showed no estrous responses (Fig. 8). The likelihood of four periods of estrus occurring following the introduction of males was thus calculated (using the Binomial distribution with $N = 7$, $P = 2/11$) as $P < 0.008$.

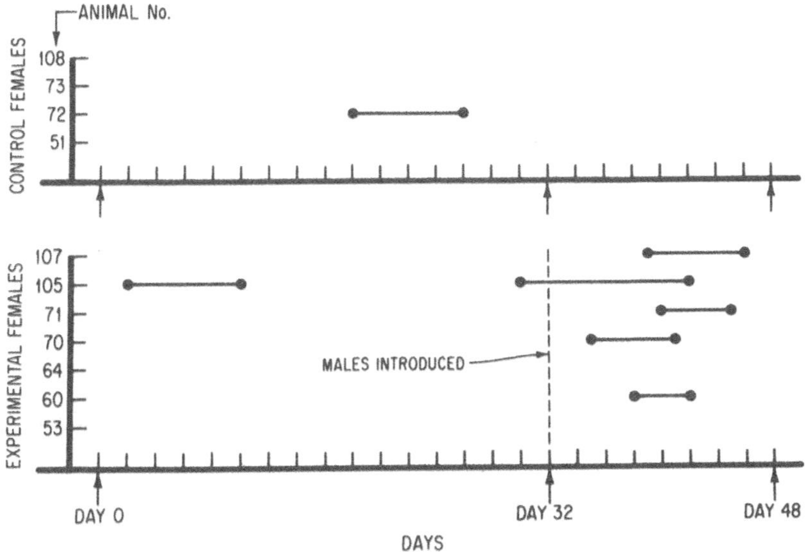

Fig. 8. The temporal pattern of the appearance of vaginal estrus in female gray opossums prior to and following the introduction of males to the experimental group. Each bar represents one period of estrus (Fadem, 1985).

No females in Group 1 exhibited a period of estrus after the males were introduced to Group 2.

Thus, pairing with males influenced the occurrence of estrus, ovulation and pregnancy in gray opossums. Although it was expected that females would show vaginal estrus and mate in a random pattern over the 32 days (ideal estrous cycle length) following pairing, this did not happen. Rather, females tended to show vaginal estrus 3 to 7 days, and to become pregnant 5 to 9 days, following pairing with a male. The 2 day difference between the appearance of vaginal estrus and the initiation of pregnancy has been previously observed (Fadem and Rayve, 1985).

Not all females were physiologically responsive to the presence of males. Approximately 54% of the females failed to produce litters and 43% of the females failed to show estrus when paired with males. Also, although they did eventually respond, three females did not become pregnant until 22-25 days following pairing. These three females may have just experienced estrus when paired with the male, and thus, may have required a longer time than females that had not just experienced estrus until estrus and ovulation could again occur.

Marsupials, like many eutherian mammals, possess well-developed vomeronasal organs (Johnson, 1977). The functional connection between the vomeronasal system and phenomena such as estrous synchrony and estrous induction is well-established for eutherian species such as the mouse (Keverne, 1983). Thus, the mechanism involved in estrous synchrony and activation in gray opossums, observed in these studies, is likely to involve the vomeronasal system - specifically, a trigger of gonadotropin secretion by pheromonal cues produced by other females and/or by males (Whitten, 1966; Bronson and Desjardins, 1974) acting through the vomernoasal system of the female. An olfactory and/or vomeronsal mechanism for estrous synchrony and acceleration might be expected since, as previously discussed, odor cues and scent-marking play an important part in social interaction in gray opossums. Work is currently in progress on the anatomy of the vomeronasal system in gray opossums and its relation to behavior in this species (Halpern, personal communication).

DISCUSSION

Although probably derived from an early common ancestor, the marsupials did not give rise to eutherian mammals (Hunsaker, 1977). Thus, the behavioral and physiological adaptations involved in chemical communication existing within these groups are likely to have evolved in a convergent fashion, allowing marsupials and eutherians to fill similar ecological niches. Because of this it has been suggested that marsupials represent the only "control" group to test hypotheses concerning the evolution of behavior within the eutherian mammals (Eisenberg and Golani, 1977).

The Didelphidae (which include the gray opossum) are of interest because they are the most ancient marsupial family and may have been the stem from which all other New World and Australian marsupials were derived (Clemens 1968; Keast, 1977). Thus, didelphid behavior, which may be little changed from that of ancestral forms, may simulate in some respects the behavior of ancient metatherians and of eutherians as well. As such, the didelphids are a particularly good model for understanding the evolution of social behavior involving chemical communication in mammals. Therefore, the findings of extensive scent-marking behavior, estrous synchrony and estrous activation in gray opossums, suggest an early origin for these phenomena in mammals.

In addition to being intrisically of interest, study of behavioral, anatomical and physiological aspects of olfaction in marsupials can contribute in a unique manner to our knowledge and understanding of the evolution of chemical communication within the Class Mammalia.

REFERENCES

Barnes, R. D., 1968, *Marmosa mitis*, a small marsupial for experimental biology, *in*: "Animal Models for Biomedical Research," Publication No. 1594, National Academy of Sciences, Washington, D.C.

Barnes, R. D., and Wolf, H. G., 1971, The husbandry of *Marmosa mitis* as a laboratory animal, *Int. Zoo Yearb.*, 11:50.

Boggs, J. R, 1969, The General Agonistic Behavior of the Mouse Opossum (*Marmosa robinsoni*), Unpubl. Master's thesis, San Diego State University, San Diego, California.

Bollinger, A., 1944, The response of the sternal integument of *Trichosurus vulpecula* to castration and to sex hormones, *J. Proc. Royal Soc. N.S.W*, 78:234.

Bollinger, A., and Whitten, W. K., 1948, The paracloacal (anal) glands of *Trichosurus vulpecula*, *J. Proc. Royal Soc. N.S.W.*, 82:3.

Bronson, F. H., and Desjardins, C., 1974, Relationship between scent-marking by male mice and the pheromone-induced secretion of the gonadotropic and ovarian hormones that accompany puberty in female mice, *in*: "Reproductive Behavior," W. Montagna, and W. A. Sadler, eds., Plenum Press, New York.

Buchmann, O. L. K., and Guiler, E. R., 1977, Behavior and ecology of the Tasmanian devil, *Sarcophilus harrisii*, *in*: "The Biology of Marsupials," F. Stonehouse, and D. Gilmore, eds., University Park Press, Baltimore.

Carter, C. S., Getz, L. L., Gavish, L., McDermott, J. L., and Arnold, P., 1980, Male-related pheromones and the activation of reproduction in the prairie vole (Microtus ochrogaster), Biol. Reprod., 23:1038.
Charles-Dominique, P., 1983, Ecology and social adaptations in didelphid marsupials: Comparison with eutherians of similar ecology, in: "Advances in the Study of Mammalian Behavior," J. Eisenberg, and D. Kleiman, eds., Special Publication No. 7, The American Society of Mammalogists.
Clemens, W. A., 1968, Origin and early evolution of marsupials, Evolution, 22:1.
Collins, L. R., 1973, "Marsupials and Monotremes," Smithsonian Institution Press, Washington.
Dobroruka, L. J., 1960, Einige Beobachtungen an Ameisenigeln, Echidna aculeata Shaw (1792), Z. Tierpsychol., 17:178.
Ebling, F. J., 1977, Hormonal control of mammalian skin glands, in: "Chemical Signals in Vertebrates," D. Muller-Schwarze, and M. M. Mozell, eds., Plenum Press, New York.
Ebling, F. J., 1948, The effect of sex hormones on the sebaceous glands of the female albino rat., J. Endocr., 5:297.
Eisenberg, J. F., and Golani, I., 1977, Communication in metatheria, in: "How Animals Communicate," T. Sebeok, ed., Indiana University Press, Bloomington.
Ewer, R. F., 1968, A preliminary survey of the behavior in captivity of the dasyrid marsupial, Sminthopsis crassicaudata (Gould), Z. Tierpsychol., 5:319.
Fadem, B. H., 1985, Evidence for the activation of female reproduction by males in a marsupial, the gray short-tailed opossum (Monodelphis domestica), Biol. Reprod., in press.
Fadem, B. H., and Cole, E. A., 1985, Scent-marking in the gray short-tailed opossum (Monodelphis domestica), Anim. Behav., 33:in press.
Fadem, B. H., and Rayve, R. S., 1985, Characteristics of the oestrous cycle and influence of social factors in grey short-tailed opossums (Monodelphis domestica), J. Reprod. Fert., 73:337.
Fadem, B. H., Trupin, G. L., VandeBerg, J. L., Maliniak, E. and Hayssen, V., 1982, Care and breeding of the gray short-tailed opossum (Monodelphis domestica), Lab. Anim. Sci., 32:405.
Fadem, B. H., and Schwartz, R., 1985, A sexually dimorphic suprasternal gland in the gray short-tailed opossum, Monodelphis domestica, J. Mammal., in press.
Farris, E. J., 1950, The opossum, in: "Care and Breeding of Laboratory Animals," E. J. Farris, ed., Wiley, New York.
Frith, H. J., and Calaby, J. H., 1969, "Kangaroos," Humanities Press, New York.
Ganslosser, V. U., 1982, Social structure and communication in marsupials, Zool. Anz. Jena., 209:294.
Gill, E. D., 1955, The problem of extinction with special reference to the Australian marsupials, Evolution, 9:87.
Glenn, E. M., and Gray, J., 1965, Effect of various hormones on the growth and histology of the gerbil (Meriones unguicalatus) abdominal sebaceous gland pad, Endocrinology, 76:1115.
Griffiths, M., 1968, "Echidnas," Pergamon Press, Oxford.
Griffiths, M., 1978, "The Biology of the Monotremes," Academic Press, New York.
Hartman, C. G., 1928, The breeding season of the opossum (D. virginiana) and the rate of intra-uterine and post-natal development, J. Morphol. Physiol., 46:143.
Hopp, S. L., and Timberlake, W., 1983, Odor cue determinants of urine marking in male rats (Rattus norvegicus), Behav. Neural Biol., 37:162.
Hughes, R. L., 1964, Effect of changing cages, introduction of the male and other procedures on the oestrous cycle of the rat, CSIRO Wildl. Res., 9:115.

Hunsaker, D., 1977, "The Biology of Marsupials," Academic Press, New York.
Hunsaker, D., and Shupe, D., 1977, Behavior of New World marsupials, in: "Biology of Marsupials," Academic Press, New York.
Johnson, J. I., Jr., 1977, Central nervous system in marsupials, in: "Biology of Marsupials," D. Hunsaker, ed., Academic Press, New York.
Johnston, R., 1975, Scent-marking by male golden hamsters (Mesocricetus auratus), II, The role of the flank gland in causation of marking, Z. Tierpsychol., 37:137.
Jurgelski, W., Jr., and Porter, M. E., 1974, The opossum (Didelphis virginiana Kerr) as a biomedical model. III. Breeding the opossum in captivity: Methods, Lab. Anim. Sci., 24:412.
Keast, A., 1977, Historical biogeography of the marsupials, in: "The Biology of Marsupials," F. Stonehouse, and D. Gilmore, eds., University Park Press, Baltimore.
Kirsch, J. A. W., and Calaby, J. H., 1977, Species of living marsupials: An annotated list, in: "The Biology of Marsupials," F. Stonehouse, and D. P. Gilmore, eds., University Park Press, Baltimore.
Kleiman, D., 1966, Scent-marking in the Canidae, Symp. Zool. Soc. Lond., 18:167.
Kumari, S., and Prakash, I., 1983, The mid-ventral gland of the Indian gerbil Tatera and soft-furred field rat, Rattus meltada, Acta Theriol., 28:425.
Langworthy, O. R., 1928, The behavior of pouch young opossums correlated with the myelination of tracts in the nervous system, J. Comp. Neurol., 46:201.
Lorenz, K., 1966, "On Aggression," Harcourt, Brace and World, New York.
Maliniak, E. and Taft, L. K., 1981, Husbandry and breeding protocol for Monodelphis domestica, in: "National Zoological Park Bulletin No.1," Dept. of Zoological Research, Washington, D.C.
Matthews, L. H., 1971, "The Life of Mammals, Vol. 2," Universe Books, New York.
McClintock, M. K., 1971, Menstrual synchrony and suppression, Nature 22:24.
McClintock, M. K., 1978, Estrous synchrony and its mediation by airborne chemical communication (Rattus norvegicus), Horm. Behav., 10:267.
McClintock, M. K., 1983, Modulation of the estrous cycle by pheromones from pregnant and lactating rats, Biol. Reprod., 28:823.
McCrady, E., Jr., 1938, The embryology of the opossum, Am. Anat. Mem., 16:1
McManus, J. J., 1970, Behavior of captive opossums, Didelphis marsupialis virginiana, Amer. Midland Nat., 84:144.
Muller-Schwarze, D., 1983, Scent glands in mammals and their functions, in: "Advances in the Study of Mammalian Behavior," J. F. Eisenberg, and D. G. Kleiman, eds., American Society of Mammalogists Special Publication No. 7.
Mykytowycz, R., and Nay, T., 1964, Studies of the cutaneous glands and hair follicles of some species of Macropodidae, CSIRO Wildl. Res., 9:200.
Parkes, A. S., and Bruce, H. M., 1961, Olfactory stimuli in mammalian reproduction, Science, 134:1049.
Quay, W. B., and Tomich, P. Q., 1963, A specialized midventral sebaceous glandular area in Rattus exulans, J. Mammal., 44:537.
Regnier, F. E., and Goodwin, M., 1977, On the chemical environmental modulation of pheromone release from vertebrate scent marks, in: "Chemical Signals in Vertebrates," D. Muller-Schwarze, and M. M. Mozell, eds., Plenum Press, New York.
Reynolds, H. C., 1952, Studies on reproduction in the opossum (Didelphis virginiana virginiana), Univ. Calif. Pub. Zool., 52:223.
Schaffer, J., 1940, "Die Hautdrüsenorgane der Säugetiere," Urban and Schwarzenberg, Berlin and Wien.

Sharman, G., 1970, Reproductive physiology of marsupials, Science, 167:1221.
Sharman, G. B., and Calaby, J. H., 1964, Reproductive behavior in the red kangaroo, Megaleia rufa, in captivity, CSIRO Wildl. Res., 9:58.
Shultze-Westrum, T. G., 1965, Innerartliche Verständigung durch Düfte beim Gleitbeutler Petaurus breviceps papuana Thomas (Marsupialia: Phalangeridae), Zeit. für vergleichende Physiologie, 50:15.
Schultze-Westrum, T. G., 1969, Social communication by chemical signals in flying phalangers, in: "Olfaction and Taste," C. Pfaffmann, ed., Rockefeller Univ. Press, New York.
Smith, M., 1980, Behaviour of the koala, Phascolarctos cinereus (Goldfuss), IV., Scent-marking, Australian Wildl. Res., 7:35.
Sokolov, W., and Skurat, L., 1966, A specific midventral gland in gerbils, Nature, 211:544.
Stonehouse, B., and Gilmore, D. P., 1977, "The Biology of Marsupials," University Park Press, Baltimore.
Strauss, J. S., and Ebling, F. J., 1970, Control and function of skin glands in mammals, Mem. Soc. Endocr., 18:341.
Streilein, K. E., 1982, Behavior, ecology and distribution of South American marsupials, in: "Mammalian Biology in South America," M. A. Mores, and H. H. Genoways, eds., Pymatuning Laboratory of Ecology, Special Publication, Linesville, Pennsylvania.
Sutherland, R. L., Evens, S. M., and Tyndale-Biscoe, C. H., 1980, Macropodid marsupial luteinizing hormone: Validation of assay procedures and changes in concentration in plasma during the oestrous cycle in the female tammar wallaby (Macropus eugenii), J. Endocrinol., 86:1.
Thiessen, D. D., 1968, The roots of territorial marking in the Mongolian gerbil: A problem of species-common topography, Behav. Res. Meth. Instrum., 1:70.
Thiessen, D. D., Blum, S. L., and Lindzey, G., 1969, A scent marking response associated with the ventral sebaceous gland of the Mongolian gerbil (Meriones unguiculatus), Anim. Behav., 18:26.
Thiessen, D. D., Owen, K., and Lindzey, G., 1971, Mechanisms of territorial marking in the male and female mongolian gerbil (Meriones unguiculatus), J. Comp. Physiol. Psychol., 77:38.
Thompson, J. A., and Pears, F. N., 1962, The functions of the anal glands of the brushtail possum, The Victorian Naturalist, 78:306.
Trupin, G. L., and Fadem, B. H., 1982, Sexual behvior of the gray short-tailed opossum (Monodelphis domestica), J. Mammal., 63:409.
Tyndale-Biscoe, H., 1973, "Life of Marsupials," American Elsevier, New York.
VandeBerg, J. L., 1983, The gray short-tailed opossum: A new laboratory animal, ILAR News, 26:9.
Walker, E. P., 1968, "Mammals of the World," Johns Hopkins Press, Baltimore.
Wemmer, C., and Collins, I., 1978, Communication patterns in two phalangerid marsupials, the gray cuscus (Phalanger gymnotis) and the brush possum (Trichosurus vulpecula), Säugetierkundliche mitteilungen, 26:161.
Whitten, W. K., 1956, Modification of the oestrous cycle of the mouse by external stimuli associated with the male, J. Endocrinol., 13:299.
Whitten, W. K., 1966, Pheromones and mammalian reproduction, Adv. Reprod. Physiol., 1:155.

OBSERVATIONS ON CHEMICAL COMMUNICATION AND ITS GLANDULAR SOURCES IN SELECTED INSECTIVORA

Walter Poduschka and Chris Wemmer

Insectivore Research Center, Wien, Austria, and
Conservation and Research Center, National Zoological
Park, Front Royal, VA 22630

INTRODUCTION

The origins of olfactory communication can be inferred from basic similarities of body plan and habitus seen in the mammal-like reptiles and morphologically conservative mammals (Eisenberg, 1981). Skin glands occur in every mammalian order of all three subclasses (Eutheria, Metatheria, Prototheria) which is ample testimony to their phylogenetic age. Vestigial dermal glands, and gland-like structures, as well as fully developed glands are found in some fish, amphibians, reptiles and birds. However, the vast number of mammalian skin glands used in communication evolved from general characteristics of the mammalian integument, namely a soft pliable slightly cornified epithelium, bearing hair and capable of sebaceous secretion over at least part of its surface. The existence of a well-developed olfactory system and nocturnal habits were also vital to the evolution of olfactory communication within the Mammalia (Starck, 1982).

The order Insectivora is a phylogenetically ancient group and therefore a most rewarding taxon for investigating the evolution of olfactory communication in mammals. It contains seven families, namely hedgehogs (Erinaceidae), shrews (Soricidae), moles (Talpidae), tenrecs (Tenrecidae), otter shrews (Potamogalidae), solenodons (Solenodontidae), and golden moles (Chrysochloridae). All families exhibit to varying degrees the morphological conservatism associated with early mammals. Specialization between families has taken place mainly in appendicular morphology and locomotion, but even in the generalized shrews anatomical and other specializations beyond the primitive mammalian condition are apparent (Pernetta, 1977). All insectivores share features associated with opportunistic, or more commonly an obligate predatory lifestyle. Vision is poorly developed at best, but hearing is acutely developed, particularly in the upper frequency range. All families exhibit extraordinary development of the nasal capsule and olfactory portion of the central nervous system. Both sound and scent therefore play important roles in communication. However, it should be emphasized that knowledge of olfactory communication within the order is still fragmentary. Only a fraction of known scent glands have been examined histologically, and no pheromones have yet been chemically identified. The purpose of this paper is to briefly review some general trends in olfactory communication within this group, and then to describe some examples of more specialized mechanisms of olfactory communication in selected insectivores.

Ubiquitous Lateral and Ventral Glands and Use of Latrines

The shrews characteristically possess a well-developed assortment of glands on the trunk. Lateral (side) glands are the most widespread of these glands, having been described in Myosorex varius (Goulden and Meester, 1978), Cryptotis parva (Kivett and Mock, 1980), Suncus murinus (Dryden and Conaway, 1967), Blarina brevicauda (Eadie, 1938; Pearson, 1946), Neomys fodiens (Crowcroft, 1957), four species of Crocidura (Goulden and Meester, 1978; Baxter and Meester, 1980), and many, if not most, species of Sorex (Eadie, 1938; French, 1980). Studies of captive shrews reveal that scent from these glands is deposited by rubbing the sides on objects and substrate, and perhaps by passive application. In two species (Suncus and Blarina) the gland has been identified as primarily sebaceous with small numbers of apocrine tubules (Eadie, 1938; Dryden and Conaway, 1967).

There have been few detailed experimental investigations of their function. A general or non sex-specific function may be inferred from the presence of lateral glands in both sexes. The lateral glands of both Blarina brevicauda and Suncus murinus are primarily sebaceous and are rubbed against objects during agonistic circumstances (Dryden and Conaway, 1967; Platt, 1976). Pearson (1946), however, noticed that the lateral and ventral glands of female Blarina are less developed in estrous than pregnant females. In Suncus the lateral gland is maintained by progesterone and testosterone (Dryden and Conaway, 1967). If lateral glands serve a general territorial function, increased secretion during pregnancy resulting from progesterone would repel intruders and minimize interspecific disturbance during a particularly vulnerable period. Lateral glands also occur in non-territorial shrews (Cryptotis parva). Clearly, comparative investigations of histology, pheromone chemistry, and experimental studies of olfactory behavior are required to clarify the social role of these glands.

Ventral glands are known from shrews and solenodons, and likely occur in tenrecs and hedgehogs. Solenodon paradoxus has a well-developed ventral gland (Fig. 1) which in the male covers an extensive area and exudes a watery, colorless secretion (Starck and Poduschka, 1981). Earlier descriptions of an oily reddish (Poey, 1861) or yellowish green secretion (Mohr, 1936) in the axillary or lateral areas undoubtedly refers to this gland, but the unusual color may have resulted from contamination. Mohr (1936) first described the peculiar courtship pattern whereby the male propels itself by its forelegs and drags its belly and outstretched hindlegs beside the female, but the probable social significance of the behavior and gland were not yet recognized. Though the gland is active year round, we noted two peaks of secretory activity. Females have a smaller gland on the sternum. Histologically, it is distinct from similar glands in the Soricidae. It consists of a 3-5mm thick lobular structure containing tubular singlelayered apocrine glands, but it lacks the encapsulated character of soricid lateral glands. In addition, myoepithelial cells are located in the walls of the secretory tubules, and the orifice opens into a hair follicle (Starck and Poduschka, 1981). The sexual role of the ventral gland in one species of shrew will be described below.

Latrine use is another ubiquitous insectivore pattern, particularly among shrews, desmans, moonrats, and tenrecs (Poduschka, 1977a). Latrines differ in details between species, but usually receive both feces and urine, and are implicated with territorial behavior. Territory holding Blarina defecate in their nests and also along the territorial boundaries of adjacent shrews; transient shrews do not exhibit the behavior (Platt, 1976). Sorex palustris was also observed to defecate on food, which deterred thievery by other shrews (Sorenson, 1962). Anal glands have been histologically

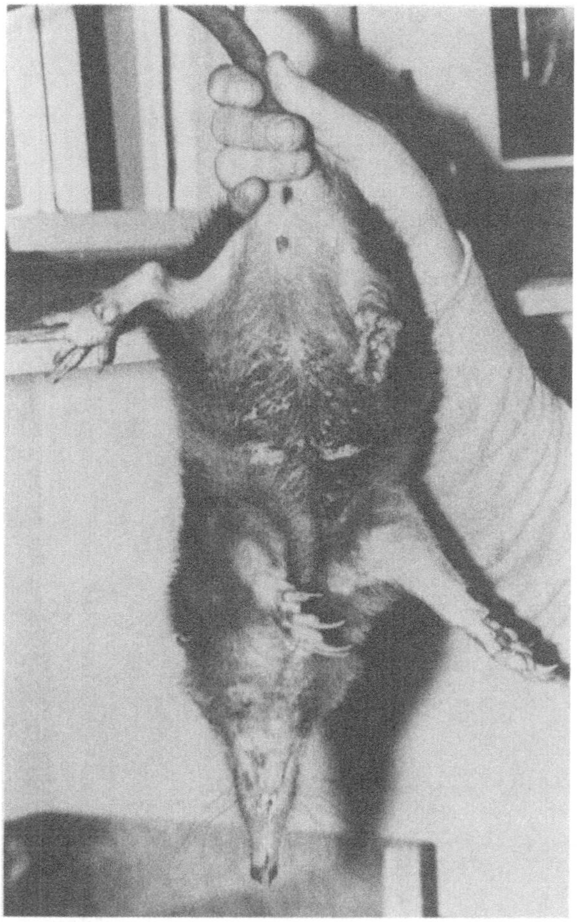

Fig. 1. Ventral side of a male Solenodon paradoxus showing the dark glandular area.

identified in several shrews (Eadie, 1938), but their presence remains to be verified in other latrine using species of insectivores. There is convincing circumstantial evidence that many mammals having this toilet habit mark their feces with anal gland secretions, but the exact role of latrines and anal glands in the Insectivora requires further study.

Anti-predator Functions

It may be hypothesized that the odor of shrews, moles and moonrats is a deterrent to some mammalian predators. Scott and Klimstra (1955), for example, concluded that moles (Scalopus aquaticus) and short-tailed shrews (Blarina brevicauda) were low on the scale of red fox food preferences because captured insectivores were often left uneaten. Shrews release a strong musky scent when disturbed (see Dryden and Conaway, 1967), but the role of scent as an anti-predator ploy is unknown. The deterrent effect of the shrew's scent does not appear to operate until the animal is captured and killed, and so it may be argued that the odor does not prevent predation. On the other hand, "prudent predators" may learn to avoid hunting prey which are not edible. Challenging experiments will be required to determine whether red foxes and other carnivores become conditioned to modify their search image for insectivore prey such as moles and shrews.

Fig. 2. Echinops telfairi encircling a horizontal branch.

Strong olfactory stimuli often release the peculiar self-anointing behavior of hedgehogs and tenrecs, which was interpreted as a means of cleaning the oral cavity and the Jacobson's organ of the olfactory stimulant (Poduschka and Firbas, 1968; Poduschka, 1977a; but c.f., Brodie, 1977).

Behavioral Convergence

Convergent evolution of behavior can be inferred from the sporadic occurrence in representatives of different families of similar or identical behavior patterns (Wemmer, 1971, 1977). Motor patterns converge evolutionarily because of fundamental similarities in body plan and movement patterns shared by many mammals. However, the function and chemical properties of convergent scent-marking patterns may differ. Within the Insectivora for example, behavior patterns similar to the ventrum dragging of Solenodon also occur in other species. Adult Setifer setosus (Tenrecidae) extend their forelegs laterally, and propel the body forward with the hindlegs while sliding the throat and sternum across the substrate. The pattern is seen in captive tenrecs placed in the living space of conspecifics, but the function of the pattern is unknown. In this species a distinct glandular area is not evident. The lesser hedgehog tenrec (Echinops telfairi) exhibits an acrobatic variation of this theme when climbing through vegetation and twigs in the cage of a conspecific. The animal supports itself like a gymnast on the horizontal bar and drags its venter and hind-quarters on the bottom and side of the branch (Fig. 2). In both cases the behavior pattern involves object-oriented contact, and thus resembles many examples of

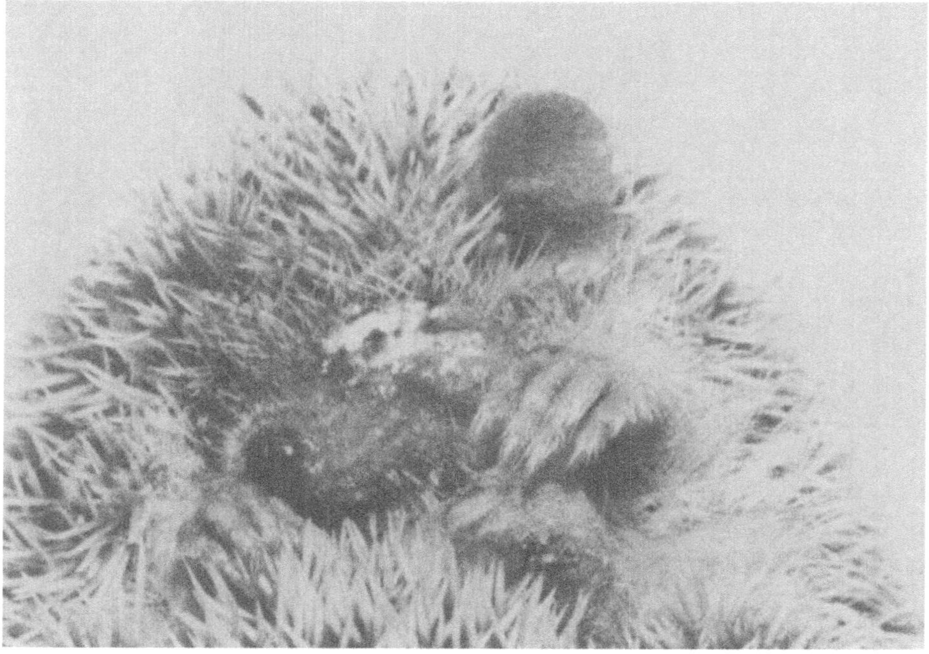

Fig. 3. Erinaceus concolor roumanicus just after death from natural causes, showing the white eyelid gland secretion.

mammalian scent marking. Yet a discrete ventral gland is not visible macroscopically. Both species are aseasonal breeders in captivity, but it is possible that periodic glandular activity would be more apparent under natural conditions in a seasonal environment (Mallinson, 1968; Scaling, 1971; Riordan, 1972; Poduschka, 1974a; Godfrey and Oliver, 1978).

The eyelid secretions of the tenrecs, hedgehogs and shrews, first described in Echinops, merit further study as convergently evolved patterns of olfactory communication. In Echinops, fright, pain, irritation, aggression, pungent odors, and sexual agitation elicit swelling and reddening of the eyelids, followed by the appearance of a white, fatty secretion from the inside of the eyelid (Poduschka, 1972). This is also accompanied by tiny watery droplets on the outside of the eyelid. When strongly stimulated, the opaque secretion covers the eyes, thus blocking vision, and passes into the nasal passages via the ductus nasolacrimalus to issue from the nostrils. In spite of the analogy to crying in humans, the glandular tissues which produce tears and this lipoid exudate are distinctive. It is perhaps a curiosity that the eyelid glands of tenrecs and hedgehogs also swell and secrete immediately before death (Fig. 3). Similar behavior has been witnessed in Nesogale talazaci and N. dobsoni (Poduschka, 1974a), as well as in shrews (von Lehmann, pers. comm.; Hutterer, 1976). Hedgehogs also exhibit the pattern to a lesser extent in response to the above mentioned stimuli.

The functional significance of this widespread insectivore behavior requires further study in its many contexts. In this connection, a few observations are pertinent. The eyelid secretion is not sex-specific, as earlier assumed by Honegger and Noth (1966), and it is not seasonal in its appearance. An olfactory function can be inferred from the strong acid taste and odor of the substance. When a male Echinops is placed in a neutral (unoccupied) cage it deposits the secretion on vertical surfaces.

Fig. 4. Female Hemiechinus auritus syriacus showing mutual sexual display.

If the male is replaced with a female the latter remains calm but exhibits searching behavior. If the first male is replaced by another male, the latter becomes agitated and runs about the cage as if searching for an intruder. Male and female Echinops thus exhibit distinctive responses to the eyelid secretion of conspecific males. Systematic experiments on reactions to this secretion by different age and sex classes should help clarify its social role.

PHEROMONES AND REPRODUCTION

The pheromonal control of reproduction, best known from investigations of domestic livestock and rodents, is regulated through a subtle but complex interplay of social and environmental factors (Bronson, 1985). Releaser effects of insectivore pheromones are most evident from studies of courtship behavior. Gould (1969) demonstrated empirically that the neck glands of Blarina brevicauda and Suncus murinus are clearly associated with reproduction. Males of both species repeatedly attempt to investigate the female olfactorily but are rebuffed by vocal threats and bites. Soon however, the area is saturated with the males' scent from repeated marking of the substrate with ventral glands and atomization of neck gland scent (Dryden and Conaway, 1967). In Suncus, the male exposes his neck gland to the female as she threatens to bite him. Her aggression gives way to receptivity as the frequency of male contact increases and culminates in mating. Dryden and Conaway (1967) hypothesized that the sebaceous secretion of Suncus lateral gland chemically facilitates the action of neck gland musk by acting as a fixative. The combined action of behavior and scent induces receptivity in a relatively short time.

Male hedgehogs produce a strong smelling urine which also appears to have the properties of a courtship pheromone (Poduschka, 1977b, 1981).

Pregnant females shun courting males by running away, but non-pregnant and unreceptive females never try to escape. In the course of the female's nocturnal meanderings the male indulges in what formerly was called a "hedgehog roundabout"; he encircles the female for hours and from a semi-erect penis he deposits in her pathway hundreds of droplets of a specialized urine-like secretion, which she inevitably treads through. The odor is quite distinguishable from normal urine and unlike the latter it leaves a whitish marking. The male's numerous and haphazard mounting attempts also transfer the substance to the female's body. It is suspected that the male's scent induces the female to remain in the courtship area, which must be fairly extensive to ensure successful mating. Because of their defensive behavior, it is not possible to monitor estrous cycles in hedgehogs by conventional methods such as vaginal smears (Poduschka, 1977b), so elucidation of the physiological correlates of scent-mediated control of reproduction must await the development of other assays. The association of urine marking and male reproductive behavior is known in a large variety of mammals (Coblentz, 1976; Smythe, 1978; Wemmer and Murtaugh, 1980), but experimental investigations of function are limited to only a few species. The insectivores unfortunately, are not among this minority.

There is suggestive circumstantial evidence that as yet undescribed primer pheromones may also regulate the reproductive cycle of some insectivores. Female hedgehogs maintained in groups, for example, exhibit synchronous prolonged estrus in which homosexual activity is pronounced and role changes frequent (Fig. 4). A long period of anestrus (18-30 months) follows which is unaffected by the presence of a male. It may be hypothesized that such a phenomenon regulates reproduction in populations in which there is a surplus of females, but controlled experiments are needed to elucidate the mechanism.

REFERENCES

Baxter, R. M., and Meester, J., 1980, Notes on the captive behavior of five species of southern African shrews, Säugetierk. Mitteil., 40:55.

Brodie, E. D., 1979, Hedgehogs use toad venom in their own defense, Nature, 268:627.

Bronson, F. H., 1985, Mammalian reproduction: An ecological perspective, Biol. Reprod., 32:1.

Coblentz, B., 1976, Functions of scent urination in ungulates with special reference to feral goats, Am. Nat., 110:549.

Crowcroft, P., 1957, "The Life of the Shrew," Max Reinhardt, London.

Dryden, G. L., and Conaway. C. H., 1967, The origin and hormonal control of scent production in Suncus murinus, J. Mammal., 19:171.

Eadie, W. R., 1938, The dermal glands of shrews, J. Mammal., 37:171.

Eisenberg, J. F., 1981, "The Mammalian Radiations, An Analysis of Trends in Evolution, Adaptation, and Behavior," University of Chicago Press, Chicago.

French, T. W., 1980, Sorex longirostris, Mammalian Species, 143:1.

Godfrey, G. K., and Oliver, W. L. R., 1978, The reproduction and development of the pigmy hedgehog tenrec, Echinops telfairi, Dodo (Jersey Wildlife Pres. Trust), 15:38.

Gould, E., 1969, Communication in three genera of shrews (Soricidae): Suncus, Blarina, and Cryptotos, Commun. Behav. Biol., 3:263.

Goulden, F. A., and Meester, J., 1978, Notes on the behavior of Crocidura and Myosorex (Mammalia: Soricidae) in captivity, Mammalia, 42:197.

Honegger, R. E., and Noth, W., 1966, Beobachtungen bei der Aufzucht von Igeltenreks Echinops telfairi Martin, Zool. Beitr., 12:191.

Hutterer, R., 1976, Über Augendrüsensekretionen und andere Phänomene bei senilen Zwergspitzmäusen, Sorex minutus Linne 1766, Säugetierk. Mitteil., 24:295.

Kivett, V. K., and Mock, O. B., 1980, Reproductive behavior in the least shrew (Cryptotis parva) with special reference to the aural glandular region of the female, Am. Midl. Nat., 103:339.
Mallinson, J. J. C., 1968, "Conservation of Madagascar Fauna in Captivity," Jersey Wildl. Pres. Trust 5th Ann. Rep., 51:54.
Mohr, E., 1936, Biologische Beobachtungen an Solenodon paradoxus Brandt in Gefangenschaft II., Zool. Anz., 116:65.
Pearson, O. P., 1946, Scent glands of the short-tailed shrew, Anat. Rec., 94:615.
Pernetta, J. C., 1977, Anatomical and behavioral specializations of shrews in relation to their diet, Can. J. Zool., 55:1442.
Platt, W. J., 1976, The social organization and territoriality of short-tailed shrew (Blarina brevicauda) populations in old field habitats, Anim. Behav., 24:305.
Poduschka, W., 1972, Augendrüsensekretionen bei Tenriciden, Film CTF 1451, Bundesstaatl. Hptst. wiss. Film, Wien.
Poduschka, W., 1974a, Fortpflanzungseigenheiten und Jungenaufzucht des Grossen Igel-Tenrek Setifer setosus (Froriep, 1806), Zool. Anz., 193:145.
Poduschka, W., 1974b, Augendrüsensekretionen bei den Tenreciden Setifer setosus (Froriep, 1806), Eichinops telfairi (Martin, 1838) Microgale dobsoni (Thomas, 1918) und Microgale talazaci (Thomas, 1918), Z. Tierpsychol., 35:303.
Poduschka, W., 1977a, Insectivore communication, in: "How Animals Communicate," T. A. Seboek, ed., Indiana Univ. Press, Bloomington and London.
Poduschka, W., 1977b, Das Paarungsvorspiel des Osteuropäischen Igels (Erinaceus e. roumanicus) und theoretische Überlegungen zum Problem männlicher Sexualpheromone, Zool. Anz., 199:187.
Poduschka, W., 1981, Paarungsappetenz bei gattungsverschiedenen Igeln: Erinaceus europaeus europaeus Linne 1758 und Hemiechinus auritus syriacus Wood 1876, Säugetierk. Mitteil., 29:72.
Poduschka, W., and Firbas, W., 1968, Das Selbstbespeichäln des Igels, Erinaceus europaeus Linne, 1758, steht in Verbindung zur Funktiondes Jacobsonchen Organes, Z. Säugetierk., 33:160.
Poey, F., 1861, Memorias sobre la historia natural de la isla Cuba, Havana, 1:23.
Riordan, D. V, 1972, Reproduction in the spiny hedgehog tenrec Setifer setosus and the pigmy hedgehog tenrec Echinops telfairi, Jersey Wildl. Pres. Trust 9th Ann. Rep., 12:14.
Scaling A., 1971, Progress report on the breeding of young hedgehog tenrec Echinops telfairi and spiny tenrec Setifer setosus, Jersey Wildl. Pres. Trust 8th Ann. Rep., 12:14.
Scott, T. G., and Klimstra, W. D., 1955, Red foxes and a declining prey population, S. Ill. Univ. Monog. Ser., No. 1.
Smythe, N., 1978, The natural history of the Central American agouti (Dasyprocta punctata), Smithson. Contrib. Zool., No. 257.
Sorenson, M. W., 1962, Some aspects of water shrew behavior, Am. Midl. Nat., 68:445.
Starck, D., 1982, Vergleichende Anatomie der Wirbeltiere III., Organe des aktiven Bewegungsapparates, der Koordination, der Umweltbeziehung, des Stoffwechsels und der Fortpflanzung, Springer-Verlag, Berlin.
Starck, D., and Poduschka, W., 1981, über die Ventraldrüse von Solenodon paradoxus Brandt, 1833 (Mammalia: Insectivora), Z. Säugetierk., 47:1.
Wemmer, C., 1971, Scent marking and anointing: Parallel evolution in mammalian behavior, Am. Zool., 11:623.
Wemmer, C., 1977, Comparative ethology of the large-spotted genet, Genetta tigrina, and related viverrid genera, Smithson. Contrib. Zool., No. 239.
Wemmer, C., and Murtaugh, J., 1980, Olfactory aspects of rutting behavior in the Bactrian camel (Camelus bactrinanus ferus), in: "Chemical Signals in Vertebrates and Aquatic Invertebrates," D. Muller-Schwarze, and R. M. Silverstein, eds., Plenum Press, New York.

PLANTAR GLANDS IN NORTH AMERICAN MUSTELIDAE

Steven W. Buskirk[1], P. F. A. Maderson[2], and Robin M. O'Connor[3]

[1]Department of Zoology and Physiology, University of Wyoming Laramie, WY 82071, [2]Department of Biology, Brooklyn College City University of New York, New York, NY 11210, [3]Alaska Department of Fish and Game, 1300 College Road, Fairbanks AK 99701

Mustelid carnivores have an extraordinary repertoire of chemical communication systems. Urine and feces are deposited in ways that suggest that they transmit intraspecific messages in several species (Lockie, 1966; Eisenberg and Kleiman, 1972). In addition, several structures have been described that serve as chemical transmitters. They include anal sacs in many species (Ewer, 1973) and abdominal glands in *Martes* and *Gulo* (Hall, 1926). For many of these structures, descriptions of the gross morphology of the transmitter organ are available. For a smaller number, a specific behavioral function of the transmitted chemical has been described or proposed. For fewer still, histological descriptions of the secretory tissues are available. Here we provide a description of structures on the feet of several species of North American mustelid carnivores that we hypothesize to function in chemical transmission. Mammal nomenclature follows Honacki et al. (1982).

We have observed structures on the plantar surfaces of hind feet of several North American mustelids, American marten (*Martes americana*), fisher (*M. pennanti*), wolverine (*Gulo gulo*) and river otter (*Lutra canadensis*). These structures appear to be homologous to each other and to those in sable (*Martes zibellina*) described by Matviev (1942). We have called these structures "plantar glands" as suggested by Schaeffer (1940) and Matviev. Four of the glands are located on the plantar surfaces of each of the hind feet (Fig. 1A). In gross appearance they vary within and between species. They may appear wart-like, resemble small inflorescences, or consist of small epidermal indentations. In American marten the gross morphology varies seasonally. During winter, when the tarsal pads are covered with pelage, the glands consist of clusters of columnar or flaky projections, yellow in color, protruding up to 4 mm beyond the skin surface. Examination of these structures with the scanning electron microscope has revealed a unique modification of the mammalian epidermis. Irregular columns of apparently cornified cells protrude in a rosette arrangement from a poorly defined gland orifice (Fig. 1B). The epidermis 2-4 mm around each gland is glabrous. Beyond that distance the epidermis is furred, and this pelage envelops the projections so that they are protected from abrasion. During summer, when the pelage covering of the plantar

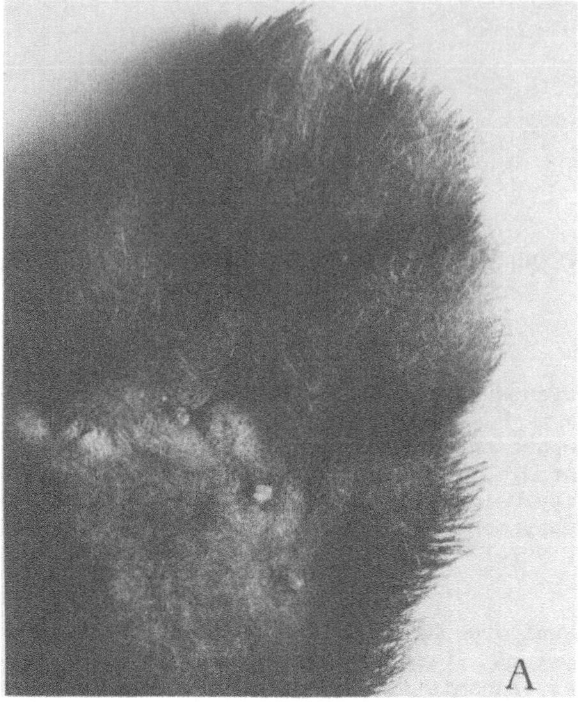
A

1mm
B

Fig. 1. Facing page. Position and gross structure of plantar glands in American marten. (A) Plantar surface of rear foot of American marten in winter pelage showing position of four plantar glands in relation to digits. Fur has been clipped to make glands visible. (B) Scanning electron micrograph of rosette-like projections from plantar gland of American marten in winter (X80).

surfaces is greatly reduced, the projections are absent, presumably as a result of abrasion. At this time, the glands appear to be indentations or pits on the plantar epidermis.

Plantar glands exhibit somewhat different morphology in river otters. Their positions correspond to those in Martes and Gulo, however, the epidermis of the plantar surface is glabrous year-round. The glands appear as disc-shaped clusters of fused columns protruding from a well-defined orifice in the epidermis (Fig. 2).

Examination of a limited amount of histological material suggests generally similar structure among the species examined, but further study will be needed to ascertain whether observed, subtle differences reflect species specificity, annual, functional changes, or possible sexual dimorphism. The plantar surface consists of a thick epidermis with conspicuous penetrations into the dermis (Fig. 3A). Each germinal invagination is associated with a column of loosely packed corneous cells. The superficial dermis is highly vascularized and the deeper

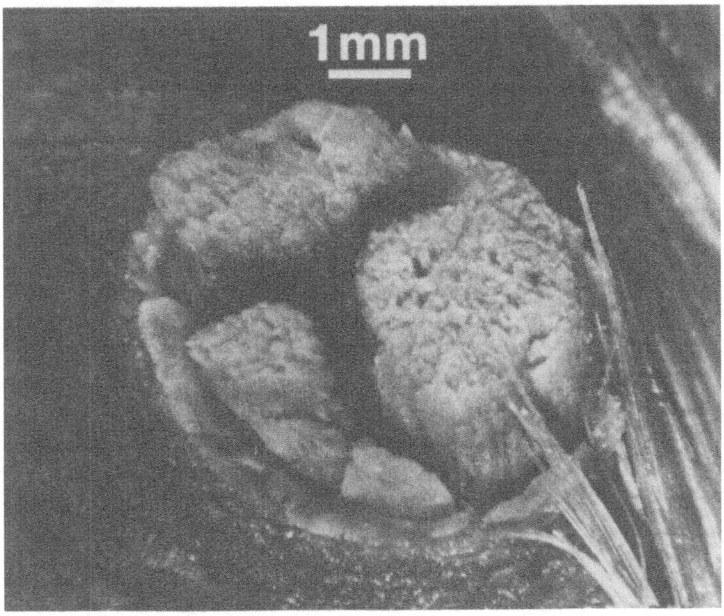

Fig. 2. Light photograph of gross superficial structure of plantar gland of American river otter. The fused columns of papillae emerging from a well-defined orifice surrounded by glabrous cornified epidermis are visible.

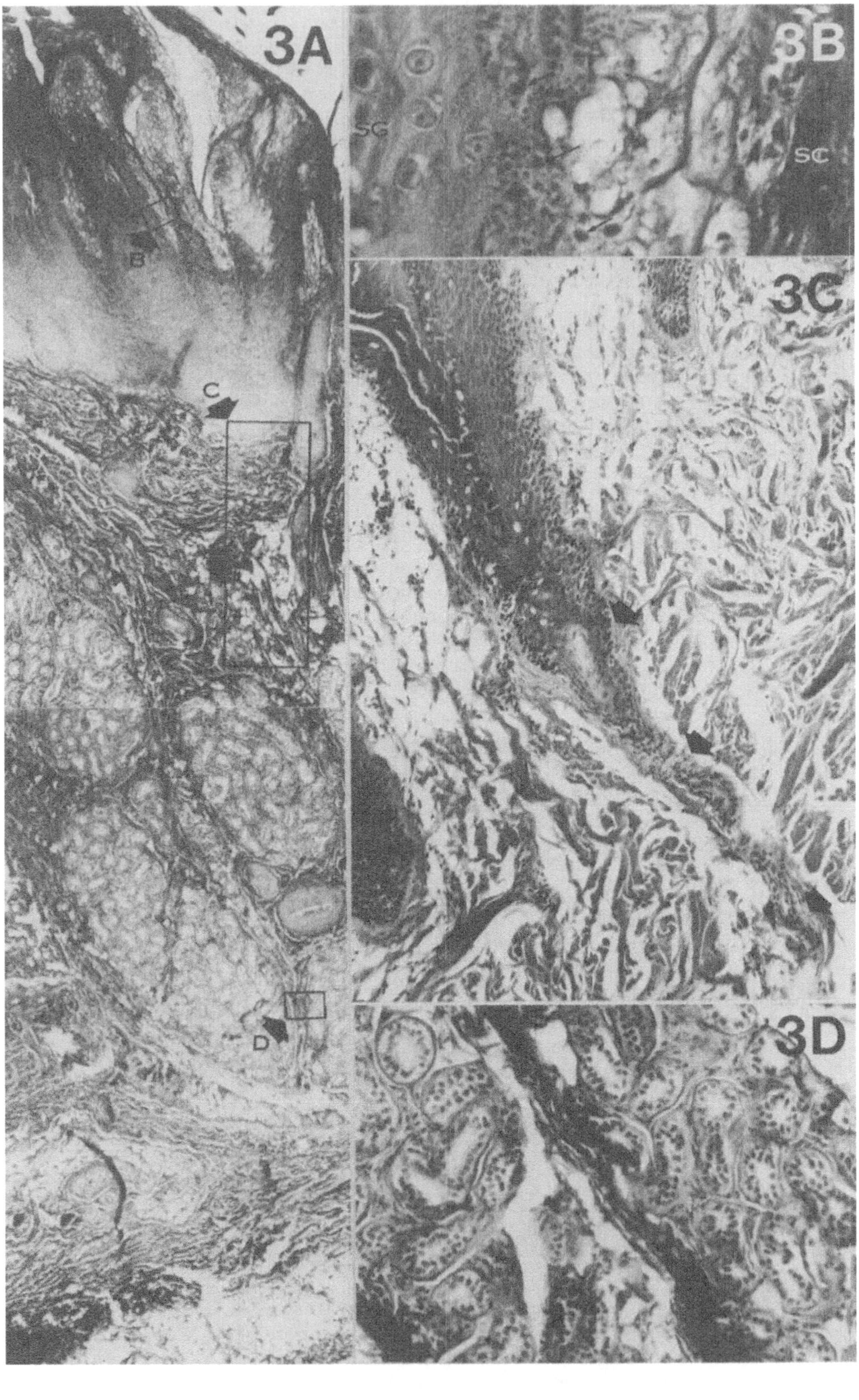
3A
B
C
D
3B
SG
SC
3C
3D

regions contain numerous acini of serous glands arranged in connective tissue-surrounded bundles. Between these bundles, and deeper to them, lie conspicuous masses of adipose connective tissue. Initial examination of hematoxylin and eosin-stained material suggests that the epidermis has a greatly exaggerated stratum granulosum. However, at higher power it becomes obvious that the conspicuous granules do not show the basophilia characteristic of keratohyalin, which is, however, present in the epidermis away from the gland. The gland granules (Fig. 3B) are eosinophilic and the more superficial cells are not conspicuously flattened as are normal mammalian epidermal corneous cells. Although further study of serial sections will provide necessary confirmation, there is a strong indication that the deep germinal invaginations connect with the serous glands by narrow tubules (Fig. 3C). Serous acini (Fig. 3D) have been observed cut at various planes, but the height of the epithelia and the presence or absence of faintly eosinophilic, amorphous material in the lumina are both highly variable. The significance of this variability cannot be adjudged until further material is available; a more complete histological study is in progress.

Superficial examination of other mustelid specimens has suggested that plantar glands are not universal among this family. Species which we have examined and which appear to lack plantar glands are _Mustela erminea_, _M. frenata_, _M. nivalis_, _M. vison_, _M. nigripes_, _Taxidea taxus_, _Enhydra lutris_ and _Mephitis mephitis_. Inasmuch as patterns of presence and absence of these glands seem to follow taxonomic lines, they may be useful in studies of systematics.

Fig. 3. Facing page. Transmission light photomicrograph of plantar glands of mustelidae. (A) Low power transmission light photomicrograph of the plantar gland of _M. americana_, sex and date of collection unknown. The gland is cut vertical to the tarsal surface and the "secretory material" is at the top of the photograph. Rectangles marked with arrows labeled "B", "C", and "D" indicate the general locations of the higher power micrographs in Figs. 3B-D. (B) High power micrograph showing the epidermis of the specimen shown in A. The stratum germinativum (SG) is separated from the innermost cells of the corneous tissues (SC) by immature cells. Those immature cells nearest the germinal layer contain large numbers of eosinophilic granules (fine arrows) which disappear or aggregate into clumps as the cells become "vacuolated." (C) Medium power micrograph showing the dermo-epidermal boundary in the tarsal gland of a female _M. pennanti_ collected during December. The broad arrows show segments of what is interpreted as a duct joining the deep invagination of the epidermis (upper left) to the serous glands (c.f., D). (D) High power micrograph of the acinar gland lying in the deep dermis of a male _M. americana_ collected in October. Although all the acinar epithelia have approximately the same height, the lumina of some contain amorphous materials, while others are empty.

We hypothesize, as did Schaffer (1940) and Matviev (1942), that plantar glands in mustelids serve as chemical transmitters. Scent deposition from other transmitters is correlated with discrete behaviors (e.g., abdominal rubbing, urination). However, no behavioral correlates of chemical transmission from plantar glands have, to our knowledge, been described. This may be because transmission from the plantar glands can be performed passively, by means of ordinary locomotion, without the use of discrete transmission behaviors. If so, plantar glands represent part of a system of chemical transmission that reduces energy expenditure of chemical signaling. A chemical signal is left on surfaces contacted by hind feet. Among wide-ranging species such as marten and wolverines, signals may be broadcast extensively without pausing to engage in discrete transmission behaviors. The adaptive value of this postulated energy savings is apparent.

ACKNOWLEDGMENTS

Materials used in this study were provided by several individuals including William Berg, Patrick Valkenburg, John Peterson and Chris Scranton.

REFERENCES

Eisenberg, J. F., and Kleiman, D. G., 1972, Olfactory communication in mammals, Ann. Rev. Ecol. Syst., 3:1.

Ewer, R. F., 1973, "The Carnivores," Cornell University Press, Ithaca, NY.

Hall, E. R., 1926, The abdominal scent gland of *Martes*, J. Mammal., 7:227.

Honacki, J. H., Kinman, K. E., and Koeppl, J. W., eds., 1982, "Mammal Species of the World: A Taxonomic and Geographic Reference," Allen Press, Inc., and Associated Systematics Collections, Lawrence, KS.

Lockie, J. D., 1966, Territory in small carnivores, Symp. Zool Soc. Lond., 18:143.

Matviev, B. S., 1942, On the skin, its glands and the sole skin glandular organs of the sable (*Martes zibellina* L.), Zool. Zhur., 21:207. (Translation.)

Schaffer, J., 1940, Die Hautdrüsenorgane der Sägetiere, Urban and Schwarzenberg, Berlin and Vienna.

THE MARKING BEHAVIOR OF MALE PRZEWALSKI'S HORSES

Lee Boyd*

Department of Physiology
Cornell University
Ithaca, NY 14853

Lonnie Kasman

Research Department
San Diego Zoo
San Diego, CA 92112

INTRODUCTION

At the time of their discovery in 1878, the distribution of Przewalski's horses (*Equus przewalskii*) was restricted to Mongolia, although they may once have ranged more widely (Mohr, 1971). Currently, Przewalski's horses are extinct in the wild; the world population of 600 animals exists wholly in zoos and all of these animals are descended from 13 founders. Przewalski's horses may be the ancestors of domestic horses (*E. caballus*), with whom they are cross-fertile. Like domestic horses they have a non-territorial harem social organization. The breeding season is from March through July, and the gestation period is 11 months.

Male equids mark the urinations and defecations of their mares with urine, or more rarely, with feces. This behavior has been reported in domestic and feral horses (Tyler, 1972; Feist and McCullough, 1976; Turner et al., 1981), Grevy's zebras (*E. grevyi*), mountain zebras (*E. zebra*), plains zebras (*E. quagga*) (Klingel, 1974), and African and Asian wild asses (*E. africanus* and *E. hemionus*) (Klingel, 1977). This behavior has not yet been described for Przewalski's horses.

Two hypotheses have been given concerning the function of marking behavior in equids. Turner et al. (1981) suggest that the stallion marks his mares' eliminations to advertise his claim to the mares. Trumler (1958) believes that the stallion is using his own eliminations to mask information about the mares' estrous status which might appear in their urine or feces.

The purposes of this study were to document marking behavior in Przewalski's horses, and to evaluate hypotheses concerning the functional significance of marking behavior.

METHODS

The study animals were 1 adult stallion and 2 adult mares, housed together in a 4 acre pasture at the Topeka Zoo's Forbes Conservation

*Present address: Dept. of Biology, Washburn University, Topeka, KS. 66621.

and Propagation Center in Pauline, Kansas. These animals were observed from January 1983 through June 1984, and from January 1985 through May 1985. A total of 293 hours of observation were made during this time. Both mares were barren in 1983, and pregnant in 1984 and 1985.

Because this was a small group of animals in a confined space, it was relatively easy to keep all individuals in sight during observations. All occurrences of urination and defecation by an individual were recorded. The time and location of each elimination were noted. In 1985 the distance of the mare to the stallion during elimination was noted. If the stallion marked the mare's elimination within 15 minutes, his marking behavior was described and the elimination was recorded as marked; otherwise the elimination was recorded to be unmarked.

RESULTS

The marking behavior of the Przewalski's stallion was similar to that described for feral horses by Turner et al. (1981). Typically, the stallion would orient toward the mare while she was urinating or defecating, approach the elimination, sniff it with his neck arched, step over it, either urinate or defecate on it, and then move away. Flehmen was often exhibited prior to stepping over the site, or while eliminating on the site. The stallion marked 9% (n = 161) of the mares' defecations. Seventy three percent (n = 15) of the time urine was used to mark the defecations rather than feces. The stallion marked 20% (n = 156) of the mares' urinations. The mares' urine was always marked with urine and never with feces.

Marking behavior was distinctly seasonal and was seen in greater frequency during months when breeding occurred (Figure 1). Marking behavior increased during the 1983 breeding season, peaking at 60% of the eliminations marked in June, and decreasing to 0 to 10% of the eliminations marked during the winter. The mean distance from the mare to the stallion during elimination was 18.9 ± 13.1 meters. She was within 5 meters of the stallion only 17% ($\bar{n}$ = 95) of the time. There was no significant difference between the distance of the mare to the stallion with regard to whether her eliminations were subsequently marked, or left unmarked ($t = 0.487$, $df = 94$, $p > 0.1$).

DISCUSSION

A mare's eliminations seem to convey important olfactory information to the stallion. In particular, a mare's urine is more likely to elicit a response than her feces. The seasonality of marking behavior is not simply a function of changing daylength, but appears to be affected by the mare's reproductive status. The frequency of marking waned in 1983 following conception. The onset of marking was delayed and then the behavior was exhibited at very low frequencies in 1984 and 1985 while the mares were pregnant. That urine appears to convey reproductive information is supported by the observation that a third of all matings (n = 6) were initiated by the stallion immediately following urination by the mare. As further evidence that urination provides information on a mare's reproductive status, Waring (1983) reports that female equids urinate more frequently and prolong the urination stance during estrus.

These data document the marking behavior of Przewalski's horses, but do not allow discrimination between the hypothesized functions of marking behavior. The harem stallion might mark a higher proportion

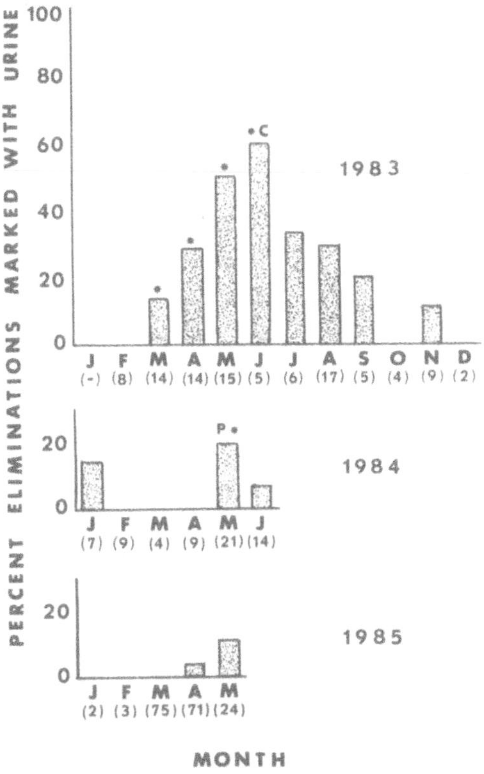

Figure 1. Percent of mares' eliminations marked with urine. Numbers in parentheses are sample sizes. Stars indicate months when breeding was seen. C = conception of 1984 foals. P = parturition.

of his mares' eliminations during the breeding season to advertise his claim to the mares when they are most attractive to other males. The harem stallion might also be differentially marking reproductively active mares' eliminations to mask attractive odors.

The frequency of marking behavior in this Przewalski's horse herd was lower than that reported by Turner et al. (1981) for feral horses. The Przewalski's stallion may be less assiduous about marking because there are no competing males to steal his mares.

Since their urine appears to contain information about reproductive status, it could be hypothesized that mares might approach the stallion before eliminating to enhance the chances that the stallion will find the urine, assess it, and breed the mares if they are in estrus. This hypothesis is not supported by the data. Mares did not approach the stallion prior to eliminating and were often at a considerable distance from him. Mares do show a preference for urinating in certain parts of the pasture, but these are broad areas rather than a discrete location that could be easily checked by the stallion. (It should be noted, however, that the four acre pasture may have been too small to show such approach.) Thus it appears to be up to the stallion to locate and respond to the urine and feces. This might be a mechanism by which mares select select stallions of superior quality if stallions who fail to find and respond to their mares' eliminations are more likely to lose their harems, and be replaced by attentive stallions.

There are some indications that females compete with one another in directing the stallion's attention toward their own urine. Social facilitation of urination is common. During the 1985 season, the mares were urinating within 5 minutes of each other 21% (n = 47) of the time. Thus the stallion may be forced to decide which elimination to visit first. Mares at the San Diego Wild Animal Park were observed to urinate simultaneously within 1 meter of each other and to urinate directly on another mare's urination within 1 minute of its deposition (pers. observation; Lee Boyd).

ACKNOWLEDGMENTS

This research was supported in 1985 by the Nixon-Griffis Fund for Zoological Research. John Sauer commented on the manuscript. Special thanks to the staff of the Topeka Zoo for their support of this project.

REFERENCES

Feist, J. D., and McCullough, D. R., 1976, Behavior patterns and communication in feral horses, Z. Tierpsychol., 41:337.
Klingel, H., 1974, A comparison of the social behaviour of the Equidae, in: "The Behavior of Ungulates and its Relation to Management," V. Geist, and F. Walther, eds., IUCN Publ. 24, Morges, Switzerland.
Klingel, H., 1977, Observations on social organization and behaviour of African and Asiatic wild asses (Equus africanus and E. hemionus), Z. Tierpsychol., 44:323.
Mohr, E., 1971, "The Asiatic Wild Horse," J. A. Allen, London.
Trumler, E., 1958, Beobachtungen an den Böhm-Zebras des Georg von Opel-Freigeheges für Tierforschung. I. Das Paarungsverhalten, Säugetierk. Mitt., 6:1.
Turner, J. W., Perkins, A., and Kirkpatrick, J. F., 1981, Elimination marking behavior in feral horses, Can. J. Zool., 59:1561.
Tyler, S. J., 1972, The behaviour and social organization of the New Forest ponies, Anim. Behav. Monogr., 5:85.
Waring, G. H., 1983, "Horse Behavior," Noyes Publications, Park Ridge, New Jersey.

CHEMICAL COMMUNICATION AMONG ASIAN ELEPHANTS

L.E. Rasmussen, M.J. Schmidt, and G.D. Daves

Oregon Graduate Center, Beaverton, Oregon
Washington Park Zoo, Portland, Oregon
Department of Chemistry, Lehigh University, Bethlehem
Pennsylvania

The Asian elephant (Elephas maximus), a mammal of recognized high intelligence with a well-organized societal structure (McKay, 1973; Eisenberg et al., 1971; Olivier, 1978; Kurt, 1974), possesses a sophisticated array of sensory apparati. These sensory systems interplay in influencing both the immediate responses of the elephant and its long-term behavior. The visual system, considered the least developed of all the elephant's sensory systems, has some color vision (Markowitz et al., 1975), visual acuity at close range (Rensch, 1959), and good night vision (Kühme, 1962); the tactile and auditory systems are well developed and relied on heavily (Eisenberg, 1980; Gadgil and Nair, 1984). Elephants produce a variety of sounds for communicative messages (Berg, 1983); recently Asian elephants have been shown to utilize low frequencies, producing infrasound (Payne et al., 1985).

The subject of the present work is the highly developed chemosensory system of the elephant; this system includes both the primary olfactory and the vomeronasal organ (VNO) systems. In such an intelligent mammal, separation of the chemosensory responses from auditory, tactile, and visual responses is difficult. A chemosensory response by the elephant depends on reception of a chemical signal, but its expression may be modified by an intricate combination of other cues, visual, tactile, and auditory, and by the physiological state and experiences of the animal.

Several years ago we began an investigation of chemosensory phenomena in the Asian elephant utilizing the largest breeding herd of Asian elephants in the western hemisphere, which is located at the Washington Park Zoo, Portland, Oregon. This herd is a unique resource for the study of elephants in captivity. The present report summarizes results obtained in a number of areas in the course of investigations during the past four years.

For the study of VNO-mediated chemosensory phenomena the elephant offers distinct advantages because of its trunk, which makes the flehmen response clearly defined and easy to identify and count. Both sexes of Asian elephants have a cigar-shaped paired vomeronasal organ located dorsal to the hard palate. Paired orifices lead into the anterior mucous-filled respiratory epithelium region, which connects in turn with the neuroepithelial area (Rasmussen, unpublished).

After initial detection of an odor (e.g., a urine puddle) by the visual or olfactory senses, bull elephants frequently and cow and sexually immature elephants infrequently, both in the wild and in captivity, contact the source with the tip of the trunk and transfer material to the openings of the VNO (Rasmussen et al., 1982). This stereotypic flehmen response, subsequent to primary olfactory detection, is an integral part of intergroup communication among Asian elephants. We have identified four specific communicating situations in which a VNO-mediated flehmen response is utilized. It is involved in the establishment and maintenance of mother-young recognition, in female-female interactions, in the mediation of male-male interactions, and especially in the responses of males to females prior to and at ovulation.

Much of our work is focused on the elucidation of the semiochemistry involved in the response of Asian bull elephants to urine of cows in estrus. The number of flehmen responses exhibited by a bull as the result of contact with female urine begins to increase about six weeks prior to ovulation (Hess et al., 1983), probably an indication that the stage of the estrous cycle and the impending receptivity of the female are being revealed to the males via specific chemical components of urine. We have exploited this dependable chemosensory response to develop a unique bioassay to guide chemical fractionation toward isolation and characterization of the responsible substance or substances. The bioassay developed has proved reliable in over 5000 double blind tests using three bulls during a period of four years.

The robustness and reliability of this semiochemical response is indicative that chemical communication may play an integral part of the mating scenario, ensuring that the bull is around and aroused at the proper time.

METHODS

Eight mature cows, several small calves, and three bulls housed in a well-designed functional facility at the Washington Park Zoo (WPZ) in Portland, Oregon, were made available for the study.

I. Bioassay for Semiochemical Responses

A. Urine sample collection. The estrous cycle of the Asian elephant has been carefully defined by determination of serum hormone levels, examination of cervical mucus, and by daily monitoring of the response of bulls to cows (Hess et al., 1983). Urine was collected from each of the eight cows during the pre-ovulatory period, during the period of peak interest by the bull (just prior to ovulation), and during anestrus.

B. Sample preparation for bioassay. The bioassay utilized 250 mL of urine or extracts or fractions equivalent to 250 mL of urine resuspended in 2 mL of acetone or methanol and added to 250 mL of buffer, saline, or urine.

C. Sample placement and controls. Two to nine samples were placed at random for each bioassay session on one of four cleaned concrete slabs available in two exercise yards, used on a rotational basis, prior to the release of one of the three regularly used bulls. Control samples, both positive and negative, were included in every bioassay session. All solvents used during the chemical procedures were bioassayed periodically as controls.

The bioassay had a built-in control: the bulls were often let out into the yard without bioassay samples being present.

D. Behavioral observations and data recording. The sixty-minute observation period began when the focal (unrestrained, solitary, free-roaming) bull entered the enclosure. The observer was positioned where a zoo visitor might stand. The bull elephants were observant of the keeper's activities but appeared to be uninterested in other observers.

A focal-animal sampling technique (Altmann, (1974) was employed and the following scoring criteria and definitions guided data recording. All occurrences of the defined selected behaviors were scored. Both the response and its frequency and duration were recorded. Primary behaviors recorded that relate to responses to test samples or other odor sources include arrested movements, hesitations in motion, sniffs (often audible), checks, placements (trunk tip placed on odor source), flehmens, aborted flehmens, peculiar flehmens named sideways-tasting-responses (STR) (after the trunk tip was placed in a sample, the trunk was curled to the side during elevation and the tip placed dorsally on the junction of the hard and soft palates), scrubs (trunk tip flattened onto the substrate in an accentuated placement and scrubbed vigorously back and forth, often preceding an STR), blows, avoidances, pauses (mild corkscrewing and shaking of trunk, terminating about 35% of flehmens), and erections. The duration of each flehmen was timed from the initial contact with the substrate or test materials until the pause, usually occurring at the termination of the flehmen. When none of the above responses were noted, the proximity (in feet) of the test bull to a test sample and the number of times he was near the sample were noted. Although the bioassay sessions were for 60 minutes, data are expressed per 30 minutes.

Bulls were observed in a similar manner during sessions when bioassay test samples were in place and when no test samples were used. In these latter sessions responses to random odor sources (e.g., urine spots from other elephants using the yard) were recorded including the number of revisitations to a particular spot.

II. Chemical Procedures for Urine Manipulation

A. Urine collection. At the defined times in the estrous cycle of a cow large volumes (50 liters or more) of urine were collected (Rasmussen et al., 1982). Post-collection bioassays of estrous urine (EU) elicited five to twenty flehmens (cf., Fig. 2). Specimens of pre-ovulatory urine were obtained during the 4-6 week period prior to ovulation. Urine samples were also obtained at the luteal, anestrous stage, during which bulls exhibited no interest toward the cow. The pH of both estrous and anestrous urine was manipulated from pH 1-12 and samples bioassayed.

B. Urine extraction. Initially the urine was extracted with hexane to remove lipids. The urine pH was then adjusted with HCl to about 5.5 (after this was determined as the pH of optimal maximum extraction) and re-extracted using dichloromethane (Table 1)(urine-to-solvent ratio 4:1) in a heavier-than-water liquid-liquid extractor (Figure 1). Control of pH during the extraction (urine increased in basicity over time) permitted 97% recovery of biological activity.

C. Preparative chromatography: chromatographic fractionation. The bioactive extract was separated first by silica gel flash chromatography (FC)(Still et al., 1978). Use of dichloromethane, benzene, or ether as an eluding solvent was equally effective; each yielded a biologically active fraction of approximately equal complexity (as seen by gc traces). This active fraction (representing about 10% of extracted organics)(Table 2) was separated further in a preparative manner by high pressure liquid chromatography (HPLC) using reverse phase 5μ octadecylsilane (C-18 Bondpak) column. Separations with high recovery of biologically active material were achieved

Table 1. Solvent Preference of Active Component(s)

Fractionation Step	Effective Solvent
Extraction from urine	DCM, Benzene
FC	DCM, Benzene, Ether
HPLC	MeOH, Acetonitrile

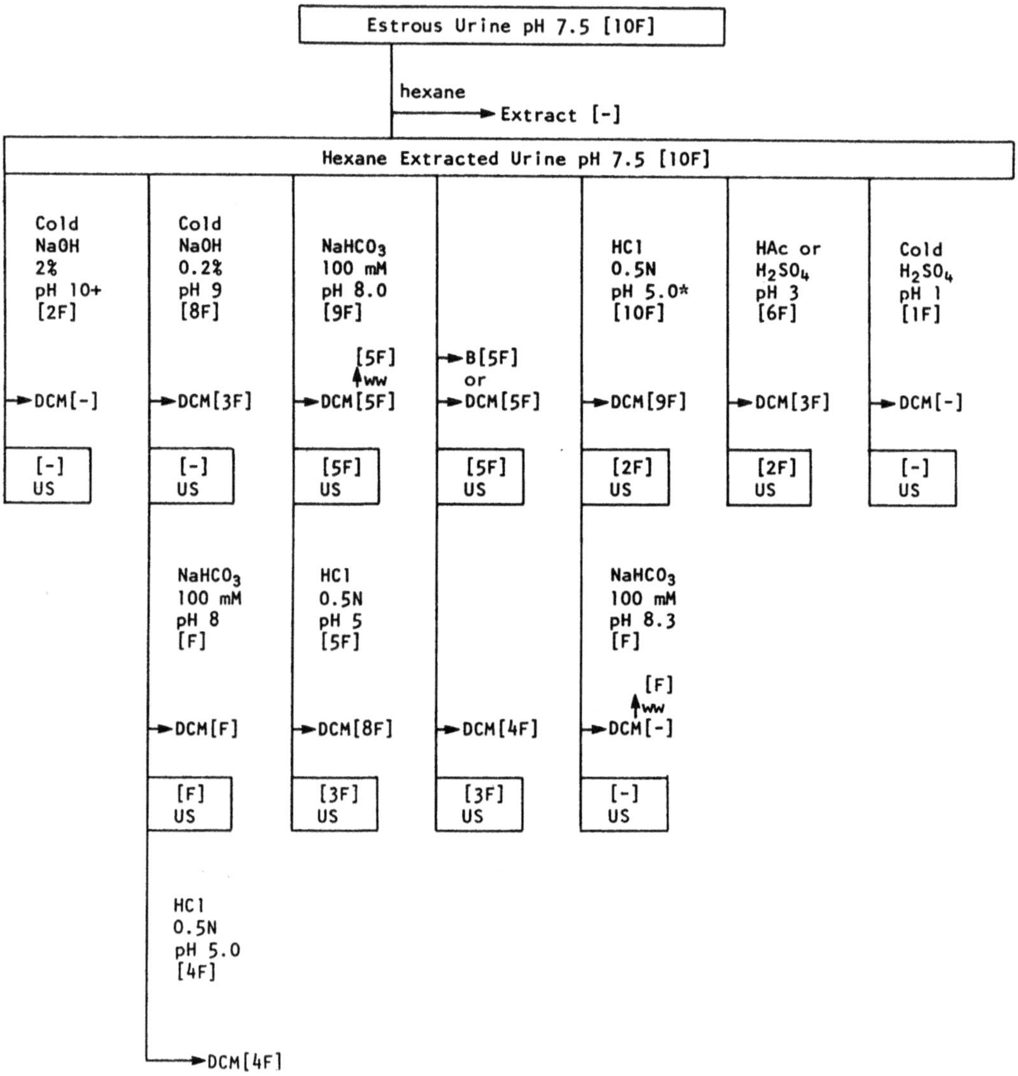

Fig. 1. Extraction scheme illustrating pH manipulations and activity of extracts. US = urine supernatant; DCM = dichloromethane extract; B = benzene extract; ww = water wash, pH 6.0. *pH optimum is between pH 4-6 for whole urine. For comparison to other extracts the elevation in urine activity by pH changes to pH 4-6 was not depicted here. Flehmen responses were elevated 10-20% by lowering pH to 4-6.

Table 2. % Recovery in Active Fractions

Active Fraction	mgm/L*	F**	% Recovery of Activity	% Reduction of Total Organics
Estrous Urine 1 Liter	50 g	10	100	0
Dichloromethane Extract	15 mg	9	97	--
Dichloromethane Eluted-FC	1.5 mg	9	90	90

*based on dried weight; ** Flehmen; liter equivalents were bioassayed

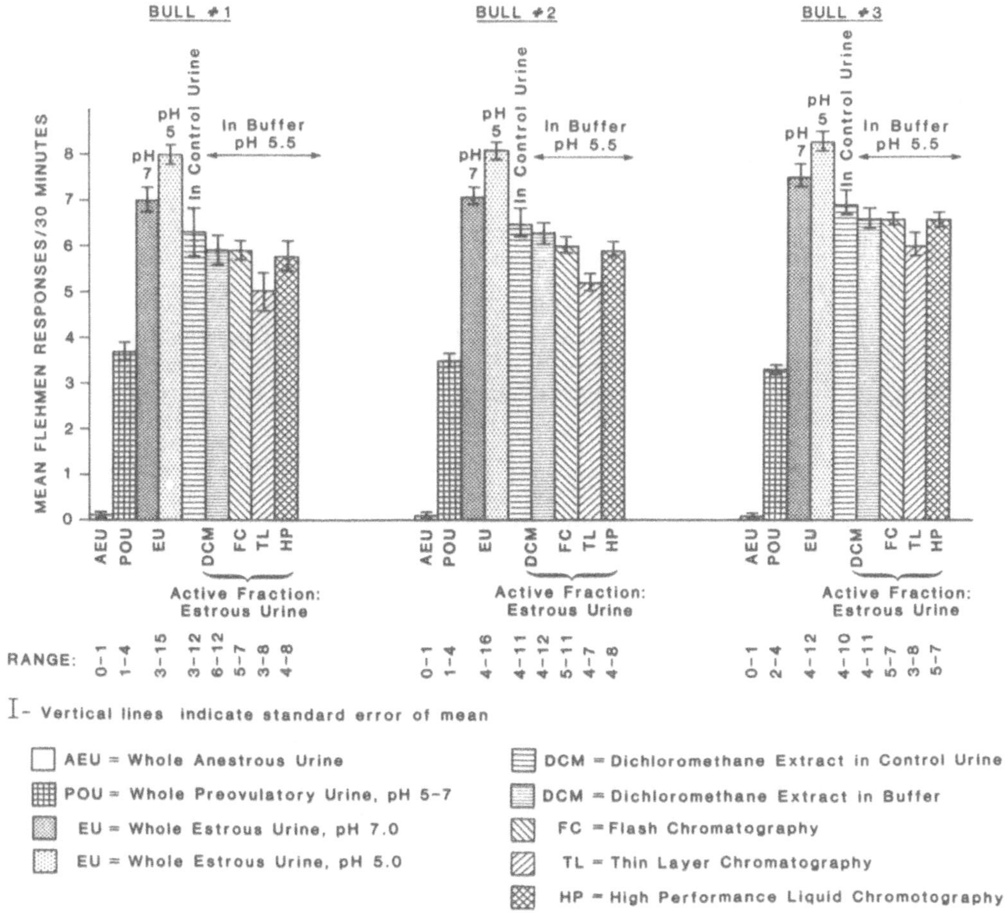

Fig. 2. Biological response data, mean flehmen responses/30 minutes of three bull elephants at Washington Park Zoo, Portland. Three types of whole urine (anestrous, pre-ovulatory, and estrous) are depicted and the active fractions of estrous urine fractionated by several chromatographic methods.

using either acetonitrile/water, 3/1 to 9/1 in an extended linear gradient; or methanol/ water, 3/1 to methanol/dichloromethane, 95/5 in an expanded gradient. The pH was maintained at 5.5 by the addition of 0.05% acetic acid to both systems (Figure 3). Elution was monitored by recording adsorption at 254 nm and 340 nm. Although the active components may not be uv-adsorbing, the uv-adsorption patterns served as a check on sample-to-sample and run-to-run variability. Fractions were analyzed by capillary column gas chromatography to establish the degree of separation attained (Figure 4).

RESULTS

I. Flehmen Responses to Urine

A. Adult female elephants. All the females of the elephant herd at the Washington Park Zoo, including the young female calf, exhibit flehmen (Figure 6); cows have been observed to exhibit flehmen responses to the urine of other cows, to human urine (male and female), to male musth urine, to female urogenital mucous secretions, to both male and female temporal gland secretions (Table 3). A maximum of 3 flehmen responses in sequence by adult females to an odor source has been observed; usually only 1 flehmen is observed. Female flehmen responses were rapid, 1-3 seconds in duration, similar to those observed in males. Data on the number and frequency of flehmens exhibited by cows have not been gathered systematically (observations: J. Haight, R. Henneous, L.E. Rasmussen, E. Michon), but it is clear that, like males, females routinely process chemosensory information (Table 3).

It is noteworthy that individual differences in trunk and lip movements occur during flehmen responses of different females. One cow (Tamba) uses the tongue to press the entry of liquid substances against the VNO (J. Haight, personal communication).

B. Calves. The three calves (2 males, 1 female) born 1982-83 at WPZ exhibit a flehmen-like response to the urine of their mother and other cows (Figure 6). The young calves were clearly transporting some urine to the openings of their VNO.

C. Adult males. 1. In response to other males: Bulls rarely exhibit flehmen to non-musth male urine (Rasmussen et al., 1984)(Figure 5). Bulls exhibit flehmen frequently in response to their own urine during musth and to the musth urine of other bulls. From October 1983 to April 1985 36 flehmens to male musth urine were recorded. Male temporal gland secretions (musth) also elicited flehmen responses (Rasmussen et al., 1984). On two occasions during field observations of several African bulls in Kruger National Park, South Africa, a bull was observed to touch the penile area of another bull and subsequently exhibit flehmen.

2. In response to females: In a wild herd adult males (living separately) can tell which females are in estrus, although the females exhibit no obvious behavior signs (Jainudeen et al., 1971; Eisenberg, 1984). Males were observed to touch urine or urine-splattered feet or the urogenital orifice of the female and subsequently place the trunk in the mouth. While this was not designated a flehmen response, it seems likely that it was. Flehmen responses by bulls to cows (in probably estrus) have been observed in Thailand work camps. Flehmen responses by African bulls to cows in a herd situation in Kruger National Park and by a bull at the Pretoria Zoo were also observed.

At the Washington Park Zoo during eight years, blood hormonal measurements indicating the periodicity of the Asian elephant estrous cycle

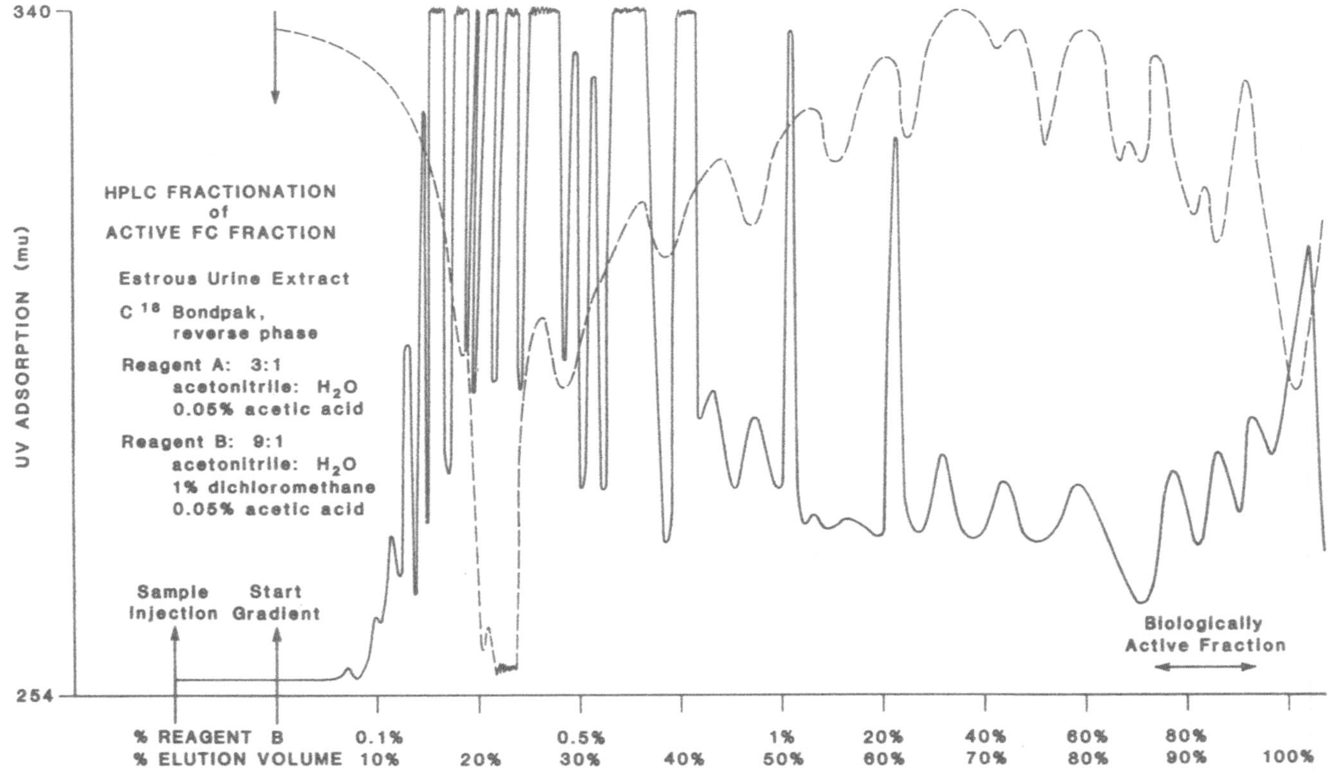

Fig. 3. High performance liquid chromatographic (HPLC) separation of components in active fraction from flash chromatography of estrous urine extract. Biologically active fraction was eluted between 65-86% of reagent B. 4-methylphenol was eluted at <0.5% reagent B. Total sample volume injected was 1 mL.

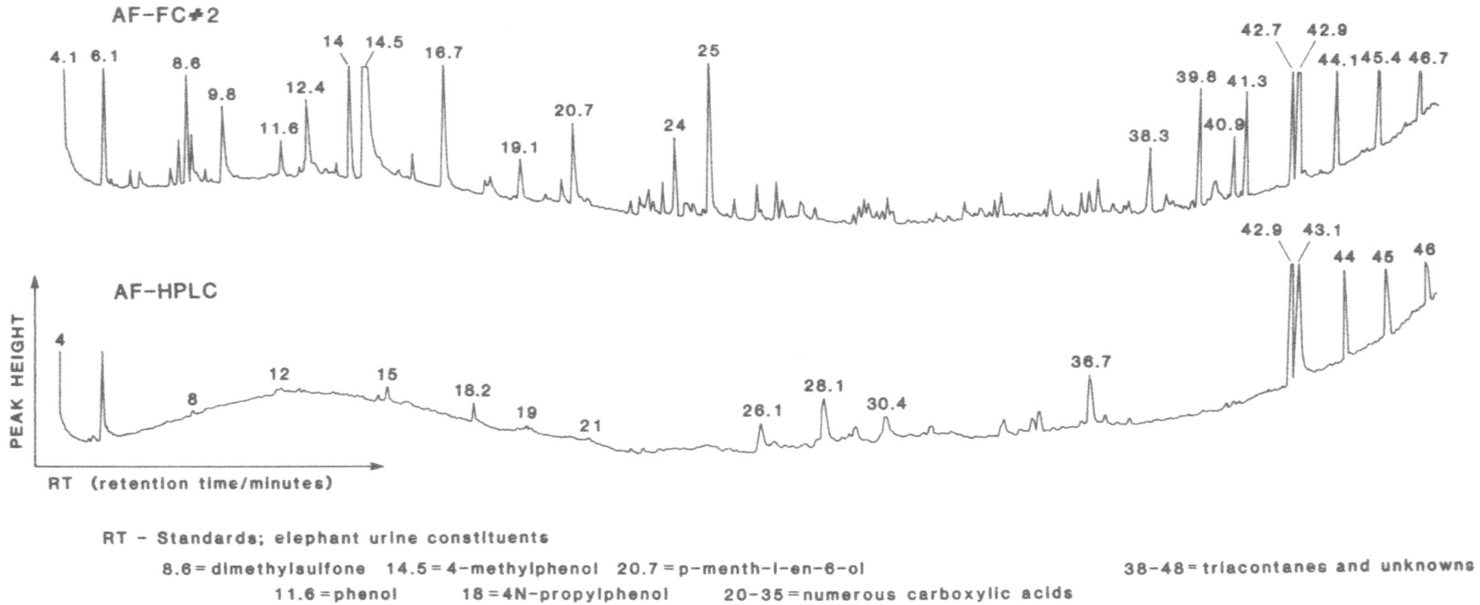

Fig. 4. Gas chromatographic separation of: AF-FC #2 (active fraction of estrous urine extract after flash chromatography and TLC), upper chromatogram; AF-HPLC (active fraction of estrous urine extract after FC, TLC, and HPLC), lower chromatogram. 1µ sample. DB-1-30N column (J&W)(0.25 µm; 0.32 mm) 30M; helium carrier gas; temperature-programmed 5%C/minute, 35°C to 330°C; retention time depicted in minutes. Solvent peak eluted prior to 4 minutes.

Table 3. Substances Eliciting Flehmens in Asian Elephants

Classification	Females	Males	Young
	Placenta: Mucus	Blood: Saliva	nt
TGS	M, F	M, F	nt
Urine: Conspecies	F Musth (M)	Musth (M) Estrous** (F) Lactating (F) Pre-Ovulatory (F)	Mother Other females
Urine: Other Species	nt	F	nt
Synthetics	nt	Novel Substances	nt

nt = not tested; F = female; M = male; ** = only high intensity response

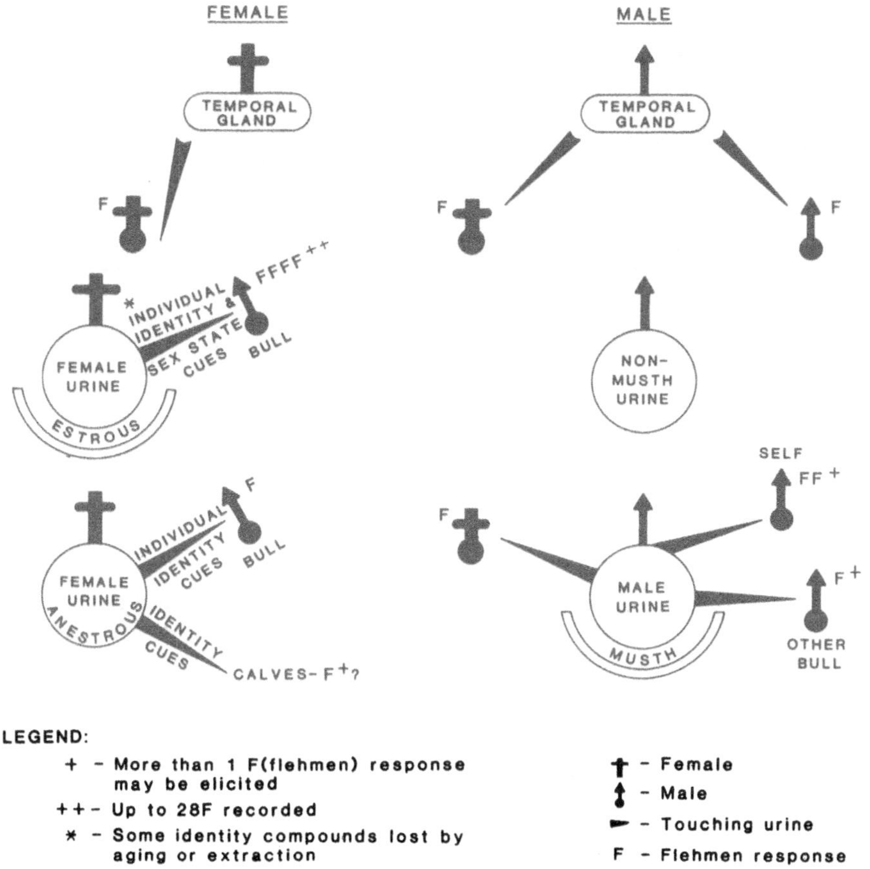

Fig. 5. Chemical communication mediated via the flehmen response in Asian elephants. Urine and temporal gland secretions both elicit flehmen responses. Emphasis of the diagram is on sex state information.

Fig. 6. Flehmen responses in Asian elephants. Upper left, female responding to male musth urine; upper right, male responding to female estrous urine; lower left, cow urinating; middle, calf touching urine with trunk tip; lower right, apparent flehmen by calf.

(Hess et al., 1983) were correlated with a daily regime of "sniff" tests during which a bull was allowed limited access to each cow for several minutes; flehmen responses and checks were recorded. The data obtained established an inverse correlation of serum progesterone levels with the male flehmen response (Hess et al., 1983).

In ten matings observed at WPZ involving three bulls and six different cows the flehmen response was utilized in the mating sequence. Prior to successful mating with a cow known to be in estrus, the experienced bulls normally exhibited 1-5 flehmens in response to the cow's urogenital area or urine. They often stimulated urination by blowing on the urogenital area of the cow. Numerous checks (4-11) and 1-5 trunk tip to palatal pits exchanges were also observed prior to the beginning of the actual mating sequence of mounting, pelvic thrusting, and ejaculation. Bull #2 previously subjected to sexual restraints exhibited fewer checks and flehmens but palatal pit exchanges were equal (J. Haight, personal communication).

Males infrequently exhibit flehmen in response to anestrous urine of cows. Some variation is observed between bulls, and the particular cow from which urine is obtained can affect the response. Female urine carries chemical signals not only concerning the estrous state but also for other factors. The bulls at WPZ have favorite and/or disliked cows. They utilize touch, sight, sound, and the chemosensory system to recognize individual cows. A "new" cow created by mixing urine from two different cows in anestrus elicited three flehmen responses, probably a response to a novel mixture.

The most dramatic exhibition of flehmen by Asian elephant bulls occurs in response to female estrous urine. Bulls exhibit flehmen more to collected estrous urine samples in a bioassay situation than when in the presence of a cow in estrus. A noteworthy exception occurred when a young female experienced her first cycle; the number of flehmen responses by the bull was high. An anestrous cow usually elicited brief check(s); a sudden increase in flehmen was observed as she entered the pre-ovulatory stage, rising dramatically just prior to ovulation.

The distinctiveness of the male response to estrous urine is the intensity of response. Males often exhibit 25-30 flehmen responses to estrous urine during a 60-minute observation period; in one instance 41 were observed. One to three responses during an observation period is typical for other odor stimuli. During 1982-1985 in observations of three bulls, 50 flehmen responses were elicited during 409 tests of anestrous urine and 995 flehmens were elicited during 165 tests of estrous urine. The ratios of flehmen responses to estrous versus anestrous urine were respectively for the three bulls, 57.5, 58.3, and 60.8 (Figure 2) The ratio of flehmen responses to estrous:anestrous urine was 7.1:0.12.

A sample of estrous urine that elicited 17 flehmens, when diluted 10-fold still elicited 7 flehmen responses; as little as 10 mL of estrous urine elicited 4 flehmens.

A fascinating observation was the failure of a bull (Packy) to exhibit flehmen either in response to the presence of his mother when she was in estrus or in response to estrous urine from his mother collected during peak estrus, whereas the dichloromethane extract of the same urine elicited the usual number of flehmen responses observed for estrous urine extracts. Another bull responded normally to urine from this cow.

The flehmen response exhibited by the 3 bulls at the WPZ show variations. The bulls have been observed for a total of 958 hours (5 different observers) in the past 4 years: Bull #3 for 207 hours, Bull #2 for 252 hours, and Bull #1 for 499 hours. For the past 2.5 years, including 575 observing hours, 1640 flehmens have been observed to urine left by prior residents of the yard, ranging from 0-37 per session.

The predominant variation from the typical flehmen response that has been observed involves a sideways twisting of the trunk and placement of the tip containing the urine on the dorsal palate at the junction of the hard and soft palates. This response (STR) has been observed 173 times. Another unusual flehmen response observed involves the use of the tongue to press materials onto the VNO, seen in one female and one male. Two bulls have been observed on three occasions each to spray a sample onto the VNO openings. Samples were often blown on if frozen or dried; sometimes a flehmen response was elicited to dry or frozen samples without prior blowing to warm or dampen the sample.

3. Cross-species responses: Bull elephants responded at low and identical levels to whole estrous and anaestrous polar bear and mare urine.

Table 4. Compounds Tested for Novel Substances Response

Compound	Bull #1 Test 1	Bull #1 Test 2	Bull #1 Test 3	Bull #1 Test 4	Bull #1 Test Y	Bull #2 Test 1	Bull #2 Test 2	Bull #2 Test 3
PHENOLS								
3-methylphenol	C	C	C	0	--	2F	F	C
4-ethylphenol	F	C	0	0	C	C	--	--
4-n-propylphenol	0	0	0	0	--	0	0	0
2-n-propylphenol	3F	P	C	0	C	P	C	C
BENZENE-RELATED								
benzoic acid	F	STR	0	--	STR	STR	C	0
para-hydroxyl benzoic acid	2F	0	0	0	C	2C	C	C
sodium benzoate	C	0	--	--	--	C	0	0
ACIDS								
valeric acid*	P	C	C	0	C	F	C	C
2-phenyl propionic acid	2C	C	0	--	--	--	--	--
phenylacetic acid	0	--	--	--	--	--	--	--
hippuric acid	0	--	--	--	--	--	--	--
p-OH phenyl acetic acid	0	--	--	--	--	--	--	--
cinnamic acid	2F	0	C	0	--	C	C	0
hexadecanoic acid	0	--	--	--	--	--	--	--
hydrocinnamic acid	3STR	0	0	0	--	C	0	0
hexanoic	2CP	--	--	--	--	--	--	--
octadecanoic	STR	0	--	--	--	--	--	--
triacontanoic acid	2F	2STR	2STR	3STR	F	STR	STR	--
hydrochloric acid	F	STR	--	--	STR	F	C	--
acetic acid	F	STR	--	--	STR	F	C	--
amino acids								
glutamic	0	0	--	--	--	--	--	--
glycine	F	C	0	--	--	--	--	--
2 methyl, 2 propyl 1,3 propandiol	2STR	2C	C	C	--	F	AF	C
menthol	AF	2C	C	0	--	P	C	C
KETONES								
5-nonanone	F	C	0	0	--	F	C	0
2-pentanone 4-hydroxyl, 4-methyl	3F ERECTION	F	P	C	0	F	C	0
acetophenone	0	--	--	--	--	--	--	--
6-methyl-5-hepten-2-one 3-pentanone	0	--	--	--	--	--	--	--
heneicosanone	0	--	--	--	--	--	--	--
butanone	0	--	--	--	--	--	--	--
ALIPHATIC HYDROCARBONS								
hexatriacontane	F	7STR	C	0	--	F	C	--
pentatriacontane	F	C	0	--	--	2C	C	--
triatriacontane	C	0	0	0	0	2C	C	--
tetracontane	2F	C	0	--	--	C	0	0
17-pentatriacontene	F	C	0	0	--	4F	--	--
octatriacontane	0	0	--	--	--	--	--	--
SULFUR COMPOUNDS								
urea	C	0	--	--	--	--	--	--
dimethyl sulfone	F	C	0	0	--	2F	F	C
ESTROGENS								
testosterone	F	F	--	--	--	C	--	--
semestrin	2F	F	2F	--	--	C	--	--
sodium estrogen sulfate	3F	F	0	--	--	3F	F	0
beta-estradiol 17-cypinate	0	--	--	--	--	0	--	--
2-methoxyl estrone	F	0	--	--	--	C	--	0
estradiol benzoate	0	--	--	--	--	0	--	--

*F = flehmen; C = check; P = place; STR = sideways testing response; AF = aborted flehmen; Test-Y = retested after one year.

However, dichloromethane extracts of these urines were inactive. Conversely, wild stallions responded to the dichloromethane extracts of estrous mare urine but not to extracts of estrous elephant urine.

4. In response to novel substances: A number of chemical substances have been presented to bulls in bioassay sessions. Most elicited no response; however, a number of synthetic compounds did elicit one or more flehmen responses. Those substances that were active at initial bioassay (Table 4) demonstrated a diminishing response upon successive re-assay. We characterize this phenomenon, a positive response during the first bioassay, followed by responses of diminishing intensity on successive tests, a novel substance response. A dramatic example of the novel substance response occurred when 4-hydroxy-4-methyl-2-pentanone was bioassayed. This compound elicited three flehmens and an erection from test bull #1 when first bioassayed. Subsequent repetitive presentations resulted in diminishing responses.

The observed habituation to novel substances has a counterpart in a slightly diminishing response by bulls over the years in the presence of a cow. It is noteworthy that no such habituation has yet been observed using estrous urine or active urine subfractions. The three bulls have been observed carefully for evidence of habituation or for indications of learned responses to bioassay materials. Caution is essential when only a few bulls are available for testing. Sample placement locations were varied daily. Duplicate samples are presented in different locations or on different days or are presented to different bulls. When samples were placed on successive days in identical locations, they elicited similar responses. On one occasion nine identical samples presented in nine different locations elicited identical responses. No evidence of learned behavior has been noted during bioassays involving 5000 samples during four years using three bulls. Periodically water samples only were placed in all areas of the concrete strips used for sample placement. No systematic checking was observed with these blanks.

II. Chemosensory Responses to Urine Fractions

Seventy bioassays of dichloromethane extracts of estrous urine re-suspended in water or saline elicited a mean of 6.2 flehmen responses per 30 minutes (total session = 60 minutes). When the extract was re-suspended in anestrous urine, a mean of 6.5 flehmen responses were elicited. No differences in responses were recorded when a dichloromethane extract from pooled estrous urine of different cows was compared with a dichloromethane extract from the urine of a single cow. Anestrous urine extracts re-presented in water, saline, or anestrous urine elicited no responses in ten bioassays. Percent recovery in the extract from pH 5.5 urine was high (Table 2).

When the bioactive dichloromethane extract of estrous urine was placed on a silica gel chromatography column and eluted with dichloromethane, ether, or benzene, the bioactivity was retained (Table 1; Figure 2). A total of 60 bioassays of these chromatographic fractions utilizing three bulls elicited 192 flehmen responses. Fractions eluted subsequently with other, more polar solvents showed no activity. Recovery was high in active fractions (Table 2). pH was manipulated several times, either to the acidic or to the basic pH; biological activity was restored when pH was readjusted to 3.5-5.5 (Figure 1).

Further separation of the active chromatographic fraction by high pressure liquid chromatography (HPLC) again yielded a single region exhibiting bioactivity. We have recorded 110 flehmens, 31 STR, and 38 places to the active fraction from 20 HPLC separations. The active HPLC fraction

Table 5. Percentages (by gc/ms) of Dominant Peaks (gc)

Compound	Retention Time (min)	Male Urine	FC Anestrous Urine	FC Estrous Urine pH 5.5	Active HPLC Fraction
Phenol	8.8	+	1.1%	1.3	--
4-methylphenol	13.9	+	70.0	75.0	--
4-ethylphenol	16.5	+	0.4	0.4	--
2-n-propylphenol	17.1	+	4.0	5.0	--
4-n-propylphenol	18.2	?	2.0	2.5	--
Total Phenols			77.5	84.2	--
Unknowns	21-29	+	++	+	--
Unknowns	42-48		+	+	++

+ = present

has no 4-methylphenol, the dominant chemical species present in the active TLC fraction and the active FC fraction (75%)(Table 5). The phenols were present in a variety of elephant urines. Concentration effects were also apparent at HPLC level; to establish if a negative bioassay resulted from the activity being located in another fraction, degradation of components, or lack of sufficient concentration, samples that elicited a check were presented in 2- or 8-fold concentration (Table 6). The partially purified HPLC chromatographic fractions exhibited less activity than those seen for intact estrous urine or the dichloromethane extract. However, the persistence of the stereotypic biological response after a series of chemical manipulations and removal of over 99% of the urine sample is remarkable and encouraging. Gas chromatographic analysis of active HPLC fractions reveals that significant sample heterogeneity remains and further fractionation will be required before one or more bioactive molecules can be identified.

DISCUSSION

We have demonstrated that Asian elephants, male and female, flehmen respond to a variety of materials, self-excreted, conspecific, non-specific or synthetic compounds, for several information-gathering reasons. These varieties of signals are then apparently processed through the vomeronasal organ system. At the WPZ the position ranking of the eight cows are clearly recognized; flehmen occurs more frequently after management-arranged changes in cow groupings. Future systematic observations may demonstrate statistically significant correlations between frequency of flehmen and rank as suggested by Pfeifer (1985) in oryx. The observed flehmens to urogenital areas of other cows, to placenta, and to temporal gland secretions suggest information on reproductive status and physiological states is being assimilated. The flehmen-like response of calves suggests the possibility of a chemo-memory imprint that aids in later recognition of relationships. We

Table 6. Concentration and Elicitation of Flehmen Response in HPLC Fractions

HPLC Sample	Response	Duration (Sec)
Fraction #9 (0.2 mL)	Brief Check	1-10
Fraction #9 (0.5 mL)	Medium Check	30-60
Fraction #9 (1.5 mL)	Flehmen	After a check, flehmen (1-2 sec)

have evidence from our organic extract bioassay data that a son recognizes his mother; in an instance of selective allo-grandmothering (Rapaport and Haight, 1985) maternal recognition can span a generation. These observations added to the variety of situations that will elicit flehmen from bulls (Figure 5) suggest that in the Asian elephant the VNO is not exclusively used for one type of communication or responsive to only one group of chemicals.

The variations in the mechanics of flehmen responses by individual elephants are noteworthy, especially the STR to the mouth roof area containing acid receptors in other mammals. The responses to dried or frozen spots and the spraying of materials onto the VNO area suggest an essential role for the moist trunk.

Why during evolution did the elephant keep its trunk, and why, with this longest and very sensitive nose, did the elephant retain the VNO system? Careful observation of the flehmen responses and examination of the types of stimuli eliciting the responses do not suggest redundancy of the VNO. Its functional survival in evolution prompts the speculation that the information received by this organ via the flehmen response has helped in survival, especially through aiding successful reproduction. Asian elephant sexes live separately, often physically large distances apart and travel moderate distances daily in their foraging. Urine spots left by a cow in the pre-ovulatory period elicit a moderate response, not as intense as the response to estrous urine but clearly elevated over response values during the long anestrous period. Initial chemo-detection of a persistent urine spot by a bull during the six-week pre-ovulatory period of the cow could induce him to remain nearby, find the cow, compete with other bulls, and perhaps successfully mate. With only a short pre-ovulation period, the probability of bull-cow encounters would be greatly reduced. The longer pre-ovulatory period increases the chances of an experienced bull receiving signals of forthcoming ovulation; these signals and his experience could result in his remaining in the proximity for the tri-yearly ovulation. Perhaps the dominant, most experienced bulls with strong genetic complements would successfully breed more frequently, producing vigorous progeny. The biohabitat of the elephants reinforces the rationale for the long estrous cycle. In his ecological environment the elephant is receiving communications through the interplay of all his sensory systems. A persistent chemical signal will aid visual and auditory clues on the presence of females. Our observations on the WPZ elephants suggest a definitive operational chemical in the alerter stages of pre-mating. The extent that pheromonal component(s) will prove functional in all stages of the mating regime is still uncertain.

Table 7. Properties of Active Component(s)

	Whole Urine	Extract	FC	HPLC
Source	Urinary filtrate or reproductive tract?	--	--	--
Potency	1. negative: anestrous 2. moderate: pre-ovulatory 3. maximal: prior to ovulation	Potency persists	Potency persists	Slightly reduced potency
Stability				
Desiccation	Stable	Stable	Stable	Stable
Heat	$60^{o}C$, 10 min	nt	nt	nt
Dialysis	<1,200 molecular weight	nt	nt	nt
Time	6 months at $20^{o}C$ and light-protected with bacteriostat	nt	nt	nt
Freeze/Thaw	Stable	Stable	Stable	Stable
pH Optimum	pH 4-6	pH 4-6	Slightly acidic	Slightly acidic
pH Reversals	(3x)	Stable	Stable	nt
Volatility	Low (wetted state) (persistent in nature)	Buffer or urine test system low volatility		
Solubility	See Table 1			

nt = not tested

The interesting observations on the variety of discrete and different chemicals eliciting the low level, non-repetitive novel substance response reinforce the multiple receiver functionality of the VNO. Sets of molecules contained in the complex urine and single molecules are both capable of eliciting flehmen. The inactivity of chemicals such as phenols, especially 4-methylphenol, identified in urines of several species (Patterson, 1967; Hradecky, 1978; Brown et al., 1978) was expected because of both their volatility and their non-uniqueness in female elephant estrous urine.

Most noteworthy is the high intensity response of the Asian bull elephant to female estrous urine. This intense response is higher than to any other substance and of higher intensity than exhibited by males of any other species to estrous females. The intensity, expressed as flehmen frequency (flehmens per unit time) of the Asian bull elephant response at estrus is at least 10-fold the response at anestrus. In the wild horse, estrous females were preferred (Turner, 1981). And in the cat, the response differential was 2-fold ($\bar{x}$ = 5.6, duration, 8.4 seconds toward estrous urine, and $\bar{x}$ = 2.8, duration, 3.3 seconds toward non-estrous urine; Verberne, 1976). Flehmen duration in the elephant exhibited no measurable difference between estrus and non-estrus; checking times to estrous samples were increased. As the duration of the flehmen response is between 1-3 seconds, differences were possibly undetected. Compared to flehmen duration reported in other species such as black-tailed deer, 20 seconds (average) (Muller-Schwarze, 1979), and giraffe, as along as 2 minutes observed (Dagg and Taub, 1970), the flehmen response in the bull elephant is of very short duration. The response to estrous urine by the male elephant is unique in its high frequency.

A concern in mammalian sex pheromone investigations, especially when studying releaser pheromones, is the ability of the animal to learn from experience and alter behavior accordingly. Therefore, a primary concern has been the observation of instances of learning. Evidence of memory can be seen in the ten deliberate revisitations per hour by a bull to the same spot. In the survival and evolution of the elephant, learning ability coupled with the dual chemosensory system have helped the elephant survive. Specifically, the elephant has an expert ability of sensory reception of tactile, auditory, and chemical stimuli, coupled to learning and memory. Johnston and Rasmussen (1984) and Meredith (1983) have suggested that the VNO might be especially important in responding to pheromones whose function is not dependent on learning. Our data on bull flehmen responses to bioassay placed samples (Figure 2) showed little difference between experienced bulls (#1 and #3) and the inexperienced one (#2). Especially if keepers were present, however, bull #2 (previously deterred from mating attempts) failed to vigorously pursue mating when allowed access to the cow. Apparently experience affected behavioral response to sexual odors despite VNO processing. Johnston and Zoharik's (1975) and Johnston and Rasmussen's (1984) results indicated that in sexually experienced male hamsters the olfactory system by itself was capable of allowing individual discrimination prior to increased sexual arousal and that an experienced male can get along without the VNO; an experienced bull elephant might also.

Our data during the past four years have suggested an unusual molecule. The properties of our active components are outlined in Table 7. The elucidation of the sensitivity of the active components to pH changes and the requirement of definite pH range for maintenance of full biological activity has been an important factor during our chemical fractionation. The urine itself is an excellent buffering system, and the recent identification of pH dependence at all stages of chemical fractionation has aided our recovery of high biological activity.

The uniqueness of the high intensity estrous urine-elicited response is further strengthened by the retention of high bioactivity in dichloromethane extract, FC fraction, and HPLC fraction. The species specificity seen at the level of dichloromethane extract suggests low level species identifiers either remain in the urine or are no longer intact. Individual recognition clues, especially mother recognition ones, apparently do not persist intact into the dichloromethane extraction suggesting the son's recognition of his mother's urine, imprinted at an early age, subsequently may function as a deterrent to mother-son matings, implementing an incest taboo.

ACKNOWLEDGMENTS

The dedication of the bioassay observers, Evie Michon, Anna Michel, Patricia Ashley, and Bonnie Comegys, is sincerely appreciated. The elephant staff's ingenuity (especially Jay Haight and Roger Henneous) in obtaining fluids and secretions and the behavioral correlate observations of palatal pit exchanges by J. Haight are recognized. I thank WPZ directors, Mr. Warren Iliff and Mr. Gene Leo, and the Friends of Washington Park Zoo for their encouraging support and Mrs. Taylor for the manuscript preparation assistance. This research was supported in part by the Medical Research Foundation of Oregon, #8240, and NIH #1 R01 HD 192 19-01.

REFERENCES

Altmann, J., 1974, Observational study of behavior, Behaviour, 49:227.
Berg, J., 1983, Vocalization and associated behaviors of the African elephant (*Loxodonta africana*) in captivity, Z. Tierpsychol., 63:63.
Brown, G., Stokke, O., and Jellum, E., 1978, Chromatographic profile of high boiling organic acids in human urine, J. Chromatogr., 145:177.
Dagg, A., and Taub, A., 1970, Flehmen, Mammalia, 34:686.
Eisenberg, J., 1980, Ecology and behavior of the Asian elephant, Elephant Suppl., 1:36.
Eisenberg, J., 1984, Recent research on the biology of the Asiatic elephant (*Elephas m. maximus*) in Sri Lanka, Spolia Zeylanica, 35:213.
Eisenberg, J., McKay, G., and Jainudeen, M., 1971, Reproductive behavior of the Asiatic elephant (*Elephas maximus maximus*), Behaviour, 38:193.
Gadgil, M., and Vijayakumaran Nair, P., 1984, Observations on the social behavior of free ranging groups of tame Asiatic elephant (*Elephas maximus* Linn.), Proc. Indian Acad. Sci. (Anim. Sci.) , 93:225.
Hess, D., Schmidt, M., and Schmidt, A., 1983, Reproductive cycle of the Asian elephant (*Elephas maximus*) in captivity, Biol. Reprod., 28:767.
Hradecky, P., 1978, "The Relationship Between the Contents of Volatile Fatty Acids in Urine and Vaginal Secretion and the Phase of the Reproductive Cycle in the Cow," Dissertation, Veterinary University, Brno, Czechoslovakia.
Jainudeen, M., Eisenberg, J., and Tilakeratne, N., 1971, Oestrous cycle of the Asiatic elephant, *Elephas maximus*, in captivity, J. Reprod. Fert., 27:231.
Johnston, R., and Zahorik, D., 1975, Taste aversions to sexual attractants, Science, 189:893.
Johnston, R., and Rasmussen, K., 1984, Individual recognition of female hamsters by males: Role of chemical cues and of the olfactory and vomeronasal systems, Physiol. Behav., 33:95.
Kühme, W., 1962, Ethology of the African elephant (*Loxodonta africanus*) in captivity, Int. Zool. Yrb., 4:113.
Kurt, F., 1974, Remarks on the social structure and ecology of the Ceylon elephant in the Yala National Park, *in*: "The Behavior of Ungulates and Its Relation to Management," V. Geist, and F. Walther, eds., 2:618.
Markowitz, H., Schmidt, M., Nadal, L., and Squier, L., 1975, Do elephants ever forget? J. Appl. Behav. Anal., 8:333.
McKay, G., 1973, The ecology and behavior of the Asiatic elephant in southeastern Ceylon, Smithsonian Contrib. Zool., 125:1.
Meredith, M., 1983, Sensory physiology of pheromone communication, *in*: "Pheromones and Reproduction in Mammals," J. Vandenbergh, ed., Academic Press, New York.
Muller-Schwarze, D., 1979, Flehmen in the context of mammalian urine communication, *in*: "Chemical Ecology: Odour Communication in Animals," F. J. Ritter, ed., Elsevier, New York.
Olivier, R., 1978, "On the Ecology of the Asian Elephant, *Elephas maximus* Linn., with Particular Reference to Malaya and Sri Lanka," Ph.D. Dissertation, Cambridge University.
Patterson, R., 1967, Possible contribution of phenolic components to boar odour, J. Sci. Food Agric., 18:8.
Payne, K., Langbauer, W., and Thomas, E., 1985, Infrasonic calls of the Asian elephant, in press.
Pfeifer, S., 1985, Flehmen and dominance among captive adult female scimitar-horned oryx (*Oryx dammah*), J. Mammal., 66:160.

Rapaport, L., and Haight, J., 1985, Some observations regarding allo-mothering in Asian elephants (*Elephas maximus*), submitted.

Rasmussen, L., Schmidt, M., Henneous, R., Groves, D., and Daves, G. D., 1982, Asian bull elephants: Flehmen-like responses to extractable components in female elephant estrous urine, *Science*, 217:159.

Rasmussen, L., Buss, I. O., Hess, D., and Schmidt, M., 1984, Testosterone and dihydrotestosterone concentrations in elephant serum and temporal gland secretion, *Biol. Reprod.*, 30:352.

Rensch, B., 1959, Trends toward progress in brains and sense organs, *Cold Spring Harb. Symp. Quant. Biol.*, 24:291.

Still, W., Kahn, M., and Mitra, A., 1978, Rapid chromatographic technique for preparative separation with moderate resolution, *J. Org. Chem.*, 43:2923.

Turner, J., 1981, Elimination marking behavior in feral horses, *Can. J. Zool.*, 58:1561.

Verberne, G., 1976, Chemocommunication among domestic cats, mediated by the olfactory and vomeronasal senses, *Z. Tierpsychol.*, 42:113.

TERRITORIAL SCENT MARKING BY TWO SYMPATRIC LEMUR SPECIES

Anne Millhollen

Department of Psychology
Fort Hays State University
Hays, KS 67601

The sifaka, *Propithecus verreauxi verreauxi*, and the ring-tailed lemur, *Lemur catta*, are diurnal, group-living, prosimian primates that inhabit the forests of Madagascar. Both species vary their group home range size with resource density (Budnitz and Dainis, 1974; Jolly et al., 1982), and vary their degree of tolerance of home range overlap between adjacent groups (Sussman and Richard, 1974). Although they demonstrate this flexibility in their ranging behavior, it appears from this study that the placement of their scent marks is territorial.

Both species were observed in the rich, 100 ha gallery forest reserve at Berenty, Madagascar, where they are sympatric. (See Jolly, 1966, and Budnitz and Dainis, 1974, for a complete description of the reserve.) They demonstrate a degree of ecological separation in that they tend to spend the majority of their time at different levels in the forest and move along different substrates (Petter, 1962; Jolly, 1966). This study demonstrates that their scent marking reflects this partitioning of the environment thus minimizing interference between the chemical messages deposited by each species.

Sifaka groups average about five animals per group at Berenty, with a more or less even number of males and females in the bisexual groups. They are territorial at Berenty. The territories have been stable for a number of years, and borders are maintained by frequent intertroop confrontations (Jolly et al., 1982). However, in northern forests, groups have exclusive use of only about 45% of their home range. Intergroup encounters do not defend this exclusive area but instead occur in the area of overlap (Richard, 1976).

Sifaka frequently scent mark both during territorial battles and in normal daily activities (Petter, 1962; Jolly, 1966; Richard, 1974). Males throat mark by rubbing the secretions from a specialized gland in the center of their throat and upper chest onto tree trunks. Both males and females urine mark by skittering up the tree leaving a trail of urine on the surface, fecal mark by releasing their feces on angled branches, and anogenital mark by rubbing the genital and/or anal region on the tree. Because the sifaka can selectively position the anal area or the genital area on the tree, what is called an anogenital mark is potentially two different messages. Unfortunately, it is difficult to identify which mark is being deposited under field conditions. All of the marks are deposited while the sifaka is vertical, which is also the posture used in locomotion. The marks are sniffed and marked again by both intra- and intergroup members.

Ring-tailed lemur groups average about 13 animals per group at Berenty, with an even number of males and females. Their home ranges have been stable for many years. They tolerate extensive boundary overlap between adjacent groups, and intergroup confrontations occur in the area of overlap (Mertl-Millhollen et al., 1979; Jolly et al., 1982). Groups studied in a gallery forest to the northwest did not allow this degree of overlap and maintained strict territorial borders (Sussman, 1974).

Ring-tailed lemurs frequently scent mark both during territorial battles and in normal daily activities (Petter, 1962; Jolly, 1966; Sussman, 1974). Males arm mark branches by dragging the antebrachial organ and associated kertainized spur that is on their forearms over the surface of the branch. They not only deposit the scent on the branch but also actually gouge the bark because of the spur (Schilling, 1974). Often the male shoulder rubs prior to arm marking; i.e., he rubs the antebrachial organ, which contains apocrine sweat glands, over the surface of a sebaceous brachial organ located just below the shoulder, thus mixing the two secretions (Montagna and Yun, 1962; Evans and Goy, 1968) and perhaps modifying the message (Mertl, 1974). Males can discriminate between the secretions produced by different individual males (Mertl, 1975). The same secretions are used in "stink fights" (Jolly, 1966). Two males in an agonistic interaction anoint their own tails with the scent, face each other and wave their tails at each other in a display that is a fine intermeshing of visual and olfactory cues (Mertl, 1976).

Both male and female lemurs genital mark horizontal and vertical surfaces. The males appear to be depositing secretions from the sebaceous glands on the surface of the scrotum and perhaps urine and feces; the females may be depositing vaginal secretions as well as labial secretions, urine and feces (Evans and Goy, 1968). Both male and female genital marks elicit a flehmen response (Mertl, 1974), and it is likely that the vomeronasal organ is involved (Bailey, 1978; Evans, 1980). All of the above lemur scent marks are frequently superimposed by group members and are sniffed and marked by animals in adjacent groups.

METHODS

To determine where the scent marks deposited by these two species were positioned, I spent over 350 hours during the months of February through July, 1975, observing two ring-tailed lemur groups, and over 100 hours, during June and July, 1975, observing one sifaka group (Mertl, 1979; Mertl-Millhollen, in preparation). Focal animal sampling (Altmann, 1974) was used to record scent marking by two males and two females from each group. The height, angle and diameter of the object marked was recorded for each mark. The position of each mark was recorded on a scale map of the forest. Home ranges for each group included any area they visited more than once during the study. Territorial borders were determined by the positions of intergroup confrontations, and adjusted according to which group, if either, retreated.

In addition, instantaneous samples (Altmann, 1974) of the activity and height in tree of the focal animal were collected every 15 minutes for the sifaka and every 3 minutes for the more active ring-tailed lemurs. Height levels were as follows: Level 1 was the ground level; Level 2 was the shrub layer, 1-3 m off the ground; Level 3 included small trees, the lower branches of larger trees and the saplings of the larger tree species, 3-7 m off the ground; Level 4 combined the closed canopy and emergent layer, more than 7 mm off the ground (Jolly, 1966; Budnitz and Dainis, 1974; Sussman, 1974).

ECOLOGICAL SEPARATION OF SCENT MARKS

Sifaka are arboreal leaf eaters that move by vertical clinging and leaping (Jolly, 1972). In this study, they ranged at all levels of the forest, spending 60% of their time in the canopy and only 3% of their time on the ground. These results are similar to those of Jolly (1966) and Richard (1974).

The sifaka were never observed to change levels in order to deposit a scent mark. They scent marked at all levels of the forest except when they were on the ground: 13% of their 174 marks were at the shrub level, 23% at the small tree level, and 64% at the closed canopy level. Thus, the vertical placement of their marks reflects their positioning in the forest for all activities.

Ring-tailed lemurs are semi-terrestrial omnivores that move quadrupedally by running and walking on horizontal branches or on the ground (Jolly, 1972). They ranged at all levels of the forest, but spent a greater proportion of their time on the ground than did the sifaka. The observations seasonally closest to those for the sifaka were those collected during 120 hours in May and June. During that period, the lemurs were observed to spend 50% of their time in the canopy level and 12% of their time on the ground. These data are comparable to those gathered by Jolly (1966) but show much less time on the ground than found by Sussman (1974) during November observations. That is likely to reflect a seasonal difference; lemurs spend more time at lower forest levels when temperatures are higher (Budnitz and Dainis, 1974).

The lemurs marked at all levels of the forest and only twice changed levels in order to mark: 16% of their 335 marks were at the ground level, 22% at the shrub level, 23% at the small tree level, and 39% in the closed canopy and emergent levels. They did somewhat more marking on ground level than would be expected from total time spent there; however, much marking occurs when they are moving from one place to another, and 26% of the activity samples that involved movement were at ground level.

Although the sifaka marked far more frequently at the canopy level and never at the ground level, the vertical distribution of marks by the two species overlapped extensively. They are flexible in their vertical use of space despite the anatomical and behavioral differences between them (Jolly, 1966). However, the ecological separation of their scent marks is dramatically increased when the difference in their choice of size of object is considered.

The size and angle of the objects marked by sifaka reflected their anatomy and preferred substrate for locomotion. The typical marking object was a vertical trunk. The average diameter of marked objects, for the 168 marks for which this was recorded, was 16.5 cm (median 15.5 cm, range 5 cm to 40 cm). All of the 135 marks for which angle was recorded were vertical.

The preferred size of object marked by the ring-tailed lemurs was a slender vertical branch or very young sapling. Of the 132 marks for which size of branch was recorded, the average diameter was 3.2 cm (median 3 cm, range .5 cm to 20 cm). Of the 315 marks for which angle was recorded, 81% were vertical and an additional 12% were at about a 45 degree angle. Thus both species marked vertical branches, but the ring-tailed lemurs marked much smaller branches than did the sifaka. These data are comparable to those of Schilling (1974). He observed an average diameter of 2.5 cm for the 100 visibly marked branches that he measured at ground level.

Petter (1962) suggested that the postural and locomotor differences between these two species causes them to use different preferred substrates. This difference is also reflected in their scent marking. There is some vertical partitioning of the forest as described above. However, the sifaka preference for moving between and also scent marking the trunks of small trees clearly separates their scent marks from those which the quadrupedal ring-tailed lemurs deposit on small saplings on the ground level or slender branches attached to the large horizontal branches on which they move. Although there is some overlap in the size of object they scent mark, and although they were observed to travel the same routes on occasion, they were only once each observed to orient to a possible marking spot by other species. Thus they maintained separate frequency channels within the forest and probably had little interspecific interference in receiving their messages.

DEMARCATION OF TERRITORIES

The sifaka group observed in this study had two males, two females and one juvenile. The group inhabited a home range of 2.5 ha adjacent to the river (Mertl-Millhollen, 1979). Eleven agonistic intergroup interactions occurred in a narrow band around the periphery. Thus home range boundaries coincided with territorial borders.

A total of 580 scent marks were recorded; 29% were deposited during the intergroup interactions. Of the remaining 409 marks, 75% were deposited in 38% of the territory, in what might be termed the battle zone--a narrow band, the width of one tree (approximately 18 m), at the perimeter of the territory. A small area of intergroup overlap in one corner accounted for 65% of the remaining marks.

In addition to depositing significantly more marks in this battle zone, the sifaka spent 54% of their time there, often seeming to patrol the area. More importantly, they marked more frequently per hour spent in the area than in the interior.

There were two ring-tailed lemur groups observed in this study, adjacent to each other occupying the same area as the sifaka group (Mertl-Millhollen, 1984, in preparation). One group consisted of 21 animals in a home range of 9.1 ha; the other group of 17 animals occupied a home range of 7.9 ha. Only 32% of their home ranges were reserved for their exclusive use. Twenty-four battles were observed. These occurred in a battle zone within their home ranges that did not demarcate the core area of exclusive use; i.e., adjacent groups entered still deeper within a group's home range. However, if an adjacent group was within this defended area, they retreated quickly when the territorial owners appeared.

A total of 1,018 scent marks were recorded. Twenty-three percent of these were deposited during battles and 353 deposited during the mating season. Of the remaining 426 scent marks, 87% of them were in the area of overlap. However, the marks were not evenly distributed. They were very sparse at the periphery of the home ranges.

Significantly more male arm marks, the often accompanying male shoulder rubs, and female genital marks were deposited in the battle zone area than in either the peripheral areas of the home range or the interior which included their core areas of exclusive use. For this analysis, the battle zone was considered an area 18 m to the inside and 18 m to the outside of a line connecting the centers of the sites where battles were observed. This is the approximate width of two trees and reflects the spread of the groups involved in a battle. If no battles were observed in an area, then the zone was positioned at the home range boundary.

Although the battle zone was only 36% of their home range areas, 76% of their marks were deposited within it. They spent 65% of their time in the zone and thus appeared to be patrolling it. They also showed a tendency to mark at a higher rate while there, but not significantly so.

The ring-tailed lemurs and the sifaka both demarcated territioral borders, the battle zones, even though their tolerance for intergroup home range boundary overlap was entirely different. The lemur pattern of marking was very difficult to observe directly in the field because of the complexity of home range overlap. There were no physical discontinuities at these borders that were obvious to people.

In an experiment that exposed a group of lemurs to marked and unmarked branches, they not only sniffed and marked the marked branch more often, but also showed a clear memory of the branch's position. If it was moved 1.5 m, they responded to it as if it were a new branch (Mertl-Millhollen, 1977). Adjacent groups were frequently observed to approach, sniff and mark a branch that had been marked earlier in the day by the adjacent group. Thus they show a clear recognition of the branches.

It is assumed that home range size and the degree of overlap between adjacent groups are modifiable traits that respond to such ecological constraints as resource density (Budnitz and Dainis, 1974; Sussman and Richard, 1974). It is possible, however, that scent marking by these two species of lemur is not modifiable. If that is the case, then one can predict a comparable pattern of deposition for the scent marks in the other areas of Madagascar where the sifaka permit more home range overlap and where the ring-tailed lemurs maintain exclusive territories. According to Gosling's economic theory (1981 and this volume), the distance from the center of the home range to the marked battle zone would be determined by the ability of the group to patrol and mark it regularly. It may therefore be positioned at or within the home range boundary.

ACKNOWLEDGMENTS

This research was supported by Grant No. 3018 from the Wenner-Gren Foundation for Anthropological Research and by a generous personal loan from Drs. Martha and Allen Neuringer. Thanks go to the Malagasy Republic for permission to do research and to the DeHeaulme family for the use of the reserve and guesthouse. Thanks also go to Dr. Peter Klopfer and the Duke University Primate Facility for help throughout.

REFERENCES

Altmann, J., 1974, Observational study of behavior: Sampling methods, Behav., 49:227.

Bailey, K., 1978, Flehmen in the ring-tailed lemur (L. catta), Behav., 65:309.

Budnitz, N., and Dainis, K., 1974, Lemur catta: Ecology and behavior, in: "Lemur Biology," R. Sussman, and I. Tattersall, eds., Plenum Press, New York.

Evans, C. S., 1980, Diosmic responses to scent-signals in Lemur catta, in: "Chemical Signals," D. Muller-Schwarze, and R. M. Silverstein, eds., Plenum Publ. Corp., New York.

Evans, C. S., and Goy, R. W., 1968, Social behaviour and reproductive cycles in captive ring-tailed lemurs (Lemur catta), J. Zool., Lond., 156:181.

Gosling, L. M., 1981, Demarcation in a gerenuk territory: An economic approach, Z. Tierpsychol., 56:305.

Jolly, A., 1972, "The Evolution of Primate Behavior," The Macmillan Co., New York.

Jolly, A., 1966, "Lemur Behavior," The University of Chicago Press, Chicago.

Jolly, A., Gustafson, H., Oliver, W. L. R., and O'Connor, S. M., 1982, Propithecus verreauxi population and ranging at Berenty, Madagascar, 1975 and 1980, Folia Primatol., 39:124.

Jolly, A., Oliver, W. L. R., and O'Connor, S. M., 1982, Population and troop ranges of Lemur catta and Lemur fulvus at Berenty, Madagascar: 1980 census, Folia Primatol., 39:115.

Mertl, A. S., 1974, "Olfactory Communication in Lemur catta," Ph.D. dissertation, The University of Oregon, Eugene.

Mertl, A. S., 1975, Discrimination of individuals by scent in a primate, Behav. Biol., 14:505.

Mertl, A. S., 1976, Olfactory and visual cues in behavioral interactions, Folia Primatol., 26:151.

Mertl, A. S., 1977, Habituation to territorial scent marks in the field by Lemur catta, Behav. Biol., 21:500.

Mertl-Millhollen, A. S., 1979, Olfactory demarcation of territorial boundaries by a primate--Propithecus verreauxi, Folia primatol., 32:35.

Mertl-Millhollen, A. S., 1984, Olfactory demarcation of territorial but not home range boundaries by ring-tailed lemurs, Amer. Zool., 24:103A.

Mertl-Millhollen, A. S., Gustafson, H. L., Budnitz, N., Dainis, K., and Jolly, A., 1979, Population and territory stability of the Lemur catta at Berenty, Madagascar, Folia Primatol., 31:106.

Montagna, W., and Yun, S. S., 1962, The skin of primates. X. The skin of the ring-tailed lemur (Lemur catta), Amer. J. Phys. Anthropol., 20:95.

Petter, J. J., 1962, Recherches sur l'écologie et l'éthologie des Lémuriens malgaches, Mém. du Mus. Nat. de l'Hist. Natur. Sér. A, 27:1.

Richard, A., 1974, Intra-specific variation in the social organization and ecology of Propithecus verreauxi, Folia Primatol., 22:178.

Schilling, A., 1974, A study of marking behaviour in Lemur catta, in: "Prosimian Biology," R. D. Martin, G. A. Doyle, and A. C. Walker, eds., University of Pittsburgh Press, Pittsburgh.

Sussman, R. W., 1974, Ecological distinctions in sympatric species of Lemur, in: "Prosimian Biology," R. D. Martin, G. A. Doyle, and A. C. Walker, eds., University of Pittsburgh Press, Pittsburgh.

Sussman, R. W., and Richard, A., 1974, The role of aggression among diurnal prosimians, in: "Primate Aggression, Territoriality, and Xenophobia," R. L. Holloway, ed., Academic Press, Inc., New York.

CHEMICAL SIGNALS IN CALLITRICHID MONKEYS - A COMPARATIVE REVIEW

G. Epple, A. M. Belcher, and A. B. Smith III

German Primate Center, Göttingen, FRG, Monell Chemical Senses Center, Philadelphia, and Department of Chemistry, University of Pennsylvania, Philadelphia, USA

INTRODUCTION

Research on chemical communication in mammals has expanded rapidly, since the classical studies by Bruce (1959) and Whitten (1956) directed attention to this field about 25 years ago. However, most of our knowledge on how chemical signals influence behavioral and endocrine events, their chemical nature and the sensory processes involved in mediating them, has been gained from the study of laboratory rodents. Research on primates is still in its infancy, partly because most primates are exotic and expensive laboratory animals, and partly because the complexity of primate behavior greatly complicates experimental studies. However, there are several groups of primates which provide clear morphological and behavioral evidence that chemical communication is of great importance in their social and sexual behaviors. One of these is the family Callitrichidae, the South American marmosets and tamarins.

Callitrichids are distributed through the neotropics from about 9°N (Panama) to about 24°S (Brazil and Bolivia) (Hershkovitz, 1977). According to Hershkovitz (1977), whose taxonomy the present paper follows, the family includes 4 genera and numerous species and subspecies. Only a few species have ever been studied in the laboratory or in the wild. Table 1 lists the species on which behavioral information concerning the use of chemical signals is available. Species of *Callithrix* and *Cebuella* are generally referred to as marmosets, species of *Saguinus* and *Leontopithecus*, as tamarins. Differences in the morphology of the lower incisors differentiate marmosets and tamarins (Hershkovitz, 1977).

The communication systems of marmosets and tamarins rely heavily on chemical signals. Of all monkeys and apes, callitrichids appear to have the most elaborately developed specialized epidermal scent glands and scent marking behaviors, suggesting that chemical communication is of great importance in this group. This report reviews some of the observational and experimental studies on scent marking, including research on the communicatory function and chemical nature of scent marks and a discussion of the roles chemical communication play in socio-sexual behavior.

Communication in social species must be discussed within the framework of the social organization of the species. Therefore, a brief account of the social system of the Callitrichidae is provided. Many

Table 1. Family Callitrichidae: Selected Species and Subspecies and their Distribution According to Hershkovitz (1977).

Genus[a]	Species and Subspecies	Common Name[b]	Distribution
Cebuella	pygmaea	pygmy marmoset	Upper Amazonia (Brazil, Columbia, Ecuador, Peru)
Callithrix	jacchus	tufted-ear marmosets	Brazil
	j. jacchus	common marmoset	Eastern Brazil
	j. penicillata	black-tufted-ear marmoset	East Central Brazil
Callithrix	argentata	bare-ear marmosets	Central Brazil, Eastern Bolivia, Paraguay
	a. argentata	silvery marmoset	Central Brazil
	a. melanura	black-tailed marmoset	Central Brazil, Bolivia, Paraguay
Callithrix	humeralifer	tassel-ear marmosets	Central Brazil
	h. intermidius	tassel-ear marmoset	Central Brazil
Saguinus	fuscicollis	saddle-back tamarins	Upper Amazonia (Brazil, Columbia, Ecuador, Peru Bolivia)
	f. fuscicollis	Lesson's saddle-back tamarin	Southeastern Columbia, Western Brazil
	f. illigeri	Illiger's saddle-back tamarin	Eastern Peru
Saguinus	labiatus	red-chested moustached tamarin	Middle Amazonia (Brazil, Bolivia)
	l. labiatus	red-chested tamarin	Western Brazil, Bolivia
Saguinus	mystax	black-chested moustached tamarin	Western Brazil, Eastern Peru
	m. mystax	moustached tamarin	Western Brazil, Eastern Peru
Saguinus	midas	Midas tamarins	Guyana, Suriname, French Guyana, Eastern Brazil
	m. midas	golden-handed tamarin	Guyana, Suriname, French Guyana, Eastern Brazil
	m. niger	black-handed tamarin	Eastern Brazil
Saguinus	leucopus	white-footed tamarin	Northern Columbia
Saguinus	oedipus	crested tamarins	Columbia, Panama, Costa Rica
	o. geoffroyi	Geoffroy's tamarin	Panama, Southeastern Costa Rica, Northwestern Columbia
	o. oedipus	cotton top tamarin	Northwestern Columbia
Leontopithecus	rosalia	lion tamarins	
	r. rosalia	golden lion tamarin	Extreme Eastern Brazil

[a]In this report, the four genera shall be abbreviated by the use of the first letter.

[b]Most common names are according to Hershkovitz (1977).

species are thought to be monogamous and to live in family groups (Epple, 1975; Kleiman, 1977a), although recent field studies have suggested that this is not the case in all species (Sussman and Kinzey, 1984). Under laboratory conditions and in the wild, each group contains a single breeding female, even if more than one adult female is present (Epple, 1967; field studies reviewed by Sussman and Kinzey, 1984). Laboratory studies have shown that most nonbreeding females do not show ovarian cyclicity. When a noncycling female is removed from the group and permanently paired with a male, cycling commences within a few days (Abbott, 1984; Epple and Katz, 1984; Evans and Hodges, 1984; French et al., 1984). If the breeding female is removed from a group, one of the other females will assume the breeding position (Epple, 1967). These findings have suggested that the breeding female suppresses ovarian cyclicity in her group mates (Abbott, 1984; Epple and Katz, 1984; Evans and Hodges, 1984; French et al., 1984). Physiological suppression of fertility among males has not been noted. In some species, however, suppression of sexual behavior has been observed with only one male regularly mating with the breeding female (Abbott, 1984; Epple, 1975; Kleiman, 1977b). Nonbreeding females, and particularly male group members, help the breeding female in the care of her offspring, usually twins.

SOURCES OF CHEMICAL SIGNALS AND CHEMOSIGNALING BEHAVIOR

All species of marmosets and tamarins studied to date possess scent glands. These macroscopically differentiated glandular pads are located in two areas, the gular-sternal region of the mid chest and the area surrounding the genitalia (scrotum, labia pudendi, suprapubic and perineal regions). In _Callithrix_ and _Cebuella_, morphologically differentiated scent glands are located on the external genitalia and the perineal skin (Epple, 1967; Christen 1974; Perkins, 1968, 1969b; Starck, 1969; Sutcliffe and Poole, 1978; Wislocki, 1930). In the present paper, these will be referred to as circumgenital glands. In many tamarins of the genus _Saguinus_ and in _L. r. rosalia_, morphologically specialized glandular skin covers the external genitalia (circumgenital gland), and a thick glandular pad, the suprapubic gland, extends rostrally across the symphysis pubis (Epple, 1967; Perkins, 1966, 1969a; Omedes et al. 1980; Wislocki, 1930). The complex of both glands will be referred to as circumgenital-suprapubic glands. Sternal glands, located above the manubrium sterni, are present in most species studied (Epple and Lorenz, 1967). In addition to specialized scent glands, microscopic concentrations of apocrine and/or sebaceous glands occur in other areas of the body such as the suprapubic, perineal and circumanal regions, in the axilla or on the muzzle. They do not form glandular fields which differ in appearance from the surrounding skin (Perkins, 1966, 1968, 1969a,b).

Histological studies of the scent glands were performed in a small number of species (_C. pygmaea_, _C. j. jacchus_, _C. j. penicillata_, _C. a. argentata_, _S. fuscicollis_, _S. o. oedipus_, _S. o. geoffroyi_). They have shown that glands located in the circumgenital and suprapubic areas generally consist of dense accumulations of very large sebaceous units and of apocrine elements. Apocrine glands predominate in the sternal area (Christen, 1974; Perkins, 1966, 1968, 1969a, b; Starck, 1969, Sutcliffe and Poole, 1978; Wislocki, 1930). It must be stressed that there are differences among species in the histological and histochemical characteristics as well as in size and morphology of the glands. In the common marmoset, for example, large sebaceous and apocrine glands are located mainly in the skin of the external genitalia, while the suprapubic area does not show morphologically differentiated glandular fields. There is no sexual dimorphism in the size of the glands (Sutcliffe and Poole, 1978). In the saddle-back tamarin, on the other hand, sebaceous and apocrine glands are found in the skin of the external genitalia and extend as a thick, deeply pigmented pad across the suprapubic area. The pad is somewhat larger in females than in males (Epple, 1980a; Perkins, 1966). In cotton top tamarins, which also possess a suprapubic gland, sex differences in gland size and gland histology are pronounced. Moreover, differences have been found between the suprapubic and genital areas of the pad in the size, glycogen content and phosphorylase activity of the sebaceous glands (Perkins, 1969a). Such differences suggest different functions for each area, a notion supported by behavioral studies on scent marking (French and Cleveland, 1984; French and Snowdon, 1981).

Systematic histological studies on the ontogenetic development of the scent glands are lacking. In infants of C. j. jacchus and S. fuscicollis no microscopic glandular specializations can be found (Epple, unpublished; Sutcliffe and Poole, 1978). In S. fuscicollis, S. l. labiatus, and C. j. jacchus, macroscopically recognizable circumgenital and suprapubic scent glands begin to develop between five and seven months of age. As the animals mature, their glands increase in size. In S. fuscicollis, they reach their full development between the ages of 18-24 months or even later (Epple, 1980), in C. j. jacchus, at the age of 12-15 months (Sutcliffe and Poole, 1978). The development of the sternal gland has been studied only in C. j. jacchus where most animals develop a small, diffuse gland at the age of 12-15 months. Fully developed, discrete glands, however, were seen only in established breeding males (Sutcliffe and Poole, 1978).

In S. fuscicollis, the morphogenetic development of the circumgenital-suprapubic glands is partially controlled by gonadal hormones. Gonadectomy prior to puberty retards their development in males and females, but does not inhibit it completely. Moreover, the sexual dimorphism in gland size is not totally dependent on gonadal hormones, since females still develop somewhat larger glands than males (Epple, 1981, 1982).

In addition to developmental processes, the social environment may influence gland size. French et al. (1984) have suggested that in cotton top tamarins, females develop more active circumgenital-suprapubic glands after being removed from their families and paired with adult males. Very similar observations were made in female S. fuscicollis (Epple, unpublished). In C. j. jacchus, there is good evidence that the full development of the sternal glands in males depends upon the individual maintaining breeding rank in its group (Epple and Lorenz, 1967; Sutcliffe and Poole, 1978).

Marmosets and tamarins possess a rich repertoire of chemosignaling behaviors. These patterns result in the application of a number of secretions and excretions to items in the environment, to the bodies of conspecifics and to their own bodies. Chemosignaling is associated closely with investigatory behaviors. Individuals of all species sniff and lick each other's bodies and scent glands in a number of sexual and social situations. Sniffing and occasional licking of scent marks deposited into the environment is also common (Christen, 1974; Epple, 1967; Heymann, 1985; Hoage, 1982; Kleiman, 1977b; Omedes, 1981; Omedes and Caroll, 1982; Stevenson and Poole, 1976; Sutcliffe and Poole, 1978).

The major sources of chemical signals appear to be the secretions of the scent glands in the circumgenital-suprapubic and sternal areas as well as urine. However, vaginal secretion, saliva, nasal secretion and possibly secretion from microscopic accumulations of skin glands also have been implicated as scent sources. In all species, the most conspicuous and most frequently seen chemosignaling behaviors are stereotyped scent marking patterns involving the circumgenital, the suprapubic and the sternal glands.

Three basic marking patterns involve the use of these glands in males and females of many species. In addition, rubbing of the entire ventral surface and rubbing of the muzzle occur quite frequently, but do not involve single, specialized scent glands. Scent marking and rubbing patterns have been given different names in the literature. For the purpose of this report they are defined as follows.

Circumgenital Marking, identical with sit rubbing, anogenital marking and anal marking (Epple, 1961; Mack and Kleiman, 1978; Moynihan, 1970; Stevenson and Poole, 1976; Sutcliffe and Poole, 1978). All species perform this lowest-intensity marking by sitting, pressing the circumgenital and circumanal areas against the substrate, and either rubbing back and forth, from side to side or in a circular motion. The monkeys apply secretions from the circumgenital glands and from the circumanal skin to the substrate, mixing them with a small amount of urine, which is discharged dropwise. In females, vaginal discharge may also be applied (Epple, 1975; French and Snowdon, 1981; Sutcliffe and Poole, 1978). Circumgenital marking is the most frequent of all marking patterns in *Cebuella*, all species of *Callithrix* and in *Saguinus*.

Suprapubic Marking, identical with pull rubbing (Moynihan, 1970). Secretions from the suprapubic glands are applied when the animal presses the suprapubic pad to the substrate, either pulling itself forward with its hands, legs dangling to either side, or pushing the body with the feet. *S. fuscicollis* also rubs the suprapubic gland against small protuberances while in a sitting position. The motor pattern is adaptable to the characteristics of the substrate. At least some species discharge urine during suprapubic marking (see Table 2). In *S. fuscicollis* (Epple, 1975) and *L. r. rosalia* (Kleiman and Mack, 1980), sit rubbing and pull rubbing are often combined into one behavior and therefore, have been recorded together as circumgenital-suprapubic marking. *S. fuscicollis* may alternate between the circumgenital and suprapubic glands several times during one marking performance and it is not possible to make a clear distinction between circumgenital and suprapubic marking (Epple, unpublished). In *S. o. oedipus* they appear to be distinct behaviors with different functions (French and Snowdon, 1981).

Sternal Marking involves rubbing of the sternal gland against items of the environment (Epple and Lorenz, 1967). Sternal marking is a variable behavioral pattern. A monkey may rub the chest against the substrate while lying flat on its stomach or press its sternal area to the item being marked while stomach and rear end are elevated. If the item to be marked is located above the surface supporting the monkey, the animal reaches it by standing on its hind legs. Sternal marking is quite rare in all species except *L. r. rosalia* where it is the most frequent marking pattern of adults (Epple and Lorenz, 1967; Kleiman and Mack, 1980; Sutcliffe and Poole, 1978 and others).

Ventral Rubbing does not involve a single glandular complex, but appears to combine the use of the circumgenital, suprapubic and sternal glands. The animal rubs the entire ventral surface over the substrate, probably mixing the secretions from different glandular areas, perhaps with the addition of secretions from glands in the unspecialized abdominal skin. This may also impregnate the ventral area with the mixture.

Muzzle Rubbing by means of lateral movements, sometimes accompanied by sneezing, appears to be a relatively common pattern among callitrichid species (Table 2). It occurs after eating or drinking when its primary function seems to be that of cleaning. However, it is also frequent prior to, during or after circumgenital and suprapubic scent marking (Coates and Poole, 1983; Epple, unpublished; French and Cleveland, 1984; Heymann, 1985;

Table 2. Occurrence and sexual dimorphism of scent marking patterns in callitrichids. + = the behavior is present; - = the behavior is absent; no symbol: data not available; u, v = a small amount of urine (u) or vaginal discharge (v) is added to the mark; ♂ > ♀ = males mark significantly more frequently than females; ♀ > ♂ = females mark significantly more frequently than males; (WQD) statement based on observations without quantitative data.

Taxon	Circumgenital Marking	Suprapubic Marking	Sternal Marking	Ventral Marking	Muzzle Rubbing	References
Cebuella pygmaea	+, u ♂ = ♀ (WQD)		+	+		Christen, 1974
Callithrix j. jacchus	+, u, v ♂ = ♀	-	+ ♂ > ♀	+	+	Box, 1977; Epple, 1976; Epple and Lorenz, 1967; Kendrick and Dixson, 1984; Stevenson and Poole, 1976; Sutcliffe and Poole, 1978
humeralifer intermedius	+, u		+	+		Rylands, 1981
a. argentata	+, u	-	+		+	Epple, 1967; Omedes, 1981
argentata melanura	+ ♀ > ♂ (WQD)	-	+			Omedes, 1981; Stevenson, 1978
Saguinus fuscicollis	+, u, v ♀ > ♂	+, u ♀ > ♂ (WQD)	+ ♂ = ♀	+	–	Epple, 1972; Epple, 1980; Heymann, 1985
midas	+, u ♂ > ♀ (WQD)	+	+			Christen, 1974; Epple, unpub.; Omedes and Carroll, 1982
o. oedipus	+, u, v. ♀ > ♂	+ ♀ > ♂	-	+	+ ♂ = ♀	Epple, 1967; French and Snowdon, 1981; French and Cleveland, 1984; Wolters, 1978
o. geoffroyi	+, u	+	+	+	+	Epple, 1967; Moynihan, 1970
leucopus	+, u ♀ > ♀ (WQD)	+ ♀ > ♂	+			Epple, unpublished
m. mystax	+, u ♀ > ♂	+ ♂ > ♀	+ ♂ > ♀		+ ♂ = ♀	Box and Morris, 1980; Heymann, 1985
l. labiatus	+, u ♀ > ♂	+, u	+		+	Coates and Poole, 1983
Leontopithecus r. rosalia	+, u ♂ = ♀	+, u ♂ = ♀	+, u ♀ = ♀	+		Epple, 1967; Mack and Kleiman, 1978

Moynihan, 1970; Sutcliffe and Poole, 1978). It may function as scent marking, resulting in the deposition of saliva, nasal discharge or of secretions from sebaceous glands located in the skin of the lips (Perkins, 1966, 1968, 1969a, b).

Most scent marking is performed on items in the environment. However, marking a social partner by means of the circumgenital and/or suprapubic glands has also been reported for several species (Christen, 1974; Epple, 1967; Moynihan, 1970; Omedes, 1981; Sutcliffe and Poole, 1978). Table 2 lists the occurrence of the scent marking patterns described above. All behaviors listed occur in both adult males and females. Quantitative sex differences in marking activity, which have been documented in some species, are also indicated in Table 2.

Several other behaviors which might play a role in chemical communication have been observed in a number of species. These patterns generally occur more infrequently than the ones described above and some of them appear to be subject to high individual variability. Rubbing of tree branches with the inside of the arms in a slow and distinct fashion has been reported for wild C. humeralifer intermedius (Rylands, 1981). No scent glands have been described in this area, but Perkins (1969a) showed that several callitrichids possess ulnar eminences, located on the wrist, in which large apocrine glands are associated with sinus hairs.

Self marking with glandular secretions and/or urine seems to be quite common. Cotton top tamarin females scent mark the heels of their feet by positioning one or both heels directly underneath the circumgenital gland when marking, keeping the gland from contact with the substrate. This behavior has been observed in a number of females in our colony, but does not appear to be displayed by all of them. Some females heel mark so frequently that their heels are stained a bright yellow (Epple, unpublished). Circumgenital marking of the tail, which is rolled into a tight curl underneath the glandular area, was observed in a S. o. geoffroyi female (Epple, 1967). Another self marking pattern observed in S. m. mystax males and females (Heymann, 1985), involves urinating onto the hand, which is held in the stream of urine while the animal urinates in a standing or sitting position. S. m. mystax males and females also rub their cheeks in urine voided by their sexual partner (Heymann, 1985). Urine washing of both hands and feet in a manner identical to that described for some prosimians and several cebid monkeys (Andrew and Klopman, 1974), has been observed in an adult male S. leucopus. However, this individual urine-washed very rarely, while the two other males and one female in our colony never showed this behavior (Epple, unpublished).

Behaviors presumably stimulating the flow or production of glandular secretions also occur. In situations of high arousal, S. fuscicollis may knead and scratch the circumgenital-suprapubic glands and/or the sternal area with one hand shortly before scent marking (Epple, 1975). Manual stimulation of the circumgenital-suprapubic gland in a similar manner occurs in S. o. geoffroyi (Moynihan, 1970), in female S. o. oedipus (Epple, unpublished), and in S. midas (Omedes and Carroll, 1980). In L. r. rosalia, chest scratching appears to serve a similar purpose (Mack and Kleiman, 1978). In addition to stimulating the scent glands these manipulations also transfer secretions to the hands, which could then transfer them to the substrate during locomotion.

FACTORS INFLUENCING CHEMOSIGNALING BEHAVIORS

Observational and experimental studies on chemosignaling behavior in callitrichids have concentrated on the most frequently displayed patterns;

circumgenital, suprapubic and sternal marking. The frequencies of display of these patterns are influenced by a large variety of physiological, social and environmental factors, indicating their importance in many areas of behavior. A small number of these factors are discussed here.

Maturational factors are of importance in determining scent marking behavior. However, the ontogenetic development of marking has been studied in only a few species. Circumgenital marking is shown at a very early age, well before the development of macroscopically recognizable scent glands. Suprapubic and sternal marking usually develop later when adult social roles are adopted. Genital marking may occasionally be shown as soon as infants begin to move around independently i.e., at approximately one month of age. It often follows immediately upon defecation and may represent a cleaning activity rather than chemosignaling behavior (Epple, 1967; French and Cleveland, 1984; Hoage, 1982; Moynihan, 1970). With increasing age, the pattern becomes more frequent. In S. o. oedipus, for example, recordable frequencies of circumgenital marking are seen in males and females at an average age of 2 months and increase during the first 20 weeks of life, when they reach adult levels in males. Female marking remains low and reaches high levels only when females assume the breeding position in their groups. Suprapubic marking is performed after 20 weeks of age and occurs at low frequencies in both sexes (French and Cleveland, 1984). In adult S. o. oedipus maintaining breeding status in their groups, sex differences are pronounced, with females showing both patterns of marking much more frequently than males (French and Cleveland, 1984; French and Snowdon, 1981).

In golden lion tamarins, circumgenital-suprapubic marking appears at 20 to 25 weeks. Sternal marking is the predominant marking pattern of adults at 38 weeks of age (Hoage, 1982; Kleiman and Mack, 1980). Infants and juveniles mark infrquently, usually displaying circumgenital-suprapubic marking. Moreover, the rate of development of marking appears to be influenced by family composition, i.e., the absence or presence of siblings (Hoage, 1982; Kleiman and Mack, 1980).

As juveniles mature, their marking activity may approach that of adults (Box, 1977; Epple, unpublished; French and Cleveland, 1984), suggesting that gonadal hormones are involved in determining the level of marking. This notion is only partially supported by experiments on the effects of gonadectomy on circumgenital-suprapubic marking in saddle back tamarins (Epple, 1978a, 1981, 1982). Adult males and females who had undergone gonadectomy as juveniles showed decreased marking, compared to control animals, under some social conditions, but not under others (Epple, 1981, 1982). However, social environment and increasing social experience had a powerful influence on scent marking. Several years after gonadectomy, most males and females, at that time living with intact mates, marked at the level of intact animals (Epple, 1981, 1982). Gonadectomy of experienced adults was without any effect on marking activity (Epple, 1978a). Thus, it appears that gonadal hormones are not essential for the development of scent marking behavior, but that they are one of several factors which influence its frequency of display.

In some species, scent marking activity appears to be influenced by the reproductive cycle, including pregnancy and birth. French and Snowdon (1981) report that circumgenital marking increases in female cotton top tamarins during the time of estrus. No changes were seen in female common marmosets at this time (Sutcliffe and Poole, 1978). In golden lion tamarins, on the other hand, female circumgenital-suprapubic marking decreases sharply during estrus, but increases during pregnancy. In this species, marking behavior shows another temporary decrease during a period of high sexual activity in midpregnancy. Following parturition, female marking activity drops to low levels. If infants survive, scent marking by the mother remains low for one to two weeks. It also decreases in the father at the time when be begins to participate in carrying the infants, i.e., approximately two weeks post partum. This decline in marking by both parents suggests that the behavior is inhibited in adults who carry small infants (Kleiman and Mack, 1980). Pregnancy also influences scent marking in other species. In C. j. jacchus, the female and her mate show increased circumgenital marking shortly before parturition (Box, 1977), and S. o. oedipus, as well as S. fuscicollis, females increase their marking during late pregnancy (Epple, 1975; c.f. Epple, 1975).

The social environment is another important factor which influences scent marking activity. In stable groups of marmosets and tamarins adult individuals scent mark at different levels. In families containing parents and adult offspring, this behavior may be inhibited in sons and/or daughters, but this seems not to be the case in all species. In S. o. oedipus families, the scent marking activity of adult daughters remains well below that of their mothers. Circumgenital marking in daughters increases if the mother dies or when the daughter is removed from the family and paired with a male. Sons, however, show circumgenital marking at the same levels as their fathers and do not show an increase after leaving the family (French and Cleveland, 1984; French et al., 1984; Tardif, 1983; Wolters, 1978). Conversely, in families of C. j. jacchus (Box, 1977) and S. fuscicollis (Epple, unpublished), sons and daughters who have reached breeding age may display circumgenital and suprapubic marking as frequently or even more frequently than their parents.

In contrast to family groups, differences in scent marking activity among individuals in groups of unrelated monkeys may be pronounced. This may be due to the fact that dominance interactions among unrelated animals are more frequent and more severe than among kin. Common marmoset groups containing nonrelated individuals establish a dominance order in which the breeding female and her pair-bonded male assume the α-position while all other group members, not related to the α-pair, are submissive (Abbott, 1984; Epple, 1967). Epple (1970) found that in such groups most circumgenital marking was shown by the α-male if the groups contained other nonrelated adult males. If the group contained several nonrelated adult females, however, most scent marking was shown by the α-female. In trios of golden lion tamarins consisting of two unrelated males and a female, sexual behavior was limited to one of the males who also showed more circumgenital-suprapubic and sternal marking than the other male. The sexually active males even increased their marking activity after the sexually inactive males were removed from the groups (Kleiman, 1978). In two male-one female trios of S. fuscicollis, no dominance relationship and no significant difference in circumgenital-suprapubic marking between the males was evident, but both marked significantly less frequently than males who lived only with one female (Epple and Katz, 1980). Castrated males, who displayed high marking activity when living with one female only, showed a severe but completely reversible reduction in scent marking while living in trios of two males and one female (Epple and Cerny, 1979).

Experiments on interactions between unfamiliar conspecifics in situations analogous to territorial defense have shown that dominance and aggression are strong determinants of scent marking activity in several species. During these experiments, established groups or pairs were given short social encounters during which they interacted with an unfamiliar male or female. Under these conditions, the dominant male and female in C. j. jacchus groups, and both members of S. fuscicollis pairs, directed a high amount of injurious aggression against the strange "intruder." In common marmosets, circumgenital marking activity following such encounters increased strongly as compared to trial free periods. This increase occurred only in the α-individual whose gender was the same as the intruder's (Epple, 1970). In saddle-back tamarins, both pair mates responded to a strange male or a strange female with contact aggression and with a strong increase in circumgenital-suprapubic marking. This response was not abolished by gonadectomy when the procedure was performed on experienced adults (Epple, 1978a). In cotton top tamarins, pairs showed a pronounced sexual dimorphism in their responses to strangers. Females showed an increase in suprapubic marking, but not in circumgenital marking, displaying both marking patterns more frequently than their mates. Males did not respond to strangers with increased scent marking, but were the only ones to display contact aggression (French and Snowdon, 1981).

The studies reviewed above show that dominance interactions influence scent marking. In stable groups, the existence of a pronounced rank order seems to be associated with high marking activity in dominant individuals and low marking activity in submissive ones. In groups in which no rank order among group members is obvious, relatively low marking activity may facilitate the stability of social relationships. Acute aggressive arousal is associated with a strong increase in marking in dominant females and, according to the species, males.

Another factor which is known to influence scent marking activity is the physical environment. In wild callitrichids, marking does not appear to be randomly distributed over the entire home range of a group. In S. fuscicollis, marking is concentrated in the center of the territory, where it is often associated with resting or insect foraging (Yoneda, 1984). In S. o. geoffroyi, Dawson (1979) observed marking in areas where the home ranges of groups overlap. Lindsay (1979), however, working on the same species, saw most scent marking along the tree branches used as regular trails within the territory. In the genera Callithrix and Cebuella, which obtain much of their diet by gouging holes into the bark of trees and consuming the exuding sap, circumgenital marking is concentrated at gouge holes (Coimbra-Filho and Mittermeier, 1977; Lacher et al., 1981; Rylands, 1981).

In captive callitrichids, scent marking behavior is also concentrated in specific locations. Common marmosets show a high amount of tree gouging under captive conditions, even though no sap can be obtained this way. Scent marking is concentrated at gouge holes and in other specific locations in the cage, such as nest boxes and food shelves (Box, 1977; Epple, 1967; Sutcliffe and Poole, 1978). The choice of marking locations is also influenced by social context and by gender (Box, 1977; Heymann, 1985; Mack and Kleiman, 1978; Sutcliffe and Poole, 1978). In a study on golden lion tamarins, for example, the monkeys performed a high percentage of their circumgenital, and particularly sternal, marking during long call displays. Several groups in a building regularly exchanged these calls, which reached the animal rooms through air vents. Scent marking during long calling was concentrated at the air vents, and males marked more frequently at these locations than females. The air vents possibly were regarded by the animals as a kind of territorial border. Marking of the feeding platform and the area around it in the process of foraging was a second major context in which this behavior occurred (Mack and Kleiman, 1978).

SIGNAL CONTENT AND CHEMICAL COMPOSITION OF SCENT MARKS

The information reviewed above implies that chemical signals are of great importance in social communication of callitrichids. However, the signal content of scent, the chemical composition of the various scent marks and the way in which chemical compounds encode communicatory signals are poorly understood. Histochemical studies on C. j. jacchus have indicated that the secretions of the circumgenital and sternal glands contain lipid and glycoprotein components (Sutcliffe and Poole, 1978). Hampton and Hampton (1975) isolated P.A.S. positive glycoprotein from the secretions of the circumgenital and suprapubic glands of S. o. oedipus. However, there were no attempts to correlate these findings with behavioral data. Our own research has concentrated on the questions raised above. The saddle-back tamarin serves as a model for these studies. However, we have recently included other species of Saguinus.

In an attempt to analyze the informational content of the scent marks, we have studied the responses of S. fuscicollis to circumgenital-suprapubic marks from different donor types and to urine (sternal marking is rare in this species and therefore was not studied). In most studies, we used choice tests of short duration during which tamarins were presented with two stimulus objects, each carrying material from a different donor type. Significant differences in the amount of contacting, sniffing and circumgenital-suprapubic scent marking directed at each stimulus object were interpreted as evidence that the animals could discriminate between the two types of scent. These studies documented the ability of the tamarins to discriminate between a number of scent types. In one series of tests, choices between stimulus objects scent marked by S. fuscicollis and by either C. j. jacchus or S. l. labiatus were presented. Choices between urine samples from these donors were also offered. The saddle-back tamarins displayed a significant preference for homospecific marks or urine samples over equivalent material from other species. Furthermore, when given a choice between the scent marks of two subspecies (S. f. fuscicollis and S. f. illigeri), the former subspecies preferred its own scent over that of S. f. illigeri (Epple et al., 1979).

When test animals were presented with choices between marks from males and from females, they preferred samples from males. This response was shown to marks from one or several donors and appears to be largely independent of the relative amounts of scent presented. The same response was shown when scents from several donors pooled in solvent were presented (Epple, 1973, 1974). Marks from intact males were preferred over those from castrates (Epple, 1979), but marks from castrated males were not discriminated from intact female marks (Epple, unpublished). When marks from males and females were presented under a screen, which made it impossible for the monkeys to contact the samples directly, the animals showed the same behavioral responses. This observation suggests that the male-female discrimination can be made on the basis of volatile constituents of the mark alone (Epple, 1978b). When voided urine only was presented, female subjects preferred urine from males over urine from females, but male subjects did not discriminate between the samples. The complex scent marks were highly preferred over voided urine from the same individual when a choice between them was given (Epple, 1978b).

Experiments using a habituation technique have shown that the tamarins are able to discriminate between the scent marks of two unknown individuals of the same sex (Epple et al., 1979). An additional experiment strongly suggested that the animals utilize scent to identify known individuals. Pairs of tamarins were allowed aggressive encounters with either a strange male or female. Each encounter was followed by a choice test, offering scent marks from both the recent opponent and from a familiar donor of the same sex. The tamarins preferentially investigated and scent-marked material from the recent opponent, even when they were tested several days after the encounter (Epple, 1974). Urine samples from two individuals presented under the same conditions, however, were not distinguished from each other (Epple, 1978b).

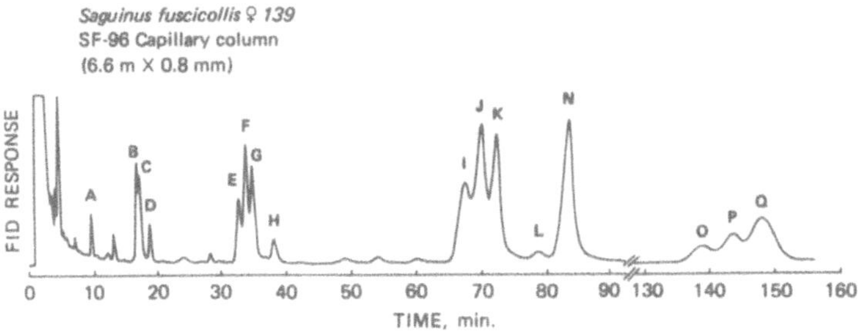

Peak	Molecular Formula	Identification
		SATURATED
A	$C_{20}H_{40}O_2$	
D	$C_{22}H_{44}O_2$	
H	$C_{24}H_{48}O_2$	
L	$C_{26}H_{52}O_2$	
		MONOENES–TYPE A
B	$C_{22}H_{42}O_2$	
F	$C_{24}H_{46}O_2$	
J	$C_{26}H_{50}O_2$	
P	$C_{28}H_{54}O_2$	
		MONOENES–TYPE B
C	$C_{22}H_{42}O_2$	
G	$C_{24}H_{46}O_2$	
K	$C_{26}H_{50}O_2$	
Q	$C_{28}H_{54}O_2$	
		DIENES
E	$C_{24}H_{44}O_2$	
I	$C_{26}H_{48}O_2$	
O	$C_{28}H_{52}O_2$	
		SQUALENE
N	$C_{30}H_{50}$	

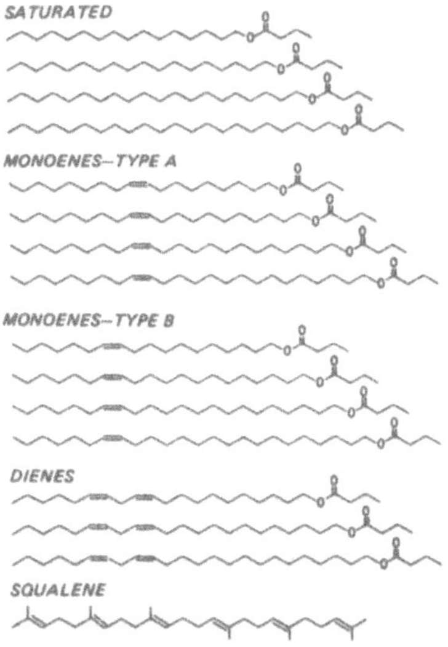

Fig. 1. The major volatile constituents of S. fuscicollis scent marks (from Smith et al., 1985, copyright April 1985 by the AAAS).

Social status appears to be encoded in the marks, either by the amount of scent deposited or by a difference in scent quality. When the monkeys received a choice between marks from unfamiliar dominant and submissive males, the scent marks from the dominant males were preferred (Epple, 1973).

The stability of the chemical signal over time also has been investigated (Epple et al., 1980). Responses to unscented stimulus objects, to objects carrying control odors and to intact scent marks of varying ages were tested. The animals showed some interest in unscented objects and in the control odors. However, they sniffed and marked objects carrying fresh conspecific scent significantly more frequently than objects carrying control scents. One day old scent marks were still investigated more intensively than controls, but the interest in the marks declined rapidly as the material aged. By three days of age, marked objects were not investigated more frequently than controls.

Our results have demonstrated that the complex circumgenital-suprapubic scent marks of S. fuscicollis are an important vehicle for communicating the identity of species, subspecies, gender, individual, social status and hormonal condition, as well as the age of the material.

Urine, which is one of the ingredients of the complex marks, carries information on species and gender, but not on individuality. Moreover, it is a much less effective stimulus of investigatory and marking behavior than the marks.

In conjunction with our behavioral studies, analytical studies were conducted to define the major volatile constituents of circumgenital-suprapubic marks. Combined gas chromatography/mass spectrometery demonstrated that these are squalene and a number of saturated and unsaturated alcohols esterified to butyric acid (Fig. 1). These components constitute 96-98% of the volatile material of the marks from males and females (Yarger et al., 1977). They are found in the scent marks as well as in secretion which can be expressed manually from the glandular pad of sedated animals. This indicates that they are secreted by the scent glands and are not part of the other ingredients of natural marks. Low concentrations of a few compounds other than the butyrates and squalene are also present in the butyrate-squalene region of the marks, and may be of importance in communication. The remaining 2-4% of the components of natural marks are of higher volatility, eluting by gas chromatography before the first butyrate ester. They represent a complex mixture, including short chain organic acids, alcohols, aldehydes, alkanes and esters. Higher concentrations of these compounds appear to be present in natural marks than in expressed glandular secretions (Belcher, unpublished). In addition, the marks contain a number of constituents which do not survive gas chromatography (Belcher, unpublished). At present, little is known about the composition of these low volatility compounds.

We have found no qualitative differences in the squalene-butyrate region between material from males and from females or between material from two subspecies, *S. f. fuscicollis* and *S. f. illigeri*. This implies that quantitative differences are important in coding information on subspecies and gender identity. To investigate this possibility, the relative concentrations of the butyrates and squalene in scent marks from males and females of two subspecies and from hybrids were determined and monitored for extended periods of time. The results show that there are differences in the relative concentrations of the butyrate esters and squalene between *S. f. fuscicollis* and *S. f. illigeri* (Fig. 2). Furthermore, they demonstrate that males of both subspecies as well as hybrids maintain a typical individual concentration profile for years (Epple et al., 1979). The profiles of females differ from those of males (Fig. 2), but maintain the same individual constancy (Belcher, unpublished). These findings imply that information on gender and on subspecies is partially encoded in the relative concentrations of these compounds.

In a recent study we have used computerized pattern recognition techniques to analyze data on the relative concentrations of squalene and most of the butyrates in a large number of scent marks from males and females of both subspecies and their hybrids. The results demonstrate that information derived solely from the butyrate-squalene concentration profiles of the marks can categorize the donors as members of one of the two subspecies and, regardless of subspecies, as males or females (Smith et al., 1985).

These mathematical differences do not necessarily provide the clues on which the animals base their discriminatory responses. To test the communicatory content of only squalene and the butyrates, scent marks from males and females were subjected to gas chromatography, and the total eluate, containing all volatile material was collected and presented to the tamarins in choice tests. The animals preferred material from males over that from females, suggesting that the information on gender had been retained. However, when the eluate was collected in two fractions, one containing only the butyrates and squalene, the other

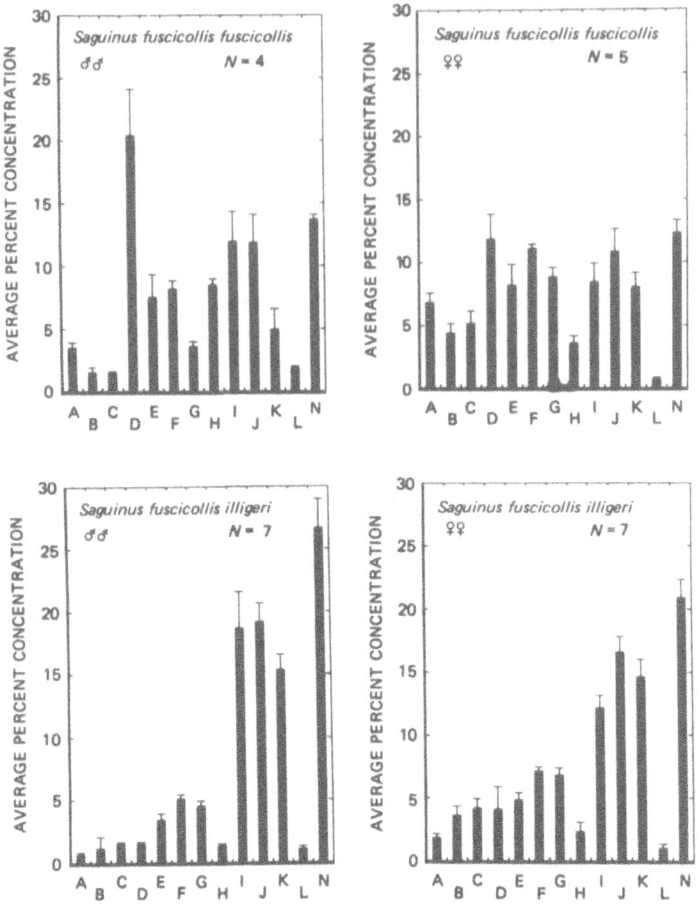

Fig. 2. Relative concentration profiles of squalene and butyrate esters in scent marks from males and females of two S. fuscicollis subspecies. (From Smith et al., 1985, copyright April 1985 by the AAAS.)

only the more highly volatile components (fixed in squalene for testing), test animals did not display a differential response (Epple et al., unpublished). Therefore, compounds from both areas of volatility are necessary to encode information on gender and possibly subspecies. From these results, we have derived the working hypothesis that specific concentration profiles of squalene and the butyrates encode part of the information relating to subspecies and gender, but that yet unknown synergists are necessary to complete it.

In addition to our studies on S. fuscicollis, we have started comparative behavioral and chemical work on other callitrichids. Gas chromatographic screening of the scent marks of several species (S. l. labiatus, S. leucopus, S. m. midas, C. j. jacchus) has suggested that the marks of these species are as complex as those of S. fuscicollis (Epple et al., 1979). There appear to be both similarities and differences among species in the volatile components of the marks. Recent studies on S. o. oedipus have shown that the circumgenital and suprapubic scent glands of this species produce squalene, cholesterol and all the butyrate esters found in S. fuscicollis. These compounds were identified by mass spectrometry in secretion expressed manually from clean glandular pads of sedated females. Squalene, cholesterol and a number of other volatile compounds as yet unidentified, can also be found in natural scent marks from the same individuals. The butyrates, on the other hand, seem to be present in such low concentrations that positive identification is not possible (Belcher et al., unpublished). Further studies are necessary to explain this finding.

CONCLUSIONS

To date, our knowledge of the social organization of wild callitrichid monkeys is very limited. It appears that these primates have a complex social system. Small, mostly territorial groups, each containing one breeding female, are the rule (Sussman and Kinzey, 1984). In captive groups, a strong pair bond between the breeding female and one of the males is evident in all species (Epple, 1975; Kleimann, 1977a, 1978 and others). If such a pair bond exists in wild marmosets and tamarins, it is likely that most offspring of the breeding female are fathered by her pair mate. The female depends on the help of her mate and other group members for the care and survival of her offspring. Group membership is somewhat flexible. Individuals may leave one group and immigrate into another, possibly changing group affiliations repeatedly. Although details on age/sex classes of dispersing individuals are unknown, it is likely that dispersal increases outbreeding and may give individuals who cannot reproduce within their natal families, a chance to find a breeding position.

It is obvious that on this social system individuals adopt different strategies, depending on their socio-sexual roles within the group. These individuals need to communicate the strategies which are necessary for the establishment and maintenance of their social roles. Considering the many sources of scent, the complexity of marking behaviors and the large number of social situations in which chemosignaling occurs, it seems likely that chemical signals mediate much of the necessary information. Our studies on _S. fuscicollis_ have shown that this species produces chemical messages which identify species, subspecies and individual, its gender and reproductive condition and may also contain information on social status and the age of the material. It is likely that other callitrichids communicate similar information. The data reviewed in this report suggest that these, and perhaps additional still unidentified chemical signals, are important in a number of behavioral contexts. Their importance in the areas of reproduction, dominance, aggression and territoriality, and in interactions with the physical environment are briefly discussed below.

A variety of reproductive functions, such as sexual behavior, ovulation, pregnancy, birth and survival of the infants, influence chemosignaling behavior in both males and females, albeit differently in different species. Events such as pair bonding, identification of the fertile phase of the ovarian cycle or of an existing pregnancy, sexual arousal and fertile mating all may be influenced by signals produced by one or both partners. For example, an increase in female scent marking during estrus (_S. o. oedipus_, French and Snowdon, 1981) may be interpreted in terms of sexual attraction to a receptive female. A decrease in female marking at the same time (_L. r. rosalia_), on the other hand, might be a means by which receptive females avoid advertising their condition to unattached males, thereby limiting sexual interaction to their bonded mate (Kleiman, 1977b). Increased scent marking by a female during late pregnancy may promote proximity to her mate as well as group cohesion, so that helpers are present at the time infants are born. Moreover, scent from a breeding female may influence the fertility of other females in the group. Epple and Katz (1984) showed that the onset of ovarian cyclicity in a S. fuscicollis female was delayed by scent from her whole family after the daughter left the family and was paired with a male. This pilot experiment, however, was performed with a single female and additional studies are needed to confirm its result.

Scent also appears to be important in mediating interactions between infants and other group members. Infant marmosets and tamarins, raised by human foster mothers, recognize the odor of their human caregivers and derive comfort from it in mildly stressful situations (Cebul et al., 1978). This suggests that, under normal social conditions, bonding between infants and caregivers is mediated by scent.

In addition to functions in the area of reproduction, there is evidence for a role of scent in aggressive behavior. In several species, scent marking behavior increases strongly in dominant individuals in connection with aggressive motivation. It appears, therefore, that high social rank and aggressive motivations are communicated either by the frequency of marking, by a specific scent signalling dominance, or both. In either case, the scents produced in aggressive marking seem to serve as olfactory threat signals. The high frequency of marking saturates the environment with the scent of an aggressive individual. Conspecifics who have experienced domination by this individual must recognize its scent and are exposed to it throughout much of their range, even in the absence of the dominant animal. In this way, the scent of the aggressor may effectively control the behavior of a submissive animal just by virtue of communicating individual identity. Should the quality of the scent be specific for a dominant animal, this effect would be further reinforced. Aggressive scent marking may function in a manner similar to other long distance threat signals. It prevents a submissive animal from challenging a dominant individual and results in avoidance of physical combat and possible injury to both. Aggressive scent marking may aid males and females to establish and maintain a high social rank within their groups, to prevent sexual competitors from challenging the pair bond between mates (Epple, 1977), and to secure their breeding privileges.

Under some ecological conditions, callitrichids are strictly territorial (Sussman and Kinzey, 1984). Therefore, scent may play a role in territorial defense and intergroup spacing (Dawson, 1979; French and Snowdon, 1981). However, scents are not necessarily a means of preventing invasion of an area. They can effectively inform conspecifics about the owners of a territory, communicating information on sex, individuality, or reproductive condition of individuals in the group. Such information can be useful to individuals who have left their own groups, either to immigrate into others or to find partners with whom to establish a new territory.

A variety of features in the physical environment of wild and captive callitrichids have a strong influence on marking activity. However, data on the functional significance of scent marking of specific locations are lacking. It is possible that under natural conditions, particularly important locations within the home range, such as regularly traveled trails or sleeping sites are scent marked. These marks might aid the whole group, or individuals which have strayed from the group, to orient themselves in a dense arboreal environment.

The marking of food resources, such as sap holes, produced by Cebuella and Callithrix species, is a frequent behavior in captivity and in the wild. Scent marking of such resources may communicate information on the nature of the resources. For example, Lacher et al., (1981) found that several groups of wild C. j. penicillata utilized the same sap holes for feeding and scent marking. Groups took turns feeding at these locations, but avoided meeting each other. The authors suggest that scent, located at sap holes, may inform other animals (intruders), about the time elapsed between uses, and therefore, about the availability of sap. It may also help to prevent the meeting of several groups at the same time, an event which would result in aggressive interactions.

In summarizing the information reviewed in this report, it is obvious that in callitrichids, chemical signals influence a large number of social and reproductive behaviors as well as the interactions of animals with their physical environment. However, our understanding of the role of chemical signals in this group of primates is still very poor and additional field and laboratory studies are needed to increase it.

ACKNOWLEDGMENTS

The long-term support of the author's research by the National Science Foundation is gratefully acknowledged. We thank Mrs. Rita Müller for the preparation of the manuscript.

REFERENCES

Abbott, D. H., 1984, Behavioral and physiological suppression of fertility in subordinate marmoset monkeys, Am. J. of Primatol., 6:169.

Andrew, R. J. and Klopman, R. B., 1974, Urine washing: Comparative notes, in: "Prosimian Biology," R. D. Martin, G. A. Doyle, and A. C. Walker, eds., Duckworth, Gloucester Crescent.

Box, H. O., 1972, Social interactions in family groups of marmosets (Callithrix jacchus), in: "The Biology and Conservation of the Callitrichidae," D. G. Kleiman, ed., Smithsonian Institution Press, Washington, D.C.

Box, H. O. and Morris, J. M., 1980, Behavioural observations on captive pairs of wild caught tamarins (Saguinus mystax), Primates, 21:53.

Bruce, H. M., 1959, An exteroceptive block to pregnancy in the mouse, Nature, 184:105.

Cebul, M. S., Alveario, M. C. and Epple, G., 1978, Odor recognition and attachment in infant marmosets, in: "Biology and Behavior of Marmosets," H. Rothe, H. J. Wolters, and J. P. Hearn, eds., Eigenverlag Rothe, Göttingen.

Christen, A., 1974, Fortpflanzungsbiologie und Verhalten bei Cebuella pygmaea und Tamarin tamarin, Fortschritte der Verhaltensforschung, Z. Tierpsychol., Suppl. 14:1.

Coates, A. and Poole, T. B., 1983, The behavior of the callitrichid monkey Saguinus labiatus labiatus, in the laboratory, Int. J. Primat., 4:339.

Coimbra-Filho, A. F. and Mittermeier, R. A., 1977, Tree-gouging, exudate-eating the "short-tusked" condition in Callithrix and Cebuella, in: "The Biology and Conservation of the Callitrichidae," D. G. Kleiman, ed., Smithsonian Institution Press, Washington, D.C.

Dawson, G. A., 1979, The use of time and space by the Panamanian tamarin, Saguinus oedipus, Folia Primat., 31:253.

Epple, G., 1967, Vergleichende Untersuchungen über Sexual- und Sozialverhalten der Krallenaffen (Hapalidae), Folia Primatol., 7:37.

Epple, G., 1974, Pheromones in primate reproduction and social behavior, in: "Reproductive Behavior," W. Montagna, and W. A. Sadler, eds., Plenum Press, New York.

Epple, G., 1975, The behavior of marmoset monkeys (Callithricidae), in: "Primate Behavior," Vol. 4, L. A. Rosenblum, ed., Academic Press, NY.

Epple, G., 1978a, Lack of effects of castration on scent marking, displays, and aggression in a South American primate (Saguinus fuscicollis), Horm. Behav., 11:139.

Epple, G., 1978b, Studies on the nature of chemical signals in scent marks and urine of Saguinus fuscicollis (Callitricidae, Primates), J. Chem. Ecol., 4:383.
Epple, G., 1979, Gonadal control of male scent in the tamarin, Saguinus fuscicollis (Callitrichidae, Primates), Chem. Sens. Flav., 4:15.
Epple, G., 1980, Relationships between aggression, scent marking and gonadal state in a primate, the tamarin Saguinus fuscicollis, in: "Chemical Signals - Vertebrates and Aquatic Invertebrates," D. Muller-Schwarze, and R. M. Silverstein, eds., Plenum Press, New York.
Epple, G., 1981, Effects of prepubertal castration on the development of the scent glands, scent marking, and aggression in the saddle back tamarin (Saguinus fuscicollis, Callitrichidae, Primates), Horm. Behav., 15:54.
Epple, G., 1982, Effects of prepubertal ovariectomy on the development of scent glands, scent marking, and aggressive behaviors of female tamarin monkeys (Saguinus fuscicollis), Horm. and Behav., 16:330.
Epple, G. and Lorenz, R., 1967, Vorkommen, Morphologie und Funktion der Sternaldrüse bei den Platyrrhini, Folia Primatol., 7:98.
Epple, G. and Cerny, V. A., 1979, Effects of castration and social change on scent-marking behavior of Saguinus fuscicollis, Folia Primatol., 32:252.
Epple, G., and Katz, Y., 1980, Social influences on first reproductive success and related behaviors in the saddle back tamarin (Saguinus fuscicollis, Callitrichidae), Int. J. Primatol., 2:171.
Epple, G. and Katz, Y., 1984, Social influences on estrogen excretion and ovarian cyclicity in saddle back tamarins (Saguinus fuscicollis), Amer. J. Primat., 6:215.
Epple, G., Golob, N. F. and Smith, A. B., III, 1979, Odor communication in the tamarin Saguinus fuscicollis. Behavioral and chemical studies, in: "Chemical Ecology: Odour Communication In Animals," F. J. Ritter, ed., Elsevier, North Holland, Amsterdam.
Epple, G., Alveario, M. C., Golob, N. F. and Smith, A. B., III, 1980, Stability and attractiveness related to age of scent marks of saddle back tamarins (Saguinus fuscicollis), J. Chem. Ecol., 6:735.
Evans, S. and Hodges, J. K., 1984, Reproductive status of adult daughters in family groups of common marmosets (Callithrix jacchus jacchus), Folia Primat., 42:127.
French, J. A. and Snowdon, C. T., 1981, Sexual dimorphism in intergroup spacing behavior in the tamarin, Saguinus oedipus, Anim. Behav., 29:822.
French, J. A. and Cleveland, J., 1984, Scent-marking in the tamarin, Saguinus oedipus: Sex differences and ontogeny, Anim. Behav., 32:615.
French, J. A., Abbott, D. H. and Snowdon, C. T., 1984, The effect of social environment on estrogen excretion, scent marking, and socio-sexual behavior in tamarins (Saguinus oedipus), Am. J. Primat., 6:155.
Hampton, J. K. and Hampton, S. H., 1975, Some expected and unexpected characteristics of reproduction in Callitricidae, Lab. Anim. Handb., 6:235.
Hershkovitz, P., 1977, "Living New World Monkeys (Platyrrhini)", The University of Chicago Press, Chicago.
Heymann, E., 1985, "Untersuchungen zur vokalen und olfaktorischen Kommunikation des Schnurrbarttamarins, Saguinus mystax mystax (Spix, 1823) (Primates, Callithrichidae)", Dissertation, Justus-Liebig-Universität, Gießen.
Hoage, R. J., 1982, Social and physical maturation in captive lion tamarins, Leontopithecus rosalia rosalia (Primates: Callitrichidae), Smithson. Contrib. Zool., 364:1.
Kendrick, K. M. and Dixson, A. F., 1984, A quantitative description of copulatory and associated behaviors of captive marmosets (Callithrix jacchus), Int. J. Primat., 5:199.

Kleiman, D. G., 1977a, Monogamy in mammals, Quarterly Rev. Biol., 52:39.

Kleiman, D. G., 1977b, Characteristics of reproduction and sociosexual interactions in pairs of lion tamarins (Leontopithecus rosalia) during the reproductive cycle, in: "The Biology and Conservation of the Callitrichidae," D. G. Kleiman, ed., Smithsonian Institution Press, Washington, D.C.

Kleiman, D. G., 1978, The Development of pair preferences in the lion tamarin (Leontopithecus rosalia): male competition or female choice? in: "The Biology and Behavior of Marmosets," H. Rothe, H. J. Wolters, and J. P. Hearn, eds., Eigenverlag Rothe, Göttingen.

Kleiman, D. G. and Mack, D. S., 1980, Effects of age, sex and reproductive status on scent marking frequencies in the golden lion tamarin, Leontopithecus rosalia, Folia Primat., 33:1.

Lacher, T. E. J., Bouchardet de Fonseca, G. A., Alves, C. Jr. and Magãlhaes-Castro, B., 1981, Exudate-eating, scent marking and territoriality in wild population of marmosets, Anim. Behav., 29:306.

Lindsay, N. B. D., 1979, A report on the field study of Geoffroy's tamarin Saguinus oedipus geoffroyi, Dodo J. Jersey Wildl. Pres. Trust, 16:27.

Mack, D. S. and Kleiman, D. G., 1978, Distribution of scent marks in different contexts in captive lion tamarins, Leontopithecus rosalia (Primates), in: "Biology and Behaviour of Marmosets," H. Rothe, H. J. Wolters, and J. P. Hearn, eds., Eigenverlag Rothe, Göttingen.

Moynihan, M., 1970, Some behavior patterns of platyrrhine monkeys. II. Saguinus geoffroyi and some other tamarins, Smithson. Contrib. Zool., 28:1.

Omedes, A., 1981, Behavioural repertoire of adult silvery marmosets Callithrix a. argentata and Callithrix a. melanura, Miscellanea Zool., 7:193.

Omedes, A. and Carroll, J. B., 1982, A comparative study of pair behaviour of four callitrichid species and the Goeldi's monkey, Dodo J. Jersey Wildl. Pres. Trust, 17:51.

Perkins, E. M., 1966, The skin of the black-collared tamarin (Tamarinus nigrocollis), Am. J. Phys. Anthrop., 25:41.

Perkins, E. M., 1968, The skin of the pygmy marmoset - Callithrix (Cebuella) pygmaea, Am. J. Phys. Anthrop., 29:349.

Perkins, E. M., 1969a, The skin of the cotton-top pinche Saguinus (Oedipomidas) oedipus, Am. J. Phys. Anthrop., 30:13.

Perkins, E. M., 1969b, The skin of the silver marmoset - Callithrix (Mico argentata), Am. J. Phys. Anthrop., 30:361.

Rylands, A. B., 1981, Preliminary observations on the marmoset Callithris humeralifer intermedius (Hershkovitz, 1977) at Dardanelos, Rio Aripuana, Mato Grosso, Primates, 22:46.

Smith, A. B., III, Belcher, A. M., Epple, G., Jurs, P. C. and Lavine, B., 1985, Computerized pattern recognition: A new technique for the analysis of chemical communication, Science, 228:175.

Stevenson, M. F. and Poole, T. B., 1976, An ethogram of the common marmoset (Callithrix jacchus jacchus): General behavioural repertoire, Anim. Behav., 24:420.

Sussman, R. W. and Kinzey, W. G., 1984, The ecological role of the Callitrichidae: A review, Am. J. Phys. Anthrop., 64:419.

Sutcliffe, A. G. and Poole, T. B., 1978, Scent marking and associated behaviour in captive common marmosets (Callithrix jacchus jacchus) with a description of the histology of scent glands, J. Zool., 185:41.

Tardif, S. D., 1984, Social influences on sexual maturation of female Saguinus oedipus oedipus, Am. J. Primat., 6:199.

Whitten, W. K., 1956, Modification of the oestrous cycle of the mouse by external stimuli associated with the male, J. Endocrin., 13:399.
Wislocki, G. B., 1930, A study of scent glands in the marmosets, especially *Oedipomidas geoffroyi*, J. Mammal., 11:475.
Wolters, H. J., 1978, Some aspects of role taking behaviour in captive family groups of the cotton-top tamarin *Saguinus oedipus oedipus*, in: "Biology and Behaviour of Marmosets," H. Rothe, H. J. Wolters, and J. P. Hearn, eds., Eigenverlag Rothe, Göttingen.
Yarger, R. G., Smith, A. B., III, Preti, G. and Epple, G., 1977, The major volatile constituents of the scent mark of a South American primate, *Saguinus fuscicollis*, Callithricidae, J. Chem. Ecol., 3:145.
Yoneda, M., 1984, Ecological study of the saddle backed tamarin (*Saguinus fuscicollis*) in northern Bolivia, Primates, 25:1.

CROSS-CULTURAL STUDIES OF TASTE AND SMELL PERCEPTION

Richard L. Doty

Smell and Taste Center
School of Medicine
University of Pennsylvania
Philadelphia, PA 19104

INTRODUCTION

There have been few quantitative cross-cultural studies of taste and smell perception in humans, despite the obvious relevance of such research to scientific and commercial enterprises, and the fact that many cultures differ in their food preparation, dietary, and personal hygienic practices. The important role of the chemical senses to the species as a whole in such diverse matters as the detection of dangerous fumes and the selection of appropriate foodstuffs suggests that the basic chemosensory mechanisms responsible for the detection and evaluation of chemical signals are essentially the same among cultures. However, genetically related sensitivities to specific odorants and tastants (e.g., androstenone and phenylthiocarbamide) vary in frequency from group to group, and likely account for at least some cross-cultural differences in taste and odor preferences and aversions. These genetically determined factors interact with the well-known influences of experience on taste and smell perception to produce a variety of chemosensory experiences.

For the purposes of the present review, a "cross-cultural" study is defined as one between or among human groups which differ on the basis of racial, geographic and/or historical factors, often reflecting differences in social customs, diet, language, artistic expression, and socio-political attitudes. "Culture" is viewed as a general and dynamic aggregate and rarely reflects mutually exclusive genetic or experiential entities.

OLFACTION

Cross-cultural Studies of Olfactory Sensitivity and Odor Preferences

A number of anthropologists evaluated the olfactory sensitivity and preferences of natives from primitive tribes in the late 19th and early 20th centuries. Unfortunately, few subjects were tested in these methodologically flawed studies. Nonetheless, they are worthy of brief mention here because of their historical interest (for additional details, see McCartney, 1968).

Lombroso and Carrara (1896/1897) presented a group of Dinkas of the Sudan (number not reported) dilutions of the oil of cloves ranging from

1:200 to 1:50,000. These authors reported that recognition did not occur for concentrations lower than 1:2000, and that three Dinkas were unable to recognize the material even at the highest concentration. No controls were tested, and Lombroso and Carrara indicated that language problems may have influenced the accuracy of their findings.

Six years later Myers (1903) used aqueous solutions of camphor to evaluate the olfactory sensitivity of a group of Murray Islanders. In addition, these individuals were asked to describe the odors of perfumes and other scents, and to indicate whether they liked or disliked them. The average olfactory acuity of the islanders was reported as being slightly higher than that of Scottish control subjects, and their likes and dislikes were noted as being quite similar (e.g., they disliked asafoetida, valeric acid, and civet, and liked musk, thyme and violet).

Rivers (1904/1905) also used a dilution series of camphor to test the olfactory sensitivity of the Todas of the Nilgiri Hills in southern India. Approximately 50 Toda men were tested, along with a few Toda women and Englishmen who were observing the activities. For a number of reasons, Rivers was unable to provide an exact comparison between the Todas and the English, but concluded that they were not superior to them and possibly inferior.

Using a Zwaardemaker olfactometer, Grijns (1906) compared the sensitivity of a small group of Javanese subjects to those of some Europeans. He concluded that the Javanese were about twice as sensitive to the three test materials -- acetic acid, ammonia and phenol. Unfortunately, these materials are poor choices for olfactory work, as at least two of them have strong trigeminal stimulative properties (acetic acid and ammonia; c.f., Doty et al., 1978) and one of them is a deadly poison (phenol). Thus, conclusions based upon them are highly suspect. A similar criticism can be leveled at Koster's (1921) study of the smell ability of the natives of the Netherlands East Indies, in which the same conclusion of greater sensitivity in the natives (compared to Dutchmen) was reached using acetic acid and ammonia.

Given such methodologic problems, no consensus can be drawn from these early cross-cultural studies. However, two modern studies (Schleidt et al., 1981; Davis and Pangborn, 1985) shed light on whether at least some cultural differences exist in olfactory perception.

In the first of these contemporary studies, Schleidt et al. (1981) sought to determine whether German, Japanese, and Italian subjects differed in their abilities to distinguish between (a) their own axillary odors, (b) the axillary odors of opposite-sexed cohabiting partners, and (c) the axillary odors of strangers, as well as in their abilities to assign such odors to correct gender categories. In addition, hedonic responses were also obtained. Twenty-four German and 25 Italian couples participated in the study, as did seven Japanese couples and 15 male and 15 female singles. Each subject wore a cotton shirt for seven consecutive nights and used no deodorants or perfume during the study. The subjects of each group were then presented with 10 shirts and asked (1) Which shirt has your own smell?, (2) Which shirt has your partner's smell?, (3) Which shirts smell male?, (4) Which shirts smell female?, and (5) Which shirts smell pleasant, which indifferent, and which unpleasant? This task was repeated three times.

One quarter to slightly more than one third of the subjects correctly reported their own odor or that of their partner on at least two of the three trials. From 20% to 64% of the subjects distinguished male from female odors, with the females of all cultures being more accurate than the males in this regard. A larger proportion of the Japanese subjects discriminiated correctly between the sexes, although it is not clear if this was due to ethnic- (or cultural-) related sex differences in odor production, odor perception, or some combination of both. The hedonic responses to male and female odors were the same in all three cultures, with male odor being judged less pleasant (and more intense) than female odor. Unlike the German women, the Japanese and Italian women classified their partners' odors more often as unpleasant than pleasant. No marked cultural variation was noted in the responses of the males, who generally judged their partners' odors as pleasant.

In the second of these studies, Davis and Pangborn (1985) established pleasantness ratings for 22 microencapsulated odorants in 24 study groups located in 20 countries (typically 30 subjects/group). Although the final analysis of these data has not yet been performed, preliminary results indicate that the odors were generally perceived as similarly pleasant or unpleasant to all of the groups tested (all intergroup correlations were greater than 0.645, with a mean of 0.864 and a median of 0.876). In a few rare instances, the pleasantness reactions to two or three odorants were idiosyncratic for a given study group.

In summary, the question as to whether olfactory sensitivity differs among persons from various cultural groups has not yet received extensive investigation. However, the recent studies by Schleidt et al. (1981) and Davis and Pangborn (1985) suggest that individuals from different cultures rate the relative pleasantness of most odors in a similar manner, implying that more similarities than differences may exist among cultures in this regard. This appears to be true for a wide variety of odors, including body odor from the axillae.

Cross-cultural Studies of Sex Differences in Odor Identification

We recently sought to determine if the sex difference in odor identification ability observed in American Blacks and Whites (Doty et al., 1984a, b; Figure 1) was also present in American Koreans and Native Japanese subjects. In this research, the University of Pennsylvania Smell Identification Test (a 40-item standardized forced-choice test with microencapsulated odors) was administered to 438 Black Americans, 1559 White Americans, 106 Korean Americans, and 308 Native Japanese. The Black and White Americans were comprised mainly of university employees and students, participants in regional health fairs, primary and secondary school students, and youngsters enrolled at summer camps and day care centers. The Korean Americans were part of a Sunday school population at a local Korean church, whereas the Japanese subjects were members of the staff and faculty of a hospital in a medium-sized Japanese metropolitan city (Kawasaki-shi).

As reported in detail elsewhere (Doty et al., 1985), the women of all four groups outperformed their male counterparts, on the average, to the same general degree (Table 1). Thus, whatever the basis of the sex difference observed in odor identification, it appears to transcend the genetic and cultural differences among these groups.

In addition to the presence of a consistent sex difference in test scores within each of the four study groups, a significant difference was observed among the groups in terms of the absolute test scores. On the average, the Korean Americans outperformed the Black and White Americans who, in turn, outperformed the Native Japanese (Table 1). All of these means differed significantly from one another, with the exception of the Black and White American ones, which did not differ (Tukey B multiple-comparison post hoc analysis; $ps < 0.01$). Although the reason(s) for these ethnic/cultural differences are not known, differential familiarity with the items is a likely factor, since several of the odors which were not easily identified by the Japanese subjects were unfamiliar to a number of

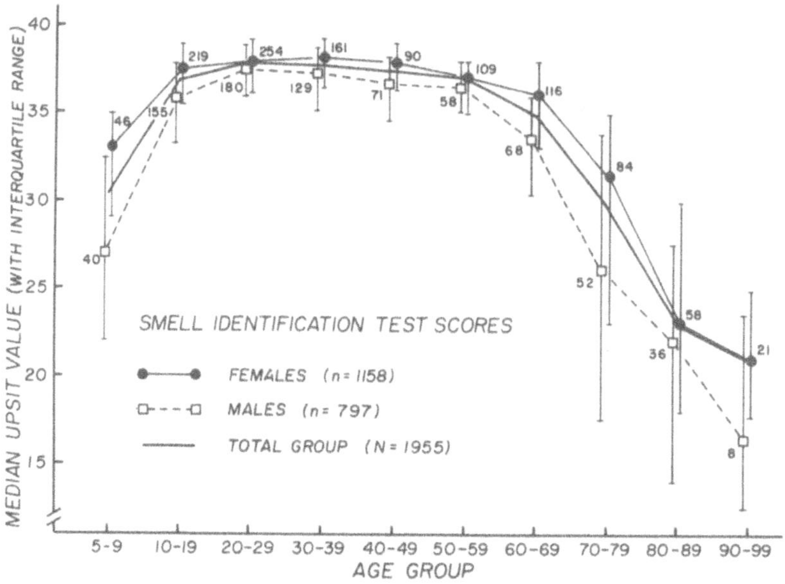

Fig. 1. Scores on the University of Pennsylvania Smell Identification Test as a function of age and gender of the subjects. Numbers by data points indicate sample sizes. From Doty et al., 1984b. Copyright 1984 by the American Association for the Advancement of Science. Reprinted by permission.

them, including "cherry," "fruit punch," "wintergreen," and "dill pickle" (see Table 2).

Despite these differences in overall test scores, the relative difficulty of the items was similar for each of the study groups. Thus, Kendall Coefficients of Concordance -- calculated for the male and female groups across the test items ranked, within each group, according to the proportion of the group correctly answering them -- were strong and statistically significant (for males, $W = 0.67$; for females, $W = 0.64$; $ps < 0.001$).

Table 1. Analysis of covariance age-adjusted means on the University of Pennsylvania Smell Identification Test for four cultural groups (covariates: age and age^2). From Doty et al. (1985).

	Males	Females	Combined
Native Japanese	29.5	31.8	31.0
Black Americans	32.4	34.0	33.2
White Americans	33.6	35.8	34.9
Korean Americans	36.6	38.0	37.3

Table 2. Percent of subjects within four ethnic/cultural groups correctly identifying items of the University of Pennsylvania Smell Identification Test (UPSIT). Note that the statistical analyses reported in the text take into account the age differences among the groups, but that the values in the table represent raw scores.

	Black Americans	White Americans	Korean Americans	Native Japanese
Banana	78.1	83.3	92.5	95.1
Bubble Gum	91.6	86.3	97.2	97.1
Cedar	63.7	76.0	90.7	74.4
Cheddar Cheese	65.8	69.9	74.8	73.4
Cherry	92.5	89.4	95.3	34.4
Chocolate	87.9	88.6	94.4	84.4
Cinnamon	85.4	82.8	98.1	92.9
Clove	71.7	81.5	69.2	55.2
Coconut	81.7	82.6	92.5	65.6
Dill Pickle	81.7	80.2	86.9	67.5
Fruit Punch	89.7	82.1	100.0	45.8
Gasoline	90.0	89.6	98.1	91.6
Gingerbread	76.3	69.1	86.0	65.9
Grape	91.8	90.4	98.1	89.6
Grass	85.8	84.0	96.3	90.6
Leather	75.6	82.3	97.2	76.9
Lemon	92.0	90.5	100.0	93.5
Licorice	87.7	90.1	98.1	73.7
Lilac	68.4	89.3	98.1	95.1
Lime	82.2	81.8	94.4	72.7
Menthol	88.6	89.4	96.3	91.2
Mint	92.9	88.5	98.1	94.8
Motor Oil	92.7	92.1	97.2	97.1
Natural Gas (Odorized)	84.0	83.6	92.5	95.5
Onion	94.1	92.4	100.0	94.5
Orange	89.3	90.2	99.1	84.4
Paint Thinner	90.4	89.7	99.1	93.2
Peach	82.0	80.6	94.4	94.5
Peanut	96.1	93.0	100.0	97.7
Pine	79.7	86.0	91.6	84.1
Pineapple	84.7	82.5	98.1	94.5
Pizza	81.5	82.2	98.1	77.6
Root Beer	90.6	86.7	99.1	76.0
Rose	85.6	83.5	93.5	81.5
Smoke	87.2	89.4	97.2	79.2
Soap	93.4	93.0	99.1	98.1
Strawberry	90.9	86.5	98.1	80.5
Terpentine	82.2	82.4	82.2	77.6
Watermelon	85.2	85.6	99.1	87.7
Wintergreen	88.4	89.5	97.2	74.4

Most of the available data on cross-cultural chemosensory differences comes from the sense of taste, rather than from the sense of smell. This is due, in part, to the traditional difficulty in presenting olfactory stimuli to subjects and to wide-spread interest in understanding the genetics of taste sensitivity, particularly to phenylthiocarbamide and related compounds. However, just as in olfaction, the data are limited and few well-controlled studies have been performed.

Cross-Cultural Studies on Taste Naming

A large number of cross-cultural studies on taste relate to how taste experiences are described in various cultures, rather than to how such cultures differ in taste perception, per se (e.g., Chamberlain, 1983; Myers, 1904; O'Mahony and Alba, 1980a, b; O'Mahony and Ishii, 1985a, b; O'Mahony and Muhiudeen, 1976, 1977; O'Mahony and Tsang, 1980). Nonetheless, as suggested by theories such as those proposed by Sapir, Whorf and others (see Kay and Kempton, 1984), language and cognition are intimately related. Clearly, words are invented to accommodate those aspects of the environment about which members of a group feel the need to communicate. Just as Eskimos and skiers classify snow into a variety of different types, perfumers, wine tasters, and taste researchers use specialized terms to describe properties of food and drink important to them in their professional activities. Thus, it is not surprising that a number of Japanese researchers and food specialists use, in addition to the traditional four basic taste categories, a fifth category termed "umami" to include monosodium glutamate and related compounds used in oriental cooking. Such a category undoubtedly reflects not only cultural experience with such chemicals, but the fact that the taste(s) of these materials are not clearly categorizable into any of the traditional taste categories. American subjects asked to sort various tastants (including monosodium glutamate) into categories of their own choosing place such materials into a separate category outside the salty one in the same way as done by Japanese subjects, suggesting they perceive them similarly (O'Mahony and Ishii, 1985; see also Schiffman and Dackis, 1975; Schiffman et al., 1975, 1980).

Cross-cultural Studies of Threshold Taste Sensitivity

Perhaps the best documented case of a genetically-determined cross-cultural difference in chemosensation is that shown for a class of bitter-tasting compounds which contain a thiocarbamide group (e.g., the thioureas, thiooxazolines, and thiothiazolines). Members of this class of goitrogenic compounds are found in edible plants of the Brassica genus, including kale, cabbage, Brussel sprouts, and turnips (Van Etten, 1969). The frequency of the "non-taster" phenotype among Caucasian populations of Western Europe and North America to phenylthiocarbamide (PTC), a member of this class of substances, is approximately 30% (Allison and Blumberg, 1959). However, this frequency is much lower among other groups that have been tested, including Black Americans (6.4%), Chinese (10.6%), and native inhabitants of highland Peru (1.9%)(Barnicot, 1959; Johnston et al., 1966; Paolucci et al., 1977). The degree of sensitivity to such substances alters, to some degree, dietary preferences. For example, Fischer et al. (1961) noted a weak, but statistically significant, negative relationship between the percentage of foods checked as disliked on a 118 item food questionnaire and taste thresholds for both quinine and 6-n-propylthiouracil (a substance closely related to PTC), but not taste thresholds for d-sucrose, sodium chloride, and hydrochloric acid. Greene (1974) presents data which suggest tasters of PTC in two Ecuadorian Andean communities in which goiter is endemic limit their ingestion of goitrogenic foodstuffs containing naturally occurring thiocarbamides.

Davis (1978) describes a group of Yucatan inhabitants who use coffee extensively and who are relatively insensitive to PTC. He argues that insensitivity to bitter substances may confer a survival advantage in that roasted coffee contains therapeutic levels of niacin, which is generally deficient in corn- or maize-based diets. Cultures with maize as the dominant dietary staple often evidence pellagra, since corn has a low tryptophan content. Interestingly, as reviewed by Katz (1982), some cultures have developed food preparation practices which help to overcome this problem. For example, a number treat maize with an alkali solution which liberates niacin from an undigestible segment and improves the amino acid quality of the digestible fraction of the food.

Cross-cultural Studies of Suprathreshold Taste Intensity and Hedonicity

Aside from differences in basal sensitivity to tastants such as PTC, differences in various suprathreshold measures of taste perception have been noted among a number of cultural groups. In one of the first such cross-cultural taste studies, Moskowitz et al. (1975) found that the pleasantness ratings given to quinine and citric acid by Indian laborers of the Karnataka region differ considerably from those given by Western and Indian medical students. Thus, the Indian laborers rated low concentrations of quinine as pleasant, whereas the medical students rated them as unpleasant. Interestingly, the Indian laborers found citric acid to be more pleasant as its concentration was increased, whereas the medical students found just the opposite. Moskowitz et al. suggest these preference differences may be the result of different dietary habits and histories.

More recently, Bertino et al. (1983) had 34 native-born Americans of European ancestry and 28 native Taiwanese residing in the United States rate the intensity (i.e., sweetness, bitterness or saltiness) and pleasantness of various concentrations of sucrose, caffeine, and sodium chloride dissolved in water. In addition, a subset of these individuals provided such ratings to cookies varying in sucrose content. In both cases, the Taiwanese assigned higher sweetness ratings to the sucrose than did the native Americans. The Taiwanese tended to rate the sucrose solutions as comparatively more pleasant, although, in the case of the cookies, this was observed only for the lower sucrose concentrations. Compared to the Americans, the Taiwanese reported the low concentrations of salt as saltier, and the high concentrations as less salty, with both high and low concentrations being rated as slightly more pleasant. In addition, there was a tendency for the Taiwanese to rate caffeine more intense, although there was no statistically significant difference between the two groups in its rated bitterness or pleasantness. A subsequent study of 30 Caucasian Americans and 30 Chinese from the People's Republic of China confirmed that Chinese subjects rate sucrose as comparatively more pleasant than American subjects (Bertino and Chan, 1985).

A number of studies of American Black and White populations suggest that they may have somewhat different taste preferences. For example, Greene, Desor and Maller (1975) found that Black children between 9 and 15 years of age evidenced a stronger preference than White children for more concentrated solutions of sucrose, lactose, and sodium chloride. Interestingly, this extensive study (which tested 311 monozygotic and like-sex dizygotic twin pairs) found heritability estimates for such preferences to be uniformly low. In a subsequent study by these same authors, no significant difference between Black and White adults was observed, suggesting that this phenomenon may disappear in adulthood (Desor et al., 1975). However, these authors noted a nonsignificant tendency for Blacks to select stronger solutions than Whites, and suggested that "Race differences in preferences for sweet and salty might be observed among adults if larger samples are tested."

Table 3. Mean hedonic and preferred frequency ratings of Black and White American military personnel for selected desserts and beverages. Modified from Meiselman (1977). Hedonic ratings based upon 9-point category scale developed by Peryam and Pilgrim (1957).

FOOD ITEM	MEAN HEDONIC RATING		MEAN PREFERRED FREQUENCY (per month)	
	Blacks	Whites	Blacks	Whites
Desserts				
Bread Pudding	5.76	5.14	8.95	5.92
Pineapple Sundae	6.37	5.83	10.49	8.12
Raisin Cookies	5.78	5.29	9.18	7.01
Raisin Pie	4.89	4.42	7.58	4.58
Sweet Potato Pie	7.36	4.53	14.60	5.23
Vanilla Wafers	6.20	5.73	10.30	7.73
Yellow Cake	5.85	5.57	8.57	6.50
Beverages				
Cherry Flavored Drink	6.36	5.55	13.40	8.72
Cherry Soda	6.38	5.71	12.87	9.26
Fruit Punch	6.84	6.03	14.69	9.69
Ginger Ale	6.48	5.93	13.06	10.30
Grape Flavored Drink	6.84	5.91	15.41	10.08
Grape Juice	6.95	6.24	16.32	12.94
Grape Lemonade	6.50	5.50	13.70	8.92
Grape Soda	6.92	5.77	16.04	9.75
Grapefruit Juice	6.41	5.66	13.90	10.54
Grapefruit Orange Juice	6.81	6.12	15.98	12.97
Grapefruit Pineapple Juice	6.46	5.57	13.96	10.05
Lemon Lime Soda	6.32	5.68	13.66	10.78
Lemonade	7.21	6.73	16.17	13.00
Lime Flavored Drink	5.91	5.28	11.06	8.09
Orange Flavored Drink	6.92	6.04	16.20	10.29
Orange Soda	7.14	6.04	16.97	11.34
Pineapple Juice	6.37	5.69	13.32	9.23
Prune Juice	4.59	3.97	7.30	4.41
Other Beverages				
Fresh Coffee	5.60	6.57	15.42	19.92
Milk	7.60	8.08	22.09	25.10
Tomato Juice	4.97	5.85	9.25	12.44

Food preference surveys suggest that such race differences are likely to be present in adults. For example, Meiselman (1977) obtained preference ratings and data of frequency of preferred use from 4,000 American military personnel for each of 378 food names. Although the subjects' region of residence or upbringing (northern or southern United States) was related to the food preferences, Blacks consistently reported stronger preferences than Whites for sweetened desserts and drinks. Mean hedonic ratings and frequency indications (how many days per month they would like to eat the food) are presented for such food items in Table 3. The greater preference by Blacks for these items contrasted with the greater preference of Whites for such drinks as milk, fresh coffee, and tomato juice. A subsequent study of 1719 Air Force personnel also found these differences, although statistical significance was present only for tomato juice in the case of the latter three items (Wyant and Meiselman, 1984).

DISCUSSION

It is apparent from this brief review that cross-cultural studies of chemosensory function are rather limited in both number and scope. Nonetheless, the available data suggest that considerable uniformity exists across cultures in the perception of tastants and odorants. Thus, the relative identifiability of odorants, as well as their perceived pleasantness, is quite similar among a number of ethnic/cultural groups. Although the Japanese use "umami" as a fifth basic taste class, American subjects similarly sort compounds of this class into a separate taste category when not constrained by verbal categories, suggesting they perceive such substances in a similar manner.

Within the framework of general perceptual similarity, however, lies evidence that cross-cultural differences are present in more subtle measures of chemosensory experience. Thus, it appears that Chinese and Black American subjects perceive several tastants as more pleasant and, in some cases, more intense than do their White American counterparts. Likewise, several lines of evidence suggest that Black Americans have a stronger preference for sweet foods and beverages than White Americans.

As with a number of sensory phenomena, the human and animal literature suggests that odor and taste preferences are a complex conglomerate of interactions between experience and inherent sensory propensities. Thus, rats can be selectively bred for high and low saccharin preferences (Nachman, 1959), although it is possible that the selection is for dams who produce sweet milk to which the infants develop an acquired preference. In general, experience seems to be prepotent for the development of human taste preferences. Thus, Greene et al. (1975) found no evidence for the heritability of the sucrose preference difference between Black and White American infants, and Beauchamp and Moran (1982, 1984) demonstrated that infants with a history of dietary exposure to sweetened water ingested relatively more sucrose-flavored water in tests at 6 months and at 2 years of age. Liking for flavors can be enhanced by pairing a relatively neutral flavor with a hedonically positive taste (Zellner et al., 1983), as well as by simply exposing a subject to the flavor (so-called "mere exposure"; see Cain and Johnson, 1978; Pliner, 1982; Rozin and Schiller, 1980; Zajonc, 1968). Interestingly, exposure of rats for 18 days postpartum to peppermint odor results in increased bulbar uptake of 2-deoxy-D-glucose in response to subsequent peppermint odor stimulation, suggesting that odor exposure may alter neurological responsiveness (Coopersmith and Leon, 1984).

The social milieu in which exposure to tastes or odors occurs is also important in facilitating the development of preferences or aversions, as evidenced in studies of the enculturation of food habits. In general, such enculturation occurs quite rapidly and depends upon a number of factors. For example, Gupta (1975) found that nonvegetarian Indian Hindus who came to the United States accepted beef in less than a year, although their vegetarian counterparts took much longer for this transition. Acceptance of Western food habits was found to be related to marital status, age, sex, length of stay in the United States, caste, and the rural-urban background of the subjects. In a similar study of first and second generation Chinese adolescent immigrant boys to Canada, Hrboticky and Krondly (1984) found that the second generation boys and the boys with more accultured patterns of language use gave higher hedonic flavor and prestige ratings to dessert, snack, and fast foods, and discriminated better between nutrient rich and poor foods (as assessed by changes in perception of health value). Interestingly, the importance of social factors is suggested even in non-human primates. Thus, chimpanzees can develop a liking for what appears to be a generally aversive tastant (chili pepper) within the context of a close social bond with a human (Rozin and Kennel, 1983).

Overall, it is clear from the present review that much more research is needed to explore cross-cultural differences in smell and taste function. Furthermore, it is apparent that the reasons for cultural differences, when present, can be quite complex. In addition to the factors mentioned above, it should be noted, in closing, that variations in the general health and nutrition of subjects may also account for some cross-cultural differences in smell and taste function. Thus, persons on sodium restricted diets come to prefer low concentrations of salt (Bertino et al., 1982), and taste preferences for sucrose can be altered by the actions of insulin (Jacobs, 1958; Mayer-Gross and Walker, 1956). As dramatically pointed out by Lepkovsky (1977), American soldiers in a World War II prison camp, who were subsisting on a ration which was 200 calories short of their basal needs, developed food preferences which differed considerably from those exhibited under normal conditions. Thus, powdered milk became the most preferred item, followed by meat. Cheese, jam, sugar and chocolate bars were not strongly preferred.

ACKNOWLEDGMENTS

I am particularly grateful to the following individuals for providing me with helpful suggestions and information related to this review; Linda Bartoshuk, Mary Bertino, Richard Davis, Solomon Katz, Sidney Mintz, Michael O'Mahony, Rose Marie Pangborn, and Paul Rozin. In addition, I thank Tom Gregor, David A. Marshall, Jeanne McSloy, R. Gregg Settle, and Teresa Anne Vollmecke for commenting on an earlier draft of the paper, and Denise Thomason for typing sections of the manuscript.

REFERENCES

Allison, A. C., and Blumberg, B. S., 1959, Ability to taste phenylthiocarbamide among Alaskan Eskimos and other populations, Human Biol., 31:352.
Barnicot, N. A., 1950-1951, Taste deficiency for phenylthiourea in African Negroes and Chinese, Ann. Eugenics, 15:248.
Beauchamp, G. K., and Moran, M., 1982, Dietary experience and sweet taste preference in human infants, Appetite, 3:139.
Beauchamp, G. K., and Moran, M., 1984, Acceptance of sweet and salty tastes in 2-year-old children, Appetite, 5:291.
Bertino, M., Beauchamp, G. K., and Engelman, K., 1982, Long-term reduction in dietary sodium alters the taste of salt, Am. J. Clin. Nutr., 36:1134.
Bertino, M., Beauchamp, G. K., and Jen, K. C., 1983, Rated taste perception in two cultural groups, Chem. Senses, 8:3.
Bertino, M., and Chan, M., 1984, Taste responses in people with Chinese or European background, Sixth Annual Meeting of the Association for Chemoreception Sciences, Sarasota, Florida, April 4-8.
Cain, W. S., and Johnson, F., Jr., 1979, Lability of odor pleasantness: Influence of mere exposure, Perception, 7:459.
Chamberlain, A. F., 1903, Primitive taste words, Am. J. Psychol., 14:146.
Coopersmith, R., and Leon, M., 1984, Enhanced neural response to familiar olfactory cues, Science, 225:849.
Davis, R. G., 1978, Increased bitter taste detection thresholds in Yucatan inhabitants related to coffee as a dietary source of niacin, Chem. Senses Flav., 3:423.

Davis, R. G., and Pangborn, R. M., 1985, Odor pleasantness judgments compared among samples from 20 nations using microfragrances. Seventh Annual Meeting of the Association for Chemoreception Sciences, Sarasota, Florida, April 24-28.

Desor, J. A., Greene, L. S., and Maller, O., 1975, Preferences for sweet and salty in 9- to 15-year-old and adult humans, Science, 190:686.

Doty, R. L., Brugger, W. E., Jurs, P. C., Orndorff, M. A., Snyder, P. J., and Lowry, L. D., 1978, Intranasal trigeminal stimulation from odorous volatiles: Psychometric responses from anosmic and normal humans, Physiol. Behav., 20:175.

Doty, R. L., Shaman, P., and Dann, M., 1984a, Development of the University of Pennsylvania Smell Identification Test: A standardized microencapsulated test of olfactory function, Physiol. Behav., 32:489.

Doty, R. L., Shaman, P., Applebaum, S. L., Giberson, R., Sikorski, L., and Rosenberg, L., 1984b, Smell identification ability: Changes with age, Science, 226:1441.

Doty, R. L., Applebaum, S. L., Zusho, H., and Settle, R. G., 1985, A cross-cultural study of sex differences in odor identification ability, Neuropsychologia, 23:667.

Fischer, R., Griffin, F., England, S., and Garn, S. M., 1961, Taste thresholds and food dislikes, Nature, 191:1328.

Greene, L. S., 1974, Physical growth and development, neurological maturation, and behavioral functioning in two Ecuadorian Andean communities in which goiter is endemic. II. PTC taste sensitiviy and neurological maturation, Am. J. Phys. Anthr., 41:139.

Greene, L. S., Desor, J. A., and Maller, O., 1975, Heredity and experience: Their relative importance in the development of taste preference in man, J. Comp. Physiol. Psychol., 89:279.

Grijns, G., 1906, Messungen der Riechschärfe bei Europäern und Javanern, Arch. Physiol., 509:517.

Gupta, S. P., 1975, Changes in the food habits of Asian Indians in the United States: A case study, Social. Soc. Res., 60:87.

Hrboticky, N., and Krondly, M., 1984, Acculturation to Canadian foods by Chinese immigrant boys: Changes in the perceived flavor, health value and prestige of foods, Appetite, 5:117.

Jacobs, H. L., 1958, Studies in sugar preference. I. The preference for glucose and its modifications by injections of insulin, J. Comp. Physiol. Psychol., 51:304.

Johnston, F. E., Hertzog, K. P., and Malina, R. M., 1966, Phenylthiocarbamide taste sensitivity and its relationship to growth variation, Am. J. Phys. Anthr., 24:253.

Kay, P., and Kempton, W., 1984, What is the Sapir-Whorf Hypothesis? Am. Anthr., 86:65.

Katz, S. H., 1982, Food, behavior and biocultural evolution, in: "The Psychobiology of Human Food Selection," L. M. Barker, ed., AVI Publishing Co., Westport, Connecticut.

Koster, S., 1921, Een vergelijkend reukonderzoek bij Inlands van Ned.-Oost Indie en Hollanders. Geneesk. T. Ned. Ind., 61:165.

Lepkovsky, S., 1977, The role of the chemical senses in nutrition, in: "The Chemical Senses and Nutrition," M. R. Kare, ed., Academic Press, New York.

Lombroso, C., and Carrara, M., 1896/1897, Contributo all'antropologia dei Dinka, Atti. Soc. Romana Antropol., 4:103.

Mayer-Gross, W., and Walker, J. W., 1946, Taste and selection of food in hypoglycemia, Brit. J. Exp. Pathol., 27:297.

McCartney, W., 1968, "Olfaction and Odours," Springer-Verlag, Berlin.

Meiselman, H. L., 1977, The role of sweetness in the food preferences of young adults, in: "Taste and Development: The Genesis of Sweet Preference," J. M. Weiffenbach, ed., DHEW Publication No. (NIH) 77-1068, Washington, D.C.

Moskowitz, H. W., Kumariah, V., Sharma, K. N., Jacobs, H. L., and Sharma, S. D., 1975, Cross-cultural differences in simple taste preferences, Science, 19:1217.
Myers, C. S., 1903, Smell. I. Olfactory acuity and discrimination of odour strengths, II. Memory and discrimination of odours, Rep. Cambridge Anthrop., Expedition to Torres Straits, 2, pt. 2, Univ. Press, Cambridge.
Myers, C. S., 1904, The taste names of primitive peoples, Brit. J. Psychol., 1:117.
Nachman, M., 1959, The inheritance of saccharin preference, J. Comp. Physiol. Psychol., 52:451.
O'Mahony, M., and Ishii, R., 1985a, A comparison of English and Japanese taste languages: Taste descriptive methodology, codability and the umami taste, Brit. J. Psychol., in press.
O'Mahony, M., and Ishii, R., 1985b, Taste sorting with Japanese and Americans: Do Americans have separate sensory concepts for salty and umami tastes? Chem. Senses, submitted.
O'Mahony, M., and Manzo Alba, M. del C., 1980a, Taste descriptions in Spanish and English, Chem. Senses, 5:47.
O'Mahony, M., and Manzo Alba, M. del C., 1980b, Taste language strategies for 'primary' and 'non-primary' stimuli in English and Spanish, I.R.C.S. Med. Sci., 8:281.
O'Mahony, M., and Muhiudeen, H., 1976, A study of non-salty type taste descriptions for NaCl in the Malay language, I.R.C.S. Med. Sci., 4:174.
O'Mahony, M., and Muhiudeen, H., 1977, A preliminary study of alternative taste languages using qualitative description of sodium chloride solutions: Malay versus English, Brit. J. Psychol., 68:275.
O'Mahony, M., and Tsang, T., 1980, A preliminary comparison of Cantonese and American-English as taste languages, Brit. J. Psychol., 71:221.
Paolucci, A. M., Ferro-Luzzi, A., Modiano, G., Morpurgo, G., and Kanashiro, V. H., 1971, Taste sensitivity to phenylthiocarbamide (PTC) and endemic goiter in the Indian natives of Peruvian highlands, Am. J. Phys. Anthr., 34:427.
Peryam, D. R., and Pilgrim, F. J., 1957, Hedonic scale method of measuring food preferences, Food Tech., 11:9.
Pliner, P., 1982, The effects of mere exposure on liking for edible substances, Appetite, 3:283.
Rivers, W. H. R., 1904/1905, Observations on the senses of the Todas, Brit. J. Psychol., 1:321.
Rozin, P., and Kennel, K., 1983, Acquired preferences for piquant foods by chimpanzees, Appetite, 4:69.
Rozin, P., and Schiller, D., 1980, The nature and acquisition of a preference for chili pepper by humans, Motiv. Emot., 4:77.
Schiffman, S. S., and Dackis, C., 1975, Taste of nutrients: Amino acids, vitamins, and fatty acids, Percept. Psychophys., 17:140.
Schiffman, S. S., McElroy, A. E., and Erickson, R. P., 1980, The range of taste quality of sodium salts, Physiol. Behav., 24:217.
Schiffman, S. S., Moroch, K., and Dunbar, J., 1975, Taste of acetylated amino acids, Chem. Sens. Flav., 1:387.
Schleidt, M., Hold, B., and Attili, G., 1981, A cross-cultural study on the attitude towards personal odours, J. Chem. Ecol., 7:19.
Van Etten, C. H., 1969, Goitrogens, in: "Toxic Constituents of Plant Foodstuffs," I. E. Liener, ed., Academic Press, New York.
Wyant, K. W., and Meiselman, H. L., 1984, Sex and race differences in food preferences of military personnel, J. Am. Diet. Assoc., 84:169.
Zajonc, R., 1968, Attitudinal effects of mere exposure, J. Pers. Soc. Psychol., 9:1.
Zellner, D. A., Rozin, P., Aron, M., and Kulish, C., 1983, Conditioned enhancement of human's liking for flavor by pairing with sweetness, Learn. Motiv., 14:338.

AUTHOR INDEX

Abbott, D. H. 425, 429, 655, 656, 661, 669, 670
Abbott, J. 408, 411, 413, 415, 417-419, 422
Abrams, G. M. 167, 171
Abravaya, J. P. 579, 584
Adams, E. 72, 76
Adams, M. A. 51, 52, 60
Adams, M. G. 32, 34
Adkins-Regan, E. 281, 282
Adler, K. 177, 188, 226, 242
Adler, N. T. 79, 84
Adron, J. W. 45, 49, 57-59
Adrian, E. D. 2, 3, 6
Agosta, W. C. 30, 32, 36, 38, 44, 561, 568, 570
Ahsan, S. N. 105, 108, 110, 111
Aida, K. 103, 112
Akhtar, M. 248, 256
Akin, G. C. 239
Aksenova, T. G. 541, 549
Alberts, J. R. 30, 36, 496, 503
Albon, S. D. 388, 394
Albone, E. S. 7, 9, 11, 14, 19, 22, 27, 28, 32, 34, 87, 96, 520, 531, 567-569
Aldridge, R. D. 243, 252, 254, 256, 287, 300
Aleinikov, P. A. 541, 550
Aleksiuk, M. 244, 245, 252, 253, 256, 258, 279, 282
Alexander, R. D. 225, 239, 397, 401, 402, 409, 413, 419, 420
Alexander, S. A. 19, 22, 23
Alleva, E. 418, 421
Allison, A. C. 359, 367, 678, 682
Almeida, O. F. X. 423, 431
Allen, B. A. 319, 320
Alsing, W. 30, 35
Altieri, R. 507, 517, 561, 568, 569
Altmann, J. 629, 644, 648, 651
Alveario, M. C. 664, 668, 669, 670
Alves, C. Jr. 662, 668, 671
Ambasta, S. P. 381, 382
Amouriq, L. 120, 122, 129
Andersen, D. C. 528-529, 530
Anderson, A. C. 370, 374, 382
Anderson, G. G. 519, 530
Anderson, H. N. 226, 242
Anderson, J. D. 178, 184
Anderson, K. K. 520, 530
Anderson, O. 173, 186, 206, 218, 342, 353, 355
Anderson, S. 548, 549
Anderson, V. H. 144, 147
Andervont, H. B. 423, 429, 441, 452
Andren, C. 262, 275
Andreoletti, G. C. 178, 186
Andreolini, F. 428
Andrew, R. J. 659, 669
Andrews, P. W. 415, 417, 418, 420, 422
Angle, J. P. 179, 184
Anokhin, P. K. 42
Ansari, A. Q. 248, 256
Anthony, R. M. 519, 530
Antunes, J. L. 167, 171
Aoki, I. 108, 111
Applebaum, S. L. 675, 676, 683
Archer, A. T. 369, 382
Archer, J. 95, 96
Argentini, M. 104, 108, 111
Arieli, Y. 46, 60
Ariëns Kappers, C. U. 357, 367
Arnold, G. P. 139
Arnold, P. 428, 430, 555, 558, 560, 600, 605
Arnold, S. J. 63, 75, 173, 174, 177-185, 187, 200, 201, 243, 256 279, 281, 282, 541, 549
Aron, M 681, 684
Aronson, L. R. 165, 169
Assmann, S. M. 443, 449, 454
Atema, J. 47, 59, 108, 110, 111, 115
Atkins, D. L. 121, 129
Attili, G. 674, 675, 684
Ayala, F. J. 224
Azulay, D. 16, 25

Bachmann, M. D. 206, 216, 217
Bacon, J. 37, 39, 44, 321
Baeder, D. H. 105, 114
Baerends, G. P. 216, 217
Bailey, K. 648, 651
Bailey, S. 30, 34
Bailie, J. H. 28, 34
Baird, C. J. 423, 431
Baker, E. C. 268, 275
Balazs, G. H. 346, 352
Baldwin, I. T. 99, 111
Ball, A. K. 127, 132, 167, 171
Ball, R. L. 243, 245, 252, 256, 257
Balthazart, J. 362, 368
Banerjee, T. K. 15, 25
Bang, B. G. 357, 358, 364, 367
Banks, L. M. 226, 242
Baranoski, L. 415, 422
Baranowitz, S. A. 16, 35
Barash, D. P. 404, 409
Barbenhenn, F. 479, 484
Barbour, R. W. 102, 112
Bard, J. 231, 240, 242, 414, 415, 422, 466, 470
Bardach, J. E. 4, 48, 49, 59, 173, 185
Barnard, C. J. 226, 240
Barnard, D. E. 195-197, 200, 202
Barnes, R. D. 579, 584, 601, 604
Barnes, V. G. 519, 530
Barnett, S. A. 441, 452, 496, 503
Barnicot, N. A. 678, 682
Barthalmus, G. L. 206, 217
Bartheld, C. S. von 128, 129, 133
Barthold, S. W. 579, 584
Bartke, A. 468-470
Bartko, J. J. 156, 172
Bartle, K. D. 79, 84
Bast, R. E. 272, 275
Bateman, A. J. 175, 185
Bateson, G. 13, 23
Bateson, P. 225, 226, 239, 397, 401, 409, 417-421
Batzli, G. O. 428, 429
Baumann, F. 262
Baxter, G. T. 262, 266, 277, 286, 300, 344, 354
Baxter, R. M. 610, 615
Beach, F. A. 7
Beall, T. 507, 514, 515, 517
Beamer, W. 468, 469
Bean, N. J. 474-476, 479, 483, 484
Beauchamp, G. K. 4, 9, 28, 34, 37, 38, 42, 231, 240, 242, 414-416, 421, 422, 466, 470, 471, 473-475, 479, 483, 485, 679, 681, 682
Becak, W. 280, 282
Beck, R. 361
Beecher, I. M. 226, 239
Beecher, M. D. 226, 239, 401, 409
Beets, M. 5
Behrends, P. 533, 539
Beiswenger, R. E. 230, 239
Bekoff, M. 402, 409
Belcher, A. M. 184, 189, 364, 368, 426, 561, 568, 570, 664-667, 671
Bell, G. 176, 185
Bellis, E. D. 206, 217
Bellringer, J. F. 434, 438
Beltramino, C. 478, 483
Belvedere, P. C. 117, 119, 121, 123, 124, 128, 130, 154, 185
Bent, A. C. 370, 372, 381, 382
Bentivegna, F. 117, 119, 121, 123, 124, 128, 130
Bentley, P. J. 289, 300, 303
Benvenuti, S. 369, 382, 383
Berg, J. 627, 644
Berger, D. F. 113
Bergink, E. W. 248, 260
Berkowitz, E. J. 156, 157, 162, 170
Berkvens, J. M. 35
Berlyne, D. E. 40, 42
Bermant, G. 244, 256
Bern, H. A. 574, 576
Bernays, E. A. 45, 60
Bernhard, R. 478, 485
Bernstein, D. T. 520, 530
Bernstein, J. W. 104, 111
Bertino, M. 679, 682
Bertram, B. C. R. 488, 503
Berruti, A. 362, 367
Beuerman, R. W. 346, 352
Beumer, R. J. 177, 189
Bhoyroo, V. D. 74, 77
Bickford, A. A. 370, 384
Bigelow, J. 479, 484
Billard, R. 124, 125, 130
Billet, F. 13, 23
Binder, M. H. 267, 277
Birch, M. C. 8, 11
Birch, M. D. 173, 174, 184, 185
Bird, S. 30, 34
Birkhead, W. S. 104, 111
Birkner, C. 520, 531
Bishop, J. E. 243, 256
Bishop, S. C. 205, 218, 262, 275
Bissinger, B. E. 305, 306, 320, 321
Blair, W. F. 185
Blanchard, F. C. 253, 256
Blanchard, F. N. 253, 256
Blaustein, A. R. 102, 111, 229-231 234, 235, 239-241, 397, 402, 408, 409, 413, 419-421, 538, 539, 541, 549, 570, 584
Blazquez, N. B. 27
Blem, C. R. 287, 300
Blest, D. A. 487, 503

Blood, D. C. 29, 36
Bloom, H. D. 119, 129
Bloom, J. L. 156, 164, 171
Blum, M. S. 7
Blum, S. L. 590, 592, 607
Blum, V. 119, 121, 125, 131
Blumberg, B. S. 678, 682
Bobek, B. 539, 540
Bock, B. C. 314, 315, 318, 320
Bock, W. J. 14, 23, 285, 286, 300
Boech, J. 184, 188
Boggs, J. R. 579, 584, 596, 604
Boisseau, C. 176, 185
Bollinger, A. 592, 595, 604
Bollinger, K. S. 562, 569
Bolton, D. M. 105, 115
Bona-Gallo, A. 280, 282
Bonnin, J. P. 124, 129
Boonstra, R. 533, 539
Boot, L. M. 423, 431
Booth, J. 346, 352
Borkoski, V. 155, 156, 169
Bossert, W. H. 135, 147, 182, 190
Bottoni, L. 178, 180, 181, 188
Bouchardet de Fonseca, G. A. 662, 668, 671
Bowen, D. V. 30, 36, 561, 568, 570
Bowman, L. E. 424, 430
Box, H. O. 658, 660-662, 669
Boyce, M. S. 562, 569
Boyd, E. M. 373, 382
Boyd, L. 626
Boyer, J. 37, 39, 43, 44
Boyse, E. A. 231, 240, 408, 411, 413-422, 466, 468-470
Boyse, J. 408, 411, 413, 415, 417-419, 422
Bradley, H. T. 323, 339
Bradshaw, S. D. 17, 23
Brady, M. K. 177, 188
Bragg, A. N. 226, 240
Brand, J. C. 38, 42
Brand, J. G. 475, 483
Brandon, R. A. 173, 185, 187
Branson, B. A. 268, 275
Brattstrom, B. H. 226, 240, 261, 275
Breckenridge, W. J. 323
Breed, M. D. 409
Breed, W. G. 424, 430
Breton, B. 124, 125, 130
Brett, B. L. H. 120, 129
Breucker, H. 159, 163, 168, 171, 172
Brinck, C. 519, 520, 530
Brock, O. G. 39, 42
Brodie, E. D. Jr. 173, 179, 185, 612, 615
Bronson, F. H. 5, 6, 10, 39, 42, 79, 84, 87, 96, 97, 174, 181, 184, 188, 190, 414, 421, 429,
Bronson, F. H. (Continued) 430, 433, 438, 440, 441, 443, 444, 450-452, 541, 559, 571, 575, 576, 603, 604, 614, 615
Brower, L. 8
Brown, C. 201
Brown, G. 642, 644
Brown, G. B. 104, 113, 115
Brown, J. H. 106, 111
Brown, J. H. 546, 550
Brown, J. L. 193, 201
Brown, K. 441, 454
Brown, R. E. 27, 34
Brown, W. S. 262, 266, 267, 276, 277, 289, 292, 295, 297, 300, 343, 352
Brownlee, R. G. 563, 568, 569
Bruce, H. M. 6, 433, 434, 436, 438, 463, 469, 571, 575, 576, 600, 606, 653, 669
Brugger, W. E. 674, 683
Brunjes, P. C. 399, 409, 480, 483
Buchmann, O. L. K. 595, 604
Buckland, F. 137, 147
Buday, J. A. 267, 277
Budnitz, N. 647-649, 651, 652
Bull, J. J. 287, 300
Bullock, A. M. 103, 113
Bunyan, P. J. 30, 34
Burghardt, G. M. 10, 39, 42, 63, 64, 67, 76, 173, 185, 245, 247, 255, 256, 259, 261, 267, 276, 279, 283, 289, 292, 299, 300, 303, 305, 306, 314, 315, 318-321, 323, 325, 326, 329, 338, 339, 341, 344, 352
Burken, R. R. 291, 300, 301
Burkholder, R. L. 18, 23
Burnett, J. W. 104, 111
Burton, T. M. 221, 224
Busack, C. A. 226, 241, 397, 411, 419, 422
Bush, M. 418, 421
Buskirk, S. A. 18, 21
Buss, I. O. 632, 645
Butler, C. G. 174, 180, 185
Butterstein, G. M. 559, 560
Buttery, R. G. 30, 35
Byers, N. 37, 44
Byers, R. E. 519, 530

Cain, W. S. 681, 682
Calaby, J. H. 587, 596, 597, 600, 605-607
Calfurni, P. 345, 352
Calhoun, J. B. 488, 503, 538, 539
Callard, I. P. 248, 257, 258
Callen, C. 345, 352
Calton, G. J. 104, 111
Camazine, B. 244-248, 251-253, 255 256, 258, 281, 282

Cameron, A. M. 104, 111
Camin, J. H. 370, 373, 383
Campbell, H. W. 63, 76
Cannon, C. W. 29, 36
Cannon, J. R. 17, 23
Capone, M. R. 533, 540
Caprio, J. 50, 59
Caracheo, F. 160, 165, 166, 171
Cardé, R. 184, 188
Carla, M. 345, 352
Carlin, B. 137, 147
Carmack, M. 30, 36, 80, 85, 468, 469
Caroom, D. 87, 96
Carpenter, C. C. 245, 256, 261, 262, 268, 276, 277
Carr, A. 346, 351, 352, 354
Carr, W. 46, 47, 58, 59
Carr, W. E. S. 48, 49, 54, 59, 568, 569
Carr, W. J. 507, 514, 517
Carrara, M. 673, 674, 683
Carroll, J. B. 656, 658, 659, 671
Carroll, R. L. 16, 23
Carroll, T. E. 345, 352
Carter, C. 305, 320
Carter, C. S. 428, 430, 546, 549, 555, 558-560, 600, 605
Case, J. 49, 59
Castro, B. M. 448, 452
Cattarelli, M. 515, 517
Cavallaro, S. A. 410
Cebul, M. S. 668, 669
Cedrini, L. 178, 185
Cerny, V. A. 661, 670
Chamberlain, A. F. 678, 682
Champlin, A. K. 423, 430
Chan, M. 679, 682
Chaney, T. B. 48, 59
Charles-Dominque, P. 583, 584, 598, 605
Charnov, E. L. 99, 101, 111
Chelazzi, G. 345, 352
Chen, L. C. 119, 130
Cherix, D. 507, 516, 517
Chichester, C. 60
Chien, A. K. 119, 130
Chiszar, D. 19, 20, 22, 25, 37, 39, 41, 43, 44, 63, 76, 263, 276, 292, 303, 305, 320, 321, 344, 352
Chiszar, J. 37, 39, 43
Chitty, D. 539
Chiu, K. W. 14, 17-19, 23, 25, 249, 259
Chohan, K. S. 127, 132, 167, 171
Christen, A. 655, 656, 658, 659, 669
Christian, J. J. 457, 461
Christiansen, E. 519, 530
Cialdini, R. B. 507, 517
Cieslak, E. S. 244, 256
Claas, B. 127, 131, 160, 167, 170
Claesson, A. 568, 569
Clancy, A. N. 476, 478, 483
Clark, D. 268, 276
Clark, D. L. 245, 258, 305, 321
Clark, L. 370, 372, 373, 375, 381, 382
Clark, R. T. 137, 148
Claus, R. 30, 35
Clausen, H. J. 262, 267, 277, 290, 292, 303, 344, 354
Clemens, W. A. 604, 605
Cleveland, J. 656-658, 660, 661, 670
Clutton-Brock, T. H. 388, 394
Coates, A. 657, 658, 669
Cobb, S. 358, 367
Coble, J. R. 479, 484
Coblentz, B. 615
Cohen, A. C. 290, 301
Cohen, J. 245, 247, 248, 252, 259, 281, 283, 292, 295, 302
Coimbra-Filho, A. F. 662, 669
Colby, D. R. 427, 431, 441, 443, 448, 452, 455
Cole, C. J. 18, 19, 23
Cole, E. A. 588-590, 595, 598, 599, 605
Cole, K. S. 126, 130
Coleman, D. L. 466, 469
Coleman, E. M. 441, 452
Colgan, P. 397, 409
Collias, E. 369, 381, 382
Collias, N. 369, 381, 382
Collins, I. 595, 607
Collins, J. W. 361, 367
Collins, L. R. 587, 595, 605
Collins, T. C. 562, 569
Colly, L. 418, 421
Colombo, L. 117, 119, 121, 123, 124, 128, 130, 154, 185
Comuzzie, D. C. 346
Conant, R. 274-276
Conaway, C. H. 610, 611, 614, 615
Cook, A. F. 151, 154
Coomes, R. M. 113
Cooper, J. C. 137, 148, 346, 355
Cooper, W. E. Jr. 19, 23, 292, 301, 315, 320, 323, 326-335, 338-340
Coopersmith, C. B. 436, 438
Coopersmith, R. 21, 23, 399, 410, 681, 682
Coppola, D. M. 79, 84, 427-429, 431, 448, 453, 458-461, 462
Coquelin, A. 476, 478, 483
Corbet, G. J. 541, 548, 549
Coues, E. 361
Courtney, R. J., Jr. 507, 514, 515, 517
Cowles, R. B. 261, 276, 299, 301

Cowley, J. J. 84, 433, 438, 453
Craig, W. 40, 43
Cramp, S. 364, 367
Crapon de Caprona, M. D. 120, 130, 155, 167, 168
Crews, D. 17, 21, 23, 24, 244-248, 250-258, 260, 263, 277, 279, 281-283, 292, 295-297, 301
Crickmer, R. 48, 59
Criddle, N. 520, 528, 530
Criddle, S. 520, 528, 530
Croll, R. P. 45, 59
Crosby, E. C. 357, 367
Crouch, G. L. 519, 530
Crow, J. F. 183, 185
Crow, R. T. 120, 122, 125, 130
Crowcroft, P. 457, 461, 610, 615
Crump, D. R. 515, 518, 520, 523, 527, 529-531, 561, 568, 569
Cuculis, J. T. 156, 171
Cuellar, H. S. 243, 257
Cupp, P. V. 194, 201
Curtassa, R. 181, 188
Cushing, B. S. 514, 517
Czaplicki, J. A. 63, 76, 292, 303, 344, 355

Dackis, C. 678, 684
Dagg, A. 644
Dahlsten, D. L. 370, 382
Dainish, K. 647, 649, 651, 652
Daly, J. W. 230, 240
Daly, M. 533, 539, 571, 575, 576
Daniel, W. W. 151, 154
Dann, M. 675, 683
Dardin, A. G. 418, 421
Darwin, C. 174, 185, 243, 257, 541, 544, 549, 579, 584
Dausset, J. 418, 421
Davenport, M. 37, 43
Daves, G. D. 628, 629, 645
Davidian, H. 359, 367
Davidson, J. M. 244, 256
Davies, J. W. 370, 374, 382
Davies, N. B. 387, 390, 394
Davis, D. D. 262, 276
Davis, D. E. 457, 461, 519, 530
Davis, L. S. 226, 240
Davis, S. F. 515, 517
Davis, R. G. 674, 675, 679, 682, 683
Davis, W. C. 179, 185
Davy, F. B. 144, 147
Dawkins, R. 176, 187, 402, 409, 414, 419, 421, 433, 436, 438
Dawley, E. M. 173, 185, 200, 201, 207, 218, 221-224
Dawson, G. A. 662, 668, 669
Dawson, J. T. 227, 229, 230, 240
Day, J. F. 113
Deck, J. 425, 431
DeFazio, A. 306, 320, 326, 327, 339
DeGroot, P. 488, 503
Delanoy, R. L. 54, 59
de la Riva, C. 426, 430, 436, 438
Delfino, G. 345, 352
Demeter, B. 37, 39, 43
Dempsey, C. H. 48, 59
Demski, L. S. 127-130, 155, 165-168, 363, 367
Deni, R. 399, 410
Dent, J. N. 189
Derby, C. D. 568, 569
Desjardins, C. 87, 96, 97, 452, 454, 468, 469, 603, 604
Desor, J. A. 679, 681, 683
Devine, M. C. 243, 245-247, 253, 254, 257, 262, 264, 276
Dewsbury, D. A. 181, 185
Dhouailly, D. 20, 23
Diamond, M. 245, 252, 253, 255-258
Dick, W. D. B. 35
Dickinson, J. A. 245, 247, 258, 292, 301
Diller, L. V. 287, 301
Dilley, S. R. 424, 429
Dills, S. S. 443, 444, 455
Distel, H. 319, 320
Dixon, A. F. 425, 431
Dixson, A. F. 658, 670
Dizinno, G. 479, 481, 482-484
Dluzen, D. 428, 430, 560
Doane, H. M. 398, 399, 402, 409, 410
Dobrogosz, W. J. 443, 444, 455
Dobroruka, L. J. 588, 593, 605
Dodd, C. K. 185
Dodson, J. J. 122, 130, 143
Dolbeer, R. A. 376, 383
Dole, J. W. 342, 352, 355
Dolphin, P. J. 248, 256
Dominey, W. 279, 282
Dominic, C. J. 424, 430
Donald, T. 370, 383
Donovan, C. A. 29, 35
Dorey, J. J. 466, 468, 469
Doty, R. L. 4, 9, 11, 27, 28, 34, 35, 173, 186, 674-676, 683
Døving, K. B. 126, 127, 130, 140-143, 147, 230, 240, 241, 519, 530
Downing, D. T. 291, 300, 301
Downs, F. 77
Drickamer, L. C. 5, 79, 80, 84, 427, 428, 430, 431, 441, 443-451, 453-455, 457, 458, 461, 462, 515-517, 533, 540, 554, 558, 560, 571, 576
Drummond, H. D. 314, 320
Dryden, G. L. 610, 611, 614, 615

D'Udine, B. 397, 410, 418, 421
Dudzinski, M. L. 90, 97
Duffy, D. C. 370, 382
Dugan, B. A. 318, 320
Dulka, J. G. 127, 130, 165, 166-168
Dullemeijer, P. 39, 43
du Mesnil du Buisson, F. 174, 189
Dunbar, J. 678, 684
Dunbar, R. I. M. 425, 430
Duncan, R. N. 137, 147
Dundee, H. A. 267, 276, 292, 301, 344, 352
Dunn, E. R. 178, 186
Dunnet, G. M. 371, 382
Dunson, W. A. 249, 260, 291, 297, 302, 303
Duvall, D. 6, 14, 19-25, 37, 39, 41, 43, 44, 262, 263, 266, 267, 276, 277, 285-289, 291-294, 296, 297, 299, 301-303, 305, 310, 320, 321, 324, 327-329, 335, 339, 343-345, 353, 354, 363, 367

Eadie, W. R. 610, 611, 615
Eaton, L. G. 105, 115
Ebbesson, S. O. E. 128, 133
Eberhard, M. J. W. 102, 111
Eberly, W. G. 260
Ebling, F. J. 6, 548, 549, 579, 584, 592, 605, 607
Eccles, J. C. 423, 430
Eccles, R. 471, 483
Edmunds, M. 109, 111
Edwards, D. 17, 23
Edwards, M. W. 230, 240
Egid, K. 231, 241, 421
Egorov, I. K. 422
Ehrenfeld, D. W. 345-347, 352-354
Eibl-Eibesfeldt, I. 87, 95, 96, 102, 111, 239, 240
Eigenmann, C. H. 174, 186
Eisenberg, J. F. 579, 584, 588, 590, 594, 595, 597, 598, 604, 605, 609, 615, 617, 622, 627, 632, 644
Eiss, I. 305, 306, 321
Ekstrom, P. 167, 168
Ellerman, J. R. 548, 549
Elton, C. S. 555, 560
Emanuel, M. E. 122, 130, 143, 147
Emlen, S. T. 243, 257, 385, 394
Endean, R. 104, 111
Engelman, K. 682
England, S. 678, 683
Enzmann, R. 520, 531
Epple, G. 4, 10, 184, 189, 364, 368, 426, 430, 561, 568-571, 576, 655-672
Erickson, R. P. 678, 684
Erlinge, S. 520, 530
Erwin, M. R. 488, 504
Estep, K. 37, 39, 43, 44
Estes, R. D. 385, 389, 392, 394
Evans, C. S. 648, 652
Evans, D. A. 7
Evans, D. M. 539
Evans, J. 519, 530
Evans, L. T. 305, 321, 324, 339, 345, 353
Evans, S. 655, 670
Evens, S. M. 600, 607
Everett, C. T. 323, 339
Evermann, J. F. 418, 421
Ewer, R. F. 588, 597-599, 605, 617, 622

Fadem, B. H. 579, 584, 588-599, 601-603, 605, 607
Fagan, J. F. 41, 43
Fahrner, K. 420, 422
Fan, Z. 519, 529, 530
Fanselow, M. S. 514, 517
Fantz, R. L. 41, 43
Farkas, S. R. 143, 147
Farner, D. S. 244, 257
Farr, J. A. 181, 186
Farris, E. J. 605
Fasolo, A. 178, 185
Fawcett, J. D. 243, 257
Feare, C. J. 370, 371, 382
Feder, H. 281, 282
Feder, M. E. 221, 224
Feiler, F. 44
Feist, J. D. 623, 626
Fekete, E. 88, 97
Fenical, W. 45, 60
Ferguson, G. W. 261, 276
Ferguson, M. M. 102, 111
Fergusson, B. 17, 23
Ferin, M. 167, 171
Fernald, R. D. 167, 168
Fernandez-Galaz, C. 478, 484
Ferro-Luzzi, A. 678, 684
Fiaschi, V. 369, 382, 383
Fiebig, E. 128, 133
Field, F. H. 30, 36, 561, 568, 570
Filsoof, M. 30, 35
Finger, T. E. 167, 168
Fink, W. L. 121, 129
Finlayson, J. S. 443, 444, 455
Finneran, L. C. 262, 276
Fiore, L. 369, 382, 383
Firbas, W. 612, 616
Fischer, R. 678, 683
Fisher, J. 195, 201
Fisher, R. A. 548, 549
Fitch, H. S. 243, 245, 252, 253, 257, 287, 288, 301, 323, 331, 339

Fitzgerald, G. J. 109, 113
FitzGerald, R. W. 546, 549, 559, 560
FitzSimons, F. W. 262, 276
Flath, R. L. 30, 36
Flavell, R. A. 420, 422
Flaxman, B. A. 15, 23, 295, 303
Fleming, A. A. 473, 484
Flier, J. 230, 240
Flood, P. F. 520, 531
Florian, M. L. E. 370, 382
Flowerdew, I. R. 539, 540
Follett, B. K. 244, 257
Ford, N. B. 243, 245-247, 252, 254, 257, 262-265, 267, 271, 272, 274-276, 292, 297, 343, 353
Forest, C. L. 159, 162, 171
Forester, D. C. 173, 186, 205, 206, 208, 210, 211, 215-218
Forrest, H. S. 104, 112
Fortune, J. E. 428, 431, 555, 559, 560
Fostier, A. 124, 125, 130
Fox, H. 243, 257
Fox, M. W. 96
Fox, V. S. 359, 363, 367
Fox, W. 244, 246, 257
Fraenkel, G. S. 269, 276
Francoer, D. 425, 431
Frank, F. 559, 560
Franken, J. J. 84
Fraser, D. F. 192, 193, 201
Fredrickson, W. T. 226, 240
Freeman, S. K. 15, 23
French, J. 27
French, J. A. 655-658, 660, 662, 667, 668, 670
French, N. R. 539, 540
French, T. W. 610, 615
Fried, J. J. 558
Friedman, D. W. 245, 257
Friess, A. E. 33, 35
Frings, H. 369, 382
Frisch, K. von 99, 101, 108, 110, 111, 507, 516, 517
Frisco, C. 119, 121, 123, 130, 154
Frith, H. J. 597, 605
Fritzch, B. 155, 167, 168
Frizel, D. E. 143, 147
Frumin, N. 63, 76
Fuchs, J. L. 63, 76
Fujimoto, K. 47, 60
Fujita, I. 127, 132
Funakoshi, M. 50, 59
Furman, J. M. 96, 97
Furudate, S. 463, 468, 469
Fusetani, N. 48, 57, 59
Gadgil, M. 627, 644
Gaines, M. S. 559, 560
Galef, B. G., Jr. 488-504
Galusha, J. C. 39, 43
Gambale, S. 90, 97
Gandolfi, G. 106, 110, 111, 120, 130
Gannon, V. P. J. 287, 288, 301
Gans, C. 13, 15, 19, 23, 24
Ganslosser, V. U. 588, 605
Gapp, D. A. 248, 257
Gardner, A. L. 579, 581, 583, 584
Gardner, J. B. 257, 262, 277
Garg, R. K. 249, 258
Garn, S. M. 678, 683
Garstka, W. R. 21, 24, 244-248, 250-258, 263, 277, 279, 281, 282, 292, 295-297, 301, 335
Garton, E. O. 528, 531
Gasco, F. 178, 186
Gavish, L. 428, 430, 555, 558, 560, 600, 605
Gehlbach, F. R. 247, 258, 262, 267, 277
Geier, S. S. 415, 422
Gentry, J. B. 464, 470
George, F. W. 281, 282
Gerd, M. A. 43
Gerell, R. 520, 530
Gergits, W. F. 173, 187, 190, 193-196, 199-202, 211, 218, 222, 224, 343, 354
Gern, W. A. 20, 262, 266, 267, 277, 344, 354
Gerzog-Thomas, D. A. 507, 512, 516, 518
Gesner, K. 136
Getz, L. L. 428, 429, 546, 549, 555, 558-560, 600, 605
Geyer, L. 575, 576
Ghoca, M. L. 507, 517
Giacoma, C. 178, 186
Giberson, R. 675, 676, 683
Gibson, A. R. 272, 275
Gill, D. E. 176, 186
Gill, E. D. 595, 605
Gillingham, J. C. 245, 247, 258, 261, 277, 292, 301, 305, 321
Gilmore, D. P. 587, 607
Glahn, J. F. 376, 383
Glen, R. M. 369, 382
Glenn, E. M. 592, 605
Glickman, S. E. 319, 321
Glissmeyer, H. R. 287, 301
Godfrey, G. K. 613, 615
Goetz, F. W. 124-126, 130, 132, 149, 154
Goh, Y. 45, 47, 48, 50, 59
Goin, O. 323, 334, 339
Golan, L. 39, 43
Golani, I. 588, 590, 594, 595, 597, 598, 604, 605

Gold, C. S. 370, 382
Golden, L. 420, 422
Goldman, A. 414, 421
Goldstein, A. 71, 76
Golob, N. F. 426, 430, 663-666, 670
Gondeck, A. R. 54, 59
Goodrich, B. 5, 28, 30, 35, 87, 91, 93, 95-97
Goodwin, M. 568, 570, 592, 595, 606
Goos, H. J. Th. 160-162, 169, 170
Gorbman, A. 129, 130, 132, 574, 576
Gordon, R. E. 206, 218
Gordon, T. A. 424, 430
Gorman, M. L. 388, 394, 519, 528, 530, 539, 540
Gorski, R. A. 476, 478, 483
Gosden, P. E. 28, 34
Gosling, L. M. 385-392, 394, 515, 517, 651, 652
Gould, E. 614, 615
Gould, S. J. 285, 295, 301
Goulden, F. A. 610, 615
Gove, D. 63, 76, 305, 306, 321, 325, 326, 339
Gower, D. B. 30, 34
Goy, J. 173, 187, 194, 195, 197-199, 202
Goy, R. W. 173, 187, 281, 282, 648, 652
Graninetti, A. 345, 352
Grant, D. L. 173, 186, 206, 218, 342, 353, 355
Grant, P. T. 49, 58, 60
Grassman, M. A. 17, 24, 347-350, 353, 355
Grau, G. A. 571, 576
Grau, H. J. 226, 240
Gravelle, K. 305, 306, 321
Graves, B. M. 14, 19, 20, 22-25, 267, 285, 288, 291-294, 296, 297, 301-303, 305, 321, 344, 363, 367
Gray, J. 592, 605
Gray, J. M. 248, 260
Green, S. 226, 240
Greenberg, N. 14, 17, 24, 306, 310, 318, 321
Greene, H. W. 306, 320, 325
Greene, L. S. 678, 679, 681, 683
Greene, P. J. 404, 409
Greenlee, R. 37, 39, 43, 44, 321
Greenwood, P. J. 99, 112
Greer, A. E. 324, 339
Gregory, E. H. 575, 576
Gregory, P. T. 243-245, 252, 253, 256, 258, 267, 277, 279, 282, 287, 302, 343, 344, 353
Gregham, G. 169
Griffin, F. 678, 683
Griffiths, M. 587, 588, 593, 594, 599, 600, 605
Grijns, G. 674, 683
Grobstein, D. 157, 168
Gromav, I. M. 548, 549
Grønneberg, T. Ø. 561, 569
Gross, M. 279, 282, 283
Grosse, D. J. 120, 129
Grossman, A. 435, 438
Groves, D. 628, 629, 645
Groves, J. D. 324, 339
Grubb, J. C. 342, 343, 353
Grubb, T. C., Jr. 173, 186, 362-364, 367, 369, 382
Grund, N. 520, 531
Guadagni, D. G. 30, 35
Guber, H. A. 19, 25
Gubernick, D. J. 405, 410
Guild, S. R. 167, 169
Guiler, E. R. 595, 604
Guillette, L. J., Jr. 246, 258, 287, 301
Gunn, D. L. 269, 276
Gupta, C. 414, 421
Gupta, S. C. 249, 258
Gupta, S. P. 681, 683
Gustafson, H. 647, 648, 652
Gutteridge, C. S. 32, 35
Gutzwiller, K. J. 20, 21, 23, 266, 267, 276, 286-289, 297, 301, 310, 320, 343, 344, 353
Gwynne, D. T. 182, 186

Hadley, N. F. 16, 24
Haigh, G. R. 424, 430
Haight, J. 632, 636, 641, 645
Hairston, N. G. 173, 186, 193, 201
Halama, L. 159, 169
Hall, E. R. 617, 622
Halliday, T. R. 177, 178, 181, 182, 186, 189, 226, 240, 243, 258, 418, 421, 575, 576
Halpern, L. R. 160, 166, 170
Halpern, M. 19, 24, 38, 39, 43, 44, 63-65, 67-69, 71, 76, 245, 247, 248, 252, 255, 259, 261-263, 267, 268, 271, 277, 281, 283, 292, 294, 295, 299, 302 305, 321, 323, 325, 339, 344, 353, 475, 483, 484
Halpern-Sebold, L. R. 127, 129, 130, 132, 156-158, 160-162, 164-167, 169-171
Halpert, A. P. 246, 258
Halstead, B. W. 103, 112
Hamilton, W. D. III. 225, 236, 240, 394, 397, 400, 408, 410, 413, 421

Hamilton, W. J., Jr. 333, 339
Hammer, U. T. 106, 111
Hammerman, D. L. 305, 321
Hammond, J. 33, 35
Hampton, J. K. 663, 670
Hampton, S. H. 663, 670
Hanken, J. 179, 186
Hansen, R. M. 528, 530
Hara, T. J. 15, 24, 49, 59, 129, 130, 158, 163, 169
Harada, S. 50, 60
Harbone, J. B. 45, 59
Harden-Jones, F. R. 136, 147
Hardy, J. W. 370, 382
Hardy, P. H. 156, 171
Harper, P. C. 359, 367
Harrington, J. E. 571, 576
Harrington, W. G. 69, 71, 73, 76
Harris, S. E. 104, 112
Harrison, K. 208, 210, 211, 218
Hart, B. L. 29, 35, 484
Hart, B. S. 548, 549
Hart, L. A. 't 123, 133
Hartman, C. G. 600, 605
Hartman, G. 119, 132
Hartwell, J. L. 113
Harvey, E. B. 571, 576
Harvey, P. H. 99, 112
Harvey, S. 80, 85
Hashimoto, Y. 47, 48, 57, 59, 60, 103, 112
Hasler, A. D. 137, 138, 147, 148, 346, 355
Hasler, J. F. 559, 560
Haslewood, G. A. D. 230, 240
Hauschka, P. V. 69, 77
Hausman, J. S. 182, 189
Hawk, H. W. 29, 35
Hawley, A. W. L. 244, 245, 252, 253, 258
Hayashi, S. 226, 240
Haynes, L. T. 48, 59
Hayes, W. R. 39, 43
Hayssen, V. 589, 599, 601, 605
Healey, M. C. 102, 112
Heap, P. F. 520, 531
Hebard, H. G. 103, 112
Heckman, S. 519, 530, 531, 562, 569, 570
Heczko, E. 108, 109, 112
Heinemann, D. 361, 367
Heiniger, H. 466, 468, 469
Heinrich, B. 376, 383
Heisermann, G. J. 248, 258
Heller, H. 166, 169
Heller, S. 262, 267, 277, 292, 302, 344, 353
Henderson, R. W. 267, 277
Hendrickson, J. R. 347-350, 353, 355
Henessy, D. F. 528, 530
Henneous, R. 628, 629, 632, 645
Henry, S. A. 222-224
Hepper, P. G. 226, 240
Herreid, C. F., II. 507, 517
Hershey, J. L. 206, 218
Hershkovitz, P. 653, 654, 670
Herskowitz, R. 324, 327-329, 335, 339, 344, 353
Hertzog, K. P. 678, 683
Hess, D. 628, 632, 636, 644
Hess, E. H. 63, 76
Hessler, C. M. 226, 242
Hesterman, E. R. 28, 35, 87, 90, 91, 93, 95-97
Heusser, H. 240
Heymann, E. 656-659, 662, 670
Hicks, C. R. 291, 302
Hidaka, I. 47, 48, 50, 54, 58-60
Highton, R. 191, 192, 201, 216, 218, 222-224
Hill, J. E. 541, 548, 549
Hiller, A. 363, 367
Hiller, U. 245, 258
Himstedt, W. 181, 186, 342, 354
Hinton, M. A. C. 541, 549
Hirth, H. F. 292, 302, 344, 354
Ho, S-m. 248, 257, 258
Hoage, R. J. 660, 670
Hodges, J. K. 655, 656, 670
Hodgson, E. S. 143, 147,
Hoffman, L. H. 246, 258
Hoffman, W. 361, 367
Hoffmeyer, I. 519, 530
Hohn, K. G. 435, 439
Hold, B. 674, 675, 684
Holl, A. 159, 169
Holland, D. 267
Holland, K. 47, 50, 59
Hollmer, A. 281, 283
Holmes, W. G. 225, 226, 240, 242, 397, 402, 404, 410, 413, 419, 421, 422
Holomuzki, J. R. 173, 186
Holzer, G. 30, 36
Homma, K. 435, 439
Honacki, J. H. 617, 622
Honda, H. 120, 122, 123, 125, 126, 130, 131
Honegger, R. E. 613, 615
Hoover, J. E. 79, 84, 445, 446, 454
Hopp, S. L. 592, 605
Hoppe, P. C. 463, 468, 469
Hooper, E. T. 548, 549
Hornbuckle, P. A. 507, 514, 515, 517
Hornby, P. J. 128-130, 165, 168
Horrall, R. M. 137, 147, 148, 346, 355

Houck, L. D. 174, 177-187, 200, 201, 205, 218, 243, 256, 541, 549
Houghton, L. E. 113
Hourston, A. S. 121, 124, 128, 132
Houston, A. 387, 394
Houston, D. C. 487, 504
Hradecky, P. 29, 35, 642, 644
Hrbacek, J. 102, 112, 239, 240
Hrboticky, N. 681, 683
Huber, G. C. 357, 367
Huber, G. E. 167, 169
Huck, U. W. 436, 438, 463, 469
Hughes, R. L. 600, 605
Humason, G. L. 296, 302
Hummel, K. P. 88, 97, 466, 469
Hunsaker, D., II. 579, 583, 584, 590, 598, 600, 604, 606
Hunt, W. A. 44
Hurley, S. S. 428, 429
Hurst, J. 420, 422
Hutchison, L. V. 359, 361, 367
Hutterer, R. 613, 615
Huxley, J. S. 389, 394
Hyatt, G. W. 387, 394
Hyder, M. 131

Iannazzi, R. 376, 383
Ichikawa, M. 127-129, 131, 132
Ikeda, M. 48, 60
Imber, M. J. 362, 367
Ina, K. 48, 58, 60
Ingersoll, D. W. 119, 120, 128, 131
Ireland, L. C. 347, 354
Isaacks, N. 568, 570
Ishii, R. 678, 684
Ito, T. 14, 24
Ivanyi, P. 414, 421, 466, 469
Iverson, S. D. 319, 321
Izard, M. K. 28, 35

Jackson, M. K. 249, 258, 291, 295, 302
Jacob, J. 15, 18, 24, 359, 362, 363, 367, 368
Jacobs, H. L. 679, 682-684
Jaeger, R. G. 173, 187, 190, 192-202, 211, 218, 222, 224, 343, 354
Jaffe, K. 507, 517
Jahan-Pawar, B. 58, 60
Jahr, C. E. 436, 438
Jainudeen, M. 627, 632, 644
Jalabert, B. 124, 125, 130
Jannett, F. J., Jr. 541, 542, 544, 546-550
Jannett, J. Z. 547, 550
Janzen, D. H. 288, 302, 370, 375, 381, 383
Jarman, M. V. 386, 394
Jarman, P. J. 385, 394
Jellum, E. 642, 644
Jemiolo, B. 30, 36, 80, 85, 428
Jen, K. C. 679, 682
Jenkinson, D. McE. 35
Jennes, L. 127, 131, 160, 167, 170
Jennings, W. G. 30, 35
Jensen, A. L. 137, 147
Joh, T. 164, 169
Johns, B. E. 96, 97
Johns, M. A. 472, 484
Johnsen, P. B. 51, 52, 60, 138-144, 146
Johnson, F., Jr. 681, 682
Johnson, J. A. 185
Johnson, J. I., Jr. 603, 606
Johnson, W. E. 464, 470
Johnston, F. E. 678, 683
Johnston, J. W. 4, 11
Johnston, R. 643, 644
Johnston, R. E. 184, 187, 431, 555, 559, 560, 590, 606, 643, 644
Johnston, R. F. 370, 382
Johnstone, A. D. F. 146, 147
Jolly, A. 647-649, 652
Joly, J. 176, 177, 185, 187
Jones, R. B. 87, 97, 515, 517
Jones, R. E. 19, 20, 22, 25, 243, 257, 287, 292, 301, 303
Jorgenson, J. W. 30, 36, 80, 85, 231, 241, 468, 469
Joseph, R. G. 195, 196, 200, 202
Josse, J. 69, 71, 73, 76
Jouventin, P. 361, 368
Joy, J. 252, 258
Jumber, J. 369, 382
Jurgelski, W., Jr. 595, 606
Jurs, P. C. 184, 189, 364, 368, 664, 666, 671, 674, 683

Kagan, M. Z. 520, 531
Kagawa, H. 125, 131
Kahmann, H. 63, 76, 325, 339
Kahn, M. 645
Kaku, T. 50, 60
Kallman, K. D. 155-157, 162, 164, 169-171
Kalra, P. S. 423, 430
Kalra, S. P. 423, 430
Kalvarsky, D. 173, 187, 193, 194, 196, 202
Kambegawa, A. 124, 133
Kamil, A. C. 40, 43, 376, 383
Kamo, N. 50, 61
Kamrin, R. P. 165, 169
Kanashiro, V. H. 678, 684
Kapoor, M. 21, 25
Kardong, K. V. 39, 44
Kare, M. R. 4, 45, 60, 369, 382

Kareem, A. M. 226, 240
Karlson, P. 255, 259
Karr, A. 347, 354
Karsay, M. A. 281, 283
Karstad, L. 370, 374, 382
Kasahara, Y. 50, 60
Kasparian, S. 357, 368
Kasumyan, A. O. 99, 105, 112
Katayama, T. 60
Katz, S. H. 679, 683
Katz, Y. 473, 478, 479, 485, 655, 661, 667, 670
Kauer, J. S. 198, 202
Kawakita, K. 50, 59
Kawamichi, T. 575, 576
Kay, P. 678, 683
Keast, A. 604, 606
Keen, W. H. 173, 187, 194, 202
Keenlyne, K. D. 287, 288, 302
Kellogg, F. E. 143, 147
Kempson, D. 539
Kempton, W. 678, 683
Kendle, E. R. 119, 131
Kendrick, K. M. 658, 670
Kennedy, J. M. 441, 454
Kennedy, J. S. 143, 147
Kennedy, R. J. 362, 368
Kennel, K. 682, 684
Kennelly, J. J. 96, 97
Kennett, D. J. 489, 491, 497-500, 504
Kerchner, M. 479, 484
Kern, M. D. 369, 381, 382
Kessel, B. 371, 372, 382
Kety, S. S. 436, 438
Keverne, E. B. 10, 368, 424, 426, 429, 430, 434, 436, 438, 443, 449, 454, 455, 472, 484
Kiddy, C. A. 29, 35
Kim, Y. S. 120, 129, 131
Kimbrough, T. D. 105, 114
Kime, D. E. 125, 131
Kimura, S. 71, 73, 77
Kimura, T. 226, 240, 354
King, B. J. 425, 531
King, M. B. 20, 21, 23, 262, 266, 267, 277, 285-289, 297, 301, 302, 310, 320, 343, 344, 353, 363, 367
Kingston, B. H. 3, 11
Kinman, K. E. 617, 622
Kinzey, W. G. 655, 667, 668, 671
Kirby, D. R. S. 421
Kirkpatrick, J. F. 623-626
Kirkpatrick, M. 175, 183, 187, 243, 259
Kirkpatrick, R. L. 559, 560
Kirsch, J. A. W. 587, 606
Kirschenbaum, D. M. 63-65, 67-69, 71, 76
Kiseleva, E. I. 240
Kittredge, J. S. 255, 259
Kittrell, E. M. W. 508, 512, 518
Kivett, V. K. 610, 616
Kiyohara, S. 47, 50, 59
Kiyohara, S. Y. 376, 382
Klauber, L. M. 266, 277, 287, 291, 292, 302
Kleeberger, S. R. 173, 187, 200, 202
Kleerekoper, H. 135, 144, 147, 148, 158, 168, 169, 350, 354
Kleiman, D. G. 579, 584, 592, 598, 606, 617, 622, 655, 656-662, 671
Klein, J. 414, 421
Kleist, S. M. 507, 517
Klemm, W. R. 29, 35
Klimstra, W. D. 611, 616
Klingel, H. 623, 626
Klopman, R. B. 659, 669
Kluge, H. A. 519, 531
Knight, L. 305, 320
Kobatake, Y. 50, 61
Koch, A. L. 346, 353, 354
Kodric-Brown, A. 546, 550
Kodrich, W. R. 287, 303
Koen, A. L. 463, 470
Koeppl, J. W. 617, 622
Kokolis, N. 104, 112
Konosu, S. 47, 48, 57, 58, 60
Koshinsky, G. D. 106, 111
Kost, T. A. 443, 444, 455
Köster, E. P. 5, 8
Koster, S. 674, 683
Koyama, Y. 128, 131
Kozlowski, G. P. 167, 171
Krakauer, I. 297, 302
Kramer, E. 143, 148
Krames, L. 507, 514, 517
Kraub, D. 520, 531
Krebs, C. J. 533, 539, 551
Krebs, J. A. 519, 531
Krebs, J. R. 99, 101, 111, 176, 187, 386, 387, 394
Kretzmann, M. B. 428, 431, 555, 559, 560
Krey, L. C. 167, 171
Kroes, R. 33, 35
Kroll, J. C. 247, 258, 262, 267, 277
Krondly, M. 681, 683
Kruczek, M. 425, 431
Krulich, L. 423, 430
Kruse, S. McK 30, 36
Kruuk, H. 388, 394
Kubie, J. L. 19, 24, 38, 39, 42-44, 63, 77, 245, 247, 248, 252, 255, 259, 261, 263, 268, 271,

Kubie, J. L., (Continued)
277, 281, 283, 292, 294,
295, 299, 302, 305, 321,
323, 325, 339, 475, 484
Kubomatsu, T. 47, 48, 54, 58,
59
Kubota, M. 77
Kuehne, R. E. 102, 112
Kühme, W. 627, 644
Kulish, C. 681, 684
Kulzer, E. 102, 112, 230, 239,
241
Kumari, S. 592, 606
Kumariah, V. 678, 679, 684
Kummer, H. 425, 430
Kumpf, D. F. 187
Kumpf, K. F. 324, 339
Kuninobu, L. S. 103, 112
Kurihara, N. 50, 61
Kuroki, T. 108, 111
Kurosumi, K. 14, 24
Kurt, F. 627, 644
Kushibe, M. 73, 77
Kux, J. 159, 163, 172
Kyle, A. L. 126-129, 131, 132,
149-151, 154, 165, 167, 169
171

Labanick, G. M. 173, 182, 187
Labat, C. P. 198, 202
Labov, J. B. 434, 436, 438,
463, 469
Labows, J. 375
Lacher, T. E. J. 662, 668, 671
Lack, D. 369, 383
Lacy, R. C. 402, 410
Ladewig, J. 29, 35, 471, 484
Lagowski, J. M. 104, 112
Laing, D. G. 99, 112, 512,
514, 517
Lam, D. M. K. 167, 171
Lam, F. L. 104, 112
Lambert, J. G. D. 122-125, 131,
133
Lamberts, R. 435, 439
Lancet, D. 13, 15, 24
Lande, R. 175, 183, 187, 243,
259
Landis, C. 44
Landmann, L. 245, 249, 259,
290, 291, 295, 299, 302
Landreth, H. F. 266, 277, 343,
354
Langbauer, W. 627, 644
Langlois, T. H. 106, 112
Langworthy, O. R. 600, 606
Lanza, B. 187
LaPointe, J. L. 17, 25, 262,
263, 277
Larsell, O. 167, 170, 357, 368
Laszlo, J. 279, 283
Laumen, J. 119, 121, 125, 131
Laurie, E. M. O. 441, 454
Lavine, B. 184, 189, 364, 368, 664,
666, 671
Law, J. H. 63, 64, 67, 77
Lawson, N. 568, 570
Lazier, C. B. 248, 256
Leahy, D. S. 105, 115
Leatherland, J. F. 156, 170
Ledедeva, N. Ye. 105, 112
Lederer, E. 561, 569
Lednor, A. J. 173, 190
Lee, C. T. 119, 120, 128, 131
Lee, K. Y. 30, 36
Lee, M. L. 79, 84
Lee, Y. C. 73, 77
Legler, J. W. 18, 25
Legrain, J. M. 103, 104, 109, 115
Leitch, A. 388, 394
Leiter, E. H. 466, 469
LeMagnen, J. 3
LeMartret, M. A. 149, 154
Lemke, J. 108, 113
Lemly, A. D. 101, 114
Lendrem, D. W. 436, 438
Lenington, S. 231, 241, 421
Leon, M. 21, 23, 399, 410, 681, 682
Leonard, J. B. 103, 112
Leontiev, A. N. 37, 40, 44
Lepkovsky, S. 682, 683
Lepri, J. J. 428, 431, 478, 483,
484, 555, 559, 560
Leuthold, K. 269, 277
Levitsky, D. A. 490, 504
Lewontin, R. C. 285, 301
Leyland, M. 319, 321
Licht, P. 280, 282
Licht, L. E. 178, 187, 241
Lidicker, W. Z., Jr. 457, 462,
546, 550, 559, 560
Likens, G. E. 221, 224
Liley, N. R. 117, 119-123, 125-132,
149, 151, 153, 154, 166, 170
173, 187
Lillywhite, H. B. 16, 20, 24, 260,
291, 295, 296, 298-300, 302,
303
Limpus, C. J. 346, 354
Lindsay, N. B. D. 662, 671
Lindsey, G. D. 519, 530
Lindstedt, K. J. 45, 60
Lindstedt, S. L. 20, 267, 344
Lindzey, G. 575, 576, 590, 592,
598, 607
Ling, L. C. 30, 35
List, J. C. 245, 259
Little, C. 13, 15, 24
Little, E. E. 48, 60
Ljungman, T. N. 291, 302
Lloyd, A. 457, 461
Lloyd, D. H. 35

Lloyd-Thomas, A. 436, 438
Lockie, J. D. 617, 622
Lomas, D. E. 434, 438
Lombardi, J. L. 443, 444, 454, 455, 463, 469
Lombroso, C. 673, 674, 683
Long, S. E. 27
Lorenz, K. Z. 40, 41, 44, 351, 354, 598, 606
Lorenz, R. 655-658, 670
Losey, G. S. 187
Lotwick, W. 401, 409
Lounsberg, D. M. 424, 430
Low, J. R., Jr. 262, 263, 271, 272, 276, 343, 353
Lowe, N. R. 69, 77
Lowry, L. D. 674, 683
Lucas, C. E. 285, 303
Ludwig, D. R. 546, 550
Luria, A. R. 37, 40, 44
Lythgoe, J. N. 129, 132, 159, 170

Macdonald, D. W. 27, 34
Macfarlane, A. 397, 410
MacFie, H. J. H. 32, 35
MacGregor, J. R. 222, 224
MacInnis, A. J. 51, 56, 58, 61
MacIntosh, T. K. 554
Mack, D. S. 657-662, 671
Mackay-Sim. A. 99, 112, 512, 514, 518
Mackenzie, D. S. 124, 132
Mackie, A. M. 45, 49, 56-59
MacLean, F. M. 262, 266, 267, 276, 289, 292, 295, 297, 300, 343, 352
MacLeod, P. 8
MacMahon, J. A. 528-530
Macrides, F. 32, 36, 38, 44, 476, 478, 483
Maderson, P. F. A. 14-25, 249, 259, 291, 295, 296, 298, 299, 302, 303
Madison, D. M. 15, 17, 18, 21, 25, 173, 187, 188, 195-197, 199, 200, 202, 203, 207, 211, 218, 261, 277, 341-343, 354, 546 549, 559, 560
Maekawa, K. 73, 77
Magãlhaes-Castro, B. 662, 668, 671
Maggio, J. C. 479, 484
Maggio, J. H. 479, 484
Magnan, P. 109, 113
Magurran, A. E. 102, 113
Maier, W. C. 426, 431
Mainardi, D. 106, 111
Makin, J. W. 398, 403, 405, 406, 410, 411
Malacarne, G. 178, 180, 181, 185, 186, 188
Malar, T. 144, 147
Malina, R. M. 678, 683
Maliniak, E. 589, 599, 601, 605, 606
Maller, O. 45, 60
Maller, O. 679, 681, 683
Mallinson, J. J. C. 613, 616
Malyukina, G. A. 105, 112
Mangold-Wernado, U. 99, 113
Mann, M. 463, 470
Manning, N. J. 125, 131
Manteyfel, Yu. B. 120, 133
Manton, M. L. 347, 351, 354
Manzo, Alba, M. del C. 678, 684
Marc, R. E. 163, 170
Marchlewska-Koj, A. 80, 85, 174, 182, 188, 425, 431, 463, 468, 469
Marconato, A. 117, 119, 121, 123, 124, 128, 130, 154
Margolis, S. E. 181, 188
Margolis-Kazan, H. 129, 132, 156-158, 160, 160-162, 164, 165, 167, 169-171
Margolis-Nunno, H. 161, 162, 164, 165, 167, 170, 171
Marker, L. 418, 421
Markl, H. 548, 550
Markowitz, H. 627, 644
Marler, P. 226, 240
Marquez, C. E. 173, 187, 194, 195, 197-199, 202
Marquez, E. M. 17, 24
Marquez, M. R. 349, 350, 353
Martin, B. 249, 259, 287, 295, 302
Martin, D. L. 198, 202
Martin, I. G. 37, 38, 42, 473, 483
Martin, R. D. 7
Martin, W. H. 303
Martinich, R. L. 119, 130
Martof, B. S. 342, 354
Martorano, R. D. 507, 514, 517
Marui, T. 50, 59
Maruniak, J. A. 87, 96, 97, 443, 452
Maser, M. D. 71, 77
Mason, E. R. 324, 339
Mason, J. R. 369, 370, 372, 375, 376, 379, 382, 383
Mason, R. T. 245, 252, 253, 255-257, 260, 279, 281, 283
Massa, R. 178, 180, 181, 188
Massey, A. 182, 427, 431, 451, 454, 457, 460, 462, 554
Mathewson, R. F. 143, 147
Mathieson, B. J. 408, 411, 413, 415, 417-419, 422
Matochik, J. A. 403, 405, 406, 411
Matsui, H. 58, 60

Matsushita, T. 47, 60
Matthews, L. H. 587, 606
Matviev, B. S. 617, 622
Maurer, B. 561, 569
Mayer-Gross, W. 682, 683
Maynard Smith, J. 99, 113, 387, 394
Mayr, E. 224
Mazder, E. 533, 540
Mazid, M. A. 60
McCall, L. 208, 210, 211, 218
McCarron, D. 262, 266, 277, 344, 354
McCartney, W. 3, 5, 8, 11, 673, 683
McClenaghan, L. R., Jr. 559, 560
McClintock, M. K. 5, 11, 79, 84, 425, 431, 600, 606
McGrady, E., Jr. 600, 606
McCullough, D. R. 623, 626
McDermott, J. L. 428, 430, 555, 558, 560, 600, 605
McDonald, J. J. 104, 113
McElroy, A. E. 678, 684
McEwen, B. S. 281, 282
McGavin, M. 196, 198, 202, 343, 354
McGlone, J. J. 514, 517
McIntosh, T. K. 79, 84, 427, 430, 431, 448, 454
McKay, D. M. 487, 504
McKay, G. 627, 644,
McKinney, T. D. 452, 454, 468, 469
McManus, J. J. 579, 583, 584, 595, 606
McMillan, V. E. 105, 113
McPartland, R. P. 463, 469
McVey, J. P. 17, 24, 349, 350, 353
Meaney, C. A. 571, 572, 576
Medina, S. R. 226, 242, 419, 422
Meester, J. 610, 615
Meisami, E. 359, 368
Meiselman, H. L. 680, 681, 683, 684
Meisner, D. H. 579, 583, 584, 595, 598
Melinkat, R. 159, 163, 168, 171, 172
Mellor, A. 420, 422
Melrose, D. R. 561, 568, 569
Meltzer, D. 418, 421
Menchel, S. 18, 25
Meredith, M. 245, 259, 305, 321, 426, 431, 434, 438, 443, 454, 471, 472, 484, 558, 560, 643, 644
Mertens, J. A. L. 369, 383
Mertl, A. S. 648, 652
Mertl-Millhollen, A. S. 648, 650-652
Merton, H. 423, 431
Metzgar, L. A. 387, 394
Meyer, D. L. 128, 133
Meyer, H. G. 156, 171
Meyer, J. H. 120, 122, 125, 131
Meyer, M. N. 548, 550
Meyers, L. E. 370, 383
Mezzalira, F. 185
Michael, R. P. 5, 363, 368
Michon, E. 632
Middendorf, G. A. 306, 320, 326, 327, 339
Mike, V. 408, 411, 413, 415, 417-419, 422
Miller, C. W. III. 292, 301
Miller, M. C. 267, 276, 344, 352
Miller, T. 39, 41, 43
Milligan, S. R. 424, 431, 433, 438
Mitchell, D. S. 29, 35
Mitra, A. 645
Mittal, A. J. 15, 25, 103, 113, 115
Mittmeier, R. A. 662, 669
Mloszewski, M. J. 386, 394
Mock, O. B. 610, 616
Modiano, G. 678, 684
Moffat, L. A. 18, 25
Mohr, E. 610, 616, 623, 626
Moncrieff, R. W. 2, 8, 11
Montagna, W. 5, 648, 652
Montgomery, R. 72, 77
Mooney, S. M. 507, 517
Moore, J. D. 397, 401, 410, 411
Moore, R. E. 188
Mora, J. M. 347, 348, 354
Mora-Novard, O. A. 478, 484
Moran, K. 37, 39, 43
Moran, M. 681, 682
Morgan, B. B. 33, 36
Morgans, T. 32, 35
Moroch, K. 678, 684
Morpurgo, G. 678, 684
Morrell, J. I. 167, 171
Morris, J. M. 658, 669
Morris, R. S. 29, 36
Morrison-Scott, T. C. 548, 549
Moschera, J. 77
Moskowitz, H. W. 679, 684
Moss, M. W. 370, 373, 383
Moulton, D. G. 4, 9, 11, 28, 34, 198, 202
Mount, R. H. 323, 339
Moynihan, M. 657-660, 671
Mozell, M. M. 6, 7, 11, 173, 246, 259
Mrosovsky, N. 346, 354
Muetterties, E. L. 96, 97
Mugford, R. A. 7, 9, 28, 34
Muhiudeen, H. 678, 684
Muir, L. 73, 77
Muller, J. F. 163, 170
Muller-Schwarze, C. 562, 568, 570

Muller-Schwarze, D. 6, 10, 11, 37, 38, 42, 44, 173, 184, 188, 246, 259, 507, 517, 519, 529-531, 561, 568, 570, 579, 584, 588, 590, 594-596, 606, 643
Mulner, O. 251, 259
Munday, K. A. 248, 256
Munro, D. F. 245, 246, 259, 262, 277
Münz, H. 127, 131, 160, 167, 170
Murphy, B. D. 105, 114
Murphy, J. B. 261, 277
Murphy, R. C. 358, 361, 368
Murphy, R. W. 329, 339
Murphy, R. X. 443, 454
Murray, K. E. 87, 93, 95, 96
Murray, L. W. 69, 77
Murtaugh, J. 615, 616
Myers, C. S. 674, 678, 684
Myers, C. W. 230, 240
Myers, G. S. 174, 186
Myers, J. H. 551, 554
Mykytowycz, R. 1, 3, 6, 10, 11, 28, 35, 87, 90, 91, 93, 95-97, 595, 606
Mylonas, N. 104, 112

Nachman, M. 681, 684
Nadal, L. 627, 644
Nagahama, Y. 124, 125, 131, 133
Nagaoko, Y. 77
Nakamura, K. 254, 259
Nakano, T. 463, 468, 469
Nathenson, S. G. 415, 422
Naulleau, G. 63, 77
Nay, T. 595, 606
Neal, E. J. 7
Neely, R. D. 519, 530
Neff, J. A. 370, 383
Nelson, K. 121, 132, 174, 188
Nelson, L. 528, 531
Nelson, R. S. 473, 484
Neusteuer, P. 99, 113
Newcombe, C. 119, 132
Newman, A. 418, 421
Newsome, G. E. 109, 113
Nichol, R. A. 436, 438
Nilaver, G. 167, 171
Nilson, G. 295, 303
Nishikawa, K. C. B. 195, 197, 202, 203
Nishimura, K. 29, 32, 35
Niswander, K. 281, 283
Noakes, D. L. G. 102, 111
Noble, E. P. 476, 478, 483
Noble, G. K. 173, 177, 188, 245-247, 255, 259, 262, 263, 267, 272, 277, 279, 283, 290, 292, 303, 323, 324, 339, 344, 354
Noble, J. F. 281, 282
Nordeng, H. 137, 148
Nordstrom, W. R. 562, 570
Norris, J. R. 32, 35
Northcutt, R. G. 127, 130, 155, 165, 167, 168, 363, 367
Nose, T. 48, 57, 59, 60
Noth, W. 613, 615
Nottebohm, F. 357, 368
Novikova, L. A. 41, 44
Novotny, M. 6, 30, 36, 80, 84, 85, 231, 241, 428, 468, 469
Nowell, N. W. 87, 97, 515, 517
Nunez, A. A. 479, 483, 484
Nussbaum, R. A. 179, 188
Nyby, J. 79, 85, 473, 479, 481-485
Nyu, N. 50, 59

O'Bleness, M. 274, 276
O'Brien, S. J. 418, 42121
O'Connell, B. 37, 39, 43, 44, 63, 76, 321
O'Connell. R. J. 30, 36, 188, 434, 438, 471, 484, 561, 568, 570
O'Connor, R. M. 18, 21
O'Connor, S. M. 647, 648, 652
Odham, G. 520, 530
O'Hara, R. K. 102, 111, 229-231, 234, 235, 239-241, 397, 409, 419, 421
Ohguchi, O. 102, 113
Ohloff, G. 561, 569
Ohsugi, T. 47, 48, 54, 58-60
Oka, Y. 127-129, 131, 132
Okada, H. 121-124, 132
Okano, K. 29, 32, 35
Oldham, R. S. 342, 355
Oliphant, L. W. 370, 383
Oliver, W. L. R. 613, 615, 647, 648, 652
Olivier, R. 627, 644
Olson, D. H. 102, 111, 235, 240
Olson, E. C. 17, 25
O'Mahony, M. 678, 684
Omedes, A. 655, 656, 658, 659, 671
O'Neill, P. 129, 132, 156-161, 167, 171
Organ, J. A. 177, 178, 188
Orians, G. H. 193, 201, 243, 259
Oring, L. W. 243, 257, 385, 394
Orlov, V. N. 548, 550
Orndorf, M. A. 674, 683
Ortleb, E. P. 345, 355
Oshima, K. 129, 132
Owen, K. 87, 97, 575, 576, 590, 598, 607
Owens, D. W. 17, 24, 346-350, 353, 355
Owen-Smith, N. 385, 386, 395
Owings, D. H. 528, 530
Ozon, R. 251, 259

Page, J. M. J. 30, 34
Page, L. M. 102, 113
Paleologou, A. M. 29, 36
Pandazis, C. 357, 368
Pandey, S. D. 424, 430
Pangborn, R. M. 674, 675, 683
Pankey, J. 226, 241, 397, 399-401
Pankhurst, N. W. 129, 132, 159, 170
Paolucci, A. M. 678, 684
Papenfuss, T. J. 177, 188
Papi, F. 369, 382, 383
Paramonova, N. P. 41, 44
Parham, J. C. 104, 112, 115
Parker, G. A. 182, 188, 387, 394, 395, 419, 422
Parker, G. H. 144, 148
Parker, W. S. 262, 277
Parkes, A. S. 6, 436, 438, 600, 606
Parks, C. R. 381, 383
Parrot, D. M. V. 433, 438
Parsons, T. S. 346, 355
Partridge, B. L. 120, 122, 123, 126-128, 132, 149, 151, 153, 154
Partridge, L. 397
Parzefall, J. 120, 132, 343, 355
Paschenko, N. I. 99, 112
Patterson, R. 642, 644
Patterson, R. G. 345, 355
Patterson, R. L. S. 4, 11, 561, 568, 569
Pawlowsky, E. N. 103, 113
Pawson, M. G. 48, 60, 143, 148
Payne, A. H. 468, 470
Payne, A. P. 181, 184, 188
Payne, K. 627, 644
Payne, T. 184, 188
Peacock, R. L. 179, 188
Peake, B. 415, 417, 418, 420, 422
Pears, F. N. 595, 607
Pearson, A. A. 167, 170
Pearson, O. P. 610, 616
Pease, L. R. 415, 422
Pease, R. 420, 422
Peck, J. N. 137, 148
Pedersen, T. 552, 554
Pelikan, J. 444, 454
Perkins, A. 623-626
Perkins, E. M. 655, 656, 659, 671
Perlmutter, A. 119, 129
Pern, U. 119, 121, 125, 131
Pernetta, J. C. 609, 616
Perrigo, G. 184, 188, 541, 550
Perrill, S. A. 323, 324, 331, 333, 335, 340
Perry, G. C. 27
Peryam, D. R. 680, 684
Peter, R. E. 128, 129, 131, 151, 154, 165, 169, 170
Peters, J. A. 346, 352
Peters, H. 552, 554
Peterson, T. T. 369, 383
Petit, G. R. 113
Petrie, M. 389, 395, 515, 517
Petter, J. J. 647, 648, 650, 652
Pfaff, D. W. 167, 171, 575, 576
Pfaffenbach, G. M. 420, 422
Pfaffman, C. 2, 3, 30, 36, 561, 568, 570
Pfeifer, S. 640, 644
Pfeiffer, W. 99, 102-105, 108-110, 113, 173, 189, 239, 241, 507, 516, 517
Pflieger, W. L. 106, 113
Phelan, R. L. 261, 276, 299, 301
Phillips, J. G. 249, 259
Pianka, E. R. 192, 203
Pigman, W. 77
Pilgrim, F. J. 680, 684
Pine, R. H. 579, 584
Pippitt, D. D. 579, 584
Pisani, G. R. 245, 259
Pitcher, T. J. 102, 113
Plasa, L. 342, 354
Platt, W. J. 610, 616
Pliner, P. 681, 684
Poduschka, W. 610, 612-616
Poey, F. 610, 616
Policard, A. 243, 260
Pollack, J. A. 333, 339
Polyakov, I. Ya. 548, 549
Pomerantz, S. M. 479, 484
Pool, T. B. 189
Poole, C. F. 30, 36
Poole, T. 37, 39, 41, 43
Poole, T. B. 655-660, 662, 663, 669, 671
Porter, M. E. 595, 606
Porter, N. 507, 517
Porter, R. H. 226, 241, 292, 303, 344, 355, 397-406
Portmann, A. 357, 368, 369, 383
Posadas-Andrews, A. 490, 504
Powell, A. J. 450, 455
Powell, R. A. 558, 560
Powers, E. B. 137, 148
Powers, J. B. 399
Powlesland, R. G. 370, 373, 375, 381, 383
Prakash, I. 592, 606
Prasad, M. 105, 108, 110, 111, 243, 260
Pratt, H. P. M. 434, 438
Preib, F. B. 216, 219
Preti, G. 4, 96, 97, 561, 570, 665, 672
Price, A. H. 17, 25, 262, 263, 277

Priesner, E. 7
Prieston, R. 102, 112
Prince, J. H. 363, 368
Pruitt, C. H. 39, 43, 63, 76
Prytula, R. E. 515, 517
Puche, H. 507, 517
Puget, A. 575, 576
Pyastolova, O. A. 235, 236, 241
Pyke, G. H. 196, 203

Quadagno, D. M. 425, 431
Quay, W. B. 22, 25, 362, 368, 541, 544, 548, 550, 592, 606
Quinn, T. P. 226, 241, 397, 411, 419, 422

Radcliffe, A. 37, 44
Radcliffe, C. W. 37, 39, 41, 43, 44, 63, 76, 263, 276
Rafuse, R. 370, 383
Rahn, H. 246, 260, 287, 303
Raisman, G. 6
Rand, A. S. 306, 314, 315, 318, 320, 321
Rand, W. M. 314, 321
Rapaport, L. 641, 645
Rapp, J. P. 463, 469
Rasmussen, K. 627-629, 632, 643-645
Rauch, J. C. 246, 260
Rausch, R. L. 274-276
Rayve, R. S. 588, 589, 595, 599, 601, 603, 605
Real, P. G. 376, 383
Redhead, T. D. 87, 97
Reddy, P. R. K. 243, 260
Reed, H. C. B. 561, 568, 569
Reed, H. D. 103, 104, 113
Reed, J. R. 105, 114
Reformato, L. S. 63-65, 67, 77
Refstie, T. 159, 171
Regaud, C. 243, 260
Regnier, F. E. 568, 570, 592, 595, 606
Reichert, S. E. 387, 395
Reid, L. D. 507, 514, 515, 517
Reidinger, R. F. 376, 379, 383
Reinert, H. K. 287, 288, 303
Reinhart, V. 29, 36
Reis, D. J. C. 169
Rensch, B. 627, 645
Ressler, R. H. 507, 517
Reucroft, P. R. 113
Reutner, T. F. 33, 36
Reyes, A. A. 415, 422
Reynolds, H. C. 581, 583, 584, 598, 601, 606
Reynolds, J. 434, 438, 449, 455
Rice, M. 568, 570, 579, 585
Rice, R. V. 71, 77
Rich, W. H. 136, 148
Richard, A. 647, 649, 651, 652
Richards, C. M. 235, 241
Richardson, F. L. 88, 97
Richardson, W. S. 329, 339
Richmond, M. E. 463, 470
Richmond, N. D. 102, 114
Ricker, W. E. 137, 148
Rieger, I. 390, 395
Rijks, J. A. 84
Riordan, D. V. 613, 616
Rissman, E. F. 428, 431, 555, 559, 560
Ritter, F. J. 6, 7, 11
Rivers, W. H. R. 674, 684
Robbins, T. 319, 321
Roberts, J. B. 260, 291, 295, 299 303
Robin, J. P. 361, 368
Robinson, C. 69, 77
Robinson, D. C. 347, 348, 354
Robinson, J. A. 156, 167, 171
Roelke, M. E. 418, 421
Rogers, N. H. 113
Rogoff, J. L. 179, 189
Rohlf, F. J. 268, 277
Romano, D. 306, 320, 326, 327, 339
Roper, T. J. 490, 504
Rose, E. A. 79, 84, 427, 430, 448, 454, 554
Rose, F. C. 235, 241
Rose, F. L. 206, 218
Rose, S. M. 235, 241
Rosenberg, H. I. 21, 25
Rosenberg, L. 675, 676, 683
Rosenberg, L. E. 571, 576
Rosenthal, G. A. 370, 375, 381, 383
Ross, P. 243, 245-247, 253, 260
Rossi, A. C. 106, 111
Rossman, D. A. 244, 260, 268, 277
Roth, G. 181, 189
Roth, J. J. 243, 257
Roth, S. I. 295, 303
Roth-Meyer, C. 425, 431
Rottman, S. J. 507, 514, 517
Rowe, F. P. 457, 461, 533, 540
Rowell, T. E. 425, 431
Rozin, P. 681, 682, 684
Roznov, V. V. 520, 531
Rubec, P. J. 119, 132
Ruddy, L. L. 105, 114
Ruibal, R. 305, 306, 321
Ruitenberg, E. J. 35
Russell, A. P. 21, 25
Russell, M. J. 5, 11, 397, 411
Ruthven, A. G. 244, 260, 273, 277
Rutten, G. A. F. M. 84
Ruttle, K. 399
Ryan, K. D. 434, 438
Rylands, A. B. 658, 659, 662, 671

Saal, F. vom 281, 283
Sackett, G. P. 226, 240, 242, 419, 422
Saglio, P. 149, 154
Sajdak, R. A. 267, 277
Sakai, D. K. 121-124, 132
Salmon, M. 387, 394
Salthe, S. N. 176, 177, 189
Sambraus, H. H. 29, 36
Samuelsen, P. 200, 203
Sanchez-Criado, J. E. 478, 484
Sandberg, A. A. 281, 283
Sandell, M. 520, 530
Sangster, A. W. 48, 59
Saplikoski, N. J. 514, 515, 518
Sar, M. 129, 131
Sargent, T. D. 40, 43
Sarig, S. 46, 60
Sarmiento, L. A. 415, 422
Satou, M. 127, 128, 131, 132
Savage, T. 216, 218
Sayler, A. 191, 203
Scalia, F. 434, 439, 482, 485
Scaling, A. 613, 616
Schadler, M. H. 463, 469, 559, 560
Schaffer, J. 579, 584, 595, 606, 617, 622
Schaller, G. B. 395, 487, 504
Scheer, B. T. 136, 148
Schell, F. M. 291, 292, 295, 303, 304
Schiffman, S. S. 678, 684
Schildknecht, H. 520, 531
Schiller, D. 681, 684
Schilling, A. 648, 650, 652
Schleidt, M. 674, 675, 684
Schmalhausen, I. I. 13, 16, 25
Schmidt, A. 628, 636, 644
Schmidt, K. P. 274, 277
Schmidt, M. 627-629, 632, 636, 644, 645
Schmidt-Nielson, K. 289, 300, 303
Schmitz, S. 479, 481, 482, 484
Schneider, D. 255, 259
Schneider, J. 479, 484
Schoffeniels, E. 362, 367, 368
Schofield, C. W. 262, 264, 272, 274, 276
Scholz, A. T. 137, 138, 147, 148, 346, 355
Schreibman, M. P. 127, 129, 130, 132, 155-162, 164-167, 169-171
Schulman, N. 64, 65, 67-69, 71, 76
Schultz, J. C. 99, 111
Schultz, T. H. 30, 36
Schultze-Westrum, T. G. 4, 588, 594, 595, 598, 601, 607
Schulze, D. H. 415, 420, 422
Schutz, F. 102, 103, 108, 109, 114
Schwanzel-Fukuda, M. 156, 167, 171
Schwartz, N. B. 434, 438
Schwartz, R. 591, 605
Schwende, F. J. 30, 36, 80, 85, 231, 241
Schwenk, K. 183, 187, 305, 319, 321, 325, 340
Scott, A. P. 124, 132
Scott, D. K. 401
Scott, H. 30, 36
Scott, J. P. 477, 484
Scott, T. G. 611, 616
Scribani, L. 64, 65, 67, 76
Scudder, K. M. 37, 39, 41, 43, 44, 263, 276, 289, 292, 294, 303, 344, 352, 355, 575, 576
Secoy, D. M. 287, 288, 301
Seghers, B. H. 108, 109, 112
Seifert, R. M. 30, 35
Selander, R. K. 464, 470
Selset, R. 126, 127, 130, 230, 240, 241
Sengupta, S. 370, 381, 383
Settle, R. G. 675, 676, 683
Sever, D. M. 173, 189, 199, 200, 203
Sexton, O. J. 345, 355
Shallenberger, R. J. 363, 364, 368
Shaman, P. 675, 676, 683
Shanfield, H. 30, 36
Sharawy, M. 249, 258, 291, 295, 302
Sharma, K. N. 679, 684
Sharma, S. D. 679, 684
Sharman, G. B. 587, 596, 600, 607
Sharp, S. 173, 187, 194, 202
Shaw, C. R. 463, 470
Shaw, E. 233, 241
Shaw, K. S. 28, 35, 87, 93, 95, 96
Sheffield, L. P. 63, 64, 67, 77
Sheffield, S. D. 428, 431, 555, 559, 560
Shepherd, G. M. 198, 202
Sherman, P. W. 225, 226, 240, 397, 402, 404, 413, 419, 421
Shibasaki, S. 14, 24
Shields, W. M. 225, 241, 397
Shillito, E. E. 533, 540
Shimizu, N. 173, 187, 193, 194, 196, 202
Shinder, T. W. 473, 484
Shine, R. 267, 277, 287, 300
Shiomi, K. 103, 112
Shire, J. G. M. 468-470
Shirer, H. W. 287, 288, 301
Shirley, S. G. 27, 36
Shiro, B. C. 448, 454, 515-517
Shlotzhauer, T. L. 211, 218
Shoop, C. R. 173, 188, 200, 202, 207, 211, 218, 342, 354
Shorey, H. H. 9, 11, 143, 147
Shuberta, H. E. 425, 431

Shupe, D. 579, 583, 584, 598, 600, 606
Shvarts, S. S. 235, 236, 241
Siegel, S. 327, 340, 544, 550
Signoret, J. P. 6, 174, 189
Sikorski, L. 675, 676, 683
Silver, W. L. 376, 383
Silverman, A. J. 156, 167, 171, 172
Silverman, R. C. 129, 132, 156-162, 167, 171
Silverstein, R. M. 6-8, 10, 37, 38, 44, 561, 563, 568-570
Simeone, J. B. 8
Simmons, K. E. L. 364, 367
Simms, D. A. 520, 528, 531
Simon, C. A. 305, 306, 319-321, 323, 326, 327, 339, 340
Simon, G. S. 173, 189, 195, 199, 203
Simonsen, L. 305, 320
Simonsen, M. 419, 422
Simpson, E. 420
Simpson, G. G. 358, 368
Simpson, K. 60
Simpson, S. J. 45, 60
Singer, A. G. 30, 32, 36, 38, 44, 561, 568-570
Sis, R. F. 29, 35
Sisemore, D. A. 226, 242
Skholl, E. D. 548, 550
Skurat, L. 541, 592, 607
Skutch, A. F. 369, 381, 383
Slaunwhite, W. R. 281, 283
Sletten, K. 230, 241
Sloan, H. E. 166, 168
Slotnick, R. M. 415, 416, 421
Smith, A. B., III. 4, 38, 42, 184, 189, 364, 368, 426, 430, 475, 483, 561, 568, 570, 663-666, 670-672
Smith, A. T. 571, 575, 576
Smith, G. C. 250, 260
Smith, H. M. 19, 20, 22, 25, 37, 39, 41, 43, 44, 63, 76, 244, 260, 289, 292, 294, 303, 344, 352, 355
Smith, J. D. 106, 107, 114
Smith, M. 595, 607
Smith, M. A. 254, 260
Smith, M. H. 464, 470
Smith, O. R. 106, 114
Smith, P. A. 534, 537, 539, 540
Smith, R. J. F. 99, 101, 103-111, 113, 114, 126, 130, 239, 241, 285, 303
Smith, W. J. 487, 504
Smolen, M. J. 558, 560
Smythe, N. 487, 504, 615, 616
Snowdon, C. T. 507, 514, 517, 655-658, 660-662, 667, 668, 670
Snyder, B. E. 120, 132
Snyder, D. H. 206, 216, 218
Snyder, G. K. 369, 383
Snyder, N. F. R. 108, 114
Snyder, P. J. 674, 683
Sober, E. 285, 303
Sokal, R. R. 268, 277
Sokolov, E. N. 37, 40, 41, 44
Sokolov, V. E. 7, 520, 531, 540, 550
Sokolov, W. 592, 607
Solomon, S. E. 17, 18, 25
Soltis, R. L. 436, 438
Soneshine, D. E. 516, 518
Sorenson, M. W. 610, 616
Sorensen, P. W. 126
Southwick, C. H. 457, 575, 576
Spaur, R. C. 41, 44
Spiro, R. G. 74, 77
Spotila, J. R. 177, 189
Springer, A. D. 127, 132, 167, 171
Squier, L. 627, 644
Sroges, R. W. 319, 321
Stabell, O. B. 159, 171, 230, 241
Stacey, N. E. 117, 120-124, 126-129, 131, 132, 149-151, 153, 154, 165-167, 170, 171, 173, 189
Stager, K. E. 361, 367, 369, 383
Stagge, B. 519, 531, 562, 570
Stalvey, J. R. D. 468, 470
Stanley, G. 87, 93, 95, 96
Stanley-Price, M. R. 391, 395
Starck, D. 609, 610, 616
Stasiak, R. H. 106, 114
Stasko, A. B. 137, 148
Staton, M. A. 101, 114
Stearns, S. C. 559, 560
Stebbins, R. C. 177, 189, 216, 218
Stecher, P. G. 105, 115
Steger, K. E. 173, 189
Stehn, R. A. 463, 470
Stein, M. 497-502, 504
Steinwascher, K. 235, 242
Stell, W. K. 127, 132, 167, 171
Sternberger, L. A. 156, 171
Stenzler, D. 108, 110, 111, 115
Sterba, G. 105, 115
Steven, D. M. 48, 59
Stevens, D. A. 507, 512, 514-516, 518
Stevens, K. 29, 36
Stevenson, M. F. 656-658, 671
Stewart, E. R. 246, 260
Stewart, K. 279, 282
Stewart, N. J. 289, 292, 294, 303, 344, 355, 575, 576
Still, W. 645
Stimac, K. 37, 39, 41, 43, 44
Stockton, M. D. 479, 484
Stoddart, D. M. 7, 9, 11, 27, 36, 341, 355, 519, 528, 529, 531,

Stoddart, D. M. (Continued)
533-535, 537-540, 579, 584
Stohrer, G. 104, 113
Stokes, G. P. 249, 260, 291, 297, 303
Stokke, O. 642, 644
Stolinski, C. 249, 259, 295, 302
Stolk, A. 122, 133
Stone, M. D. 286, 300
Stonehouse, B. 587, 607
Stoner, D. 370, 383
Storez, R. A. 178, 189
Strauss, J. S. 579, 584, 592, 607
Streilein, J. W. 418, 421
Streilein, K. E. 579, 584, 588-590, 607
Strupp, B. J. 490, 504
Stumpf, W. E. 127, 129, 131, 160, 167, 170
Sugiura, K. 104, 115
Sugiwaka, K. 121-124, 132
Sugowdz, G. 87, 93, 95, 96
Sullivan, D. S. 527, 529
Sullivan, T. P. 515, 518-520, 527, 529, 531
Summers, R. G. 103, 112
Sun, H. H. 45, 60
Sussman, R. W. 647-649, 651, 652, 655, 667-669, 671
Sustarsic, D. L. 463, 469
Sutcliffe, A. G. 655-660, 662, 663, 671
Sutherland, R. L. 600, 607
Sutterlin, A. M. 48, 49, 61
Suzuki, N. 105, 115
Svare, B. 463, 470
Svendsen, G. E. 562, 570
Sweatman, H. P. A. 181, 186,
Swigar, A. A. 561, 568, 569
Switz, M. G. 5, 11
Sype, W. E. 242
Szalo, G. 295, 303
Szybalski, W. 370, 383

Taft, L. K. 589, 606
Takahashi, F. T. 255, 259
Takashima, Y. 47, 60
Taleisnik, S. 478, 483
Tamarkin, L. 423, 431
Tamura, T. 45, 47, 48, 50, 59
Tanaka, R. D. 51, 56, 58, 61
Tanaka, Y. 60
Tanner, W. M. 18, 23
Tanzer, M. L. 69, 71, 73, 77
Tardif, S. D. 661, 671
Tarver, M. 173, 187, 194, 195, 197-199, 202
Taub, A. 643, 644
Taub, F. B. 193, 203
Tavolga, W. N. 120, 122, 126, 133
Taylor, E. H. 329, 340
Taylor, S. 305, 320
Taylor, S. V. 63, 76
Tedford, B. L. 361, 367
Teipner, C. L. 528, 531
Teeter, J. H. 142-144, 146, 147
Telle, H. J. 488, 491, 504, 505
Teller, M. N. 104, 115
Templeton, A. R. 223, 224, 273, 277
Tepper, V. J. 402, 405
Terman, C. R. 424, 431
Teyssedre, C. 181, 189
Thaler, H. T. 408, 411, 413, 415, 417-419, 422
Thapliyal, J. P. 249, 258
Thau, R. 167, 171
Theisen, B. 159, 171
Thiessen, D. D. 10, 508, 512, 518, 568, 570, 575, 576, 579, 585, 590, 592, 598, 607
Thines, G. 103, 104, 109, 110, 115
Thisted, R. 181, 183, 184, 187
Thomas, E. 627, 644
Thomas, E. S. 274-276
Thomas, L. 231, 240, 242, 408, 411, 413-419, 422, 466, 470
Thomas, R. F. 305, 321
Thomas, R. L. 515, 517
Thomas, S. 48, 59
Thommesen, G. 126, 127, 133, 163, 171, 230, 240
Thompson, J. A. 595-598, 607
Thompson, K. 5, 11
Thompson, W. J. P. 370, 383
Thornhill, R. 182, 189, 279, 283
Tilakeratne, N. 632, 644
Tilley, S. G. 182, 183, 187, 189, 205, 210, 216, 218
Tilson, E. F. 479, 484
Timberlake, W. 592, 605
Timms, A. M. 144, 147
Tinbergen, N. 40, 44
Tinkle, D. W. 287, 303
Toch, E. 425, 431
Todd, J. H. 48, 59, 173, 185
Tokarz, R. R. 245, 247, 248, 252, 253, 255, 256, 258, 281, 282
Tomich, P. Q. 592, 606
Tracy, C. R. 342, 355
Trainer, D. O. 370, 374, 382
Trauth, S. 335
Traynier, R. M. M. 143, 148
Tristram, D. A. 173, 189, 195, 196, 198, 200, 203, 211, 218, 342, 355
Trivers, R. L. 101, 115, 210, 218, 398, 411, 418, 422
Truitt, R. V. 262, 278
Trumler, E. 623, 626
Trupiano-Duvall, J. 324, 327-329, 335, 339, 344, 353

Trupin, G. L. 579, 584, 589, 595, 599, 601, 605, 607
Trussell, P. C. 370, 382
Tsang, T. 678, 684
Tsumagari, M. 50, 60
Tsushima, J. 48, 61
Tucker, D. 105, 115, 376, 382
Turk, A. 4, 11
Turner, J. 42
Turner, J. W. 575, 576, 623-626
Twitty, V. C. 173, 179, 185, 186, 189, 206, 218, 219, 341, 342, 353, 355
Tyler, S. J. 623, 626
Tyndale-Biscoe, H. 583, 585, 595, 598, 600, 607

Ueda, K. 124, 127-129, 131-133
Uhazy, L. S. 51, 56, 58, 61
Ulibarri, C. 399
Utsumi, K. 29, 32, 35
Uzzell, T. 178, 189

Vagvolgyi, A. 245, 255, 259
Valentine, G. L. 559, 560
VandeBerg, J. L. 589, 599, 601, 605, 607
Vandenbergh, J. G. 5, 6, 9, 11, 27, 36, 84, 427-429, 431, 433, 439, 441, 443, 444, 448, 451-455, 457-463, 469, 478, 484, 554, 555, 558, 560, 571, 576
Van Den Berk, J. 519, 531
Vandenbussche, E. 110, 115
van den Hurk, R. 122-125,131, 133, 156, 157, 162, 170
van der Lee, S. 423, 431
Van Etten, C. H. 678, 684
van Oordt, P. G. W. J. 123, 133
Vasilieva, V. S. 520, 531
Vaticon, M. D. 478, 484
Vellano, C. 178, 180, 181, 186, 188
Verberne, G. 642, 645
Vernet-Maury, E. 515, 518, 528
Verrell, P. A. 178, 179, 181, 182, 186, 189, 190
Vial, J. L. 216, 219
Vigouroux, M. 517
Vijayakumaran Nair, P. 627, 644
Vinogradova, O. S. 37, 44
Vinogradova, V. M. 120, 133
Vinyard, G. L. 101, 115
Vitt, L. J. 19, 23, 292, 301, 315, 320, 323, 326-335, 338-340
Voronin, L. G. 37, 40, 44
Vrba, E. S. 285, 295, 301
Vrtis, V. 541, 544, 550

Waagen, G. N. 18, 25
Wacher, T. 392
Wade, G. N. 248, 260
Wade, M. J. 183, 185, 190
Wahlert, G. von, 14, 23
Waite, J. H. 69, 77
Wakabayashi, R. 77
Wake, D. B. 177, 179, 188, 190
Walcott, C. 173, 190
Waldman, B. 102, 104, 108-110, 115, 225-236, 238, 242, 397
Wallace, R. B. 415, 422
Walker, E. P. 589, 595, 607
Walker, J. W. 682, 683
Walker, S. E. 127, 132, 167, 171
Wallace, R. A. 248, 260
Wallace, R. L. 287, 301
Wallis, J. 425, 426, 431
Wallraff, H. G. 369, 383
Walsh, T. 37, 39, 43
Walter, D. G. 346, 354
Walther, F. R. 386, 388, 392, 395
Walton, I. 136, 148
Waltz, E. C. 488, 505
Wand, M. A. 344, 352
Wand, B. U. 363, 367
Warburg, M. R. 291, 303
Ward, A. L. 528, 530
Ward, H. B. 136, 137, 148
Ward, P. 488, 505
Ware, G. C. 28, 34
Warham, J. 363, 368
Waring, G. H. 624, 626
Waring, C. H. 29, 36
Warnock, N. 37, 44
Warwick, R. T. T. 359, 367
Wasden, R. E. 507, 514, 515, 517
Wasilewski, W. 554
Wassersug, R. J. 226, 242
Wasylik, A. 370, 383
Watanabe, K. 125, 126, 128, 133
Watanabe, Y. 239, 242
Watkins, J. F. 247, 258, 262, 267, 277
Watson, J. D. 423, 432
Watson, M. R. 69, 77
Watson, P. 179
Weatherall, J. A. 69, 77
Weiner, B. E. 19, 25
Wedekind, H. P. 155, 171
Weiss, E. H. 420, 422
Welch, J. F. 87, 95, 97
Weldon, P. J. 99, 115, 279, 283, 285, 291, 292, 295, 303, 304, 518
Wellborn, S. 344, 352
Wellington, J. L. 37, 38, 42, 471, 473-475, 483, 485
Wells, K. D. 192, 203
Wells, K. E. 41, 44
Welty, J. C. 369, 381, 383

Wemmer, C. 595, 607, 612, 615, 616
Wenzel, B. M. 4, 10, 13, 357, 359, 361, 362, 364, 367-369, 383
Werner, J. K. 173, 187, 200, 202
Wertz, P. W. 291, 300, 301
Wesley, A. L. 226, 242
West, J. R. 561, 568, 569
West, L. B. 235, 242
West Eberhard, M. J. 337, 340, 400
Westerberg, H. 140-143, 145, 147, 148
Westlake, G. F. 144, 147
White, D. M. 401, 402, 405
White, F. N. 246, 260
Whitear, M. 103, 113, 115
Whiteman, C. E. 370, 384
Whiting, A. M. 335, 340
Whitney, G. 473, 479, 481-485
Whitsett, J. M. 443, 444, 454, 455
Whitten, W. K. 4-6, 10, 174, 181, 190, 423, 432, 433, 439, 468, 469, 571, 576, 595, 600, 603, 604, 607, 653, 672
Whittier, J. M. 243, 245, 252, 253, 257, 260
Whitworth, M. R. 572, 573, 575, 577
Wieland, W. 105, 114
Wiens, J. A. 361, 367
Wiesler, D. 30, 36, 80, 85
Wigmore, S. W. 488-491, 493-498 504
Wilcoxon, H. C. 63, 76
Wilde, W. S. 63, 77
Wildt, D. E. 418, 421
Williams, G. C. 99, 101, 104, 115, 285, 304
Williams, M. 233, 242
Williams, S. 179, 190
Williamson, N. B. 29, 36
Wills, G. D. 226, 242
Wilson, E. O. 4, 11, 118, 133, 135, 147, 182, 190, 229, 242
Wilson, J. D. 281, 282
Wilson, M. 468, 469
Wilson, M. I. 533, 539
Wilson, S. R. 468, 469
Wilz, I. 520, 531
Wimberger, P. H. 370, 384
Winans, S. S. 434, 439, 482, 485
Windholz, M. 105, 115
Winer, B. J. 327, 340
Winkler, C. A. 418, 421
Winokur, R. M. 18, 25
Wisby, W. J. 137, 147, 148
Wise, D. R. 433, 438, 453
Wishlow, W. P. 120, 131
Wislocki, G. B. 655, 672
Witkin, J. W. 167, 172
Witzell, W. N. 346, 355
Wolf, H. G. 601, 604
Wolff, J. O. 434, 439
Wolff, P. R. 450, 455
Wolters, H. J. 658, 661, 672
Wong, K. L. 15, 18, 23
Wood, H. B., Jr. 113
Woods, J. E. 206, 219
Wright, A. A. 262, 278
Wright, A. H. 262, 278
Wright, D. E. 18, 25
Wright, P. J. 29, 36
Wright, R. H. 143
Wrobel, D. J. 173, 190
Wu, H. M. H. 226, 242, 419, 422
Wuensch, K. L. 533, 540
Wuttke, W. 435, 439
Wyant, K. W. 681, 684
Wyche, C. J. 102, 113
Wyrick, M. 226, 241, 397, 399-401, 403, 411
Wysocki, C. J. 37, 38, 42, 231, 242, 415, 416, 421, 422, 466, 470-476, 478-480, 484, 485

Yahr, P. 575, 577
Yamada, K. M. 15, 25
Yamaguchi, K. 127, 132
Yamaguchi, M. 415, 417, 418, 420, 422
Yamamoto, M. 126, 133, 158, 172
Yamashita, S. 50, 60
Yamazaki, F. 125, 126, 129, 133
Yamazaki, K. 231, 240, 246, 408, 411, 413-422, 466, 468-470
Yang, S. Y. 464, 470
Yao, P. 67-69, 77
Yapp, W. B. 13, 25
Yarger, R. G. 561, 570, 665, 672
Yoneda, M. 662, 672
York, D. S. 306, 319, 320
Yoshii, K. 50, 61
Young, G. 125, 131
Young, R. S. 519, 530
Yuhara, M. 29, 32, 35
Yun, S. S. 648, 652

Zahavi, A. 243, 260, 388, 395, 488, 505
Zahorik, D. 643, 644
Zajonc, R. 681, 684
Zar, J. H. 552, 554
Zayas, Z. A., 408, 411, 413, 415, 417-419, 422
Zeiske, E. 159, 163, 168, 171, 172
Zellner, D. A. 681, 684
Zemstova, M. I. 41, 44

Ziegler, I. 104, 112
Ziegler, M. 520, 531
Zinkevich, E. P. 7, 520, 531, 541, 550
Ziswiler, V. 15, 18, 24, 359, 362, 367
Zivin, J. A. 156, 172
Zlatkis, A. 30, 36
Zohar, Y. 124, 125, 130
Zotterman, Y. 2, 3, 11
Zucker, I. 473, 484
Zusho, H. 675, 676, 683

SUBJECT AND TAXONOMIC INDEX

<u>Acheilognathus</u> <u>lanceolatus</u>, 120, 122
<u>Achillea</u> <u>millefolium</u>, 372, 373, 374, 376, 377, 379
<u>Acomys</u> <u>cahirinus</u>, 226, 397, 400-404, 406-409
Adaptation, 41, 299, 300,
- defined, 285, 286
- rattlesnake skin lipids, 289-298

Adrenal glands, 79-84
Adrian, E. D., 2, 3, 6
<u>Aepyceros</u> <u>melampus</u>, 386
Afghan pika, 575
African antelopes, 385
African catfish, 125
African elephants, 632
African wild ass, 623
Aggregation chemosignals
- snakes, 262, 266-267, 289-295

Aggression and chemosignals
- lizards, 333, 334
- in mice, 476-479
- and pika familiarity, 571-575
- and pika mating, 571-575

<u>Agkistrodon</u> <u>contortrix</u>, 288
<u>Agkistrodon</u> <u>contortrix</u> <u>mokeson</u>, 288
Agouti, 487
<u>Agrimonia</u>, 377
<u>Agrimonia</u> <u>paraflora</u>, 372-374, 376, 377, 379
Agrimony, 373
Alarm pheromones (<u>see</u> <u>also</u> alarm signals)
- analgesia, 514
- anti-pathogenic function, 104
- in anuran tadpoles, 238, 239
- blood odors, 507, 513, 514
- chemistry in fishes, 104, 105
- ecology, 514-517
- evolution, 514-517

Alarm pheromones (Continued)
- fishes, 99-110
- fright reactions, 515
- habituation to, in fishes, 110
- kairomone, 507
- population density, 514
- in prey-predator interactions, 507-517
- sensitization to, in fishes, 110

Alarm signals, 507-517 (<u>see</u> <u>also</u> alarm pheromones)
- analgesia, 514
- attraction, 510
- avoidance, 510
- blood odors, 507, 513, 514
- ecology, 514-517
- evolution, 514-517
- exploration, 512, 513
- fright reactions, 515
- gerbil activity suppression, 508-510
- gerbil hesitation, 510
- physiological arousal, 512
- population density, 514

Alarm substance in fishes, 99-110 (<u>see</u> <u>also</u> alarm pheromones)
- alarm substance cells, 99-110
- of darters, 99-101
- defensive behavior, 99-110
- and epidermal damage, 99-101, 103-107
- evolution, 99-110
- and fright reaction, 102
- in ostariophysan fishes, 99-110
- predator deterrent, 104, 108-110
- sacciform cells, 99, 100
- in schools, 101, 102
- selection on receiver, 107-110
- selection on sender, 100-107

Alarm signal theories, 100-107, 507, 514-517
Alliaria, 377
Albatrosses, 358, 362, 365
Albone, E. S. 7, 9
Alcelaphus buselaphus, 386
Alcids, 358
Alfalfa, 572
Alle alle, 358
Alliaria officinalis, 374, 377, 379
Altruism, 413
Amago salmon, 120
Ambrose Monell Foundation, 4
Ambystoma annulatum, 177
Ambystoma dumerili, 178
Ambystoma gracile, 178
Ambystoma jeffersonianum, 178
Ambystoma laterale, 178
Ambystoma macrodactylum, 178, 181
Ambystoma maculatum, 177
Ambystoma mexicanum, 178
Ambystoma opacum, 177
Ambystoma texanum, 177
Ambystoma tigrinum, 178
Ambystomatidae, 176, 177, 198
American martin, 617, 619
  plantar glands, 619
American opossums, 588
American toad, 226
Americans, 678
  of European ancestry, 679
  soldiers, 682
Ammocrypta beani, 109
Amniotes, 17, 19, 20
Amphibians, 14-16, 173-184, 341, 342, 351, 352, 609
  chemical landmarks, 342
  chemosensory homing, 342, 343
  chemosensory orientation, 342, 343
  habitat odors, 342
Anal and fecal secretion marking in house mice, 88, 89
  chemistry, 92-95
  dominance effects, 91, 92
  effects of own versus others odors, 90-92, 95
  headspace analysis, 93-95
  heart rate responses, 94, 95
  gas chromatographic and mass spectrometric analysis, 93, 94
  spatial effects, 89, 90
Anas platyrhynchos, 362
5α-androst-16-en-3β-ol, 568
5α-androst-16-en-3-one, 4, 568
Aneides, 191
Aneides aeneus, 194, 206, 216
Anguilla anguilla, 103, 129
Anguilla japonica, 48
Animal Communication by Pheromones, 9
Annelid cuticle collagens, 73-75
22nd Annual Symposium on Biology of the Skin, 5
Anolis, 248, 313, 318
Anolis trinitatus, 306
Anoptichthys antrobius, 103
Anoptichthys jordani, 103, 104
Antarctic fulmar, 358, 361
Antarctic petrel, 359
Antechinus flavipes, 595, 597
Antelope, 385, 392
Anthracosaurian amphibian, 16, 19
Antilocapra americana, 568
Anuran chemosignals and kin recognition, 225-239
  bile salts, 230
  development, 230-235
  familiarity, 230-235
  fitness effects, 235-239
  growth regulating factors, 235-238
  habituation, 233
  histocompatability responses, 231
  kin directed growth regulating factors, 236-239
  learning, 233
  maternal effects, 230, 231
  phenotype matching, 239
  phenotypic template recognition, 233, 234
  population specific substances, 230
  preferences and aversions, 226-229
  sensory accommodation, 230
  sensory mechanisms, 226-229
  sibling groups, 228-230
  signal characterization, 229, 230
  steroids, 230
  toxins, 230
Anurans, 102, 174, 226-239, 342, 343
  chemical signaling, 227-239
  evolution of social behavior, 225-239
  kin recognition, 225-239
  social structure, 225-239
  tadpole schools, 232, 233
  tadpoles and larvae, 226-239
Apes, 653
Apodemus flavicollis, 535
Apodemus sylvaticus, 533-539
  comparative social behavior, 533-539
  conspecific chemical signals, 533-539

Apodemus sylvaticus (Continued)
heterospecific chemical signals, 533-539
population variation, 537, 538
sex chemosignal discrimination, 536, 537
Appetitive behavior, 40, 41
Apterygiformes, 364
Apteryx, 364
Arthropods, 375
Arvicola, 548
Aquatic and marine snakes, 343
Aquatic turtles, 345-352
homing and orientation, 345-352
social odors, 345, 346
Arius felis, 103
Ashy storm-petrel, 362
Asian elephant, 627-643
Asian elephant chemical signals, 627-643
acid responses, 638
aliphatic hydrocarbon responses, 638
benzene-related responses, 638
calve responses, 632
estrogen responses, 638
estrous signals, 637, 642, 643
evolution, 641
female responses, 632
interspecific responses, 637-639
ketone responses, 638
male responses, 632-637
maternal recognition, 641
musth, 632, 636
phenol responses, 638
of placenta, 640
response habituation, 639
responses to urine fractions, 639, 640
semiochemical bioassay, 628
sulfur compound responses, 638
temporal gland exudates, 632, 635, 640
urine chemistry, 629-632
urine Flehmen responses, 632-639
urogenital mucus, 632
Asian wild ass, 623
Aspredininae, 103
Association for Chemoreception Sciences, 7
Aster, 377
Aster praealtus, 376, 377
Aster prenanthoides, 376, 377
Astyanax mexicanus, 103
Atlantic salmon, 136, 140-142, 159
Auburn symposium, 4
Auk, 358
Aulacomys, 548
Aves, 341, 357, 364, 369, 381, 528
Avoidance chemosignals
and crop protection, 519, 520, 527-530
mustelid anal gland compounds, 519-529
mustelid avoidance by gophers, 519-529
mustelid 3,3-dimethyl-1,2-dithiolane, 520, 523-528
mustelid 3-propyl-1,2-dithiolane, 520, 523-528
rodent predator avoidance, 519-529
snakes, 266
Ayu, 120
Azadirachta indica, 381

Bacteria, 373, 374, 381
Bacterial commensalism, 28
Badgers, 388, 528
BALB/cByJ mice, 463-466, 468
Bank swallows, 226
Bank vole, 519, 551, 552, 554
population density effects, 551-554
puberty influencing chemosignals, 551-554
Barbarea vulgaris, 377
Bardach, J. E., 4
Bare-ear marmosets, 654
Bartholin's Gland
(see Cattle)
(see Major vestibular gland of cattle)
Bathygobius soporator, 120, 122
Bathystoma rimator, 48
Beach, F. A., 6
Beagle, 30
urinary headspace semiochemistry, 30
Beauchamp, G. K., 4, 9
Beaver, 519, 561-569
chemosignal behavioral functions, 561
field study, 561-569
responses to castoreum fractions, 561-569
Beets, M., 5
Belontiids, 119, 120, 128
Birch, M. C., 8
Bird chemical signals, 357-366
(see also Procellariiform chemical signals)
anal glands, 362, 363
in feeding homing and communication, 369
Harderian gland, 363

Bird chemical signals (Continued)
  and nest construction, 371-373
  and plant fumigant discrimination, 371-373
  plant fumigant use, 369-381
  plant volatile chemical discriminations, 376-381
  preen substance, 362
  sebaceous glands, 362, 363
  secretory products, 362, 363
  uropygial preen gland, 15, 362, 363
Bird chemoreception, 357-366, 369-381
  and nest fumigant use, 369-381
  plant volatile discrimination, 376-381
Bird olfaction
  ecology, 357-366
  vomeronasal system absence, 357
  nasal cavity structure, 357
  sensory mucosa, 358
  conchae, 358, 359
  mitral cell counts, 359, 360
  bulb volume, 359, 360
  (see also Procellariiform olfaction)
  responses to prey odors, 360-362
  and foraging, 361, 362
  terminal nerve absence, 357, 363
  evolution of, 364-366
  plant volatile discriminations, 376-381
Bird reproduction, 363, 364
  and chemical signals, 363, 364
  nest fumigant functions, 369, 370, 381
  parasite and pathogen control, 369, 370, 373, 374, 381
Birds, 10, 19, 101, 173, 174, 216, 226, 280, 281, 357-366, 369-382, 609
  nasal salt gland, 358
Bitterling, 122
Black Americans, 675-681
Blackbanded darters, 109
Black-chested moustached tamarin, 654
Black goby, 119
Black-handed tamarin, 654
Black mustard, 373
Black-tailed deer, 42, 561, 568, 643
  alert reaction of, 42
Black-tailed marmoset, 654
Black-tufted-ear marmoset, 654
Blandfordimys, 548
Blarina, 610
Blarina brevicauda, 610, 611, 614
Blennius pavo, 119, 121, 125
Blenny, 119
Blind cave characins, 104, 109
Blind cave fishes, 103
Blood plasma
  cattle, 29
Blood worm, 65, 73
Bluegill, 104
Bluehead chub, 100
Blum, M. S., 7
Boa constrictor, 293, 299
Bobcat, 528
Bolitoglossa engelhardti, 179
Bolitoglossa flavimembris, 179
Bolitoglossa franklini, 179
Bolitoglossa macrinii, 177
Bolitoglossa morio, 179
Bolitoglossa occidentalis, 179
Bolitoglossa resplendens, 179
Bolitoglossa rostrata, 179
Bolitoglossa subpalmata, 178
Bonnethead shark, 142-144, 146
Bower birds, 369
Brachydanio rerio, 104, 106-108, 119, 122-124
Brain-pituitary-gonadal axis, 155-168
Brassica, 678
Brassica nigra, 373, 377, 379
Broad-headed skink, 327, 335
Bronson, F. H., 6, 10
Brower, Lincoln P., 8
Brown bullhead, 142
Brown snakes, 344
Brown trout, 125
Bruce effect, 433, 434
  mouse urine esterase genetics, 463-469
Bruce, Hilda M., 6, 433
Brush possum, 595
Brussel sprouts, 678
Buffalo, 386
Bufo, 230
Bufo americanus, 226-238
Bufo boreas, 235, 342
Bufo bufo, 238
Bullhead, 48
Bunocephalus bicolor, 105
Burghardt, G. M., 10
2-sec-Butyldihydrothiazole, 30

C57BL/10J mice, 463, 465, 466, 468
C57BL/6J mice, 463, 465, 466, 468
C57BL/KsJ mice, 463, 465, 466, 468
C57BL/6ByJ mice, 463, 464, 465, 466, 468

Cabbage, 678
Caecobarbus geertsi, 103, 104
Calamoichthys calabricus, 103
California vole, 428, 429, 555
Callichthyidae, 105
Callitrichidae, 653
Callithrix, 653, 655, 657, 662, 668
Callithrix argentata, 654, 658
Callithrix argentata argentata, 654, 655, 658
Callithrix argentata melanura, 654, 658
Callithrix humeralifer, 654, 658
Callithrix humeralifer intermidius, 654, 658, 659
Callithrix jacchus, 654
Callithrix jacchus jacchus, 425, 654-656, 658, 661, 663, 666
Callithrix jacchus penicillata, 654, 655, 668
Callitrichids, 653-669
  anal marking, 657
  anogenital marking, 657
  apocrine gland, 655-659
  breeding system, 653, 655
  butyrates in scent marks, 666, 667
  chemical signal computer pattern analysis, 664-667
  chemical signal sources, 655-659
  cholesterol in scent marks, 666, 667
  circumgenital glands, 655-659
  circumgenital marking, 657-658
  circumgenital-suprapubic gland complex, 655-659
  comparative chemical signals, 653-659
  epidermal scent glands, 653
  food resource marking, 668, 669
  genital glandular glands, 655-659
  gular-sternal glandular pads, 655-659
  hormones and marking, 660
  individual recognition, 663, 664
  muzzle rubbing, 657-659
  nasal discharge, 659
  physical environment and scent marking, 662
  pull rubbing, 657
  reproduction and scent marking, 661, 662, 667
  saliva, 659
  scent and aggression, 668
  scent gland development, 656
  scent gland histology, 655, 656
  scent glands, 655-659
  scent and infant-adult interactions, 668
  scent mark chemistry, 662-667
  scent mark information content, 662-667
  scent mark volatile constituents, 664-667
  scent marking, 653, 655-659
  scent marking development, 660, 661
  scent and territoriality, 668
  sebaceous glands, 655-659
  self marking, 659
  sit rubbing, 657
  social determinants and scent marking, 661, 662, 667
  social organization, 653, 655
  squalene in scent marks, 666, 667
  sternal glands, 655-659
  sternal marking, 657, 658
  suprapubic glands, 655-659
  suprapubic marking, 657, 658
  urine washing, 659
  ventral rubbing, 657, 658
Caluromys philander, 598
Campostoma anomalum, 107
Canary, 357
Canis latrans, 528
Cape petrel, 358, 361
Carassius auratus, 103
Caretta caretta, 349
Carnivores, 390, 587
Carnivorous fishes, 46
  amino acid feeding stimulants, 48, 49, 56
  chemosensitivity and feeding, 46-58
Carp, 48
Cascades frogs, 229
Castor canadensis, 519, 561
Castoreum, 561-569
  chemistry, 561
  fractionation, 562-564
  multicomponent activity, 567-569
  response to fractions, 564-567
  in scent mounds, 562

Casuariformes, 364
Catfish, 99, 102, 104
Catostomus commersoni, 106, 107
Cats, 507, 513, 642
Cattle chemical signals, 28-34
anogenital region anatomy, 32-34
anogenital region histology, 32-34
anogenital region semiochemistry, 32-34
estrous cycle and semiochemicals, 32-34
headspace analysis, 30
major vestibular (Bartholin's) gland, 32, 33
microbial commensals, 32
sebaceous glands of vulva skin, 32, 34
semiochemistry, 28-34
sweat glands, 32, 34
urinary headspace gas chromatography, 31
vagina, 29, 32
vestibular glands, 32
vestibule, 29
Caucasians
North American, 678, 679
Western European, 678
Cave fishes, 104
Cavity nesters, 370
CBA/J mice, 463-466, 468
Cebuella, 653, 655, 657, 662, 668
Cebuella pygmaea, 654, 655, 658
Cephalopods, 360
Cervus elaphus, 388
Cetaceans, 13, 17, 361
vocalization, 13
Chaca chaca, 105
Channel catfish, 48
Char, 126
Characidae, 174
Characoidei, 103
Cheetah, 418
Chelonia, 351
Chelonia mydas, 346, 347, 349
chemosignal learning, 347
Chelydra serpentina, 345
Chemical Senses, (The), 8
Chemical Senses and Flavour (Chemical Senses), 8
Chemical signals
alarm chemosignals, 507-517
amphibian and reptile homing, 341-352
amphibian and reptile orientation, 341-352
in aquatic animals, 341
in Asian elephants, 627-643
beaver castoreum fractions, 561-569
in callitrichids, 653-669
chemical complexity, 6
computer pattern analysis, 364, 568, 664, 665, 667
dietary effects, 398-400
and dispersal in water, 135-146
in eukaryotes, 15
in fish reproduction, 117-129
in gray opossums, 588-604
habitat odors, 342
history of recent study, 1-11
in humans, 5, 674, 675
individual recognition, 400-406, 413-420
in insectivores, 609-615
interactive, 568
major histocompatability gene complex, 414-417
mammalian ovarian function, 423-429, 433-437
in marsupials, 587-600
in monotremes, 587, 588, 593, 594, 599, 600
mustelid anal glands, 520-527
mustelid plantar glands, 617-622
opossum sternal scent glands, 579, 583
and pika breeding facilitation, 571-575
predator avoidance, 519-529
in Przewalski's horse, 623-626
puberty effects, 426-429
and rat diet selection, 487-503
rodent conspecific and heterospecific, 533-539
rodent individual, 536
rodent puberty-influencing, 441-452, 457-461, 551-554
rodent population variation, 537, 538
rodent reproduction, 79-84
rodent species recognition, 547
rodent sex discrimination, 536, 537
rodent urinary, 424-429, 441-452
rodent vomeronasal organ, 471-483
in salmonid homing, 136-142
sea turtle imprinting, 346-352
and sexual selection, 541-548
spiny mouse kin recognition, 397-409
trees, 99

Chemical signals (Continued)
vole familial chemosignals, 555-559
Chemical signals and feeding in fishes, 45-58
diet selection, 50, 57, 58
feeding deterrents, 45, 50, 51, 58
feeding enhancers, 45, 50, 51, 58
feeding stimulants, 45, 50, 51, 58
Chemical signals and kin recognition, 226
in anurans, 226-239
in ground squirrels, 226
major histocompatibility gene complex, 413-420
mouse mating preferences, 417, 418
in rats, 226
in salmon, 226, 230
in spiny mice, 226
Chemical signals and salamander nesting, 205-217
chronological age of eggs, 215-217
clutch recognition, 211, 217
egg recognition, 210, 211, 217
individual recognition, 206-213, 217
nest odors, 211-213, 217
oophagy, 216, 217
philopatry and homing to the nest, 205, 206, 217
species recognition, 213, 214, 217
Chemoattractant properties of earthworm wash, 67
amino acid profile, 68, 69, 73-75
carbohydrate component, 72-75
glycoprotein component, 72, 73
imino acid, 73, 74
molecular mass, 73-75
pH, 67
periodate effects, 72, 73
temperature variation, 67, 71
Chemoreception Abstracts, 8
Chemosensory orientation and homing
amphibians, 341-343
birds, 369
reptiles, 341, 343-346
sea turtles, 346-352
Chilodus punctatus, 105
Chimpanzee, 425, 426, 682
Chinese, 678
American, 681
Canadian immigrant boys, 681
of the People's Republic, 679
Chrosomus neogaeus, 107
Chrysemys picta, 345
Chrysochloridae, 609
Chrysophrys major, 47, 48, 58
Cichlids, 119, 120
Cinquefoil, 373
Clam, 47
Clarias lazera, 125
Classical conditioning, 137
Claytonia virginica, 373
Clemmys, 351
Clemmys insculpta, 345
Clethrionomys glareolus, 519, 535, 551, 554
Clinostomus funduloides, 107
Clupea harengus, 48
Clupea harengus pallasi, 121, 124
Cod, 48, 127, 143
Coelenterates, 360
Coho salmon, 137, 138, 226, 397, 419
Colubrid, 244, 263
Colubridae, 262
Columba livia, 358
Columbidae, 362
Commission on Chemoreception of the International Union of Physiological Sciences, 4
Common marmoset, 654, 660-662
Common shiner, 102, 108, 109
predator specific fright reaction, 109
Communication
alarm signals, 507-517
assessment, 392, 575
display, 487
formalized, 487, 503
manipulative, 176
ritualized, 42, 487, 503, 516
specialized, 487, 503, 516
unspecialized, 487, 503
Competition (see Territoriality)
Compleat Angler, (The), 136
Computer pattern analysis, 364, 568, 664-667
Conditioning, 11
Conium maculata, 373, 374
Connochaetes taurinus, 389
Constrictor constrictor, 297
Consummatory behavior, 40, 41
Coral gobie, 103
Corvus frugilegus, 362
Corynopoma riisei, 121
Cotton top tamarin, 654, 656, 659, 660, 662
Coturnix coturnix japonica, 226
Couesius plumbeus, 107

Cow headspace volatiles
aldehydes, ketones, carboxylic acids, phenols, sulfides and hydrocarbons, 32
Coyotes, 528
Craig-Lorenz-Tinbergen Approach, 40
Crested newt, 180, 181
Crested tamarins, 654
Crocidura, 610
Crocodile, 101, 363
Crocodylus acutus, 314
Crotalus, 266, 343, 528
Crotalus atrox, 261
Crotalus horridus, 262, 266, 267, 287, 288, 295
Crotalus horridus horridus, 287, 288
Crotalus viridis, 262, 266, 344
Crotalus viridis oreganus, 287
Crotalus viridis viridis 286, 288, 296-299
Crural system in monotremes, 594, (see also Monotreme chemical signaling)
sexual dimorphism, 594
Crustaceans, 360, 568
Cryptotis parva, 610
Cuculidae, 362
Cyclostomes, 357
Cynomolgus monkey, 425
Cynops, 177
Cyprinid, 104
Cyprinoids, 122
Cyprinus carpio, 48

Dab, 49
Damaliscus korrigum, 392
Dandelions, 572
Danio malabricus, 104
Daption capense, 361
Darters, 99-102, 108-110
Dasycercus, 595
Dasyuridae, 595
Dasyurus, 598
Daucus carota, 372-374, 376, 377
DBA/2J mice, 463-466, 468
Dear-enemy recognition, 195, 196
Dedicated motor patterns, 39-42
to sensory function, 41, 42
Dehydro-exo-brevicomin, 30
Dendrotriton bromeliacia, 179
Dendroaspis polylepis, 262
Desmans, 610
Desmognathus, 191
Desmognathus auriculatus, 206
Desmognathus fuscus, 178, 194, 206-209, 213, 214
Desmognathus monticola, 178, 194
Desmognathus ochrophaeus, 173, 178, 180, 181, 183, 205-209, 213, 214, 216
Desmognathus santeetlah, 213
Desmognathus wrighti, 179
Diamondback rattlesnake, 261
Didelphidae, 579, 588, 595, 614
Didelphids, 590
Didelphis, 579
Didelphis virginiana, 579, 581-583, 588, 595
Diadophis punctatus arnyi, 344
Dietary effects on odors, 6
17α, 20β-dihydroxy-4-pregnen-3-one, 124, 154
Dimethyl disulfide, 38, 568
Dinkas, 673, 674
Diomedeidae, 358
Diplodactylus, 21
Dipsosaurus, 19
Diving-petrels, 358, 363, 365, 366
Dogs, 29
Domestic boar, 4
Doty, R. L., 4, 9
Dover sole, 49, 58
Dual olfactory system, 434, 482, 483
(see also Olfactory system)
(see also Vomeronasal system)
Dutchmen, 674
Duvall, D., 6

Earthworm, 63-75
chemoattractant for garter snake feeding, 63-75
chemoattractant wash, 64-78
cuticle collagen, 69-75
Earthworm cuticle collagen, 70
(see also Chemoattractant properties of earthworm wash)
annelid, 73-75
chemical characteristics, 64-75
chemoattractant in garter snake feeding, 70, 71
molecular mass, 73-75
Eastern massasaugas, 287
Ebling, F. G., 6
Echidna, 587, 593, 594
cloacal scent marking, 593, 594
Echinops, 613, 614
Echinops telfairi, 611
Ecology of Vertebrate Olfaction, (The), 9
Ecto-parasites, 373, 381
Ecuadorian Andeans, 678
Eel, 48, 103, 129, 159

Elaphe obsoleta, 295
Elapidae, 262
Elephas maximus, 627
Electric gymnotiformes, 105
Elicited motor patterns, 37-39
Enhydra lutris, 621
Ensatina, 177, 198
Ensatini eschscholtzii, 177
Epple, G., 4, 10
Equids, 623
Equus africanus, 623
Equus caballus, 623
Equus hemionus, 623
Equus grevyi, 623
Equus przewalskii, 623
Equus quagga, 623
Equus zebra, 623
Eretmochelys imbricata, 346
Erigeron, 372, 373, 377, 379
Erinaceidae, 609
Erinaceus concolor roumancus, 613
Ermine, 519
Escherichia coli, 373, 374
Estrogen, 32
Etheostoma exile, 109
Etheostoma nigrum, 109
Etheostoma olmstedi, 100
Etheostoma swaini, 109
Ethology, 1, 40
Etiocholanolone glucuronide, 119, 121, 123, 154
Eumeces, 323, 324, 329, 331, 336, 338
Eumeces egregius, 323
Eumeces fasciatus, 323, 324, 326-335
Eumeces inexpectatus, 323, 324, 326, 329-334, 336
Eumeces laticeps, 323-338
Eumeces multivirgatus, 323
Eumeces obsoletus, 323, 324
Eumeces septentrionalis, 323
Eupomacentrus leucostictus, 46
Euproctus, 177
Europeans, 674
European Chemoreception Research Organization, 5, 8
European minnow, 104, 105
European rook, 362
European starling, 371
European wild rabbit, 1, 2, 90, 95
  feces, 87
  odoriferous anal glands, 87
  territoriality, 87
Eurycea bislineata, 178
Eutheria, 609
Eutherians, 587, 588, 590, 594
Evans, D. A., 7
Evolution, 173-184
  adaptation, 285, 286, 300
  assessment, 252, 392, 575
  and bird chemosignaling and olfaction, 364-366
  convergent, 358
  cross-cultural human taste and smell, 673-684
  demography, 460, 461
  garter snake chemical communication, 244, 255
  garter snake heterogamety, 280, 281
  garter snake W chromosome, 280
  insectivore behavioral convergence, 612, 613
  kin recognition, 397-409
  kin recognition and fitness, 235-239
  life history tactics in salamanders, 192
  mating preferences, 417, 418
  mouse urine esterase genetics, 463-469
  and opossum scent glands, 579-583
  optimal outbreeding, 225, 418-420
  parental investment theory, 210
  population density effects, 457-461
  procellariiform olfaction, 364-366
  r- and K-selection, 192
  recognition alleles, 414-420
  reproductive isolation, 221-224, 329-334
  rodent chemosignal population variation, 537, 538
  rodent pregnancy block, 433-437
  rodent scent glands and sexual selection, 541-548
  rodent urinary chemosignals, 443
  sex-ratio bias, 253
  sibling species and hybridization avoidance, 221-224
  Thamnophis radix complex, 272-275
  and variation in salamander courtship pheromones, 183, 184
  vole familial chemosignals, 559
Excessive construction, 15
Exploratory behavior, 305-319, 344
  and alarm pheromones, 512, 513
  iguanas, 305-319
  in nesting iguanas, 314-318
  novel environments, 326, 327
  reptilian curiosity, 319

Eye movement
ocular orientation, 41

Familial chemosignals (see also Vole puberty influencing chemosignals)
beaver family odor, 562
in pine voles, 555-559
Fasciatus group of Eumeces, 329-334
Fathead minnows, 101, 104-106, 126
Fat-tailed marsupial mouse, 597, 598
scent-marking social aspects, 598, 599
Feces, 29
cattle, 29
mice, 87-95
Feeding, 45
chemical signals and, 45-58
chemoreceptor sensory tuning in fishes, 45, 46
earthworm wash and garter snake, 63-75
Ferrets, 507, 513, 514, 520
Finescale dace, 106
First International Symposium on Olfaction and Taste, 3, 4
Fish alarm substance cells, 103-107
abrasion hypothesis, 105-107
breeding cycle and, 105-107
clavate cells, 103-107
club cells, 103-107, 516
dorsal and pectoral spine poison glands, 103, 104
endocrine control, 105-107
epidermal, 103-107
mechanical damage, 103-107
in ostariophysans, 103-107
seasonal occurrence, 105-107
toxic secretions, 103, 104
venoms and mechanical disruption, 104
Fisher, 617
Fishes
alarm pheromone evolution, 99-110
carnivorous, 45, 46
central nervous system chemosensory projections, 127, 128
chemosensory orientation and homing, 135-146
development of fright reaction, 109, 110
evolution of chemical signaling, 99-110, 117, 118
fright reaction, 102, 105, 106-110
gonopore, 122, 123
herbivorous, 45, 46
olfactory bulb, 126, 127
olfactory receptors, 126
olfactory and reproductive system development, 155-168
olfactory terminal fields, 127, 128
parental care, 102
reproductive pheromones, 117-129
schools, 101, 102, 108, 109
unspecialized skin toxins, 104
Fish olfactory and reproductive systems (see also Olfactory system)
development in platyfishes, 155-168
heredity in platyfishes, 155-168
Fish orientation and homing
anemotaxis, 143
circling and turning behavior, 144
facial (VII) nerve, 142
handedness in turning behavior, 144
home stream selection, 138-140
klinotaxis, 136
maxillary barbels, 142
odor imprinting, 137, 138
olfactory hypothesis, 137
parent stream concept, 136, 137
responses to sign-stimuli, 136
reversing anemomenotaxis, 143
rheotaxic responses, 138, 141, 142, 145, 146
role of chemical cues, 135-146
in salmon, 136-142
spawning migrations, 136
stereotypic swimming pattern, 136
taste receptive fields, 142
tropotaxis, 136, 146
zig-zag swimming, 138, 143, 146
Fish reproductive pheromones
from anal fin appendages, 121, 125
androgen metabolites, 123-125
caudal glands, 121
chemical identity, 123, 124
cloacal mouthing, 123
control of production and release, 122-126

Fish reproductive pheromones
(Continued)
control of responsiveness, 128, 129
estrogen metabolites, 123-125
etiocholanolone glucuronids, 119, 121, 123
female attractors, 119
female pheromones, 119-124
gonadal origin, 121-123
gonadotropin effects, 124, 125
kidney, 125, 126
male attractors, 120
male pheromones, 119, 124-126
neural response mechanisms, 126
ovarian fluid, 122, 123
ovarian stimulators, 119
progestogens, 124
prostaglandins, 126
reproductive hormones, 124-126, 128, 129
reproductive steroids, 124-126, 128, 129
seasonal variation, 124
from secondary sex structures, 121
skin origin, 121
social attraction, 119
sources and sites, 121-123
steroid glucuronides, 125
in urine, 121
Fish schooling and genotype structuring, 102
Fitness, 101, 102, 174, 175, 180-184, 238, 243, 337, 388, 434
reproductive success, 385
Five-lined skinks, 344, 345
Fixed action pattern, 39, 40
Fleabane, 373
Flehmen, 29, 628, 629, 632-639
in cattle, 29
in elephants, 627, 629, 632-643
in ring-tailed lemurs, 648
Flightless land birds, 364
Foraging
chemocommunication and rat diet selection, 487-503
rat chemocommunication and distant diet selection, 487-503
central place foraging, 488
information centers, 488
social cues and orientation, 492-495
olfactory signals, 495-496, 499, 500
rat dietary familiarity, 496-499
contextual cues, 500-502
prey-predatory chemosignals, 507-517
predatory-prey-odor preferences, 513
predator avoidance by prey, 519-529
Fowl mites, 383
Fox, 507, 513-515, 528
Frillfin goby, 120, 122
Frog, 226-239, 251, 342, 343
Fugu pardalis, 47, 48, 58
Fulmars, 358, 362, 365
Fulmarus glacialis, 359
Fulmarus glacialoides, 361
Fundulus heteroclitus, 48

Gadus morhua, 48, 127, 143
Gambusia affinis, 104
Garter snake, 63, 64, 66, 67, 70-75, 243-255, 261-275, 292, 325, 344, 475
annual reproductive cycle, 244-255
body size, 243
chin rubbing, 245
cloacal plug, 243
courtship behavior, 245, 246
dermal vitellogenin receptor, 249, 250
gametogenesis in garter snakes, 244
hibernation, 253
lipoprotein receptor, 250
renal sex segment, 243
reproduction, 243-255
reproductive pheromones, 243-255
vascular bed of Rauch, 246-249
vitellogenin female pheromone, 246-252
Garter snake pheromone trails, 63
aggregation, 263
contact chemoreception, 268, 269
den location, 266, 267
directionality cues, 268-272
evolution, 272
gravid female aggregation, 267
klinotaxis, 269
lipoprotein, 263
male responses, 268-271
in non-sexual aggregation, 266, 267
predation, 263

Garter snake pheromone trails (Continued)
  prezygotic reproductive isolation, 263, 271-275
  reproduction, 263, 266
  seasonal changes, 264, 265
  species differences, 271-275
  tongue flicking, 261
  trail orientation, 268-272
  tropotaxis, 269
  vitellogenin, 263
  vomeronasal system, 261
Garter snake reproductive pheromones, 245-255, 261-275 279-282
  assessment, 252
  and courtship behavior, 245, 246
  epidermal percolation, 248-250, 281
  female attractiveness, 246-255, 261-275, 279-282
  hinge region, 249, 254, 255
  liver lipoprotein, 250-252, 280, 281
  mate selection, 252, 253
  reproductive hormones, 247-248, 250, 252
  and sexual mimicry, 279-282
  of she-males, 280, 282
  tongue flicking, 245, 261, 279
  trails, 245-252
  vascular bed of Rauch, 246-249
  vitellogenin, 246-252, 281
  vomeronasal system, 245-255
Gaviiformes, 364
Gazella thomsoni, 386
Gekkonidae, 19
Gelada baboons, 425
Geoffroy's tamarin, 654
Geomyidae, 519
Georgian diving-petrel, 358
Gerbils, 399, 507-516, 571, 575
  alarm chemosignals, 507-517
  odors and predatory preferences, 513
Germans, 674, 675
Geum canadense, 374, 377, 379
Geum virginianum, 374
Giant danio, 104
Ginglymostoma cirratum, 143
Giraffe, 643
Glands
  antelope pedal, 392
  beaver anal, 561, 565, 566
  beaver castor sacs and castoreum, 561-569
  bird anal, 362, 363
  bird harderian, 363
  bird nasal salt, 358
  bird sebaceous, 362, 363
  bird preen uropygial, 362, 363
  cattle anogenital, 32-34
  caudal in fishes, 121
  defined, 15
  elephant temporal, 632, 635, 640
  evolution of in tetrapods, 14-22
  fisher tarsal, 621
  gazelle antorbital, 386
  gecko caudal, 21
  gray opossum scent, 594, 595
  gray opossum suprasternal scent, 591-593, 594-599
  holocrine epidermal, 14, 22
  house mouse anal, 87-95
  house mouse sex, 87-95
  insectivore anal, 610, 611
  insectivore eyelid secretion, 613, 614
  insectivore lateral, 610
  insectivore neck, 614
  insectivore scent, 609-615
  insectivore ventral, 610
  lizard generation, 19, 22
  lizard preanal organ, 18, 19
  marmoset scent, 655-659
  marsupial cloacal (anal), 588
  marsupial frontal and sternal, 588
  marsupial scent marking, 588
  marten acinar, 621
  microtine anal, 541, 544, 546, 548
  microtine comparative morphometry, 541-548
  microtine posterolateral, 541, 542, 544, 546, 548
  microtine preputial, 541, 542, 544, 546, 548
  microtine proctodeal, 542
  microtine scent, 541-548
  microtine sebaceous, 541
  microtine vesicular, 542, 544, 546, 548
  monotreme scent, 593, 594
  morphometry and sexual selection, 541-548
  mustelid abdominal, 617
  mustelid anal, 520-527, 617
  mustelid plantar, 617-622
  opossum sebaceous (holocrine), 580-582
  opossum sudoriferous (apocrine), 580-582
  physiologically controlled, 18
  pika cheek, 571, 572, 575

Glands (Continued)
plethodontid cephalic, 178, 179, 182
plethodontid mental (submandibular), 178-180, 200
rabbit anal, 87
rabbit chin, 2
rabbit inguinal, 2
river otter plantar, 619
salamander abdominal, 178-184
salamander cloacal, 177-184, 197-200
salamander genial, 179, 180
salamander granular, 200
salamander submandibular region, 200
salamander vent, 199, 200
skink cloacal, 326, 328, 335, 336
skink urodaeal, 335, 336
snake nucho-dorsal, 21
spermatheca, 199, 200
tamarin scent, 655-659
of tetrapod integument, 18
Virginia opossum cutaneous scent, 579
Virginia opossum sternal scent, 579-583
Glandulocaudine fishes, 121, 174
Glechoma hederacea, 374, 377, 379
Glycera debranciata, 65, 73
Goatfish, 47
Gobiodon quinquestrigatus, 103
Gobius jozo, 119, 121, 124, 128
Goldfish, 103, 120, 123-129, 149-153, 165
sex pheromones, 149
ovulatory pheromone, 149
Golden hamster, 561, 568, 590
Golden-handed tamarin, 654
Golden lion tamarin, 654, 660-662
Golden moles, 609
Goldenrod, 373
Gonadotropin, 156
Gonopodium, 157
Goodrich, B. S., 5
Gopher snake, 528
Gophers, 519
Gopherus agassizi, 345
Gordon Conference on the Chemical Senses, 4
Gray kangaroos, 595, 597
Gray opossum, 588-593, 595-604
(*see also* Gray short-tailed opossum)
Gray opossum chemical signaling, 588
chest marking, 589
comparative aspects, 590
estrous synchrony, 601
evolutionary aspects, 604
female activation by male chemosignals, 601-603
and female reproduction, 600-604
flank marking, 589
functional significance, 590, 592
head marking, 589
hip marking, 589
hormonal aspects, 592, 593
priming effects, 600
scent glands and marking, 594-599
scent marking, 589
scent marking ethology, 589
sebaceous glands, 592
sex and age relationships, 589, 590
sexual dimorphism, 592
social aspects, 599
suprasternal scent gland, 591-593
Gray short-tailed opossum, 588
(*see also* Gray opossum)
Great tits, 386
Greater kudu, 386
Grebes, 364
Greek valerian, 373
Green frog, 342
Green iguana, 306, 309, 312
(*see also* Iguanas)
chemosensory ethology of, 305-319
field exploration, 314-318
laboratory exploration, 306-313
Green salamanders, 216
Green sea turtle, 346, 347, 350
Grevy's zebra, 623
Ground skinks, 344, 345
Ground squirrels, 226, 419
Group selection, 235, 236
Guinea pigs, 471-475, 480
learning and chemoinvestigation, 473-475
Gulf darter, 109
Gulo, 617, 619
Gulo gulo, 617
Guppy, 120, 123, 125
female pheromone, 120, 121
Gustation, 45-47, 135-146
(*see also* Taste)
and feeding in fishes, 45, 46
and fish orientation, 135-146
in humans, 673, 681, 682
Gymnotidae, 103

Habituation, 27, 110, 233, 327, 639, 663
Hag, 361

Hagfish, 49
Haldane, J. B. S., 15
Hamadryas baboons, 425
Hamster, 30, 418, 472
female attractant pheromone, 38
semiochemistry of vaginal secretion, 30
vaginal discharge, 38
Haplochromis burtoni, 120
Harlequin fish, 110
Hartebeest, 386, 387, 389, 391
Hawaiian crab, 47
Hawaiian goatfish, 47
Hawksbill sea turtle, 346
Headstanders, 105
Hedgehogs, 609, 610, 612-615
Hemiechinus auritus syriacus, 614
Heptan-2-one, 30
Herbivores, 587
Herbivorous fishes
chemosensitivity and feeding, 46-58
plant amino acid feeding stimulants, 56-58
Herring, 121, 128
larvae, 48
Highland Peruvians, 678
Homing
amphibians, 341-343
birds, 369
reptiles, 346-352
salmon, 136-142
sea turtles, 346-352
Homology, 18
Horse, 623-625, 642
House mouse, 87-95, 423, 424, 427-429, 441-452, 457, 461, 554 (*see also* Mice)
anal and fecal marking, 88, 89
anal gland secretion, 87-95
anal glands, 88, 93-95
BALB/c strain, 91, 92
chemosignals and male aggression, 87
chemosignals and territoriality, 87, 95
coagulating glands, 87, 93-95
feces odors, 87-95
plagues in Australia, 87
preputial glands, 87, 93-95
wild and laboratory, 88, 91, 92
House sparrow, 360, 381
Human olfaction, 5, 673-677, 681, 682
Human pheromones, 5, 674, 675
menstrual cycle synchrony, 5
Human smell and olfaction, 673
axillary odor discrimination, 674, 675
cross-cultural odor sensitivity and preference, 673-675, 681, 682
cross-cultural sex differences in odor identification, 675-677, 681, 682
genetics versus experience, 681, 682
human pheromones, 674, 675
Human taste and gustation, 673, 681, 682
cross-cultural suprathreshold intensity and hedonicity, 679-681
cross-cultural taste naming, 678
cross-cultural threshold sensitivity, 678, 679
genetics versus experience, 681, 682
Humans, 397, 404, 434, 673
axillary sweat and odors, 30, 674, 675
cross-cultural defined, 673
cross-cultural taste and perception, 673, 678-682
cross-cultural smell perception, 673-677, 681, 682
genetically related taste and smell sensitivities, 673
3α-Hydroxy-5α-androst-16-ene, 4
cis-4-Hydroxydodec-6-enoic acid lactone, 568
Hypomesus olidus, 121-124
Hypoxanthine-3(N)-oxide, 104
Hypsiprymnodon moschatus, 595

Ictalurids, 119
Ictalurus catus, 104
Ictalurus melas, 104
Ictalurus natalis, 48, 104
Ictalurus nebulosus, 142
Ictalurus puntactus, 48
Iguanas, 307-319
(*see also* Green iguana)
Iguana iguana, 306, 319
Iguanidae, 19, 305
Iguanids, 323
Illiger's saddle-back tamarin, 654
Impala, 386
Inclusive fitness, 238, 400, 413, 418-420
(*see also* Individual fitness)
(*see also* Fitness)
mating preferences, 417, 418
recognition alleles, 414-420
and vole reproduction, 558
Indians, Asian
Hindus, 681

Indians, Asian (Continued)
laborers of the Karnataka region, 679
medical students, 679
Individual fitness, 101, 102, 174, 175, 180-184, 433
Individual recognition, 400-406
Apodemus sylvaticus, 536
in callitrichids, 663, 664
comparative rodent, 536
familiarity and learning, 406-409
and familiarity in pika, 571
genetic basis for, 413-420
major histocompatibility gene complex, 413-420
mating preferences, 417, 418
Microtus agrestis, 536
in pika, 571
recognition alleles, 414-420
rodent pregnancy blockage, 437
rodent self- and non-self discrimination, 539
in Saguinus fuscicollis, 663-665
spiny mice, 400-406
systems, 418-420
Indian field mouse, 424
Insectivora, 609, 612
Insectivore chemical signals
anal glands, 610, 611
anti-predator functions, 611, 612
apocrine gland component, 610
eyelid secretion, 613, 614
and glandular sources, 609-615
lateral glands, 610
latrine use, 610, 611
neck glands, 614
primers, 615
and reproduction, 614, 615
releasers, 614
scent-marking behavioral convergence, 612-614
sebaceous gland components, 610
self-anointing behavior, 612
urine, 614, 615
ventral glands, 610
Insectivores, 587, 609-615
(see also Insectivore chemical signals)
Insects, 2, 118, 174, 182, 184, 279, 516, 568
social, 404, 516
Italians, 674, 675
8th International Congress on Animal Reproduction and Artificial Insemination, 6
International Congress of Physiology, 3
International Society of Chemical Ecology, 8
International Summer Course on Odour Perception, 5
3rd International Symposium on Comparative Reproduction, 6
Invertebrate, 507
Investigatory behavior, 37-42
reflexive, 40-42
Iowa darters, 109, 110
Isovaleric acid, 568
Isoxanthopterin, 104

Japanese, 674, 675, 678
native, 675-677
Japanese Association of Taste and Smell, 4
Japanese quail, 226, 419
Javanese, 674
Johnny darter, 109
Johnston, J. W., 4
Journal of Chemical Ecology, 8

Kale, 678
Kangaroos, 594
Kare, M. R., 4
Kemp's Ridley sea turtle, 347, 350
nasal morphology, 347
Keratinocytes, 15, 16, 18
Keverne, E. B., 10
Kingston, B. H., 3
Kin Recognition,
development in anurans, 230-235
familiarity, 230-235, 400, 401
and fitness, 225, 226, 230-235
genetic recognition system, 234
histocompatibility responses, 231
major histocompatibility gene complex, 413-420
phenotype matching, 239, 402-406
phenotype template recognition, 233, 234
recognition alleles, 408, 409, 413, 414, 418-420
recognition systems, 418-420
selective benefits, 235-239
spiny mice, 397-409
templates in anurans, 239
in vertebrates, 226
Kin selection, 101, 102, 225, 235, 236, 397

Kin selection (Continued)
familiarity, 400, 401
fitness effects, 235-239
genetic relatedness, 225
individual recognition, 400-406
major histocompatability gene complex, 413-420
phenotype matching, 402-406
self-matching, 404
Kinship, 226
Kiwis, 364
Koala, 596-598
sternal gland chest marking, 596, 597
Korean Americans, 675, 676, 677
Köster, E. P., 5, 8

Lacerta muralis, 20
Lacertilian, 19
Lagodon rhomboides, 48
Lagurus luteus, 519, 529
Lamium purpureum, 372, 373, 377, 379
Lampropeltis getulus, 262
Leach's Storm-Petrel, 362-364
Learning, 6, 27, 233, 346, 347
diet selection and sociality, 487-503
familiarization, 400-404
imprinting defined, 351
and rat dietary chemo-communication, 487-503
and rodent vomeronasal system, 471-483
sea turtle orientation, 343-352
ungulate scent matching, 389, 390
LeMagnen, Jacques, 3
Lemmings, 541
Lemur catta, 647
Lemurs, 647-651
heterospecific scent-marking ecological separation, 649, 650
sympatric heterospecific battle zone, 650, 651
teritorial scent marking in sympatric heterspecifics, 647-651
Leontopithecus, 653
Leontopithecus rosalia, 654
Leontopithecus rosalia rosalia, 654-659, 667
Lepidochelys kempi, 346
Lepidochelys olivacea, 346, 348
Lepomis macrochirus, 104
Leptotyphlopidae, 262
Leptotyphlops, 267
Leptotyphlops dulcis, 262
Lepus americanus, 520
Lesser hedgehog tenrec, 612
Lesson's saddle-back tamarin, 654
Life history
mouse lifespan, 437
rodent demography, 460, 461
rodent population density, 457-461
and vole familial chemosignals, 559
Limanda limanda, 49
Lion tamarins, 654
Little auk, 358
Lizard reproduction (see also Skink pheromones)
iguana nesting, 314-318
skink ethological isolation, 329-334
skink head coloration, 327, 333, 334
Lizards, 16, 17, 19, 248, 305-319, 323-338, 343
preanal organ, 18, 19
iguanas, 305-319
iguana nesting, 314-318
Eumeces pheromones, 323-338
skinks, 323-338
Loggerhead sea turtle, 346, 350
Long-tailed weasel, 520, 528
Lonicera japonica, 374
Loons, 364
Loricarridae, 103
Loricarid catfishes, 105
Louse, 373
Lumbricus terrestris, 63, 64, 73-75
Lungless salamanders, 191
Lutra canadensis, 617
Lygosominae, 324
Lynx rufus, 528

Macaca facicularis, 425
Macaca mulatta, 5
Macaques, 419
Mackerel, 102
MacLeod, P., 8
Macronectes giganteus, 361
Macropisthodon, 254
Macropus, 598
Macropus cangaru, 595, 596
Macropus giganteus, 597
Macropus robustus, 596
Macropodidae, 595
Madtoms, 103
Magellan diving-petrel, 358
Major histocompatability gene complex, 413-420, 466
genetics, 413, 414, 466
H-2 alleles, 414-417, 419, 420
and individual chemosensory recognition, 414-420
mating preferences, 417, 418

Major histocompatability gene
complex, (Continued)
in mice, 413-420
and mouse urine odors, 414-416
odor constitution, 416
pregnancy blocking odors, 414
recognition alleles, 414-420
Major vestibular (Bartholin's)
gland of cattle, 32, 33
Malapterurus electricus, 105
Mallard duck, 362
Mammalia, 604, 609
Mammal-like reptiles, 609
Mammalian chemical signals
alarm substances, 507-517
in Asian elephants, 627-643
in callitrichid monkeys, 653-669
endocrine effects, 423, 424
estrous cycle, 424, 426
in gray opossum, 588-604
individual recognition, 536
inhibitory, 423-429
in insectivores, 609-615
in marsupials, 587, 588, 594-600
in monotremes, 587, 588, 593, 594, 599, 600
mustelid anal glands, 520-527
and ovarian function, 423-429
ovarian repression, 423-424
pika breeding facilitation, 571-575
predator avoidance, 519-529
priming pheromones, 428
Przewalski's horse, 623-626
puberty accelerating, 427, 555
puberty suppressing, 426-429, 441-452, 551-559
rodent conspecific and heterospecific, 533-539
rodent population variation, 537, 538
and rodent vomeronasal system, 471-483
scented trap studies, rodents, 533-539
sex discrimination, 536, 537
Mammalian chemosignals and kin recognition
familiarity, 400, 401
phenotype matching, 402-406
self-matching, 404
spiny mice, 397-409
Mammalian ovarian function
Bruce effect, 433
chemosignal repression, 423-424
in elephants, 637
endocrine regulation, 423, 434, 435
in gray opossums, 600-603
in insectivores, 614, 615
pika breeding facilitation, 571-575
puberty-influencing chemosignals, 426-429, 433, 441-452, 457-461
rodent pregnancy blocking pheromones, 433-437
Mammalian reproduction, 9
and callitrichid chemosignals, 653, 655, 660, 661, 667
and chemosignals, 423-429, 433-437, 441-452, 551-554, 555-559
elephant chemosignals, 632, 637, 640, 642, 643
endocrine regulation, 423
in gray opossums, 592, 593, 600-604
microtine sexual selection, 541-548
in monotremes, 594
and opossum sternal scent glands, 579-583
pika breeding facilitation, 571-575
Przewalski's horse, 624, 625
rodent vomeronasal organ, 472, 473-476
Mammalian semiochemistry, 9
Mammals, 19, 27, 99, 118, 167, 173, 174, 181, 226, 280, 281, 341, 359, 363, 365, 385-393, 433, 434, 471, 482, 528, 533, 538, 541, 555, 579, 590, 604, 609, 613, 653
Margosa, 381
Marine catfishes, 103
Marine invertebrates, 360
Marine turtles, 346
Marmosa, 579, 595, 598
Marmosa domestica, 595
Marmosa robinsoni, 595, 597
Marmosets, 653, 655, 661, 667, 668
apocrine gland, 655-659
chemical signal sources, 655-659
circumgenital glands, 655-659
circumgenital-suprapubic gland complex, 655-659
comparative chemosignals, 653-669
epidermal scent glands, 653
food resource marking, 668, 669
genital glandular pads, 655-659

Marmosets, (Continued)
gular-sternal glandular pads, 655-659
hormones and marking, 660
individual recognition, 663, 664
muzzle rubbing, 657, 659
physical environment and scent marking, 662
reproduction and scent marking, 660, 661, 667
scent and aggression, 668
scent and infant-adult interaction, 668
scent glands, 655-659
scent gland development, 656
scent gland histology, 655, 656
scent marking, 653, 655-659
scent mark chemistry, 662-667
scent mark information content, 662-667
scent marking development, 660, 661
sebaceous glands, 655, 659
sociality and scent marking, 661, 662, 667
sternal glands, 655-659
sternal marking, 657, 658
suprapubic glands, 655-659
ventral rubbing, 657, 658
Marsupial chemical signaling, 587, 588
anal glands and cloacal marking, 595
cloacal (anal) glands, 588
feces, 594, 595
frontal and sternal glands, 588
mother-young interactions, 599, 600
neonate migration to the pouch, 600
neonatal vomeronasal system, 600
saliva and oral marking, 594, 595, 597, 598
scent marking, 588
scent marking social determinants, 598
sternal glands and marking, 596
urine marking, 594, 595, 598
Marsupialia, 579, 583, 587
Marsupials, 583, 587-589, 593-595, 597-600, 603, 604
Martes, 617, 619, 622
Martes americana, 617, 621
plantar glands, 621
Martes pennanti, 617, 621
tarsal glands, 621
Martes zibellina, 617
Martin, R. D., 7
Masticophis taeniatus, 262
Mate choice, 175, 397
female choice, 243
individual recognition, 400-406, 413-420
kin recognition, 397, 406-409, 413-420
major histocompatability gene complex, 413-420
mating strategies, 385
opossum sternal scent gland, 579, 583
optimal outbreeding, 418-420
pheromones in mate selection, 252, 253
pika chemosignals, 571-575
and relatedness, 418-420
reproductive assessment, 252
rodent scent glands and sexual selection, 542, 547
sex-ratio bias, 253
sexual mimicry, 279
McCartney, W., 3, 5, 8
McClintock, Martha, 5
Meadow vole, 428, 429
Megaleia rufa, 595
Menacanthus, 373
Menidia menidia, 49
Mephitis mephitis, 621
Meriones libycus, 572
Meriones unguiculatus, 10, 507, 508, 568, 572
Mesocricetus auratus, 561, 568
Mesozoic, 14, 17
Metatheria, 587, 609
Metatherians, 579
3-methylbutyl methyl sulfide, 30
17α-methyltestosterone, 106
Mice, 2, 6, 30, 79, 87-95, 104, 413-420, 423, 424, 427, 434-437, 441, 452, 457, 461, 463, 469, 471-482, 537-539 (see also House mouse)
inhibitory chemosignals, 423-429
learning and aggression chemosignals, 476-479
learning and reproductive chemosignals, 475, 476
learning and ultrasonic vocalization chemosignals, 479-482
ovarian function, 423-429
semiochemistry and aggression, 30
urinary headspace semiochemistry, 30
urinary pheromones, 441-452

Michael, R. P., 5
Microtidae, 551, 554
Microtinae, 541
Microtine rodents, 433, 541
scent gland morphometry and sexual selection, 541-548
Microtine scent glands and sexual selection, 541-548
and mate choice, 547
and territoriality, 547
Microtus, 541, 546-548
(see also Microtine rodents)
Microtus afganus, 543, 545, 548
Microtus agrestis, 519, 533-539
comparative social behavior, 533-539
conspecific chemical signals, 533-539
heterospecific chemical signals, 533-539
population variation, 537, 538
sex chemosignal discrimination, 536, 537
Microtus arvalis, 519, 541, 543-545, 548
Microtus brandti, 543, 545
Microtus californicus, 428, 429, 543, 545, 546, 548, 555
Microtus chrotorrhinus, 542, 543, 545, 546, 548
Microtus evoronensis, 543-545
Microtus fortis, 543, 545, 546
Microtus gregalis, 543, 545
Microtus gud, 543, 545
Microtus juldaschi, 543, 545
Microtus kirgisorum, 543, 545
Microtus longicaudus, 543, 545
Microtus maximowiczii, 543, 545
Microtus mongolicus, 543, 545
Microtus montanus, 527, 542-548
Microtus ochrogaster, 429, 543-546, 555
Microtus pennsylvanicus, 429, 543-546, 548
Microtus pinetorum, 428, 429, 519, 543, 545, 546, 555
Microtus richardsoni, 542, 543, 545-548
Microtus rossiaemeridionalis, 543-545, 548
Microtus socialis, 543, 545
Microtus townsendii, 533
Microtus transcaspicus, 543, 545
Midas tamarins, 654
Milk, 29
Mimicry, 279
sexual, 279
Minnows, 99, 104
Miopithecus talapoin, 424
Misgurnus anguillicaudatus, 120, 122, 125
Moles, 609, 611
Moncrieff, R. W., 2, 8
Monell Chemical Senses Center, 4
Mongolian gerbil, 10, 568, 590
Mongooses, 388
Monkey, 437, 653
Monodelphis, 579, 589
Monodelphis domestica, 588, 589
Monotreme chemical signaling, 587, 588
alveolar gland, 594
crural system, 594
evolutionary aspects of, 604
heel spur, 594
mother-young interaction, 599, 600
neonate pouch migration, 600
scent glands and marking, 593, 594
venom, 594
Monotremes, 587-589, 593, 594, 599 600
breeding biology, 594
Montagna, W., 5
Moonrats, 610, 611
Mosquito fish, 104
"Mother Carey chicken," 361
Moulton, D. G., 4, 9
Mountain dusky slamander, 205
chemical ecology of nesting, 205-217
Mountain zebra, 623
Mouse opossum, 595, 596
Mouse pregnancy blocking pheromones, 2, 463-469
chemistry, 463, 467, 468
endocrine effects, 463, 466, 467
genetics, 463-469
social factors, 463-469
urinary esterases, 463-469
Mouse puberty-influencing chemosignals, 441-452
dispersal, 450
generation time, 451
grouped female urine, 441, 448-451
on highway islands, 457-458
male urine, 441-445
population density effects, 457-461
pregnant or lactating female urine, 441, 445-447
urine from females in estrus, 441, 447, 448
urinary, 441-452, 457-461
Mouse urinary chemosignals, 30, 32, 79, 441-452
effects of adrenalectomy, 79, 80

Mouse urinary chemosignals, (Continued)
effects of ovariectomy, 79
environmental effects, 442-446
esterase genetics, 463-469
estrous females, 441, 447, 448
evolutionary functions, 443, 444, 457-461
female puberty acceleration, 79
female puberty delay, 79, 84
gas chromatography, 81, 83
grouped females, 441, 448-451
headspace analysis, 79
hormonal involvement, 442, 443, 448
male, 441-445
olfactory system, 449
ovarian cycle synchrony, 79
population density effects, 457-461
pregnant or lactating female, 441, 445-447
seasonal variation, 444
social effects, 442, 447, 448
timing effects, 442
urethra, 450
variation across estrous cycle, 80, 82-84
volatile components, 79-84, 442, 443, 446
vomeronasal system, 449
Moustached tamarin, 654
Mozell, M. M., 6, 7
Mugford, R. A., 7, 9
Muller-Schwarze, D., 6, 10
Multiple sensory modalities, 181
Mummichog, 48
Murid rodents, 397, 433
Muridae, 541
Muroid rodents, 399
Murray islanders, 674
Mus, 399, 451
Mus booduga, 424
Mus domesticus, 463
Mus musculus, 87-95, 399, 423, 429, 441, 451, 457-461, 533, 535
(see also Mice)
(see also House mouse)
Muskrat, 519
Mustela, 528
Mustela erminea, 519, 520, 621
Mustela frenata, 621
Mustela nigripes, 621
Mustela nivalis, 519, 621
Mustela putorius furo, 507, 514
Mustela vison, 621
Mustelidae, 520, 522, 525, 527-529, 617, 621, 622
plantar glands, 18
Mustelid anal gland secretion, 520-529
chemistry, 520, 523-527
3,3-dimethyl-1,2-dithiolane, 520, 523-528
3-propyl-1,2-dithiolane, 520, 523-528
Mustelid plantar glands, 617-622
morphology and histology, 617-622
and chemosignals, 617
as chemical transmitters, 617, 619
in marten, 619
in river otter, 619
serous components, 621
and scent deposition, 619
Mykytowycz, R., 3, 5, 6, 10
Myosorex varius, 610
Myrmecobius fasciatus, 595
Myxine glutinosa, 49

Naked sand darter, 109
Nannostominae, 105
NATO Conference on Chemical Ecology: Odour Communication in Animals, 7, 10
Natricinae, 244, 254
Natrix, 254
Natural history
of salamander territoriality, 191, 192
garter snake reproductive pheromones, 244-246
in prairie rattlesnakes, 286, 287
bird plant fumigant use, 369, 370, 381
Natural selection, 174-176, 180-184, 243, 255, 451, 547
(see also Sexual selection)
and adaptation, 285, 286, 300
and kin recognition, 234, 235
r- and K-selection, 192
Neal, E. J., 7
Nearest-neighbor analysis, 231-233
Neomys fodiens, 610
Nereis virens, 65, 71, 73, 74
Nerodia, 244
Nerodia cyclopion floridana, 297
Nerodia sipedon, 262
Nesogale dobsoni, 613
Nesogale talazaci, 613
Netherlands East Indie's Natives, 674
Netherlands Organization for Applied Scientific Research, 7

New Guinea long-beaked echidna, 587, 594
sweat glands, 594
Newt, 177, 180, 181, 342
Nocomis biguttatus, 107
Nocomis leptocephalus, 100, 107
Non-olfactory chemical communication
monotremes, 594
salamanders, 177-180
Northern fulmar, 359-361, 363
Northern pocket gopher, 519
Norway rats, 399, 487
Notomys alexis, 424
Notophthalmus, 178
Notophthalmus viridescens, 179, 181, 182, 200, 206
Notropis chiliticus, 107
Notropis cornutus, 102, 107, 108
Notropis lutrensis, 107
Notropis venustus, 107
Notropis whipplei, 107
Noturus flavus, 104
Noturus spp., 103
Novotny, M., 6
Nucleus olfactoretinalis, 155, 159, 160-161, 163, 164, 166
in fishes, 155-168
(see also Nervus terminalis or terminal nerve)
effects of lesions on development, 165-166
Nurse shark, 143
Nuthatches, 370, 381

Oberhäutchen, 19, 249, 250
Oceanitidae, 358
Oceanodroma homochroa, 362
Oceanodroma leuchorhoa, 362
Oceanites oceanicus, 361
Ochotona princeps, 571
Ochotona rufescens, 575
Odocoileus hemionus columbianus, 561, 568
Odor image, 28, 34, 561, 567-569
beaver castoreum fraction, 561, 567-569
Gestalt, 568
Odor imprinting
in Asian elephants, 640, 641, 643
defined, 351
in fish orientation and homing, 135-146
habitat odors in turtles, 350
in prairie rattlesnakes, 294
rodent individual recognition, 437
in salmon, 135-146, 346
in sea turtles, 346-352
Odor mixtures, 568
Olfaction in Mammals, 9
Olfaction and Odours, 8
Olfactory block to pregnancy, 433-437
adaptive significance, 433, 434, 436, 437
brain mechanisms, 434-436
evolution, 433, 434, 436, 437
familiarity, 435, 436
neural bases, 434
neuroendocrine bases, 434, 435
noradrenergic pathways, 436
in rodents, 431-437
urine esterase genetics, 463-469
vomeronasal mediation, 434
Olfactory system, 14, 17, 63, 143, 146, 255, 261, 482
alarm signaling in fishes, 99-110
in birds, 357-360, 369-381
bird mitral cells, 359, 360
bird olfactory bulbs, 358-360
bird olfactory conchae, 358, 359
bird sensory mucosa, 358
bird volatile cue detection, 357-366, 369-381
development in platyfishes, 155-168
in elephants, 627
evolution in tetrapods, 14
and familiarity, 575
fish feeding, 45, 46
fish olfactory bulbs, 126-129, 158, 160, 161, 166
fish olfactory nerve, 140, 141
fish neuroendocrine system, 117-129, 155-168
in fish orientation, 140-142
and fish reproductive pheromones, 126-129
and fish reproductive system development, 155-168
and goldfish spawning, 149-154
in humans, 5, 673-677, 681, 682
insectivore nasal capsule, 609
non-sensory epithelium, 159
odor imprinting in marine turtles, 17
opossum scent discrimination, 583
and predator avoidance, 519-529
preoptic area, 127-129
and rat diet selection, 487-503
in rodent pregnancy blockage, 434, 436
sensory epithelium in platyfishes, 156-159, 162, 163, 166
spiny mice, 399

Olfactory system, (Continued)
telencephalon, 127-129, 158
terminal fields in fishes, 127, 128
turtle nasal morphology, 346, 347
Oligochaete, 73
Olive Ridley sea turtle, 348
sand smelling, 348
Oncorhynchus, 136
Oncorhynchus kisutch, 137, 226
Oncorhynchus masou, 120
Oncorhynchus nerka, 102
Oncorhynchus rhodurus, 120
Ondatra zibethica, 519
Opheodrys aestivus, 262
Ophidians, 261, 263, 325
Opossum, 579-583
Optimal foraging theory, 40, 196
central place foraging, 488
information centers, 488
Optimal outbreeding, 225
genetic relatedness, 225
phenotype matching, 225
Oregon Regional Primate Research Center, 5
Orientation and homing, 37-42
amphibian, 341-343
birds, 369
chemical landmarks, 342
magnetic fields, 345, 346
odor imprinting, 346-352
reflexes, 40-42
reptiles, 341, 343-346
salmon, 136-142
turtles, 346-352
Ornithonyssus sylviarum 373, 374
Ornithorhynchus anatinus, 587
Ornithorhynchidae, 587
Orthopristis chrysopterus, 49
Oryctolagus cuniculus, 1, 87
Oryx, 386, 392
Oryx beisa, 386, 392
Osprey, 362
Ostariophysan fishes, 103-107, 239
(see also Alarm substance cells)
Otter shrews, 609

Pacific herring, 121, 128
Pagodroma nivea, 361
Pandion haliaetus, 362
Papio hamadryas, 425
Parental investment, 210, 397-400
in salamanders, 210
nepotism, 397
Parkes, A. S., 6
Parus major, 386
Parupeneus porphyreus, 47
Passer domesticus, 381
Passeriformes, 357
Passerines, 370, 381
Patterson, R. L. S., 4
Pearl dace, 106
Pelecanoididae, 358
Pelecanoides georgicus, 358
Pelecanoides magellani, 358
Pencil fishes, 105
Penguins, 358, 364, 366
z-2-penten-1-yl acetate, 30
Pentylacetate, 30
Perca flavescens, 104
Perch, 104
Percidae, 99
Percina nigrofasciata, 109
Peromyscus leucopus, 533
Peromyscus maniculatus, 533
Peromyscus maniculatus bairdii, 424
Petaurus breviceps, 598
Petrels, 358, 365
Petromyzon marinus, 49
Pfaffmann, C. 2
Phalanger, 598
Phalanger gymnotis, 595, 597
Phalangeridae, 595
Phascolarctos cinereus, 595, 596
Phascolomyidae, 595
Phenotype matching, 225, 234, 239, 402-406
in anurans, 234, 239
major histocompatability gene complex, 413-420
mating preferences, 417, 418
and optimal outbreeding, 418-420
recognition alleles, 408, 409, 414-420
recognition systems, 418-420
and relatedness, 418-420
in spiny mice, 402-406
Phenylacetic acid, 568
Pheromones, 8
airborne, 118, 207, 221-224, 327, 335
alarm, 507-517
aphrodisiac, 174
in beaver castoreum fractions, 567, 568
in callitrichid monkeys, 653, 669
and chin scraping in salamanders, 199
concept, 6
contact, 327, 335
elephants, 627-643
evolution, 14
and goldfish spawning, 149-154
injection in salamanders, 177-180

Pheromones, (Continued)
in insectivores, 609-615
in lizards, 323-338
in mammalian ovarian function, 423-429, 433-437, 441-452
and mammalian reproduction, 9
primer, 5, 28, 118, 128, 174, 182, 215, 428, 433-435, 441-452, 457-461, 615
releaser, 118, 128, 433, 614, 643
rodent puberty-influencing, 441-452, 457-461, 551-554, 555-559
and rodent vomeronasal organ, 471-483
in salamander courtship, 173-184
and salamander reproductive isolation, 221-224
and salamander territorial defense, 191-200
sex attractant, 174
in snakes, 245-255, 261-275, 279-282
signaling, 39, 433
urinary, 433, 441-452, 457-461, 463-469
waterborne, 118, 135-146, 227, 229, 345
Phoxinus phoxinus, 104, 105
Physiological ecology, 14
Physiologically controlled glands
amphibians, 18
avian uropygial gland, 18
chelonian musk glands, 18
crocodilians, 18
desert tortoise, 18
mammalian skin glands, 18
marine turtles, 18
non-serpentinform lizards, 18
Pig, 6, 174, 514, 561, 568
C_{19}-Δ16 steroid pheromone, 30
Pigeon, 358, 369
Pigfish, 49
Pigtail macaques, 226
Pikas, 571-575
Pimephales notatus, 107
Pimephales promelas, 101, 104-107, 126
Pimephales vigilax, 107
Pine vole, 428, 429, 555-559
familial chemosignals and reproduction, 555-559
Pinfish, 48
Pinus resinosa, 562
Pinus sylvestris, 562
Pirhana, 103
Pituophis, 528
Pituophis melanoleucus, 262
Pitymys subterraneus, 551, 552, 554
Plaice, 49, 58
Plains zebra, 623
Plant nest fumigants, bird's,
bacterial control, 373, 374, 381
head-space analysis, 376
in passerines, 370, 381
in starlings, 371-381
mono- and sesquiterpenes, 375
pathogen control, 373, 374, 381
species' selected, 371-373
volatile chemical profiles, 375, 381
Plantago major, 377, 379
Plants, 369-382
Platyfish, 127, 129, 155-168
(see also olfactory system)
Platypus, 587, 594
neck glands, 594
foot pad sweat gland, 594
Plecoglossus altivelis, 120, 122, 123
Plethodon, 191, 192, 194, 221, 223
Plethodon aureolus, 221, 222
Plethodon cinereus, 191-200, 206, 217, 222, 342, 343
chemical ecology of, 191-200
Plethodon glutinosus, 192, 193, 200, 221, 222
Plethodon hoffmani, 179, 193
Plethodon jordani, 178, 193, 195, 221-224, 342, 343
Plethodon kentucki, 222
Plethodon neomexicanus, 179
Plethodon shenandoah, 192, 193, 196, 200
Plethodon teyahalee, 222, 223
Plethodontidae, 176, 177, 191, 198, 199, 221
Plethodontids, 177-182, 206, 207, 211
Plethodon vehiculum, 179
Plethodon yonahlossee, 178
Pleuronectes platessa, 49, 58
Plotosus lineatus, 103
Pocket gopher, 520-529
Podicipediformes, 364
Podophthalmus vigil, 47
Poecilia chica, 120
Poecilia melanogaster, 120
Poecilia reticulata, 120
Poecilia sphenops, 120
Poeciliidae, 120, 155
Poison hemlock, 373
Polar bear, 637
Polemonium reptans, 372-374, 377
Pollution, 352

Polychaete, 73
Polypteriformes, 103
Pondsmelt, 121
Potamogalidae, 609
Potentilla recta, 373, 377, 379
Potentilla reptans, 379
Prairie deer mouse, 433
Prairie rattlesnake pheromones
birthing rookeries, 287-298
chemistry, 298, 299
and cutaneous waterloss, 20, 289-291
and den location, 20, 286-298
lipid, 285
odor imprinting, 294
odor trailing, 39
proto-integument, 289-298
semiochemistry, 298, 299
skin lipids, 291-298
Prairie rattlesnakes, 266, 285-300
birthing rookeries, 287-298
den location, 286-298
neonatal behavior, 286-291
proto-integument functions, 289-298
skin lipid pheromones, 291-298
skin shedding, 294, 295
Prairie vole, 428, 429, 555
Preti, G., 4
Prey-predator interactions
alarm chemosignals, 507-517
chemosignals and crop protection, 519, 520, 527-530
pocket gophers and mustelids, 519-529
Priesner, E., 7
Procellariidae, 358
Procellariiform chemical signals, 357-366
anal glands, 362, 363
evolution of, 364-366
feces, 363
harderian gland, 363
homing, 363
mate recognition, 363, 364
and reproduction, 363, 364
responses to prey chemicals, 361, 362
sebaceous glands, 362, 363
secretory products, 362, 363
stomach oil, 363
uropygial preen gland, 362, 363
Procellariiform olfaction, 357
bulb volume, 359, 360
conchae, 358, 359
ecology, 357-366
evolution of, 364-366
and foraging, 361, 362
homing, 363
mate recognition, 363, 364
mitral cell counts, 359, 360
nasal cavity structure, 357
responses to prey odors, 360-362
sensory mucosa, 358
terminal nerve absence, 357, 363
vomeronasal system absence, 357
Procellariiformes, 357-366
taxonomy, 358
Primary defense, fish, 109
Primates, 184, 226, 425, 426, 434, 647, 653, 682
callitrichids, 653-669
humans, 673-682
inhibitory chemosignals, 424-426
lemurs, 647-651
ovarian function, 424-426
Prions, 358, 365
Prionotus carolinus, 49
Pronghorn antelope, 568
Propithecus verreauxi verreauxi, 647
Prosimians, 647
Prostaglandins
fish reproductive pheromones, 124, 126
goldfish spawning behavior, 149-154
Proteus anguineus, 342
Protoadaptation
versus preadaptation, 15
rattlesnake skin lipids, 289-298
snake skin shedding, 294, 295
Prototheria, 587, 609
Przewalski's horse, 623-626
harem social structure, 623
Przewalski's horse scent marking, 623-626
fecal, 623, 624
functions, 623
marking behavior, 624, 625
urinary, 623, 624
urinary fecal marking, 624, 625
urinary reproductive information, 624, 625
urinary social facilitation, 625
Psammomys obesus, 572
Pseudacris triseriata, 342
Pseudemys scripta, 314, 345
Pseudoeurycea brunnata, 179
Pseudoeurycea goebeli, 179

Pseudoeurycea rex, 179
Pseudomonas aeroginosa, 373, 374
Pseudopleuronectes americanus, 49
Pteridophytes, 17
Pterophyllum scalare, 119
Puberty influencing chemosignals
(see Mouse puberty-influencing chemosignals)
(see Vole puberty-influencing chemosignals)
Puffer fish, 47, 48, 58
lip taste receptors, 47
Puffinus griseus, 358
Puffinus pacificus, 363
Pygmy marmoset, 654

Q/K ratio, 106

Rabbits, 28, 249, 359
semiochemistry of scent gland and fecal pellets, 30
Rainbow trout, 49, 104, 119, 122, 143, 159
Raisman, G., 6
Rana arvalis, 235
Rana cascadae, 229-231, 234, 235
Rana clamitans, 342
Rana sylvatica, 231-234
Rana temporaria, 233
Ranunculus, 377, 379
Rasbora heteromorpha, 110
Rat kangaroo, 595
Rats, 21, 29, 79, 167, 226, 359, 413, 425, 433, 472, 473, 487-503, 514, 528, 575
chemocommunication and diet, selection, 487-503
maternal pheromone, 21
sociality and diet selection, 487-503
Rattlesnakes, 39, 266, 287, 288, 343, 528
Rattus norvegicus, 226
Reciprocal altruism, 101
Red-backed salamander, 191, 206
Red-chested moustached tamarin, 654
Red-chested tamarin, 654
Red-dead nettle, 373
Red deer, 388
Red fox, 30, 611
semiochemistry of male urine, 30
Red kangaroos, 595-597
Red sea bream, 47, 48, 58
Red pine, 562
Red-sided garter snakes, 279
Red-spotted newt, 179, 206
Redundancy, 566, 569
Regina septemvittata, 262, 268-271
Reproductive isolation
hybridization, 221-224
role of chemical signals in, 221-224
in salamanders, 200, 221-224
and sibling species, 221
Reptile integument, 14
β-keratin, 16
development, 20
histology in snakes, 295-298
lipid housing, 16
mechanical protection, 16
Oberhäutchen, 19, 22
odor trails, 18
pheromone expression in snakes, 249, 280, 286, 289-298
rattlesnake proto-integument, 289-291
snake cutaneous waterloss, 290, 291
snake skin shedding, 294, 295
squamate β-layer, 19, 22
and squamate epidermatoglyphics, 292
Reptiles, 10, 14-17, 19, 173, 174, 341, 343, 346, 352, 359, 363, 365, 528, 609
cloaca, 17
fecal pellet, 17
olfactory homing and orientation, 343-346
urine, 17
Response guided strategy, 28, 34, 567-569
Rhabdophis, 254
Rheiformes, 364
Rhesus monkey, 5
Rhinichthyes cataractae, 107
Rhipocephalus phoenix, 45
Rhodeus ocellatus, 120, 122
Ringneck snake, 344
Ring-tailed lemur, 647-651
antibrachial organ, 648
apocrine sweat glands, 648
arm scent marking, 648, 651
fecal scent marking, 648
Flehmen, 648
genital scent marking, 648, 651
interspecific scent marking, 648-650
scent marking battle zone, 650, 651
scent marking and territoriality, 647-651
sebaceous brachial organ, 648
shoulder rubbing, 651
stink fights, 648

Ring-tailed lemur, (Continued)
urine scent marking, 648
vaginal and labial secretion, 648
Riparia riparia, 226
Ritter, F. J., 7
River otter, 617, 619
plantar glands, 619
Rock dove, 358-360
Rodent chemical signals, 533-539
and comparative social behavior, 533-539
estrous synchrony, 2, 28
heterospecific versus conspecific, 533-539
individual discrimination, 536
population variation, 537, 538
pregnancy influencing, 2, 555, 571, 575
scent glands and sexual selection, 541-548
self and non-self discrimination, 539
sex chemosignal discrmination, 536, 537
trap odors, 533, 534
Rodent puberty-influencing chemosignals, 441-452, 551-554, 571
(see Mouse puberty-influencing chemosignals)
(see Vole puberty-influencing chemosignals)
Rodent scented-trap studies, 533-539
Rodents, 79, 397, 399, 423, 424, 429, 433, 442, 507, 514, 515, 519, 533, 534, 551, 555, 614, 653
Rook, 362
Rose bitterling, 120
Russell, M. J., 5
Russian physiologists, 40-42
Russian reflexologists, 40-42

Sable, 617
Saddle-back tamarins, 654, 655, 662, 663
Saguinus, 653, 655, 657, 663
Saguinus fuscicollis, 184, 364, 561, 654-659, 661-663, 666-668
(see also Tamarins)
butyrates in scent marks, 666, 667
chemical signal computer pattern analysis, 664-667
cholesterol in scent marks, 666, 667
scent mark volatile constituents, 664-667
squalene in scent marks, 666, 667
Saguinus fuscicollis fuscicollis, 654, 663, 665
Saguinus fuscicollis illigeri, 654, 663, 665
Saguinus labiatus, 654
Saguinus labiatus labiatus, 654, 656, 658, 663, 666
Saguinus leucopus, 654, 658, 666
Saguinus midas, 654, 658, 659
Saguinus midas midas, 654, 666
Saguinus midas niger, 654
Saguinus mystax, 654
Saguinus mystax mystax, 654, 658, 659
Saguinus oedipus, 654
Saguinus oedipus geoffroyi, 654, 655, 658, 659, 662
Saguinus oedipus oedipus, 654, 655, 657-661, 663, 666, 667
Salamander chemical signaling, 173-184
abdominal glands, 178
in aggression, 173
chemosensory receptors, 198-200
chin tapping, 198-200
cloacal exudates, 197-200
cloacal glands, 177-184, 197-200
in courtship, 173-184
courtship pheromone delivery, 176-184
egg recognition and clutch fidelity, 210-213
evolution of, 173-184
fecal pellets, 197-200
and female receptivity, 180-184
genial glands, 179, 180
granular gland, 200
individual recognition, 197
individual and species recognition, 173
Jacobson's organ, 198
and mating success, 180-184
nares, 198
nasolabial cirri, 195, 196, 198
nasolabial cirri tapping, 198-200
nasolabial grooves, 194, 198
and nesting behavior, 205-217
and parental care, 205-217
philopatry and homing to the nest, 206
phylogenetic occurrence of courtship pheromones, 174-184
plethodontid cephalic glands, 178, 179, 182

Salamander chemical signaling, (Continued)
plethodontid mental (submandibular) glands, 178, 180, 200
predator repellent and poisons, 173
premaxillary or mandibular teeth, 177-180
and reproductive isolation, 221-224
submandibular gland, 200
substrate marking, 196, 197
vent gland, 199-200
Salamander reproduction, 176-184
and cloacal exudates, 178, 179
courtship, 176
fungal infestation and oophagy, 216
mate location, 176
and mating success, 180-184
nesting behavior, 205-217
parental investment, 205-217
pheromone delivery, 177-179
pheromones in pre-mating reproductive isolation, 200
sperm storage, 176
sperm transfer, 176
spermatheca, 176, 182, 183, 199, 200
spermatophore, 176, 180, 181
Salamander territoriality, 191-200
chemical signaling, 191-200
competition, 192, 193
defined, 193
foraging success, 192
natural history, 191, 192
and optimal foraging, 196
resource periodicity, 192
Salamanders
brooding behavior, 206, 208, 210
chemical signals and nesting, 205-217
chemosensory receptors, 198-200
chin tapping, 198-200
competition among, 192, 193
dear enemy recognition, 195-197
evolution of courtship pheromones, 173-184
fat storage organ, 194
granular glands, 200
Jacobson's organ, 198
life history tactics, 192
migration and homing, 173
nares, 198
nasal chamber, 198
nasolabial cirri, 195, 196, 198
nasolabial cirri tapping, 198-200
nasolabial grooves, 194, 198
olfactory cues and clutch recognition, 215, 216
parental investment, 205-217
pheromonal functions, 200
pheromones and territorial advertisement, 191-200
premaxillary or mandibular teeth, 177-180
and reproductive isolation, 221-224
reproductive patterns in aquatic and terrestrial forms, 176-180
sensory-feedback hypothesis, 216
submandibular region, 200
territoriality, 191-200
vent gland, 199-200
Salamander territorial pheromones, 191-200
chemosensory receptors, 198-200
chin tapping, 198-200
cloacal exudates, 197-200
cloacal glands, 197-200
dear enemy recognition, 195, 196
fecal pellets, 197-200
nares, 198
nasolabial cirri tapping, 198-200
and nasolabial grooves, 194, 198
receiver behavior, 195-198
and territorial advertisement, 194
vent gland, 199-200
<u>Salamandra</u>, 177, 342
Salamandridae, 176, 177
<u>Salmo</u>, 136
<u>Salmo alpinus</u>, 126, 230
<u>Salmo gairdneri</u>, 49, 119, 120, 122-124, 143, 159
<u>Salmo salar</u>, 136, 159, 230
<u>Salmo trutta</u>, 125
Salmon, 137, 138, 226, 230, 346, 397, 419
Salmon homing, 136-142
instream orientation, 138-140
near shore orientation, 137, 140-142, 145
oceanic phase, 137, 145
sensory adaptation, 146
stream phase, 137
Salmonoids, 122
Sandworm, 65, 73, 74
<u>Sarcophilus</u>, 595, 598
<u>Sarcophilus harrisii</u>, 595
<u>Sarotherodon mossambicus</u>, 125

Scalopus aquaticus, 611
Sceloporus occidentalis, 345
Scent marking, 385-393
  antelope scent glands, 385, 386
  antelope skin glands, 385, 386
  beaver mud piles, 561-569
  beaver scent mounds, 561-569
  callitrichids, 653, 655-659
  competitor assessment, 385, 387, 388, 392
  cost-benefit analysis, 390
  echidna cloacal, 593, 594
  economics of, 385-393
  in fat-tailed marsupial mouse, 598, 599
  gazelle antorbital gland, 386, 392
  golden hamster, 590
  in gray opossums, 588, 589, 591-593, 594-599
  individual recognition, 388-390
  in insectivores, 609-615
  koala chest marking, 596, 597
  mammal mating strategies, 385
  marmosets, 653, 655-659
  in marsupials, 588, 595-598
  marsupial personal-space marking, 592
  Mongolian gerbil, 590
  in monotremes, 593, 594
  and mustelid plantar glands, 622
  opossum sternal scent glands, 579, 581, 583
  optimal territory size, 392, 393
  pedal glands, 392
  in pika mating, 571-575
  Przewalski's horse fecal and urinary, 623-626
  rabbit chinning, 1
  rabbit dung hills, 1
  rabbit enurination, 1
  rabbit tail-flagging, 1
  salamander chin scraping, 199
  scent matching, 388-390, 392, 393
  status advertisement, 385, 387, 388, 392
  in sugar glider, 598
  tamarins, 653, 655-659
  and territoriality, 385-393
Schreckstoff, 99
  in ostariophysan fishes, 99
Schultze-Westrum, T. G., 4
Scincella lateralis, 324, 327, 329
Scinidae, 336, 338
Scincinae, 324
Scolopacidae, 362
Scomber scombrus, 102
Scophthalmus maximum, 49
Scorpionflies, 182
Scotch pine, 562
Scots, 674
Sea lamprey, 49
Sea robin, 49
Sea turtles, 346-348, 350, 351
  chemical orientation, 346-352
  juveniles, 346-352
  migrations, 346
  nasal morphology, 346, 347
  sand smelling, 347, 348
  seasonal cycle, 346
  visual sea-finding, 346
Secretion
  mechanical expression, 18
  physiological expression, 18
Semiochemicals, 4, 14, 20, 21, 28, 341, 343
  beaver castoreum, 561-569
  bioassays, 28
  of cattle, 27-34
  and crop protection, 519, 520, 527-530
  estrous-associated, 28
  facultative controlled release, 18
  gastrointestinal, 17
  headspace analysis, 30-32
  interactive, 568
  lipid, 20
  microtine scent glands and sexual selection, 541-548
  mustelid 3,3-dimethyl-1,2-dithiolane, 520, 523-528
  mustelid 3-propyl-1,2-dithiolane, 520, 523-528
  predator avoidance, 519-529
  uncontrolled release, 17, 19
Semiochemistry
  beaver castoreum, 561-569
  in callitrichids, 664-667
  computer pattern analysis, 364, 568
  defined, 27
  in elephants, 628-640
  gas chromatographic analysis, 520-527
  Gestalt as odor image, 568
  headspace analysis, 30
  insect, 27
  mammalian, 27-34
  mustelid anal gland, 520-527
  odor image approach, 28, 34, 561, 567-569
  radioimmunoassay analysis, 30

Semiochemistry (Continued)
response guided strategy, 28, 34, 567-569
Semotilus atromaculatus, 107
Senecio obvatus, 374
Sensory processes
motor patterns dedicated to, 41, 42
Sensory tuning, 47-48
Setifer setosus, 612
Sex-linked P gene, 155-157
Sexual selection, 174-176, 180-184, 243, 333, 334, 336-338, 389
(see also Natural selection)
in antelope territorial scent marking, 389
characters suggesting, 544
microtine polygyny, 542, 546, 547
and microtine scent glands, 541-548
and monogamy, 546
and opposum scent glands, 579-583
and reproductive isolation, 336-338
in salamander courtship pheromones, 176-184
in skink pheromones, 323, 329-338
Shearwaters, 358, 361, 362, 365
Shiner, 102, 108, 109
Shorey, H. H., 9
Short-necked clam, 47
Short-tailed shrew, 611
Short-tailed vole, 538
Short-tailed weasel, 520
Shrews, 609-611, 613
Sifaka, 647-651
anogenital scent marking, 647
chest scent marking, 647
fecal scent marking, 647
interspecific scent marking, 648-650
scent marking battle zone, 650, 651
scent marking and territoriality, 647-651
throat scent marking, 647
urine scent marking, 647
Signoret, J. P., 6
Silver dollar, 103
Silverside, 49
Silverstein, R. M., 6-8, 10
Silvery marmoset, 654
Simeone, J. B., 8
Sistrurus catenatus catenatus, 287
Sitta canadensis, 370
Sitta carolinensis, 370
Skin glands (see glands)
Skink pheromones
cloacal glands, 323, 326, 335, 336
cloacal odors, 328, 331, 335
conspecific discriminations, 324
contact and airborne, 327, 335
dorsal, 336
egg and offspring identification, 324
ethological isolation, 329-334, 336-338
evolution of, 323-325, 329, 338
female, 335, 336
interspecific odor discrimination, 329-331
male agonism, 333, 334
proctodaeum, 335
and reproduction, 327-329
reproductive isolation, 329-334, 336-338
scent trails, 323, 334, 335, 337
skin odors, 326, 331, 335
snout-dipping, 324
social discrimination, 323
urodaeal glands, 335-337
Skinks, 323-338
prey odors, 328, 329
Slender bitterling, 120
Smell, 673
(see also Olfactory system)
Smell identification test, University of Pennsylvanis, 675-677
Sminthopsis, 598
Sminthopsis crassicaudata, 597, 598
Smith, A. B. III, 4
Smooth newt, 177, 181
Snake aggregation
gravid female, 266, 267
in neonatal snakes, 289-298
non-sexual, 266, 267
in prairie rattlesnakes, 285
prairie rattlesnake birthing rookeries, 287-298
to reduce waterloss, 267
thermal benefits, 288
Snake reproduction
aggregation thermal benefits, 288
annual reproductive cycle in garter snakes, 244-255
gametogenesis in garter snakes, 244
garter snakes, 243-255
garter snake courtship behavior, 245, 246
mating balls, 279-281
pheromone trails, 263-266

Snake reproduction (Continued)
rattlesnake birthing rookeries, 287-298
renal sex segment, 243
reproductive hormones, 250-252
reproductive pheromones, 243-255
sexual mimicry, 279-282
she-males, 280-282
vitellogenin, 246-252, 281
Snakes, 17, 19, 243, 261-275, 285-300, 305, 326, 343, 344, 528
aggregation, 262, 344
birthing rookeries in prairie rattlesnakes, 287-298
combat dancing, 261
cutaneous waterloss, 290, 291
den location, 262, 286-298
dermal poison glands, 254
dispersal, 343, 344
hinge region vacuolated cells, 21
integumentary mesos layer, 21
mating balls, 279-281
nuchodorsal glands, 21
odor trailing behavior, 262, 291, 295, 344
rattlesnake skin lipids, 289-298
reproduction, 243-255, 261-275, 279-282
reproductive pheromones, 243-255
skin shedding, 295
social behavior, 261-275
Thamnophis radix complex, 272-275
Sniffing behavior, 41
Snow petrel, 359, 361, 363
Snowshoe hares, 515, 520, 527
Social behavior, 261-275
beaver family odors, 562
beaver scent marking, 561-569
and callitrichid chemical signals, 653-659
chemocommunication and rat diet selection, 487-503
chemosignal social facilitation, 626
comparative, in rodents, 533-539
and elephant chemosignals, 627-643
gray opossum chemical signals, 588-604
marsupial chemical signals, 587, 588, 594-600
monotreme chemical signals, 587, 588, 593, 594, 599, 600
opossum sex discrimination, 579, 583
pika mate choice, 571-575
Przewalski's horse, 623
rodent sex discrimination, 536, 537
and rodent odors, 533-539
snakes, 261-275
spiny mice, 397-409
and vole familial chemosignals, 559
Social learning of diet selection, 487-503
chemical cues, 495, 496, 499, 500
contextural cues, 500-502
dietary familiarity, 496-499
foraging orientation, 492-495
multiple messages, 491-492
rat demonstrators, 488-490
rat observers, 488-490
and rat olfactory communication, 487-503
Sociobiology, 4, 225, 261, 262, 275, 434, 436
cooperative behaviors, 225
in garter snake pheromone trails, 261-275
Sockeye salmon, 102
Sokolov, V. E., 7
Sole, 49, 58
Solea solea, 49, 58
Solenodon paradoxus, 611
Solenodons, 609, 610, 612
Solenodontidae, 609
Solidago canadense, 377
Solidago rugosa, 372-377, 379
Solidago ulmiflora, 374, 377
Songbirds, 369
Sonora, 267
Sonora semiannulata (episcopa), 262
Sooty shearwater, 359-361
Sorex, 610
Sorex araneus, 535
Sorex minutus, 535
Sorex palustris, 610
Soricidae, 609, 610
South American didelphids, 588
Southern giant petrel, 359, 361
Spermophilus tridecemlineatus, 226
Sphenisciformes, 364
Sphyrna tiburo, 142
Spiny mice, 226, 397-409
(_see also_ Spiny mouse chemosignals)
chemosignals in kin recognition, 397-409
kin recognition, 397-409
Spiny mouse chemosignals
dietary effects, 398-400

Spiny mouse chemosignals
(Continued)
familiarity, 400-406
individual signatures, 400-406
maternal pheromone, 399
mother's milk, 398
parent-infant recognition, 397-400
phenotype matching, 402-406
self-matching, 404
sibling recognition, 401, 402
urine and feces, 398, 399
Spiny mouse kin recognition, 397-409
familiarity, 400, 401
individual olfactory signatures, 400-406
ontogeny, 403
parent-infant recognition, 397-400
phenotype matching, 402-406
recognition alleles, 408, 409
self-matching, 404
sibling recognition, 400-407
Spring beauty, 373
Squamate reptile, 19, 21, 295, 323, 325, 344
neonate skin-shedding, 20
skin lipids and waterproofing, 20
skin shedding, 20, 21
Squid, 58
Staphylococcus epidermis, 373, 374
Starlings, 369-382
chemosensory electrophysiology, 376
plant nest fumigant use, 371-373
plant use ecology, 381
plant volatile discrimination, 376-381
Stoat, 519, 520, 524, 528
Stoddart, D. M., 7, 9
Stonecat, 104
Storm-petrels, 358, 361, 365
Streptococcus aurealis, 373, 374
Strike-induced chemosensory searching, 39
Storeria dekayi, 262, 344
Sturnus vulgaris, 371
Suckers, 99, 106
Sugar glider, 594, 595, 598, 601
scent-marking social aspects, 598
Suncus, 610, 614
Suncus murinus, 610, 614
Sus scrofa, 561, 568
Swamp wallaby, 596
Swiss albino mice, 463
Switz, M. G., 5
Swordtail, 163
SWR/J mice, 463-468
Synapsid, 14
Syncerus caffer, 386

Tachyglossidae, 587
Tachyglossus aculeatus, 587
Taiwanese natives residing in the United States, 679
Talopoin monkey, 424, 425
Tamarin, 364, 425, 561, 653, 655, 661, 663, 667, 668
apocrine glands, 655-659
chemical signal computer pattern analysis, 664, 665-667
chemical signal sources, 655-659
circumgenital glands, 655-659
circumgenital-suprapubic gland complex, 655-659
comparative chemical signals, 653-669
cotton top, 654, 656, 659, 660, 662
epidermal scent glands, 653
genital glandular pads, 655-659
gular-sternal glandular pads, 655-659
hormones and marking, 660
individual recognition, 663, 664
muzzle rubbing, 657-659
physical environment and scent marking, 662
reproduction and scent marking, 660, 661, 667
scent and aggression, 668
scent and infant-adult interactions, 668
scent glands, 655-659
scent gland development, 656
scent gland histology, 655, 656
scent marking, 653, 655-659
scent mark chemistry, 662-667
scent marking development, 660, 661
scent mark information content, 662-667
scent mark volatile constituents, 664-667
sebaceous glands, 655-659
self marking, 659
social determinants and scent marking, 661, 662, 667

Tamarin, (Continued)
sternal glands, 655-659
sternal marking, 657, 658
suprapubic glands, 655-659
suprapubic marking, 657, 658
urine washing, 659
ventral rubbing, 657, 658
Tantilla gracilis, 262
Tapes japonica, 47
Taraxacum officinale, 374, 572
Taricha, 178, 179, 351
Taricha rivularis, 205, 342
Tasmanian devil, 595
Tassel-ear marmosets, 654
Taste, 45, 305, 319, 325
(see also Gustation)
and feeding in fish, 45
in humans, 673, 681, 682
lizard tastebuds, 305, 319
Taxidea taxus, 528, 621
Teleosts, 119, 123, 124, 149, 154, 159, 166
Tenebrio molitor, 307
Tenrecidae, 609, 612
Tenrecs, 609, 610, 612, 613
Terrestrial turtles, 345, 346
fecal pellets, 345, 346
homing and orientation, 345, 346
social odors, 345, 346
Territoriality
beaver mud piles, 561-569
beaver scent marking, 561-569
beaver scent mounds, 561-569
cost of competition, 387
defense costs, 390-393
individual recogiition, 388-390
and insectivore latrine use, 610, 611
and lemur group home range size, 647, 651
male competition, 386
optimal territory size, 392, 393
owner advantage, 386, 387, 389
pika scent marking and familiarity, 571-575
resource defense, 385, 390, 392
and resource density, 647, 651
in salamanders, 191-200
in scent gland dimorphism, 547
scent marking, 385-393
scent marking economics, 390-393
and scent marking in sympatric heterospecific lemurs, 647-651
scent matching, 388-393
Tesselated darter, 100
Testudo hermanni, 345
Tetrapod, 13, 14
Thamnophis, 63, 244-246, 251, 254, 265, 267, 270, 275, 325, 344
Thamnophis brachystoma, 273, 275
Thamnophis butleri, 245, 262, 264, 267-271, 273-275
Thamnophis couchi, 262
Thamnophis cyrtopsis, 273, 274
Thamnophis elegans, 262
Thamnophis equus, 254, 273
Thamnophis melanogaster, 245, 252 253, 254
Thamnophis marcianus, 262, 265, 267, 273, 274
Thamnophis proximus, 262, 264, 268-271
Thamnophis radix, 245, 262, 272-275
Thamnophis radix complex, 274
Thamnophis radix haydeni, 274
Thamnophis radix radix, 274
Thamnophis sirtalis, 244, 245, 250-254, 262, 264, 267-271 344
Thamnophis sirtalis parietalis, 21, 279, 282
Theropithecus gelada 425
Thiessen, D. D., 10
Thomomys talpoides, 519
Thompson, K., 5
Thomson's gazelle, 386, 388
Tiger, 538
Tilapia 125
Tilapia zillii, 46, 51, 53-55, 57, 58
plant chemical feeding stimulants, 52-58
Timber rattlesnakes, 266, 287 295
Toad, 226-239, 343
Todas, 674
p-toluidine, 30
Tongue flicking, 39, 41, 63, 65-67, 70, 72, 292-294
ethology of, 306-319
exploratory, 305-319
garter snakes, 63, 64
iguanas, 305-319
rattlesnakes, 39
in skinks, 323-326
types, 306, 307, 309, 312
Topi, 392
Tortoises, 343, 345
Tragelaphinae, 392
Tragelaphus strepsiceros, 386
Trees, 99, 519
Trichosurus, 598
Trichosurus caninus, 595

Trichosurus vulpecula, 592, 595, 597
Trigeminal system
in birds, 369
Triturus, 177
Triturus cristatus, 178, 180, 181
Triturus vulgaris, 177, 181
Tubenoses, 358
Tufted-ear marmoset, 654
Turbot, 49
Turk, A., 4
Turnips, 678
Turtles, 17, 343, 345
Typhlopidae, 262
Typhlops, 267
Typhlops pusillus, 262

Ultrasonic vocalization and chemosignals in mice, 479-481
Uma, 19
Ungulates, 390
Urine
bladder, 29
cattle, 27-34
elephant buffering capacity, 643
in elephant semiochemistry, 628-640
European wild rabbit, 2
guinea pigs, 38
male mouse esterases, 463-469
mice, 30, 32, 79-84, 433-437
odors and estrous synchrony, 6
odors and pregnancy blockage, 6, 433-437
puberty-influencing chemosignals, 441-452, 457-461, 551-554
in Przewalski's horse marking, 623-626
rodent chemosignals, 433-437, 474
Urodele, 342
Uta, 19

Vandenbergh, J. G., 6, 9
Vegetation, 369-382
Veronica, 377
Vertebrates, 15, 173, 174, 226, 279-282, 503, 507
Viola, 373, 374
Violet, 373
Vipera aspis, 262
Vipera berus, 262
Viperidae, 262
Viperids, 287
Virginia, 267
Virginia opossum scent glands, 579-583
bacteria, 583
dewlap, 581
gross morphology, 579-582
scent marking, 579, 581, 583
sex recognition or advertisement, 579
sexual dimorphism, 579-583
sternal, 579-583
sternal integument gross morphology, 581, 582
sternal integument histology, 580, 582, 583
Virginia opossums, 579-582, 588, 595, 597, 598, 601
sternal scent gland morphology, 579-583
Virginia striatula, 262
Vitellogenin, 246-252, 254, 255, 272, 281
(see also Garter snakes)
(see also Garter snake reproductive pheromones)
Vole, 519, 527-529, 535-539, 541, 542, 544, 548
(see also Microtine rodents)
Vole puberty influencing chemosignals, 551-554, 555-559
(see also Familial chemosignals)
and dispersal, 558
familial chemosignal repression, 555-559
female urine influences, 551-554
male odor effects, 555-559
ovarian suppression, 551-554
population density effects, 551-554, 558, 559
Vomeronasal system, 19, 38, 41, 63, 65, 71, 73, 245, 255, 261, 279, 292, 305, 319, 323-325, 344, 346
absence in birds, 357
in elephants, 627, 640-643
enzyme system in snakes, 71
extinction, 474, 475
in garter snake mating, 245
in garter snake pheromone trails, 261
in guinea pig chemoinvestigation, 473-475
iguana nucleus sphericus, 319
in iguana tongue flicking, 305, 319
lesion effects, 472
in marsupial reproduction, 603, 604
in mouse aggression, 476-479

Vomeronasal system, (Continued)
in mouse reproduction, 475, 476
in mouse ultrasound, 479-482
in neonatal marsupials, 600
organ removal effects, 473
reinforcement of investigation, 38
in ring-tailed lemurs, 648
in rodents, 471-483
rodent learning, 473-482
rodent neuroanatomy, 471, 472
rodent pregnancy blockage, 434, 436
and rodent social behavior, 471-483
in salamanders, 198
sea turtle, 346
in skink tongue flicking, 323-326
in snake tongue flicking, 261, 279, 344
stimulation as reinforcement, 471-483
in vole reproduction, 558
Vulpes, 528
Vulpes fulva, 507, 514
Vulture, 487

Wallabia bicolor, 596
Water snake, 344
Weasels, 515, 519, 520, 522, 525, 526, 528, 529
Wedge-tailed shearwaters, 363-365
Wenzel, Bernice M., 4, 10
Western fence lizard, 345
Western red-bellied newt, 205
White Americans, 676, 677, 679-681
White catfish, 104
White flounder, 49
White-footed tamarin, 654
Whitten, W. K., 4, 6, 10
Wild carrot, 373
Wild mice, 433, 451, 457-461
Wildebeest, 389
Wild hopping mouse, 424
Wilson, E. O., 4
Wilson's storm-petrel, 361, 362
Wolverine, 617, 622
Wood frog, 231
Wooly opossum, 598
Woodmice, 538

Xenopus, 251
Xiphophorus helleri, 120, 163
Xiphophorus maculatus, 127, 155

Yamame salmon, 120
Yarrow, 373
Yellow-footed marsupial mouse, 597
Yellow vole, 519, 529
Yucatan inhabitants, 679

Zaglossus bruijni, 587
Zebra danios, 104, 106, 108, 110
Zebrafish, 119
Zinkevich, E. P., 7
Zoological Society of London, Symposium on Olfaction in Mammals, 7
Zoological Society of Southern Africa, Symposium on Animal Communication, 7
Zottermann, Y., 2, 3

GPSR Compliance

The European Union's (EU) General Product Safety Regulation (GPSR) is a set of rules that requires consumer products to be safe and our obligations to ensure this.

If you have any concerns about our products, you can contact us on ProductSafety@springernature.com

In case Publisher is established outside the EU, the EU authorized representative is:

Springer Nature Customer Service Center GmbH
Europaplatz 3
69115 Heidelberg, Germany

Batch number: 10838292

Printed by Printforce, the Netherlands